physica status solidi (a)

ISSN 0031-8965 ✕ VOL. 77 · NO. 1 · MAY 1983

AKADEMIE-VERLAG · BERLIN

THEORIA CUM PRAXI

Classification Scheme

1. Structure of Crystalline Solids
   1.1 Perfectly Periodic Structure
   1.2 Solid-State Phase Transformations
   1.3 Alloys. Metallurgy
   1.4 Microstructure (Magnetic Domains See 18; Ferroelectric Domains See 14.4.1)
   1.5 Films
   1.6 Surfaces
2. Non-Crystalline State
3. Crystal Growth
4. Bonding Properties
5. Mössbauer Spectroscopy
6. Lattice Dynamics. Phonons
7. Acoustic Properties
8. Thermal Properties
9. Diffusion
10. Defect Properties (Irradiation Defects See 11)
    10.1 Metals
    10.2 Non-Metals
11. Irradiation Effects (X-Ray Diffraction Investigations See 1 and 10)
12. Mechanical Properties (Plastic Deformations See 10)
    12.1 Metals
    12.2 Non-Metals
13. Electron States
    13.1 Band Structure
    13.2 Fermi Surfaces
    13.3 Surface and Interface States
    13.4 Impurity and Defect States
    13.5 Elementary Excitations (Phonons See 6)
        13.5.1 Excitons
        13.5.2 Plasmons
        13.5.3 Polarons
        13.5.4 Magnons
14. Electrical Properties. Transport Phenomena
    14.1 Metals. Semi-Metals
    14.2 Superconductivity. Superconducting Materials and Devices
    14.3 Semiconductors
        14.3.1 Films
        14.3.2 Surfaces and Interfaces
        14.3.3 Devices. Junctions (Contact Problems See 14.3.4)
        14.3.4 High-Field Phenomena, Space-Charge Effects, Inhomogeneities, Injected Carriers (Electroluminescence See 20.3; Junctions See 14.3.3)
    14.4 Dielectrics
        14.4.1 Ferroelectrics
15. Thermoelectric and Thermomagnetic Properties
16. Photoconductivity. Photovoltaic Effects
17. Emission of Electrons and Ions
    17.1 Field Emission Microscope Investigations
18. Magnetic Properties
    18.1 Paramagnetic Properties
    18.2 Ferromagnetic Properties
        18.2.1 Ferromagnetic Films
    18.3 Ferrimagnetic Properties
    18.4 Antiferromagnetic Properties

(Continued on cover three)

Author Index

K. Abdelmoula K5
N. R. Agamalyan 107
Md. Iqbal Ahmed 643
R. Alcala 625
A. A. Alekhin 51
A. S. Aleksandrov 725
V. A. Aleshin K37
S. Alexandrova 721
S. Amelinckx 249, K1, K9
A. N. Anagnostopoulos 595
C. Ance 241
G. A. Andreev 793
M. A. Angadi 685
M. Anglada 283
V. E. Antonov 71, K23
T. E. Antonova 71, K23
I. Apostol 679
A. G. Arias K185
K. Arlauskas 387
D. A. Aronov K41
P. V. Ashrit 685

C. M. van Baal 521
V. Hari Babu 643
I. L. Baginskii K99
A. A. Bahgat 809
M. Balarin 181
M. Balaskó K87
M. Balčiūnienė 463
C. Ballesteros K29
D. Basu 121
A. K. Batra 399
I. T. Belash 71, K23
A. M. Belyantsev 437
C. Bernhardt 225
B. P. Bespalov K13
J. Betko 331
R. P. Bhardwaj 347
K. Biedrzycki K95
J. Bonnafe K69
O. Yu. Borkovskaya K129
O. Breitenstein K107
B. Bridge 655
A. W. Brinkman 535
K. D. Bronsema 249
U. Brückner 545
V. N. Brudnyi K33
G. Brun 669
E. Burzo 127
A. B. Bykov 107

F. Calderón K185
G. Campet 603
V. Carbayo K69

J. Červenák 323, K125
Sh. Chadraabal 709, K53
M. M. Chandra 715
R. Chaoui K189
R. K. Chaudhary K161
Y. Chen 625
T. A. Cherepanova 35, 561
M. I. Chertanov K173
J. Chrzanowski K95
J. I. Chyi 741
J. Claverie 603
B. Cochrane 269
A. J. Cohen 619
D. Conte 603
D. Coquillat 669
I. Cosma 127
S. Cristoloveanu K155
C. Cruz 225

R. Deblieck 249
P. Delavignette 291
E. Demkovicsová 785
S. Deonarine K199
M. Dinescu 679
N. L. Dmitruk K129
J. L. Dormann 611
W. Drabik 513
A. Dragūnas 463
M. N. Drozdov 437
A. Dubon 309
V. P. Dudkevich K37
D. van Dyck 249

G. Eichhorn 195
V. F. Elesin 725
Yu. F. Eltsev 759

M. K. Fayek 809
E. G. Fesenko K37
F. Fiedler K75
D. W. Field 733
V. I. Fineberg K129
G. Le Flem 603
C. P. Foley 491
V. Ya. Fritsberg 369

A. P. Gaevskis 369
F. Găiseanu K59
P. Gaworzewski 571
Yu. M. Gerasimov 107
V. Yu. Gertsman 97
T. Gessner 133
D. Ghose 121

Z. Gierak 775
B. Gillot 339
A. A. Gladyshchuk 765
K. Gmucová 323
O. A. Golikova 449
A. L. Golovin K91
Yu. I. Golovko K37
S. I. Golovkova 375
M. A. Gonzalez K69
P. J. Goodhew 269
P. M. Grangeon 309
V. P. Gribkovskii 765
G. I. Gudzenko 115
F. Guiu 283
J. Gunji K45
A. M. Gurvich 375

H. Haefner 527
P. Hagenmuller 603
K. Hehl 207
U. Heinecke 225
S. Heinemann 153
K. Herrmann 175
I. Hevesi 679
T. Hoshikawa K121
R. Hrabański K143
T. Huang K193
A. Hübler K17

R. M. Imamov K91

C. Jäger K167
Mukesh Jain 67
A. Jalyschko 175
G. G. Janezic 619
W. Jeleński 393
F. Jemmali 339
S. Jetschke 207
J. Jimenez K69
Th. Jung K203
G. Juška 387

O. A. Kaibyshev 97
N. P. Kalmykova 803
A. A. Kaminskii 107, K173
T. Kanadani K49
K. N. Kang K155
H. Karge 201, 207
S. B. Karmohapatro 121
S. B. Kaschieva 699
Y. Kawai K103
M. P. Kazeko 725
Sh. Kh. Khannanov 97
L. G. Khvostantsev 161
H. Kido K121
M. Kittler 139
A. Kiveris 471

R. Klabes 553
W. Kleinn K149
V. A. Klimov 793
H. Klose 233
M. Koizumi K121
B. T. Kolomiets 803
R. V. Konakova K125, K129
M. Konč 785
Z. Korczak 497
P. Kordos K125
E. G. Kostsov K99
W. Kowalsky K75
A. S. Koziejowska 35, 561
G. P. Kramar 483
A. I. Kravchenko 375
E. W. Kreutz 583
G. Kühnel 167
S. P. Kuila 477
K. Ram Kumar 467
K. T.-Y. Kung 355
N. Kuramochi 633
D. B. Kushev 709, K53

C. Lacour 309
H. B. Lal 405
T. Lalinský 323
J. van Landuyt 249, K1
A. Lehr 775
M. L. Levitan 455
M. D. Lin 741
M. S. Lin 741
E. Linke 545
R. N. Litovskii 443, 699
B. X. Liu 355
J. Llopis K29
M. M. Lokshin 443
J. C. Lou 741
C. W. Lung 81
I. Lupsa 127
V. S. Lysenko 443, 699

R. Mach K179
J. Mahy K1
K. Makariūnas 463
K. Mäkinen 263
R. M. Mallya 87
C. Manolikas 595
A. Mansingh 399
D. Marton K17
S. C. Mathur 399
M. Maurin 669
E. E. Meerson K99
A. N. Men 455
B. A. Men 455
R. G. Mendiratta 693
K. Měřínský 331
A. Mertens 233

J. J. Métois K5
L. Michailovits 679
I. N. Mihailescu 679
V. V. Mikhailin 375
H. Mizubayashi 633
S. V. Moharil 413
V. B. Molodkin 115
Y. Momose K45
J. W. Moroń 775
M. Morvič K125
R. N. Mukherjee 799
V. M. Mukhortov K37
Vl. M. Mukhortov K37
G. O. Müller K179
M. Müller K167

K. K. Nagpaul 347
L. Nanai 679
G. K. Narula 693
A. N. Nazarov 699
E. Nebauer K203
N. R. Nedev 699
D. F. Neely 491
T. H. Nguyen K135
M.-A. Nicolet 355
G. Nihoul K5
A. Niklas 393
N. A. Nikolaev 161
M. Nogues 611
V. G. Novitskii 115

G. Oelgart 527
T. Ogawa K103
M. Ohkubo K45
M. Ohta K49
S. Okuda 633
V. M. Orera 625
K. L. Ovanesyan K173
A. N. Ozernoi 63

B. Pałósz 11
D. Panknin 553
Ya. I. Panova 483
D. Papadopoulos 595
V. V. Parashchuk 765
M. Pasemann 133
N. D. Patel 655
A. G. Petrosyan K173
H. K. Phan K135
P. K. C. Pillai 693
P. Pipinys 471
J. Piqueras K29
B. Pistoulet 669
V. P. Plakhty 505
E. K. Polychroniadis 291
E. G. Ponyatovskii 71, K23
A. S. Popov 709, K53
A. I. Potapov K33
H.-H. Potthoff 215
F. Poulin 535
L. Pranevičius K171
M. Prasad 715
N. G. Puttaswamy 87

J. K. Quamara 347

Ramani 87
A. V. Krishna Rao K161
K. V. Rao K161
T. S. Rao K63
U. V. Subba Rao 643
V. I. Rashupkin 71, K23
H. Ratajczak 785
N. M. Ravindra 241
K. N. Reddy 643
P. V. Reddy K63
R. Reetz 233
A. Remón K29
U. Rensch 195
R. Reulke 527
B. Rheinländer 527, K111
A. Rimeika 471
V. D. Rodriguez 625
A. L. Rouff K193
V. V. Rubinov K41
Yu. V. Rud K33
T. E. Rudenko 699

A. Sakakibara K49
A. Sakalas K171
S. Samatov 449
K. K. Saraf 381
S. E. Sarkisov 107
M. P. Sathyavathiamma 87
M. Satyam 467
S. Sche 741
A. Schlachetzki K75
K. Schmalz 571
B. Schmidt 133
H. G. Schneider 201
K.-W. Schröder 139
H. K. Sehgal 67
V. N. Semkin 421
S. C. Sen 799
M. Serginov K33
V. I. Shakhovtsov K129
A. U. Shareef 381
A. K. Sharma K81
B. L. Sharma 347
L. A. Shebanov 369
M. F. Sherbakova 443
M. Shimada K121
V. Kh. Shpunt 803

A. V. Sidorin 679
W. Siegel 167
P. Slančo 785
O. P. Smirnov 505
E. A. Smorgonskaya 803
H. Sodolski 749
A. P. Srivastava 381
O. P. Srivastava 405
D. N. S. Srivastwa 799
M. A. Starikov 759
R. Stegmann 527
M. Subotowicz 497
A. Subrahmanyam 45
K. N. Subramanian 361
B. Sujak K95
E. V. Sviridov K37
H. Syhre 553
A. Szekeres 721

T. Taguchi K115
Ng. Tang 233
T. L. Tansley 491
J. C. Tedenac 669
V. Ten 175
G. van Tendeloo 249, K9
J. Teuho 263
V. G. Thiessen 71
J. C. Thuillier 649
I. Thurzo 323
T. Tima 785
V. A. Timofeeva 107
Yu. A. Tkhorik K125
A. Tomas 611
I. Tomáš K87
J. W. Tomm 175
J. Toruń 513
A. K. Tripathi 693
B. Tuck K189
I. E. Türe 535

A. M. Vainrub K13
R. Z. Valiev 97
A. N. Vasilev 375
G. Vértesy K87
Z. Vértesy K87
Yu. I. Vesnin 759
R. M. Vlasova 421

A. V. R. Warrier 67
D. N. Waters 655
B. Weber 201
H.-H. Wehmann K75
G. A. Wiegers 249
E. Wieser 225, 553
U. Winter K139
J. Wollbrandt 545
J. Woods 535
T. Wosinski K107

L. Y. Xiong 81

G. P. Yablonskii 765
H. Yamada K49
M. Yamada K49
G. Yu. Yashin 421
A. E. Yunovich 175

I. L. Zaitsevskii K125
S. P. Zakovryashin 759
P. Zaumseil K139
A. Zentko 785
J. Zhang 649
A. K. Zhetbaev 63
E. Ziegler 167
G. Zubauskas K171
V. V. Zubritskii 765

physica status solidi (a)
applied research

Volume 77 · Number 1 · Pages 1 to 426, K1 to K98, and A1 to A10

May 16, 1983

PSSA 77(1) 1–426, K1–K98, A1–A10 (1983)

ISSN 0031-8965

AKADEMIE-VERLAG · BERLIN

Subscriptions and orders for single copies should be directed

in the GDR:

to the Postzeitungsvertrieb or
to the Akademie-Verlag, DDR-1086 Berlin, Leipziger Str. 3 – 4;

in the other socialist countries:

to a book-shop for foreign language literature or
to the competent news-distributing agency;

in the FRG and BERLIN (WEST):

to a book-shop or
to the wholesale distributing agency Kunst und Wissen, Erich Bieber OHG, D-7000 Stuttgart 1, Wilhelmstr. 4 – 6;

in the other Western European countries:

to Kunst und Wissen, Erich Bieber GmbH, CH-8008 Zürich, Dufourstr. 51;

in USA and CANADA:

to Verlag Chemie International, Inc., Plaza Centre, Suite E, 1020 NW 6th Street, Deerfield Beach, FL 33441, USA;

in other countries:

to the international book- and journal-selling trade,
to Buchexport, Volkseigener Außenhandelsbetrieb der Deutschen Demokratischen Republik, DDR-7010 Leipzig, Postfach 160, or
to the Akademie-Verlag, DDR-1086 Berlin, Leipziger Str. 3 – 4.

Editorial Note:

"physica status solidi (a)" undertakes that an original paper *accepted for publication* before the 23th of any month will be published within 50 days of this date unless the author requests a postponement. In special cases there may be some delay between *receipt* and *acceptance* of a paper due to the review and, if necessary, revision of the paper.

Schriftleiter und verantwortlich für den Inhalt:
Professor Dr. Dr. h.c. P. Görlich, DDR-1020 Berlin, Neue Schönhauser Str. 20 bzw. DDR-6900 Jena, Schillbachstr. 24.

Verlag:
Akademie-Verlag, DDR-1086 Berlin, Leipziger Str. 3 – 4; Fernruf: 2236221 und 2236229; Telex-Nr.: 114420; Bank: Staatsbank der DDR, Berlin, Kto.-Nr.: 6836-26-20712.

Chefredakteur:
Dr. H.-J. Hänsch.

Redaktionskollegium:
Prof. Dr. E. Gutsche, Dr. H.-J. Hänsch, Dr. H. Lange, Dr. S. Oberländer.

Anschrift der Redaktion:
DDR-1020 Berlin, Neue Schönhauser Str. 20; Fernruf: 2823380.

Veröffentlicht unter der Lizenznummer 1620 des Presseamtes beim Vorsitzenden des Ministerrates der Deutschen Demokratischen Republik.

Gesamtherstellung:
VEB Druckerei „Thomas Müntzer", DDR-5820 Bad Langensalza.

Erscheinungsweise:
Die Zeitschrift „physica status solidi (a)" erscheint jeweils am 16. eines jeden Monats. Jährlich erscheinen 6 Bände zu je 2 Heften.

Bezugspreis:
Je Band 200,– M zuzüglich Versandspesen (Preis für die DDR: 130,– M).

Bestellnummer dieses Bandes: 1085/77.

Printed in the German Democratic Republic.
AN (EDV) 20735

Contents

Review Article

B. Pałósz — Reasons for Polytypism of Crystals of the Type MX_2 (I) 11

Original Papers

T. A. Cherepanova and A. S. Koziejowska
Sublattice Models of the Binary Crystal Growth (I) 35

A. Subrahmanyam
Effect of DC Field and X-Irradiation on the Optical Absorption and Thermoluminescence of NaCl Single Crystals 45

A. A. Alekhin — The Bloch Lines Dynamics in Permalloy Films 51

A. K. Zhetbaev and A. N. Ozernoi
Mössbauer Study of Radiation Effect of the $^{181}Ta(p, n)^{181}W$ Reaction . . . 63

Mukesh Jain, H. K. Sehgal, and A. V. R. Warrier
The Variation of Optical Energy Gaps with Composition in $Pb_{1-x}Hg_xS$ Alloy Films 67

V. E. Antonov, T. E. Antonova, I. T. Belash, E. G. Ponyatovskii, V. I. Rashupkin, and V. G. Thiessen — The Pd–Ni–H System at High Pressure 71

C. W. Lung and L. Y. Xiong
The Dislocation Distribution Function in the Plastic Zone at a Crack Tip 81

Ramani, M. P. Sathyavathiamma, N. G. Puttaswamy, and R. M. Mallya
Mössbauer Effect Study of the Thermal Decomposition in Hydrogen of Homogeneously Mixed Ferrous Nickel Oxalates with Different Iron to Nickel Ratios 87

R. Z. Valiev, V. Yu. Gertsman, O. A. Kaibyshev, and Sh. Kh. Khannanov
Non-Equilibrium State and Recovery of Grain Boundary Structure (I) . . 97

A. A. Kaminskii, V. A. Timofeeva, N. R. Agamalyan, A. B. Bykov, S. E. Sarkisov, and Yu. M. Gerasimov
Luminescence Properties of Nd^{3+} Ions in Disordered $Na_5YSi_4O_{12}$ Crystals . . 107

G. I. Gudzenko, V. B. Molodkin, and V. G. Novitskii
Dynamical Diffraction in Alloys. Influence of Long-Range Order 115

D. Ghose, D. Basu, and S. B. Karmohapatro
Sputtering of Germanium by 15 to 35 keV Noble Gas Ions 121

E. Burzo, I. Lupşa, and I. Cosma
On the Magnetic Behaviour of NiCuSi Solid Solutions 127

T. Gessner, M. Pasemann, and B. Schmidt
Calculation of Neutron-Induced Defect Clusters in Silicon and Comparison with TEM Investigations 133

M. Kittler and K.-W. Schröder
Determination of Semiconductor Parameters and of the Vertical Structure of Devices by Numerical Analysis of Energy-Dependent EBIC Measurements 139

S. Heinemann On the Stationary Behaviour of Magnetic Flux Embraced by Asymmetric Two-Junction SQUIDs 153

L. G. Khvostantsev and N. A. Nikolaev
Electroresistance of Cerium at Phase Transitions under Hydrostatic Pressure up to 8.0 GPa. *P–T* Diagram of the Cerium Phase Transitions 161

W. Siegel, E. Ziegler, and G. Kühnel
Impurity Parameters of Undoped and Zn-Doped p-Type GaP with Two Acceptor Levels Determined by Hall Measurements 167

J. W. Tomm, Karin Herrmann, A. Jalyschko, A. E. Yunovich, and V. Ten
Direct Comparison of Photo- and Electroluminescence in $Pb_{1-x}Sn_xTe$ Diode Lasers 175

M. Balarin Zum Einfluß einer Temperaturabhängigkeit des Frequenzfaktors auf die Form und Breite von Glow-Peaks 181

U. Rensch and G. Eichhorn
Investigations on the Structure of MOCVD AlN Layers on Silicon 195

H. G. Schneider, H. Karge, and B. Weber
Characterization of Heteroepitaxial Silicon Layers on Sapphire by Rutherford-Backscattering (RBS) Analyses 201

S. Jetschke, H. Karge, and K. Hehl
Anisotropic Effects in N^+-Implanted $LiNbO_3$ 207

H.-H. Potthoff Velocities of Dislocation Groups in Very Thin Neutron-Irradiated Copper Single Crystals Measured by Slip Line Cinematography 215

U. Heinecke, C. Cruz, E. Wieser, and C. Bernhardt
Structural Transformations and Connected Changes of Magnetic Properties by Mechanical Treatment of $SrFe_{12}O_{19}$ 225

H. Klose, A. Mertens, R. Reetz, and Ng. Tang
Ion Beam Contacting of Silicon (I) 233

C. Ance and N. M. Ravindra
Departure of Hydrogen from a-Si:H. A Spectroscopic Study 241

R. Deblieck, G. A. Wiegers, K. D. Bronsema, D. van Dyck, G. van Tendeloo, J. van Landuyt, and S. Amelinckx
Clustering and "Phase Ordering" Transitions in Molybdenum (III) Sulphide, $Mo_{2.065}S_3$ 249

J. Teuho and K. Mäkinen
A Study of Non-Periodic Antiphase Boundaries in Long-Period Ordered $Au_3Zn(H)$ (I) 263

B. Cochrane and P. J. Goodhew
TEM Images of Faceted Bubbles and Voids 269

M. Anglada and F. Guiu
Temperature and Strain Rate Dependence of the Cyclic Flow Stress of Molybdenum Single Crystals 283

E. K. Polychroniadis and P. Delavignette
Accurate Characterization of Subgrain Boundaries in a TEM 291

C. Lacour, A. Dubon et P. M. Grangeon
Aspect aléatoire de la formation des composés par frittage réactif 309

I. Thurzo, K. Gmucová, J. Červenák, and T. Lalinský
GaAs–Anodic Oxide Interface Examination by Deep-Level Transient Current Spectroscopy 323

J. Betko and K.Měřínský
Geometrical Magnetoresistance Effect in Semiconductors with Mixed Conductivity. Determination of the Electron Hall Mobility in Semi-Insulating GaAs 331

B. Gillot and F. Jemmali
Electrical Conduction Mechanism in Al-, Cr-, and Zn-Substituted Magnetites over the Temperature Range from 77 to 1000 K 339

R. P. Bhardwaj, J. K. Quamara, K. K. Nagpaul, and B. L. Sharma
The Non-Isothermal Depolarization Current Studies in Kapton-H Thermoelectrets 347

K. T-Y. Kung, B. X. Liu, and M-A. Nicolet
Study of Ni–Nb System by Ion Mixing 355

K. N. Subramanian
Deformation Structure in LiF Single Crystals 361

L. A. Shebanov, V. Ya. Fritsberg†, and A. R. Gaevskis
Crystallographic Properties and Superconductivity of Solid Solutions of the $BaBi_xPb_{1-x}O_3$ System 369

S. I. Golovka, A. M. Gurvich, A. I. Kravchenko, V. V. Mikhailin, and A. N. Vasilev
Enhancing and Quenching of Intrinsic Luminescence and Characteristic Features of Calcium Tungstate Phosphorescence in the Presence of Lanthanides 375

A. U. Shareef, K. K. Saraf, and A. P. Srivastava
Spontaneous Electric Current from Polyvinyl Alcohol Induced by the First Heating Run 381

G. JUŠKA and K. ARLAUSKAS
Features of Hot Carriers in Amorphous Selenium 387

A. NIKLAS and W. JELEŃSKI
X-Ray Luminescence of YAG:Nd^{3+} 393

A. K. BATRA, S. C. MATHUR, and A. MANSINGH
Dielectric Dispersion in Triglycine Sulphate Single Crystals 399

H. B. LAL and O. P. SRIVASTAVA
Electrical Transport Studies in Copper Mercuric Iodide 405

S. V. MOHARIL A Phenomenological Model for Supralinearity and Sensitization in LiF-TLD 100 413

V. N. SEMKIN, R. M. VLASOVA, and G. YU. YASHIN
Absence of 3.39 μm Photoconductivity in the Quasi-One-Dimensional Organic Semiconductor $Cs_2(TCNQ)_3$ 421

S hort Notes

J. MAHY, J. VAN LANDUYT, and S. AMELINCKX
Electron Diffraction Evidence for Superstructures in $TaTe_4$ and $NbTe_4$. . . K1

K. ABDELMOULA, J. J. MÉTOIS, and G. NIHOUL
Reconstructed Surface Structure of (111) Gold Platelets, Observed by High Resolution Techniques K5

G. VAN TENDELOO and S. AMELINCKX
Real Space Evidence for the Occurrence of Microdomains in "Disordered" Cu_3Pd K9

B. P. BESPALOV and A. M. VAINRUB
Thermal Conductivity of Pressed Pellets of TCNQ Complexes. Manifestation of Amorphous Structure K13

A. HÜBLER and D. MARTON
Energy Dependence of the Si–SiO_2 Interface State Cross-Section Measured by the TSC Method K17

V. E. ANTONOV, T. E. ANTONOVA, I. T. BELASH, E. G. PONYATOVSKII, and V. I. RASHUPKIN
Superconductivity of Pd–Au–H Solid Solution K23

A. REMÓN, J. PIQUERAS, J. LLOPIS, and C. BALLESTEROS
Cathodoluminescence and Optical Absorption in 20 kV Electron Irradiated MgO Single Crystals K29

V. N. BRUDNYI, A. I. POTAPOV, YU. V. RUD, and M. SERGINOV
Electrical Properties of Electron-Irradiated GaAs and $ZnSiAs_2$ Solid Solutions K33

V. M. Mukhortov, Yu. I. Golovko, V. A. Aleshin, E. V. Sviridov, Vl. M. Mukhortov, V. P. Dudkevich, and E. G. Fesenko
Ferroelectric Phase Transition in Heteroepitaxial Films of (Ba, Sr) TiO_3 . . K37

D. A. Aronov and V. V. Rubinov
On the Minimum Coefficient of Light Absorption by Free Carriers under the Conditions of Exclusion in "Pure" Semiconductors with Non-Injecting Contacts K41

Y. Momose, M. Ohkubo, and J. Gunji
Interaction of Mechanically Rubbed Aluminium with Polar Organic Vapors Found by Measurements of Optically Stimulated Exoelectron Emission under the Influence of Applied Electric Potential K45

M. Ohta, T. Kanadani, A. Sakakibara, H. Yamada, and M. Yamada
Effects of Grain Boundary on Age Hardening and Reversion in Al–Zn Alloys K49

A. S. Popov, D. B. Kushev, and Sh. Chadraabal
A Model for Photoconductivity of Fe-Doped GaP K53

F. Găiseanu On the Diffusion Profile of Boron in Silicon at High Concentrations K59

P. V. Reddy and T. S. Rao
A Study of the Semiconducting Properties of Mixed Mn–Mg Ferrites . . . K63

J. Jimenez, M. A. Gonzalez, V. Carbayo, and J. Bonnafe
On the Thermal Evolution of the Photoconductivity in Bulk Fe-Doped Semi-Insulating InP K69

W. Kowalsky, H.-H. Wehmann, F. Fiedler, and A. Schlachetzki
Optical Absorption and Refractive Index near the Bandgap for InGaAsP . . K75

A. K. Sharma Thickness Dependence of the Thermoelectric Power of Tellurium Films K81

G. Vértesy, I. Tomáš, Z. Vértesy, and M. Balaskó
Temperature Hysteresis of Demagnetized Domain Structures K87

A. L. Gogolin and R. M. Imamov
Investigation of the X-Ray Scattering Intensity for the Laue-Case Diffraction under Total-External-Reflection Conditions K91

K. Biedrzycki, J. Chrzanowski, and B. Sujak
Some Electrical and Optical Effects Observed in Polycrystalline Ice at 200 K K95

Pre-Printed Titles

of papers to be published in the next issues of physica status solidi (a) and physica status solidi (b) A1

Systematic List

Subject classification: Corresponding papers begin on the following pages (pages given in italics refer to the principle subject classification):

1	*11*
1.1	*11*, *291*
1.2	*161*, *225*, *369*, *K95*
1.3	*71*, *115*, *249*, *263*, *K1*, K23, *K49*
1.4	*97*, 133, *263*, 269, *291*, *309*, *K1*, *K9*, *K37*
1.5	*139*, *195*, *201*, 355
1.6	*K5*, *K91*
2	*241*, *355*, *387*
3	*35*, *195*
5	*63*, *87*, 225
9	*K13*, *K59*
10	11, 97
10.1	*81*, *115*, *215*, 269, *283*, K49
10.2	241, *361*, 405
11	63, *121*, *133*, *207*, 215, *233*, *269*, K29, *K33*
12.1	81, 161, *283*, K49
13.3	323, *K17*
13.4	*167*, *181*, *375*, *393*, 413, K33, *K53*, *K69*
14	K95
14.1	*161*
14.2	71, *153*, *369*, *K23*
14.3	*167*, *331*, *339*, *K33*, *K63*
14.3.2	139, *323*
14.3.3	139, *381*
14.3.4	*233*, 347, *387*, *K41*
14.4	*347*, *405*
14.4.1	*399*, *K37*
15	*405*, *K63*, *K81*
16	*421*, *K53*, *K69*
17	*K45*
18	*127*
18.2	71, 127, *K87*
18.2.1	*51*
18.3	*225*
20	K95
20.1	*45*, *67*, *107*, 207, *K29*, *K41*, *K75*
20.3	*45*, *107*, *175*, *375*, *393*, *413*, *K29*

21	63, 71, 81, 115, 283, 355, K45, K49
21.1	51, 71, 87, 115, 127, 215, 263, 269, 355, K9
21.1.1	51, 87
21.2	K49
21.4	161
21.6	263, K5, K23
22	175, 181, 249, K1
22.1.1	121, K41
22.1.2	133, 139, 201, 233, 241, K17, K59, K91
22.1.3	387, K81
22.2	195
22.2.1	323, 331, K33
22.2.2	167, K53, K69
22.2.4	K75
22.4.4	67
22.5.2	45, 361, 413
22.5.4	405
22.6	K17, K29
22.8	107, 309, K33
22.8.1	207, 369, 375, K37, K63
22.8.2	225, 339, 393, K87
22.9	347, 381, 399, 421, K13

Contents of Volume 77 Continued on Page 429

Review Article

phys. stat. sol. (a) 77, 11 (1983)

Subject classification: 1 and 1.1; 10

Institute of Physics, Warsaw Technical University[1])

Reasons for Polytypism of Crystals of the Type MX_2[2])

I. Polytypism and Structure Behaviour of Inorganic Crystals: General Problems

By

B. Pałosz

Contents

1. Introduction

2. Organization of crystal structure of layered materials

3. Polytypism or polytypisms

4. Polytypism — basic definitions

5. Basic problems in polytypism

5.1 Reasons for creation of faults

5.2 Ordering of faults — creation of polytypes — theories of polytypism

5.3 Stability of polytypic structures

5.4 Physical properties of polytypic crystals

6. Summary

References

1. Introduction

Progress in investigations of the structure of materials is usually accompanied by discoveries of new phenomena and new physical properties of materials. Observation and discoveries of some phenomena is possible when specific experimental techniques or "tools" with high precision are used, other phenomena are discovered during detailed analysis of commonly known experimental observations or they may be predicted on the basis of theoretical considerations. New experimentally discovered phenomena are frequently, at the beginning, classified as "exceptional", valid for one compound, group of materials, etc. Further studies frequently show, however, that numerous "exceptional phenomena", which seem to be governed by new specific physical laws are, in fact, typical for materials. Moreover, most of them may be described by "classical laws" — they are governed by commonly known general physical laws.

The phenomena which were first regarded as exceptional and seemed unimportant

[1]) ul. Koszykowa 75, 00-662 Warszawa, Poland.

[2]) Dedicated to the memory of Prof. K. Boll-Dornberger.

for the formulation of general physical theories are for example: liquid crystals, modulated structures, order–disorder structures and others; they were investigated extensively in recent years.

The phenomenon which seems to be typical for solids, but is usually regarded as a phenomenon exceptional for solids, is structural polytypism of inorganic crystals. Polytypism, being modification of polymorphism, was first found by Baumhauer in 1912 [16] and until about 1940 was presented as a phenomenon valid for SiC only. Real progress in investigations of polytypism of inorganic crystals began in 1951 when Frank presented the screw dislocation theory of crystal growth. According to this theory polytypism should appear in crystals as typical growth phenomenon when a screw dislocation mechanism operates during crystallization. Theoretical predictions of the theory of Frank were partly confirmed in numerous experiments performed for various materials, e.g. CdI_2, PbI_2, SnS_2, SiC, etc. The appearance of numerous structural modifications, both periodic and disordered, in crystals is not understandable from the point of view of classical crystallography: this difficulty in explaining the reasons for polytypism was circumstantiated in the theory of Frank by regarding the polytypic modifications of structure of crystals as "faulted" structures formed by a destruction of a given correct structure.

From classical crystallography it follows that solids should form crystals with a lattice periodic in three dimensions: minimal structural elements of crystals should be unit cells repeated periodically in three dimensions. A new general theory of construction and description of structure of inorganic crystals was presented in 1956 by Dornberger-Schiff [9] — the theory of OD structures. This theory is based on the assumption that the structure of crystals is constructed of two-dimensional structural elements, layers, being structural units. According to OD theory disordering in an arrangement of layers should appear in some crystals, moreover ordered structures (they may be described basing on classical crystallography) and disordered ones may be, in some cases, regarded as equivalent structures. The method of interpretation of the structure of crystals presented in the OD theory was not, however, used in investigations of polytypism: the theory of OD structures and theories of polytypism were developed independently one from the other since now. In 1966 the first complex review of polytypism of crystals was presented by Verma and Krishna in [17]. In this book (practically a handbook) experimental data on polytypism of about ten various materials were summarized and the theories proposed to explain reasons for polytypism were discussed. The number of materials known now as polytypic is a few times higher than 15 years ago. The frequency of appearance of the term "polytypism" in literature increases continuously but, it should be stressed that this term is used for notation of a number of phenomena which are, in fact, phenomena differing from classical polytypism defined for SiC. One may say that numerous modifications of polytypism occur in inorganic crystals, while classical polytypism is observed in a relatively small number of crystals of compounds. Observation and discovering of polytypism of single crystals is possible by using typical experimental techniques of X-ray and electron diffraction analysis but, in some cases, it may be not observed when standard procedure of structural analysis is used, especially when polytypism is not expected. Discovering polytypism in powdered materials is very difficult and, it seems that in materials which do not exist in the form of single crystals polytypism if any may not be discovered.

The main purpose of investigations of polytypism developed extensively in recent years is the determination of the laws governing the structural behaviour of polytypic crystals. The problem of reasons for polytypism of crystals might be resolved uniquely by determination of quantitative relations between some physical parameters and structure of polytypic crystals. Most of investigations performed in the field of polytypism have qualitative importance while quantitative calculations are performed occassionally only.

It is clear now that one definition of polytypism and one unique theory of polytypism do not exist. The problems appearing in investigations of polytypism may be divided into two main groups: (i) general problems: geometrical rules of organization of layered structure of crystals and mechanism of growth of crystals, (ii) detailed problems characteristic of different modifications of polytypism: stoichiometric crystals, doped crystals, alloys, etc., (iii) detailed questions concerning structural behaviour of crystals of one material: conditions of polytypic growth, stabilization of crystal structure, applications, etc.

The unusual structural behaviour of polytypic crystals — if compared to classical polymorphs — which is not understandable from the point of view of classical crystallography cannot be uniquely described by using definitions of classical theories of solids. For dozens of years of studies on polytypism of crystals numerous special methods were elaborated for description, structural analysis, and measurement of various effects observed in polytypic crystals. One may say that a special "vocabulary" is used for the presentation of polytypism. A consequence of this specialization is that polytypism is a phenomenon not commonly known and may seem not well understandable for the whole of physicists. Some problems of those which are usually presented as "problems of polytypism" may be directly defined on the basis of commonly known definitions and laws used in solid state physics. Translation of "polytypic problems" into the language commonly used for the description of physical phenomena seems to be important for further progress in the explanation of polytypism.

One of the purposes of this review is to present polytypism as the phenomenon typical for inorganic crystals with emphasis on the problems important for some general theories of solids. This review is divided into two parts: Part I concerns general problems of polytypism of various materials, in Part II polytypism of crystals of the type MX_2 — compounds CdI_2, PbI_2, $CdBr_2$, SnS_2, and crystals of the system Ti–S — is discussed in detail. Thus, while in Part I the problems which should be solved for an explanation of polytypism are formulated, in Part II "tools" used in the analysis of the phenomenon are presented in the case of MX_2 crystals.

2. Organization of Crystal Structure of Layered Materials

A great number of data on the structure of crystals brings about the need for continuous development of existing methods for description and classification of structure [1 to 8]. Crystals of periodic structure (perfect crystals) may be described uniquely by lattice constants and the positions of atoms in a unit cell. However, such a procedure for the description of a structure is not suitable in general. Besides a unit cell, for each structure one can define a number of different structural units as the basic elements. For a crystal chemist, for example, the most important information is the spatial arrangement of atoms (individual atom coordination) and not the lattice constants [8]. The classification of structure of inorganic crystals based on various structural units — atoms, molecules, sheets, frameworks, etc. — was developed by Lima de Faria and Figueiredo [5, 7]. It is clearly seen from this classification that most of inorganic crystals (both elements and compounds) may be described as composed of layers. The classification proposed by Lima de Faria is based on the analysis of known (identified) exactly periodic structures and classifies them with regard to structural units, coordination polyhedra of elements, etc. A more complete classification of the layered structures of materials was proposed by Dornberger-

Schiff in 1956 [9] and is developed continuously as the OD theory of solids [10 to 12]. The problem of organization of crystal structure is in the OD theory the ordering of two-dimensional structural units — packing of layers (similarly as in the classification of Lima de Faria). The classification proposed by Lima de Faria and Figueiredo [5, 7] is very useful for the analysis of stability conditions of a periodic arrangement of layers: Lima de Faria found some basic rules governing the construction of periodic layered structures [13]. However, a number of real crystals exhibit a certain disordering of the crystal lattice. For layered structures the typical deviation from periodicity of the lattice is one-dimensional disordering in the periodic arrangement of layers. The appearance of disordering (creation of stacking faults) is predicted by the theory of OD structures. From the OD theory it follows that by assuming (i) a given set of layers (layers may be composed of a number of different elements) as basic units and (ii) a certain type of interaction between the units: the resulting structure composed of layers is not in every case periodic but, in some cases is expected to be non-periodic. The tendency for disordering of OD structures is a direct consequence of the energetical equivalence of different arrangements of a given set of layers. It is evident that the term "unit cell" looses its sense in the case of disordered structures. The description of non-periodic structures requires the introduction of some specific parameters for the notation of disordering. A special system of notation for OD structures was elaborated by Dornberger-Schiff [10 to 12]. The system is rather complicated and is not commonly used by physicists but by a not numerous group of crystallographers collaborating with Dornberger-Schiff.

The results of theoretical investigations of OD structures are fully confirmed by experiment. A tendency to form disordered structures, predicted according to the theory, was confirmed for a number of materials, elements and compounds [11, 12]. One may expect that an increasing number of crystals for which classical forms of classification and notation are not satisfied will cause, in consequence, growth of interest in the concept of OD structures. The success of the OD theory in explaining of disordering in some layered materials confirmed fully the assumptions of the theory. It seems to be baseful to focus the attention of the physicists on the fact that a number of materials have a natural tendency for disordering. Thus, the structure of new synthetized materials should be initially analyzed on the basis of OD theory. It seems nearly sure that a number of materials form crystals with stacking faults but they are not observed because of the lack of a detailed analysis of their structure (typical examples of such "hidden" structural behaviour of materials are modulated structures, which can be easily discovered, although not during standard structural analysis [14, 15]).

Structure and behaviour of materials with stacking faults were investigated extensively in recent years. The main reason for the interest in materials with disordered structure is the influence of stacking faults on physical properties of crystals. Usually, stacking faults appear only incidentally, disturbing a given (basic) crystal lattice in a random manner, e.g. crystals of metals, alloys, silicon, etc. In crystals of not numerous group of materials, stacking faults appear systematically exhibiting a tendency for ordered (periodic) arrangement of stacking faults within the basic lattice. The OD theory gives us the possibility to classify inorganic crystals into such in which stacking faults may form and those in which they are geometrically forbidden. However, it is known that in real crystals the ability for creation of stacking faults is realized in some of the isostructural crystals only.

The application of the phenomenon of the creation of stacking faults needs the determination of the rules governing the process of generation of these faults. This problem is now the subject of a separate branch of solid state physics — polytypism.

The term "polytypism" appeared nearly half a century before the OD theory was formulated; in the beginning this term was used to denote a special kind of polymorphism — in one dimension only — which was first observed in crystals of SiC in 1912 by Baumhauer [16]. From the point of view of the present state of investigations of the structure of crystals it is clearly seen that the general problem of the organization of the structure of layered materials (including polytypic) is contained in the OD theory of solids. Nevertheless, here conservation of the term "polytypism" is intentional because of the historical aspect as well as because of the extraordinary variety of different structural modifications observed in polytypic crystals. Moreover, the subject of the investigations of polytypism is not only the formulation of the geometrical rules governing the construction of structure but, first of all, it is the analysis of methods for controlling the process of formation of polytypic structures.

A great number of structural modifications of polytypic crystals required the introduction of a special vocabulary for polytypism, of new forms for notation and classification of structures and development of methods of structural analysis. In 1966 the first, fundamental work (practically a handbook) of Verma and Krishna on "Polymorphism and Polytypism in Crystals" appeared [17]. However, 15 years since this book was published the experimental data presented here are to a high degree not up-to-date. The number of materials which are known to be polytypic and the number of modifications of earlier known polytypic materials are far larger than in 1966. After 1966 a number of review articles has appeared; the subject of one of these articles is usually one polytypic material or a selected aspect of investigations of a given group of polytypic crystals, e.g. SiC [18, 19], metal dichalcogenides [20], ZnS [21 to 23], CdI_2 [22 to 27], alloys [28], micas [29], etc.

The general question of polytypism "why does polytypism appear in crystals" has not been answered satisfactorily by none of the numerous theories of polytypism presented up to date. In this review the main problems and questions which appeared in earlier studies of polytypism of crystals are formulated. In Sections 3 and 4 the definitions of "reasons" and "results" analyzed in investigations of polytypism are presented; a detailed study of literature on polytypism has shown that frequently "reasons" and "results" are confused and, in consequence, a number of problems

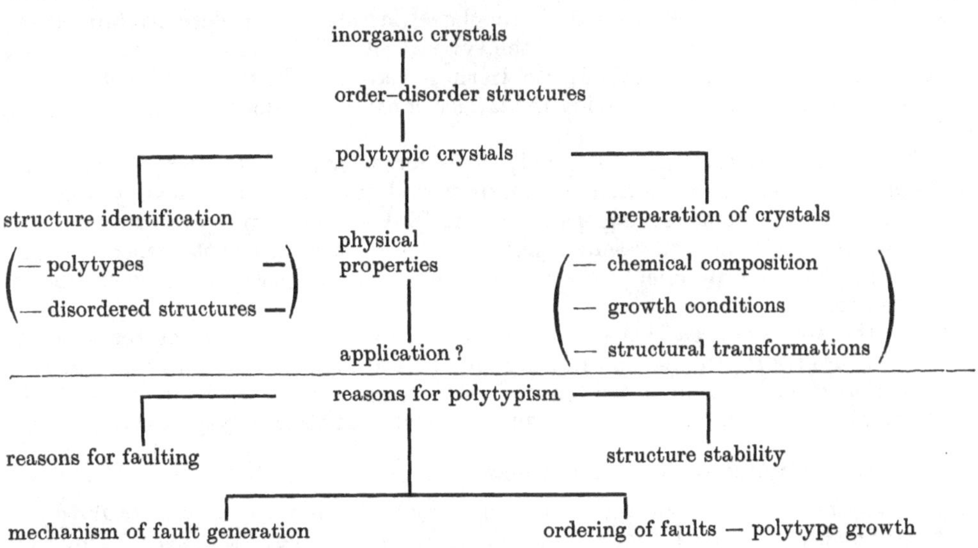

Fig. 1. Diagram of basic problems in the investigations of polytypism of inorganic crystals

appearing in polytypism are not well defined. The general problems of polytypism and some "physical aspects" of the phenomenon are discussed in Section 5 on the basis of the definitions presented in Sections 3 and 4. A diagram of some basic problems in the investigations of polytypism is presented in Fig. 1.

3. Polytypism or Polytypisms

The classical definition of polytypism was formulated by Verma and Krishna [17]: "The ability of a substance to crystallize into a number of different modifications, in all of which two dimensions of the unit cell are the same, while the third is a variable integral multiple of a common unit". The correct and consistent use of the above definition is irremissible for a clear formulation of problems which must be resolved to understand the phenomenon itself. The consequences of the above definition are that:

1. polytypic modifications should be composed of identical structural units, layers, and may differ in stacking of layers only;
2. each layer should have the same chemical composition, equal to the stoichiometry of the crystal;
3. each pair of adjacent layers is equidistant.

Thus, it is evident that "pure polytypism" is possible only in crystals with well defined chemical formula. However, the term "polytypism" is frequently used for description of phenomena similar to polytypism but not fulfilling the above definition. The differences in "polytypisms" lie mainly in the interpretation of reasons (rules) of polytypism. Here we present some examples of the use of the term "polytypism":

1. Classical polytypism — SiC, ZnS, CdI_2 [17], GaSe [30], etc. In these materials constructive layers: Si–C, I–Cd–I, Ga–Se–Se–Ga, may occupy several equivalent positions — individual layers (of elements) forming the sequence are packed in close manner in SiC and CdI_2 and in other way in GaSe. Constructive layers are close packed — individual constructive layers differ in orientation with respect to the given crystallographic axis. Nearest coordination of individual atoms is the same for the whole crystal.
2. Stoichiometric crystals with two types of nearest coordination — TaS_2, $TaSe_2$, MoS_2, etc. Coordination of metal atoms may change from octahedral to trigonal prismatic [20].
3. Non-stoichiometric crystals with the basic lattice of the close packing type: Si–Al–O–N [31] (the basic lattice is of the type of SiC), Ti–S [32] (with basic lattice isostructural to CdI_2), alloys [28, 33], etc. In these materials the structure of crystals — stacking of layers of the basic lattice and lattice constants — depends on the chemical composition of crystals.
4. Magnetic materials, e.g. Fe–S [34]. The ferro- and antiferrimagnetic "polytypic modifications" differ in their magnetic period but not in the stacking of layers.
5. Non-layered structures, e.g. ganophyllite [35] — basic units are columns. The columns have the same equivalent construction and may shift with respect to each other in the direction parallel to the column axis. The basic modifications are monoclinic and triclinic.

From the above comparison it follows that in every case when the term "polytypism" is used some further information on the phenomenon should be given. It is evident that the reasons for stoichiometric polytypism must differ from those for the non-stoichiometric one or from those due to which magnetic polytypism appears.

4. Polytypism — Basic Definitions

Layers The basic unit of polytypic structure may be composed of a number of simple, monoatomic layers. In such composite layers (constructiv layers) the layers of individual elements may be stacked in many ways. In polytypic crystals the stacking of constructive layers is in every case of a close packing type; the basic lattice of a crystal is composed of hexagonal layers:

1. ZnS, SiC — the structure is based on the close packed lattice of hexagonal layers of anions (e.g. sulphur); metallic layers (Zn) are hexagonal layers, too, and are placed in tetrahedral voids between the anion layers. The structure of a crystal is described uniquely by a sequence of layers of one type [17].

2. GaSe — the basic layer is composed of two Ga and two Se hexagonal layers in the arrangement Ga–Se–Se–Ga. This arrangement possesses a symmetry plane dividing the slab: A–β | β–A, A–γ | γ–A, etc. into two equivalent mirror parts [30]. For a given sequence of metallic layers the positions of anion layers may be different.

Thus, the condition of close packing, which is contained indirectly in the definition of polytypism (see Section 3) should be fulfilled by basic, constructive layers and not by individual layers of elements.

Stacking faults Frequently polytypism is presented as a process of generation and ordering of stacking faults in crystals. If we take into account that stacking faults are usually regarded as a "damage" of the proper structure (similarly as defects, dislocations, etc.), then polytypism will appear simply as the result of desctruction of a good crystal. Together with the introduction of a stacking fault into a crystal structure some irregularities should appear in the lattice: lattice constants should change, local symmetry should be destroyed, etc. This is, however, not a correct description of faulting observed in polytypic crystals.

Stacking faults represent arrangements of layers different from those assumed as the correct or faultless ones, but it should be stressed that not one arrangement only can possibly be assumed as faultless. Detailed analysis of stacking faults as defects in close packing of layers was presented by Hirth and Lothe [36]. These faults, defects are observed in various materials (both elements and chemical compositions) and appear incidentally under not well defined conditions. Stacking faults as changes in the stacking of layers in polytypic crystals appear systematically, frequently under well-defined conditions. Stacking fault defects are created together with dislocations and may appear with twinning etc., while fault arrangements of layers formed in polytypic crystals may pass through the whole volume of a crystal, i.e. without generation of dislocations. Thus, the polytypic structure may fulfil the condition of close packing at every point of a crystal. It should be noted that polytypism is observed in crystals of mixed chemical composition (e.g. compounds, alloys), while stacking fault defects are rather characteristic of crystals of elements, e.g. metals, silicon, etc. For materials which possess one well-defined structural modification as basic, the term "stacking fault" is adequate for each arrangement of layers different from the basic one. If several modifications form simultaneously with similar probability, then the definition of the given arrangement as "fault" is not in every case unique. In this paper the term "fault arrangement" and not "stacking fault" is usually used to present polytypism.

Formation and motion of dislocations in crystals during crystallization and phase transformations were presented frequently as the "creative" factors in polytypism (see Section 5.2). It seems useful to separate the reasons for creation of fault arrangements of layers in polytypic crystals and the mechanism by which stacking faults are introduced to the lattice. Fault arrangements of layers may be regarded as "correct structure", whereas dislocations may accompany the process of creation of various arrangements of layers during crystallization or transformation of crystals.

Polytypes and disordered structures. Polytypic structures may be divided into two groups; (i) polytypes — structures with well-defined unit cells; period in the direction perpendicular to the layer plane is determined by the number of layers forming the repeated arrangement of layers: $c = nc_0$, n number of layers within a period, c_0 minimal interlayer distance, and (ii) disordered structures for which stacking of layers is not repeated periodically. The structure of each polytype is described uniquely by the sequence of layers within a period, for disordered modifications only approximate characterization of a structure is possible. In the case of disordering, the structure may be characterized by a certain set of parameters, e.g. types of faults, density and distribution of a given type of faults, etc.

The classicication of polytypic structures for polytypes and disordered structures is, however, not necessary. Typical methods for the identification of polytypes are based on X-ray or electron diffraction methods [17]. These methods make it possible to observe interference maxima as separated reflexions of up to a few hundred layers in a unit cell. For higher polytypes the overlapping reflexions form continuous lines along the c^*-direction observed on photographs of the reciprocal lattice of a crystal [37]. Thus, the disordered modifications may, in fact, be regarded as periodic structures and are not observed experimentally because of the above experimental limitation. An application of the "lattice imaging" technique with the use of high resolution electron microscopy for layered materials made possible the direct observation of the stacking of layers in a number of materials, e.g. SiC [38, 39], Friauf-Laves phases [40], etc. However, this method is limited to a crystal thickness of up to few dozens nm. The real period in a bulk material may be far higher.

The difficulties in the identification of polytypes rise rapidly with the increase of the number of layers in a unit cell. Identification of polytypes composed of more than 100 layers in a unit cell appear to be an entirely separate problem. The satisfaction of solving the problem of identification of a polytype by the proper choice of one from thousands of possible hypothetical polytypes is, certainly, very high but the importance of such detailed information for polytypism is rather problematic.

Hundreds polytypes of various compounds were identified but disordered structures were analyzed occassionally only. This seems a little strange if we take into account that polytypes appear in a low number of crystals while disordered structures form the main group of polytypic modifications for a number of polytypic materials.

5. Basic Problems in Polytypism

The basic questions formulated in investigating polytypism are related to the possibility of controlling the polytypic structure of crystals. It is usually believed that polytypism is not a "quantitatively governed phenomenon", thus, it is impossible to predict the structure of a growing crystal. The question of controlling polytypism was, in fact, understood as the question of controlling the stacking of layers within a period of polytypes. This seems an improper way of understanding polytypism. Polytypism is not only the creation of polytypes, but is a very complex problem of the organization of the structure of layered materials (see Section 2). The problem of formation of polytypes of a period of hundreds of Ångstroems is, certainly, fascinating but is not representative for the phenomenon and seems to be not the most important subject for research and applications. Most papers on polytypism are original papers in which only some aspects are discussed, but systematic investigations and general problems are presented occassionally, however. The present state of investigation of the phenomenon can be considered collection and classification of facts. In spite of a great number of information on structure of polytypes, dislocations, physical behaviour, etc. our knowledge about the reasons and laws governing polytypism is not satisfying. To understand polytypism systematic investigations should be performed. Some problems are of greatest importance — a tentative list of the basic problems of polytypism is discussed in Sections 5.1 to 5.4: 5.1 — reasons for creation of fault arrangements of layers; 5.2 — reasons for ordering, generation of polytypes and disordered structures, mechanism of growth; 5.3 — stability conditions of polytypic structures, role of growth conditions and chemical composition of materials on a structure of polytypic crystals; 5.4 — relation between structure of crystals and their physical properties.

A theory of polytypism should resolve the above problems. In particular it should explain why polytypism appears only in some of the isostructural crystals of different compounds. Thus, while the OD theory makes possible to select the inorganic structures which may exhibit polytypism, the theory of polytypism should enable us to predict in which of them polytypism will appear.

5.1 Reasons for creation of faults

In elaborating the problem presented earlier, stacking faults were not clearly distinguished from "defects" and "fault arrangements of layers" being not defects. In a number of papers the role of "stacking faults in the growth of polytypic structures" was indirectly assumed equivalent to the role of structural defects of the type stacking faults. As discussed in Section 4 this immediately suggests that polytypism is not the phenomenon which is governed by quantitative physical laws (except, perhaps, for statistics). Such an assumption does not seem profitable for further studies of polytypism. If possible, regularities in arrangement of layers in polytypic crystals should be found and it seems that it should be assumed that faults appear in the lattice free of defects and dislocations.

5.1.1 Description of faults

In the beginning of the analysis of the reasons for faulting, sequences free of faults must be defined. The polytypes, which appear most frequently in polytypic crystals (usually two or three) are assumed as the most stable modifications, e.g. 2H and 3C in ZnS, 2H and 4H in CdI_2, 4H, 6H, 15R in SiC, etc. Faulting is usually defined by comparison of a given faulted sequence with one of the basic sequences. Thus, the same arrangement of layers may be defined as "faulty" in more than one way. Polytypic structures may be described in a number of symbolic notations, e.g. ABC, Hägg, Zhdanov, hc, etc. — these notations are described in [17]. Consequently, faulting may be defined on the basis of each of these notations. For analysis of faulting appearing in polytypic crystals the choice of a given notation is very important — for the analysis of regularities in the stacking of fault layers and for the analysis of the mechanism of their introduction into the structure. The method of description of faulting suggests the interpretation of the reasons (mechanism) for faulting. Below, we present some examples of simple faulting in the ABC, Hagg, and hc notations:

ABC sequence: A B C A B C | ↓ ↓ ↓ ↓ ↓ B C A B C | ↓ ↓ ↓ ↓ B C A B ...

creation of faults may be interpreted as a set of shifts of layers in the basic sequence ABCABCA ... (designed by arrows).

Hägg sequence: + + + + + $\overset{\downarrow}{-}$ + + + + + $\overset{\downarrow}{-}$ + + + ...

faults appear due to a change in the direction of stacking of atoms on (+) or (—) lines (see zigzag sequence [17]).

hc sequence: c c c c [h h] c c c [h h] c c ...

faulting causes the introduction of layers in the arrangement different from the basic one (here c is basic, h is fault).

Notations of faulting are usually chosen depending on the mechanism proposed by the authors: the ABC and Hägg notations are usually used in dislocation theories of growth, hc notation for thermodynamic considerations [17]; the problem of notation of faulting is discussed in detail in Part II of this paper, dealing with MX_2 structures.

Most information on the fault arrangements of layers existing in real crystals is derived from the analysis of the organization of the structure of polytypes, i.e. structures representing ordered arrangements of layers. This is because the analysis of disordering — disordered structures seem to be more convenient as a source of information on faulting — is rather complicated and time consuming and, detailed studies of disordering were performed occassionally only, e.g. [41 to 51]. Description of disordered structures needs the use of a number of parameters and such a description contains in every case approximate information on the "average structure"; identification (description) of polytypes is far easier and acceptable. Using classifications performed for polytypes it should be remembered that conclusions about faults formed for polytypes may appear not complete (representative) for a description of the structural behaviour of polytypic crystals; polytypes can appear in crystals only in the case when a given faulted sequence fulfils the geometrical conditions of construction of a unit cell (cf. Section 5, Part II), while such a limitation is not valid for disordered structures.

5.1.2 Classification of faults

Two ways to introduce faults in a crystal structure are possible: (i) faults are created during crystallization — growth faults and (ii) faults appear by a change of an initially formed basic sequence in a crystal — deformation (transformation) faults. Such a classification was first proposed for stacking fault defects, e.g. [36], and was frequently respected for dislocation theories of polytypism [18, 19, 21, 26, 41, 52 to 54]. The above-mentioned division of faults is, of course, simple and very suggestive but is not suitable (and frequently not true) for faults in polytypic crystals. In the simplest monoatomic structures of close packed hexagonal layers two different arrangements of layers are only possible — h and c. In every case one of them is "correct", the second is "faulty". In complex polytypic structures, when neither c nor h represents the basic structure, the problem of faulting is far more complicated — the hc notation does not ensure a unique classification of faults. If several sequences are acceptable as the faultless ones, the problem of a choice of one as basic to calculate the faulting cannot be resolved in a unique way. Thus, it seems speculative to decide at which stage the given fault was introduced in a crystal merely on the basis of geometrical analysis of the "destruction" of a given, faultless structure.

The problem of determining the stage at which faults form is not current for phase transformations. In this case faults are (i) created, (ii) rearranged, or (iii) reduced in number as a result of transformation. Any transformation takes place in solid state and is, in every case, accompanied by rearrangements of layers or parts of layers.

Two groups of factors should be analyzed as possible reasons for faulting: (i) faults appear incidentally due to any disturbance in equilibrium growth (and so, under unspecified conditions), (ii) creation of faults is necessary to stabilize the structure. Determination of reasons for faulting created during crystallization cannot be, in general, unique: the faults formed during structural transformation may be regarded as the arrangements of layers formed according to thermodynamics — in order to form a more stable structure.

From the analysis of regularities in the arrangement of layers in polytype cells it was found that some of fault arrangements geometrically possible for a given basic lattice appear in crystals far more frequently than others [19, 26, 52, 55]. (Possibly similar regularities occur in disordered structures.) The fault sequences appearing most frequently in a given crystal may be regarded as the faults energetically favourable — among those geometrically possible. The problem of stabilization of faults may be qualitatively analyzed by using the concept of stacking fault energy (s.f.e.). Calculation of the parameter s.f.e. was proposed first by Hirth and Lothe [36] to estimate the internal energy of stacking faults in close packed structures. This concept of s.f.e. was accepted and developed for polytypic structures by Tiwari et al. [56, 57] and used for identification of polytypes [58, 59]. Values of s.f.e. of a structure — which may be assumed as a measure of the internal energy of faulting — are estimated comparing a given stacking of layers with one assumed as basic. The number of pairs of layers separated by 1, 2, 3, ... layers in arrangements differing from those in the basic stacking is a measure of s.f.e.. It was found that polytypes are characterized by the lowest value of s.f.e. possible [19, 26, 58, 59]. This leads to the important conclusion that faults are introduced in crystals not exactly incidentally, i.e. faulting causes an increase in the internal energy of a crystal as low as possible. This fact suggests that the process of faulting is indeed controlled by thermodynamics. The s.f.e. theory does not explain the reasons for faulting but makes it possible the predict the results of faulting, giving specific "rules of choice" for faulting. Note that this is possible only in the case when faults are not energetically equivalent. It seems that the s.f.e. concept may in future establish the basis for classification of faults in polytypic structures.

5.1.3 Stacking faults in theories of polytypism

Various classifications of faults are possible by assuming that polytypic crystals grow by a dislocation mechanism. The generally acknowledged concept for generation of faults, and of periodic ordering of faults, was proposed by Frank [60]. According to this theory, faults appear during the growth process due to disturbances in the front of crystallization. Stacking faults originate from the parent dislocation through buckling and slip (cracking) of an otherwise perfect thin platelet. Presence of dislocations is, in the theory of Frank, the reason for the creation of faults and for polytypism. This mechanism itself, however, cannot be accepted as the reason for generation of only some fault arrangements in crystals and not as the reason for polytypism. The main discrepancies between predictions of the theory of Frank and experiment are as follows: (i) all the faults possible to form through dislocations should appear with similar frequency in crystals, in real crystals some of the theoretically predicted faults (hypothetical) are observed, (ii) faults (and polytypism) should appear in each crystal isostructural to a polytypic crystal, moreover, polytypism should appear in crystals of elements, (iii) a number of faults which are observed in crystals are not predicted by this mechanism, e.g. polytypes formed when the polytype 2H is the parent structure; hexagonal and rhombohedral polytypes are formed when the basic is polytype 4H, etc. [17, 61].

The epitaxial theory of Vand and Hanoka [62] is based on the assumption that the reason for polytypic growth of crystals is the appearance of a foreign nucleus on a front of crystallization; polytypes form by epitaxial growth due to the nucleus. However, epitaxial growth does not explain the appearance of only some faults in the real crystals. Thus, while in the theory of Frank well defined arrangements of layers should appear during the spiral growth, in the case of epitaxial growth no regularities are predicted.

The dislocation model of polytypic growth of crystals was used successfully to explain the formation of polytypes in ZnS and $Ti_{1.2}S_2$ crystals grown from vapour: Mardix et al. suggested the mechanism of periodic slip process for growth of polytypes of ZnS, when stacking faults were generated due to thermal fluctuations or mechanical stresses in the initially perfect crystals [63 to 66]. They found that the angles of tilt calculated theoretically agree very well with those measured experimentally for ZnS [63, 64]. Similar correlations were observed in $Ti_{1.2}S_2$ crystals [67, 68]. It was found also that the dislocation structure observed on the surface of crystals may be related to the polytypic structure of crystals assuming that the structure is formed during growth or through structure transformations [18, 25, 69 to 72]. The relations between dislocations and stacking faults found experimentally, however, cannot confirm the assumption that stacking faults appear because of dislocations. It seems that the growth process as well as a phase transformation may force the creation of dislocations — it is possible that the reason for dislocations is the formation of stackings of layers, faults, favourable energetically in given conditions. The phase transformations going on in crystals during annealing are, certainly, governed by thermodynamics, thus it seems to be clear that transformations cannot depend merely on the presence of dislocations in the lattice.

Dislocation theories seem to be a kind of "geometrical throught experiments", with no well-defined physical assumptions. Thus, it is evident that a number of quantitative relations observed in real crystals between structure and thermodynamic conditions of their formation cannot be explained together with the assumption of the "creative" role of dislocations in polytypism of crystals.

In the thermodynamic theory of Jagodzinski [73] the problem of reasons for faulting was not clearly formulated.

From the above it follows that the problem of reasons for faulting in polytypic crystals and a lack of polytypism in a number of crystals isostructural to polytypic crystals was practically omitted in the "classical" theories of polytypism. Several problems seem to be most important to solve before the general theory of polytypism will be formulated: (i) mechanism of generation of layers in arrangements which are not basic, (ii) rules governing the frequency of occurrence and distribution of faults in crystals, and (iii) stabilization of faults in a crystal lattice. The ordering of faults in periodic arrangements seems to be of secondary importance for an explanation of polytypism (cf. statistical theories, Section 5.2.3).

A separate problem is the formation of faults in non-stoichiometric crystals. If faults appear together with a change of chemical composition (and only in this case) then the reasons for faulting may be directly connected with the presence of "foreign" particles in the given basic lattice free of faults for stoichiometric composition, e.g. Ti–S, see Part II. The term "stacking fault" seems to be non-adequate in the case of mixed crystals: a change in the arrangement of layers may be related to the creation of a chemically different phase in the given parent material. This phase (the given arrangement of layers) may be characterized by its own internal structure having interatomic distances different from those of the basic lattice — the additional atoms may form their own coordination polyhedra, etc. Thus, it seems that the reasons for faulting and the mechanism of generation of faults in stoichiometric, non-stoichiometric, and in mixed crystals should be analyzed separately (c.f. Section 5.3.2). Pure polytypism is possible in stoichiometric crystals of compounds only (cf. Section 3).

5.2 Ordering of faults — creation of polytypes — theories of polytypism

Most of the existing theories of polytypism are simply theories of polytype formation: disordered structures are regarded as destroyed (faulted) polytypes. Nearly all the results obtained in the field of polytypism were limited to two main theories of polytypism: the dislocation theory of Frank [60] and the disorder theory of Jagodzinski [73]. These theories were analyzed in a number of original and review papers, e.g. [18, 19, 25, 26, 74]. In recent years several reports appeared on a theory of polytypism based on statistical considerations [75, 76].

5.2.1 *Dislocation theories*

The dislocation theory was first presented by Frank in 1951 [60]. A number of dislocation theories developed later are based on the assumption of Frank: formation of polytypic structures is attributed to the spiral growth mechanism. Polytypes, in these theories, may be created at any stage of crystallization by growth of a crystal around the screw dislocation. Thus, polytypism is a surface phenomenon; structure of a polytype is related to the pitch of the parent screw dislocation while the unit cell height of a polytype is equal to the Burgers vector of a dislocation for hexagonal structures and is three times larger for rhombohedral polytypes. Various modifications of dislocation theory differ in the definition of the origin of the parent dislocation and the mode of generation of faults. Frank assumed that a dislocation may originate through buckling and slip of a perfect thin platelet of the parent crystal; according to Vand and Hanoka [62] growth of a polytype is initiated by nucleation of epitaxial growth of layers on a front of crystallization; Rai [74] and Dubey [52] assumed that the structure of a polytype is determined by the initial seed, a microtwin appearing on the crystal surface.

To overcome the difficulties in the application of the theory of Frank, Pandey, and Krishna proposed the faulted matrix model of spiral growth of polytypic crystals [19, 26, 53]. They considered separately the generation of faults and the mechanism ordering them.

Most recently Kuhlmann-Wilsdorf et al. [54] suggested that polytypism can be caused by giant steps generated due to a radial impurity gradient in the crystals.

Dislocation theories may all be assumed as general theories of a mechanism of growth of layered materials. Thus, there are no reasons to assume the dislocation theory as valid for polytypic crystals only, so, according to dislocation theories, polytypism should appear in each crystal isostructural to a polytypic crystal and in every case when the structure of a crystal is of OD type, cf. Section 2.

The assumption that the structure of the spiral ledge (structure of a polytype) depends on dislocations only led many investigators in the past to an incorrect interpretation of reasons for polytypism in various materials, e.g. $CdBr_2$, PbI_2 [77, 78]. It is clear now that the problem of the reasons for polytypism of crystals cannot be resolved completely assuming that polytypism is a growth phenomenon.

5.2.2 *Disorder theory of Jagodzinski*

The theory of Jagodzinski which is based on equilibrium thermodynamics, is usually presented as counterpart to the theory of Frank. Jagodzinski assumed that each sequence of hexagonal layers in close packing has the same value of the internal energy. The stable polytypic structures should have, according to equilibrium thermodynamics, minimum free energy, so they should have maximum entropy. The stability of polytypic modifications depends, according to the theory, on the total entropy of the structure, consisting of configurational and vibrational parts. Assuming that configurational entropy increases with an increase of the number of faults in crystals and simultaneously vibrational entropy decreases, Jagodzinski found that the most stable polytypic modifications are: polytypes of exactly periodic arrangement of layers and some specific disordered structures with well defined degree of disordering. This theory is based on the qualitative analysis of simple thermodynamical rules, however, it is slightly speculative for polytypism: (i) faultless sequences are not defined uniquely — they may be chosen with no limitation, (ii) disordering may be measured in various ways — the value $\alpha = 0.12$ suggested by Jagodzinski cannot be assumed as the value characteristic of all crystals [79]. It is possible that by a selection of conditions (i) and (ii) each polytypic structure — polytypes and disordered structures — may be shown to be energetically favourable assuming the criteria of the theory of Jagodzinski. It should be stressed that reasons for faulting are not well defined in disorder theory, thus this theory should be regarded as general theory of ordering of any structural faults appearing in a crystal lattice. According to the theory of Jagodzinski polytypism should appear in all materials the structure of which is composed of structural elements, here layers, which may be arranged in more than one way, cf. OD structures, Section 2.

It should be noted that the procedure of calculating the free energy of crystals which is based on the analysis of qualitative thermodynamical considerations was earlier used to show that in real crystals a certain amount of disorder in the lattice should exist, see [81].

Independently of the above, there is one more reason to assume that the criterion of maximum entropy used by Jagodzinski is not appropriate for polytypic crystals: Wahab and Trigunayat [80] estimated the specific surface free energy and the entropy contributions to the polytypes and they found that the stacking fault energy is twelve orders of magnitude higher than the entropy contribution.

5.2.3 Statistical theories

The main purpose of the theories of Frank and of Jagodzinski was to explain the way in which layers are ordered in crystals to form polytype cells. It should be remembered, however, that higher polytypes with no disordering appear in crytals occasionally only, while most of crystals have a disorder type structure. The mechanism of formation of disordered structures was analyzed in the last few years in theories based on statistical considerations.

It may be assumed that even in the case when no quantitative physical rules govern the ordering of polytypic structures the statistical relations are valid. Thus, one may expect that even in the case when the faults present in crystals will be distributed randomly according to statistics polytypes may appear as "statistical fluctuations". It is, however, known that faults appearing in crystals are created depending on a number of physical factors (cf. Section 5.3). The resulting structure of crystals may be regarded as formed according to (i) physical rules quantitatively determining arrangement of nearest layers and (ii) statistical rules determining the distribution of layers ordered in simple stable sequences within a crystal. The main difficulty in applying statistics to the theory of polytypism depends on the use and availability of experimental data on (i) interactions between layers, (ii) growth conditions, (iii) chemical composition, etc., to the theoretical calculations. The theories based on statistics presented up to date concern the simplest polytypic structures, i.e. structures comparable to close packing of hexagonal layers of one type of atoms. Uppal et al. [75] presented computer simulation of polytypes using the model of a one-dimensional spin-1/2 Ising chain. In this model the spins $\pm 1/2$ of the Ising chain were substituted by two types of layers, h and c. Short-range and long-range interactions were introduced in the model: short-range interactions were defined (and employed by model potential calculations) with respect to growth conditions; long-range interactions were related to the elastic interactions in a lattice. In this model the short-range interactions are responsible for faults introduced to the structure, long-range interactions represent the tendency to form the basic, faultless sequences (h or c). Parameters of the model were simulated for crystallizations from a melt and from vapour phase. It was found that using the Ising model of a statistically governed process of ordering of layers pure polytypes of only several layers in a cell may form while the sequences which are unit cells of higher polytypes appear occasionally. Simultated structures were usually of the type (simple polytype + disorder).

Kozielski and Tomaszewicz [76] analyzed the statistical mechanism of formation of disordered structures: 2H, 4H, 6H, 10H + disorder. Structures of this type were observed in crystals of solid solutions of ZnS (with CdS, ZnSe [82]) and in doped ZnS crystals (e.g. Al [83]). To calculate models of crystal structure Kozielski assumed that (i) density of faults in a structure (layers h intermixed in a c-sequence) and (ii) interactions between h and c layers are the two main parameters limiting the random distribution of faults (here, h-layers). It was shown that theoretically calculated structural models were in good agreement with the structure of real crystals. It should be noted, however, that the model proposed by Kozielski was confirmed for crystals grown from a melt (crystallization by the Bridgman method at argon pressure of about 100 atm), while crystals of similar chemical composition but obtained by chemical transport [84, 85] have far more complex structure. Moreover, the above procedure for calculating models of polytypic structures cannot be directly related to the stoichiometric crystals: in mixed and doped ZnS crystals analyzed by Kozielski faults appear together with foreign atoms added to the basic ZnS lattice. It is clear that these foreign chemical components change the interactions within the crystal lattice and, in this way, influence the stability of various arrangements of layers.

The calculations of layered models of polytypic structures performed by Uppal and Kozielski show that by assuming a statistical mechanism of ordering of layers in crystals, higher polytypes should not appear in macrodomains in crystals. The importance of these calculations depends on the possibility to simulate experimental conditions in quantitative calculations of the structural behaviour of polytypic crystals: in the model of Kozielski the density of faults (here layers h) may be controlled experimentally by the concentration of foreign atoms in a crystal: in calculations of Uppal et al. short-range interactions between layers in fault arrangements are dependent on growth conditions. Quantitative calculations of the role of growth conditions on stacking of layers in crystals seem to be important for an understanding of laws governing polytypism of crystals and for their controlled crystallization.

It seems to be possible that the process of statistical distribution of faults in crystals cooperates with the mechanism of growth proposed by Frank — this was suggested by Uppal et al. [75]. In such "cooperative" process the structure of the parent dislocation should depend on growth conditions and, polytypes should appear in the case when the screw dislocation mechanism operates during crystallization (cf. Part II, Section 8).

5.3 Stability of polytypic structures

It is well known that different arrangements of one set of structural elements (atoms, ions, molecules, etc.) have different internal energies. The internal energy of a given structure depends, in general, on thermodynamic conditions — changing the thermodynamic parameters of a system we change the internal energy of the structure and, simultaneously, the free energy. This relation may explain structural transformations observed in a number of materials during annealing, quenching, or pressing of solids. If a given assembly of structural components may be arranged in more than one way — each arrangement having its specific internal energy, then according to equilibrium thermodynamics the structure with the lowest free energy will be stable. Interstructural transformations are frequently observed in crystals but the number of different structural modifications of a given material — polymorphs — is usually two or three; the given polymorph is stable under well defined thermodynamic conditions. In polytypic crystals, however, the number of structural modifications which may coexist under the same conditions is practically infinite. This anomaly may be explained in two ways: (i) all the modifications of polytypic crystals are equivalent; (ii) the factors which influence the stacking of layers in the structure fluctuate during formation of a crystal and the resulting structure of the crystal is, in general, not stable (metastable).

To resolve the problem of stability of a polytypic structure precise calculations of free energy should be performed. However, calculation of the internal energy of various polytypic modifications is at least very difficult especially in the case of long-period polytypes and for disordered structures.

It was shown earlier that a qualitative analysis of simple energetical relations describing the interactions between elements (atoms or ions) can be very effective to understand the stability of selected structural modifications of the given compound, e.g. [1 to 6, 13, 86]. Such qualitative analysis may explain the stability of basic polytypes but not a great variety of polytypic modifications; it is clear that interactions of a range of chemical bonds cannot stabilize structures consisting of hundreds of layers (cf. theory of Jagodzinski, Section 5.2.2 and of Uppal et al., Section 5.2.3).

It is usually assumed that the main difference between polymorphism and polytypism is in structure stability: polymorphs differ significantly in their internal energies while polytypes coexisting in one crystal are assumed to be structural modifications energetically equivalent or nearly equivalent [17]. Values of internal energies were estimated for some polytypic modifications by Uppal et al. [75] (for ZnS), by Bind [88] (for SiC), and by Wahab and Trigunayat [80] (for MX_2). The differences in internal energies of various structures calculated by Uppal seem to be negligible for thermodynamically dependent stabilization of polytypic structures. The calculations of Bind, however, gave the possibility to predict the direction of inter-polytypic transformations taking place during annealing of SiC crystals.

The results of investigations of the structural behaviour of polytypic crystals reported up to date give rise to the conclusion that stacking of layers in crystals depends directly on some internal and external parameters of the process of formation of a polytypic structure: (i) different structures (basic polytypes and fault arrangements) are created in different conditions with specific abundance; (ii) the structure of crystals may be transformed during annealing or quenching and not under pressure treatment of crystals — assuming that polytypic structures of a given compound have the same density phase transformations should not occur under pressure; (iii) different polytypic modifications have different values of parameters of selected physical properties.

The effect of temperature on polytypism of crystals was analyzed most frequently both for crystallization and for phase transformations. Results presented by various authors are, however, not congruent and sometimes contradictory. Thus, it seems to be clear that thermal stabilization of crystal structure is not the only important factor for structure stabilization of polytypic crystals. Apart from the effect of temperature, the role of impurities and chemical composition of crystals in stabilization of polytypic structures was frequently analyzed.

5.3.1 *Effect of temperature*

Temperature is the parameter of crystallization and phase transformation which in most cases plays the dominant role in structure formation. Usually polymorphic modifications may transform into each other with a change of temperature. Interphase transitions are usually reversible; different modifications may coexist at the temperature of a phase transition. Thus, for polymorphic crystals the phase transition temperature divides the whole range of temperature into two parts: above the transition point one modification is stable, below — the second. For polytypic crystals, however, such a division of polytypic structures is, in general, impossible, e.g.:

SiC The polytype 3C was classified by Knippenberg [89] as metastable in the whole range of temperatures. Bind [88, 90] calculated the range of "thermal stability" equal to 456 to 2287 °C while Shaffer [18] classified 3C as low-temperature modification. The polytype 6H was defined by Bind [88, 90] as low-temperature modification, by Knippenberg [89] as high-temperature structure. The temperature of phase transition cannot be defined uniquely for these crystals: 3C transforms to 6H above 1800 °C [89], 2H transforms to 3C at 1400 to 1600 °C, and to 6H above 2400 °C [91], polytypes 2H, 3C, 4H, 15R all transform to 6H at 1500 to 2830 °C [92].

ZnS The polytype 2H is usually assumed as high-temperature structure, 3C as low-temperature one; the transition temperature is assumed 1020 °C [17]. However, assuming the above thermal stability a number of experimental facts cannot be interpreted uniquely, e.g. the structure of the type 2H + disorder transforms to 3C during annealing at 400 to 750 °C [21], the crystals obtained by the Bridgman method at about 1800 °C have the structure of the type 3C + disorder [93].

CdI_2 The polytype 2H is assumed the low-temperature modification, 4H the high-temperature one. Thus, it is strange that the polytype 2H and disordered structures transform to 4H both during annealing and quenching [87, 94, 95]. The polytype 2H transforms during annealing at 50 °C to the structure of the type 2H + disorder [87] but this structure cannot be assumed as the most stable one at 50 °C: the crystals of CdI_2 grown at 50 °C from a solution are highly polytypic — the structure 2H + disorder appears occasionally only [96]. The CdI_2 crystals grown at −5 and −15 °C have nearly exclusively the structure 4H or 4H + disorder [97].

From the above it follows that thermal stability of polytypic structures cannot be defined uniquely. It is clear that the effect of temperature on crystallization and on phase transformations of crystal should be analyzed separately: (i) temperature is merely one of the growth parameters — the resulting structure is formed as a result of cooperation of a number of factors; (ii) crystallization occurs on a crystal surface (in the inter-phase region: solid–vapour, melt, solution), the phase transformation occurs in solid state; (iii) during annealing of crystals layers tend to rearrange and form the structure of lowest free energy, but the resulting structure can possibly be not the most stable one — transformation needs a given activation energy, thus, the resulting structure may be just a metastable phase; (iv) during crystallization the growth parameters may fluctuate and a number of various phases, which are not related to "average growth conditions", may appear.

To conclude, temperature is one important factor influencing arrangement of layers in polytypic crystals, but the effect of temperature on formation and stabilization of polytypic structures should be considered together with other factors. A good definition for "thermal stabilization" of polytypic structures is hard to find and the meaning of this term is in general questionable.

5.3.2 Stoichiometry of crystals — effect of foreign particles on polytypism of crystals

The possibility of influencing the polytypic structure of crystals by controlling their chemical composition was analyzed for ZnS, SiC, CdI_2, Ti–S, and some other materials. It was found that some particles influence the polytypic structure strongly while for others the structure of crystals is insensitive. The role of chemical composition of crystals on their polytypic structure was analyzed for three cases: (i) the basic lattice of a stoichiometric composition is doped by foreign atoms; (ii) the extra atoms added to the basic lattice are the same as the components of the parent material, e.g. Ti–S; (iii) some atoms of the parent material are substituted by other atoms, e.g. in solid solutions.

Case (i): Crystals of SiC doped by Al were analyzed by Hayashi [98] who found that the polytypes 4H, 6H, and 15R are stable in different ranges of Al concentration 0.3% and more; 0.00 to 0.05 and 0.1 to 0.3; 0.05 to 0.1%, respectively. These relations, however, were not confirmed by Knippenberg [89] who applied a different procedure of crystallization than that used by Hayashi [98]. Thus, as we can see, atoms of Al added to the lattice of SiC stabilize polytypes, but it should be stressed they do not influence polytypism of the crystals, i.e. they do not generate faults. It was found by Lundqvist [99] that atoms of Mg, Mn, and Fe introduced into the crystals of SiC showed no correlation with their polytypic structure.

The effect of atoms of Al on the polytypic structure of ZnS crystals is different from that found in SiC. It was found by Kozielski [83] that the number of polytypic modifications appearing in crystals of ZnS doped by Al increases significantly in comparison with pure ZnS crystals. So, in the case of ZnS foreign atoms in the lattice (here Al) cause promotion of polytypism of crystals — they cause generation of a number of faults in the lattice.

The effect of doping presented above may be related to a disturbance of electrical equilibrium of the crystal lattice. It is then possible that some types of lattice defects, e.g. vacancies, may cause a similar result to that obtained by doping. Taking this into account the possibility of influencing the structure of crystals by introduction of different structural defects should be considered: it seems important to correlate the polytypic structure behaviour of crystals with the method of their formation: a given process of formation of crystals may be connected with the generation of one type of defects.

Case (ii): The correlation of chemical composition of crystals with their structure was observed for a number of materials: Ti–S, Si–Al–O–N, alloys, etc. (cf. Section 3). It should be remembered, however, that in the case when different structural modifications have different chemical compositions "polytypism" has its special meaning. The polytypic modifications which are formed in these materials are usually periodic structures: the given modification is stable within the given composition range. In some cases the crystallographic period of the basic lattice (e.g. SiC type lattice for Si–Al–O–N, CdI_2 — for Ti–S) may be related to the specific distribution of chemical components of the crystal, e.g. in polytypes of Si–Al–O–N [31], Ti–S [100]. The special type of polytypism was found by Tronc and Huber [32] in crystals of $Ti_{1.2}S_2$, in which a number of polytypic modifications — both polytypes and disordered structures — were observed. The effect of extra titanium atoms introduced in the TiS lattice was not well defined, but it is probable that polytypism in these crystals is connected with the different distribution of extra atoms in the basic lattice of TiS_2 (cf. Sections 6.5 and 7.3, Part II).

Case (iii): The effect of substitution of Zn by Cd and S by Se or Te in solid solutions of ZnS was analyzed by Kozielski [82], and Pałosz et al. [84, 85]. It was found that with an increase of the percentage of Cd atoms in crystals grown by the Bridgman method [82] (from 0 to about 7%) the number of layers h in the lattice of ZnS increases from 7 to 100% (structure changes from the type 3C + disorder to pure 2H). In this case hexagonality (%h) of crystals is closely related to chemical composition. The crystals obtained by the Bridgman method have all disordered structure; pure polytypes are not observed beside the basic polytypes 2H and 3C.

In the crystals Zn–Cd–S obtained by chemical transport distinct relations between chemical composition and structure of crystals were not found [84, 85]. In this case apart from disordered structures higher polytypes were observed. It should be noted that in the case of chemical transport the resulting structure of crystals depends to a high degree on the temperature of crystallization [85].

The effect of substitution of I atoms by Br or Cl in crystals of CdI_2 on their polytypic structure was analyzed for crystals grown from the vapour phase [101] and from a solution [96]. It was found that in crystals grown from the vapour phase and from solutions the number of faults increases in comparison to pure CdI_2 crystals: (i) the structure of crystals grown from the vapour phase is always based on the polytype 4H (4H + disorder): (ii) a small amount of Br atoms (1%) influences the initial structure of crystals of CdI_2 grown from solutions noticeably, but the next Br atoms added to the lattice do not influence their structure. In this case a great variety of polytypic modifications is formed; the number of higher polytpes is similar as in pure crystals of CdI_2.

In investigations performed up to date the possibilities for influencing the polytypic structure of crystals by modulation of their chemical composition were utilized to a small degree only. It is clear from the present state of these investigations that foreign particles introduced to the crystal lattice influence their polytypic structure in many ways: (i) they may stabilize the given structural modification by reducing the number of faults or (ii) may cause generation of faults in crystals by "destabilization" of the given structure. Thus, special interest should be put on the elimination of impurities which, as it was found, may activate polytypism noticeably [102]. Moreover, it should be remembered that the effects of impurities (or foreign atoms) and of structural defects may be comparable and the separation of these two effects can be in some cases impossible.

A number of factors — mechanical, geometrical, chemical, thermodynamical, etc. — may influence the formation and stability of polytypic structures. The effect of only an inconspicuous part of them on polytypism of crystals was studied in earlier investigations. Discrepancies between various interpretations of the reasons for polytypism presented by numerous authors are frequent in literature in the field of polytypism. The main reason for such "non-adequateness" of similar experiments may be the following: (i) a complete list of factors important for polytypism is not known and various authors may overlook some of them; (ii) the effects of different factors on polytypism of crystals should be separated, but in many cases such separation is impossible. So, erroneous interpretations of experimental observations are very probable: it was found that cooperation of growth parameters, temperature of crystallization and solvent used for crystallization of CdI_2 from a solution, leads to very complicated phase diagrams [96]. In this case the effect of temperature cannot be described uniquely: if different solvents are used for crystallization the effect of temperature may be different.

The resolution of the problem of completing the list of factors important for formation and stabilization of polytypic structures is necessary for controlling the structure of polytypic crystals and thus, important for their application.

5.4 Physical properties of polytypic crystals

Physical properties of a crystal depend both on its chemical composition and on its structure (lattice constants and defect structure). Most of materials may crystallize in several modifications, each of them having well defined physical parameters. Physical properties of the given modification may be modified by (i) defects present in the crystal or (ii) by its chemical composition. The discovery of polytypism in crystals gives us new possibilities of preparation of materials of desired properties: the number of polytypic modifications of the polytypic material of one chemical composition, each of them of different physical parameters, is practically infinite. The interpretation of the relations between structure and physical properties observed in polytypic crystals depend on the "type of polytypism": (i) in stoichiometric crystals (pure polytypism) a change of physical parameters depends directly on crystal structure, i.e. on the stacking of layers; (ii) the properties of non-stoichiometric crystals may depend both on chemical composition and on structure, i.e. on distribution of chemical components in the lattice and on stacking of layers of the basic lattice; (iii) in doped crystals the relation between the real structure of a crystal and its properties seems to be most complex: dopands may influence the stacking of layers, defect structure and may determine some properties themselves.

The detailed analysis of the relations structure–physical properties makes possible a direct calculation of physical properties as a function of simple structural parameters describing the arrangement of layers. In polytypic crystals as the only parameter necessary to describe the structure may be chosen the parameter describing the one-dimensional ordering of structural elements of one type — a layer. One may expect that a direct description of the role of ordering of layers in the determination of physical properties of crystals will enable a verification of various theories of solids and, possibly, application of polytypism.

The possibility of controlling of physical properties of polytypic crystals by controlling their structure was reported for some materials, e.g.:

SiC The energy band gap changes from 2.2 to 3.1 eV when hexagonality of crystals changes from 0 to 50%h [89].

ZnS Birefringence ($n_e - n_0$) changes linearly with an increase of hexagonality of crystals: from 0 for 0%h to 24×10^{-3} for 100%h [103].

PbI_2 Raman spectra differ for different polytypic structures [104]. Resistivity changes from 8.5×10^{13} for 2H to 5.66×10^{7} Ωcm for disordered structures [105].

SnS_2 The energy band gap changes from 0.92 to 2.18 eV for polytypes 24H and 2H [106].

CdI_2 The dielectric constant of crystals changes from about 20 for 4H to 150 for higher polytypes [107, 108]. Luminescence spectra are shifted from 500 to 560 nm for 2H and 6H, respectively [109].

The above relations were not investigated extensively and are now regarded as supplementary to the current studies of polytypism. It seems that further relations may be easily discovered.

The theoretical investigations of general relations between physical properties and structure of polytypic crystals were performed by Geilikman [110] for MX type crystals. He calculated theoretically elastic moduli for some simple polytypic modifications and estimated the role of the arrangements of layers in h and c sequences in the determination of elastic properties of crystals. Theoretical calculations of elastic moduli for 2H and 3C — differing noticeably — are in good agreement with experimentally measured values. It seems that considerations of quantitative relations between elastic properties and structure of crystals may be used for the description of relations between structure and various physical parameters of crystals.

It might to be interesting to correlate quantitative calculations of vibrations of polytypic crystals with the theories of polytypism which are based on the assumption that vibrations of the crystal lattice are the factors influencing an ordering of layers in crystals: theory of Jagodzinski (see Section 5.2.2); of Uppal et al. (Section 5.2.3), theory of Peibst [111].

Relations between structure and physical properties observed in polytypic crystals were usually interpreted with the assumption that physical parameters of a crystal depend on stacking of layers of its structure. It seems that an alternative interpretation of these relations is possible: "physical parameters of crystals — their properties — determine the structure of crystals". For such an interpretation it should be assumed that the stackings of layers formed in a crystal are stable for given values of internal and external physical parameters of a material and these parameters are determined by the conditions of the process of formation of the crystal structure. Such an interpretation was suggested by Knippenberg for relations between energy band gap and structure of crystals of SiC [89] and by Komura and Kitano [112] for crystals of Mg-base Friauf-Laves phases. Knippenberg suggested that the energy of electrons (configuration of electron bands) may determine stacking of layers in polytype cells of SiC. Komura found that the structure of crystals of alloys Mg–Cu–Ni, Mg–Zn–Ni, Mg–Cu–Zn, etc. depends on electron–atom ratio e/m — different stackings of layers are stable within different ranges of values e/m. It was found that stacking of layers does not depend on the chemical composition of crystals.

It seems to be probable that the structure of polytypic crystals depends on the physical parameters of materials; this assumption is equivalent to the assumption that polytypism is an energy governed phenomenon. The physical parameters of crystals — e.g. electron energy bands, vibrations — may be influenced by (i) some thermodynamical factors (temperature, pressure, etc.) of the process of formation of the crystal structure as well as by (ii) chemical factors (impurities, dopands, stoichiometry). One may expect that the relations between structure and physical properties of crystals may be correlated with the conditions of stability of polytypic structures. Further studies of the problem seem to be important for the performance of controlled growth of polytypic crystals, i.e. for application of polytypism.

6. Summary

1. Polytypism appearing in inorganic crystals should be regarded as a property of a crystal lattice and not as the property of the compound: this follows from the OD theory of solids of Dornberger-Schiff. Polytypism of crystals of a given compound can appear merely under specific conditions of formation of a crystal structure: crystals of a given compound may be polytypic in some cases, in others they may occur in one structural modification — they are monotypic. The relatively small number of materials which are known as being polytypic, results from the fact that polytypism is not a well and commonly known phenomenon, so, it is not an expected and researched one in standard structural analysis. The great number of known polytypic modifications in some "most polytypic" materials is the result of detailed studies of their structure performed just for discovering these structures. The lack of information on polytypism of numerous materials having crystals isostructural to polytypic crystals does not mean that these materials cannot be polytypic.

2. Polytypism is a phenomenon depending on a number of energetic factors which influence the stacking of layers in crystals. The arrangements of layers formed in real crystals are those which are energetically favourable under the given conditions. It seems that accidental factors which may disturb the equilibrium of the crystal lattice — defects, dislocations, etc. — are not important for the promotion of polytypism of crystals — they are of secondary importance in the determination of the polytypic structure of crystals influencing merely the mechanism of structure formation.

3. The process of ordering of layers in periodic structures — polytypes — and in disordered sequences seems to be dependent on energetic and geometrical factors connected with the mechanism of crystal growth.

4. The limited possibilities of controlling the polytypic structure of crystals restricts further progress in explanation and in application of polytypism. A number of various factors influence the structure of polytypic crystals but only a little part of them was analyzed up to date. Systematic studies on the role of factors important for the stability of the crystal structure and on the process of polytypic growth of crystals are necessary to complete the theory of polytypism.

5. The most exciting problem in polytypism is the creation of crystals having a structure composed of unit cells formed of hundreds of layers. The question of the possibility of controlling the growth of a desired polytype is seen frequently in literature. On the basis of the present state of investigations of polytypism one may conclude that polytypes are formed incidentally in the crystals and stacking of layers in polytype cells cannot be controlled. Nevertheless, the question of the possibility of controlling structure and periodicity of polytypes may be replaced by the question of the possibility of influencing the ordering of layers in desired sequences: basic sequences (simple polytypes) and in faulted sequences of a desired type. The structure of a polytype is usually described by lattice constants but, it may be described as well by the parameters used for disordered structures, i.e. (i) types of basic and faulted arrangements of layers in a cell, (ii) density of faults, (iii) their distribution in the crystal, etc. It seems that structural parameters of these types can be controlled in the processes of growth or structural transformations of crystals. This problem is discussed in detail for MX_2 crystals in Part II of this paper.

In memory of Prof. K. Boll-Dornberger

More than 25 years have passed since the OD theory was first presented by Prof. K. Boll-Dornberger. This theory which was written to determine the rules governing the formation of layer crystal lattices in solids appeared very helpful in the explanation of various structural behaviour of solids found unexplicable on the basis of classical crystallography. One of the non-classical phenomena predicted by the OD theory was polytypism of inorganic crystals. One may have expected that this theory would have been adopted by investigators of polytypism — a phenomenon observed in some materials before the OD theory was published. Unfortunately, however, in most of the papers published in the field of polytypism the OD theory was not mentioned.

I am interested in investigating polytypism of inorganic crystals. I found a number of valuable indications for my studies of detailed problems of polytypism by analyzing the theory of OD structures.

By dedicating this paper to the memory of Prof. K. Boll-Dornberger I would like to express my gratitude for the help I have found in the papers of the author of the OD theory.

References

[1] F. C. Frank and J. S. Kasper, Acta cryst. **11**, 184 (1958).
[2] S. Iida, J. Phys. Soc. Japan **12**, 222 (1957).
[3] A. L. Loeb, J. Solid State Chem. **1**, 237 (1970).
[4] P. A. Beck, Z. Krist. **124**, 101 (1967).
[5] J. Lima de Faria and M. O. Figueiredo, Garcia d Orta, Sér. geol. (Lisboa) **2**, (2), 69 (1978).
[6] C. J. J. van Loon, J. Solid State Chem. **10**, 195 (1974).
[7] L. Lima de Faria and M. O. Figueiredo, J. Solid State Chem. **16**, 7 (1976).
[8] E. Parthé, Acta cryst. **B36**, 1 (1980).
[9] K. Dornberger-Schiff, Acta cryst. **9**, 593 (1956).
[10] K. Dornberger-Schiff, Kristall und Technik **14**, 1027 (1979).
[11] K. Dornberger-Schiff, Acta cryst. **A38**, 483 (1982).
[12] H. Grell and K. Dornberger-Schiff, Acta cryst. **A38**, 49 (1982).

[13] J. LIMA DE FARIA, Z. Krist. **148**, 1 (1978).
[14] P. M. DE WOLFF, Acta cryst. **A30**, 777 (1974).
[15] A. JANNER and T. JANSSEN, Phys. Rev. B **15**, 643 (1977).
[16] H. BAUMHAUER, Z. Krist. **50**, 33 (1912).
[17] A. R. VERMA and P. KRISHNA, Polymorphism and Polytypism in Crystals, Wiley, New York 1966.
[18] P. T. B. SHAFFER, Acta cryst. **B25**, 477 (1969).
[19] D. PANDEY and P. KRISHNA, Phil. Mag. **31**, 1133 (1975).
[20] J. A. WILSON and A. D. YOFFE, Adv. Phys. **18**, 193 (1969).
[21] W. L. ROTH, Physics and Chemistry of II-VI Compounds, North Holland Publ. Co., 1967.
[22] G. C. TRIGUNAYAT and G. K. CHADHA, phys. stat. sol. (a) **4**, 9 (1971).
[23] G. C. TRIGUNAYAT, phys. stat. sol. (a) **4**, 281 (1971).
[24] B. PRASAD, phys. stat. sol. (a) **38**, 11 (1976).
[25] G. C. TRIGUNAYAT and A. R. VERMA, in: Crystallography and Crystal Chemistry of Materials with Layered Structures, Ed. F. LEVY, Reidel, Dordrecht 1976.
[26] D. PANDEY and P. KRISHNA, Phil. Mag. **31**, 1113 (1975).
[27] M. R. TUBBS, phys. stat. sol. (b) **49**, 11 (1972).
[28] H. SATO, R. S. TOTH, and G. HONJO, J. Phys. Chem. Solids **28**, 137 (1967).
[29] J. ZUSSMAN, Bull. Soc. Franc. Minéral. Crist. **102**, 5 (1979).
[30] J. C. J. M. TERHELL and W. C. VAN DER VLEUTEN, Mater. Res. Bull. **11**, 101 (1976).
[31] A. HENDRY and K. H. JACK, Inst. Phys. Conf. Ser. **36**, 157 (1977).
[32] E. TRONC and M. HUBER, J. Phys. Chem. Solids **34**, 2045 (1973).
[33] B. I. NIKOLIN and N. N. SHEVCHENKO, Scripta metall. **14**, 467 (1980).
[34] L. A. MARUSAK and L. N. MULAY, Phys. Rev. B **21**, 238 (1980).
[35] D. A. JEFFERSON, Acta cryst. **A34**, 491 (1978).
[36] J. P. HIRTH and J. LOTHE, Theory of Dislocations, McGraw-Hill Publ. Co., New York 1968.
[37] V. VAND and J. L. HANOKA, Mater. Res. Bull. **2**, 241 (1967).
[38] S. R. SINGH and G. SINGH, Acta cryst. **A36**, 779 (1980).
[39] N. W. JEPPS, D. J. SMITH, and T. F. PAGE, Acta cryst. **A35**, 916 (1979).
[40] Y. KITANO, Y. KOMURA, and I. T. KAJIWARA, Trans. Japan. Inst. Metals **18**, 39 (1977).
[41] H. JAGODZINSKI, Acta cryst. **2**, 201, 208, 298 (1949).
[42] M. FARKAS-JAHNKE, Acta cryst. **B29**, 407, 413 (1973).
[43] B. PALOSZ, Acta cryst. **A33**, 172 (1977).
[44] D. PANDEY, S. LELE, and P. KRISHNA, Proc. Roy. Soc. **A369**, 435, 451, 463 (1980).
[45] D. PANDEY and P. KRISHNA, J. Phys. D **10**, 2057 (1977).
[46] H. MULLER, Neues Jahrb. Miner. Abh. **84**, 43 (1952).
[47] J. SINGER, Acta cryst. **16**, 601 (1963).
[48] A. EBINA and T. TAKAHASHI, J. appl. Phys. **38**, 3079 (1967).
[49] T. MINAGAWA, Acta cryst. **A33**, 687 (1977).
[50] R. K. JAIN and G. O. TRIGUNAYAT, Acta cryst. **A26**, 463 (1970).
[51] M. J. KOZIELSKI, J. Crystal Growth **30**, 86 (1975).
[52] M. DUBEY, phys. stat. sol. (a) **29**, 213 (1975).
[53] D. PANDEY and P. KRISHNA, Mater. Sci. Engng. **26**, 53 (1976).
[54] D. KUHLMANN-WILSDORF, D. PANDEY, and P. KRISHNA, Phil. Mag. A **42**, 527 (1980).
[55] R. SRINIVASAN and V. PARTHASARATHI, Z. Krist. **137**, 296 (1973).
[56] R. S. TIWARI, A. K. RAI, and O. N. SRIVASTAVA, Phys. Rev. B **9**, 5155 (1974).
[57] R. S. TIWARI, A. K. RAI, and O. N. SRIVASTAVA, phys. stat. sol. (a) **31**, 419 (1975).
[58] B. PRASAD and O. N. SRIVASTAVA, J. appl. Cryst. **4**, 516 (1971).
[59] V. K. AGRAWAL, J. appl. Cryst. **6**, 44 (1973).
[60] F. C. FRANK, Phil. Mag. **42**, 1014 (1951).
[61] J. I. HANOKA, K. VEDAM, and H. K. HENISH, in: Crystal Growth, Ed. H. S. PEISER, Pergamon Press, Oxford 1967.
[62] V. VAND and J. I. HANOKA, Mater. Res. Bull. **2**, 24 (1967).
[63] S. MARDIX, Z. H. KALMAN, and I. T. STEINBERGER, Acta cryst. **A24**, 464 (1969).
[64] E. ALEXANDER, Z. H. KALMAN, S. MARDIX, and I. T. STEINBERGER, Phil. Mag. **21**, 1237 (1970).
[65] S. MARDIX and I. T. STEINBERGER, J. appl. Phys. **41**, 5339 (1970).

[66] I. T. Steinberger, E. Alexander, Y. Brada, Z. H. Kalman, I. Kiflawi, and S. Mardix, J. Crystal Growth **13/14**, 285 (1972).
[67] J. J. Legendre, M. Huber, and M. Sauvage, phys. stat. sol. (a) **40**, K101 (1977).
[68] J. J. Legendre and M. Huber, Acta cryst. **A36**, 1031 (1980).
[69] M. R. Tubbs, Acta cryst. **B27**, 857 (1971).
[70] V. K. Agrawal, Crystal Lattice Defects **8**, 7 (1978).
[71] V. K. Agrawal, phys. stat. sol. (a) **57**, 143 (1980).
[72] V. K. Agrawal, J. Crystal Growth **49**, 580 (1980).
[73] H. Jagodzinski, Neues Jahrb. Miner. Abh. **3**, 49 (1954).
[74] K. N. Rai, phys. stat. sol. (a) **8**, 271 (1971).
[75] M. K. Uppal, S. Ramasesha, and C. N. R. Rao, Acta cryst. **A36**, 356 (1980).
[76] M. Kozielski and A. Tomaszewicz, Bull. Acad. Pol. Sci., Sér. Math. Astr. Phys. **26**, 749 (1978).
[77] B. Pałosz and J. Przedmojski, Z. Krist. **151**, 31 (1980).
[78] B. Pałosz, Z. Krist. **153**, 51 (1980).
[79] R. K. Jain and G. C. Trigunayat, Acta cryst. **A26**, 463 (1970).
[80] M. A. Wahab and G. C. Trigunayat, Indian J. pure appl. Phys. **18**, 577 (1980).
[81] A. J. Dekker, Solid State Physics, MacMillan & Co., London 1958.
[82] M. J. Kozielski, J. Crystal Growth **30**, 86 (1975).
[83] M. J. Kozielski, Bull. Acad. Pol. Sci., Sér. Math. Astr. Phys. **24**, 367 (1976).
[84] W. Pałosz, Ph. D. Thesis, Warsaw Techn. Univ., 1979.
[85] W. Pałosz, M. J. Kozielski, and B. Pałosz, J. Crystal Growth **58**, 195 (1982).
[86] M. Kogachi and K. Katada, J. Phys. Chem. Solids **36**, 501 (1975).
[87] T. Minagawa, J. appl. Cryst. **12**, 57 (1979).
[88] J. M. Bind, Mater. Res. Bull. **14**, 1173 (1979).
[89] W. F. Knippenberg, Philips Res. Rep. **18**, 161 (1963).
[90] J. M. Bind, Mater. Res. Bull. **13**, 91 (1978).
[91] G. A. Bootsma, W. F. Knippenberg, and G. Verspui, J. Crystal Growth **8**, 341 (1971).
[92] P. Krishna, R. C. Marshall, and C. E. Ryan, J. Crystal Growth 8, 129 (1971).
[93] B. Pałosz and J. Przedmojski, Acta phys. Polon. **A47**, 435 (1975).
[94] T. Minagawa, J. appl. Cryst. **11**, 243 (1978).
[95] R. M. Arora and G. C. Trigunayat, J. low-Temp. Phys. **23**, 231 (1976).
[96] B. Pałosz, J. Crystal Growth **52**, 969 (1981).
[97] S. Gierlotka and B. Pałosz, EMCG 82 – ME, Prague 1982.
[98] A. Hayashi, J. Miner. Soc. Japan **4**, 363 (1960).
[99] D. Lundqvist, Acta chem. Scand. **2**, 177 (1948).
[100] G. A. Wiegers and F. Jellinek, J. Solid State Chem. **1**, 519 (1970).
[101] S. D. Sharma and K. Mehrotra, J. Electrochem. Soc. **126**, 325 (1979).
[102] P. C. Jain and G. C. Trigunayat, J. Crystal Growth **48**, 107 (1980).
[103] O. Brafman and I. T. Steinberger, Phys. Rev. **143**, 501 (1966).
[104] W. M. Sears, M. L. Klein, and J. A. Morrison, Phys. Rev. B **19**, 2305 (1979).
[105] M. Rao and O. N. Srivastava, J. Phys. D **11**, 919 (1978).
[106] S. Acharya and O. N. Srivastava, phys. stat. sol. (a) **56**, K1 (1979).
[107] A. M. Fernandez and O. N. Srivastava, J. appl. Cryst. **8**, 645 (1975).
[108] A. M. Fernandez and O. N. Srivastava, J. appl. Cryst. **10**, 32 (1977).
[109] I. M. Bolesta and I. K. Gloskowskaya, Ukr. fiz. Zh. **23**, 691 (1978).
[110] M. B. Geilikman, phys. stat. sol. (b) **91**, 725 (1979).
[111] H. Peibst, Z. phys. Chem. (Leipzig) **223**, 193 (1963).
[112] Y. Komura and Y. Kitano, Acta cryst. **B33**, 2496 (1977).

(Received June 2, 1981)

Original Papers

phys. stat. sol. (a) 77, 35 (1983)

Subject classification: 3

Computer Center, Latvian State University, Riga[1])

Sublattice Models of the Binary Crystal Growth

I. The Growth Kinetics of Ordering Binary Crystals from Solution and Melt

By

T. A. Cherepanova and A. S. Koziejowska

In the frames of a lattice model the general regularities of the ordering binary crystal growth is considered. Discussion concerns the applicability of models with one lattice and two sublattices used to describe the growth of an ionic-type crystal from the solution. The growth kinetics of crystals of GaAs-type from the melt is studied by means of a sublattice model. The {111}-face polarity effect is shown.

В рамках решеточной модели изучены общие закономерности роста упорядочивающихся бинарных кристаллов. Обсуждается вопрос применимости моделей с одной решеткой и двумя подрешетками для описания роста кристалла ионного типа из раствора. С помощью подрешеточной модели изучена кинетика роста кристаллов типа GaAs из расплава. Описан эффект полярности граней {111}.

1. Introduction

Lately a wide range of investigations was done to study descriptive methods of multicomponent monocrystal growth kinetics in the frames of lattice models, based on Monte Carlo computer simulation [1 to 6] and kinetic equations [7 to 15]. In [16] the kinetic equations of the growth of a crystal with substitution of components were obtained, based on the introduction of sublattices enclosed one in the other. Let us dwell now on the question of the applicability of such a description in the case of growth of an ordering α–β binary crystal from solution and melt.

2. Description of the Model

Let us examine a simple cubic (type NaCl) lattice, all the sites of which are occupied by α- and β-particles belonging to solid or liquid phases, and the solvent particles. In the solid phase the diffusion processes are assumed to be so slow in comparison with the crystal growth rate (or in comparison with the time of the numerical experiment) that they could be neglected, so that the solid component particles α and β are "frozen" in the crystal.

Describing the solution we shall consider two utmost models. One of them is based on the assumption of a complete intermixing of the component particles α and β and the solvent particles, so that the probability to discover the particle of sort ν ($\nu = \alpha, \beta$) is equal to C_0^ν, the concentration of ν-particles in the solution (moreover $C_0^\alpha + C_0^\beta < 1$). Another model of a solution [17] is based on the introduction of two sublattices — a cation one (I) for α-particles and an anion one (II) for β-particles. The particles of α-type may occupy only the sites of the cation sublattice, while β-particles only the sites of the anion sublattice. The solvent particles are distributed equiprobably over both sublattices of the liquid phase, which corresponds to the condi-

[1]) Boulevard Raina 29, Riga, USSR.

tion of a complete intermixing of α-particles and the solvent particles in the cation sublattice, β-particles and the solvent particles in the anion sublattice.

For a (100) face the kinetic equations in the models under consideration in mean field approximation, taking into account only configurations without overhangs, have the form:

1. Model with a complete intermixing of components in the solution,

$$\frac{1}{W_{10}^{\nu}} \frac{\mathrm{d}[{}_pC_1^{\nu}(z)]}{\mathrm{d}t} = \exp\left(\frac{\Delta\mu_{\nu}}{kT}\right) [{}_qC_1^s(z-1) - {}_pC_1^s(z)] - \\ - \frac{{}_pC_1^s(z) - {}_qC_1^s(z+1)}{{}_pC_1^s(z)} \, {}_pC_1^{\nu}(z) \times \\ \times \exp\left[\theta_{\nu} - 4\,\frac{\varphi_{\nu\nu\,q}C_1^{\nu}(z) + \varphi_{\alpha\beta\,q}C_1^{\xi}(z)}{kT} - \\ - \frac{\varphi_{\nu\nu\,q}C_1^{\nu}(z-1) + \varphi_{\alpha\beta\,q}C_1^{\xi}(z-1)}{kT\,{}_qC_1^s(z-1)}\right];$$

$$\nu, \xi = \alpha, \beta; \quad \nu \neq \xi\,, \quad p, q = \mathrm{I}, \mathrm{II}; \quad p \neq q\,. \tag{1}$$

2. Model with a complete ordering of the component distribution over the sublattices in the solution

$$\frac{1}{W_{10}^{\nu}} \frac{\mathrm{d}[{}_pC_1^{\nu}(z)]}{\mathrm{d}t} = \exp\left(\frac{\Delta\mu_{\nu}}{kT}\right) [{}_qC_1^{\xi}(z-1) - {}_pC_1^{\nu}(z)] - \\ - [{}_pC_1^{\nu}(z) - {}_qC_1^{\xi}(z+1)] \exp\left[\theta_{\nu} - \varphi_{\alpha\beta}\,\frac{4\,{}_qC_1^{\xi}(z) + 1}{kT}\right];$$

$$\nu, \xi = \alpha, \beta; \quad \nu \neq \xi\,, \qquad p, q = \mathrm{I}, \mathrm{II}; \quad p \neq q\,. \tag{2}$$

Here ${}_pC_1^{\nu}(z)$ is the concentration of the solid ν-component in the p-sublattice in atomic layer z, and ${}_pC_1^s(z) = {}_pC_1^{\alpha}(z) + {}_pC_1^{\beta}(z)$; the quantity $\varphi_{\nu\xi}$ is the difference in the interaction energy of the nearest neighbour ν and ξ particles in liquid phase and crystal, $\Delta\mu_{\nu}/kT$ the supersaturation of the ν-component in the liquid phase, W_{10}^{ν} the probability per time unit for a liquid phase particle to attach to a crystal. We assume that the value W_{10}^{ν} is independent of the distribution configuration of the nearest neighbour particles of different sorts. The entropy factor θ_{ν} difference in the first (1) and the second (2) models is $\ln C_0^{\nu}$, which associates with the account of the fact how the configuration part of the solution entropy influences the chemical potential of solution. The short-range order parameters in solid phase and interface zone usually are expressed through the concentrations of components in the sublattices. For instance, the concentration of solid–solid bonds of ν–ξ particles in a layer z is

$$_{\mathrm{I\,II}}C_{11}^{\nu\xi}(z) = {}_{\mathrm{I}}C_1^{\nu}(z)\, {}_{\mathrm{II}}C_1^{\xi}(z)\,, \qquad {}_{\mathrm{I\,II}}C_{11}^{\nu\nu}(z) = {}_{\mathrm{I}}C_1^{\nu}(z)\, {}_{\mathrm{II}}C_1^{\nu}(z)\,,$$

$$C_{11}^{\nu\xi}(z) = {}_{\mathrm{I}}C_1^{\nu}(z)\, {}_{\mathrm{II}}C_2^{\xi}(z) + {}_{\mathrm{II}}C_1^{\nu}(z)\, {}_{\mathrm{I}}C_1^{\xi}(z); \qquad \nu, \xi = \alpha, \beta; \quad \nu \neq \xi\,. \tag{3}$$

As shown earlier [8, 9, 14, 16], the crystallization kinetics of binary alloys may be described in the frames of a pair approximation of the distribution function by means of introducing into consideration one lattice (not divided into sublattices). In the case of a complete intermixing of components in solution the system of kinetic equations, at such description, has the form [8, 9]

$$\frac{1}{W_{10}^{\nu}} \frac{\mathrm{d}C_1^{\nu}(z)}{\mathrm{d}t} = C_0^{\nu} C_{01}^{fs}(z, z-1) - C_{01}^{f\nu}(z+1, z) \exp(\theta_{\nu})\, Q_{\nu,s}(z, z-1)\, [Q_{\nu}(z)]^4\,,$$

$$\frac{1}{W_{10}^{\nu}} \frac{\mathrm{d}C_{11}^{\nu\nu}(z)}{\mathrm{d}t} = C_0^{\nu} C_{01}^{f\nu}(z)\, C_{01}^{fs}(z, z-1)\, [C_0^{f}(z)]^{-1} - 2C_{01}^{f\nu}(z+1, z)\, C_{11}^{\nu\nu}(z) \times$$

$$\times \exp\left(-\frac{\varphi_{\nu\nu}}{kT} + \theta_\nu\right) [C_1^{\nu}(z)]^{-1}\, Q_{\nu,s}(z, z-1)\, [Q_\nu(z)]^3 ,$$

$$\frac{1}{W_{10}^{\nu}} \frac{\mathrm{d}C_{11}^{\nu\nu}(z, z-1)}{\mathrm{d}t} = C_0^{\nu} C_{01}^{f\nu}(z, z-1) - C_{01}^{f\nu}(z+1, z)\, C_{11}^{\nu\nu}(z, z-1)\, [C_1^{\nu}(z)]^{-1} \times$$

$$\times \exp\left(-\frac{\varphi_{\nu\nu}}{kT} + \theta_\nu\right) [Q_\nu(z)]^4 , \tag{4}$$

$$\frac{\mathrm{d}C_{11}^{\alpha\beta}(z)}{\mathrm{d}t} = [C_0^{\alpha} C_{01}^{f\beta}(z)\, W_{10}^{\alpha} + C_0^{\beta} C_{01}^{f\alpha}(z)\, W_{10}^{\beta}]\, C_{01}^{fs}(z, z-1)\, [C_0^{f}(z)]^{-1} -$$

$$- C_{11}^{\alpha\beta}(z) \exp\left(-\frac{\varphi_{\alpha\beta}}{kT}\right) \{C_{01}^{f\alpha}(z+1, z)\, [C_1^{\alpha}(z)]^{-1} \exp(\theta_\alpha)\, Q_{\alpha,s}(z, z-1) \times$$

$$\times [Q_\alpha(z)]^3\, W_{10}^{\alpha} + C_{01}^{f\beta}(z+1, z)\, [C_1^{\beta}(z)]^{-1} \exp(\theta_\beta)\, Q_{\beta,s}(z, z-1) \times$$

$$\times [Q_\beta(z)]^3\, W_{10}^{\beta}\} ,$$

where

$$Q_\nu(z) = C_{11}^{\nu\nu}(z)\, [C_1^{\nu}(z)]^{-1} \exp\left(-\frac{\varphi_{\nu\nu}}{kT}\right) + C_{11}^{\alpha\beta}(z)\, [2C_1^{\nu}(z)]^{-1} \exp\left(-\frac{\varphi_{\alpha\beta}}{kT}\right) +$$

$$+ C_{01}^{f\nu}(z)\, [2C_1^{\nu}(z)]^{-1} ,$$

$$Q_{\nu,s}(z, z-1) = \left[C_{11}^{\nu\alpha}(z, z-1) \exp\left(-\frac{\varphi_{\nu\alpha}}{kT}\right) + C_{11}^{\nu\beta}(z, z-1) \times\right.$$

$$\left. \times \exp\left(-\frac{\varphi_{\nu\beta}}{kT}\right)\right] [C_1^{\nu}(z)]^{-1} ,$$

and, for example, $C_{01}^{fs}(z, z-1)$ is a concentration of bonds of the nearest neighbour liquid (f) and solid (s) particles belonging to layer z and $z-1$, respectively, $C_{01}^{f\nu}(z+1, z)$ is the concentration of bonds of the nearest neighbour liquid (f) particles and solid particles of sort ν, belonging to layers $z+1$ and z, respectively.

Relations (3) allow to compare formally the solutions of kinetic equations for the three offered models (1), (2), (4). In particular, from (3), taking into account the conditions ${}_{\mathrm{I}}C_1^{\alpha} + {}_{\mathrm{I}}C_1^{\beta} = {}_{\mathrm{II}}C_1^{\alpha} + {}_{\mathrm{II}}C_1^{\beta} = 1$, we have a reverse relationship,

$${}_{\mathrm{I}}C_1^{\alpha} = \{1 - C_{11}^{\alpha\alpha} + C_{11}^{\beta\beta} + [(1 - C_{11}^{\alpha\alpha} + C_{11}^{\beta\beta})^2 - 4C_{11}^{\alpha\alpha}]^{1/2}\}/2 ,$$

$${}_{\mathrm{II}}C_1^{\alpha} = \{1 - C_{11}^{\alpha\alpha} + C_{11}^{\beta\beta} - [(1 - C_{11}^{\alpha\alpha} + C_{11}^{\beta\beta})^2 - 4C_{11}^{\alpha\alpha}]^{1/2}\}/2 . \tag{5}$$

In accordance with (5) the following correlations take place: $({}_{\mathrm{I}}C_1^{\alpha} + {}_{\mathrm{II}}C_1^{\alpha})/2 = C_1^{\alpha}$, $({}_{\mathrm{I}}C_1^{\beta} + {}_{\mathrm{II}}C_1^{\beta})/2 = C_1^{\beta}$. Here C_1^{ν} is the ν-component concentration in the volume of the solid phase ($\nu = \alpha, \beta$) and $C_1^{\nu} = C_{11}^{\nu\nu} + C_{11}^{\alpha\beta}/2$. At the same time, models (1), (2), (4) are based on the same assumptions that the crystal growth process is the result of the accomplishment of random elementary events of the particle phase transitions on a face surface with determined transition probabilities. Monte Carlo computer simulation is a precise solution of the above-formulated problem, and (1), (2), (4) are approximate descriptions.

General regularities for the binary ordering crystal growth from solution were studied in [4] by the Monte Carlo method. It is shown that in a wide region of supersaturations there grows an ideally ordered (as regards the α- and β-component distributions over the lattice sites) crystal. Increase in supersaturation results in the destruction of the order by the formation of domains. Inside the domains the ordered distribution of α- and β-components remains, while on the boundaries this order is disturbed. With increasing growth rate the characteristic size of domains decreases, which leads to a complete disordering of the component distribution in the solid phase.

3. Calculations for a Model System

In the given paper the results of computer simulation by the Monte Carlo method were used as a criterion for the applicability of models with two sublattices (1) and one lattice (4). In computer simulation the condition of a complete intermixing of components in the solution was given. To carry out a Monte Carlo computer simulation, a face (100) was chosen, containing 20×21 particles. Periodic boundary conditions were given. The length of the random-number generator used here was $\approx 5 \times 10^7$ Determination of the process statistic characteristics at a stationary growth regime was made every $N_0 = 15 \times 10^4$ trials. In solution of kinetic equations and computer simulation the following values of the system parameters were given: $\varphi_{\alpha\alpha}/kT = \varphi_{\beta\beta}/kT = -1$, $\varphi_{\alpha\beta}/kT = 1$; $W_{10}^{\alpha} = W_{10}^{\beta} = W_{10}$; $\Delta\mu_\alpha = \Delta\mu_\beta = \Delta\mu$. The parameter θ in the model of an "ordered" solution (2) is equal to 3.693. The parameter θ in models with a complete intermixing of components in the solution ((1) and (4)) is equal to 3. In the simulation the dimensionless growth rate R was determined as an average number of particles added to a crystal in unit time, related to one surface site: $R = \Delta M (N\tau W_{10})^{-1}$, where ΔM is the increase in the number of particles in a crystal during N trials, τ the time interval during which not more than one possible elementary event can take place; $\tau = [\max(W_{10}^{\nu} + W_{01}^{\nu})]^{-1}$. In the frames of kinetic equations (1), (2), and (4) the value of a dimensionless growth rate R was computed according to

$$R = \sum_{z,\nu} \left\{ \frac{\mathrm{d}[{}_{\mathrm{I}}C_1^{\nu}(z)]}{\mathrm{d}t} + \frac{\mathrm{d}[{}_{\mathrm{II}}C_1^{\nu}(z)]}{\mathrm{d}t} \right\} \frac{1}{W_{10}N(z)}, \tag{6}$$

where $N(z)$ is the number of sublattices, the sites of which belong to layer z.

Fig. 1 shows how the concentrations $C_{11}^{\nu\nu}$ of bonds of the nearest neighbour particles of the same sort, forming in a crystal during the growth process, depend on super-

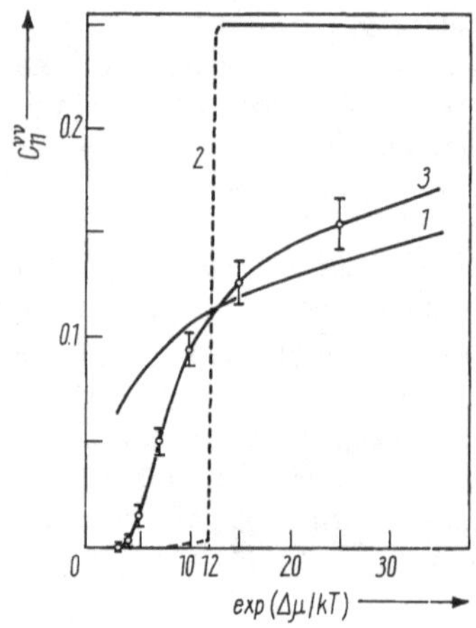

Fig. 1. Dependence of bond concentrations $C_{11}^{\nu\nu}$ of the nearest-neighbour particles of the same sort in a crystal on supersaturation $\Delta\mu/kT$ (complete intermixing of components in solution). (1) one-lattice model (4), (2) model with two sublattices (1), (3) Monte-Carlo computer simulation

saturation $\Delta\mu/kT$ in the case of complete component intermixing in the solution. Curve 1 corresponds to a model with one lattice (4), curve 2 to a model with two sublattices (1), curve 3 represents the results of Monte Carlo computer simulation. In the region of relatively small supersaturations ($\exp(\Delta\mu/kT) \leqq 4$), where an ordered (NaCl-type) α–β crystal grows, the sublattice model is in better agreement with the computer simulation results, than the model considering one lattice. With the increase in supersaturation the concentrations $C_{11}^{\nu\nu}$ rise, which corresponds to an ever-growing disordering in the component distribution over the sublattices. Consequently for the values $\exp(\Delta\mu/kT) > 8$ the kinetics of the short-range order formation in a crystal is better described by kinetic equations (4) of binary alloy crystallization.

In the description of the crystal growth in the frames of a sublattice model, the kinetic order–disorder transition occurs in a very narrow region of values $\Delta\mu/kT$ (in the case of the considered model system at $\exp(\Delta\mu/kT) = 12$). The value of bond concentration $C_{11}^{\nu\nu}$ obtained in the limits of low values of the growth rate ($R \to 0$) from equations (4), corresponding to the description of binary alloy crystallization, is non-zero. This leads, as is seen from Fig. 2, to the difference in values of the component equilibrium concentrations in the solution in the sublattice model and Monte Carlo computer simulation, on the one hand (curve 2 and 3, respectively), and in the model with one lattice, on the other hand (curve 1).

Fig. 2 presents the dimensionless growth rate dependence on supersaturation of components in the solution, $\Delta\mu/kT$. Curve 4 corresponds to the model of an "ordered" solution (2). The results shown in Fig. 2, agree with the above-mentioned derivations concerning the regions of correspondence of different models with the computer simulation results, as the growth rate is determined by the short-range order at the interface. Here the kinetic order–disorder transition in a sublattice model leads to a non-

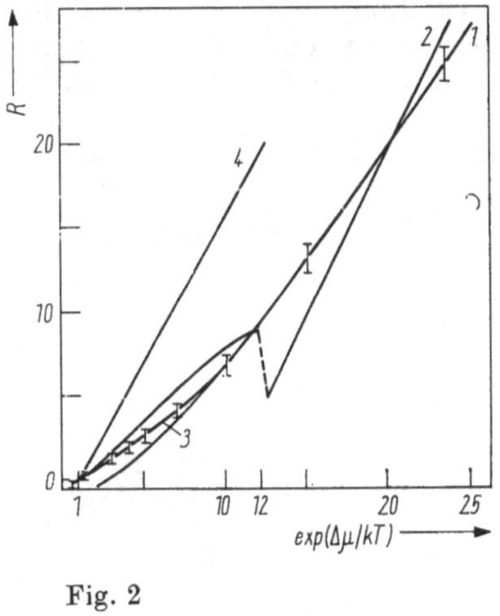

Fig. 2

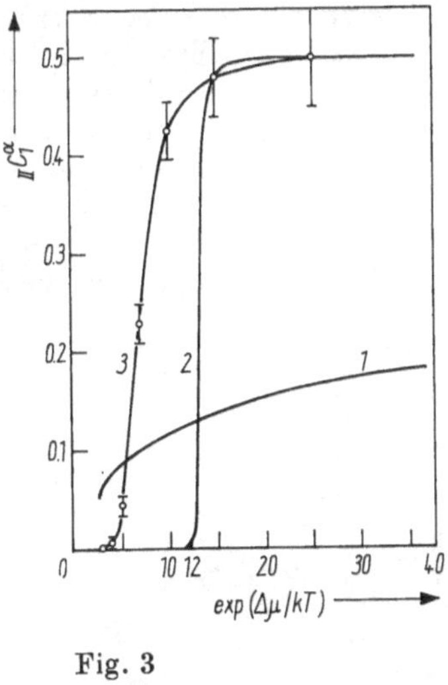

Fig. 3

Fig. 2. Dependence of the normal dimensionless growth rate R on supersaturation $\Delta\mu/kT$. (1) one-lattice model (4), (2) model with two sublattices (1), (3) Monte-Carlo computer simulation, (4) model of an "ordered" solution (2)

Fig. 3. Dependence of the α-component concentration ${}_{II}C_1^{\alpha}$ in the β-sublattice of a crystal on supersaturation $\Delta\mu/kT$. (1) one-lattice model (4), (2) model with two sublattices (1), (3) Monte-Carlo computer simulation monotonous change (by jump) in the growth rate with the supersaturation increasing in the region $\exp(\Delta\mu/kT) = 12$ of a kinetic transition (curve 2 in Fig. 2). If the short-range order in the atomic layers of the solid phase at the interface zone determines the growth rate, then the long-range order formed in a crystal presents information on the mechanism of the kinetic order-disorder transition.

Fig. 3 shows the dependence of the α-component concentration ${}_{\text{II}}C_1^{\alpha}$ in the β-sublattice of the crystal on supersaturation. The correspondence of curves to the models under consideration and methods of description is the same as in Fig. 1. As follows from the Monte Carlo computer simulation results, disordering in the component distribution over the solid phase is realized through the formation of domains. In the region of relatively small supersaturations ($\exp(\Delta\mu/kT) \leqq 4$) a characteristic size of domains d_0 in the plane of a growing crystal face is comparable with its size l and the value ${}_{\text{II}}C_1^{\alpha} \approx 0$. With the increase in $\Delta\mu/kT$ the size d_0 decreases. When $\exp(\Delta\mu/kT) > 10$, it becomes smaller than l and the concentration ${}_{\text{II}}C_1^{\alpha} \approx 0.5$, while the bond concentration $C_{11}^{\nu\nu}$ may be still considerably smaller than 0.25, the value corresponding to a complete intermixing of components in the sublattices. Fig. 4 presents the dependences of the characteristic domain size d_0 on supersaturation and concentration of bonds $C_{11}^{\nu\nu}$, obtained in computer simulation. Thus, as the kinetic equations of binary alloy crystallization (model with one lattice) do not describe the formation of clusters with an ordered distribution of components, the results of the solution of the kinetic equation in a model with sublattices for the parameter of long-range order forming in the crystal growth process correspond better to the Monte Carlo computer simulation results, than the results in a model with one lattice. This conclusion is evident and we could infer it a priori. Nevertheless we stressed it especially in order to underline the difference between one-lattice and sublattice models, and consequently the validity of the choice of the model for describing the multicomponent crystal growth.

Let us consider now the peculiarities of the order–disorder transition in the interface zone crystal–solution. Fig. 5 a and b display the distributions $C_{11}^{\nu\nu}/C_{11}^{ss}$ of the number of bonds of nearest-neighbour particles of the same sort per one solid–solid bond, over the atomic layers of the interface zone and the crystal for different values $\Delta\mu/kT$. Fig. 5a corresponds to a model with two sublattices (1), Fig. 5b to a model with one lattice (4). Here $C_1^s(z) = [{}_{\text{I}}C_1^s(z) + {}_{\text{II}}C_1^s(z)]/2$; $C_{11}^{ss}(z) = C_{11}^{\alpha\alpha}(z) + C_{11}^{\alpha\beta}(z) + C_{11}^{\beta\beta}(z)$. In Fig. 5a curve 1 corresponds to the value $\exp(\Delta\mu/kT) = 4$, curve 2 to the value $\exp(\Delta\mu/kT) = 12$, and curve 3 to values $\exp(\Delta\mu/kT) > 12$. For $\exp(\Delta\mu/kT) <$

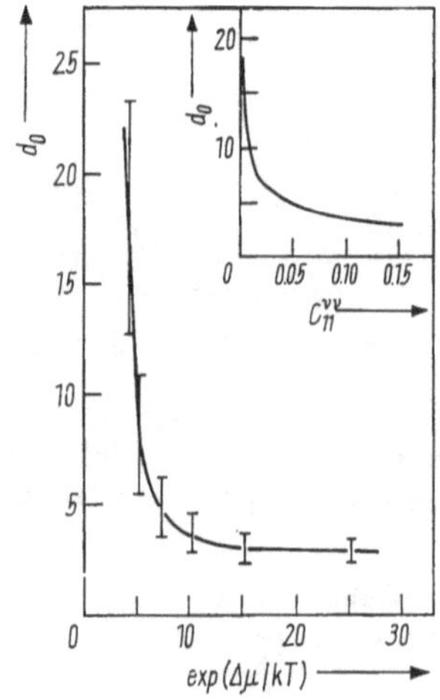

Fig. 4. Dependence of the characteristic domain size d_0 with an ordered component distribution on supersaturation $\Delta\mu/kT$ and on the bond concentration $C_{11}^{\nu\nu}$ of the nearest-neighbour particles of the same sort in the crystal. Results of Monte-Carlo computer simulation

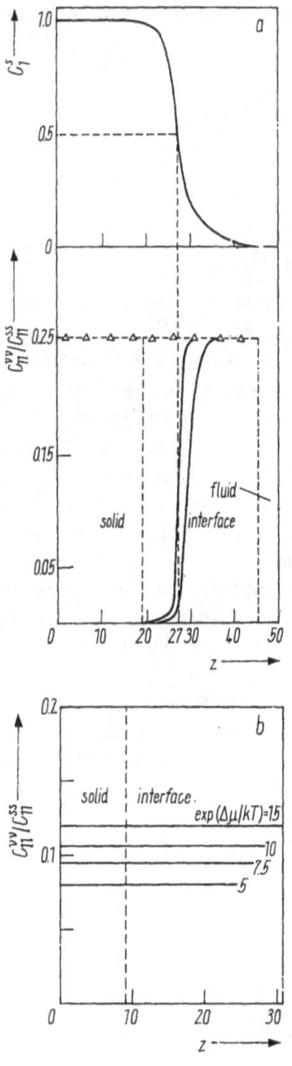

Fig. 5. a) Distribution of the number C_{11}^{vv}/C_{11}^{ss} of bonds of the nearest-neighbour particles of the same sort over the atomic layers of interface zone and crystal — model with two sublattices (1). (1) exp $(\Delta\mu/kT) = 4$, (2) 12, (3) > 12. b) Distribution of the number C_{11}^{vv}/C_{11}^{ss} of bonds of the nearest-neighbour particles of the same sort over the atomic layers of interface zone and crystal — one-lattice model (4)

< 12 the region of disordering includes only that part of the interface zone, where $C_1^s(z) < 0.5$. When exp $(\Delta\mu/kT) > 12$ it covers the whole interface zone leading to a complete disordering in the component distribution over the growing crystal. It is evident now that in a sublattice model the increase of supersaturation results in disordering of the component distribution over sublattices, which starts at the beginning of the interface zone (where the values of the solid particle concentrations in the layer are small: $C_1^s(z) \ll C_0^f(z)$).

The interface zone cross-section computed from the kinetic equations for a model with one lattice (4) is shown in Fig. 5b. Different from the results shown in Fig. 5a here the value of the short-range order parameter at the given supersaturation is the same over the whole interface zone. In other words, when $\Delta\mu/kT$ increases, disordering in the component distribution taken place over all the interface zone width equally.

Consequently, the computer simulation and the solution of the kinetic equation point out that the sublattice models describe successfully the crystal growth kinetics with an ordered distribution of components in the solid phase. The processes of this kind concern, for example, the growth of crystals of GaAs type from the melt.

4. GaAs-Type Crystal Growth from the Melt

Let us consider how the kinetic coefficients of growth of α–β crystals of GaAs type depend on the growing face orientation. Here in the melt we give the condition of complete intermixing of gallium and arsenic components, that is we assume the mixing heats in a sublattice model of liquid phase $\omega/kT > (\omega/kT)_c = -0.693$, where the value $(\omega/kT)_c$ is estimated from the Bethe approximation in statistical thermodynamics.

The set of kinetic equations for concentrations in the gallium $\left({}_{\mathrm{I}}C_1^{\alpha}(z)\right)$ and arsenic $\left({}_{\mathrm{II}}C_1^{\beta}(z)\right)$ sublattices, describing a normal growth, has the form

$$\frac{1}{W_{10}^{\nu}}\frac{\mathrm{d}[{}_pC_1^{\nu}(z)]}{\mathrm{d}t} = C_0^{\nu}{}_pC_0^{f}(z)\left[\frac{{}_qC_1^{s}(z-1) - {}_pC_1^{s}(z)}{{}_pC_0^{f}(z)}\right]^{m'} - {}_pC_1^{\nu}(z)\left[\frac{{}_pC_1^{s}(z) + {}_qC_1^{s}(z+1)}{{}_pC_1^{s}(z)}\right]^{m''} \times \exp\left[\theta_\nu - n\frac{\varphi_{\alpha\beta}}{kT}\,{}_qC_1^{\xi}(z) - m'\frac{\varphi_{\alpha\beta}}{kT}\right], \tag{7}$$

where $\nu = \alpha, \xi = \beta$ for $p = \mathrm{I}$ and $\nu = \beta, \xi = \alpha$ for $p = \mathrm{II}$. Here m', m'' are the number of z-layer particles having nearest neighbours in layers $z - 1$ and $z + 1$ respectively, n is the coordination number in a layer. For a (100)-face $n = 0$, $m' = m'' = 2$; for a (110)-face $n = 2$, $m' = m'' = 1$. For a (111)-face $n = 0$, $m' = 1$, $m'' = 3$ or $m' = 3$, $m'' = 1$ (if between the layers $z - 1$ and z values $m' = 1$, $m'' = 3$ than between the layers z and $z + 1$ values $m' = 3$, $m'' = 1$, between the layers $z + 1$ and $z + 2$ again $m' = 1$, $m'' = 3$ and so on). When $n = 0$ (faces (100) and (111)), in a chosen order of the atomic layer numeration, in the even-numbered layers only the sites of the first sublattice (I) occur, while in the odd-numbered layers — the sites of the second sublattice (II) only. The entropy factors: $\theta_\nu = \ln C_0^{\nu} + (n + m' + m'')\,\varphi_{\alpha\beta}/2kT_{\mathrm{eq}}$; here T_{eq} is an equilibrium temperature and $\nu = \alpha$ or β. To solve equation (7) the following system parameter values were assigned: $W_{10}^{\alpha} = W_{10}^{\beta} = W_{10}$, the phase transition enthalpy per one bond $\varphi_{\alpha\beta} = 21797$ J/mol. The value of a dimensionless growth rate was computed according to (6).

Fig. 6 presents the dependence of the normal dimensionless growth rate on temperature for a stoichiometric melt composition $C_0^{\alpha} = 0.5$. Here the parameter $\theta_\alpha = \theta_\beta = 2.775$. Curves 1 and 2 in Fig. 6 present the growth kinetics of the faces (110) and (100), respectively; curve 3 the growth kinetics of the faces $(111)_{\mathrm{Ga}}$ and $(111)_{\mathrm{As}}$. The

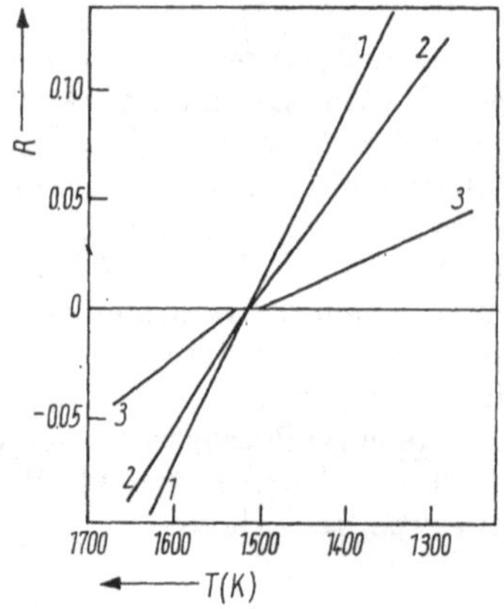

Fig. 6

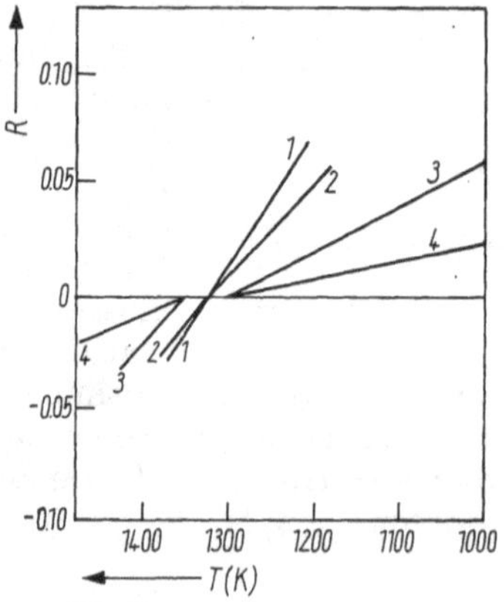

Fig. 7

Fig. 6. GaAs. Dependence of the normal dimensionless growth rate R on temperature for a stoichiometric melt composition ($C_0^{\alpha} = 0.5$). (1) growth of face (110), (2) growth of face (100), (3) growth of faces $(111)_{\mathrm{Ga}}$ and $(111)_{\mathrm{As}}$ (7)

Fig. 7. GaAs. Temperature dependence of the normal dimensionless growth rate R for a nonstoichiometric melt composition. Curves (1), (2), (3), (4) growth of faces (110), (100), $(111)_{\mathrm{As}}$, $(111)_{\mathrm{Ga}}$, respectively, for the melt composition $C_0^{\alpha} = 0.2$. Curves (1) and (2), (3) and (4) growth of faces (110), (100), $(111)_{\mathrm{Ga}}$, $(111)_{\mathrm{As}}$, respectively, for the melt composition $C_0^{\alpha} = 0.8$ (7)

melting temperature of a GaAs crystal at $C_0^\alpha = 0.5$ was given equal to 1511 K. The value of critical supercooling for the growth of the (111)-face is different from zero and equals ≈ 12 K. As is evident from Fig. 6, at stoichiometric melt composition the model reveals no difference in the growth kinetics of polar faces $(111)_{Ga}$ and $(111)_{As}$.

When in the melt gallium or arsenic is in excess, the polarity of the direction ⟨111⟩ influences the growth of GaAs crystals depending on growth conditions. Fig. 7 shows how the normal dimensionless growth rate of faces (110), (100), $(111)_{Ga}$, $(111)_{As}$ depends on temperature for the melt compositions $C_0^\alpha = 0.2$ (curves 1, 2, 4, 3, respectively) and $C_0^\alpha = 0.8$ (curves 1, 2, 3, 4, respectively). At $C_0^\alpha = 0.2$ the parameters $\theta_\alpha = 2.352$, $\theta_\beta = 3.738$, at $C_0^\alpha = 0.8$ the parameters $\theta_\alpha = 3.738$, $\theta_\beta = 2.352$. The melting temperature of GaAs crystals for both these melt compositions was given equal to 1323 K. As is obvious from computation results, with the excess of arsenic in the melt the kinetic coefficient of the $(111)_{As}$-face is greater than the kinetic coefficient of the $(111)_{Ga}$-face. Vice versa, with the excess of gallium in the melt the kinetic coefficient of the $(111)_{Ga}$-face is greater than the kinetic coefficient of the $(111)_{As}$-face. The {111}-face polarity effect is associated with the fact that in the case of non-stoichiometric melt composition the configurational parts of the entropy in the expressions for the liquid phase chemical potentials of α- and β-components are different. The computation results presented are in qualitative accord with the experimental data of investigations concerning the primary growth direction of GaAs crystals from the melt [18 to 20].

It should be noted that in some papers [18, 19] the influence of the ⟨111⟩ polarity on GaAs-crystal growth is explained by the kinetics of the layer growth over the {111} surface owing to the attachment of separate atoms and atom complexes formed in melt. It was assumed that the melt contains two types of such complexes: arsenic complexes each consisting of one As atom and three Ga atoms, and gallium complexes each consisting of one Ga atom and three As atoms. In the present paper, in terms of a simple model assuming the condition of complete intermixing and absence of atom complexes in the melt, we managed to explain the face polarity effect in the crystal growth kinetics.

Thus, as follows from computation results shown in Fig. 6 and 7 and experimental data, the growth forms of GaAs crystals are determined by the {111}-faces.

References

[1] A. A. Chernov and J. Lewis, J. Phys. Chem. Solids **28**, 2185 (1967).
[2] A. A. Chernov, Uzpekhi fiz. Nauk **100**, 277 (1970).
[3] T. A. Cherepanova, A. V. Shirin, and V. T. Borisov, in: Industrial Crystallization, Plenum Press, New York 1976 (p. 113).
[4] T. A. Cherepanova, J. P. van der Eerden, and P. Bennema, J. Crystal Growth **44**, 537 (1978).
[5] T. A. Cherepanova and V. F. Kiselev, Kristall und Technik **14**, 545 (1979).
[6] T. A. Cherepanova, in: Crystal Growth, Vol. 13, Izd. Nauka, Moscow 1980 (p. 143) (in Russian).
[7] T. A. Cherepanova, Dokl. Akad. Nauk SSSR **238**, 162 (1978).
[8] T. A. Cherepanova, phys. stat. sol. (a) **58**, 469 (1980).
[9] T. A. Cherepanova, phys. stat. sol. (a) **59**, 371 (1980).
[10] H. Pfeiffer and W. Haubenreisser, phys. stat. sol. (b) **101**, 253 (1980).
[11] H. Pfeiffer, J. Crystal Growth **52**, 350 (1981).
[12] Y. Saito and H. Müller-Krumbhaar, J. chem. Phys. **74**, 713 (1981).
[13] D. E. Temkin, J. Crystal Growth **52**, 299 (1981).
[14] T. A. Cherepanova, J. Crystal Growth **52**, 319 (1981).
[15] G. T. Didrikhsons and H. Pfeiffer, phys. stat. sol. (a) **71**, 169 (1982).

[16] T. A. CHEREPANOVA, in: Proc. XI. Internat. Mineralogical Association Meeting Crystal Growth, Izd. Nauka, Moscow 1980 (p. 145) (in Russian).
[17] M. TEMKIN, Acta Phys.-Chim. USSR 20, 105 (1946).
[18] V. A. PRESNOV, V. A. SELIVANOVA, and S. S. HLUDKOV, in: Crystal Growth, Vol. 6, Izd. Nauka, Moscow 1965 (p. 275) (in Russian).
[19] E. N. BOYKO, L. A. BORISOVA, and K. E. MIRONOV, in: Crystal Growth, Vol. 7, Izd. Nauka, Moscow 1967 (p. 246) (in Russian).
[20] YU. B. BOLKHOVITYANOV, in: Poluprovodnikovye plenki dla mikroelektroniki, Izd. Nauka, Novosibirsk 1977 (p. 170) (in Russian).

(Received November 8, 1982)

phys. stat. sol. (a) 77, 45 (1983)

Subject classification: 20.1 and 20.3; 22.5.2

Department of Physics, Birla Institute of Technology and Science, Pilani[1])

Effect of DC Field and X-Irradiation on the Optical Absorption and Thermoluminescence of NaCl Single Crystals[2])

By

A. Subrahmanyam[3])

Optical absorption and thermoluminescence of NaCl single crystals X-irradiated at 300 and 373 K in presence of moderately large dc fields ($\approx$ 0 to 15 kV/cm) are reported. With dc field the F-band peak shifts to lower wavelength side and the overall colour center concentration is observed to increase. An attempt is made to analyse the shift in the F-band.

Untersuchungsergebnisse der optischen Absorption und Thermolumineszenz von NaCl-Einkristallen, röntgenbestrahlt bei 300 und 373 K unter mäßig hohen Gleichspannungsfeldern (0 bis 15 kV je cm), werden mitgeteilt. Unter Gleichspannungsfeldern verschiebt sich das Maximum der F-Bande zu längeren Wellenlängen und es wird beobachtet, daß sich die Gesamtkonzentration der Farbzentren erhöht. Es wird versucht, die Verschiebung der F-Bande zu analysieren.

1. Introduction

Radiation-induced colouration in alkali halides has been studied extensively by many workers. It is well established that the colourability of the crystal strongly depends upon the previous history of the specimen, like electrical, mechanical, thermal, etc. treatments to which the crystal is subjected prior to or during irradiation. For example, ac field treatment on KBr crystals before X-irradiation yielded interesting results both in optical [1] and dielectric properties [2]. However, a similar treatment with dc field was found to be ineffective. Hence, it is thought that if dc field treatment has been carried out while the crystal is being X-irradiated, it may be possible to study the space charge effects [3] along with the local distribution of colour centers, in particular the F-center. Preliminary data along these lines are found to be quite interesting and have been already reported [4, 5]. The present paper deals with more details of the optical absorption and thermoluminescence data on NaCl single crystals (as-cleaved and quenched) coloured with X-rays at room temperature and at elevated temperature in the presence of a dc field ($\approx 0-15$ kV/cm). An attempt is made to analyse these data.

The details of the experimental method are the same as those given in [4] and [5]. The data reported here are found to be reproducible both on laboratory grown and commercially obtained (Harshaw chemical company) crystals.

2. Results

Fig. 1 presents the optical absorption coefficient (α) as a function of wavelength (λ) of 'as-cleaved' NaCl crystals X-irradiated at room temperature (300 K) under different

[1]) Pilani 333031, Rajasthan, India.

[2]) This work forms part of the Ph. D. Thesis submitted by the author to the Indian Institute of Technology, Kharagpur, India.

[3]) Present address: Department of Physics, Indian Institute of Technology, Madras 600036, India.

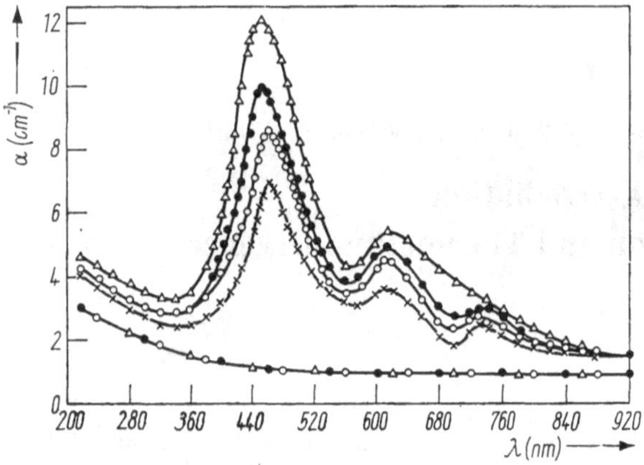

Fig. 1

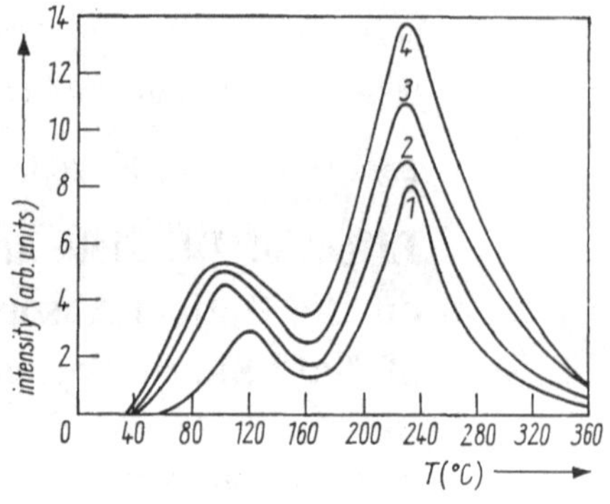

Fig. 2

Fig. 1. Optical absorption coefficient (α) as a function of wavelength for NaCl single crystals X-irradiated at 300 K under different dc fields. × 0, ○ 3.5, ● 8.4, △ 15.5 kV/cm

Fig. 2. Thermoluminescence (TL) curves of NaCl single crystals on X-irradiated at 300 K under different dc fields. (1) 0, (2) 3.5, (3) 8.4, (4) 15.5 kV/cm

dc fields. It may be noted that the background absorption of these crystals has very low values down to 400 nm below which α slightly increases with decreasing wavelength. The X-irradiated crystal exhibits prominent F-, R-, and M-bands at 460, 610, and 730 nm, respectively. The dc field seems to produce a slight shift of about 20 nm towards higher energy side in the F-band peak position along with an increase in the overall colour center concentration. However, at a field strength of 15.5 kV/cm the M-band ceases to appear.

Thermoluminescence (TL) glow curves of these (field treated) crystals are presented in Fig. 2. Application of a dc field (during irradiation) seems to shift the low temperature glow peak from 395 to 377 K and increases the TL light output considerably. The high temperature glow peak is practically uneffected.

Table 1

Summary of results on NaCl

dc field (kV/cm)	F-band peak position (nm)	F-center concentration N_F ($10^{16}/cm^3$)	new band peak position (nm)	new band colour center concentration ($10^{16}/cm^3$)	(K)		thermoluminescence peak temperature intensity (arb. units)
0	460	3.2	—	—	395	507	3810
3.5	460	2.4	460	3.0	377	504	5420
8.4	460	3.1	450	3.2	377	504	7160
15.5	460	3.4	440	5.1	377	504	8800
NaCl (quenched)							
0	465	7.0	—	—	423 to 433		9200
2.5	465	6.3	460	5.0			
4.8	465	7.4	440	7.7			
6.4	465	8.6	440	9.8			
8.7	465	9.4	440	10.2	423		10300

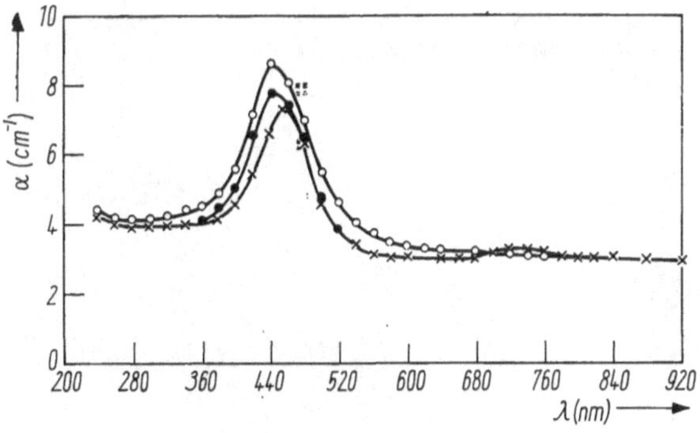

Fig. 3

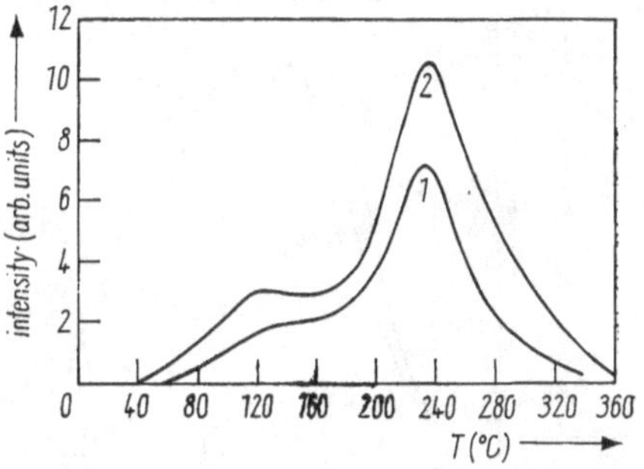

Fig. 4

Fig. 3. α vs. wavelength for NaCl single crystalsX-irradiated at 373 K under different dc fields. × 0, ● 4.5, and 7.9 kV/cm

Fig. 4. TL curves of NaCl single crystals on X-irradiation at 373 K under dc field. (1) 0 and (2) 7.6 kV/cm

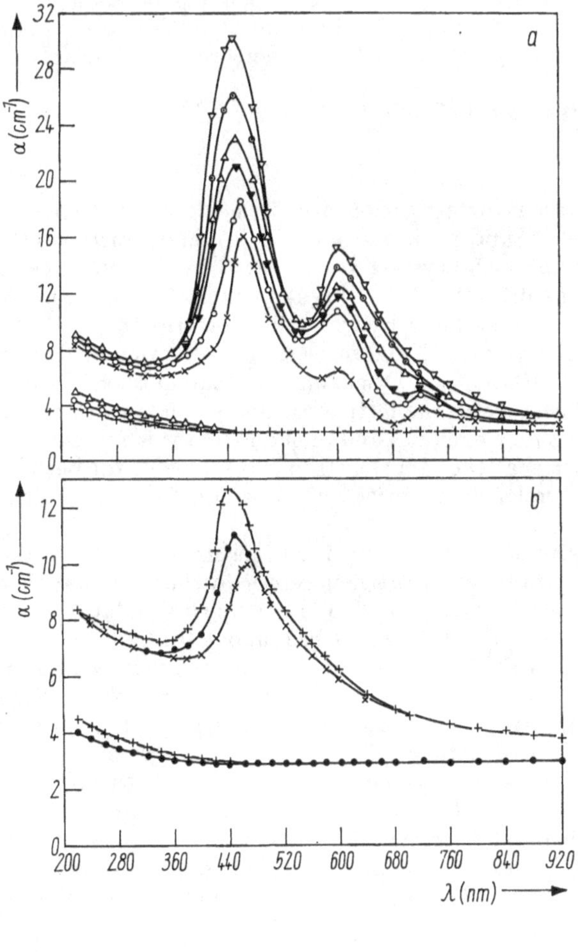

Fig. 5. α vs. wavelength for quenched NaCl single crystals X-irradiated at a) 300 and b) 373 K under different dc fields. × 0, ○ 2.5, ● 4.7, ▼ 4.8, △ 6.4, + 8.2, ⊙ 8.7, △ 14.8 kV/cm

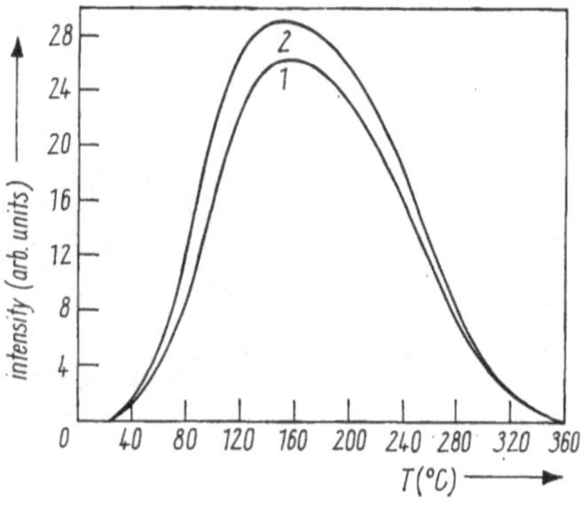

Fig. 6. TL curves of quenched NaCl crystals on X-irradiation at 300 K under dc fields. (1) 0 and (2) 8.7 kV/cm

When X-irradiation has been carried out in the presence of a dc field on NaCl crystals at an elevated temperature (373 K), there has been a considerable reduction in the colour center concentration; R- and M-bands have practically smoothed out (Fig. 3). However, with dc field qualitative results could still be observed (Fig. 3, 4).

When similar measurements as mentioned above have been carried out on these crystals quenched from about 1045 K, the behaviour is similar except for the fact that the colour center concentration is greater in the corresponding bands and the M-band disappears at a relatively low dc field of 6.4 kV/cm in optical absorption measurements (Fig. 5a and b) and the glow peaks have been broadened widely in thermoluminescence (Fig. 6).

A summary of these results is presented in Table 1.

3. Discussion

Interesting observations of the present investigation are (i) a slight but constant shift (towards higher energy) in the F-band peak position and (ii) an overall increase in the colour center concentration in NaCl crystals X-irradiated in the presence of a moderately large dc field. A similar shift in the F-band peak position has been reported in coloured alkali halides subjected to large hydrostatic pressures [6 to 8]; by electron irradiation of KCl and NaCl crystals [9], and in plastically deformed coloured KBr crystals [10]. It is proposed by Maisch and Drickamer [7] that in alkali halides hydrostatic pressure results in a shrinkage in the lattice parameter, in turn affecting the energy level scheme in which the first excited band of the F-center is replaced by a new K′-band level. It is also shown that the same electron is responsible for both F- and K′-bands. It is obvious that the relative intensities of these bands depend on the transition probability.

Assuming that the new band observed in the present investigation is similar to the F-band in all respects, the lattice parameter of the new center in these crystals has been calculated to be 2.7068×10^{-1} nm by Ivey-Mollow empirical relations. In order to produce this shift the required hydrostatic pressure is about 10.1325×10^{8} Pa [7]. Though it is quite unlikely that a dc field of a few kV/cm can produce a fairly constant stress of this magnitude, still the lattice parameter at some preferential locations may shrink because of impurities experiencing the electric field. It is well known that impurities will diffuse to dislocations where the extra charge at the dislocation lines is being compensated by space charge [11, 12], so that both impurity centers and intrinsic anion vacancies will be linked to the space charge region around the dislocations. In general, NaCl and KCl crystals are known to contain $(OH)^-$ impurities [13] pinned to dislocations present in the crystal. (The absorption band due to this impurity is reported to lie at 185 nm.) X-irradiation in presence of a dc field probably facilitates the dissociation of the impurity from the dislocation region and also may assist in the creation of O^{--} which is locally compensated by the surrounding negative ion vacancies. This model being analogous to the Z-center, it may be assumed that the anion impurity is responsible for the extra attractive (coulombic) force required to shrink the lattice. Electrons trapped by such negative ion vacancies are likely to show an absorption band as observed in the present investigation. In other words, the presence of mobile $(OH)^-$ impurities in the presence of a dc field modifies the excited state of the F-center such that it lies very near to the conduction band. Consequently, the concentration of the space charge existing at the dislocation region seems to increase. This argument is supported by band calculations [15] and dielectric measurements on these crystals [16].

An attempt has been made to resolve the composite band in the field-treated NaCl crystals. The experimentally observed composite band for a particular field value has been enhanced in intensity by multiplying the band (arbitrarily) with different numbers from 1 to 3. The regular F-band has been subtracted from the enhanced composite bands which resulted in a number of graphs. From these graphs, the band at 440 nm having a good bell shape and Smakula number (a_s) [17] which is approximately equal to that of the regular F-band, has been designated as the new band (Fig. 7a, b). The suitable numbers with which the composite band has been multiplied are shown in respective figures in brackets.

The formation of the new absorption band is attended by a reduction in the strength of the F-band under the same dosage of X-irradiation, as compared to that obtained without any dc field; this is expected since negative ion vacancies in the case of irradiation under dc field are being used up in the formation of new centers. The results indicate further that the formation and stability of the new center are dependent on the temperature and the strength of the dc field during irradiation. Thus, at elevated temperature both F- and new centres are produced to a less extent by the same dosage of irradiation. At higher fields, when the electrolytic current through the crystal is comparatively large, the new centers appear to be destroyed.

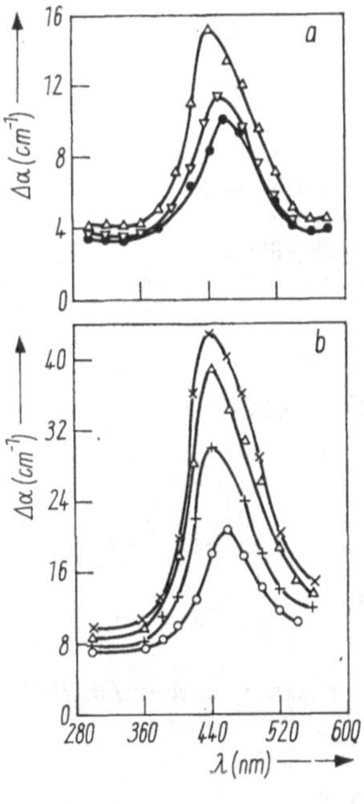

Fig. 7. Δα vs. wavelength for the new band in a) as cleaved and b) quenched NaCl crystals (ir- radiated at 300 K) under different dc fields. ○ 2.5 (2), ● 3.5 (2), + 4.8 (2), △ 6.4 (2.25), ▽ 8.4 (1.75), × 8.7 (2.25), △ 15.5 (1.75) kV/cm. Numbers in brackets relate to those which have been used for enhancing the composite band

With this limited data, it is not possible to identify the possible reason for the disappearance of M-bands in the field-treated crystals.

The overall increase in the colour center concentration with dc field may be attributed to the decrease in the recombination rate of the generated carriers because of preferential movements in presence of the dc field.

The thermoluminescence (TL) glow peaks observed in the present investigation in 'as-cleaved' NaCl single crystals (X-irradiated at 300 K) at 395 and 507 K agree well with values reported by Mariani and Alvarez Rivas [18] (the slight discrepancy is due to the difference in heating rates). It is quite likely that the presence of dc field affects the first-stage colouration (or the energy levels of the existing anion vacancies) leading to a shift in the low temperature glow peak. The second TL peak at 507 K is presumed to be due to destruction of aggregates of F-centers; such aggregates have been found to be formed in X-irradiated NaCl after it was heated to about 400 K [19]. The concentration of these aggregate centers is directly dependent on the initial concentration (at 300 K) of F-centers. These aggregates, in general, are linked to the dislocation regions. Such being the reason with dc field, the high temperature glow peak is practically uneffected; however, the TL light output has been enhanced because of the dissociation of aggregates and increase in recombination efficiency. It is interesting to note that the TL pattern has not been changed with increase in the colour center concentration [18].

In conclusion, X-irradiation of NaCl single crystals in presence of a moderately large dc field (i) seems to produce a K'-band — the energy level of which lies in the conduction band and (ii) the overall colour center concentration increases because of the decrease in the recombination rate of the generated carriers.

Acknowledgements

The author is greatly indebted to Prof. K. V. Rao and Prof. H. N. Bose for their valuable guidance and helpful discussions during the course of the present work.

References

[1] S. Govinda, Solid State Commun. **16**, 297 (1975).
[2] S. Govinda, Physica (Utrecht) **79B**, 192 (1975).
[3] A. Papadika, J. Phys. Chem. Solids **28**, 650 (1967).
[4] A. Subrahmanyam and K. V. Rao, phys. stat. sol. (a) **52**, K147 (1979).
[5] A. Subrahmanyam and K. V. Rao, Indian J. pure appl. Phys. **17**, 469 (1979).
[6] I. S. Jacobs, Phys. Rev. **93**, 993 (1954).
[7] W. G. Maisch and H. G. Drickamer, J. Phys. Chem. Solids **5**, 328 (1958).
]8] A. D. Brothers and D. W. Lynch, Phys. Rev. **164**, 1124 (1967).
[9] L. C. Arsenovici and P. D. Townsend, Phil. Mag. **25**, 381 (1972).
[10] R. Chang, Phys. Rev. **138**, A839 (1965).
[11] K. L. Kleiwer, J. Phys. Chem. Solids **27**, 705, (1966).
[12] K. L. Kleiwer and J. S. Coehler, Phys. Rev. **140**, A1226, A1241 (1965).
[13] J. Rolfe, Phys. Rev. Letters **1**, 56 (1958).
[14] H. W. Etzel and D. Patterson, Phys. Rev. **112**, 1112 (1958).
[15] P. G. Dawber and I. M. Parker, J. Phys. C **3**, 2186 (1970).
[16] A. Subrahmanyam, phys. stat. sol. (a) **69**, 773 (1982).
[17] J. J. Markham, Solid State Phys., Suppl. **8**, 21 (1966).
[18] D. F. Mariani and J. L. Alvarez Rivas, J. Phys. C **11**, 3499 (1978).
[19] K. V. Rao and J. Sharma, Physica (Utrecht) **28**, 653 (1962).

(Received January 6, 1983)

phys. stat. sol. (a) 77, 51 (1983)

Subject classification: 18.2.1; 21.1; 21.1.1

State University of the Mordov ASSR, Saransk[1])

The Bloch Lines Dynamics in Permalloy Films

By

A. A. ALEKHIN

A theoretical calculation of the overall structure of the cross-tie wall is made. The wall static and dynamic parameters (including the cross and circle Bloch lines radii) are found making use of the energy minimization of the wall unit cell. The velocity of the circle Bloch line motion and the annihilation field are calculated. The results of calculation are compared with experimental data.

Проведен теоретический расчет всей структуры доменной границы с поперечными связями. Используя минимизацию энергии элементарной ячейки границы, найдены статические и динамические параметры границы (в том числе и радиусы круговой и крестообразной линий Блоха.) Вычислены скорость движения круговой блоховской линии и поля аннигиляции. Результаты расчета сравниваются с экспериментальными данными.

1. Introduction

Domain walls with cross-ties are of great interest for their application in a cross-tie memory (CM) [1 to 3]. Fast response of such CM depends on the velocity of the circle Bloch line (CBL) motion. The structure of static insulated CBL was first calculated in [4] but the consideration such as this does not make it possible to go to the CBL dynamics since a domain wall (DW) is not taken into account. A similar calculation was presented for a cross Bloch line (Cr.BL) [5]. The CBL velocity estimation was first done in [6, 7] where the model of two infinitely extended Néel segments of opposite polarity separated by CBL was used.

The aim of this paper is to make a theoretical calculation of the structure of DW with cross-ties, to define the velocity of the CBL motion and the Bloch line annihilation field. The domain wall is periodic in the Z-direction, therefore, a unit cell with length l should be considered (Fig. 1). The cross-tie is taken as a strip with a width equal to the Cr.BL diameter ($2R_2$): segments of opposite polarity are separated by a strip with a width equal to the CBL diameter ($2R_1$). Considering such a complex DW as a wall with cross-tie we tried to build a simple model limiting ourselves to a minimum of variation parameters necessary. The cell energy is estimated from the expression $E = E_0 + \mathscr{E}$, where E_0 is the energy of "a solid cell" (without strips and BL), $\mathscr{E}$ takes into account the energy of strips, CBL, Cr.BL, and their interaction energies. The examination is made at the film median plane. The magnetization vector $\boldsymbol{j}$ in the domains on both sides of DW deviates from the Z-axis by an angle φ_0 ($\sin \varphi_0 = H/H_K = C_\infty$, H_K is the anisotropy field) if the external field H is applied along the X-axis. Thus favourably and unfavourably magnetized segments of DW appear. The circle Bloch lines aspire to balance the system, removing towards the length increase of the DW segments polarized by a field. A DW with cross-ties is a Néel wall of alternate polarity, therefore, the magnetization distribution there must be described by a nucleus and a lengthy offshoot ("a tail"). The magnetization distribution over the film thickness, for the thickness under consideration (20 to 70 nm), may be thought

[1]) 430000 Saransk, USSR.

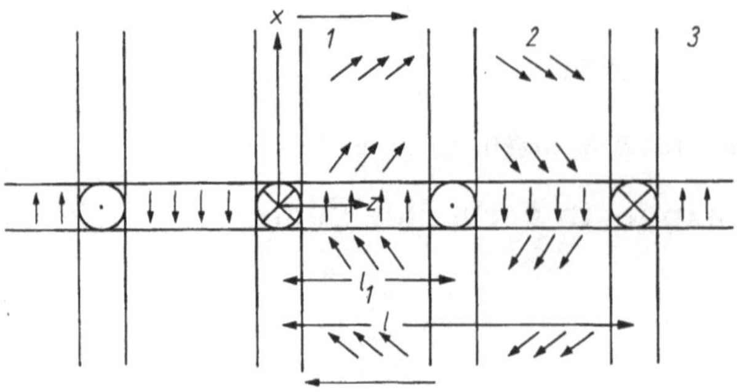

Fig. 1. Geometric model of a cross-tie wall

to be homogeneous. The magnetization change in the direction normal to the DW plane occurs much faster than along the wall, and so one can take the magnetization distribution for the favourably oriented segments the same as for the Néel DW [8],

$$\frac{j_x^{(1)}}{j} = (1 - C_1)(1 + d_1^2x^2)^{-1} + (C_1 - C_\infty)(1 + \nu_1^2x^2)^{-1} + C_\infty\,, \tag{1}$$

and for unfavourably oriented segments one obtains [6]

$$\frac{j_x^{(2)}}{j} = -(1 - C_2)(1 + d_2^2x^2)^{-1} - (C_2 + C_\infty)(1 + \nu_2^2x^2)^{-1} + C_\infty\,, \tag{2}$$

where d_i, ν_i, C_i are variation parameters characterizing the nucleus and "the tail" of DW; x is measured in units of the film half-thickness D. Expression (1) of [8] is a generalization for the case of an available external field of distribution suggested by Feldkeller on the basis of the analysis of the Néel DW structure observed in experiment [9]. Let as call DW with strips and Bloch lines "a real wall" and under "a basic part" of DW we understand a wall where areas occupied by strips and Bloch lines are left out of account. We characterize "the real wall" by nine parameters: d_i, ν_i, C_i, λ_i, and β_1,$\beta_l = l_i/l$ $[i = 1,2]$. Heıe $\lambda_i = R_1/D$. Before examination such a complex real wall" with inhomogeneous content it is necessary to calculate a simpler wall. It is common knowledge that it is easier to complicate gradually the solved problem than to solve a complex one anew. Therefore, at a first stage we calculate "the solid wall" (without strips and Bloch line) with a prescribed value of period l, find its static and dynamic parameters, define the force acting on CBL on the side of "the solid wall" reforming in the external field.

2. "The Solid Wall" Model

"The solid cell" consists of two segments with lengths l_1 and l_2, respectively. The energy of such a cell is defined by the expression

$$E_0 = 2Dl\sigma_0 = \sum_{l=1}^{2} [E_K^{(i)} + E_H^{(i)} + E_S^{(i)} + E_A^{(i)}] + W(l_1)\,, \tag{3}$$

where the terms are the anisotropy energy, the segment Zeeman, magnetostatic, and exchange energies and the energy of the cell interaction with the rest DW segments (i is the segment number, σ_0 the energy of the cell surface unit). Having put (1) and (2) into the known expression for anisotropy, Zeeman, and exchange energies one obtains for the cell first segment

$$\sigma^{(1)}_{K,H} = \sigma^{(1)}_K + \sigma^{(1)}_H = \tfrac{1}{2}\pi KD\,[(1 - C_1)^2 d_1^{-1} + (C_1 - C_\infty)^2 \nu_1^{-1} + 4(1 - C_1)(C_1 - C_\infty)(d_1 + \nu_1)^{-1}]\,,$$

$$\sigma^{(1)}_A = 2\pi AD^{-1}\left\{(3\nu_1 - d_1) - \frac{4C_1 d_1^2\,[(C_1 - C_\infty)\,d_1 + (1 - C_1)\,\nu_1]}{(1 - C_1)(C_1 - C_\infty)(d_1^2 - \nu_1^2)} + \right.$$
$$\left. + \frac{n_1\,[(C_1 - C_\infty)d_1^2 + (1 - C_1)\,\nu_1^2 + (1 - C_\infty)\,d_1\nu_1\sqrt{(1 + C_\infty)\,2^{-1}}]\,(\sqrt{b_1 + n_1} - \sqrt{b_1 - n_1})}{(1 - C_1)(C_1 - C_\infty)(d_1^2 - \nu_1^2)^2}\right\},$$

where K is the anisotropy constant, A the exchange constant, $n_1 = [b_1^2 - 8(1 + C_\infty)\,d_1^2\nu_1^2]^{1/2}$, $b_1 = (1 + C_1)\,d_1^2 + (2 - C_1 + C_\infty)\,\nu_1^2$. The energy of the cell interaction with the rest DW segments has the form

$$W(l_1) = \sum_{i=1}^{3} (W_{1,2i+1} + W_{2,2i+2}) + 2\sum_{i=0}^{3} W_{1,2i+2} + W_{\mathrm{d}}\,. \tag{4}$$

The sums have been calculated from the expression $W = -\int \boldsymbol{j}\cdot\boldsymbol{H}_{\mathrm{S}}\mathrm{d}V$, where H_{S} is a magnetostatic field calculated as a gradient of a magnetostatic potential $\Phi(\boldsymbol{r})$

$$\Phi(\boldsymbol{r}) = \int \varrho(\boldsymbol{r}_1)\cdot|\boldsymbol{r} - \boldsymbol{r}_1|^{-1}\,\mathrm{d}V_1\,.$$

Here $\varrho(\boldsymbol{r}_1) = -\operatorname{div}\boldsymbol{j}$ is the volume density of "magnetic charges". The following expression is obtained for the magnetostatic self-energy of the first segment:

$$\begin{aligned}(\sigma^{(1)}_{\mathrm{S}})_{\mathrm{S}} = {} & \pi^2 j^2 D\,[(1 - C_1)^2 d_1^{-1} + (C_1 - C_\infty)^2 \nu_1^{-1} + \\ & + 4(1 - C_1)(C_1 - C_\infty)(d_1 + \nu_1)^{-1}] + \\ & + 4\pi j^2\alpha_1^{-1} D\,\{(1 - C_1)^2 d_1^{-3}\,[F(d_1, d_1) - F(d_1, d_1, \alpha_1)] + \\ & + (C_1 - C_\infty)^2 \nu_1^{-3}\,[F(\nu_1, \nu_1) - F(\nu_1, \nu_1, \alpha_1)] + \\ & + (1 - C_1)(C_1 - C_\infty)(d_1 + \nu_1)(d_1\nu_1)^{-2}\,[F(d_1, \nu_1) - F(d_1, \nu_1, \alpha_1)] - \\ & - \alpha_1^2 J_1(\alpha_1)\}\,.\end{aligned} \tag{5}$$

The following notations are introduced here:

$$\begin{aligned}F(d, \nu, x) = {} & 2\varphi(d, \nu, x) - \int_0^\infty (a_0^2 + t^2)^{-1} \times \\ & \times \ln|(\sqrt{t^2 + 1 + x^2} + 1)(\sqrt{t^2 + 1 + x^2} - 1)^{-1}|\,\mathrm{d}t\,,\end{aligned}$$

$$\varphi(d, \nu, x) = \begin{cases}(a_0\sqrt{b_0^2 - a_0^2})^{-1}\operatorname{arctg}(\sqrt{b_0^2 a_0^{-2} - 1}\,; & b_0 > a_0\,,\\ (a_0\sqrt{a_0^2 - b_0^2})^{-1}\ln|(a_0 + \sqrt{a_0^2 - b_0^2})\,b_0^{-1}|\,; & a_0 > b_0\,,\end{cases}$$

$$\begin{aligned}J_1(x) = {} & 2(1 - C_1)^2\,[d_1(4 - d_1^2)]^{-1}\,(f(x) - \varphi(d_1, d_1, x)] + \\ & + 2(C_1 - C_\infty)^2\,[\nu_1(4 - \nu_1^2)]^{-1}\,[f(x) - \varphi(\nu_1, \nu_1, x)] + \\ & + 2(1 - C_1)(C_1 - C_\infty)(d_1 + \nu_1)\,[(d_1 + \nu_1)^2 - d_1^2\nu_1^2]^{-1} \times \\ & \times [f(x) - \varphi(d_1, \nu_1, x)]\,,\end{aligned} \tag{6}$$

$$a_0 = (d^{-1} + \nu^{-1})\,, \qquad b_0^2 = 1 + x^2\,, \qquad f(x) = x^{-1}\operatorname{arctg} x\,,$$

$$\alpha_i = \frac{l_i}{D}\,, \qquad i = 1, 2\,.$$

For the second segment $\sigma^{(2)}_{K,H}$, $\sigma^{(2)}_A$, and $(\sigma^{(2)}_{\mathrm{S}})_{\mathrm{S}}$ are obtained by means of substituting the energies of the first segment d_1 for d_2, ν_1 for ν_2, C_1 for C_2, C_∞ for $(-C_\infty)$ and $J_1(\alpha_1)$ for $J_2(\alpha_2)$ (by the same substitution we obtain $J_2(x)$ from $J_1(x)$). The energy of the interaction of the first segment with the nearest favourably oriented segments is

$$\begin{aligned}\sigma_{1,2i+1} = {} & 4\pi j^2 D\alpha_1^{-1}\,\{2(i\alpha)^2 J_1(i\alpha) - (i\alpha+\alpha_1)^2 J_1(i\alpha+\alpha_1) - \\ & - (i\alpha-\alpha_1)^2 J_1(i\alpha-\alpha_1) + (1-C_1)^2\, d_1^{-3}\times \\ & \times [2F(d_1,d_1,i\alpha) - F(d_1,d_1,i\alpha+\alpha_1) - F(d_1,d_1,i\alpha-\alpha_1)] + \\ & + (C_1-C_\infty)^2\, \nu_1^{-3}\,[2F(\nu_1,\nu_1,i\alpha) - F(\nu_1,\nu_1,i\alpha+\alpha_1) - \\ & - F(\nu_1,\nu_1,i\alpha-\alpha_1)] + (1-C_1)\,(C_1-C_\infty)\,(d_1+\nu_1)\,(d_1\nu_1)^{-2}\times \\ & \times [2F(d_1,\nu_1,i\alpha) - F(d_1,\nu_1,i\alpha+\alpha_1) - F(d_1,\nu_1,i\alpha-\alpha_1)]\}\,; \\ i = {} & 1, 2, 3. \end{aligned} \tag{7}$$

The expression for $\sigma_{2,2i+2}$ is obtained from $\sigma_{1,2i+1}$ by the substitution mentioned above (d_1 for d_2 and so on), $i = 1, 2, 3$. It is taken into account that $\sigma_{23} = \sigma_{12}$, $\sigma_{25} = \sigma_{14}$, $\sigma_{27} = \sigma_{16}$, and $\sigma_{29} = \sigma_{18}$. The cell interaction with the distant DW segments ($i = 4, 5, 6 \ldots$) was taken as dipole–dipole. The magnetic dipole moment $\boldsymbol{M}$ of the sample was determined by analogy with the electrical dipole moment,

$$\boldsymbol{M} = \int_V \boldsymbol{r}\varrho\,\mathrm{d}V\,, \tag{8}$$

where $\boldsymbol{r}$ is radius vector drawn from the origin to the "magnetic charge" element. Having put the distributions (1) and (2) into (8) and having integrated, we obtain the following expressions for the linear density of the magnetic moment of the first and the second segments:

$$\begin{aligned} m_1 &= \frac{M_1}{l_1} = 2\pi jD^2\,[(1-C_1)1\,d_1^{-1} + (C_1-C_\infty)\,\nu_1^{-1}]\,, \\ m_2 &= \frac{M_2}{l_2} = -2\pi jD^2\,[(1-C_2)\,d_2^{-1} + (C_2+C_\infty)\,\nu_2^{-1}]\,, \\ M_i &= M_x\,, \qquad M_y = M_z = 0\,; \qquad i = 1, 2\,. \end{aligned} \tag{9}$$

The expression of closed form for the energy of dipole–dipole interaction is

$$\begin{aligned} W_\mathrm{d} = {} & m_1^2 l^{-1}\beta_1^2 \sum_{i=4}^{\infty} [i(i^2-\beta_1^2)]^{-1} + m_2^2 l^{-1}\beta_2^2 \sum_{i=4}^{\infty} [i(i^2-\beta_2^2)]^{-1} - \\ & - m_1\,|m_2|\, l^{-1}\,\beta_1\beta_2 \sum_{i=4}^{\infty} (2i+1)\,[i(i+1)\,(i+\beta_1)\,(i+\beta_2)]^{-1}\,. \end{aligned} \tag{10}$$

The total magnetostatic energy of the cell surface unit can be written as

$$\begin{aligned} \sigma_\mathrm{S} &= \beta_1\sigma_\mathrm{S}^{(1)} + \beta_2\sigma_\mathrm{S}^{(2)}\,; \\ \sigma_\mathrm{S}^{(1)} &= (\sigma_\mathrm{S}^{(1)})_\mathrm{S} + \sum_{i=0}^{3}\sigma_{1,2i+2} + \sum_{i=1}^{3}\sigma_{1,2i+1} + \sigma_\mathrm{d}^{(1)}\,. \end{aligned}$$

In the case of the limiting transition to the unipolar Néel DW ($\alpha_1 \to \infty, \beta_1 \to 1, \beta_2 \to 0$) one obtains for σ_S

$$\begin{aligned} \lim \sigma_\mathrm{S} = {} & \pi^2 j^2 D\,[(1-C_1)^2\,(d_1+2)^{-1} + (C_1-C_\infty)^2\,(\nu_1+2)^{-1} + \\ & + 4(1-C_1)\,(C_1-C_\infty)\,d_1\nu_1(d_1+\nu_1+d_1\nu_1)^{-1}\,(d_1+\nu_1)^{-1}]\,. \end{aligned}$$

This expression is in agreement with the value of $\sigma_\mathrm{S}^\mathrm{N}$ obtained in [8].

As the CBL radius is of the order of hundred ångstrom and the period l is of some μm in the "solid wall' model the CBL size (and consequently, the strip size) was not taken into account. To determine a static structure of such a wall with the prescribed value of C_∞ field the cell energy σ_0 is minimized over seven parameter (d_i, v_i, C_i, and β_1, $i = 1, 2$) at a fixed period ($l = 5, 10$ μm). With the field increase the cell energy decreases. To investigate the CBL motion one needs the DW dynamic parameters. To find them C_∞ is prescribed and for β_1 (0.5 to 1) the energy σ_0 is minimized over six parameters (d_i, v_i, and C_i, $i = 1, 2$). As was mentioned above the DW segments of opposite polarity are separated in the cell by a strip whose width is equal to the CBL diameter. Therefore, one should consider the CBL strip motion in the domain wall as the motion of the united system. Let us call this system "effective" CBL. The "effective" CBL moves in a DW along the Z-axis. The "effective" CBL–DW system has one degree of freedom since the wall is stationary and the "effective" CBL motion is considered as a translational motion of its centre of mass (the centre of mass co-ordinate is l_1). The force acting on the "effective" CBL on the side of the "solid wall" reforming in the external field has the form $F_0 = -\mathrm{d}E_0/\mathrm{d}l_1 = -2D\,\mathrm{d}\sigma_0/\mathrm{d}\beta_1$. Let us give the equation a different form,

$$\sigma_0 = (\beta_1\sigma_A^{(1)} + \beta_2\sigma_A^{(2)}) + (\beta_1\sigma_{K,H}^{(1)} + \beta_2\sigma_{K,H}^{(2)}) + \\ + (\beta_1\sigma_S^{(1)} + \beta_2\sigma_S^{(2)}) = \sigma_A + \sigma_{K,H} + \sigma_S\,.$$

A general idea on the character of the effect of magnetostatics on the "effective" CBL motion can be obtained from the plot of σ_S as a function of β_1 (Fig. 2). The slope of the curve at each point is equal to the force linear density taken with the sign minus $f_S = -\mathrm{d}\sigma_S/\mathrm{d}\beta_1$. With β_1 increase the value f_S decrease, being positive (accelerating force), then goes though zero, and becomes a stopping power. Consequently, the effect of magnetostatics on the CBL motion is more complicated than it was suggested in [7]. The plots of σ_0 and σ_A as a function of β_1 are of the same form as for σ_S as a function of β_1. The force linear density $f_{K,H} = -\mathrm{d}\sigma_{K,H}/\mathrm{d}\beta_1$ over the whole region of β_1 values from 0.5 to 1 remains positive.

Thus the importance of the "solid wall" calculation is that basing on the distributions (1) and (2) we calculated precisely the main part of DW and ascertained the main features of its dynamic structure. We determined the force acting on the CBL on the side of "the solid wall" reforming in the external field, considered the contributions to this force, analysed the role of these contributions and consequently, cleared up the mechanism of the "effective" CBL motion in the domain wall. Basing on the results of the "solid wall" calculation we considered "the real wall".

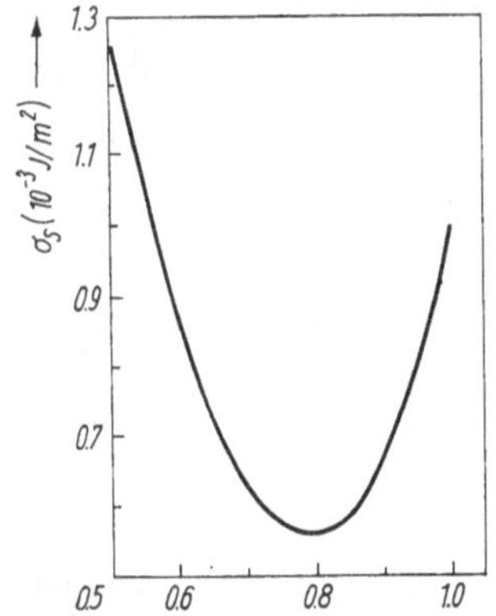

Fig. 2. The magnetostatic energy of "the solid wall" as a function of CBL position $C_\infty = 0.4$, $2D = 37$ nm, $H_K = 0.2$ kA/m

3. "The Real Wall"

As was mentioned above a DW with cross-ties is characterized by nine parameters which describe the nucleus and "the tail" of the wall and the CBL, Cr.BL, and strip dimensions as well. The cell energy is estimated from the expression $E = E_0 + \mathcal{E}$, where E_0 is determined in the "solid wall" model and $\mathcal{E}$ takes account of the energy of CBL, Cr.BL, strips, and their interaction energy. CBL structure and energy are determined as in [6], but the line radius R_1 is taken as a variation parameter. The trial function $C(\varrho_1)$ that satisfies the boundary conditions $C(0) = 1$, $C(1) = 0$ has the form

$$C(\varrho_1) = (1 - \varrho_1)\,(1 + a_1\varrho_1)\,, \tag{11}$$

where a_1 is a variation parameter, $C(\varrho_1)$ the projection of the magnetization vector on the film plane, and ϱ_1 the distance from the CBL centre in the R_1 line radius. While calculating Cr.BL we proceed from the fact that the magnetization vectors are at conjugated equiangular hyperboles. The origin is taken in their common centre. The circle with radius R_2 touching the hyperboles apices is drawn from this centre. Since the magnetization distribution in Cr.BL differs by the sign of the j_x-component only, one obtains the same expressions for the exchange and magnetostatic energy of the surface charges of Cr.BL as for CBL. The x- and z-axes divide the circle with radius R_2 into four equal parts whose volume charges have alternate signs ($\varrho_{\text{I}} > 0$, $\varrho_{\text{II}} < 0, \ldots$). We write down the volume density of the Cr.BL "magnetic charge" in cylindrical coordinates,

$$\varrho = -\operatorname{div}\boldsymbol{j} = j\left[\frac{\sqrt{1 - C^2(r)}}{r} + \frac{C(r)}{\sqrt{1 - C^2(r)}}\,\frac{\mathrm{d}C(r)}{\mathrm{d}r}\right]\sin 2\varphi\,.$$

We face some difficulties when calculating the magnetostatic energy of the Cr.BL volume charges by the usual method, but since the Cr.BL energy is only a small part of the cell total energy, it will not be an appreciable error to estimate as follows. We average the charge volume density in each of the four parts of Cr.BL and then calculate the magnetostatic self-energy of each part and the energy of interaction of these parts with each other. The trial $C(\varrho_2)$ is taken in the same form as for CBL (11), but with the parameters a_2, $\varrho_2 = r/R_2$. The strips can be considered as the walls inside a wall (see Fig. 1). Let us call a strip containing CBL the first strip (S_1) and a strip with Cr.BL the second one (S_2). The strips are divided by lines parallel to the x-axis and passing through the middle of the strips into parts S_1', S_1'', S_2', and S_2''. The following magnetization distributions are chosen for them:

$$\begin{aligned}
\tilde{j}_x &= jC_\infty - (z - l_1)\,R_1^{-1}(j_x^{(1)} - C_\infty j)\,; && (l_1 - R_1) \leqq z \leqq l_1\,,\\
\tilde{j}_x &= jC_\infty + (z - l_1)\,R_1^{-1}(j_x^{(2)} - C_\infty j)\,; && l_1 \leqq z \leqq (l_1 + R_1)\,,\\
\tilde{j}_x &= jC_\infty - (z - l)\,R_2^{-1}(j_x^{(2)} - C_\infty j)\,; && (l - R_2) \leqq z \leqq l\,,\\
\tilde{j}_x &= jC_\infty + (z - l)\,R_2^{-1}(j_x^{(1)} - C_\infty j)\,; && l \leqq z \leqq (l + R_2)\,.
\end{aligned} \tag{12}$$

Since the volume density of "magnetic charges" decreases slowly with x increase one can calculate the magnetostatic energy of the strips making use of the mean value of the volume charge density. For the S_1 region, for example, the mean charge density is

$$\bar{\varrho}_I = (R_1 - L_1)^{-1} \left\{ \frac{1}{2} [\bar{j}_x^{(1)}(L_1) - \bar{j}_x(R_1)] + j \left[(L_1 - R_1) R_1^{-1} (1 - C_\infty^2)^{1/2} - \right.\right.$$

$$\left.\left. - R_1^{-1} \int_{R_1}^{L_1} \left[1 - \left(\frac{j_x^{(1)}}{j} \right)^2 \right]^{1/2} dx \right] \right\},$$

where $L_1 = D/v_1$ is the strip length, $\bar{j}_x^{(1)} = j_x^{(1)} - jC_\infty$. An arbitrarily chosen strip length does not have a great effect on the results of the calculation, since a region with length L_1 makes the main contribution to magnetostatic energy. While calculating the strip exchange energy we proceed from the well-known expression

$$\mathscr{E}_A = Aj^{-2} \int_V \left[\left(\frac{\partial \boldsymbol{j}}{\partial x} \right)^2 + \left(\frac{\partial \boldsymbol{j}}{\partial z} \right)^2 \right] dV .$$

It is clear from the qualitative considerations that the main contribution to $\mathscr{E}_A$ is made by the $/\partial \boldsymbol{j}/\partial z)^2$ term, i.e. $\mathscr{E}'_A \approx Aj^{-2} \int_V (\partial \boldsymbol{j}/\partial z)^2 \, dV$, that is corroborated by calculation. The magnetostatic energy of the strip interaction with the main DW part and with each other is found by means of the mean density of "magnetic charges". The integrals obtained depend on L_1 and λ_i ($i = 1, 2$). They were calculated by the Korobov method 10]. The cell total energy is given as follows:

$$E = E'_0 + \mathscr{E}_1 + \mathscr{E}_2 ,$$

where $\mathscr{E}_1$ is the CBL total energy and $\mathscr{E}_2$ the Cr.BL total energy. E'_0 is a functionof all nine parameters and $\mathscr{E}_1$ depends on λ_1 and a_1, and $\mathscr{E}_2$ depends on λ_2 and a_2. The parameter a_1 is in $\mathscr{E}_1$ only, therefore, while minimizing over a_1 it is sufficient to consider the condition $d\mathscr{E}_1/da_1 = 0$. The task is to define the value $\mathscr{E}_1$ which is minimum over a_1 for a prescribed λ_1. Having a number of values of the energy $\mathscr{E}_1$ which are minimum over a_1 one can approximate $\mathscr{E}_1$ by polynominals in λ_1. As a result $\mathscr{E}_1(\lambda_1, a_1)$ is transformed into $\mathscr{E}'_1(\lambda_1)$ which is minimum over a_1. $\mathscr{E}_2$, the energy of Cr.BL, has an analo-

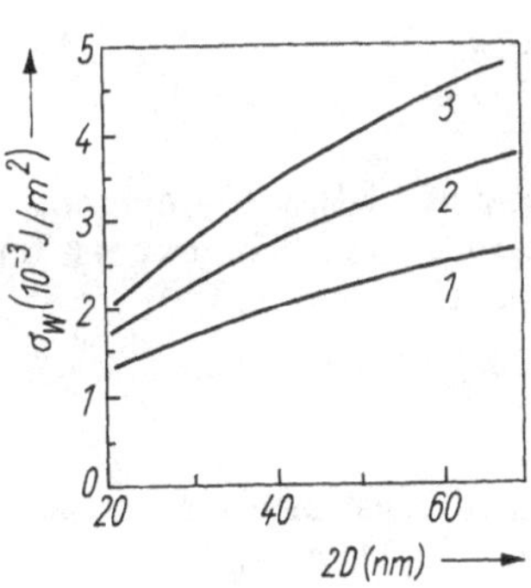

Fig. 3.

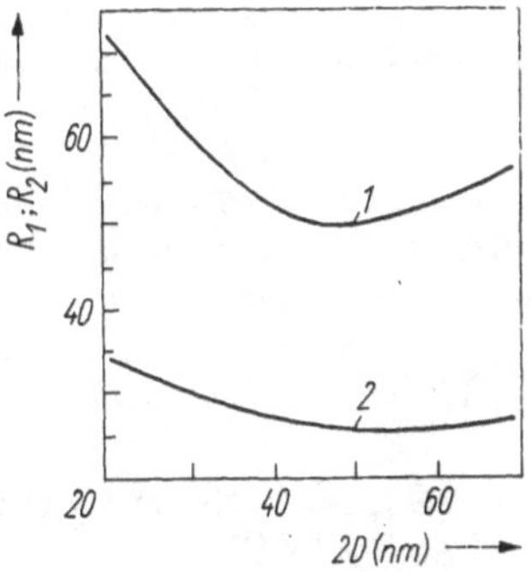

Fig. 4.

Fig. 3. The energy of the domain wall surface unit as a function of film thickness. Curve (1) shows the energy of the cross-tie wall according to the results of the present paper. Curves (2), (3) show the Néel wall energy according to the results of [18] and [8], respectively

Fig. 4. The Bloch lines radii as a function of film thickness: (1) CBL; (2) Cr.BL gous transformation. Minimizing the cell energy E over d_i, ν_i, C_i, λ_i, and α ($i = 1, 2$) with $\beta_1 = 0.5$ and $C_\infty = 0$ one obtains the period l and other parameters of the static structure. In the further consideration the value l is considered to be constant. The DW period decreases with film thickness increase (20 to 70 nm) in agreement with experiment [11]. An increase of the l value with further increase of film thickness [12] is not expected from our model. In the thickness interval 20 to 30 nm the change of l is 870 nm and for 60 to 70 nm only 150 nm. The polarity alternation in DW with cross-ties decreases greatly the domain wall energy, which is seen from a plot of the DW energy as a function of the film thickness (Fig. 3, curve 1).

It is known that stray fields compress DW, therefore, the wall length must increase with decreasing stray field value. In accordance with this we have found from our calculations that the nucleus length in DW with cross-ties is larger than that in Néel walls. Since Cr.BL contains surface and volume charges (moreover, the stray fields of the central part of the wall intensity the stray fields of the Cr.BL charges and attenuate stray fields of the CBL surface charges), then as a result of the compressing effect of the stray fields the Cr.BL radius R_2 will be smaller than the CBL radius R_1 (Fig. 4). The quantitative relationship of R_1 and R_2 values was influenced by the strip sizes taken to be $2R_1$ and $2R_2$. The energy of the strip with Cr.BL is larger than that of the strip with CBL by 40%.

4. The Definition of the DW Dynamic Parameters and the Velocity of the "Effective" CBL Motion

We describe the "effective" CBL motion by Newton's second law. The equation of motion of such a system with the initial condition $t = 0$, $\beta_1 = 0.5$, $v = 0$ has the form

$$\frac{\mathrm{d}\,Mv}{\mathrm{d}t} = F + \mathcal{F}_1\,, \tag{13}$$

where $F = -\mathrm{d}E/\mathrm{d}l_1 = -2D\,\mathrm{d}\sigma_\mathrm{w}/\mathrm{d}\beta_1$ is the force acting on the "effective" CBL on the side of "the real wall" reforming in the external field. v is the velocity of its motion. $\mathcal{F}_1$ is a resistance force that is expressed by the Rayleigh function which is, in its turn, expressed by the energy dissipated in unit time in the volume of CBL and the strip. M is the mass of the "effective" CBL. The energy dissipated in unit time in the volume of the moving CBL is [6]

$$\frac{\mathrm{d}\mathscr{E}_1}{\mathrm{d}t} = -\alpha_0(\gamma j)^{-1}\,v^2 \int\limits_V \left(\frac{\mathrm{d}j}{\mathrm{d}z}\right)^2 \mathrm{d}V = -\alpha_0 j(2\gamma A)^{-1}\,\mathscr{E}_A^0 v^2\,,$$

where α_0 is an attenuation constant, $\mathscr{E}_A^0$ the exchange energy of CBL, γ the magnetomechanical ratio. One has the similar relationship for a strip as well. The expression of closed from for the loss power in the "effective" CBL has the form

$$\frac{\mathrm{d}\mathscr{E}}{\mathrm{d}t} = -2R_0 = -\alpha_0 j\mathscr{E}_A(2\gamma A)^{-1}\,v^2\,, \tag{14}$$

where $\mathscr{E}_A = \mathscr{E}_A^0 + 2\mathscr{E}'_A$, $\mathscr{E}'_A$ is the strip exchange energy. The resistance force acting on the moving "effective" CBL is expressed by the Rayleigh function R_0.

$$\mathcal{F}_1 = -\alpha_0 j\mathscr{E}_A(2\gamma A)^{-1}\,v\,. \tag{15}$$

Since the mass M depends on the "effective" CBL position, (13) can be written as

$$M\left(\frac{\mathrm{d}v}{\mathrm{d}t}\right) = 2Df - bv - v^2 l^{-1}\left(\frac{\mathrm{d}M}{\mathrm{d}\beta_1}\right), \tag{16}$$

where $f = F/2D$ is the force linear density, $b = \alpha_0 j \mathscr{E}_A (2\gamma A)^{-1}$. The mass of the surface unit of the "effective" CBL, m, was taken proportional to the mass obtained by Doring's formulae [13]. The value $K(m)$ was introduced as a proportionality coefficient with values 1, 3, 5, 10. Similarly the coefficient $K(\alpha_0)$ was introduced for considering the attenuation change. With $\alpha_0 = 0.01$ taken as the initial value the coefficient $K(\alpha_0)$ has the following values: 0.5, 1, 1.5, and 2. To define the forces F and $\mathscr{F}_1$ it is necessary to known the dynamic structure parameters of DW. Assume the strip and CBL have displaced by some value $\Delta\beta_1$. If we "fix" the system and consider then all the parameters to be in the equilibrium state, the angle of rotation of the magnetization vector in the DW basic part will be much smaller than the mean angle of rotation in the strip. Therefore, the wall basic part reforms much faster and it tends to the state corresponding to the energy minimum at every point of the strip position. Moreover, the dynamic strip characteristics will be defined less exactly than the dynamic properties of the DW basic part, the definite influence of which on the "effective" CBL motion is shown in "the solid wall" model. Proceeding from this we separate β_1 into parts $\beta_1 = 0.5 + K\,\Delta\beta_1$ in the range from 0.5 to 0.98 and minimize the cell energy E over d_i, v_i, C_i, and λ_i $(i = 1, 2)$ for every value of C_∞ at given value of β_1. As a result of this minimization we obtain values of these dynamic parameters, with the help of which we define σ_w, f, and b. (Volume charges appearing in moving CBL are left out of account.)

One can obtain a qualitative notion about the character of the "effective" CBL motion from the potential curve, the plot of σ_w as a function of β_1 (Fig. 5). The tangent of the angle of rotation in every point is equal to the force linear density f taken with the minus sign. In the point where σ_w is minimum f is equal to zero. f will be an accelerating force on the left handside of this point and a retarding one on the right-hand side. The potential well depth decreases with increasing field C_∞ and the minimum position displaces to the right to $\beta_1 = 1$. It is clear from Fig. 5 that the full energy of the wall $(\sigma_w)_{min}$ corresponding to the balanced state, decreases with the increase of external magnetic field. This may be explained by the fact that when applying the magnetic field along the x-axis the magnetization vector at the domains on both sides of DW deviates from the z-axis by the angle φ_0. As the result the full angle to which the magnetization vector revolves in DW, decreases in half of the cell and increases in the other. Thus favourably and unfavourably magnetized segments of DW appear. The "effective" CBL moves in a way that the length of segments with the highest energy decreases, and that with the least energy increases; that is why with increasing field the structure of a domain cross-tie wall more and more approaches the Néel wall structure, for which full energy of the balanced structure σ_w decreases with the increase of external field. Solving numerically equation (16) one finds the displacement value $\Delta\beta_1$, the "effective" CBL motion time and the value of its velocity. The

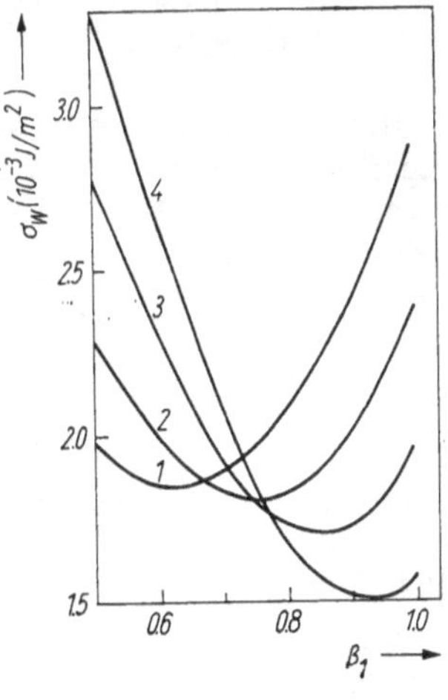

Fig. 5. The energy of the domain wall surface unit as a function of the CBL position for different field values: (1) $C_\infty = 0.1$, (2) 0.2, (3) 0.3, (4) 0.4; $2D = 37$ nm, $H_K = 0.2$ kA/m

mean velocity of motion is defined from the relationship $\bar{v} = l/2t_{an}$, where t_{an} is the Bloch line annihilation time. These mean velocities have values of 10^2 to 10^3 m/s in agreement with experiment [1, 14, 15]. The Bloch line annihilation occurs at a field $C_\infty = 0.4$; this value is in the range $(0.3 \leqq C_\infty \leqq 0.5)$ determined in [12]. (It should be noted that for $K(m) > 10$ annihilation takes place in the field $C_\infty = 0.3$.) It should be noted that the domain structure reforming time and the "effective" CBL coercive force are not taken into account. The dependence of the mean velocity of the "effective" CBL on the attenuation factor for different values of $K(m)$ is given in Fig. 6. For the film 37 nm thick, for $C_\infty = 0.4$, $H_K = 0.4$ kA/m, $K(\alpha_0) = 1$, and $K(m) = 1$, the "effective" CBL mean velocity is equal to 1524 m/s. If $K(m) = 10$, then $\bar{v} = 559$ m/s, if $K(m) = 300$, then $\bar{v} = 98.5$ m/s. One should note that for small values of $C_\infty \leqq 0.3$ the "effective" CBL mean velocity increases almost linearly with C_∞, whereas the $\bar{v}$ value decreases slowly with increasing H_K. Consequently, basing on the experimental data [1, 2, 14 to 16] one can make a reasonable choice of $K(m)$ and $K(\alpha_0)$ values. It follows from the calculation that DW period, nucleus, and tail length, CBL and Cr.BL radii, and annihilation time decrease with increasing anisotropy constant but f increases a little. For the film thickness of 37 nm the DW energy increases by 40% with the anisotropy constant increasing by eight times. The thickness dependence of the mean velocity of the CBL motion depends on many factors. For the unambiguous solution of this problem it is necessary to determine experimentally the dependence of attenuation factor α_0, circular Bloch line coercitivity H_B, domain structure rearrangement time, retarding field of the "magnetization ripple", and the coefficient $K(m)$ on the film thickness. If we assume $K(\alpha_0)$, H_B and $K(m)$ to be the same for different thicknesses (20 to 70 nm), then our calculation will give the "effective" CBL mean velocity increase with film thickness (Fig. 7). But the experimental data of [14] give the inverse dependence of the mean velocity on the film thickness. Feldkeller has pointed out, however, that the velocity of the CBL motion in thin films ($2D \leqq 20$ nm) is smaller than that in thick ones [17]. But it is necessary to note the following: with the increase of film thickness (for $K =$ const) and K (for $2D =$ const) the period of the DW decreases, that is why the time of annihilation of Bloch lines and not the average velocity of CBL motion characterizes high-speed memory. It is impossible to deduce the correct conclusions about high-speed memory from the fact that $\bar{v}$ increases or decreases. In fact our calculation shows that with the increase of K (for just the same film thickness) $\bar{v}$ decreases; with the increase of film thickness (for $K =$ const) $\bar{v}$ increases, while in both cases the annihilation time decreases. It is necessary to note that the annihilation time of Bloch lines in DW with balanced wall period is less than the annihilation time in DW with unbalanced period.

It appears from the above that there is a good qualitative and quantitative agreement between the theoretical calculation using our model of the cross-tie wall and the experiment. It testifies the fact that the parameters chosen for the model description

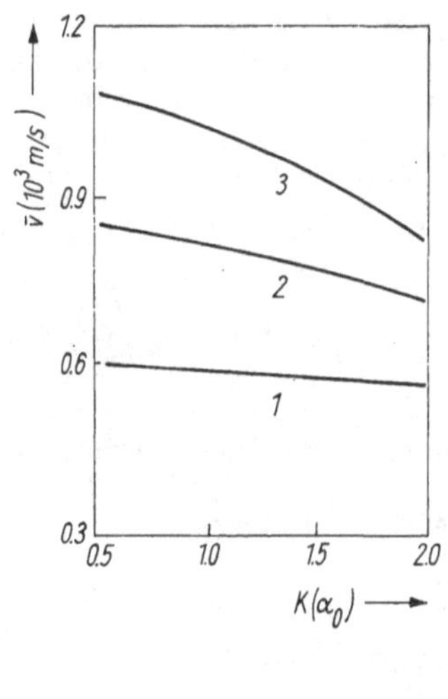

Fig. 6. The "effective" CBL mean velocity as a function of the coefficient $K(\alpha_0)$ for different values of $K(m)$. (1) $K(m) = 10$; (2) 5; (3) 3; $2D = 37$ nm, $H_K = 0.2$ kA/m

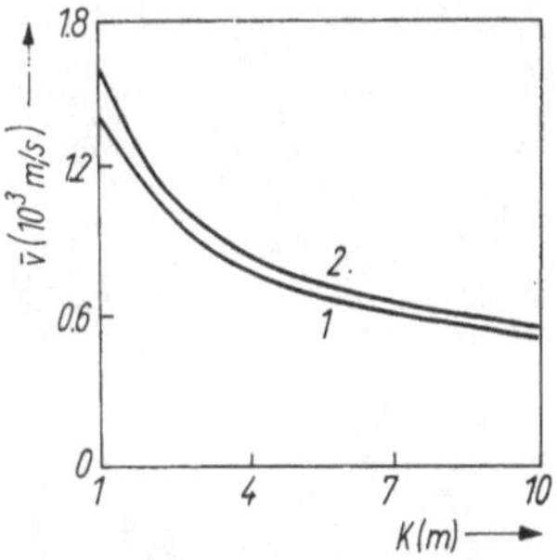

Fig. 7. The "effective" CBL mean velocity as a function of the coefficient $K(m)$ for different values of film thickness. (1) $2D =$ $= 37$, (2) 60 nm; $C_\infty = 0.4$; $K(\alpha_0) = 1$, $H_K = 0.2$ kA/m

influence greatly the character of the CBL motion in the domain wall. One can expect that the model complication and improvement (first of all the strip "liberation") will lead to better results.

Acknowledgements

The author is greateful to Dr. R. V. Telesnin for his interest in the paper, thanks are also due to Dr. A. G. Shishkov for helpful discussions.

References

[1] L. Schwee, H. Irons, and W. Anderson, IEEE Trans. Magnetics **10**, 564 (1974); **12**, 608 (1976).
[2] L. Schwee, W. Anderson, Y. Lin, and R. Lee, J. appl. Phys. **49**, 1831 (1978).
[3] R. Sery, IEEE Trans. Magnetics **11**, 29 (1975).
[4] E. Feldkeller and H. Tomas, Phys. kondens. Materie **4**, 8 (1965).
[5] H. Bäurich, phys. stat. sol. **16**, K39 (1966).
[6] A. A. Alekhin, O. A. Vinogradov, and V. V. Sergeev, phys. stat. sol. (b) **71**, K123 (1975).
[7] A. A. Alekhin, O. A. Vinogradov, and V. F. Tschischin, Proc. VI. All-Union Coll. Thin Magnetic Films, Saransk 1979 (p. 78) (in Russian).
[8] O. A. Vinogradov, Izv. Acad. Nauk SSSR, Ser. fiz. **31**, 370 (1967).
[9] E. Feldkeller, Z. angew. Phys. **15**, 206 (1963).
[10] N. M. Korobov, Dokl. Akad. Nauk SSSR **124**, 1207 (1959); **132**, 1009 (1960).
[11] H. Rubinstein and R. Spain, J. appl. Phys. **31**, 30S (1960).
[12] W. Bürger, phys. stat. sol. (a) **13**, 429 (1972).
[13] W. Doring, Z. Naturf. **3a**, 373 (1948).
[14] A. G. Shishkov, E. N. Ilicheva, N. G. Kanavina, O. S. Kolotov, and R. V. Telesnin, phys. stat. sol. (a) **50**, K101 (1978).
[15] A. G. Shishkov, E. N. Ilicheva, N. G. Kanavina, and O. S. Kolotov, Fiz. tverd. Tela **18**, 2134 (1976).
[16] O. Bostanjoglo and Th. Rosen, phys. stat. sol. (a) **57**, 561 (1980).
[17] E. Feldkeller, Z. angew. Phys. **13**, 257 (1960).
[18] H. Riedel and A. Seeger, phys. stat. sol. (b) **46**, 377 (1971).

(Received January 10, 1983)

phys. stat. sol. (a) 77, 63 (1983)
Subject classification: 5; 11; 21

Institute of Nuclear Physics, Academy of Sciences of the Kazakh SSR, Alma-Ata[1])

Mössbauer Study of Radiation Effect of the ^{181}Ta(p, n)^{181}W Reaction

By
A. K. ZHETBAEV and A. N. OZERNOI

The influence of irradiation and subsequent annealing upon Mössbauer spectrum parameters of tantalum-181 in metallic tantalum is investigated. Probable reasons of Mössbauer line broadening after irradiation are discussed.

Исследовано влияние облучения и последующего отжига на параметры мессбауэровского спектра тантала-181 в металлическом тантале. Приводится обсуждение вероятных причин уширения мессбауеровской линии после облучения.

1. Introduction

As it is known [1 to 3], the Mössbauer effect can give interesting information on nature and properties of radiation defects in metals. Vogl [1] and Petry et al. [2] pointed out the possibility of tracing interaction kinetics of cobalt (iron) impurity atoms with defects in aluminium irradiated by neutrons and electrons. The appearance of an additional line in the spectrum is assigned to a dumb-bell configuration formation from matrix interstitial and ^{57}Co impurity atoms.

Zhetbaev [3] suggests to use the nuclear reactions for selective assignment of primary knocked-on atom Mössbauer spectra. The essence of this technique is that during neutron or charged particle irradiation nuclear reactions take place which can lead to Mössbauer radionuclide formation in the target. During this process recoil nuclei receive kinetic energy sufficient for their displacement from lattice sites. Therefore, the maximum damaged zone of the target is labelled by the Mössbauer radioactive nuclei. Results on the Fe–C system [4] have revealed a high ability of the method. It has been shown that under the effect of ionizing radiation the process of graphitization with formation of α-iron clusters around the primary knocked-on atom proceeds.

The present paper, which essentially continues the previous investigations on the ^{57}Fe [3, 4] isotope for selective assignment of the states of primary knocked-on atoms, summarizes the results of Mössbauer spectroscopy of ^{181}Ta in proton irradiated metallic tantalum samples.

2. Experimental Technique

The experiments were carried out on metallic tantalum (99.975% pure). The foil, 12 μm thick, was rolled down to a thickness of 3 μm in several stages with intermediate annealing at 2000 °C in 10^{-8} Torr vacuum and then, annealed at 2100 °C for 6 h. One part of the foil was then used as a cyclotron target. The other served as an absorber for Mössbauer spectra measurements.

[1]) 480082 Alma-Ata, USSR.

The targets were irradiated by protons of 10 MeV energy in vacuum in an isochronous cyclotron. The tantalum nuclear reaction with protons led to radioactive ^{181}W nuclei formation in the target which populate the ^{111}Ta 6.2 keV Mössbauer level after decay. The reaction ^{181}Ta(p, n)^{181}W has the maximum of the excitation function with a cross-section of 100 mb at the proton energy of about 10 MeV [5], and the competitive reactions, which could contaminate the target with foreign radio-isotopes at the given energy, have negligibly small cross-sections. There is a possibility of using the ^{181}Ta(d, 2n) ^{181}W reaction with a cross-section six times larger than that of the (p, n) reaction. However, the ^{181}Ta(d, p)^{182}Ta reaction giving about 10% of the (p, n) reaction yield proceeds simultaneously. This determines the choice of the energy and the type of bombarding particles.

The tantalum foil was fastened with the help of a joint washer to a water-cooled holder, the wall thickness of which in the zone of the beam hit did not exceed 0.5 mm. The current density of the proton beam has been maintained at 1 μA cm^{-2} during the experiment. The integral irradiation dose was 10^{18} cm^{-2}. The samples have been kept for 10 days at room temperature before measurements.

Mössbauer spectra were measured on a NP-255 spectrometer at room temperature with a source moving at constant acceleration in standard transmission geometry. The velocity calibration was carried out with the help of a laser interferometer; γ-radiations of 6.2 keV were detected by a proportional counter filled with argon–methan mixture.

3. Results and Discussion

Contrary to a previous report [6], in which the attempt to use the cyclotron tantalum target as a Mössbauer source failed, we succeeded in observing a resonance absorption at the 6.2 keV level of ^{181}Ta with a ^{181}W(Ta) source immediately afetr proton irradiation without intermediate annealing. The emission Mössbauer spectrum is represented in Fig. 1, curve a, which shows a very broadened single line, $W = 0.44(2)$ mm s^{-1}, with isomer shift characteristic of the ^{181}W(Ta) source. Such a source can be also prepared by diffusion of ^{181}W into tantalum as, for example, in [6].

The foil target was also studied as an absorber and ^{181}W(W), which was obtained using the technique described in [7], served as a source. After irradiation of the sample the Mössbauer line has been broadened $(W = 0.30(2)$ mm s$^{-1})$ almost by a factor of

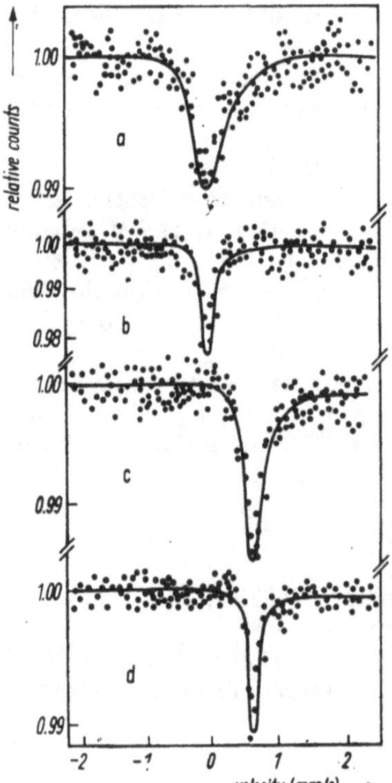

Fig. 1. Mössbauer spectra of tantalum foil irradiated by protons: (a), (b) emission spectra before and after annealing with metallic tantalum absorber, respectively; (c), (d) absorption spectra before and after annealing with the ^{181}W (W) source, respectively

two in comparison with the initial one while the isomer shift did not change significantly.

One can see that the Mössbauer spectrum of the irradiated sample has a different linewidth depending upon whether the investigated sample is a source or an absorber. Such a difference is ascribed to by the fact that the emission spectrum was registered only from the nuclei which underwent direct collision with the bombarding protons. It should be expected that the maximum possible matrix radiation damage under the given conditions takes place around these nuclei. In the case of the absorption spectrum the information is obtained from the whole sample where the average defect concentration over its volume cannot be so high as in local regions of the primary knocked-on atoms. No similar phenomena have been observed in experiments using $^{56}Fe(d, p)^{57}Fe$ reactions in metallic iron and stainless steel [8]. This result can be understood if one takes into consideration that the γ-resonance at the 6.2 keV level of ^{181}Ta is quite different in its nuclear characteristics from ^{57}Fe. Owing to the great difference in nuclear radii of the ground and excited states and also in large quadrupole momenta of both states, ^{181}Ta possesses a very high sensitivity to insignificant changes of the local surrounding of resonance atoms which is observed in our experiments.

After annealing of the cyclotron target at 2000 °C in 10^{-8} Torr the linewidths in absorption and emission spectra (Fig. 1, curves b and d) become practiclly equal: $W = 0.16(1)$ mm s^{-1} which shows a recovery of the matrix structure to the initial state.

According to Sauers data [6] the Mössbauer linewidth increases linearly with increasing impurity concentration with a slope $\mathrm{d}W/\mathrm{d}C = 1380$ mm s^{-1} and this increase is connected with the lattice parameter change associated with the introduction of interstitial impurities, i.e. interstitial oxygen, nitrogen, or hydrogen atoms, into metallic tantalum, at least, at low concentrations $\leqq 0.1$ at%. The valence in this process has secondary meaning. Assuming that the line broadening is mainly caused by interstitial atoms near the Mössbauer atom, we have estimated the average concentration of these atoms near the knocked-on atom after irradiation as 0.02 and 0.01 at%, respectively, from linewidth changes after irradiation and annealing.

An atom which underwent nuclear transformation and received a definite recoil energy (knocked-on) will migrate in the target as if it was implanted by an ion accelerator ("inner implantation"). Therefore, the spatial distribution of the elemtary defects produced by this atom will be directly connected with the known distribution of projected ion ranges.

As quantitative characteristics of the elementary defect distribution profile in the initial damage made by an atom with energy E one can choose the function $S_\mathrm{D}(E, x)$ in such a way that $S_\mathrm{D}(E, x)\,\mathrm{d}x$ is the mean energy transferred to atoms with the initial energy E in the projected range interval $(x, x + \mathrm{d}x)$. Then the average number of Frenkel pairs $\mathrm{d}N$ produced by an atom with the energy E in the target layer $(x, x + \mathrm{d}x)$ will be proportional to S_D,

$$\mathrm{d}N \approx \frac{S_\mathrm{D}(E,x)\,\mathrm{d}x}{E_\mathrm{d}},$$

where E_d is the threshold energy of atomic displacement from the lattice site.

Zelenskii et al. [9] calculated the function $S_\mathrm{D}(x)$ for iron irradiated by iron ions and showed that the defect concentration in the damaged zone is significantly higher than over the projected range. A similar situation should be expected for ^{181}W recoil atoms arising in the tantalum lattice as a result of the (p, n) nuclear reaction. This expectation has been realized by the data we have obtained from an evaluation of the concentration of interstitial atoms in proton-irradiated tantalum.

4. Conclusions

(i) The influence of irradiation and subsequent annealing upon Mössbauer spectrum parameters of ^{181}Ta in metallic tantalum has been investigated for the first time.

(ii) It has been found that distortion of the crystal structure near the knocked-on atom is high.

(iii) The ^{181}W(Ta) source of a cyclotron target obtained after annealing has the linewidth $W = 0.16(1)$ mm s^{-1}. This is the best among known values for this matrix which is obviously ascribed to a small impurity content introduced into the matrix during its doping with the ^{181}W isotope according to the ^{181}Ta(p, n)^{181}W reaction.

References

[1] G. Vogl, J. Physique 6, 165 (1977).
[2] W. Petry and G. Vogl, Phys. Rev. Letters 45, 1862 (1980).
[3] A. K. Zhetbaev, J. Physique 41, 1 (1979).
[4] Sh. Sh. Ibragimov, A. K. Zhetbaev, and M. F. Vereshchak, in: Radiation Defects in Metals, Ed. Sh. Sh. Ibragimov, Izd. Nauka, Alma-Ata 1981.
[5] D. Brune and J. J. Schmidt (Ed.), Handbook on Nuclear Activation Cross-Section, IAEA, Vienna 1974.
[6] C. Sauer, Z. Phys. 222, 439 (1969).
[7] A. K. Zhetbaev, A. N. Ozernoi, B. G. Akhmetova, and M. K. Akchulakov, Zh. eksper. teor. Fiz., Pisma 8, 45 (1982).
[8] J. Christiansen, Nuclear Phys. A99, 345 (1967).
[9] V. F. Zelenskii, I. M. Neklyudov, L. S. Ozhigov, E. A. Reznichenko, V. V. Rozhkov, and T. P. Chernyaeva, Some Problems of Radiation Technology Physics, Naukova Dumka, Kiev 1979.

(Received November 30, 1982)

phys. stat. sol. (a) 77, 67 (1983)

Subject classification: 20.1; 22.4.4

Department of Physics, Indian Institute of Technology, Delhi[1]) *(a)*
and Solid State Physics Laboratory, Ministry of Defence, Delhi[2]) *(b)*

The Variation of Optical Energy Gaps with Composition in $Pb_{1-x}Hg_xS$ Alloy Films

By

MUKESH JAIN (a), H. K. SEHGAL (a), and A. V. R. WARRIER (b)

Studies of optical properties of flash evaporated $Pb_{1-x}Hg_xS$ films, in the composition range $0.07 \leq x \leq 0.93$, show that the alloy films are formed by alloying of PbS with α-HgS. The lead rich films show an indirect gap and a direct optical gap whereas the mercury rich alloy films show only the direct optical gap. The optical gap increases linearly in polycrystalline films with increasing Hg concentration. The single crystal films, grown at 100 °C ($0.07 \leq x \leq 0.50$) have indirect and direct optical band gaps; the optical gap although increasing linearly with x has values smaller than the films grown at 25 °C.

Untersuchungen der optischen Eigenschaften von schnell aufgedampften $Pb_{1-x}Hg_xS$-Schichten im Zusammensetzungsbereich $0{,}07 \leq x \leq 0{,}93$ zeigen, daß die Legierungsschichten durch Legierung von PbS mit α-HgS gebildet werden. Die bleireichen Schichten zeigen eine indirekte Bandlücke und eine direkte optische Bandlücke, während die quecksilberreichen Legierungsschichten nur das direkte optische Gap zeigen. Die optische Bandlücke in polykristallinen Schichten nimmt linear mit steigender Hg-Konzentration zu. Die Einkristallschichten, bei 100 °C gewachsen mit $0{,}07 \leq x \leq 0{,}50$ besitzen eine indirekte und eine direkte optische Bandlücke, die optische Bandlücke besitzt kleinere Werte als die der bei 25 °C gewachsenen Schichten, obgleich sie linear mit x zunimmt.

1. Introduction

Earlier optical investigations carried out on $Pb_{1-x}Hg_xS$ films, grown by the electroless technique [1], report a direct optical band gap which varies with compositions in the range $x = 0.0$ to 0.33. The electroless technique could not be used for growing films with higher concentration of mercury [2]. In addition, it was difficult to grow films even with $x > 0.15$ with reproducible properties. In order to understand the properties of the alloy films over the complete composition range, we have prepared films by flash evaporation of homogeneous mixture of PbS and HgS. In this communication we report the study of optical absorption in the $Pb_{1-x}Hg_xS$ films in the composition range $x = 0.07$ to 0.93.

2. Experimental

The starting mixtures for film deposition were prepared by bubbling H_2S gas through mixtures of aqueous solutions of $N/2$ lead acetate and $N/2$ mercurie chloride mixed in appropriate molar proportions to obtain desired values of x. Films were grown on KBr single crystal substrates, maintained at different temperatures, by flash evaporation of the PbS–HgS mixtures.

The transmittance and reflectance of these films were measured in the wavelength

[1]) New Delhi 110-016, India.

[2]) Lucknow Road, Dehli 110-007, India.

range 0.4 to 2.5 μm using a Cary-14 spectrophotometer and a Varian reflectance attachment. The absorption coefficient for a particular photon energy was calculated from the measured transmittance and reflectance spectra using the formula [3]

$$\alpha d = \ln\left[\frac{(1-R)^2}{2T} + \sqrt{\frac{(1-R)^4}{4T^2} + R^2}\right],$$

where d is the thickness of films.

3. Results and Discussion

The wavelength dependence of the optical absorption coefficient for $Pb_{1-x}Hg_xS$ ($0.07 \leqq x \leqq 0.93$) alloy films grown at 25 °C is shown in Fig. 1. Structural investigations [4] of these films indicate that the films in the range $0.07 \leqq x \leqq 0.50$ were polycrystalline whereas films containing a higher concentration of mercury were found to be amorphous. The polycrystalline films show a uniphase f.c.c. structure. The lattice parameters for the polycrystalline films were observed to increase with x (see inset in Fig. 1). It may be seen from the optical absorption data (Fig. 1) that the absorption coefficient rises slowly with increase in the photon energy followed by an abrupt steep rise. The initial slow increase in α is more predominant in Pb rich films. For large value of x this portion appears to be non-existent. Optical absorption of films grown at 100 °C is shown in Fig. 2. The films over the complete range are of f.c.c. structure; the films in the range $0.07 \leqq x \leqq 0.50$ were single crystalline whereas those with concentration $0.50 < x \leqq 0.93$ were polycrystalline. The lattice parameter with concentration is shown as inset in Fig. 2. The optical absorption, as may be seen, is qualitatively similar to that of the films grown at 25 °C.

The initial slow rise followed by a steep increase in α with photon energy indicates the presence of an indirect and a direct band gap in the Pb rich films. An abrupt increase in α suggests the presence of only a direct band gap. The interpretation for indirect and direct optical gaps was obtained from curves of $\sqrt{\alpha h\nu}$ versus energy and α^2 versus energy plotted in the appropriate regions. α-HgS is known to have a direct optical gap of 2.1 eV [5] at the centre of the Brillouin zone. PbS is an ionic semiconductor in which the maximum of the valence band and the minimum of the conduction bands are located at the boundary of the Brillouin zone along the [111] direction. The direct optical gap of PbS is 0.41 eV [6]. Scanlon [7, 8] has shown from optical

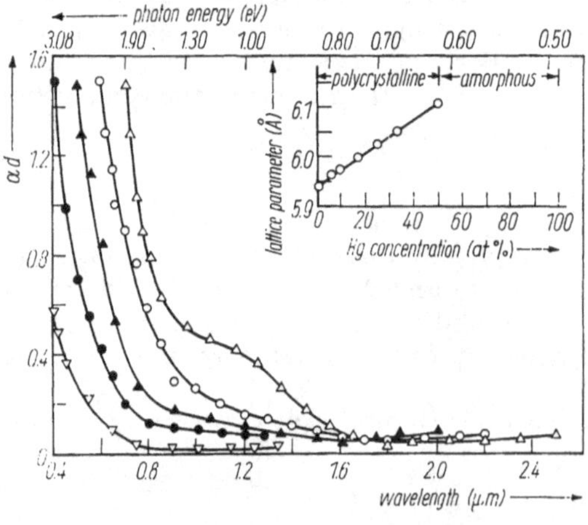

Fig. 1. Room temperature optical absorption spectra for $Pb_{1-x}Hg_xS$ thin films (x = △ 0.07, ○ 0.25, ▲ 0.50, ● 0.75, ▽ 0.91). The inset shows the variation of the lattice parameter as a function of Hg concentration

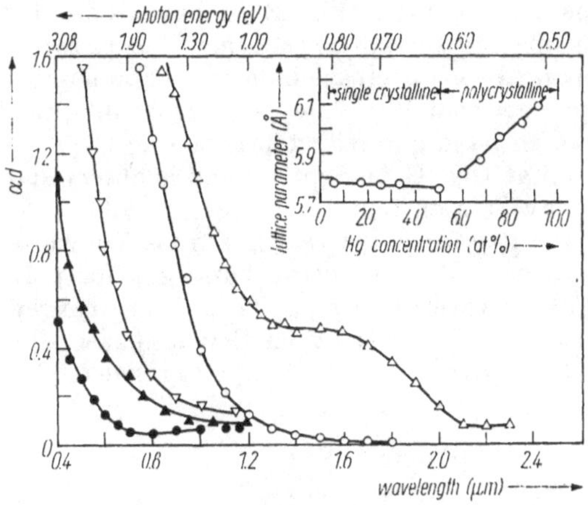

Fig. 2. Spectral variation of absorption coefficient for $Pb_{1-x}Hg_xS$ thin films grown at 100 °C ($x = \triangle$ 0.07, ○ 0.25, ▽ 0.34, ▲ 0.50, ● 0.67). The inset shows the variation of the lattice parameter as a function of Hg concentration in the films

measurements that PbS also has an indirect gap of 0.37 eV. This shows that in the uniphase $Pb_{1-x}Hg_xS$ alloy films, the lead rich films qualitatively retain the optical behaviour typical of PbS, whereas the films with higher concentration of mercury follow the optical behaviour of α-HgS. This type of optical behaviour has also been observed in alloy semiconductors like GaAsP [9], GaAlAs [10], etc.

The variation of direct and indirect gaps with composition for films grown at 25 and 100 °C is shown in Fig. 3. It may be seen that the direct optical gap in films grown at 25 °C shifts towards higher energies almost linearly with mercury concentration ($0.07 \leqq x \leqq 0.50$) in the polycrystalline films. It may be seen from Fig. 3, that even for a small concentration of Hg, the band gap shifts appreciably towards higher energies from the PbS value. The large shift in the band gap for small concentration of Hg is due to the large difference in the band gaps between PbS and α-HgS; the behaviour is similar to that observed in HgCdTe alloys [11]. The indirect band gaps of ≈ 0.7 eV show a small increase with mercury concentration. In case of single crystal films ($0.07 \leqq x \leqq 0.50$), grown at 100 °C, the variation of band gaps with mercury concentration was observed to be qualitatively similar to those of films

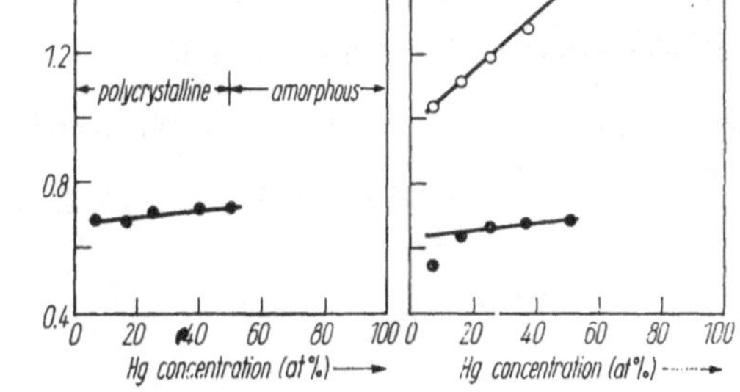

Fig. 3. Optical gap as a function of mercury concentration in $Pb_{1-x}Hg_xS$ grown at a) room temperature (25 °C) and b) 100 °C substrate temperature. The dashed curve in Fig. 3a is for films grown at 100 °C. ○ direct optical gap, ● indirect optical gap

grown at 25 °C. The single crystal films are observed (Fig. 3) to have somewhat lower values of direct optical gaps as compared to the polycrystalline films grown at 25 °C. The decrease in optical gap accompanied with decrease in lattice parameter, in the single crystal films, could be due to the fact that PbS shows a decrease in optical gap with contraction of its lattice (decrease of band gap with temperature).

In the case of polycrystalline films grown at 100 °C, the optical gap is observed to increase with mercury concentration. It may be seen that the direct optical gap for these films lies on the foil of the direct band gap for polycrystalline films (shown by dashed curve in Fig. 3a). Therefore, it can be said that optical band gap for polycrystalline films in the range $0.07 \leqq x \leqq 0.93$ varies linearly with mercury concentration. These observations supports the structural investigations that uniphase alloy films of $Pb_{1-x}Hg_xS$ can be formed by flash evaporation in the complete range $0.07 \leqq x \leqq 0.93$.

4. Conclusion

The present investigation shows that uniphase thin films of $Pb_{1-x}Hg_xS$ have an indirect and a direct optical gap in the Pb rich films ($0.07 \leqq x \leqq 0.50$), whereas the optical gap is direct in case of mercury rich films ($0.50 < x \leqq 0.93$). The band gap shifts towards higher energy with increase in Hg concentration in the films, suggesting that PbS alloys with α-HgS-structure have formed.

Acknowledgement

The authors are grateful to Dr. S. C. Jain, Director, Solid State Physics Laboratory, Delhi, for his encouragement and providing facilities in his laboratory.

References

[1] N. C. SHARMA, D. K. PANDYA, H. K. SEHGAL, and K. L. CHOPRA, Thin Solid Films 62, 97 (1979).
[2] N. C. SHARMA, D. K. PANDYA, H. K. SEHGAL, and K. L. CHOPRA, Thin Solid Films 59, 157 (1979).
[3] D. L. GREENWAY and G. HARBEKE (Ed.), Optical Properties and Band Structure of Semiconductors, Pergamon Press, Oxford 1968 (p. 9).
[4] MUKESH JAIN, H. K. SEHGAL, and A. V. R. WARRIER, XV. Ann. Conf. Electron Microscopy Soc. of India, Varanasi, December 16 to 18, 1982.
[5] G. A. SOLER, J. P. AICARDI, and J. P. LEYRIS, phys. stat. sol. (b) 105, 249 (1981).
[6] N. KH. ABRIBOSOV, V. F. BANKINA, L. V. PORETSKAYA, L. E. SHELIMOVA, and E. V. SKUDNOVA, Semiconducting II–VI, IV–VI, and V–VI Compounds, Plenum Press, New York 1969 (p. 109).
[7] W. W. SCANLON, J. Phys. Chem. Solids 8, 423 (1959).
[8] W. W. SCANLON, Phys. Rev. 109, 2, 47 (1958).
[9] D. LONG, Semiconductors and Semimetals, Vol. 1, Ed. R. K. WILLARDSON and A. C. BEER, Academic Press, New York 1966 (p. 150).
[10] A. BALDERESCHI, E. HESS, K. MASCHKE, H. NEUMANN, K.-R. SCHULZE, and K. UNGER, J. Phys. C 10, 4709 (1977).
[11] D. LONG and J. L. SCHMIT, Semiconductors and Semimetals Vol. 5, Ed. R. K. WILLARDSON and A. C. BEER, Academic Press, New York 1970 (p. 175).

(Received September 17, 1982)

phys. stat. sol. (a) 77, 71 (1983)

Subject classification: 1.3; 14.2; 18.2; 21; 21.1

Institute of Solid State Physics, Academy of Sciences of the USSR, Chernogolovka[1])

The Pd–Ni–H System at High Pressure

By

V. E. Antonov, T. E. Antonova, I. T. Belash, E. G. Ponyatovskii,
V. I. Rashupkin, and V. G. Thiessen

Studies on crystal structure, superconducting and magnetic properties of hydrogen solid solutions in the Pd–Ni alloys with 5, 20, 40, 60, and 80 at% Ni are presented. For solutions on the base of alloys containing 20 to 80 at% Ni the phenomenon of dissolution in poor and rich nickel phases is revealed. Hydrogenation of the samples is performed at hydrogen pressures up to 7 GPa and temperatures up to 350 °C.

Изучены кристаллическая структура, сверхпроводящие и магнитные свойства твердых растворов водорода в сплавах Pd–Ni с 5, 20, 40, 60 и 80 at% Ni. Обнаружено явление распада растворов на базе сплавов, содержащих от 20 то 80 at% Ni, на обедненную в обогащенную никелем фазы. Гидрирование образцов проводилось при давлениях водорода до 7 GPa и температурах до 350 °C.

1. Introduction

In recent years much attention has been focussed on the superconducting properties of hydrogen solid solutions in palladium alloys, and a number of new results are of interest (see, for instance, review [1]). When treating the data in question, hydrogen was implied to be an interstitial component of the alloys that occupies interstices in the f.c.c. metal lattice and only increases its parameter. It seemed somewhat disturbing, however, that the superconducting properties of these Pd–Me–H solutions charged with hydrogen by different techniques (implantation, electrolytically, under high pressure) were sometimes found to differ drastically. Research into one of the systems of solutions of this type [2], Pd–Cu–H, charged with hydrogen under high pressure, has shown [3] that at $0.5\ \text{GPa} \leqq P_{H_2} \leqq 2\ \text{GPa}$ and $T > 220\ °\text{C}$ there occurs an irreversible phase transformation accompanied by a tetragonal distortion of the f.c.c. metal sublattice and, most probably, caused by its atomic ordering.

A new phase transformation revealed in the Pd–Cu–H system suggests that for other Pd–Me–H systems as well a structural investigation, particularly, in conjunction with studies on other physical properties, may yield some interesting results. The Pd–Ni–H system seemed to be promising for such an investigation because of the cardinal difference among the concentration dependences of the temperature of transition into the superconducting state, T_k, obtained for this system in cases of sample hydrogenation by implantation [1] or electrolytically [4] and under high pressure [5, 6]. The alloys with 5, 20, 40, 60, and 80 at% Ni are under study in the present work.

2. Experimental

The ingots were melted in an induction furnace in vacuum from electrolytical palladium and nickel. After a 10 h homogenization at 1100 °C and water-quenching, these ingots were rolled to strips ≈ 0.1 mm thick, then subjected to stress-relief annealing

[1]) 142432 Chernogolovka, Moscow District, USSR.

in vacuum at 1100 °C for 5 min and again quenched in water. The samples were cut from these polycrystalline strips.

Hydrogen compression to pressures of ≈ 2 GPa was performed in "piston-cylinder"-type chambers, to pressures of ≈ 7 GPa in "toroid"-type chambers by the method suggested in [7]. Petrol was used to create a hydraulic gate. The temperature in those chambers was measured with an accuracy to ± 2 and ± 10 K, respectively, and the pressure to ± 0.03 and ± 0.3 GPa. Measurements of the electrical resistance were conducted by a conventional four-probe technique, the Curie points were determined by a differential transformer method [8].

The Pd–Ni–H samples aimed at further investigation at atmospheric pressure were produced by exposing at fixed values of hydrogen pressure ($P_{H_2} \leqq 7$ GPa) and temperature (200 °C $\leqq T \leqq$ 350 °C). The high-pressure chamber was then rapidly cooled, the pressure lowered to atmospheric, the samples were removed from the chamber and later on kept in liquid nitrogen. In case of the "piston-cylinder"-type chamber the quenching was done down to ≈ -20 °C, and in case of the "toroid"-type chamber down to ≈ -50 °C (such quenching has been shown to be sufficient to prevent hydrogen losses from the proper samples at atmospheric pressure). The hydrogen content in the samples was determined with an accuracy of 5% by decomposing them into metal and molecular hydrogen in a pre-evacuated container of known volume at $T \leqq 500$ °C.

The magnetization of the samples was measured with 5% error in a pulsed magnetic field up to 5 T by the induction method [9] in the temperature range 4.2 to 200 K, the pulse duration being ≈ 0.01 s. Samples for the magnetic measurements were couples of strips $6 \times 1 \times 0.2$ mm^3 in size, the total weight being ≈ 20 mg. An X-ray study was carried out by a phototechnique at $T = 90$ K using a DRON-1 diffractometer with CuKα radiation, the samples being in the form of discs ≈ 5 mm in diameter. The temperature of transition to the superconducting state was determined by the induction method.

3. Results and Discussion

The phase diagrams of the Pd–H and Ni–H systems are similar. Both the systems reveal in the T–P_{H_2} diagrams a line of isomorphous phase transformation, $\gamma_1 \rightleftarrows \gamma_2$, ending in a critical point, the $(P_{H_2})_{cr}$ value for the Ni–H system being higher than that for the Pd–H system by approximately three orders of magnitude [10 to 12]. Nickel and palladium form continuous disordered solid solutions with f.c.c. lattice [13]. It was shown in [6] that at room temperature the pressure of the $\gamma_1 \rightarrow \gamma_2$ transition increases monotonously with nickel content of the alloys from $P_{H_2} \approx 1.8$ kPa for the Pd–H solutions to ≈ 0.6 GPa for the Ni–H solutions. The data of the present work agree with these results of [6] and demonstrate that as palladium being doped with nickel, the line of the $\gamma_1 \rightleftarrows \gamma_2$ transformation is found to be shifted rigidly towards higher pressures.

3.1 Electrical resistance of Pd–Ni alloys under high hydrogen pressure

The isotherms presented in Fig. 1 were taken during stepwise increase and decrease in pressure with an exposure in each point for a certain period of time, $\Delta\tau$, up to the termination of the temporal resistance drift, $R(\tau)$, caused by the diffusional nature of the formation of Pd–Ni–H solid solutions. There are final values of R given in Fig. 1. As one can see from the figure, all the $R(P_{H_2})$ isotherms plotted possessed distinct anomalies corresponding to the $\gamma_1 \rightarrow \gamma_2$ and $\gamma_2 \rightarrow \gamma_1$ transitions.

The shape of the $R(P_{H_2})$ dependences for the Pd–Ni alloys under study remained nearly unchanged with temperature varying within 100 °C $\leqq T \leqq$ 350 °C. At $T \gtrapprox 250$ °C typical values of $\Delta\tau$ were $\lessapprox 30$ min in the ranges of the $\gamma_1 \rightarrow \gamma_2$ and

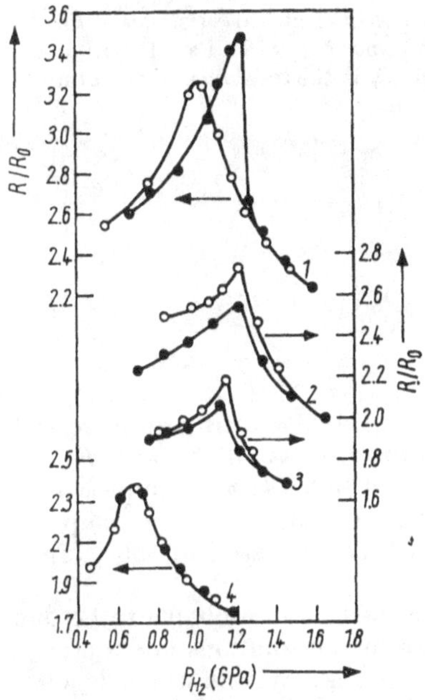

Fig. 1. Electiical resistance isotherms ($T = 250$ °C) of Pd–Ni alloys in hydrogen atmosphere. (1) 100, (2) 80, (3) 60, (4) 40 at% Ni. ● increasing pressure, ○ decreasing pressure. R_0 is the resistance of the samples under normal conditions

$\gamma_2 \rightarrow \gamma_1$ transitions and $\lesssim$ 5 min in the single-phase regions. At lower temperatures these values increased rapidly and at 100 °C reached hours. A typical phase diagram where cations of maxima on the $R(P_{H_2})$ dependences are plotted, can be found in Fig. 2.

The T–P_{H_2} phase diagram of the $Pd_{60}Ni_{40}$–H sys-tem (Fig. 2) apart from the curve of the $\gamma_1 \rightleftarrows \gamma_2$ transformation, has a line of Curie points, T_C for the γ_1-phase (the γ_2-phase is paramagnetic in the temperature and pressure range studied). In an inert medium the Curie point of $Pd_{60}Ni_{40}$ alloy increases approximately linearly with pressure, the slope being $(dT_C/dP)_{in} \approx 2.0$ K GPa^{-1} [14]. A considerable and monotonous decrease of the Curie point of the alloy in hydrogen atmosphere is, thus, caused by an incrcrease of the hydrogen solubility in the γ_1-phase of the $Pd_{60}Ni_{40}$–H solution at increasing pressure (at $T = 100$ °C and $P_{H_2} = 0.2$ GPa the concentration of the solution reaches the hydrogen-to-metal atomic ratio $n = 0.05 \pm 0.03$). The drift time for the Curie point of the $Pd_{60}Ni_{40}$–H γ_1-solution did not exceed 15 to 20 min after changing the hydrogen pressure.

As a whole, it can be said that the phase diagrams for the Pd–Ni–H solutions in question are completely analogous to those for Ni–H [12] and Pd–H [10] solutions. Further investigation has shown, however, that in Pd–Ni–H solutions there occur phenomena that are not observed in Ni–H and Pd–H solutions: in case of the alloys with 20, 40, 60, and 80 at% Ni exposed at $T \gtrsim 250$ °C and $P_{H_2} > (P_{H_2})_{\gamma_1 \rightarrow \gamma_2}$ (this region is shaded in the T–P_{H_2} diagram of the $Pd_{60}Ni_{40}$–H system, Fig. 2) over time periods exceeding considerably the period required for their saturation with hydrogen,

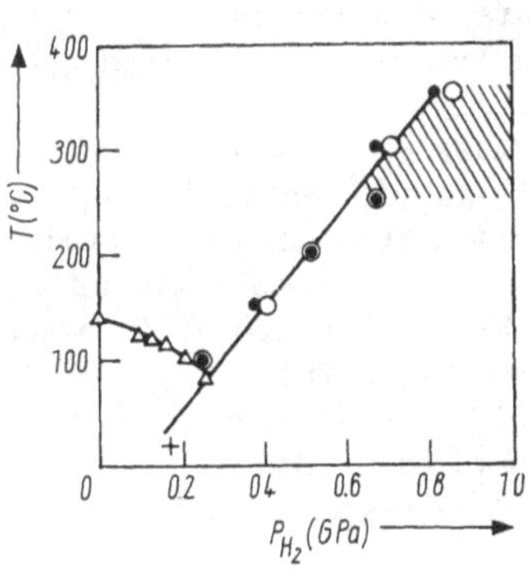

Fig. 2. T–P_{H_2} phase diagram of the $Pd_{60}Ni_{40}$–H system. ● pressures of the $\gamma_1 \rightarrow \gamma_2$ transition, ○ $\gamma_2 \rightarrow \gamma_1$, + pressure of the $\gamma_1 \rightarrow \gamma_2$ transition according to [6], △ Curie points

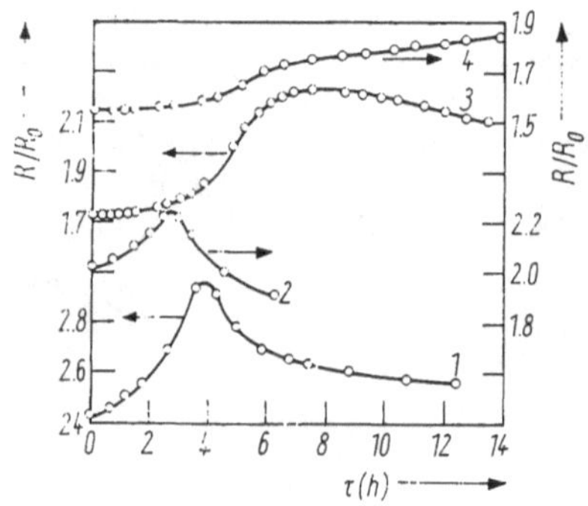

Fig. 3. Time dependences of electrical resistance of Pd–Ni alloys at $T = 350$ °C and $P_{H_2} = 2$ GPa. (1) 80, (2) 60, (3) 40, (4) 20 at% Ni. R_0 is the resistance of the samples under normal conditions

there appears a relatively slow temporal drift of resistance. The drift rate increased and decreased with temperature. Typical $R(\tau)$ dependences at 350 °C and $P_{H_2} =$ 2 GPa are given in Fig. 3. The curves were obtained as follows: the samples were preexposed at $P_{H_2} = 2$ GPa and $T = 200$ °C up to the termination of the temporal resistance drift ($\Delta\tau \lessapprox 1$ h), and the temperature was then increased to 350 °C in a jump-like manner.

The anomalous behaviour of electrical resistance seemed most probable to be due to new irreversible phase transformations taking place in the solutions under study. As to the temperature ≈ 250 °C at which the transformations start and which depended weakly on both pressure and composition of the Pd–Ni alloys, its magnitude is likely to be determined by kinetic reasons.

3.2 X-ray data

At first we studied the constitution of the Pd–Ni–H samples which underwent no irreversible phase transition at high temperatures and were to be ordinary γ-solutions.

All the Pd–Ni–H samples subjected to an X-ray study in the present work were exposed to $T \approx -20$ °C for some minutes in the course of manipulations at atmospheric pressure. This temperature was low enough to brake the hydrogen exchange between the samples and the environment but, at the same time, it was high enough for redistribution of hydrogen within the samples to occur and their phase composition to come to an agreement with the metastable (i.e. at a fixed total content of hydrogen in the solution) T–c diagram during these some minutes (see [10, 15]). According to such diagrams, at atmospheric pressure and $T \lessapprox -20$ °C the Pd–H and Ni–H samples with $n_{\gamma_1}^{max} < n < n_{\gamma_2}^{min}$ consist of a mixture of the γ_1- and γ_2-phases of concentrations $n_{\gamma_1}^{max}$ and $n_{\gamma_2}^{min}$, and those with $n \geqq n_{\gamma_2}^{min}$ constitute single-phase solutions (for Pd–H solutions $n_{\gamma_1}^{max} \lessapprox 0.008$, $n_{\gamma_2}^{min} \approx 0.6$ (see [10]), and for Ni–H solutions $n_{\gamma_1}^{max} \lessapprox 0.02$, $n_{\gamma_2}^{min} \approx 0.7$ [16]). Thus, to treat the data obtained at atmospheric pressure correctly we needed investigating such a separation for the Pd–Ni–H solutions under study.

The Pd–Ni–H samples aimed at the given series of measurements were produced by exposing at fixed P_{H_2} and $T \leqq 350$ °C during 1 h, that was sufficient for hydrogenation of the Pd–Ni alloys to be completed, but was much less than the period of time required for the phase transformation to occur, observed in the γ_2-solutions on the base of alloys with 20 to 80 at% Ni (see, for instance, Fig. 3). The X-ray study has shown that the $Pd_{60}Ni_{40}$–H, $Pd_{20}Ni_{80}$–H, and Ni–H solutions with $0.2 \lessapprox n \lessapprox 0.6$ consist of a mixture of the two phases with a f.c.c. metal lattice, their parameters, a, denoted with closed circles in Fig. 4, being independent of the hydrogen content in the sample. In case of the samples with $n \gtrapprox 0.6$ no lines of phases with smaller a-value were observed on X-ray patterns.

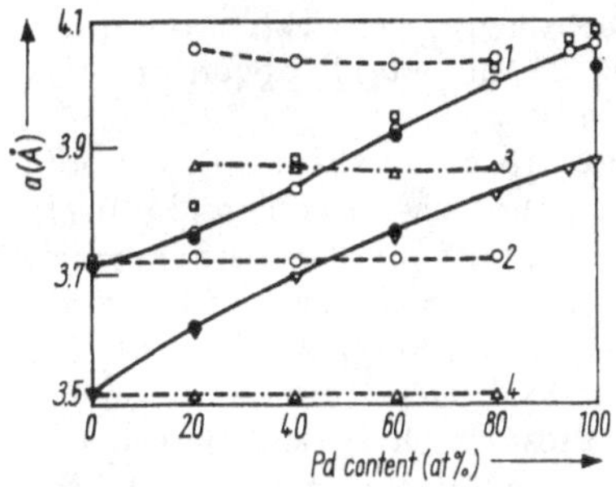

Fig. 4. Parameters a of the f.c.c. metal sublattice for different phases of Pd–Ni–H solutions at $T = 90$ K and atmospheric pressure. ▽ initial Pd–Ni samples, □ after exposure at $T = 200$ °C and $P_{H_2} = 2$ GPa for 5 h, ○ after exposure at $T = 350$ °C and $P_{H_2} = 2$ GPa for 10 h ((1) phases poor in nickel, (2) rich in nickel), △ the same samples after annealing in vacuum at $T = 300$ °C for 2 ((3) phases poor in nickel, (4) rich in nickel), ● phases of concentrations $n_{\gamma_1}^{max}$ and $n_{\gamma_2}^{min}$ h (see text)

As is seen from Fig. 4, the smaller values of a-parameters for the two-phase samples are close to those for corresponding alloys with no hydrogen. It is reasonable to assume them to be parameters of γ_1-phases of composition $n_{\gamma_1}^{max}$, whereas greater parameters to relate to γ_2-phases of composition $n_{\gamma_2}^{min}$. The $a(n_{\gamma_1}^{max})$ and $a(n_{\gamma_2}^{min})$ values obtained for the Ni–H solutions are in line with the data of [12, 16].

To evaluate the changes in a-parameters of Pd–Ni–H γ_2-solutions with hydrogen concentration, a series of samples was prepared by exposure at $P_{H_2} = 2$ GPa and $T = 200$ °C for 2 h. Their parameters, a^{200}, are shown with squares in Fig. 4. The hydrogen concentration in the alloys with 0,5, 20, 40, 60, 80, and 100 at% Ni was $n = 0.99$, 1.03, 1.12, 1.20, 1.16, 1.21, and 1.09, respectively. The a^{200}-values obtained for the Pd–H and Ni–H γ_2-solutions of the concentrations mentioned are in line with the data of [17] and [12].

The above results enabled us to go on to elucidating the question what happens to the Pd–Ni–H γ_2-solutions in the process of a prolonged exposure in hydrogen atmosphere at $T \gtrsim 250$ °C. For this purpose a number of samples were produced by exposing at $T = 350$ °C and $P_{H_2} = 2$ GPa for 10 h. The data of X-ray investigation of such samples are denoted in Fig. 4 with open circles. According to these data, the samples based on nickel, palladium, and $Pd_{95}Ni_5$ alloy were ordinary γ_2-solutions, for which $a^{200} > a^{350} > a(n_{\gamma_2}^{min})$ (the a^{350}-value for the Ni–H solution is not shown in order to simplify the figure). A decrease in a^{350}-value compared to a^{200} is due to the fact that at $P_{H_2} = 2$ GPa the hydrogen solubility in all the metals under study is smaller at 350 than at 200 °C.

In case of samples based on alloys with 20, 40, 60, and 80 at% Ni the situation was quite different. They proved to consist of three phases with f.c.c. metal lattices. The value of the parameter a for one of the phases lay between the values $a(n_{\gamma_2}^{min})$ and a^{200} for a corresponding alloy (these points are connected with the solid line in Fig. 4) and coincided with the a^{350}-value for the same alloy charged with hydrogen at $P_{H_2} = 2$ GPa and $T = 350$ °C for 1 h. Thus, one of the three phases observed in the investigated samples was a residue of the Pd–Ni–H γ_2-solution which had not enough time to transform into a new phase state during the exposure at $P_{H_2} = 2$ GPa and $T = 350$ °C. The parameters a of the other two phases depended weakly on the concentration of the starting Pd–Ni alloys and corresponded to the a^{350}-parameter for palladium containing a few atomic percent of nickel (these points are linked with dashed curve 1 in Fig. 4) and that for nickel (curve 2).

In the light of these data, the most probable process occurring at a prolonged exposure of the alloys with 20 to 80 at% Ni at $T \gtrsim 250$ °C and $P_{H_2} > (P_{H_2})_{\gamma_1 \to \gamma_2}$, is believed to be a dissolution of hydrides of these alloys to a hydride of nearly pure nickel and a hydride of palladium with some atomic percent of nickel. The kinetics of the dissolution, however, remains relatively sluggish even at 350 °C, as is seen already from the curves of electrical resistance presented in Fig. 3. The kinetics inertia is confirmed by a X-ray examination as well. For instance, a detectable amount of the γ_2-phase of the $Pd_{60}Ni_{40}$–H solution was retained in the samples even after exposing them at $P_{H_2} = 2$ GPa and $T = 350$ °C for 48 h.

A considerably lower diffusion mobility of the metal atoms in the Pd–Ni–H solutions compared to that of hydrogen atoms proved, on the other hand, to be useful to confirm, in addition, the dissolution of these solutions at high pressure and increased temperature. An annealing in vacuum for 2 h of the samples on the base of the alloys with 20, 40, 60, and 80 at% Ni, produced by exposure at $P_{H_2} = 2$ GPa and $T = 350$ °C for 10 h, resulted in a complete extraction of hydrogen from them, but the samples retained their three-phase nature. The phases which in the presence of hydrogen had $a^{200} > a^{350} > a(n_{\gamma_2}^{\min})$ were found to have now parameters a coinciding with those for corresponding Pd–Ni alloys with no hydrogen (∇ in Fig. 4). The values of the parameters which we used earlier to ascribe to phases poor in nickel (curve 1 in Fig. 4) decreased to those for such alloys with no hydrogen (curve 3). The same happened in the case of the phases considered to be hydrogen solutions on nickel base (curves 2 and 4, respectively).

After a further annealing of these samples in vacuum at $T > 400$ °C the alloys, as revealed by the X-ray examination, recovered the single-phase homogeneous state with starting values of parameters of their f.c.c. lattices. The recovery of this initial state occurred also after plastic deformation of the dissolved samples with no hydrogen at room temperature. Such a behaviour of the samples demonstrates the fact that under normal conditions the dissolution to phases rich and poor in nickel is not an equilibrium property of Pd–Ni alloys themselves. It is the triple Pd–Ni–H solutions with a high hydrogen content which are found to dissolve.

3.3 Superconductivity

The available data concerning the behaviour of superconducting properties of hydrogen solid solutions in Pd–Ni alloys containing up to ≈10 at% Ni are in line with each other [1, 4 to 6]: over this concentration range the replacement of palladium by nickel causes a monotonous decrease in the maximum attainable temperature, T_k^m, of transition of Pd–Ni–H solutions to the superconducting state from ≈8.8 K for palladium to ≈1 to 2 K for the alloys with nickel atomic fraction $x_{Ni} \approx 0.1$. For the alloys with a higher nickel concentration no such an agreement is observed. The studies on the $Pd_{80}Ni_{20}$ alloy charged with hydrogen by implantation at helium temperature [1] and on the $Pd_{75}Ni_{25}$ alloy hydrogenated electrolytically at $T \approx 200$ K [4] have shown them to possess no superconductivity at $T \gtrsim 1.1$ K. The results obtained in the case of hydrogenation of Pd–Ni alloys under a high hydrogen pressure and room temperature [5, 6] were quite different: on the dependence $T_k^m(x_{Ni})$ a deep minimum at $x_{Ni} \approx 0.1$ was followed by a maximum ($T_k^m \approx 10$ K) at $x_{Ni} \approx 0.2$, and then the T^m values decreased sharply. It should be noted that the $T_k(n)$ dependences for the alloys with $x_{Ni} \approx 0.2$ exhibited a chaotic behaviour [6]. In the light of the arguments of the previous section, a divergence in the data of [1, 4] and [5, 6] can be straightforwardly explained if one assumes that in the Pd–Ni–H γ_2-solutions with $x_{Ni} \gtrsim 0.2$ investigated in [5, 6] there occurred a dissolution to phases rich and poor in nickel, and that the jumps on the temperature dependences of electrical resistance, registered in those works, corresponded to the transition of the nickel-poor phases to the superconducting state.

The dependences $T_k(n)$ for the Pd–Ni alloys with 5, 20, and 40 at% Ni derived in the present paper are given in Fig. 5. The Pd–Ni–H samples were produced by exposing for 10 h at $P_{H_2} \leqq 7$ GPa and $T \leqq 350$ °C (with the exception of solutions on the base of $Pd_{80}Ni_{20}$ and $Pd_{60}Ni_{40}$ alloys with maximum n-values, see below). Typical dependences $\Delta(T)$ of the signal of disbalance of an ac bridge in the range of transition

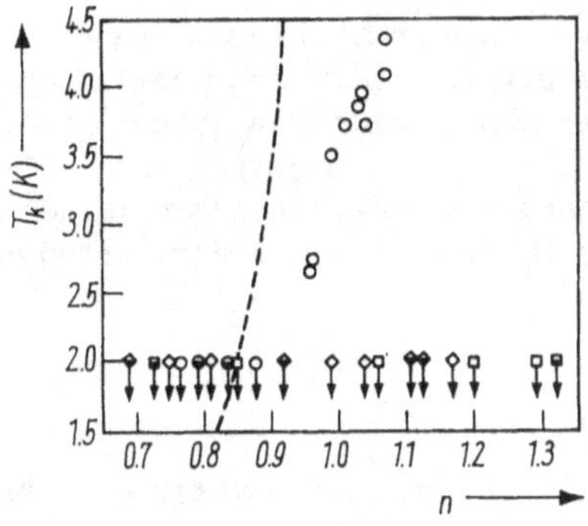

Fig. 5. Superconducting temperature T_k vs. hydrogen content n for hydrogen solutions in Pd-Ni alloys with ○ 5, ◇ 20, and □ 40 at% Ni. The symbols with arrows stand for samples possessing no superconductivity at $T \geqq 2$ K. The half-blackened symbols concern samples where dissolution to phases poor and rich in nickel was observed. Dashed line represents the $T_k(n)$ dependence for the Pd-H solutions [1]

to the superconducting state for the $Pd_{95}Ni_5$-H solutions are demonstrated in Fig. 6. Relatively narrow steps on the $\Delta(T)$ dependences corresponding to such a transition are indicative of a homogeneous hydrogen distribution over the volume of the samples. The T_k-values were estimated by locations of the midpoint of these steps.

As is seen from Fig. 5, the $T_k(n)$ dependence for the $Pd_{95}Ni_5$-H γ_2-solutions is markedly shifted in respect to a similar dependence for Pd-H solutions (dashed line) towards higher n-values. At $T \geqq 2$ K no transition to the superconducting state has been observed for the $Pd_{80}Ni_{20}$-H and $Pd_{60}Ni_{40}$-H γ_2-solutions (empty symbols in Fig. 5). Thus, our data for the Pd-Ni-H γ_2-solutions having undergone no separation to nickel-poor and nickel-rich phases agree with the data of [1, 4 to 6] and contradict the data of [5, 6].

A surprising result was the absence of superconductivity at $T \geqq 2$ K in case of $Pd_{80}Ni_{20}$-H and $Pd_{60}Ni_{40}$-H samples dissolved to phases rich and poor in nickel (half-blackened symbols in Fig. 5), for even by the most understated estimates among the samples under study there were to be found such ones for which the hydrogen concentration in the nickel-poor phase was sufficient to make it a superconductor with $T_k \gg 2$ K. This effect might be due to suppression of superconductivity in particles of the nickel-poor phase by the surrounding paramagnetic medium owing to their small sizes.

Nevertheless, from our point of view, the most probable cause responsible for the appearance of superconductivity in the Pd-Ni-H samples with $x_{Ni} \gtrapprox 0.2$ studied in [5, 6] seems to be the dissolution with precipitation of particles of the phases poor in nickel. Indeed, the γ_2-solutions of hydrogen in alloys with $x_{Ni} \geqq 0.2$ do not possess superconductivity at $T \geqq 2$ K even at the n-values exceeding considerably those reached in [5, 6] (see Fig. 5). But it is difficult to expect, at the same time, that one more new transformation may occur in those solutions at high hydrogen pressure and

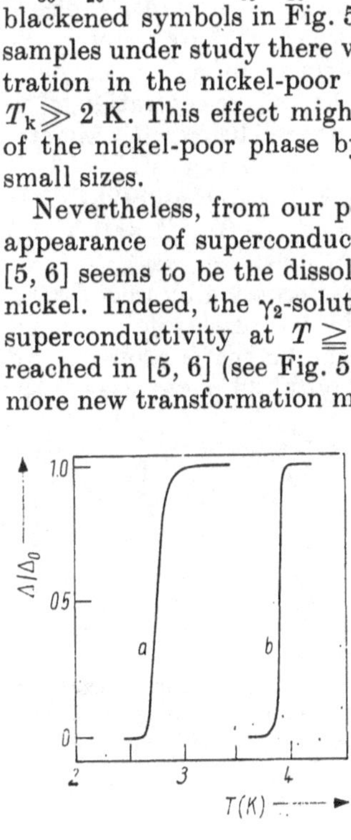

Fig. 6. Temperature dependences of the signal of disbalance Δ of an ac bridge in the range of transition to the superconducting state for $Pd_{95}Ni_5$-H solutions with (a) $n = 0.96$ and (b) 1.04. Δ_0 is the magnitude of the signal for samples in the normal state

room temperature. In any case, our experiments have shown that the exposure of the $Pd_{80}Ni_{20}$ and $Pd_{60}Ni_{40}$ alloys, pre-charged with hydrogen at an increased temperature, at $P_{H_2} = 7$ GPa and room temperature for 48 h resulted neither in a change of the phase composition of the solutions formed at an increased temperature nor in the appearance of superconductivity (it is just the manner used to produce samples with a maximum hydrogen content, the data on superconductivity of which are presented in Fig. 5).

3.4 Magnetic measurements

The last that we are going to consider in our work is the effect of hydrogen on the magnetization of Pd–Ni alloys. The Pd–Ni–H samples were produced by exposure for 2 h at $T \leqq 300$ °C and $P_{H_2} \lesseqgtr 2$ GPa. The X-ray examinations of the samples thus obtained have revealed no dissolution of the γ_2-phases to the nickel-rich and nickel-poor phases. Thus, only ordinary γ-solutions were under investigation. In [18] it has been shown that the behaviour of the magnetic properties of Pd–Ni alloys is well described by equations of the theory of weak itinerant ferromagnetism [19, 20]. In the present work we have used these equations to treat the experimental data for the Pd–Ni–H samples.

Ferromagnetic ordering in the Pd–Ni alloys arises at a nickel concentration of 2.5 at%, further the Curie points, T_C, and the spontaneous magnetization, σ_0, at $T = 0$ K increase monotonously up to $T_C^{Ni} = 631$ K and $\sigma_0^{Ni} = 0.616\mu_B$/atom [21, 18], where μ_B is the Bohr magneton. The $Pd_{60}Ni_{20}$ and $Pd_{60}Ni_{40}$ alloys have been chosen for investigation. The $\sigma_0(n)$ dependences for these alloys are exhibited in Fig. 7. In full analogy with that done for the Ni–H solutions in [16], in the present work for the $Pd_{80}Ni_{20}$–H and $Pd_{60}Ni_{40}$–H solutions it has been shown that at atmospheric pressure their γ_2-phases with hydrogen concentrations down to $n_{\gamma_2}^{min}$ are found to be paramagnetic at $T \geqq 4.2$ K, and the samples with $\sigma_0 \neq 0$ consist of a ferromagnetic γ_1-phase of concentration $n_{\gamma_1}^{max}$ and a paramagnetic γ_2-phase of concentration $n_{\gamma_2}^{min}$. Thus, the hydrogen concentration at which the $\sigma_0(n)$ dependences, being approximately linear in the $(\gamma_1 + \gamma_2)$ two-phase regions, intersect the $\sigma_0 = 0$ axis is $n_{\gamma_2}^{min}$. As is seen from Fig. 7, the $n_{\gamma_2}^{min}$ values estimated in this way for the $Pd_{80}Ni_{20}$–H and $Pd_{60}Ni_{40}$H solutions are close to those for the Pd–H ($n_{\gamma_2}^{min} \approx 0.6$ [1]) and Ni–H($n_{\gamma_2}^{min} \approx 0.7$ [16]) solutions.

Concluding we note that the suppression of ferromagnetic ordering on introducing hydrogen to Pd–Ni alloys is in agreement with the prediction of the rigid d-band model that earlier proved to be a good one to describe the magnetic properties of hydrogen solid solutions on the base of f.c.c. and h.c.p. alloys of 3d metals (see [15]).

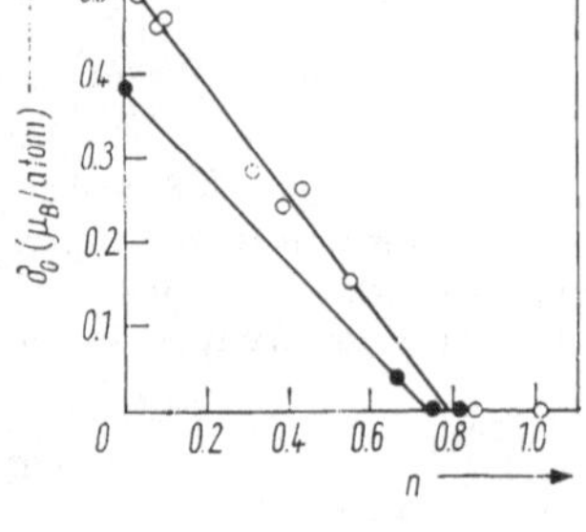

Fig. 7. Spontaneous magnetization σ_0 at $T = 0$ K in dependence on hydrogen content n in $Pd_{80}Ni_{20}$ (●) and $Pd_{40}Ni_{60}$ (○) alloys. The magnetization is given in μ_B/atom of the Pd–Ni alloy

Acknowledgements

The authors wish to thank V. F. Degtyareva for the equipment for X-ray measurements, K. A. Peresada, A. N. Grachev, and A. I. Amelin for their assistance in preparing and conducting the expreriments, as well as V. G. Glebovskii in whose group the palladium–nickel alloys were melted.

References

[1] B. Stritzker and H. Wühl, in: Topics in Applied Physics, Vol. 29, Ed. G. Alefeld and J. Völkl, Springer-Verlag, 1978 (p. 243).
[2] V. E. Antonov, I. T. Belash, E. G. Ponyatovskii, and V. I. Rashupkin, Zh. eksper. teor. Fiz., Pisma **31**, 451 (1980).
[3] V. F. Degtyareva, V. E. Antonov, I. T. Belash, and E. G. Ponyatovskii, phys. stat. sol. (a) **66**, 77 (1981).
[4] L. Sniadower, L. Dumoulin, P. Nedellec, and J. P. Burger, in: Metal–Hydrogen Systems, Ed. T. N. Veziroğlu, Pergamon Press, Oxford 1982 (p. 357).
[5] T. Skośkiewicz, High Temp.–High Press. **7**, 684 (1975).
[6] T. Skośkiewicz, phys. stat. sol. (a) **48**, K165 (1978).
[7] I. T. Belash and E. G. Ponyatovskii, USSR Patent 741105.
[8] G. T. Dubovka and E. G. Ponyatovskii, Fiz. Metallov i Metallovedenie **33**, 540 (1972).
[9] L. Jacobs and P. Lawrence, Rev. sci. Instrum. **28**, 713 (1958).
[10] E. Wicke and H. Brodowsky, see [1] (p. 73).
[11] E. G. Ponyatovskii, V. E. Antonov, and I. T. Belash, Dokl. Akad. Nauk SSSR **230**, 649 (1976).
[12] V. E. Antonov, I. T. Belash, and E. G. Ponyatovskii, Dokl. Akad. Nauk SSSR **233**, 1114 (1977).
[13] M. Hansen and K. Anderko, Constitution of Binary Alloys, McGraw-Hill Publ. Co., Inc., New York/Toronto/London 1958.
[14] H. Fujiwara, H. Kadomatsu, K. Ohishi, and Y. Yamamoto, J. Phys. Soc. Japan **40**, 1010 (1976).
[15] E. G. Ponyatovskii, V. E. Antonov, and I. T. Belash, Uspekhi fiz. Nauk **137**, 663 (1982).
[16] V. G. Thiessen, V. E. Antonov, I. T. Belash, B. K. Ponomarev, and E. G. Ponyatovskii, Dokl. Akad. Nauk SSSR **242**, 1390 (1978).
[17] J. E. Schirber and B. Morosin, Phys. Rev. B **12**, 117 (1975).
[18] B. K. Ponomarev and V. G. Thiessen, phys. stat. sol. (b) **88**, K139 (1978).
[19] E. P. Wohlfarth, J. appl. Phys. **39**, 1061 (1968).
[20] D. M. Edwards and E. P. Wohlfarth, Proc. Roy. Soc. **A303**, 127 (1968).
[21] E. O. Wollan, Phys. Rev. **167**, 461 (1968).

(Received December 27, 1982)

phys. stat. sol. (a) 77, 81 (1983)

Subject classification: 10.1; 12.1; 21

International Centre for Theoretical Physics, Trieste[1]) *and Institute of Metal Research, Academia Sinica, Shenyang*[2])

The Dislocation Distribution Function in the Plastic Zone at a Crack Tip

By

C. W. Lung and L. Y. Xiong[3])

The dislocation distribution function in the plastic zone at a crack tip is calculated with the help of a modified BCS-type crack model. It is found that opposite sign dislocations appear on the slip plane immediately ahead of the crack tip. These negative dislocations seem to enter the crack tip and blunt it, resulting in a dislocation free zone at the crack tip.

Die Versetzungsverteilungsfunktion in der plastischen Zone an einer Rißspitze wird mit Hilfe eines modifizierten BCS-Rißmodells berechnet. Es wird gefunden, daß Versetzungen mit entgegengesetztem Vorzeichen auf der Gleitebene unmittelbar über der Rißspitze auftreten. Diese negativen Versetzungen scheinen in die Rißspitze einzutreten und sie stumpf zu machen, was zu einer versetzungsfreien Zone an der Rißspitze führt.

1. Introduction

The problem of dislocation distribution in the plastic zone at a crack tip has received much interest recently due to its importance to elastic plastic fracture of materials, which was first studied by Bilby et al. (BCS-type crack model) [1]. They described the plastic zone by an array of dislocations co-planar with the crack and derived the same relationship between the crack size (c), the plastic zone size ($a-c$), and the applied stress (σ^∞) as Dugdale had done macroscopically before [2]. Recently, Kobayashi and Ohr [3] investigated crack propagation and the structure of the plastic zone formed ahead of a crack with the help of in situ electron microscope fracture experiments of the b.c.c. metals molybdenum and tungsten. They found that a part of the plastic zone immediately ahead of the crack tip is free of dislocations. Weertman [4] proposed that perfect dislocations are not emitted directly from the crack tip, and that pairs of partial or perfect dislocations of opposite sign are created on the slip plane immediately ahead of the crack tip. Dislocations of one sign move away from the tip region leaving dislocations of opposite sign piled up against the crack tip. When the applied stress is sufficiently large the piled up dislocations can enter the crack tip and blunt it.

It should be noted that the dislocation density function in the original BCS-type crack model monotonically decreases with the distance from the crack tip, and all the dislocations are of positive sign. This picture is not consistant with the recent experimental results and with the physical pictures of Weertman. The present paper shows that, with the help of a modified BCS-type crack model, opposite sign dislocations appear on the slip plane immediately ahead of the crack tip. These negative dislocations enter the crack tip and blunt it, resulting in a dislocation free zone at the crack tip.

---

[1]) P.O. Box 586, 34100 Trieste, Italy.

[2]) Wenhua Road, Shenyang, The People's Republic of China (permanent address).

[3]) Present address: Istituto di Fisica, Politecnico di Milano, 20133 Milano, Italy.

2. Small-Scale Yielding near Cracks

Unlike the original BCS-type crack model, we treated this problem within a homogeneous and continuous system. Acturely, we only consider the plastic zone which does not include both the crack and plastic zone in a whole system to avoid discontinuity and inhomogeneity, because the crack plane is a free surface and the resistance to motion of the dislocations in the crack region (σ_0) is different from that in the plastic zone (σ_1).

At first, we do not consider crack blunting. Suppose one of the dislocations with Burgers vector of magnitude b is located at a distance x from a crack of length $2c$ (Fig. 1). In this case the resultant shear stress on it is zero when the system is in equilibrium [1, 5]. This condition leads to the following integral equation for the density function $D(x)$ of the dislocation:

$$A \int_0^a \frac{D(x')\,\mathrm{d}x'}{x - x'} - \sigma_1 + \sigma^\infty + \sigma^\mathrm{c} = 0\,, \tag{1}$$

where $A = Gb/2\pi(1 - \nu)$, G is the shear modulus, b is Burgers vector. The first term is the interactions between this dislocation and the others. σ_1 is the frictional stress for the dislocation motion (which is taken as the yield stress), and σ^∞ is the applied stress field taken as constant, and σ^c is the elastic stress field due to crack. The image stresses at the crack tip can be neglected due to its small radius of curvature [6]. For small scale yielding, one may write

$$\sigma^\mathrm{c} = \frac{K}{(2\pi x)^{1/2}}, \qquad K = \sigma^\infty(\pi a)^{1/2}\,. \tag{2}$$

Let $u = x'/a$, $w = x/a$, where a is the plastic zone size, u and w are dimensionless parameters, then from (1)

$$A \int_0^1 \frac{D(u)\,\mathrm{d}u}{w - u} - \sigma_1 + \sigma^\infty + \sigma^\mathrm{c} = 0\,. \tag{3}$$

This singular integral equation can be solved by finite Hilbert transformation [7]. We get

$$D(u) = \frac{\sigma_1}{\pi^2 A}\left(\frac{1-u}{u}\right)^{1/2} (1 - \alpha)\,\pi \left[1 - \frac{1}{4\sqrt{1-u}} \ln\left(\frac{1 + \sqrt{1-u}}{1 - \sqrt{1-u}}\right)\right], \tag{4}$$

where $\alpha = \sigma^\infty/\sigma_1$.

The conditions for it to exist lead to the relation (Appendix)

$$\frac{a}{c} = \left[\frac{2\sqrt{2}\alpha}{\pi(1-\alpha)}\right]^2. \tag{5}$$

As expected, $a \to 0$, when $\alpha = 0$; that is $\sigma_1 \to \infty$ (infinite resistance to dislocation motion) or $\sigma^\infty \to 0$ (no applied stress on dislocations). Again, $a \to \infty$, that is, yielding spreads across the infinite plate, when $\alpha \to 1$; this means $\sigma^\infty \to \sigma_1$ (applied stress tends to resistance σ_1 to dislocation motion) (Fig. 2).

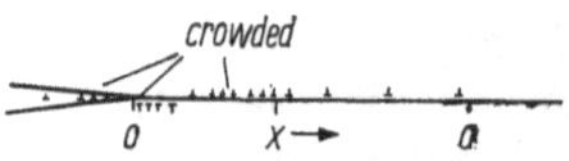

Fig. 1. Schematic figure of dislocation in the plastic zone at a crack tip

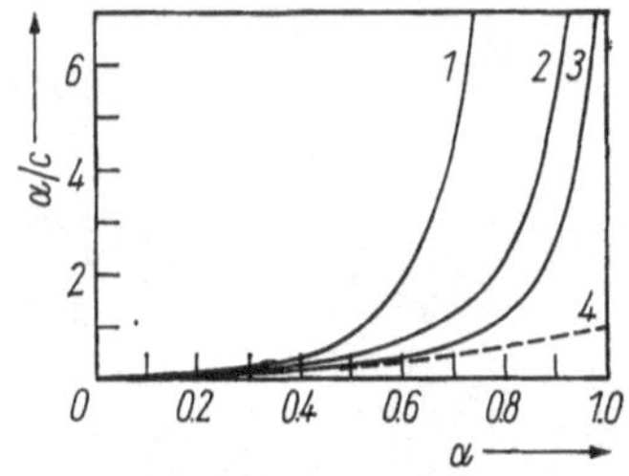

Fig. 2. Relationship between the plastic zone size a/c and the applied stress $\alpha = \sigma^\infty/\sigma_1$.

(1) $\frac{a}{c} = \left[\frac{2\sqrt{2}\,\alpha}{\pi(1-\alpha)}\right]^2$; (2) BCS: $\frac{a}{c} = \sec\left(\frac{\pi}{2}\alpha\right) - 1$; (3) $\alpha = \dfrac{\pi}{2\sqrt{2}\,n + \arccos\dfrac{2n-1}{2n+1}}$;

(4) $\frac{a}{c} = \alpha^2$

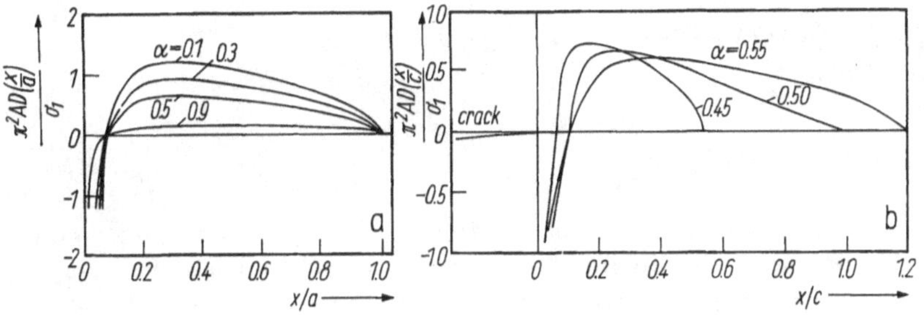

Fig. 3. Dislocation distribution in the plastic zone under small scale yielding, a) x/a as unit, b) x/c as unit

Unlike the original BCS-type crack model, $D(u)$ can change sign from $u = 0$ to $u = 1$. It results from the solution of (1). This means, there are opposite sign dislocations existing in the plastic zone simultaneously as shown in Fig. 3a. In this type of σ^c, $D(u) = 0$ at $u_0 = 0.08$, and it seems independent on α. That is, the dislocations always change their sign at $x_0 = 0.08a$.

If we use the length of crack as unit, $D(u)$ may be changed into $D(x/c)$, (Fig. 3b). It is shown that the higher the applied stress level, the larger the area under the distribution function curve; and then, the larger the COD of the cracked material.

The fact that $D(u)$ is negative in (4) means that the sign of the Burgers vector of the dislocations relative to that of dislocations which create a positive stress field is negative.

3. Elastic–Plastic Stress Fields

In this case, the plastic zone is larger. If the size of plastic zone at fracture is not very much less than the crack length, (2) is too simple to describe the stress field near the crack tip accurately. The following expression may be better:

$$\sigma^c = \frac{\sigma^\infty}{\sqrt{1 - \left(\frac{c}{x+c}\right)^2}} - \sigma^\infty . \tag{6}$$

As above, the singular integral equation is solved by the finite Hilbert transformation. Then, we get the distribution function of dislocation in the plastic zone

$$D(u) = E(\alpha) \sqrt{\frac{1-u}{u}} - F(\alpha) \frac{u+n}{\sqrt{u(u+2n)}} \times \\ \times \ln \frac{[\sqrt{2n} + \sqrt{(2n+u)(1-u)}]^2 + u^2}{u(2n+1)}, \qquad (7)$$

where

$$E(\alpha) = \frac{\sigma_1}{\pi^2 A} \left\{2\alpha \left[\operatorname{arctg} \sqrt{2n} - \frac{\pi}{2}\right] + \pi\right\},$$

$$F(\alpha) = \frac{\alpha\sigma_1}{\pi^2 A},$$

$$n = \frac{c}{a}.$$

The condition similar to (A6) leads to the relation

$$\alpha = \frac{\pi}{2\sqrt{2n} + \arccos\left(\frac{2n-1}{2n+1}\right)}. \qquad (8)$$

As expected, $n = c/a \to 0$ when $\alpha = 1$; that is, when $\sigma^\infty \to \sigma_1$, $a \to \infty$ and yielding spreads across the infinite plate. Again, $n \to \infty$ when $\alpha \to 0$ that is, when $\sigma_1 \to \infty$ (infinite resistance to dislocation motion) or when $\sigma^\infty \to 0$, $a \to 0$ (no plastic zone).

Then, we may get the size of plastic zone as a function of various applied stress level (α). It is compared to that of the BCS-type crack model in Fig. 2.

The distribution function of dislocations in the plastic zone as a function of distance from the crack tip is given in Fig. 4a and b. Similar to the small scale yielding case, the dislocation distribution function increased from minus infinity passing through zero to positive value. After reaching a maximum, it goes down to zero at the border of the plastic zone. It seems that u_0, the root of $D(u) = 0$, also nearly constant with a little departure due to various value of α this means, $u_0 = \varkappa/a \approx$ const.

Opposite sign dislocations will attract each other and become closer and closer. Ultimately, they will annihilate. As for the crack case, part of the negative dislocations

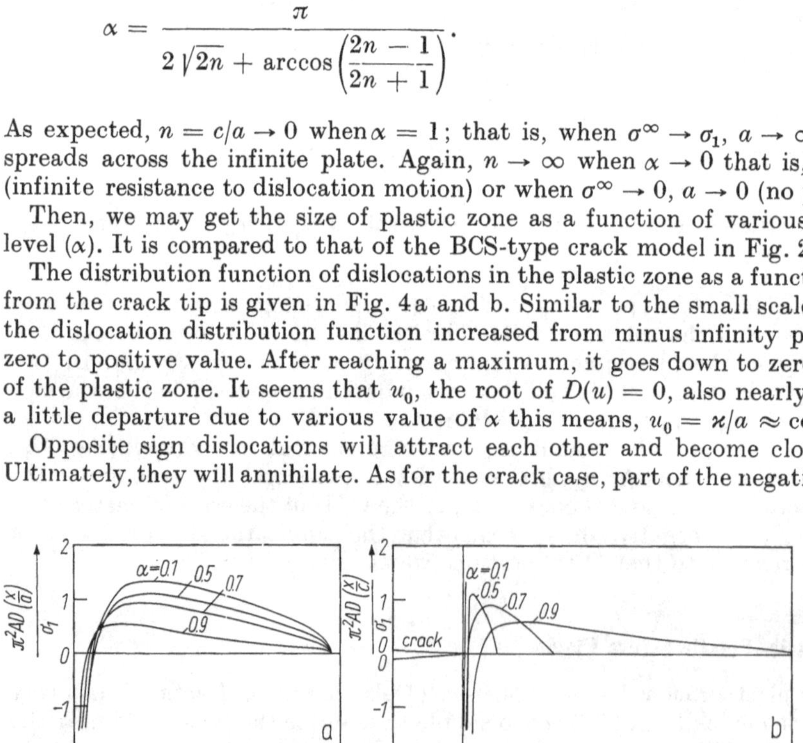

Fig. 4. Dislocation distribution in the plastic zone under elastic plastic stress field a) x/a as unit, b) x/c as unit

will move to enter the crack tip and blunt it by consuming some of the energy available for creation of new surface. The process will continue till an energy balance is reached between the energy gain from annihilation and the energy consumed by dislocation movement and new surface creation. Therefore, it is possible that a part of dislocation free zone to be left over after blunting. Dislocations of positive sign are also created and move away from the crack tip region settling down to its equilibrium position on the slip plane.

Positron lifetime and Doppler broadening lineshape parameter have been measured in the plastic zone at the crack tip by Jiang et al. [8]. It is believed that by using the collimation technique to improve the accuracy, the positron annihilation technique used in this direction will promote our understanding about this problem in the future.

Appendix

$$A \int_0^1 \frac{D(u)\, \mathrm{d}u}{w - u} - \sigma_1 + \sigma^\infty + \sigma^c = 0\,. \tag{A1}$$

Let

$$\sigma_1 - \sigma^\infty = P\,,$$

$$\sigma^c = \frac{\sigma^\infty \sqrt{c}}{\sqrt{2x}} = \frac{\sigma^\infty \sqrt{c/a}}{\sqrt{2w}}\,, \tag{A2}$$

change variables

$$u = \tfrac{1}{2}\,(\xi + 1)\,, \qquad w = \tfrac{1}{2}\,(\eta + 1)\,, \tag{A3}$$

we get

$$A \int_{-1}^{+1} \frac{f(\xi)}{\xi - \eta}\, \mathrm{d}\xi = -P + \frac{\sigma^\infty \sqrt{c/a}}{(\eta + 1)^{1/2}}\,. \tag{A4}$$

Solving the above singular integral equation by finite Hilbert transformation [7], we get for the boundary condition that $f(+1)$ is finite,

$$f(\xi) = -\frac{1}{\pi^2 A} \left(\frac{1 - \xi}{1 + \xi}\right)^{1/2} \int_{-1}^{+1} \left(\frac{1 + \eta}{1 - \eta}\right)^{1/2} \frac{1}{\eta - \xi} \left[-P + \frac{\sigma^\infty \sqrt{c/a}}{(\eta + 1)^{1/2}}\right] \mathrm{d}\eta\,, \tag{A5}$$

where the solution would fulfil the Lipschitz condition in $-1 < \eta \leqq 1$ and $f(\infty) = 0$. But, the solution would not fulfil the Lipschitz condition at $\eta = -1$ and $f(\xi = -1)$ would be infinite [9]. In fracture mechanics, it is just the case at the crack tip. However, we had better let $f(-1)$ to be infinite of lowest power which leads to a sufficient and necessary condition, then

$$\int_{-1}^{+1} \left(\frac{1 + \eta}{1 - \eta}\right)^{1/2} \left[P - \frac{\sigma^\infty \sqrt{c/a}}{(\eta + 1)^{1/2}}\right] \mathrm{d}\eta = 0\,. \tag{A6}$$

From this equation, we may get the plastic zone size as a function of α and c

$$\frac{a}{c} = \left[\frac{2\sqrt{2}\,\alpha}{\pi(1 - \alpha)}\right]^2. \tag{A7}$$

In all these formulae the Cauchy principal values of the integrals are to be taken to avoid divergence at singular points.

Similarly, we get the solution for the elastic plastic stress fields.

Acknowledgements

The authors would like to thank Prof. A. Salam, the International Atomic Energy Agency, and UNESCO for hospitality at the International Centre for Theoretical Physics, Trieste, where most of this work was finished during their stay in the summer of 1982.

References

[1] B. A. BILBY, A. H. COTTRELL, and K. H. SWINDEN, Proc. Roy. Soc. **A272**, 304 (1963).
[2] D. S. DUGDALE, J. Mech. Phys. Solids 8, 100 (1960).
[3] S. KOBAYASHI and S. M. OHR, Phil. Mag. A **42**, 763 (1980).
[4] J. WEERTMAN, Phil. Mag. A **43**, 1103 (1981).
[5] C. W. LUNG, Preprint of the Internat. Centre for Theoretical Physics, IC/79/72. Trieste 1979; Sci. Sinica **23**, 441 (1980).
[6] J. FRIEDEL, Dislocations, Pergamon Press, 1964 (p. 47).
[7] L. G. CHAMBERS, Integral Equations, Internat. Textbook Co., Ltd., 1976 (p. 133).
[8] J. JIANG, L. Y. XIONG, C. W. LUNG, C. K. JI, N. Q. LIU, and S. Y. WANG, 6th Internat. Conf. Positron Annihilation, PA, USA, Ed. P. G. COLEMAN, S. C. SHARMA, and L. M. DIANA North-Holland Publ. Co., 1983 (p. 469)
[9] V. I. SMIRNOV, Kurs visshei matematiki, Vol. III-2 (p. 27); Vol. IV-1, Gos. izdat. techniko-teoreticheskoi literatury., 1952 (p. 57)

(Received December 28, 1982)

phys. stat. sol. (a) 77, 87 (1983)
Subject classification: 5; 21.1; 21.1.1

Department of Physics, Bangalore University[1] *(a)*
and Department of Metallurgy, Indian Institute of Science, Bangalore[2] *(b)*

Mössbauer Effect Study of the Thermal Decomposition in Hydrogen of Homogeneously Mixed Ferrous Nickel Oxalates with Different Iron to Nickel Ratios

By
RAMANI (a), M. P. SATHYAVATHIAMMA (a), N. G. PUTTASWAMY (a), and R. M. MALLYA (b)

Mössbauer effect and X-ray measurements are carried out on product samples of the thermogravimetric analysis (TGA) and isothermal decomposition in hydrogen of homogeneously mixed ferrous nickel oxalates with different iron to nickel ratios. The formation of Fe–Ni alloy is obtained at considerably lower temperatures (≈ 300 °C) in each case. The Fe–Ni alloys obtained shift from iron-rich to nickel-rich composition as the nickel ratio in the mixed metal oxalates is increased. The formation of Fe–Ni Invar from mixed metal oxalate with Fe:Ni = 1:1 is indicated in the early stages but not from those with Fe:Ni = 2:1 or 64:36.

An Produktproben von homogen verteilten Eisen–Nickeloxalaten mit unterschiedlichem Eisen–Nickel-Verhältnis nach thermogravimetrischer Analyse (TGA) und isothermem Zerfall in Wasserstoff werden Mößbauereffekt- und Röntgenmessungen durchgeführt. In allen Fällen wird die Bildung der Fe–Ni-Legierung bei beträchtlich niedrigeren Temperaturen (≈ 300 °C) erhalten. Die erhaltenen Fe–Ni-Legierungen verschieben sich von der eisenreichen zur nickelreichen Zusammensetzung, wenn das Nickelverhältnis in dem Metall-Mischoxalat erhöht wird. Die Bildung der Fe–Ni-Invar-Legierung aus dem Metall-Mischoxalat mit Fe:Ni = 1:1 wird in frühen Zuständen beobachtet, jedoch nicht aus Oxalaten mit Fe:Ni = 2:1 oder 64:36.

1. Introduction

In preparing ferrites from metal oxides long durations of heating at high temperatures are required in order to sinter and cause interdiffusion of metal ions to give homogeneous ferrites of uniform properties. Metal oxalates form the starting material for the production of ultrafine mixed metal oxides such as ferrites. During the treatment of the oxalates other intermediate products like oxides and carbides could appear. Co-precipitation of the metal oxalates has often resulted in advantages in subsequent processing for nearly homogeneous ferrite formation.

Halsey and Pritchard [1] and Gallagher and Kurkjian [2] have shown that when ferrous oxalate is heated in nitrogen and air, respectively, iron oxide forms in bulk or in fine form. Earlier work [3, 4] by the present authors on thermal decomposition of ferrous oxalates in hydrogen shows that fine particles of iron can be obtained at lower temperatures and that sometimes it is accompanied by formation of fine particles of Fe_3O_4 and Fe_3C under certain conditions.

Often during co-precipitation, depending on experimental details, the metal ions could precipitate separately causing phase separatione, .g. if pH is not carefully controlled. Sometimes, due to local experimental changes minor hydroxides could also precipitate. In the present work homogeneous co-precipitation of the oxalates is carried

[1]) Bangalore 560056, India.
[2]) Bangalore 560012, India.

out to ensure mixing of the two metal oxalates at atomic levels. Metal oxalates with different molar ratios are prepared. The thermal and structural changes occurring during the heating of Fe–Ni oxalates in hydrogen atmosphere are followed using thermogravimetry, Mössbauer spectroscopy, and X-ray diffractometry.

2. Experimental Method

Ferrous and nickel salt solutions are mixed to give iron to nickel molar ratios of 2:1, 1:1, 1:10 and also 64:36, 1:3, and 1:19. The mixed oxalates are simultaneously co-precipitated from an oxalate solution maintained at a sufficiently low pH ($pH < 5$) to prevent hydroxide precipitation. The precipitation is carried out in an inert atmosphere to prevent oxidation of ferrous ions. The precipitated oxalates are filtered, washed thoroughly with distilled water, and dried.

2.1 Thermogravimetric analysis

Thermogravimetric analysis (TGA) of 0.2 g of ferrous nickel oxalate is carried out in hydrogen using a modified type of thermobalance based on that developed by Vasudevamurthy et al. [5]. The linear rates of heating employed are 0.5 and 2 K/min.

2.2 Isothermal decomposition

Isothermal decomposition of mixed metal oxalates are carried out in a flowing hydrogen atmosphere between 200 and 500 °C for different periods of time ranging from 15 min to several hours in two types of configurations:

1. *Vertical experiments*: Initially, isothermal experiments have been carried out taking the samples in a 5 cm long glass "culture tube" mounted vertically inside the reaction tube. The reaction tube could be maintained at the desired temperature by a vertically aligned tubular furnace which could be moved up and down.

2. *Horizontal experiments*: Experiments have been carried out by spreading the sample as a thin layer in a porcelain boat mounted horizontally in an all glass reaction tube. The reaction tube is covered by a tubular furnace mounted on wheels which could be moved horizontally away from the heated zone. The boat is mounted such that it is in the centre of the furnace when it covers the reaction tube. Pure (O_2-free) dry H_2 gas is passed over the sample. The spreading of the sample in the horizontal arrangement also helps in allowing more of the sample surface to come in contact with the hydrogen. The gases evolved during the decomposition are swept away by the hydrogen gas flow much faster in the horizontal than in the vertical experiments. The gas flow is suitably monitored using bubblers.

Table 1
Isomer shift and quadrupole splitting of undecomposed mixed metal oxalates

sample	Fe:Ni ratio	δ (mm/s)	ΔE (mm/s)
ferrous oxalate	—	1.08	1.71
Fe–Ni oxalates	2:1	1.08	1.71
	1:1	1.08	1.71
	1:3	1.20	1.71
	1:10	1.20	1.71
	1:19	1.20	1.71

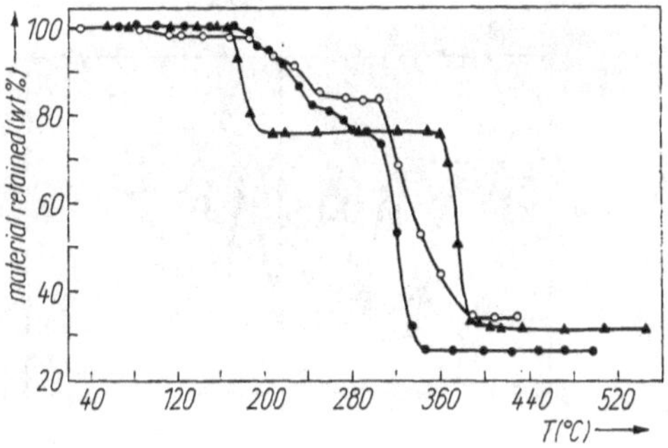

Fig. 1. TGA of nickel oxalate (○), ferrous oxalate (▲), and ferrous nickel oxalate (Fe:Ni = = 1:10) (●) in hydrogen at 2 K/min

Table 2

Mössbauer measurements of the products of mixed Fe–Ni oxalates with Fe:Ni = 1:10 in vertical experiments

T (°C)	time (h)	unde-composed	central	Fe_3O_4	Fe–Ni	percentage absorption of Fe–Ni (positive extreme peak)	
199	18	p	p	—	—	—	
232	102	—	—	—	p	Fe-rich	0.36
						Ni-rich	0.26
250	1	p	p	—	—	—	—
	1 (80 K)	p	p	—	—	—	—
	12	p	p	—	p	Ni-rich	0.56
254	2	p	p	—	—	—	—
	4	p	p	—	p	Ni-rich	0.28
						Fe-rich	0.11
	4 (80 K)	p	p	—	p	Ni-rich	0.49
	12	p	p	—	p	Ni-rich	0.9
						Fe-rich	0.2
	48	—	—	—	p	Ni-rich	0.6
						Fe-rich	—
273	1	p	p	—	p	Ni-rich	0.2
	4	p	p	—	p	Ni-rich	0.5
	4 (80 K)	p	p	—	p	Ni-rich	0.6
	12	—	—	—	p	Ni-rich	1.1
300	0.25	p	p	ps	p	Ni-rich	1.05
	0.25 (80 K)	p	p	p	p	Ni-rich	1.14
	0.5	p	p	ps	p	Ni-rich	0.7
	0.5 (80 K)	p	p	p	p	Ni-rich	0.7
	7	—	—	—	p	Ni-rich	0.8
						Fe-rich	0.4
	12.5	—	—	—	p	Ni-rich	1.1
						Fe-rich	0.4
499	14.5	—	—	—	p	Ni-rich	2.0
						Fe-rich	0.52

p present.
ps traces indicated.

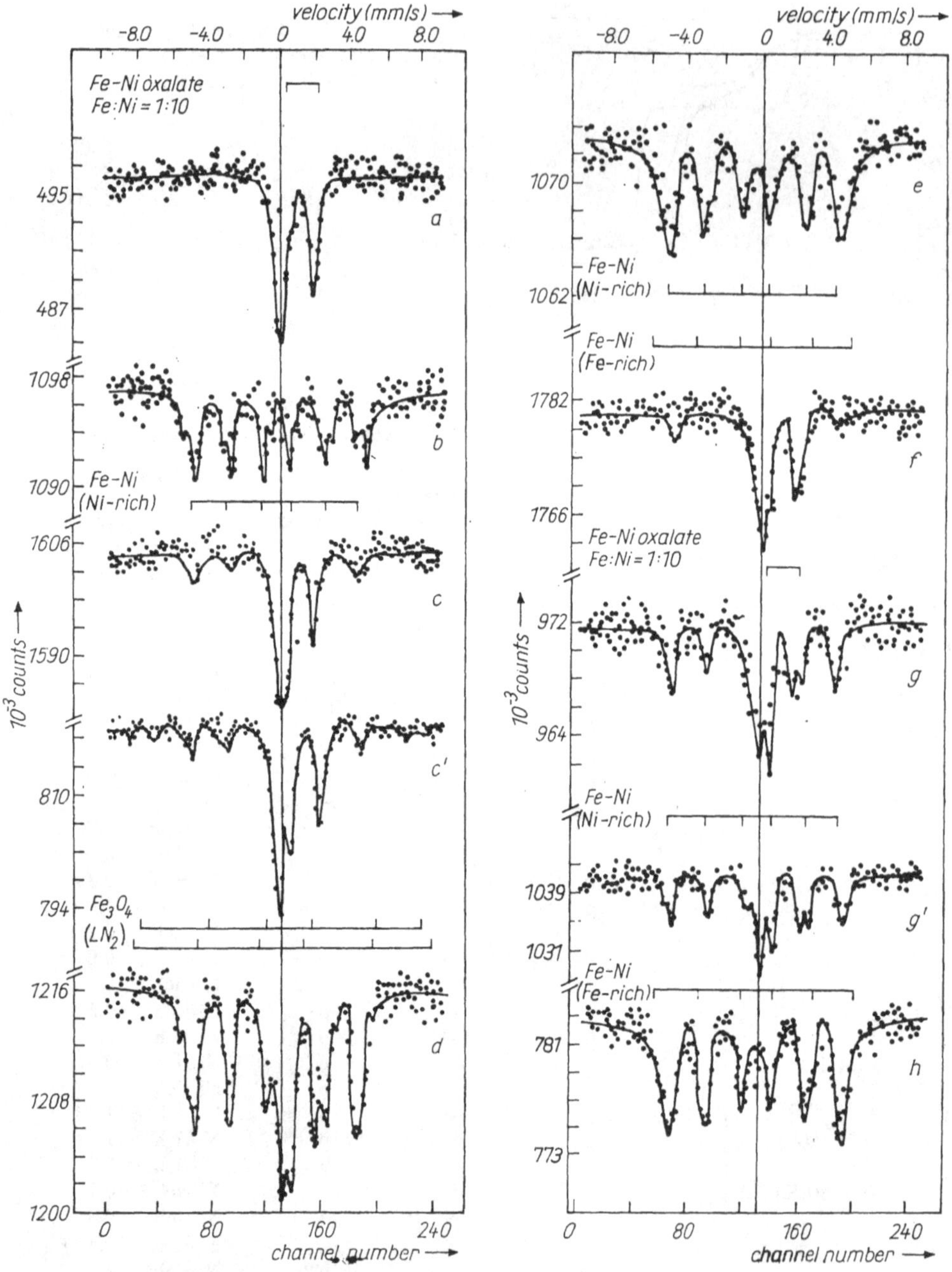

Fig. 2

Before heating the sample all the air in the reaction tube is driven off by H_2 displacement to prevent explosions. The reaction product is very pyrophoric and the sample should be exposed to atmosphere extremely slowly and in controlled stages. After the desired reaction period the product is initially cooled in a current of H_2, then the H_2 flow is stopped, and the product retained for about 12 h at a slight over-

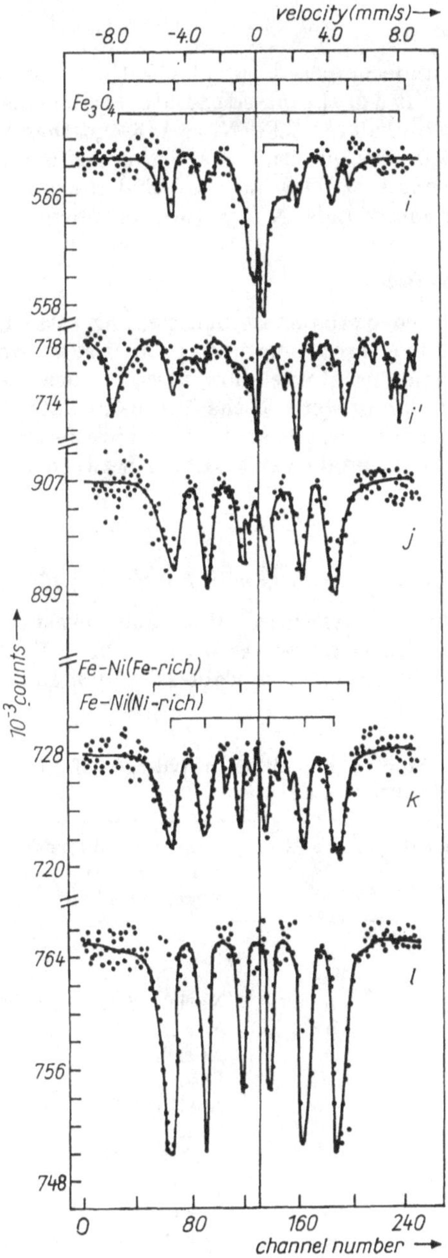

Fig. 2. Mössbauer spectra of ferrous nickel oxalate with Fe:Ni = 1:10 heated in hydrogen in vertical experiments. (a) at 199 °C for 18 h, (b) 232 °C for 102 h, (c) 254 °C for 4 h, (c′) LN_2 spectrum of (c), (d) 254 °C for 12 h, (e) 254 °C for 49 h, (f) 273 °C for 1 h, (g) 271 °C for 4 h, (g′) LN_2 spectrum of (g), (h) 273 °C for 12 h, (i) 300 °C for 0.5 h, (i′) LN_2 spectrum of (i), (j) 307 °C for 1 h, (k) 305 °C for 12.5 h, (l) 499 °C for 14.5 h

pressure of H_2. Air is then permitted to diffuse in through a capillary tube for a day. Next, more air is admitted through a 4 mm tube for a day and then air is allowed to pass directly. Then the sample is taken out and quickly transferred to a container. This controlled method of exposure avoided pyrophoricity of the product and prevented oxidation as confirmed by Mössbauer spectra and ESCA studies of these samples.

The decomposition experiments are carried out in a reducing atmosphere of hydrogen for all the samples, the different temperatures at which the specimens are maintained are chosen suitably from the reaction zones obtained by thermogravimetric analysis.

2.3 Mössbauer measurements

Mössbauer absorbers are prepared from the decomposed samples and the room and liquid nitrogen temperature measurements are carried out using a 10 mCi/^{57}Co (Pd) source in most of the cases and with a 25 mCi/^{57}Co(Rh) source with some samples. A krypton filled proportional detector is employed to detect 14.4 keV γ-rays.

The spectra are computer fitted whenever necessary. In certain cases areas are calculated by addition of counts under the peak. The samples are examined by X-ray powder diffractometry using a PW1050 type Philips diffractometer.

3. Results and Discussion

3.1 Undecomposed mixed metal oxalates

The results of Mössbauer measurements of undecomposed samples with different Fe:Ni ratios are listed in Table 1. The isomer shift δ of the mixed oxalates has increased with increasing nickel content in the mixed oxalate at Fe:Ni = 1:3 and then it remains constant with further increase of nickel. The quadrupole splitting ΔE is not changed and is equal to that of normal ferrous oxalates. This indicates that the mixed metal oxalates are solid solutions and not new compounds. X-ray studies confirm the same.

3.2 Products of TGA of mixed metal oxalates

The two-stage decomposition in H_2 of the mixed oxalates at high heating rates is resolved at slow heating rates (0.5 to 2 K/min) into a three- four-stage pattern as shown in Fig. 1. The first stage is dehydration. The further stages show that, besides the decomposition of nickel oxalate, the decomposition of some of the ferrous oxalate in the mixed oxalate is taking place at lower temperatures than that of pure ferrous oxalate in hydrogen. The weight loss in the mixed metal oxalate occurs between 170 and 300 °C.

3.3 Products of isothermal decomposition in H_2

3.3.1 Mössbauer measurements of mixed metal oxalates with Fe:Ni = 1:10

The results of room temperature (RT) (unless stated otherwise) Mössbauer measurements of products of the decomposition of mixed metal oxalates with Fe:Ni = 1:10 decomposed in vertical and horizontal experiments are listed in Tables 2 and 3. Möss-

Table 3

Mössbauer measurements of the products of the decomposition in hydrogen of mixed Fe–Ni oxalate (Fe:Ni = 1:10) in horizontal experiments

T (°C)	time (h)	undecomposed	central	Fe–Ni	percentage absorption of Fe–Ni (positive extreme peak)
200	24	p	p	—	
250	1	p	p	—	
	4.5	—	—	p	Ni-rich 2.05
	25	—	—	p	Ni-rich 2.0
300	0.5	—	p	p	Ni-rich 1.2
					Fe-rich —
	1	—	—	p	Ni-rich 0.6
					Fe-rich —
	2	—	—	p	Ni-rich 1.18
					Fe-rich —
	24	—	—	p	Ni-rich 11.5
480	0.5	—	—	p	Ni-rich 1.9
					Fe-rich —
486	4	—	—	p	Ni-rich 2.7
					Fe-rich —
490	1	—	—	p	Ni-rich 2.1
					Fe-rich —

p present.

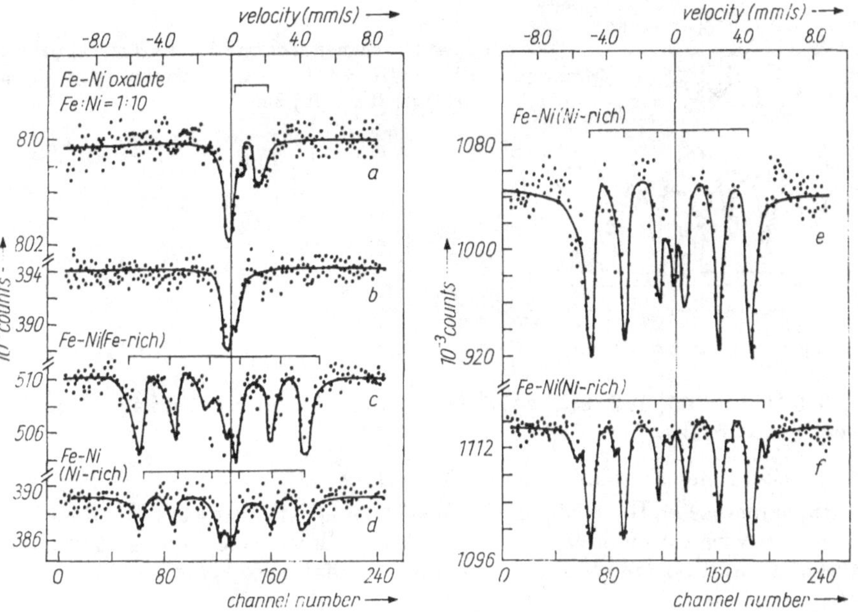

Fig. 3. Mössbauer spectra of ferrous nickel oxalate with Fe:Ni = 1:10 heated in hydrogen in horizontal experiments. (a) at 200 °C for 24 h, (b) 250 °C for 1 h, (c) 250 °C for 25 h, (d) 300 °C for 0.5 h, (e) 300 °C for 24 h, and (f) 490 °C for 1 h

bauer spectra of these samples are given in Fig. 2 and 3, respectively. (i) In the beginning stages of the decomposition, undecomposed peaks with central contributions indicative of superparamagnetic fine particles are obtained in the RT Mössbauer spectra as indicated for the samples obtained at 199 °C for 18 h (Fig. 3), 250 °C for 1 h, 254 °C for 2 h. (ii) As decomposition proceeds, along with the above products an Fe–Ni alloy starts appearing as indicated for the samples obtained at 254 °C for 4 and 12 h (Fig. 2, curves c and d), 273 °C for 1 and 4 h (Fig. 2, curves f and g), and 300 °C for 0.5 h (Fig. 2, curve i). (iii) Fe–Ni peaks with varied amounts of Ni-rich and Fe-rich contributions are obtained in the Mössbauer spectra as in the samples obtained at 232 °C for 102 h (Fig. 2, curve b), 254 °C for 48 h (Fig. 2, curve e), 307 °C for 1 h (Fig. 2, curve j), 305 °C for 12.5 h (Fig. 2, curve k), and at 499 °C for 14.5 h (Fig. 2, curve l).

The trend as shown by Fig. 2 and Tables 2 and 3 is that at low temperatures of decomposition, both Fe-rich and Ni-rich Fe–Ni alloys are obtained with extremely fine particles being formed. At higher temperatures (>270 °C) Ni-rich Fe–Ni is dominant. When the temperature is still higher or the period is too long, then both Fe-rich and Ni-rich Fe–Ni are obtained. Fe_3O_4 is formed transitorily and is present in a fine particle form and an indication to fine particles of Fe–Ni is also obtained through liquid nitrogen Mössbauer measurements (Fig. 2, curve k).

In horizontal experiments the Fe–Ni alloy formation is also inferred from the Mössbauer spectra. These reactions being faster, Fe_3O_4 traces are not observed in any RT spectrum. Also the Ni-rich contribution to the Fe–Ni alloy formation is dominant as seen in Table 3. In samples heated at 200 °C for 24 h and 250 °C for 1 h, peaks of the undecomposed sample with a slight central contribution are obtained in Mössbauer spectra (Fig. 3, curves a and b). Mössbauer spectra of samples treated at 250 °C for

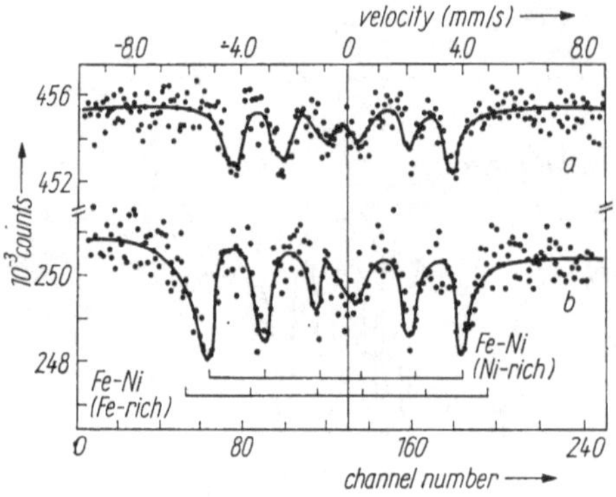

Fig. 4. Mössbauer spectra of ferrous nickel oxalate with Fe: Ni = 1:19 heated in hydrogen. (a) at 360 °C for 0.5 h, (b) 360 °C for 1 h

4 and 25 h, 300 °C for 0.5, 1, 2, and 24 h, 480 °C for 0.5 and 4 h, and 490 °C for 1 h clearly reveal Fe–Ni peaks.

3.3.2 *Mixed metal oxalates with Fe: Ni = 1:19*

As the Ni ratio is increased, the peaks in the Mössbauer spectra are very broad indicating that many hfs with different H_{int}-values of Fe–Ni alloy corresponding to different compositions of Fe and Ni are obtained. Fig. 4 gives Mössbauer spectra of the sample

Table 4

Mössbauer measurements of the products of mixed Fe–Ni oxalates with Fe:Ni = 1:1 and 2:1 in vertical experiments

T (°C)	time (h)	undecomposed	central	Fe_3O_4	Fe–Ni	percentage absorption of Fe–Ni (positive extreme peak)
Fe:Ni = 1:1						
270	4	p	p	—	ps	
	12	p	p	—	Invar	
	12 (80 K)	p	p	—	Invar	
300	0.5	p	p	—	—	
	1	p	p	p	—	
	2	p	ps	p	—	
	4	p	—	p	Invar	
	10	—	—	p	Invar	
	10 (80 K)	—	—	p	Invar	
307	11.33	—	—	—		
Fe:Ni = 2:1						
244	3	p	p	—	—	
250	9.5	p	p	—	—	
270	10	p	p	—	—	
294	0.5	p	p	—	—	
297	1	p	p	—	—	
300	2	p	p	—	—	
	4	ps	—	p	p	Fe-rich 3.3
	15	—	—	p	p	Fe-rich 4.17
333	6	—	—	p	p	Fe-rich 3.0

p present.
ps traces indicated.

Table 5
Mössbauer measurements of the products of mixed Fe–Ni oxalates with Fe:Ni = 1:1, 2:1, and 64:36 in horizontal experiments

T (°C)	time (h)	unde-composed	central	Fe–Ni	percentage absorption of Fe–Ni (positive extreme peak)
Fe:Ni = 1:1					
280	4.5	—	p	p	Fe-rich 2.4
300	1	p	p	p	Invar
	2	—	p	p	Fe-rich 3.6
	3	—	—	p	Fe-rich 7.0
320	2	—	p	p	Fe-rich 3.7
Fe:Ni = 2:1					
300	1	p	p	—	
	2	p	p	p	Fe-rich 2.0
340	0.5	p	p	p	Fe-rich 2.2
	1	—	—	p	Fe-rich 6.2 Ni-rich (slight)
Fe:Ni = 64:36					
270	12	p	p	—	
340	20	—	—	p	Fe-rich 4.4
360	1	—	p	p	(slight Invar tendency) 4.9
	8	—	—	p	Fe-rich 7.0 Ni-rich 4.0
	24	—	—	p	Fe-rich 1.0 Ni-rich 0.4
400	1	—	p	p	Fe-rich 5.0
480	7	—	—	p	(slight Invar tendency 1.6)

p present.

decomposed at 360 °C for 0.5 and 1 h. Even the sample decomposed at 360 °C for 24 h gives a similar spectrum.

3.3.3 *Mixed metal oxalates with Fe:Ni = 1:1, 2:1, and 64:36*

The results are tabulated in Tables 4 and 5. Mainly Invar is formed from samples with Fe:Ni = 1:1. Samples with Fe:Ni = 64:36 decompose in a manner similar to that of the 2:1 samples. In the decomposition products of this even at 480 °C no Invar is obtained. The results are discussed in detail elsewhere [6].

The X-ray diffraction patterns confirm Mössbauer measurements. At lower Ni concentrations Fe–Ni peaks are prominent on the Fe side and vice versa.

4. Conclusions

The TGA of mixed metal oxalates indicates that the decomposition temperature of ferrous oxalate is lowered in the homogeneous mixed oxalates in comparison to the decomposition of ferrous oxalate alone in hydrogen. The decomposition together with reduction occurs over a range of temperature (180 to 340 °C). Initially, nickel fine particles seem to be formed by the preferential decomposition of some of the nickel oxalates and subsequently an alloy is formed by further decomposition of the rest of the mixed oxalate. The changed stoichiometry of the alloy formed, with respect to the starting stoichiometry in the mixed metal oxalates, indicates this. The formation of Invar in 1:1 mixed oxalate rather than in 2:1 samples also points to the same. The Fe–Ni alloy formed is either Fe-rich or Ni-rich depending on the Fe:Ni ratio. In 1:10 samples, Ni-rich Fe–Ni is dominant. Fe_3O_4 formation is obtained only transitorily in vertical experiments, and this also in very fine form. Fe_3C is totally absent in the mixed metal oxalate decomposition in hydrogen. The size of the alloy particles produced is a function of time and temperature of the decomposition and an indication of fine particles of Fe–Ni is obtained from Mössbauer measurements at liquid nitrogen temperature.

Acknowledgement

The work at Bangalore University has been supported by a research grant from the University Grant Commission (UGC).

References

[1] M. J. HALSEY and A. M. PRITCHARD, J. Chem. Soc. A **80**, 2878 (1968).
[2] P. K. GALLAGHER and C. R. KURKJIAN, Inorg. Chem. **5**, 214 (1966).
[3] RAMANI, M. P. SATHYAVATHIAMMA, N. G. PUTTASWAMY, and R. M. MALLYA, Nuclear Phys. and Solid State Phys. (India) 19C, p. 411 (1976).
[4] RAMANI, M. P. SATHYAVATHIAMMA, N. G. PUTTASWAMY, and R. M. MALLYA, Nuclear Phys. and Solid State Phys., Symp. New Delhi (India) (1980).
[5] A. R. VASUDEVAMURTHY, BHARADWAJ, and R. M. MALLYA, Chem. and Ind., London, 1956 (p. 300).
[6] RAMANI, M. P. SATHYAVATHIAMMA, N. G. PUTTASWAMY, and R. M. MALLYA, Proc. Internat. Conf. Appl. Mössbauer Effects, Jaipur (India) 1981, Indian Nat. Sci. Acad. (p. 494).

(Received December 17, 1982)

phys. stat. sol. (a) 77, 97 (1983)

Subject classification: 1.4; 10

Aviation Institute, Ufa[1])

Non-Equilibrium State and Recovery of Grain Boundary Structure

I. General Analysis, Crystallogeometrical Aspects

By

R. Z. Valiev, V. Yu. Gertsman, O. A. Kaibyshev, and Sh. Kh. Khannanov

The non-equilibrium state of high-angle grain boundary structure caused by the presence of grain boundary defects of dislocation nature is analysed. Such a state is characterized by the presence of long-range elastic fields of the boundary and/or by mechanical instability of the grain boundary structure. The creation of a non-equilibrium grain boundary structure and its recovery to an equilibrium one is considered within the framework of the crystallogeometrical approach for the case of interaction between lattice dislocations and grain boundaries.

Проанализировано неравновесное состояние структуры большеугловых границ зерен, обусловленное присутствием зернограничных дефектов дислокационной природы. Такое состояние характеризуется наличием у границы зерен дальнодействующих упругих полей и механической неустойчивостью зернограничной структуры. Образование неравновесной структуры границ зерен и её возврат к равновесной рассмотрены в рамках кристаллогеометрического подхода для случая взаимодействия решеточных дислокаций с границами зерен.

1. Introduction

In recent years considerable progress has been reached in understanding the high-angle grain boundary (GB) structure. However, most achievements are related to perfect GB's possessing an equilibrium structure. A thermodynamic equilibrium structure is a structure with minimum free energy at the given crystallographic parameters of the GB. In real materials, however, the GB structure can be far from perfect just as crystalline lattices in real materials are imperfect due to various lattice defects. Such an imperfect GB structure may be defined as a non-equilibrium one since it must possess a higher energy at the same crystallographic parameters. The transition of a non-equilibrium GB structure into a more stable state is the process of GB structure recovery. It is to be noted that a thermodynamic non-equilibrium GB structure can be mechanically stable just as some lattice defects are stable, though their presence increases the lattice energy.

By now experimental evidence has been obtained that various kinds of treatment result in the fact that GB's acquire a non-equilibrium structure and exhibit properties different from those of equilibrium GB's [6 to 11]. This indicates the importance of the development of general conceptions of a non-equilibrium grain boundary structure and its recovery to an equilibrium one. At present it is difficult to give a complete conception of the non-equilibrium state of GB's embracing all the possible distortions of their structures. Nevertheless even now it is possible to analyse the non-equilibrium

[1]) Ufa 450025, USSR.

GB structure taking into account mainly defects of dislocation character since considerable progress has been attained in the study of such defects. The present paper is an attempt to analyse creation, development, and recovery of the non-equilibrium structure of high-angle GB's. The first part deals with crystallogeometrical aspects, in the second part the energetic analysis is given.

2. Physical Basis

We consider a special case of the non-equilibrium GB state, namely a non-equilibrium state caused by defects of dislocation nature only. GB's having mechanically stable structures without long-range elastic fields will be called equilibrium GB's. (Such GB's are often called true GB's [16, 24].) In other words, the interaction forces between the elements of an equilibrium GB structure are mutually balanced, which determines its mechanical stability, and the equilibrium GB divides the crystals not deforming them elastically. Such an equilibrium state does not necessarily correspond to a thermodynamic equilibrium state since at fixed crystallographic GB parameters there may exist several structures corresponding to true GB's and the one with minimum free energy will be in thermodynamic equilibrium. However, each true GB is in relative equilibrium compared with GB's having mechanically unstable structures and long-range elastic fields. Further on we shall use the notion of equilibrium GB in this very sense.

The energy of a non-equilibrium GB and that of an equilibrium one creating the same misorientation far from the boundary differ in the values of elastic field energy and interaction energy. This does not mean that if two GB's have different values of energy one of them is necessarily non-equilibrium since the energies of these GB's may differ simply due to differences in their crystallographic parameters. The GB energy is known to depend on grain misorientation parameters and on the boundary plane. E.g., a special GB is in a sense more in equilibrium than a random one (and in general each GB is a non-equilibrium one compared with the perfect lattice). However, using the term "non-equilibrium GB" we imply only that such a boundary has uncancelled elastic fields and/or that its structure is mechanically unstable.

To represent a non-equilibrium GB we use the method of imaginary cuts as it has been done for lattice defects — translational and rotational dislocations [12 to 14].

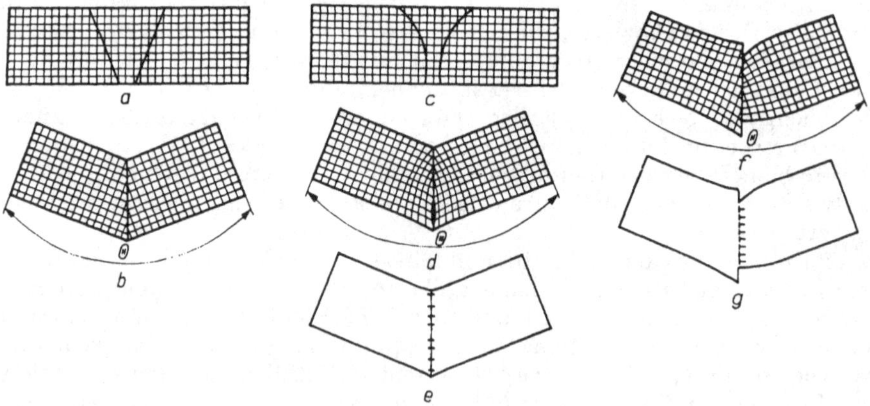

Fig. 1. Schemes showing the ways of introduction of equilibrium high-angle grain boundary and non-equilibrium ones (see text). (Networks of fine lines do not show a crystalline lattice but are given here only to show elastic deformations)

Fig. 1 a, b show how to obtain an equilibrium GB: cut surfaces are joined together without deformation. If the cut is done as shown in Fig. 1 c the deformation of cut surfaces will be necessary for joining them together which creates long-range elastic fields of the boundary (Fig. 1 d). The non-equilibrium GB in Fig. 1 f is obtained in another way: after cutting along the plane of equilibrium GB (see Fig. 1 b) cut surfaces are joined together with deformation. Certainly all the possible variants of the introduction of a non-equilibrium GB state are not reduced to these two schemes which are given here only as examples. The ways of obtaining non-equilibrium GB's are greatly analogous to the introduction of Somigliana dislocations [12 to 14]. In each of these schemes (Fig. 1 d, f) the whole boundary surface has a singularity in stresses and strains. Such a representation of the non-equilibrium state of GB structure seems to be the most general and with respect to it the schemes of Balluffi and co-workers [15, 16] of introduction of grain goundary dislocations (GBD's) by means of cuts along the GB plane may be regarded as particular cases.

The representation of non-equilibrium GB structure given above is related to continuum. However, according to modern conceptions [1 to 4] GB's have crystallographically ordered structures. Therefore, discrete distortions of their structures — GBD's and GBD complexes — may be regarded as sources of elastic fields. Fig. 1 e, g show schematically GBD complexes creating elastic distortions of the same character as in Fig. 1 d, f, respectively. (The introduction of a continuous distribution of infinitesimal dislocations would be completely equivalent to the continuum conception.) In all the cases given in these figures the GB release from elastic fields (GB structure recovery) can occur by carrying GBD's out from GB, besides, in the example in Fig. 1 e homogeneous distribution of GBD's in GB is possible leading to their mutual annihilation. Of course, these examples do not exhaust all the possible ways of non-equilibrium GB structure recovery since the given schemes do not exhaust all the cases of non-equilibrium states of a GB structure. In general recovery mechanisms are likely to be determined by the concrete type of non-equilibrium GB structure.

Since we always proceed from the dislocation description of the GB structure, it is necessary to specify the terminology. We consider GBD's having Burgers vectors which are DSC lattice translations [1] only. In general we adhere to the terminology of grain boundary defects of dislocation nature introduced previously [15 to 17] with making some refinements. The classification adopted by us is as follows: GB dislocation defects are divided into two groups — either they may be a part of equilibrium GB structure or they may be non-equilibrium. The first group includes intrinsic (or structural or secondary) GBD's (IGBD's) which are a requirement of the boundary geometry, e.g., they provide the deviation of grain misorientation from the special one. IGBD elastic fields are mutually cancelled. Therefore, IGBD's do not create long-range stresses and GB's with IGBD's are equilibrium ones. IGBD's have DSC Burgers vectors [1 to 4]. Besides, non-equilibrium defects — extrinsic GBD's (EGBD's) — may exist in grain boundaries. EGBD's may evidently be generated directly in the boundary but the main way of their generation is the interaction between lattice dislocations and GB's. From the purely geometrical viewpoint a lattice Burgers vector in the boundary may be represented as a sum of primitive DSC translations [1]. Therefore, lattice dislocations in the boundary may split into GBD's with DSC vectors [4, 15, 17 to 19]; these GBD's are EGBD's. The dislocation with lattice Burgers vector located in the GB will be called trapped lattice dislocation (TLD). (The passage of LD through the boundary is a special case of the process of TLD dissociation into a complex of EGBD's: a dislocation of orientational misfit [3] also belonging to EGBD's is left in the GB after LD passage.) EGBD's are non-equilibrium defects since they have uncancelled elastic fields unlike IGBD's while having DSC Burgers vectors like IGBD's.

It should be noted that there is no doubt concerning the applicability of the GBD conception to coincidence GB's and near-coincidence ones, but this problem is rather debatable in the case of random GB's. For random GB's DSC periods become so small that the GBD Burgers vectors may be regarded as close to zero. Without discussing the problem of the physical significance of dislocation conceptions for such GB's we shall note that many processes in any GB's may be described, at least formally, by means of GBD's.

Now we consider a concrete case of creation and recovery of non-equilibrium GB structure.

3. Crystallogeometry of Interaction between Lattice Dislocations and Grain Boundaries

In real conditions the creation of the non-equilibrium GB structure, i.e. the appearance of long-range stress fields of the boundary and/or loss of mechanical stability of its structure may be caused by trapping lattice dislocations by the boundary during plastic deformation or during GB movement through the deformed matrix. TLD's interact with the equilibrium GB structure. We proceed from the model of TLD dissociation into discrete EGBD's [17 to 19]. In our opinion the dissociation model is more preferable for crystallogeometrical analysis than the model of continuous delocalization of TLD cores [20 to 22] since

(i) it is more suitable for a formal description;

(ii) it allows to consider concrete atomic rearrangements in the GB structure;

(iii) it has better experimental verification which will be outlined later.

The analysis of creation and recovery of a non-equilibrium GB structure in the case of interaction between LD's and GB's leads in fact to the problem of dislocation absorption by GB's.

3.1 The analysis of dislocation rearrangements in grain boundaries

During lattice slip LD's accumulate in the plane of separation of two neighbouring crystals and grain misorientation changes since TLD's create some rotation of the crystalline lattice. However, since each dislocation creates an inhomogeneous rotation this misorientation change is generally inhomogeneous which leads to a non-equilibrium state of the GB structure, i.e. to the presence of uncancelled stresses. The rearrangement of GB structure is necessary for the transition into the equilibrium state (GB structure recovery). This rearrangement may consist either in carrying EGBD's (the products of TLD dissociation) away from GB's or in the formation of such EGBD configuration that GB has not long-range elastic fields and internal forces in the boundary plane are mutually balanced. In the latter case the homogenization of lattice rotation at the boundary will occur and EGBD's will "turn into" IGBD's. In case the total Bugers vector of TLD's is zero complete annihilation of EGBD's and recovery of crystallogeometrical parameters of the pre-deformational GB structure are possible after TLD dissociation [8, 23].

If the total Bugers vector of TLD's is not zero, a network of EGBD's is left in the boundary after any possible annihilation and recovery to the original (pre-deformational) structure is impossible due to rotation created by EGBD's and a new equilibrium GB configuration must be formed.

The formation of a low-angle dislocation boundary (polygonization) is an example of such a process. Long-range elastic fields of dislocations forming a boundary are mutually compensated. It is impossible, however, to represent the recovery of a high-angle boundary structure to an equilibrium state as the alignment of localized LD's in the boundary into a compensated network.

Firstly, the equilibrium configuration of the given set of LD's in the given boundary plane may not exist. E.g., a single-row wall of edge dislocations will always have long-range elastic fields if orientations of their Burgers vectors do not coincide with the normal to the boundary plane.

Secondly, even if creation of an equilibrium network of LD's in the given plane is possible (see, e.g., Fig. 2a), such a configuration is not in equilibrium in a high-angle boundary: each LD in GB dissociates into a complex of EGBD's and due to the irregularity of the total EGBD network each EGBD which is not in the centre of the complex is affected by unbalanced forces of other EGBD's. Therefore, the configuration in Fig. 2b is unstable and should be rearranged. In this case an equilibrium configuration will be represented by a regular wall of EGBD's (Fig. 2c) formed by climb of TLD dissociation products. Elastic field compensation in the case shown in Fig. 2c is better than in Fig. 2a. Thus, a new equilibrium GB represents in fact an original high-angle boundary with a superimposed low-angle one, the latter being defined in a DSC lattice, but not in an atomic crystal lattice.

It is to be noted that not only the configuration of edge EGBD's with Burgers vectors normal to the boundary plane can be in equilibrium, but also EGBD's with other Burgers vector orientations may create no long-range stress fields at GB. E.g., GBD's in Fig. 3 in some cases have no long-range stresses, i.e. they are IGBD's as it is shown in [16, 24]. If there are two EGBD systems with different orientations of Burgers vectors parallel to the boundary plane, then the creation of an EGBD network forming twist misorientation without long-range elastic fields is possible e.g., this may be a system of two mutually perpendicular rows of screw GBD's).

The general way of analysis of non-equilibrium GB structure recovery seems to be the search for a stable EGBD configuration with the best compensation of their stress fields. EGBD rearrangement is necessary for the establishment of the equilibrium state and during the whole period of this rearrangement the GB structure will be a non-equilibrium one.

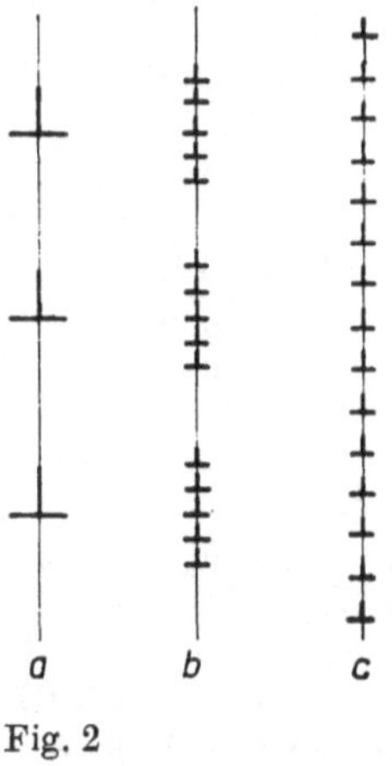

Fig. 2

Fig. 3

Fig. 2. a) Lattice dislocations trapped by high-angle grain boundary; b) TLD's dissociated into EGBD's; c) equilibrium EGBD network

Fig. 3. A possilbe equilibrium structure with EGBD Burgers vectors parallel to the boundary plane

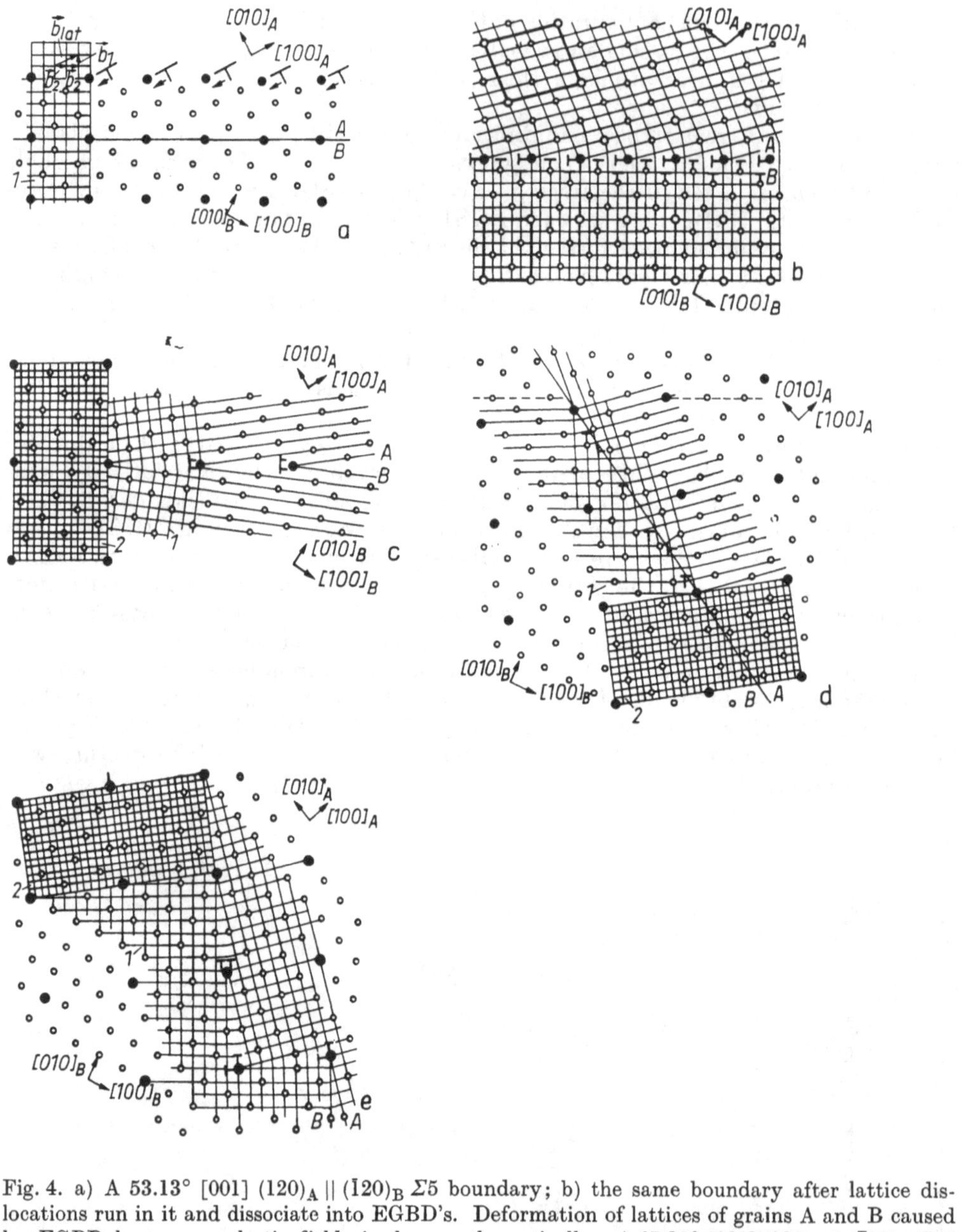

Fig. 4. a) A 53.13° [001] $(120)_A \parallel (\bar{1}20)_B$ $\Sigma 5$ boundary; b) the same boundary after lattice dislocations run in it and dissociate into EGBD's. Deformation of lattices of grains A and B caused by EGBD long-range elastic fields is shown schematically; c) 67.38° [001] $(230)_A \parallel (\bar{2}30)_B$ $\Sigma 13$ boundary resulting from deformed $\Sigma 5$ boundary (see Fig. 4b) after EGBD $\boldsymbol{b}_2$ going out and EGBD $\boldsymbol{b}_1$ alignment into an equilibrium wall. GBD's are shown by $\Sigma 5$ DSC lattice extra half-planes; d) 67.38° [001] $(8\bar{1}0)_A \parallel (470)_B$ $\Sigma 13$ boundary resulting from deformed $\Sigma 5$ boundary after boundary plane migration and alignment of EGBD's $\boldsymbol{b}_1$ and $\boldsymbol{b}_2$ into equilibrium configuration. EGBD's are shown by $\Sigma 5$ DSC lattice extra half-planes. The original boundary position is shown by dashed line; e) a $\Sigma 13$ boundary resulting from deformed $\Sigma 5$ boundary after formation of EGBD $\boldsymbol{b}_2$ vertical walls. Large black circles represent coincidence sites, 1 DSC lattice for $\Sigma 5$, DSC lattice for $\Sigma 13$

Suppose there is a symmetrical $\Sigma 5$ tilt boundary (Σ is the inverse density of coincidence sites) between crystals with cubic lattices (Fig. 4a). As a result of slip in each (010) plane of grain A LD's run in the boundary where each of them dissociates into EGBD's defined from the DSC lattice, $\boldsymbol{b}_1$ and $2\boldsymbol{b}_2$. In terms of the crystal lattice of grain A the dissociation reaction is given by

$$[100]_A \rightarrow \tfrac{1}{5}\,[120]_A + 2\,\tfrac{1}{5}\,[2\bar{1}0]_A\,. \tag{1}$$

Fig. 4b shows the scheme of changes of grain boundary structure: EGBD's $\boldsymbol{b}_1$ determine the lattice rotation of grains A and B, while EGBD's $\boldsymbol{b}_2$ create the deformation so that grain A becomes extended and grain B becomes compressed along the boundary. It should be noted that during running in the boundary each LD creates a GB step. Therefore, the GB plane changes and the GB plane in Fig. 4b is in fact rotated with respect to the boundary plane position of Fig. 4a. For simplicity Fig. 4a shows that each dislocation of grain A runs in the boundary in its own plane, however, the structure in Fig. 4b may be formed if the same number of dislocations run in the boundary not simultaneously and along any given set of planes and if the distribution of their dissociation products into a uniform EGBD network will occur. GB release from long-range elastic fields may take place in several ways:

(i) The equilibrium GB structure results from EGBD $\boldsymbol{b}_2$ going out from the boundary (to a triple point or free surface) by means of slip and from EGBD $\boldsymbol{b}_1$ alignment to a wall without long-range elastic fields (Fig. 4c). EGBD $\boldsymbol{b}_2$ slip will result in sliding along the boundary. (Since GB steps are associated with GBD's [3, 19] then, firstly, the GBD slip rate is limited by these steps and, secondly, slip of GBD's with steps in their cores will cause boundary migration alongside with sliding.) The resultant boundary is a $\Sigma 5$ boundary with a superimposed dislocation one. The misorientation around the $[001]_{A/B}$ axis created by EGBD $\boldsymbol{b}_1$ wall and calculated by means of the well-known Frank formula (cf. [13]) is equal to 14.25°. Together with the original boundary misorientation a new boundary misorientation angle of 67.38° results. On the other hand, the boundary in Fig. 1c is a $\Sigma 13$ tilt boundary (its misorientation angle is just 67.38°) and there are no GBD's in the DSC lattice for $\Sigma 13$ shown in this figure. Thus $\boldsymbol{b}_1$ dislocations in this boundary are "imaginary" ones: they exist in the original DSC lattice only and there are no dislocations in the new GB structure. One should note that during lattice dislocation absorption true GBD's may be left in GB provided the original DSC lattice is not destroyed.

(ii) Fig. 4d shows the boundary structure resulting from boundary rotation (migration) to the position where the total EGBD Burgers vector is perpendicular to the boundary plane. Without considering the GB migration mechanism we only note that the resulting dislocation configuration has no long-range stresses. A low-angle asymmetrical tilt boundary corresponds to it and its misorientation angle is 14.25°. The new boundary is an asymmetrical $\Sigma 13$ tilt boundary and there are again no GBD's in the $\Sigma 13$ DSC lattice.

(iii) The equilibrium GB structure may also result from the formation of vertical EGBD $\boldsymbol{b}_2$ walls (Fig. 4e). It is again an asymmetrical $\Sigma 13$ tilt boundary and on an average the GB plane is the same as in Fig. 4d, but this boundary is divided into facets passing along the most densely packed CSL planes.

The simple example considered here does not evidently exhaust all the possible ways of recovery of the non-equilibrium GB structure formed during interaction between LD's and GB. But it confirms the general conclusion that complete LD absorption in GB consists in the formation of a stable GB structure without long-range elastic fields. It should be noted that LD absorption in GB must lead to the change of the boundary crystallographic parameters.

The above-given conceptions refer to single boundaries without taking into account their joining with other GB's. The GB structure recovery in polycrystals may be supposed to occur in an analogous way though it is to be noted that in polycrystals this process may be complicated by the necessity of coordinated changes of structures of all the boundaries having common joints. Therefore, additional reactions in triple points of grains may be expected to occur alongside with the mechanisms described above.

3.2 Experimental evidence

Experimental data obtained by now (mainly by means of transmission electron microscopy) give sufficient evidence for the model conceptions proposed in this work. Direct observations of TLD dissociation and alignment of resulting DSC-GBD's into equilibrium IGBD networks were made in near-coincidence boundaries [15, 17 to 19, 25 to 27]. IGBD changes are sure to lead to a change of GB parameters. Jones et al. [28] were evidently the first to indicate the change of parameters of GB's containing non-annihilated EGBD networks. They even proposed to use the term "GB recrystallization" instead of the term "GB structure recovery" due to parameter changes compared with the pre-deformational state. Many works have been published [28 to 32] with examples of creation of regular EGBD networks in twin boundaries. Diffraction experiments carried out in a number of cases showed the difference from exact twin misorientations.

A special case of the proposed model is polygonization, i.e. formation of low-angle boundaries. An LD network without long-range stress fields is an equilibrium configuration for low-angle GB's. Horsewell [33] observed LD incorporation into a low-angle boundary during creep. This attests to a change of misorientation angle at that boundary and also confirms the proposed model of GB structure recovery.

In random GB's it is impossible to observe LD dissociation directly by electron microscopy. During in-situ heating the phenomenon of TLD "spreading" consisting in a continuous widening and fading of electron microscopic contrast of TLD images up to complete disappearance of contrast was discovered [5 to 7, 34, 35]. The problem of the mechanism of such TLD "spreading" is still debatable — either it is associated with delocalization of LD core in the boundary or is caused by dissociation of these LD's into EGBD's. However, it should be noted that the disappearance of electron microscopic TLD images alone cannot be a criterion of complete GB structure recovery since recent experiments [8 to 10] showed that GB properties differ from equilibrium ones during some definite period of time after TLD spreading.

4. Summary

The modern data concerning the structure and properties of crystalline interfaces lead to the conclusion that the GB structure in real materials greatly depends on the thermomechanical history of the specimen. In real polycrystals the GB structure may prove to be far from perfect, i.e. it may be non-equilibrium thermodynamically. A non-equilibrium GB structure is characterized by the fact that stress fields of these defects are not completely compensated and as a result a non-equilibrium GB must have an elevated energy compared with an equilibrium one only in case one takes into account the imperfections of GB structure which are of dislocation nature only.

The possible ways of creation and recovery of non-equilibrium GB structure are considered for the case of interaction between LD's and GB's. A non-equilibrium state of GB structure is created when GB contains either TLD's or their dissociation products — EGBD's. The recovery of this non-equilibrium structure consists in the formation of a stable EGBD configuration without long-range elastic fields, i.e. EGBD's "turn" in fact into IGBD's. In the process of GB structure recovery crystallographic GB parameters are generally changed and this process may include EGBD going out from GB, EGBD annihilation, EGBD reactions with the elements of equilibrium GB structure sliding along the boundary, and GB migration.

References

[1] W. Bollmann, Crystal Defects and Crystalline Interfaces, Springer-Verlag, Berlin 1970.
[2] Internat. Coll. Grain Boundaries in Metals, Grenoble 1974; J. Physique **36**, C4 (1975).
[3] A. N. Orlov, V. N. Perevezentsev, and V. V. Rybin, Grain Boundaries in Metals, Metallurgizdat, Moscow 1980 (in Russian).
[4] V. M. Kosevich, V. M. Ievlev, L. S. Palatnik, and A. I. Fedorenko, Structure of Intercrystalline and Interphase Boundaries, Metallurgizdat, Moscow 1980 (in Russian).
[5] H. Gleiter, Progr. Mater. Sci. **25**, 125 (1981).
[6] P. H. Pumphrey and H. Gleiter, Phil. Mag. **32**, 881 (1975).
[7] R. A. Varin, phys. stat. sol. (a) **52**, 347 (1979).
[8] R. Z. Valiev, V. Yu. Gertsman, and O. A. Kaibyshev, phys. stat. sol. (a) **61**, K95 (1980).
[9] R. Z. Valiev, V. Yu. Gertsman, and O. A. Kaibyshev, Fiz. tverd. Tela **22**, 2213 (1980).
[10] M. W. Grabski, R. Z. Valiev, J. W. Wyrzykowski, and W. Łojkowski, Res. Mechanica Letters **1**, 489 (1981).
[11] R. Z. Valiev and O. A. Kaibyshev, Dokl. Akad. Nauk USSR **258**, 92 (1981).
[12] J. D. Eshelby, Solid State Phys. **3**, 79 (1956).
[13] J. Friedel, Dislocations, Pergamon Press, 1964.
[14] V. A. Likhachev and R. Yu. Khairov, Introduction to the Theory of Disclinations, University Press, Leningrad 1975 (in Russian).
[15] R. W. Balluffi, Y. Komem, and T. Schober, Surface Sci. **31**, 68 (1972).
[16] J. R. Hirth and R. W. Balluffi, Acta metall. **21**, 929 (1973).
[17] D. J. Dingley and R. C. Pond, Acta metall. **27**, 667 (1979).
[18] R. C. Pond and D. A. Smith, Phil. Mag. **A36**, 353 (1977).
[19] T. P. Darby, R. Schindler, and R. W. Balluffi, Phil. Mag. **A37**, 245 (1978).
[20] P. H. Pumphrey and H. Gleiter, Phil. Mag. **30**, 593 (1974).
[21] H. Gleiter, Scripta metall. **11**, 305 (1977); Phil. Mag. **A36**, 1109 (1977).
[22] W. Łojkowski, H. O. K. Kirchner, and M. W. Grabski, Scripta metall. **11**, 1127 (1977).
[23] R. Z. Valiev, O. A. Kaibyshev, and Sh. Kh. Khannanov, phys. stat. sol. (a) **52**, 447 (1979).
[24] J. R. Hirth, Metallurg. Trans. **3**, 3047 (1972).
[25] W. Bollmann, B. Michaut, and G. Sainfort, phys. stat. sol. (a) **13**, 637 (1972).
[26] W. A. T. Clark and D. A. Smith, J. Mater. Sci. **14**, 776 (1979).
[27] L. M. Clarebrough and C. T. Forwood, phys. stat. sol. (a) **58**, 597 (1980).
[28] A. R. Jones, P. R. Howell, and B. Ralph, Phil. Mag. **35**, 603 (1977).
[29] P. R. Howell, J. O. Nilsson, and G. L. Dunlop, Phil. Mag. **A38**, 39 (1978).
[30] L. E. Svensson and G. L. Dunlop, Canad. Metall. Quart. **18**, 39 (1979).
[31] G. L. Dunlop and J. O. Nilsson, Mater. Sci. Engng. **42**, 273 (1980).
[32] G. Wirmark, J. O. Nilsson, and G. L. Dunlop, Phil. Mag. **A43**, 93 (1981).
[33] A. Horsewell, Phil. Mag. **A38**, 747 (1978).
[34] R. Z. Valiev and O. A. Kaibyshev, phys. stat. sol. (a) **44**, 477 (1977).
[35] W. Łojkowski and M. W. Grabski, Scripta metall. **13**, 511 (1979).

(Received February 14, 1983)

phys. stat. sol. (a) **76**, 777 (1983)

Subject classification: 20.1 and 20.3; 22.8

A. V. Shubnikov Institute of Crystallography, Academy of Sciences of the USSR, Moscow[1])

Luminescence Properties of Nd^{3+} Ions in Disordered $Na_5YSi_4O_{12}$ Crystals

By

A. A. KAMINSKII, V. A. TIMOFEEVA, N. R. AGAMALYAN, A. B. BYKOV, S. E. SARKISOV, and YU. M. GERASIMOV

Spectral and luminescence investigations of disordered $Na_5YSi_4O_{12}:Nd^{3+}$ crystals with high ionic conductivity are carried out. Absorption and luminescence spectral analysis allow to refine some details of crystal structure of this superionic conductor and to determine crystal-field splitting schemes of Nd^{3+} ions' $^4F_{3/2}$, and $^4I_{9/2-15/2}$ manifolds. Such luminescence intensity characteristics, as radiative lifetime of metastable $^4F_{3/2}$ state, intermanifold luminescence branching ratios, and effective cross-section of transitions, connected with intensive bands in $^4F_{3/2} \rightarrow {}^4I_{11/2}$ and $^4F_{3/2} \rightarrow$ $\rightarrow {}^4I_{13/2}$ channels are also determined.

Проведены спректрально-люминесцентные исследования разупорядоченных кристаллов $Na_5YSi_4O_{12}:Nd^{3+}$ с высокой ионной проводимостью. Анализ спектров поглощения и люминесценции позволил уточнить детали кристаллической структуры этого суперионного проводника и построить схемы расщепления мультиплетов $^4F_{3/2}$ и $^4I_{9/2-15/2}$ ионов Nd^{3+}. В работе также определены такие интенсивностные характеристики люминесценции активатора, как излучательное время жизни метастабильного состояния $^4F_{3/2}$, межмультиплетные коэффициенты ветвления люминесценции и эффективное поперечное сечение переходов, связанных с интенсивными полосами в каналах $^4F_{3/2} \rightarrow {}^4I_{11/2}$ и $^4F_{3/2} \rightarrow {}^4I_{13/2}$.

1. Introduction

The special way to improve the generation parameters in trivalent rare-earth ion (TR^{3+}) stimulated emission (SE) of luminescing dielectric materials was shown in the beginning of the sixties [1, 2], and was connected with the search and creation of disordered crystals. A special peculiarity of this new class of laser media is that TR^{3+} ions enter the composition of activator centres (AC) in a variety of structures. This results in the broadening of absorption spectra, consisting of broad bands, providing a substantial increase in the coefficients of light energy utilization from pumping sources. The energy migration of electron excitation over the levels of the centres close by their structure, also improves the SE excitation characteristics. Now some methods of creation of disorder and multicentres in crystals with TR^{3+} ions are known. The most effective one among them is the synthesis of mixed systems on the basis of compounds or components with different crystallochemical properties [3]. Thus, in yttrium–fluoride (CaF_2–YF_3–TRF_3) systems with trifluoride concentration of 5 wt%, complicated coordination polyhedra of cations are created, in which F atoms occupy positions differing from those usual ($\frac{1}{4}, \frac{1}{4}, \frac{1}{4}$) in CaF_2-type crystal [4]. In a solid solution of $YScO_3$ type (Y_2O_3–Sc_2O_3) the disorder from the point of view of TR^{3+}-centre formation is connected with the difference of ionic radii, energies, and radial extension of 4f orbitals of Y^{3+}, Sc^{3+}, and TR^{3+}. In such systems maximal disorder takes place

[1]) Leninskii Prospekt 59, 117333 Moscow, USSR.

when the component ratio is 1:1 [5]. Also such components are known, e.g. $K_5 \cdot Nd(MoO_4)_4$, in which disorder, stipulating a variety of AC of TR^{3+} ions is already "inherent" in the crystal structure itself [6].

A list of references from nearly 20 years of SE and luminescence study of disordered dielectric crystals with TR^{3+} ions is given in [7]. The data accumulated over such a period show that compounds with ionic conductivity can be also attributed to the disordered luminescing crystals. The present work was carried out in this connection. It deals with the study of luminescence and absorption properties of Nd^{3+} ions in the superionic conductor $Na_5YSi_4O_{12}$, where structural disorder is determined by the behaviour of the alkaline ions [8]. The problems of the investigations carried out were: study of AC properties of Nd^{3+} ions in $Na_5YSi_4O_{12}$ and determination of their level energies; temperature and activator concentration dependences of luminescence lifetime of the metastable $^4F_{3/2}$ state, and also determination of all basic luminescence intensity characteristics, including the cross-section of the transitions between $^4F_{3/2}$ and $^4I_{11/2,13/2}$ states, to estimate the possibility of SE excitation in such a crystal. This work is continuing investigations of luminescence properties of yttrium silicates, started in [9].

2. Crystals and Experimental Technique

Crystals of $Na_5TRSi_4O_{12}$ (TR = Sm to Lu, Y, and Sc) type with high ionic conductivity by Na^+ cations were grown at first under hydrothermal conditions in 1969 [10]. The structure peculiarity of this class of Na, TR silicates are plane twelve-member rings ($Si_{12}O_{36}$), which are connected in the framework by single TR-polyhedra. According to structural investigations compounds $Na_5TRSi_4O_{12}$ (TR = Sc, Y, and Lu) have in first approximation the common inversive $R\bar{3}c$ space group [11, 12]. However, according to X-ray analysis data for $Na_5LuSi_4O_{12}$ [13] and $Na_5YSi_4O_{12}$ [14] crystals, there

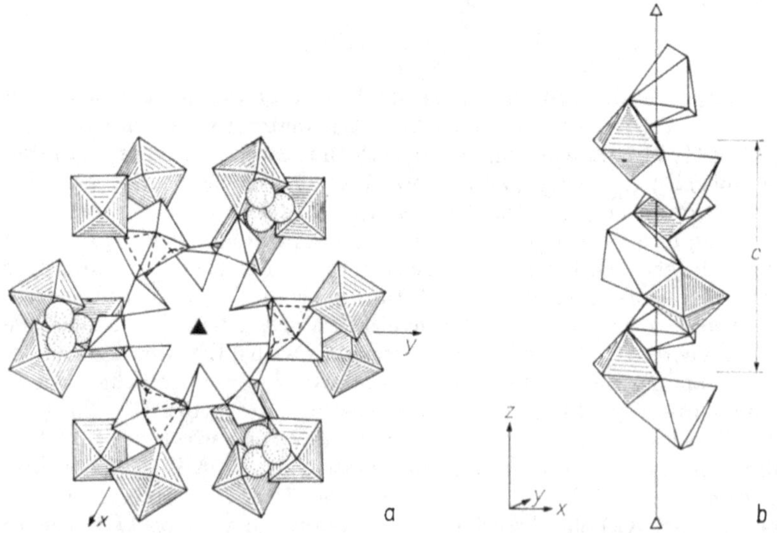

Fig. 1. $Na_5YSi_4O_{12}$ crystal structure fragment [13]: a) right and left "screws" from TR_1 and TR_2 octahedra around the twelve-member ring of Si tetrahedra. Circles denote Na atoms; b) right screw axis 3_1 from the alternating TR_1 (shaded) and Na octahedra. Polyhedra are connected by common edges

were some reflections contradicting the c-plane in the $R\bar{3}c$ group symbol, common for the given compounds. Structure refinement was carried out within the acentric R32 space group, which in its turn is under the $R\bar{3}c$ group [13, 14]. Rhombohedral unit cell parameters of $Na_5YSi_4O_{12}$ are: $a = 13.41$ Å, $\alpha = 150° 36'$ (in hexagonal coordinates $a = 22.03$ and $c = 12.62$ Å) and $z = 18$. In the acentric structure model TR^{3+} ions occupy two crystallographic positions, arranging near right 3_1 and left 3_2 screw axes with octahedral cavity-cells filled by Na^+ ions or being empty, respectively (Fig. 1). In the case of the centrosymmetric model TR^{3+} ions occupy one crystallographic position, arranging near the sliding c-plane.

Single crystals of $Na_5YSi_4O_{12}$ were grown from the flux with partial substitution of Y^{3+} ions by Nd^{3+}. Crystallization occurred from fluor-oxide flux with $Na_5YSi_4O_{12}$ concentration about $\approx 40\%$ mass, by slow cooling of the solution from 950 to 800 °C [15]. Other phases, such as $NaYSi_3O_9$, $NaYSiO_4$, and $Na_3YSi_2O_7$ were observed in our flux system together with the given $Na_5YSi_4O_{12}$ compound. $Na_5YSi_4O_{12}:Nd^{3+}$ was grown at rotating a Pt rod with 6 mm diameter, introduced into the bulk of the solution (standard Pt crucibles with 175 ml capacity were used as crystallizers) using the crystallization centre localization method [16]. Due to the high initial value of flux concentration, during the whole time of crystallization (3 to 6 days) at the cooling rate of 0.5 K/h, we were able to obtain a quite large quantity (10 to 20 g) of transparent $Na_5YSi_4O_{12}$ crystals of about ≈ 3 mm size. The introduction of Nd_2O_3 into the flux instead of Y_2O_3 up to their equal (1:1) molar ratio in the solution, practically did not change the phase formation in the Na_2O–Y_2O_3–SiO_2 solvent system. However, higher content of Nd_2O_3 (1:2) in the solution resulted in the narrowing of the crystallization region of $Na_5YSi_4O_{12}:Nd^{3+}$ and further to its total disappearance. As it seems, this can be connected with the incomplete "isostructure" between Y^{3+} and Nd^{3+} ions in the compound [17]. Growth experiments gave a concentration series of $Na_5(Y_{1-x}Nd_x)\cdot Si_4O_{12}$ with x varying from 0.01 to ≈ 0.2. The concentration value of Nd^{3+} ions in $Na_5YSi_4O_{12}$ crystals was determined using a transmission electron microscope "Hitachi-H-500 H" equipped with a special cell for the investigation of large samples and an X-ray spectrometer "Micro-X 7000", allowing to carry out microanalysis by energy dispersion. The spectra obtained were treated using a computer programme.

Absorption and luminescence spectra of $Na_5(Y_{1-x}Nd_x)Si_4O_{12}$ crystals were investigated with grating DFS-12 and SDL-1 spectrometers as well as with SF-8 and "Specord 75-IR" spectrophotometers. Luminescence excitation of Nd^{3+} ions was accomplished by KGM-12-100 type tungsten-iodine lamps and ILA-120 Ar laser. Prism monochromator ZMR-3, connected with a photomultiplier FEU-28 and broadband oscilloscope S1-16 were used to measure the luminescence decay time. A stroboscopic ISSh-100-3 type lamp was used as luminescence excitation source ($\tau_{exc} \approx 10$ μs). All spectroscopic measurements were carried out in "isotropic" approach using microcrystals with layers of different thickness placed in special specimen cells [9].

3. Spectroscopic Properties of Superionic $Na_5YSi_4O_{12}:Nd^{3+}$ Crystal

The studied absorption and luminescence bands of Nd^{3+} ions, corresponding to $J \to J'$ transitions fall within the optical transparency region of $Na_5YSi_4O_{12}$ crystals spreading from ≈ 0.4 to ≈ 6 μm. The structural disorder of $Na_5YSi_4O_{12}$ crystal appears in the inhomogeneous broadening of absorption and luminescence lines of the activator, the width of which changes slightly (in 1.5 times) with temperature changing from 77 to 300 K. Such a behaviour is stipulated by the statistical character of Na^+ ion arrangement (as a result of their mobility) near Nd^{3+} polyhedra. Our spectroscopic measurements also answered the question on the number of the crystallographic positions

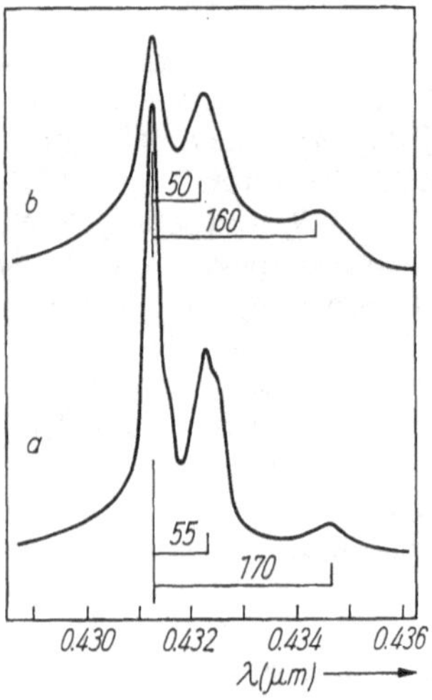

Fig. 2. Absorption spectra of Nd^{3+} ions in $Na_5YSi_4O_{12}$ crystal, corresponding to the $^4I_{9/2} \rightarrow {}^2P_{1/2}$ transition at (a) 77 and (b) 300 K. The marks denote splitting of the ground $^4I_{9/2}$ state

occupied by $Y^{3+}(Nd^{3+})$ ions in $Na_5YSi_4O_{42}$. A number of lines at 77 K, e.g. in the absorption $^4I_{9/2} \rightarrow {}^2P_{1/2}$ (Fig. 2) and luminescence ($^4F_{3/2} \rightarrow {}^4I_{9/2}$) (Fig. 3) spectra, have doublets, which supports the acentric model of trigonal structure of $Na_5YSi_4O_{12}$: Nd^{3+} crystals. Because the distance between the components of the observed doublets is nearly equal to the inhomogeneous linewidth, and doublet structure as well as lineshape are independent of activator concentration, we analysed the crystal-field splitting scheme of Nd^{3+} ions in $Na_5YSi_4O_{12}$ within the quasi-centre model [18]. The results of this analysis (accuracy of measurements is ± 10 cm^{-1}) are listed in Table 1. It should be noted that in these experiments the main attention was paid to the manifolds the

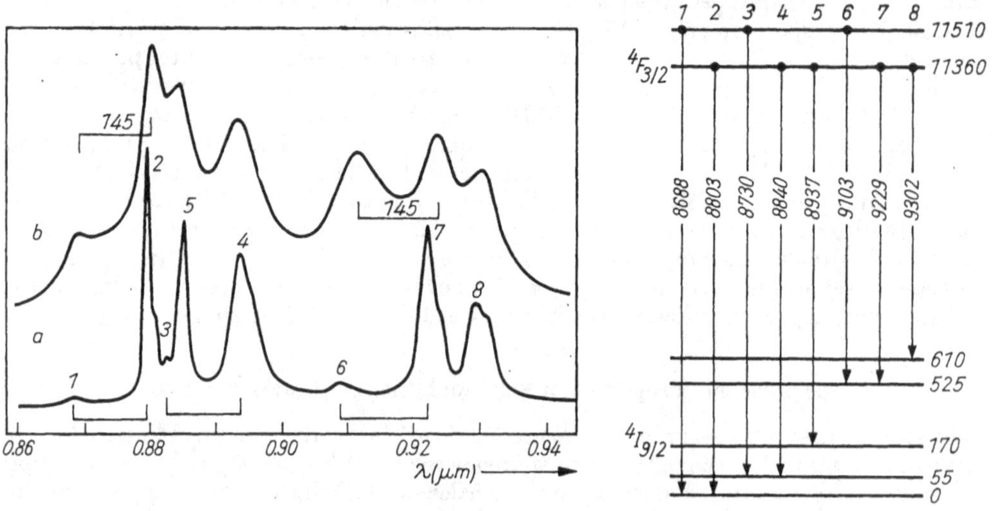

Fig. 3. Luminescence spectra and at (a) 77 and (b) 300 K ($^4F_{3/2} \rightarrow {}^4I_{9/2}$ channel) and crystal-field splitting scheme of $^4F_{3/2}$ and $^4I_{9/2}$ (in the lower part) manifolds at 77 K of Nd^{3+} ions in $Na_5YSi_4O_{12}$ crystal. Energy of levels is given in cm^{-1} and transitions between them in Å. Lines in the spectrum (a) and transitions in the scheme are denoted by corresponding numbers to facilitate their comparison. Marks in the spectra denote splitting of the metastable $^4F_{3/2}$ state

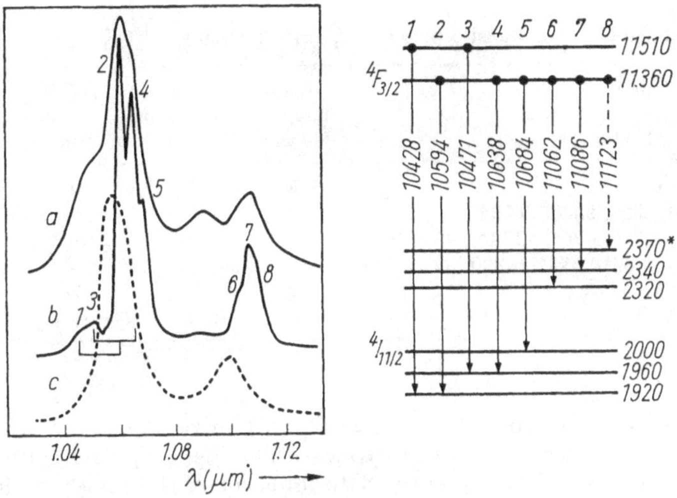

Fig. 4. Luminescence ($^4F_{3/2} \rightarrow {}^4I_{11/2}$ transition) spectra at (a) 300 and (b) 77 K and crystal-field splitting scheme ($T = 77$ K) on the right-hand side of $^4F_{3/2}$ and $^4I_{11/2}$ manifolds of Nd^{3+} ions in $Na_5YSi_4O_{12}$ crystal and also in $Na_3YSi_3O_9$ crystal ($T = 77$ K). Asterisk denotes energy value needing refinement

luminescence, transitions between which are potential for SE excitation at these wavelengths. Spectra of these luminescence channels are given in Fig. 4 and 5, where the crystal-field splitting schemes of the corresponding manifolds are also shown.

To determine the intensity characteristics of emission $^4F_{3/2} \rightarrow {}^4I_{J'}$ channels of Nd^{3+} ions in $Na_5YSi_4O_{12}$ crystal, intermanifold luminescence branching ratios ($\beta_{JJ'}$) from the metastable $^4F_{3/2}$ state were measured and concentration and temperature dependences of the $^4F_{3/2}$ state lifetime ($\tau_{lum}(C)$ and $\tau_{lum}(T)$) were studied. The $\beta_{JJ'}$ values were

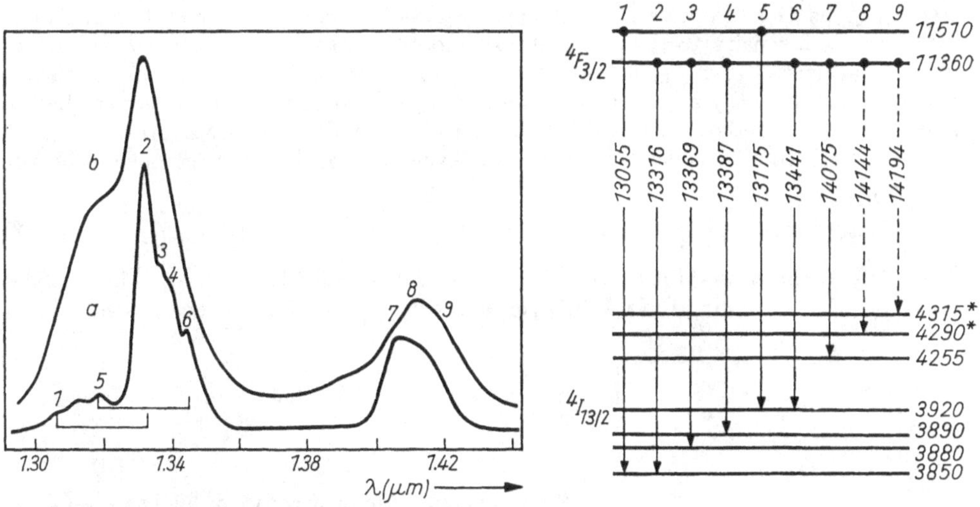

Fig. 5. Luminescence ($^4F_{3/2} \rightarrow {}^4I_{13/2}$ transition) spectra at (a) 77 and (b) 300 K and crystal-field splitting scheme (in the lower part) of $^4F_{3/2}$ and $^4I_{13/2}$ manifolds of Nd^{3+} ions in $Na_5YSi_4O_{12}$ crystal. All notations are the same as in Fig. 4

Table 1

Splitting manifold of the Nd^{3+} ion quasi-centre in $Na_5YSi_4O_{12}$ crystal at 77 K

manifold	energy of levels (cm^{-1})	number of levels		ΔE
		theor.	exper.	(cm^{-1})
$^4I_{9/2}$	0, 55, 170, 525, 610	5	5	610
$^4I_{11/2}$	1920, 1960, 2000, 2320, 2340, 2370*)	6	6	450*)
$^4I_{13/2}$	3850, 3880, 3890, 3920, 4255, 4290*), 4315*)	7	7	465*)
$^4I_{15/2}$	5760, 5795, 5895*), 6240, 6305, 6320, 6370, 6390	8	8	630
$^4F_{3/2}$	11360, 11510	2	2	150
$^2P_{1/2}$	23180	1	1	—

*) Refinement is needed.

found from the spectroscopic quality parameter $X = \Omega_4/\Omega_6$ (where Ω_4 and Ω_6 are phenomenological parameters of intensity according to the Judd-Ofelt approach) [19], which was measured using the method based on the determination of the line strength ($s_{JJ'}$) ratio between absorption transitions $^4I_{9/2} \rightarrow {}^2P_{1/2}$ and $^4I_{9/2} \rightarrow {}^4I_{15/2}$ [7, 19]

$$X = 1.23 s(^4I_{9/2}, {}^2P_{1/2})/s(^4I_{9/2}, {}^4I_{15/2}) .$$

For the determined value of $X \approx 0.5$, the following values of $\beta_{JJ'}$ were found according to [7]: $\beta_{J,9/2} = 0.34$, $\beta_{J,11/2} = 0.54$, $\beta_{J,13/2} = 0.12$, and $\beta_{J,15/2} = 0.006$.

The most important intensity characteristics of a luminescence transition for the estimation of the possibility for SE excitation at its wavelength is the cross-section σ_e. In the case of disordered media, among which are $Na_5YSi_4O_{12}:Nd^{3+}$ crystals with inhomogeneously broadened and overlapped lines in the spectra, it is appropriate to use its effective peak value [7]

$$\sigma_e^{eff} = \frac{\lambda^2 A_{JJ'}}{8\pi n^2 \, \Delta\nu_{lum}^{eff}}, \quad (1)$$

where $\Delta\nu_{lum}^{eff}$ is the effective luminescence $^4F_{3/2} \rightarrow {}^4I_{J'}$ bandwidth equal to the ratio of its integral intensity to the peak intensity at the wavelength λ, n is the refractive index of the crystal, and $A_{JJ'}$ is the probability of intermanifold spontaneous radiative transition, equal to β_{JJ}/τ_{rad}. The behaviour of $\tau_{lum}(N_0)$ shown in Fig. 6, was studied to determine the τ_{rad} value. At the activator content at which cross-relaxation processes did not quench luminescence from the $^4F_{3/2}$ levels, τ_{lum} and τ_{rad} are connected by the relation [7]

$$\tau_{rad} = \tau_{lum}(1 + \sum_{J'} W_{JJ'}/\sum_{J'} A_{JJ'}) , \quad (2)$$

where $W_{JJ'}$ is the probability of non-radiative transitions $^4F_{3/2} \rightarrow {}^4I_{15/2}$. The analysis of the $\tau_{lum}(T)$ behaviour for $Na_5YSi_4O_{12}$ crystal with low Nd^{3+} ion concentration shows

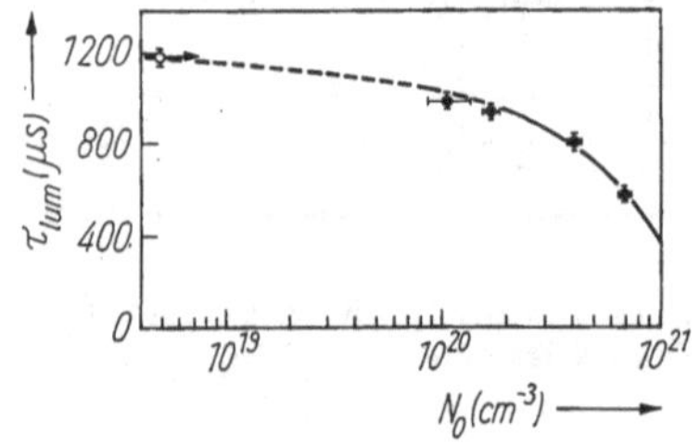

Fig. 6. Concentration dependence of lifetime $\tau_{lum}(N_0)$ of $Na_5YSi_4O_{12}:Nd^{3+}$ crystal at 77 (○) and 300 K (●)

that $W_{JJ'} \approx 0$. Thus, the measurements of luminescence decay carried out resulted in the value $\tau_{rad} \approx 1200$ μs for the $^4F_{3/2}$ state.

The values of τ_{rad}, $\beta_{JJ'}$, and $\Delta\nu_{lum}^{eff}$ found allowed to estimate by formula (1) the value of σ_e^{eff} at the wavelengths of the intensive peaks of two luminescence bands $^4F_{3/2} \rightarrow$ $\rightarrow {}^4I_{11/2}$ (1.0594 μm) and $^4F_{3/2} \rightarrow {}^4I_{13/2}$ (1.3320 μm). The results of the calculations carried out are listed in Table 2. More accurate values of σ_e must be given from orientational measurements of the luminescence parameters. According to the value of σ_e^{eff} the crystal studied can be assigned to the materials with medium gain coefficient, e.g. close to the laser glass of GLS-7 type [20].

At present we are planning to carry out investigations of the spectroscopic properties of $Na_5(Y_{1-x}Nd_x)Si_4O_{12}$ crystals under the conditions of ionic conductivity.

Table 2

Spectral and luminescence parameters of basic $^4F_{3/2} \rightarrow {}^4I_{11/2}$ and additional $^4F_{3/2} \rightarrow {}^4I_{13/2}$ luminescence channels of Nd^{3+} ions in $Na_5YSi_4O_{12}$ crystal

parameter	$^4F_{3/2} \rightarrow {}^4I_{11/2}$		$^4F_{3/2} \rightarrow {}^4I_{13/2}$	
	77 K	300 K	77 K	300 K
λ (μm)	1.0594	1.0594	1.3316	1.3320
$\beta_{JJ'}$	0.54	0.54	0.12	0.12
$\Delta\nu_{lum}^{eff}$ (cm^{-1})	≈125	≈240	≈110	≈210
σ_e^{eff} (10^{-20} cm^2)	≈2.4	≈1.2	≈0.9	≈0.5

Acknowledgements

The authors thanks Dr. B. K. Sevastianov for support and Dr. B. A. Maksimov for discussions.

References

[1] A. A. Kaminskii and V. V. Osiko, Izv. Akad. Nauk SSSR, Ser. neorg. Mater. **1**, 2040 (1965).
[2] Kh. S. Bagdasarov, Yu. K. Voronko, A. A. Kaminskii, V. V. Osiko, and A. M. Prokhorov, Kristallografiya **10**, 746 (1965).
[3] Yu. K. Voronko, A. A. Kaminskii, V. V. Osiko, and A. M. Prokhorov, Izv. Akad. Nauk SSSR, Ser. neorg. Mater. **2**, 1161 (1966).
[4] V. B. Aleksandrov and L. S. Garashina, Dokl. Akad. Nauk SSSR **189**, 307 (1969).
[5] Kh. S. Bagdasarov, A. A. Kaminskii, A. M. Kevorkov, L. Li, A. M. Prokhorov, T. A. Tevosyan, and S. E. Sarkisov, Dokl. Akad. Nauk SSSR **224**, 798 (1975).
[6] A. A. Kaminskii, S. E. Sarkisov, J. Bohm, P. Reiche, D. Schultze, and R. Uecker, phys. stat. sol. (a) **43**, 71 (1977).
[7] A. A. Kaminskii, Laser Crystals, Their Physics and Properties, Springer-Verlag, Berlin/Heidelberg/New York 1981.
[8] B. A. Maksimov, B. V. Merinov, V. S. Borovkov, A. K. Ivanov-Shits, and N. V. Belov, Kristallografiya **24**, 265 (1979).
[9] A. A. Kaminskii, N. R. Agamalyan, S. E. Sarkisov, L. N. Demyanets, A. N. Lobachev, and G. A. Emelchenko, Izv. Akad. Nauk SSSR, Ser. neorg. Mater. **16**, 2185 (1980).
[10] B. A. Maksimov, B. N. Litvin, V. V. Ilyukhin, and N. V. Belov, Kristallografiya **14**, 498 (1969).
[11] B. A. Maksimov, Yu. A. Kharitonov, and N. V. Belov, Dokl. Akad. Nauk SSSR **213**, 1072 (1973).
[12] B. V. Merinov, B. A. Maksimov, V. V. Ilyukhin, and N. V. Belov, Dokl. Akad. Nauk SSSR **248**, 1108 (1979).

[13] B. V. MERINOV, B. A. MAKSIMOV, YU. A. KHARITONOV, and N. V. BELOV, Dokl. Akad. Nauk SSSR **240**, 81 (1978).
[14] B. A. MAKSIMOV and N. V. BELOV, Dokl. Akad. Nauk SSSR **261**, 623 (1981).
[15] V. A. TIMOFEEVA, Rost kristallov iz rastvorov–rasplavov, Izd. Nauka, Moscow 1978.
[16] V. A. TIMOFEEVA and A. B. BYKOV, Progr. Crystal Growth Charact. **2**, 377 (1979).
[17] A. A. KAMINSKII, V. V. OSIKO, S. E. SARKISOV, M. I. TIMOSHECHKIN, E. V. ZHARIKOV, J. BOHM, P. REICHE, and D. SCHULTZE, phys. stat. sol. (a) **49**, 305 (1978).
[18] A. A. KAMINSKII, Zh. eksper. teor. Fiz. **58**, 407 (1970).
[19] R. D. SHANNON, B. E. TAYLOR, T. E. GIER, H.-Y. CHEN, and T. BERZIZS, Inorg. Chem. **17**, 958 (1978).
[20] A. M. PROKHOROV (Ed.), Spravochnik po lazeram, Vol. 1, Izd. Sovetskoe Radio, Moscow 1978 (p. 330).

(Received February 15, 1983)

phys. stat. sol. (a) 77, 115 (1983)

Subject classification: 1.3 and 10.1; 21; 21.1

Institute of Metal Physics, Academy of Sciences of the Ukrainian SSR, Kiev[1])

Dynamical Diffraction in Alloys

Influence of Long-Range Order

By

G. I. GUDZENKO, V. B. MOLODKIN, and V. G. NOVITSKII

Within the framework of an earlier proposed method the dynamical diffraction of X-rays, neutrons, electrons, and other charged particles by ordered alloys are considered theoretically. Anomalous transmission, pendellösung patterns, intensity jumps near the absorption edges for fundamental and superlattice reflections, and channeling of fast particles are discussed. The influence of the long-range order and other characteristics of alloys on these dynamical effects are clarified.

С помощью ранее предложенного метода рассматривается теория динамического рассеяния рентгеновских лучей, нейтронов, электронов и других заряженных частиц в упорядочивающихся сплавах. Обсуждаются аномальное прохождение, маятниковые решения, скачки интенсивности у краев поглощения для структурных и сверхструктурных отражений, а также каналирование быстрых частиц. Выясняется влияние параметра дальнего порядка и других характеристик сплава на эти динамические эффекты.

1. Introduction

Within the framework of our method [1 to 3] the infinite system of equations of the dynamical theory in the case of a crystal with large irregularities can be solved approximately in terms of the perturbation theory, since the energy is large in comparison with the potential. In this case the system of equations of the dynamical theory is reduced to a set of two equations (two-beam case) for two strong Bragg waves; it has the same form as in the case of a perfect crystal. However, this set contains now the effective potential (V^{eff}) which depends on the crystal parameters and takes into account the correction due to the influence of weak Bragg (ΔV^{B}) and diffuse (ΔV^{D}) waves, so that

$$V_{\boldsymbol{h}}^{\text{eff}} = \overline{V}_{\boldsymbol{h}} + \Delta V_{\boldsymbol{h}}^{\text{B}} + \Delta V_{\boldsymbol{h}}^{\text{D}}\,, \tag{1}$$

where $\overline{V}_{\boldsymbol{h}}$ is the Fourier component of the potential averaged over the defect configurations and thermal displacements. In turn, the diffuse correction consists of terms due to the thermal displacements (ΔV^{T}), static distortions (ΔV^{S}), scattering factor fluctuations (ΔV^{F}), and a term caused by the non-additive contribution of static distortions and scattering factor fluctuations,

$$\Delta V_{\boldsymbol{h}}^{\text{D}} = \Delta V_{\boldsymbol{h}}^{\text{T}} + \Delta V_{\boldsymbol{h}}^{\text{F}} + \Delta V_{\boldsymbol{h}}^{\text{SF}}\,.$$

Using the approximate expressions for the scattering factors [4], we have estimated these corrections in the following situations:

1. Slow neutron diffraction by the NiAl alloy. Taking into account various Ni isotopes, the effective potential and corrections have the following magnitudes:

$$|\overline{V}_{\boldsymbol{h}}| = 1.5 \times 10^{11}\ \text{cm}^{-2}\,, \qquad |\Delta V_{\boldsymbol{h}}^{\text{B}}| = 4.0 \times 10^{6}\ \text{cm}^{-2}\,,$$
$$|\Delta V_{\boldsymbol{h}}^{\text{T}}| = 3.5 \times 10^{7}\ \text{cm}^{-2}\,, \qquad |\Delta V_{\boldsymbol{h}}^{\text{F}}| = 3.8 \times 10^{9}\ \text{cm}^{-2}\,,$$
$$|\Delta V_{\boldsymbol{h}}^{\text{S}}| = 4.0 \times 10^{5}\ \text{cm}^{-2}\,, \qquad |\Delta V_{\boldsymbol{h}}^{\text{FS}}| = 5.0 \times 10^{6}\ \text{cm}^{-2}\,.$$

[1]) pr. Vernadskogo 36, 252680 Kiev 142, USSR.

Table 1

electron energy (eV)	μ_0/C (10^{15} cm^{-2})	μ_h/C (10^{15} cm^{-2})	p_0/C (10^{15} cm^{-2})	p_h/C (10^{15} cm^{-2})	V_0^{Ge} (10^{15} cm^{-2})	V_0^{Si} (10^{15} cm^{-2})
10^1	60	52	−72	−63	41	13
10^2	19	16	−23	−20	41	13
10^3	6.0	5.2	− 7.2	− 6.3	41	13
10^4	1.9	1.6	− 2.3	− 2.0	41	13
10^5	0.6	0.5	− 0.7	− 0.6	41	13
10^6	0.1	0.1	− 0.15	− 0.15	41	13
					χ_0^{Ge}	χ_0^{Si}
X-rays	(10^{-8} cm^{-2})	(10^{-8} cm^{-2})	(10^{-8} cm^{-2})	(10^{-8} cm^{-2})	(10^{-8} cm^{-2})	(10^{-8} cm^{-2})
	4.5	4.1	− 5.4	− 4.9	1300	800

2. Electrons and X-rays (GeSi). Table 1 gives $\mu_h = \text{Im}\,(\Delta V_h^F)$ and $p_h = \text{Re}\,(\Delta V_h^F)$, divided by $C = C_{Ge}C_{Si}$, for various electron energies.

It can be stated now that in solid solutions with arbitrary concentrations and scattering factor differences of the components the main dynamical effects will take place. Indeed, our estimations show that the condition

$$|\Delta V_h^D| \ll |\overline{V}_h| \tag{2}$$

is satisfied. According to [2], this means that the strong Bragg peaks and the weak diffuse background will be observed in the experiment. The same prediction was made in the kinematical theory [5]. However, the influence of the crystal defects on the lineshape and on the integrated intensity will now be different. Moreover, a new possibility arises to study defects by their influence on the coherent effects.

2. The Borrmann Effect

Two strong Bragg waves are modulated so that if the absorption power is concentrated at the atoms, absorption of one of them will be higher than normal, while the other will exhibit anomalous transmission (Borrmann effect). Interesting experimental possibilities appear at the study of the integral characteristics of anomalously transmitted X-rays (ATXR) for the fundamental and superlattice reflections. In order to illustrate these possibilities let us consider a stoichiometric binary alloy with a simple cubic β-brass structure, consisting of the components A and B. If this alloy is completely ordered, the interference absorption coefficient μ_h^i of the anomalously transmitted wave in case of the fundamental reflections may be represented as follows:

$$\mu_h^i = \mu^A(\cos\theta)^{-1}\,\{1 - P\varepsilon_h^A \exp[-M_1^A(\boldsymbol{h})]\} + \\ + \mu^B(\cos\theta)^{-1}\,\{1 - P\varepsilon_h^B \exp[-M_2^B(\boldsymbol{h})]\}\,. \tag{3}$$

where μ^α (α = A, B) is the photoelectric absorption coefficient, θ the angle of incidence, P the polarization factor, ε_h^α the local absorption coefficient, $\boldsymbol{h}$ the diffraction vector, M_1^A the thermal Debye-Waller factor for atoms A occupying the first sublattice, and M_2^B the same factor for the B-type atoms occupying the second sublattice. Supposing that $|\varepsilon^\alpha - 1| \ll 1$, M_1^A, $M_2^B \ll 1$; $P = 1$ (3) simplifies to

$$\mu_h^i \cos\theta = \mu^A M_1^A(\boldsymbol{h}) + \mu^B M_2^B(\boldsymbol{h})\,. \tag{4}$$

It is possible now to determine experimentally M_1^A and M_2^B by measuring the integrated intensities of anomalously transmitted wave fields for two wavelengths λ, λ' near the K absorption edges of the alloy components. However, this possibility is limited by the necessity of taking into account in (3) and (4) the corrections due to the effective absorption caused by the diffuse scattering. This difficulty can be avoided using the experimental method of Datsenko et al. [6]. In this method the integrated intensity ratios (jumps) are measured near the K absorption edges, excluding thus μ_D owing to its equality at both sides of the K-edge. Plotting then the depth dependence of these ratios, the increments

$$A^A = \Delta\mu^A(\cos\theta)^{-1} M_1^A(\boldsymbol{h})\,; \qquad A^B = \Delta\mu^B(\cos\theta)^{-1} M_2^B(\boldsymbol{h}) \tag{5}$$

may be measured. Here $\Delta\mu^\alpha$ are the jumps of the photoelectric absorption coefficients. Now (5) enables one to determine the Debye-Waller factors $M_{1,2}^\alpha$.

In the case of superlattice reflections each wave field is localized on the atoms of the corresponding sublattice. This means that, generally speaking, the anomalous transmission is absent. However, if the coefficients μ^α differ significantly, e.g. $\mu^B \ll \mu^A$, the Borrmann effect for superlattice reflections becomes possible. The interference absorption coefficient will now, instead of (3), take the form

$$\mu_{\boldsymbol{h}'}^{i} = \mu^A(\cos\theta)^{-1}\{1 - P\varepsilon_{\boldsymbol{h}'}^A \exp[-M_1^A(\boldsymbol{h}')]\} + \\ + \mu^B(\cos\theta)^{-1}\{1 + P\varepsilon_{\boldsymbol{h}'}^B \exp[-M_2^B(\boldsymbol{h}')]\}\,. \tag{6}$$

Here the diffraction vector $\boldsymbol{h}'$ corresponds to the superlattice reflection. Under the conditions which lead to (4), (6) is also simplified to

$$\mu_{\boldsymbol{h}'}^{i}\cos\theta = \mu^A M_1^A(\boldsymbol{h}') + \mu^B[2 - M_2^B(\boldsymbol{h}')]\,. \tag{7}$$

This equation shows that the Borrmann effect for the superlattice reflection can take place only if $\mu^B \ll \mu^A$.

If the considered alloy is not completely ordered, i.e. if the order parameter $\eta \neq 1$, additional distortions of the crystal lattice appear along with the thermal vibrations and other deviations from periodicity. The effective absorption is modified correspondingly, nevertheless μ_D can again be excluded by measuring integrated intensity jumps. In this case for the fundamental reflections

$$\begin{aligned} A_{\boldsymbol{h}}^A &= \Delta\mu^A(2\cos\theta)^{-1}\{D_1^A(\boldsymbol{h}) + D_2^A(\boldsymbol{h}) + \eta[D_1^A(\boldsymbol{h}) - D_2^A(\boldsymbol{h})]\}\,, \\ A_{\boldsymbol{h}}^B &= \Delta\mu^B(2\cos\theta)^{-1}\{D_1^B(\boldsymbol{h}) + D_2^B(\boldsymbol{h}) + \eta[D_2^B(\boldsymbol{h}) - D_1^B(\boldsymbol{h})]\} \end{aligned} \tag{8}$$

and for the superlattice reflections

$$\begin{aligned} A_{\boldsymbol{h}'}^A &= \Delta\mu^A(2\cos\theta)^{-1}\{2 - 2\eta + \eta[D_1^A(\boldsymbol{h}') + D_2^A(\boldsymbol{h}')] + [D_1^A(\boldsymbol{h}') - D_2^A(\boldsymbol{h}')]\}\,, \\ A_{\boldsymbol{h}'}^B &= \Delta\mu^B(2\cos\theta)^{-1}\{2 + 2\eta - \eta[D_1^B(\boldsymbol{h}') + D_2^B(\boldsymbol{h}')] + [D_1^B(\boldsymbol{h}') - D_2^B(\boldsymbol{h}')]\}\,. \end{aligned} \tag{9}$$

If the generalized Debye-Waller factors D_γ^α are such that $D_1^\alpha = D_2^\alpha = D^\alpha$, then these equations take, respectively, the form

$$A_{\boldsymbol{h}}^\alpha = \Delta\mu^\alpha(\cos\theta)^{-1} D^\alpha(\boldsymbol{h})\,, \tag{10}$$

$$A_{\boldsymbol{h}'}^\alpha = \Delta\mu^\alpha(\cos\theta)^{-1}\{1 - \eta[1 - D^\alpha(\boldsymbol{h}')]\}\,. \tag{11}$$

These equations enable one to determine the generalized Debye-Waller factor and the order parameter by experimental analysis of both fundamental and superlattice reflections.

In case of neutron diffraction the usual absorption is absent. Nevertheless the Borrmann effect may take place owing to the effective absorption which is due to the diffuse scattering by fluctuating atomic factors. For the fundamental reflection the anomalous transmission of the field concentrated between the atoms is determined by the effective local absorption $(1 - \varepsilon)$,

$$\mu^- = \mu_0 - \mu_{\boldsymbol{h}} \approx (1 - \varepsilon)\,\mu_0 \approx 10^{-4}\mu_0\,. \tag{12}$$

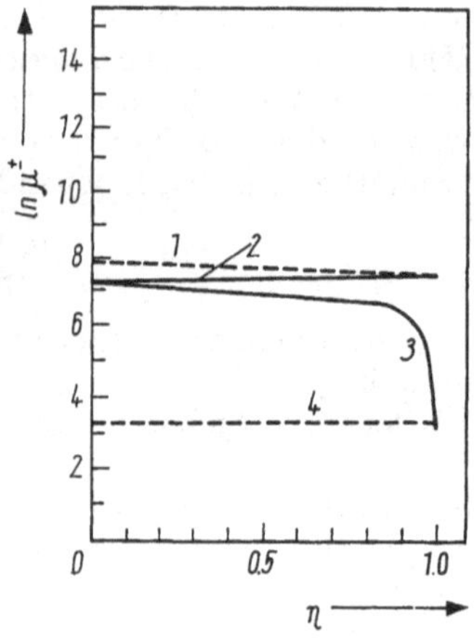

Fig. 1. Long-range order dependence of the effective absorption in a stoichiometric alloy NiAl, taking into account various Ni isotopes. (1) Absorbed fundamental field; (2) absorbed superlattice field; (3) transmitted superlattice field; (4) transmitted fundamental field

The absorption of the other wave field which is localized at atoms, is determined by

$$\mu^+ = \mu_0 + \mu_h \approx 2\mu_0 . \tag{13}$$

So the large difference between μ^+ and μ^- is due to the highly local character of the effective absorption [3, 7, 8].

For the superlattice reflections at complete disorder, when the diffuse background is maximum, the Borrmann effect is absent because the absorption conditions for both wave fields do not differ. At the same time in case of a completely ordered stoichiometric alloy these conditions may become different, if only one sublattice contains different isotopes. The wave field, localized at the other sublattice, will show anomalous transmission. Fig. 1 shows the dependence of the effective absorption on the order parameter in NiAl taking into account various Ni isotopes.

3. Pendellösung Interference Patterns

In experiments on wedge-like crystal samples the pendellösung interference patterns can be observed. The distance between the intensity maxima in case of a non-perfect crystal is given by [3, 9]

$$t_h = 2\pi(\gamma_0\gamma_h)^{1/2} \, [\lambda(\overline{V}_h + P_h) \sin \alpha_0]^{-1} , \tag{14}$$

where $\overline{V}_h = \overline{V}_{-h}$, α_0 is the angle of the wedge, γ_0 and γ_h are the cosine functions of the angles between the normal to the crystal surface and transmitted and reflected beams, respectively.

Taking into account only the correction due to the scattering factor fluctuations in the case of a stoichiometric binary alloy with simple cubic β-brass structure, (14) gives for the fundamental reflections

$$t_h = 2\pi(\gamma_0\gamma_h)^{1/2} \, \{\lambda[2N(c_A V_h^A + c_B V_h^B) + (c_A c_B - \tfrac{1}{4}\eta^2) \, \Phi_h^F] \sin \alpha_0\}^{-1} \tag{15}$$

and for the superlattice reflections

$$t_{h'} = 2\pi(\gamma_0\gamma_h)^{1/2} \, \{\lambda\eta[N(V_{h'}^A - V_{h'}^B) + (c_B - c_A) \, \Phi_{h'}^F] \sin \alpha_0\}^{-1} . \tag{16}$$

Here c_α is the concentration of the component α, V_h^α the atomic scattering factor, η the long-range order parameter, N the number of crystal cells, and Φ a factor which remains in the correction if the dependence on c_α and η is written down explicitly. It is easily seen from (14) to (16) that, neglecting the corrections ΔV^D, the distance between the fundamental pendellösung maxima depends on the concentrations of constituting atoms and their isotopes and does not depend on the order parameter.

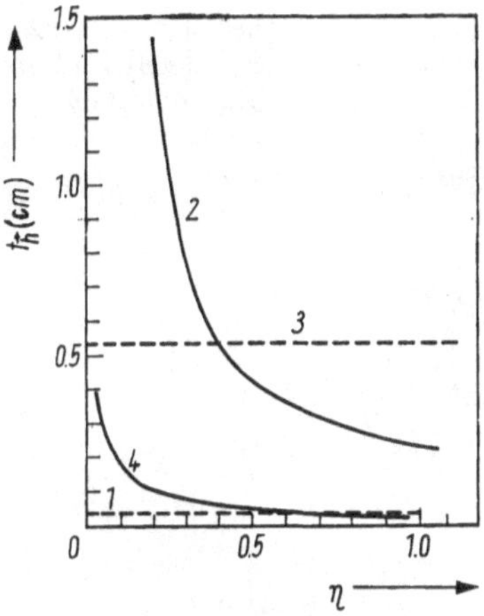

Fig. 2. Long-range order dependence of the distance between the pendellösung maxima in case of neutron diffraction; (1) and (2) distance for the fundamental and superlattice reflections, respectively, in a NiAl alloy; (3) and (4) the same quantities in a TiAl alloy

At the same time the distance $t_{h'}$ depends on the order parameter and does not depend on concentration (see Fig. 2). This is due to the fact that the fundamental reflection is determined by the averaged Fourier component of the potential, and the superlattice reflection is determined by the difference between the Fourier components averaged over each sublattice.

An interesting phenomenon can be predicted if one of the alloy components has a negative scattering factor (e.g. Ti in case of neutron diffraction). Then, according to (15), (16), the distance between the superlattice fringes $t_{h'}$ may become much less than t_h (effect of pattern inversion). In this situation the superlattice interference maxima may be observed.

Consider now a completely disordered alloy, taking into account static and thermal displacements. In this case

$$t_h = 2\pi(\gamma_0\gamma_h)^{1/2}\left\{\lambda\left[2N(c_A V_h^A + c_B V_h^B)\exp(-D_h) + T\Phi_h^T + \right.\right.$$
$$\left.\left. + c_A c_B \Phi_h^F + c_A c_B\left(\frac{1}{V}\frac{dV}{dc_A}\right)\Phi_h^S\right]\sin\alpha_0\right\}^{-1}, \tag{17}$$

where V is the crystal volume. When the distortions are weak enough, this expression takes the form

$$t_h = 2\pi(\gamma_0\gamma_h)^{1/2}\exp\left\{\frac{1}{2}\langle(hU_S^T)^2\rangle_T^B + \frac{1}{2}c_A\sum_{s'}(h\cdot U_{s'})^2 - \frac{1}{24}c_A\sum_{s'}(h\cdot U_{s'})^4 - \right.$$
$$\left. - \frac{1}{8}c_A\sum_{s'}[\Delta(h\cdot U_{s'}^T)^2]^2 - \frac{1}{4}c_A\sum_{s'}\Delta(h\cdot U_{s'}^T)^2(h\cdot U_{s'})^2\right\}(\lambda N\,|V_h^B|\sin\alpha_0)^{-1}. \tag{18}$$

Here $\langle(h\cdot U_s^T)^2\rangle_T^B$ is the averaged squared thermal displacement of an atom B, $\Delta(h\cdot U_s^T)^2$ is the change of this quantity caused by the appearance of an atom A in the s-site. The terms proportional to h^4 take into account the non-additivity of static and thermal displacements [3, 5, 9]. As long as pendellösung patterns may be observed in experiment with great accuracy [10 to 12], (16) to (18) provide a useful possibility for the interpretation of experimental data.

4. Diffraction of Highly Energetic Particles

When the particle energy increases sufficiently, a large number of strong Bragg waves may be excited in the vicinity of the Ewald sphere. If this sphere is tangent to a plane

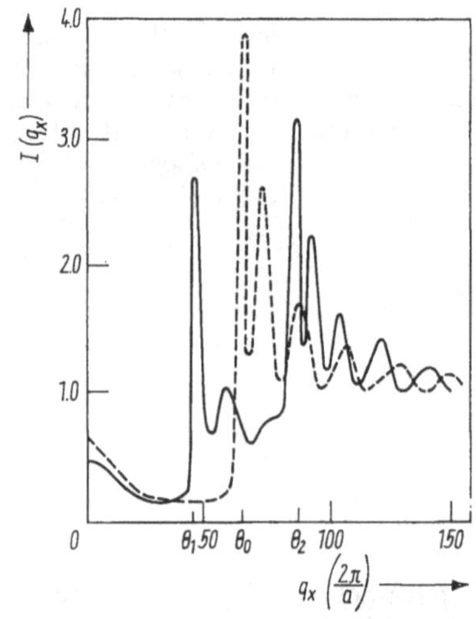

Fig. 3. Angle dependence of the depth-averaged nuclear reaction yield; – – – completely ordered stoichiometric AuCu alloy; ——— completely disordered alloy. These curves were obtained for the binary Kronig-Penney model potential in [14]

or a line in the reciprocal lattice, all the waves with the wave vectors lying in this plane (line) will be excited, which results in axial or planar channeling, respectively [13]. Neglecting non-coherent effects the problem of planar channeling in a crystal with two sublattices is equivalent to the problem of evolution in a one-dimensional periodic potential with two maxima in the same unit cell [2]. These two barriers are determined by the averaged potential of atoms populating the corresponding sublattice. The difference between the maxima is largest in a completely ordered alloy, while in the case of complete disorder both maxima become equal. Such a behaviour of the interplanar potential leads to the splitting of the critical channeling angle. This effect is illustrated in Fig. 3, which gives the angle dependence of the nuclear reaction yield in an ordered alloy. It is seen from Fig. 3 that the critical channeling angle θ_0 in the completely disordered alloy is split in two angles, θ_1 and θ_2, when the order increases. These two angles correspond to different maxima of the effective potential for each sublattice and are quite simply related to η [14]. This enables one to determine the long-range order parameter from experimental data on planar channeling.

References

[1] G. I. GUDZENKO and V. B. MOLODKIN, Preprint IMP, No. 75.4, Kiev 1975.
[2] V. B. MOLODKIN, Metallofizika 2, 3 (1980).
[3] G. I. GUDZENKO, V. B. MOLODKIN, and V. G. NOVITSKII, Preprint IMP, No. 77.4, Kiev 1977.
[4] J. B. FORSYTH and M. WELLS, Acta cryst. 12, 412 (1959).
[5] M. A. KRIVOGLAZ, Theory of X-Ray and Thermal-Neutron Scattering by Real Crystals, Plenum Press, New York 1969.
[6] L. I. DATSENKO, Ukr. fiz. Zh. 24, 577 (1979).
[7] P. H. DEDERICHS, phys. stat. sol. 23, 377 (1967).
[8] V. B. MOLODKIN and E. A. TIKHONOVA, Fiz. Met. Mater. 24, 385 (1967).
[9] G. I. GUDZENKO and V. B. MOLODKIN, Preprint IMP, No. 75.6, Kiev 1975.
[10] S. S. SCHILSTEIN and V. A. SOMENKOV, in: Proc. XIII. Conf. Application of X-Rays, Tchernogolovka 1982.
[11] N. M. OLEKHNOVICH, A. L. KARPEY, V. M. SKIDAN, and L. D. PUZENKOVA, ibid. (p. 199).
[12] N. D. RARANSKII and G. S. BOYADYAN, ibid. (p. 181).
[13] YU. KAGAN and YU. V. KONONETS, Zh. eksper. teor. Fiz. 58, 226 (1970).
[14] V. B. MOLODKIN and V. G. NOVITZKY, Preprint IMP, No. 79.8, Kiev 1979.

(Received December 30, 1982)

phys. stat. sol. (a) 77, 121 (1983)

Subject classification: 11; 22.1.1

Saha Institute of Nuclear Physics, Calcutta[1])

Sputtering of Germanium by 15 to 35 keV Noble Gas Ions

By

D. GHOSE, D. BASU, and S. B. KARMOHAPATRO

Sputtering yields of Ge bombarded by 15 to 35 keV $^{40}Ar^+$, $^{84}Kr^+$, and $^{132}Xe^+$ ions are measured. The data are discussed on the basis of Sigmund's linear collision cascade theory taking into account the different reduced nuclear stopping cross-section functions as suggested by different authors. For Ar and Kr ions the theory with the stopping cross-section due to Wilson et al. agrees reasonably well with the experiment. The yields for Xe ions tend to increase from the theoretical curve. The cause of the discrepancy is discussed in the light of the energy deposition in the loaded target with projectile ions and the spike effect due to heavy ion irradiation, which are not included in the simple cascade theory.

Die Sputterausbeute von mit 15 bis 35 keV $^{40}Ar^+$-, $^{84}Kr^+$- und $^{132}Xe^+$-Ionen bombardierten Ge wird gemessen. Die Werte werden anhand der Sigmundschen Theorie der linearen Stoßkaskade unter Berücksichtigung von verschiedenen reduzierten Kernbremsquerschnittsfunktionen verschiedener Autoren diskutiert. Für Ar- und Kr-Ionen stimmt die Theorie mit Bremsquerschnitten von Wilson u. a. vernünftig mit dem Experiment überein. Die Ausbeuten für Xe-Ionen scheinen die theoretischen Kurven zu übersteigen. Der Grund für diese Diskrepanz wird mit der Energiedepositron in dem mit Projektilionen beladenen Target und dem Spike-Effekt, der von der Bestrahlung mit schweren Ionen herrührt, diskutiert, die nicht in der einfachen Kaskadentheorie enthalten sind.

1. Introduction

Recently Andersen and Bay [1] presented data on sputtering yields for different ion-target combinations as a function of energy of the incident ions based on a systematic search in the literature. The survey reveals that for Ge the sputtering yield data in the keV energy range are scarce. The prominent ones are those due to Southern et al. [2] who measured the sputtering yield of Ge by Ar^+ ion bombardment in the energy range 1 to 5 keV and due to Sommerfeldt et al. [3] who extended the measurements from 14 to 34 keV. It is known that the semiconductor materials develop an amorphous phase at the surface at doses required for sputtering experiments. Therefore, measurement of sputtering yield for such targets is important for a comparison with Sigmund's multiple collision theory of sputtering [4], one of the basic assumptions of which is that the target is random. In the present paper, we have measured the sputtering yield of Ge by $^{40}Ar^+$, $^{84}Kr^+$, and $^{132}Xe^+$ ions in the energy range 15 to 35 keV and compared the results with Sigmund's theory of sputtering [4] with nuclear stopping cross-sections of various approximations.

2. Experimental

The sputtering experiments were carried out in a Bernas-type electromagnetic isotope separator [5] which is reinstalled recently in the new institute campus at Bidhan Nagar. Before bombardment, the Ge crystal was mechanically polished by 0.3 μm alumina powder in the Minimet polisher (Buehler Ltd., USA) followed by cleaning

[1]) Sector 1, Block AF, Bidhan Nagar, Calcutta 700064, India.

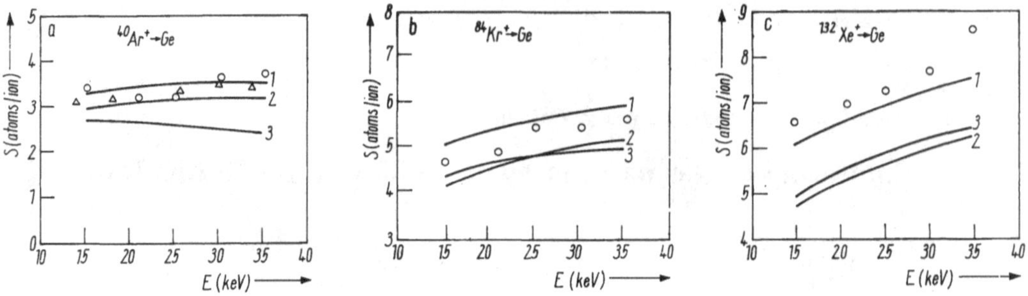

Fig. 1. Sputtering yield S of Ge by a) $^{40}Ar^+$, b) $^{84}Kr^+$, and c) $^{132}Xe^+$ ion bombardment as a function of ion energy E. The solid curves represent the sputtering yield according to the theory of Sigmund [4] with the stopping cross-sections after (1) [7], (2) [8], and (3) [9], respectively; ○ present results, △ after [3]

in an ultrasonic cleaner. Ions from the separator were incident normally on the target surface with a current density of 0.3 to 1.3 mA/cm^2 depending on the ion species and energy. The residual gas pressure during bombardment was 2.5×10^{-5} Torr. The sputtered atoms were collected on a hemispherical copper cup placed before the target. The collector cup was also maintained at a negative potential with respect to the target for suppressing the secondary electrons produced by the impact of energetic ions. The ion dose was measured on the target by a current integrator. In each run, the target was bombarded with a fixed dose of about 2.7×10^{19} ions/cm^2. Measuring the increase in weight of the collector cup which was of the order of milligrams after the bombardment and knowing the integrated beam current at the target, the sputtering yield was determined. The estimated accuracy is within $\pm 10\%$ which is typical for similar sputtering experiments.

3. Results and Discussion

Fig. 1 a to c shows the sputtering yield of Ge under the bombardment of $^{40}Ar^+$, $^{84}Kr^+$, and $^{132}Xe^+$ ions normally incident on the target as a function of ion energy in the range 15 to 35 keV. The data of Sommerfeldt et al. [3] for $Ar^+ \rightarrow Ge$ are also included in Fig. 1a which agree well with the present data.

Sigmund [4] derived a thick-target sputtering yield formula by applying general transport theory to the atomic collision cascade developed in the amorphous target due to ion bombardment,

$$S = \Lambda \alpha N S_n(E)\,, \tag{1}$$

where S is the sputtering yield; $\Lambda = 0.042\, N^{-1}\, U_0^{-1}$ Å^{-2}; N the target atomic density; U_0 the surface binding energy; α a dimensionless quantity depending on the mass ratio between the target atom mass M_2 and the impinging ion mass M_1; $S_n(E)$ the nuclear stopping cross-section given by the following expression:

$$S_n(E) = 4\pi a Z_1 Z_2 e^2 \frac{M_1}{M_1 + M_2}\, s_n(\varepsilon)\,, \tag{2}$$

where $a = 0.8853 a_0 (Z_1^{2/3} + Z_2^{2/3})^{-1/2}$ is the Thomas-Fermi screening radius; $a_0 = 0.529$ Å; Z_1 and Z_2 the atomic numbers of the incident ion and targe tatom, respectively; $s_n(\varepsilon)$ the reduced nuclear stopping cross-section, and ε the dimensionless energy variable defined by Lindhard et al. [6] (hereafter referred to simply as LSS).

For covalent crystals, Sigmund [4] assumed a spherical surface barrier with the surface binding energy U_0 equal to the measured cohesive energy per atom, U_c ($U_c = 7.63$ eV for Ge) and the bulk binding energy $V = 2U_c$. This implies that the Λ factor of (1) must be multiplied by the factor $\beta = 0.805$ [4]. However, in the analysis of sputtering yields of amorphous semiconductors, Sigmund [4] found that the yield values can well be predicted by neglecting V and putting $\beta = 1$.

Another important parameter for the determination of S is the reduced nuclear stopping cross-section $s_n(\varepsilon)$. Sigmund [4] used the Thomas-Fermi value of $s_n(\varepsilon)$ tabulated by LSS [7] for comparison of the experimental data with the theory. However, it is known from range measurements that the LSS data [7] overestimate the stopping cross-section at low energies, $\varepsilon \lesssim 0.3$. In the present experiment ε values lie between 0.3 to 0.02 depending on the ion–target combination and the energy E (for Ar–Ge, $\varepsilon = 0.132$ to 0.308; for Kr–Ge, $\varepsilon = 0.042$ to 0.099; and for Xe–Ge, $\varepsilon = 0.02$ to 0.047). More accurate values of $s_n(\varepsilon)$ at low values of ε can be obtained from the recently derived expressions of the reduced nuclear stopping power function by Wilson et al. [8] (hereafter referred to simply as WHB) and Kalbitzer et al. [9] (hereafter referred to simply as KOGF). The analytic expression of $s_n(\varepsilon)$ according to WHB [8] is

$$s_n(\varepsilon) = \frac{A \ln (B\varepsilon)}{B\varepsilon - (B\varepsilon)^{-C}}\,; \tag{3}$$

where $A = 0.56258$, $B = 1.1776$, and $C = 0.62680$; while the empirical form of $s_n(\varepsilon)$ due to KOGF [9] is

$$s_n(\varepsilon) = \frac{1.7\varepsilon^{1/2} \ln (\varepsilon + e)}{1 + 6.8\varepsilon + 3.4\varepsilon^{3/2}}\,, \tag{4}$$

where e is the base of natural logarithm.

We have calculated S values using the LSS data of $s_n(\varepsilon)$ as well as those obtained from (3) and (4) (Fig. 1 a to c).

Blank and Wittmaack [10] measured the sputtering yields of Si bombarded with Ar and Xe ions and found that their data agree well for the stopping cross-section after WHB [8]. Their data also revealed that KOGF data [9] underestimate the stopping cross-section at energies $\varepsilon > 0.1$. So, we may reasonably assume that in the energy range investigated, the theoretical yield values are properly represented by the curve WHB.

Fig. 1 a to c shows that the measured yield values are always higher than that predicted by the theory after WHB. Although the theoretical yield curve (WHB) lies approximately within the range of the experimental error for Ar^+ and Kr^+ ions, the discrepancy is more pronounced in case of Xe^+ ions. The possible reasons for the observed high values of the sputtering yield are (i) the increase in energy deposition near the surface due to implanted projectile ions and (ii) the spike effect in the dense collision cascades for heavy ion irradiation.

The first mechanism can affect the sputtering yield if the concentration of the implanted ions is of appreciable amount and also if there is a large difference in mass and atomic number between the target and projectile ions as observed in the sputtering of Si by Xe^+ ions [10, 11]. In the case of simple saturation, the concentration of the projectile ions near the surface is $1/S$. Therefore, for targets with not too low sputtering yield like Ge in the present experiment, this effect is expected to be negligible.

The second mechanism, i.e. the existence of the energy spike might be the appropriate explanation for the observed deviation of the experimental data from the theory.

Sigmund's theory of sputtering [4] is based on the linear collision cascade approximation where the collision density due to impact of energetic ions is sufficiently small so that in each collision the struck atom is assumed to be stationary. This approximation is valid only when the mass of the incident ion and the target atom are not too large. But with increasing mass of incident ion and the target atom, the energy density in the cascade region increases; as a result the majority of atoms within a certain cascade volume are instantaneously in motion and collision between moving atoms can no longer be neglected. Such a situation is described as a spike which results in non-linear effects in sputtering yield [12, 13]. Sigmund [14 to 16] proposed that the spike event may occur when the mean energy θ per atom deposited by the incident ion within a collision cascade volume is larger than a certain minimum value θ_{min}. From radiation damage studies in Si and Ge by low energy heavy ion bombardment, Thompson et al. [17, 18] observed that the energy spike becomes important in determining the number of displaced atoms per incident ion whenever the energy density within the collision cascade exceeds ≈ 0.1 eV/atom. The above value is very close to the heat of melting U_m of the respective solids ($U_m = 0.5$ eV for Ge). Thus for Si and Ge, θ_{min} may be taken to be approximately equal to U_m. This value of the upper energy limit of θ_{min} is assumed to be valid in the case of sputtering since both the radiation damage and sputtering results from cascades of atomic collisions. Sigmund [14 to 16] showed that the spike effect is determined by two quantities, namely, the effective maximum energy density θ_0 in the central core of the spike and the time constant τ for decay of the spike. The spike component of the sputtering yield is appreciable when the following two conditions are simultaneously satisfied: θ_0 is greater than θ_{min} and τ is greater than τ_0, the slowing down time of the projectile ion. Fig. 2 shows the spike parameters for Ge bombarded with Xe^+ ions in the energy range 10 to 40 keV where θ_0, τ_0, and τ are calculated for $m = 1/3$, from (3), (4), and (9) of [14], respectively, where m is the strength of the power law approximation to the interatomic potential. It appears from Fig. 2 that in the energy range investigated a significant contribution of the energy spike effect to the sputtering yield of Ge by Xe^+ ions may be expected. Similar plots of spike parameters in the case of Ar^+ and Kr^+ ions incident on Ge show that the development of the spike is also possible for these combinations, though, however, less pronounced than that for $Xe^+ \rightarrow$ Ge.

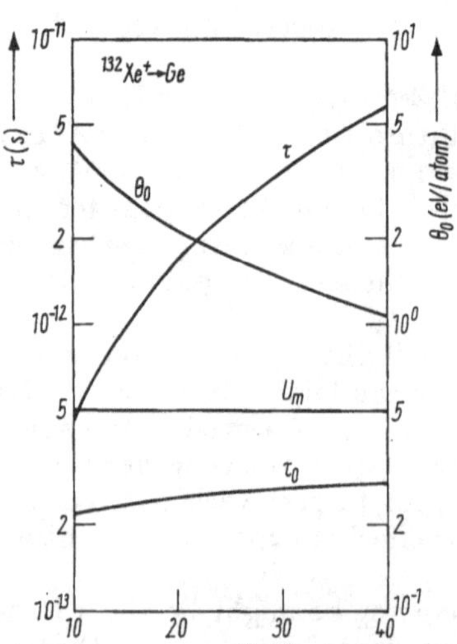

Fig. 2. Spike parameters vs. incident ion energy for Ge bombarded with Xe^+. U_m is the heat of melting of Ge

Thus for low energy heavy ion bombardment, the spike mechanism may be considered as a process parallel to the collision cascade leading to the enhancement of sputtering yield values. It may be noted that in metals the spike mechanism is operative when θ_0 exceeds the surface binding energy which is usually taken as the respective sublimation energy [15]; whereas unlike metals the present data for Ge can be explained assuming θ_{min} as the heat of melting rather than the sublimation energy or the cohesive energy, the importance of the former in spike phenomena was demonstrated by Thompson et al. [17, 18] in radiation damage studies of Si and Ge.

Despite all the assumptions entering the sputtering theory, there is a reasonable agreement between experiment and the simple cascade theory with a suitable stopping cross-section, in the case of Ar and Kr bombardment. For Xe, the yield tends to deviate from the theoretical curve indicating the possibility of existence of the spike effects. In addition, the effect of increase in energy deposition due to the loading of the target with projectile ions, if any, can be ascertained by studying the dose dependence of sputtering yield. For such experiments, surface defects, microstructure, impurities, and the topographical changes during irradiation are to be considered carefully [15], which may be helpful in establishing a valid expression for sputtering yield from a well defined but real surface.

Acknowledgement

The authors thank Mr. M. K. Chakraborty for operating the isotope separator.

References

[1] H. H. Andersen and H. L. Bay, in: Sputtering by Ion Bombardment, Ed. R. Behrisch, Institute of Physics, University of Aarhus, 1980 (Chap. 4).
[2] A. L. Southern, W. R. Willis, and M. T. Robinson, J. appl. Phys. **34**, 153 (1963).
[3] H. Sommerfeldt, E. S. Mashkova, and V. A. Molchanov, Phys. Letters A **38**, 237 (1972).
[4] P. Sigmund, Phys. Rev. **184**, 383 (1969).
[5] R. Bernas, Thèse de Docteur des Sciences, Université de Paris, 1954.
[6] J. Lindhard, M. Scharff, and H. E. Schiøtt, Kong. Danske Vid. Selsk., mat.-fys. Medd. **33**, No. 14 (1963).
[7] J. Lindhard, V. Nielsen, and M. Scharff, Kong. Danske Vid. Selsk., mat.-fys. Medd. **36**, No. 10 (1968).
[8] W. D. Wilson, L. G. Haggmark, and J. P. Biersack, Phys. Rev. B **15**, 2458 (1977).
[9] S. Kalbitzer, H. Oetzmann, H. Grahmann, and A. Feuerstein, Z. Phys. **A278**, 223 (1976).
[10] P. Blank and K. Wittmaack, J. appl. Phys. **50**, 1519 (1979).
[11] P. Blank and K. Wittmaack, Radiat. Eff. **27**, 29 (1975).
[12] H. H. Andersen and H. L. Bay, J. appl. Phys. **46**, 2416 (1975).
[13] H. L. Bay, H. H. Andersen, W. O. Hofer, and O. Nielsen, Nuclear Instrum. and Methods **132**, 301 (1976).
[14] P. Sigmund, Appl. Phys. Letters **25**, 169 (1974); **27**, 52 (1975).
[15] P. Sigmund, in: Inelastic Ion-Surface Collisions, Ed. N. H. Tolk, J. C. Tully, W. Heiland, and C. W. White, Academic Press, Inc., New York 1977 (p. 121).
[16] P. Sigmund, in: Sputtering by Ion Bombardment, Ed. R. Behrisch, Fysisk Institut, Odense Universitet, 1980 (Chap. 2).
[17] D. A. Thompson, R. S. Walker, and J. A. Davies, Radiat. Eff. **32**, 135 (1977).
[18] D. A. Thompson and R. S. Walker, Radiat. Eff. **36**, 91 (1978).

(Received September 28, 1982)

phys. stat. sol. (a) 77, 127 (1983)

Subject classification: 18; 18.2; 21.1

National Center of Physics, Bucharest[1]) *(a)*
and Department of Physics, Polytechnical Institute, Cluj-Napoca[2]) *(b)*

On the Magnetic Behaviour of NiCuSi Solid Solutions

By

E. Burzo (a), Ileana Lupsa (b), and I. Cosma (b)

Results are presented of magnetic measurements performed on $Ni_uCu_{1-u-x}Si_x$ alloys with $u = 0.61$ and 0.91, in the temperature range (4.2 to 1200) K and fields up to 50 kG. The saturation magnetization decreases by increasing silicon content. The reciprocal susceptibility obeys a modified Curie-Weiss law: $\chi = \chi_0 + C(T - \theta)^{-1}$. Around the Curie points the initial susceptibility and low field magnetizations follow asymptotic power laws: $\chi \propto t^{-\gamma}$ and $M \propto t^{\beta}$, where $t = |T - T_c| \times T_c^{-1}$, $\beta = 0.33 \pm 0.01$, and $\gamma = 1.40 \pm 0.02$. By plotting the magnetizations of NiCu and NiCuSi alloys, as function of the electron concentration of the diamagnetic matrix, an universal behaviour is evidenced. It is emphasized that the magnetic properties of the NiCu system can be regarded as a peculiar case of a more general-type behaviour mainly determined by the composition of the diamagnetic matrix.

Nous présentons les résultats des mesures magnétiques sur les alliages $Ni_uCu_{1-u-x}Si_x$ avec $u = 0,61$ et respectivement $u = 0,91$, dans les domaines des témpératures (4,2 à 1200) K et champs jusqu'a 50 kG. L'aimantation a saturation décroit si le contenu en silicium est plus grand. L'inverse de la susceptibilité magnétique suit une loi de Curie-Weiss: $\chi = \chi_0 + C(T - \theta)^{-1}$. Autour des points de Curie, la susceptibilité initiale et l'aimantation dans les champs faibles suivent des lois asymptotiques $\chi \propto t^{-\gamma}$ et $M \propto t^{\beta}$ ou $t = |T - T_c|\, T_c^{-1}$, $\beta = 0,33 \pm 0,01$ et $\gamma = 1,40 \pm 0,02$. En représentant les aimantations des alliages NiCu et NiCuSi en function de la concentration des electrons de conduction de la matrice diamagnétique (CuSi), on évidence un comportement universel. Les propriétés magnétiques du system NiCu peuvent être considerés comme un cas particulier d'un comportement plus general, determiné essentiellement par la composition de la matrice diamagnétique.

1. Introduction

The magnetic measurements performed on $Ni_{1-x}Cu_x$ alloys show that the magnetization M_s decreases almost linearly with increasing copper content up to $x \approx 0.50$. The extrapolation of M_s-values gives zero moments for $x \approx 0.55$. In the composition range $x \geqq 0.50$, a deviation from the linear dependence is evidenced [1, 2]. In an initial stage, the above data were analysed in the band model [3]. Later on, a number of experimental results showed that this model cannot describe all the magnetic properties of CuNi alloys [4]. Then, the local environment model [4, 5] or coherent potential approximation [6, 7] was used. In the local environment model one supposes that the magnetic moment of a nickel atom in NiCu alloys is the same as in pure metal, if in their nearest neighbourhood there is at least a critical number N_c of similar atoms and is zero if this number is smaller than N_c.

In order to check the general validity of the above models previously we analysed the magnetic properties of $Ni_uCu_{1-u-x}Si_x$ alloys, having a constant nickel content, namely $u = 0.7$ and 0.8, respectively [8, 9]. Copper contributes one electron and silicon four electrons to the conduction band. Thus, by varying the composition of the CuSi

---

[1]) P. O. Box 5206 Bucharest, Romania.

[2]) Cluj-Napoca, Romania.

matrix, a change in the number of conduction electrons takes place. In both cases the magnetization decreases linearly with increasing silicon content, the electron concentration, respectively. These data suggest that the local environment model may be used to analyse only the magnetic behaviour of nickel in a simple matrix, as copper.

Further to these studies, we analyse the magnetic behavour of $Ni_uCu_{1-u-x}Si_x$ alloys in two extreme composition ranges. These cover the region with rich Ni content ($u = 0.91$) and also the composition range, where the onset of ferromagnetism is observed ($u = 0.61$).

The samples were studied by magnetic measurements in a large temperature range. Attention has been given to the analysis of the magnetic behaviour close to the Curie points.

2. Experimental

The samples were prepared by melting the constituents in an arc furnace. These were thermally treated at 1100 °C for three days and then quenched. By fast cooling we expect that the constituent atoms are randomly distributed in the lattice. Chemical

Table 1

The composition (in at%) of $Ni_uCu_{1-u-x}Si_u$ alloys

samples	1	2	3	4	5	6	7
Ni	60.88	61.76	62.86	91.07	91.11	90.86	91.39
Cu	32.08	32.61	34.15	1.87	3.79	5.84	7.77
Si	7.04	5.63	2.98	7.06	5.10	3.30	0.84

analyses were also made. The final composition of the samples are listed in Table 1.

The X-ray measurements show that the compounds crystallize in a f.c.c.-type structure.

The magnetic measurements were performed between 4.2 and 1200 K. Below room temperature the data were obtained with an "Oxford Instruments"-type equipment in fields up to 50 kG. Above room temperature, a Faraday-type balance was used. In this temperature range the samples were sealed in quartz tubes under vacuum. The paramagnetic data were obtained at two values of magnetic field, both at heating and cooling the samples. No presence of ferromagnetic impurities or oxidation phenomena were observed.

3. Experimental Results

The magnetization isotherms obtained at 4.2 K are plotted in Fig. 1. The spontaneous magnetizations were determined by computer analysis of the experimental data, according to the relation $M = M_s[1 - (a/H)] + \chi_0' H$. The a values decrease as the Curie point of the alloy is smaller. The Pauli contributions χ_0' are greater as the silicon content increases. For example, in case of sample 1 $\chi_0' \approx 6 \times 10^{-6}$ m³/kmol was obtained and for sample 3 $\chi_0' \approx 0.9 \times 10^{-6}$ m³/kmol was determined.

The thermal variation of spontaneous magnetization for selected alloys is plotted in Fig. 2. The data show that for both series of samples, the saturation magnetization and the Curie temperatures decrease as the silicon content is greater. The curvature radius of $M(T)/M(0)$ versus T/T_c graphs increases as the Curie points are smaller. This fact reflects the diminution of the exchange interactions in alloys.

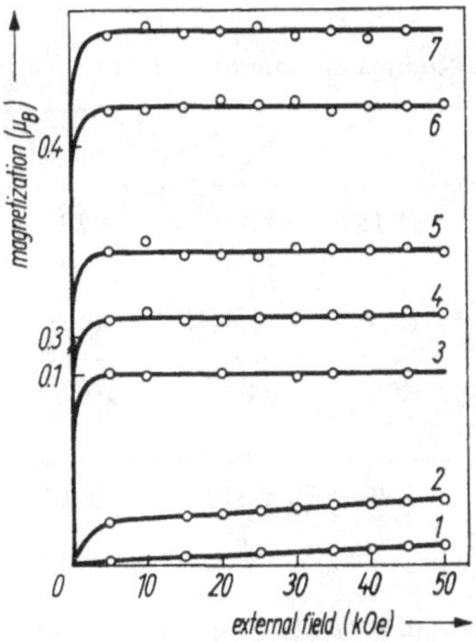

Fig. 1. The magnetization isotherms at 4.2 K for $Ni_uCu_{1-u-x}Si_x$ alloys

The thermal variation of reciprocal susceptibility is plotted in Fig. 3. The data may be analysed according to a modified Curie-Weiss law

$$\chi = \chi_0 + C(T - \theta)^{-1} . \tag{1}$$

We denoted by C the Curie constant, θ is the paramagnetic Curie temperature and χ_0 a Pauli paramagnetic term. The C-, θ-, and χ_0-values were determined by the least squares fit of the experimental data. These are listed in Table 2.

We note that χ_0' values determined from the analysis of magnetization isotherms at 4.2 K are greater than those obtained from susceptibility measurements. This behaviour was observed in a number of alloys [10]. Probably in the ferromagnetic range the χ_0-values are enhanced by the magnetic interactions.

The effective nickel moment μ_{eff} decreases as the silicon content is greater, cf. Table 2.

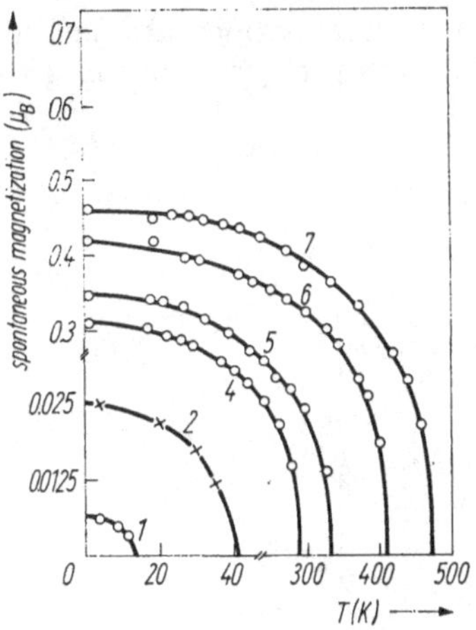

Fig. 2

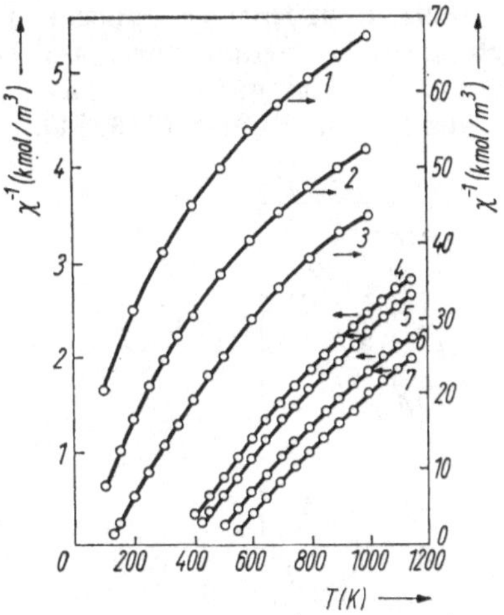

Fig. 3

Fig. 2. Thermal variation of spontaneous magnetization

Fig. 3. Thermal variation of reciprocal susceptibility

Table 2

Parameters C, χ, θ, and μ_{eff} obtained from paramagnetic measurements

samples	1	2	3	4	5	6	7
Curie constant C (10^{-3} m^{3}K/kmol)	0.367	0.792	1.181	2.092	2.279	2.563	2.670
χ_0 (10^{-6} m^3/kmol)	1.17	1.03	0.89	0.85	0.79	0.79	0.80
paramagnetic Curie temperature θ (K)	17	50	117	330	365	440	510
effective Ni moment μ_{eff} (μ_B/atom)	0.6206	0.921	1.094	1.209	1.261	1.339	1.363

Interesting information may be obtained in the temperature range close to the Curie points. Previously it was evidenced that in the temperature range $10^{-4} < t < 10^{-1}$, where $t = |T - T_c| T_c^{-1}$, the magnetic behaviour is gouverned by asymptotic power laws [11, 12]. For initial susceptibility and low-field magnetization, the following relations are given:

$$\chi \sim t^{-\gamma} \qquad (T \to T_c^-, H \to 0) \tag{2}$$

$$M \sim t^{\beta} \qquad (T \to T_c^+, H \to 0) . \tag{3}$$

For a Heisenberg ferromagnet the theoretical predictions give critical exponents $\gamma = 1.40$ and $\beta = 0.36$, respectively [11, 12].

We plotted in Fig. 4 ln $(1/\chi)$ as a function of ln t for samples 4 and 7. A linear variation is obtained. As seen in Fig. 4, the critical exponents are close to the theoretical predictions.

The low-field magnetizations as functions of t in a logarithmic scale are plotted in Fig. 5. The data agree with the prediction of relation (3). The critical exponents are $\beta = 0.33 \pm 0.01$, close to those expected in the Heisenberg ferromagnet.

Although the Curie temperatures and the saturation magnetizations are smaller than those of nickel, the magnetic behaviour in the critical temperature range is similar to that observed in Ni metal [13, 14].

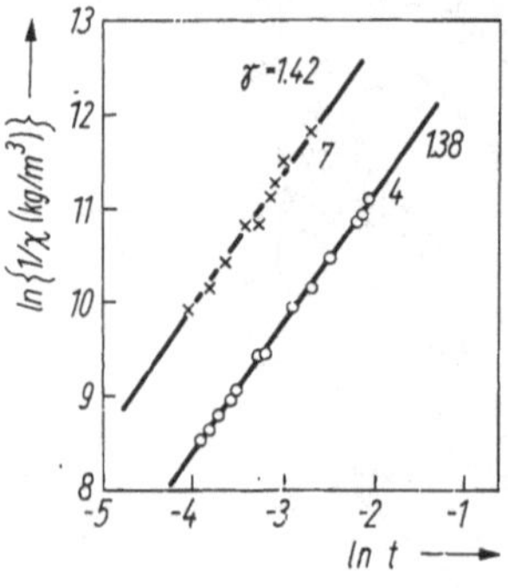

Fig. 4

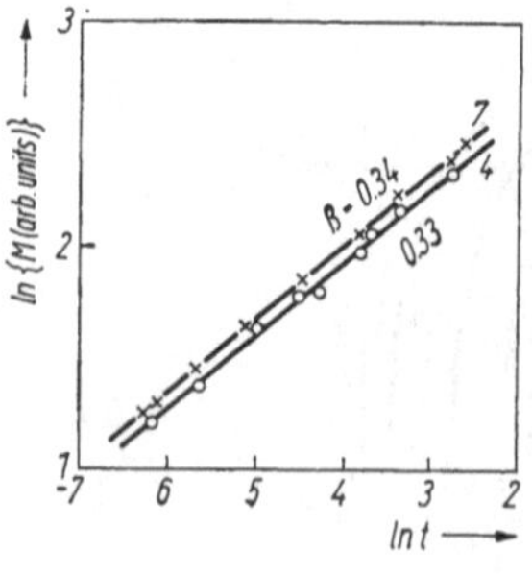

Fig. 5

Fig. 4. The variation of ln $(1/\chi)$ as a function of ln t, for samples 4 and 7

Fig. 5. The dependence of ln M on ln t, for samples 4 and 7

4. Discussion

The present data show that even if we keep constant the nickel content, the magnetizations and the Curie temperatures change strongly by substituting copper by silicon. In Fig. 6 and 7 we plotted the M_s- and T_c-values as functions of the electron concentration of the diamagnetic matrix. On the same graphs some data obtained in case of NiCu [1, 2], $Ni_{0.3}Cu_{0.8-x}Si_x$ [8], and $Ni_{0.7}Cu_{0.3-x}Si_x$ [9] are also presented. For samples having different nickel content, but the same conduction electron concentration, identical M_s- and T_c-values are observed. In the limit of experimental errors all the data show a similar type of behaviour. The magnetic data obtained in case of NiCuAl alloys [15] have also the same trend. We conclude that the electron concentration of the diamagnetic matrix determines mainly the magnetic behaviour of ternary NiCu-based alloys. The NiCu system appears as a peculiar case of this general behaviour, where the copper matrix contributes an electron number equal to the atomic fraction of copper to the conduction band.

The similarities between the magnetic behaviour of ternary NiCu-based alloys and NiCu ones are also evidenced by other types of experiments. The ferromagnetic resonance measurements performed on the NiCuAl system [15] show that the g-values, $g = 2.19 \pm 0.01$, are not dependent on composition and equal to those obtained in Ni or NiCu alloys [16]. The magnetic behaviour of NiCuAl [17] or NiCuSi alloys close to the Curie points resembles that evidenced in Ni or the NiCu system [13, 14].

The above facts suggest that the magnetic behaviour of NiCuSi and NiCu may be analysed in a common manner. The model which seems to describe better their physical properties is the coherent potential approximation. The XPS spectra [18] of the NiCu alloys are in a good approximation given as superpositions of the density of states of the two elements. The Ni and Cu bands seem to retain their identity over the entire composition range. Probably in more complex systems as NiCuSi the CuSi effective medium shares a common conduction band.

Although the local environment effects are present in ternary systems they do not seem to play an important role in describing the magnetic behaviour of NiCu-based

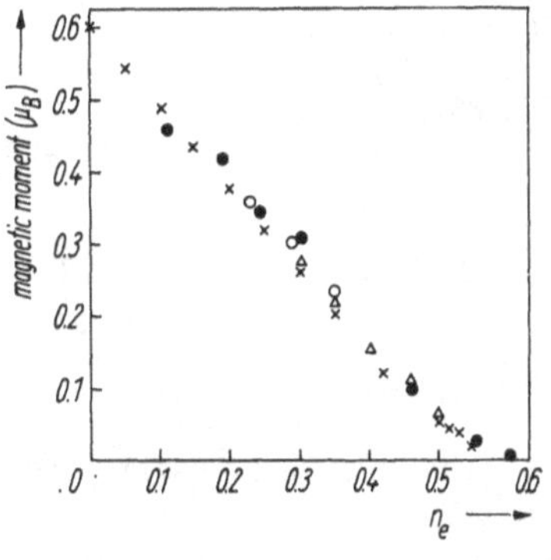

Fig. 6

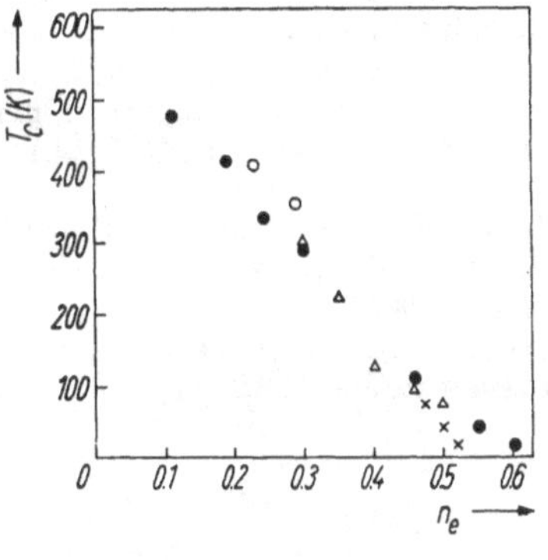

Fig. 7

Fig. 6. The dependence of magnetization for the studied samples (●) as a function of electron concentration of the diamagnetic matrix. The data for NiCu (×) [1, 2] $Ni_{0.7}Cu_{0.3-x}Si_x$ (△) [8], and $Ni_{0.8}Cu_{0.2-x}Si_x$ (○) [7] are also plotted

Fig. 7. The dependence of the Curie temperatures as functions of electron concentration of the diamagnetic matrix, for the studied samples (●). Some data for NiCu (×), $Ni_{0.7}Cu_{0.3-x}Si_x$ (△), and $Ni_{0.8}Cu_{0.2-x}Si_x$ (○) are also plotted alloys, except in the composition range where the onset of ferromagnetism takes place. As in case of NiCu alloys, a deviation from a linear behaviour of M_s- and T_c-values is observed, for a contribution of the diamagnetic matrix greater than $n_e = 0.48$. For the NiCu system — in this case — the moments are associated with Ni-rich regions on an atomic scale [19, 20], which due to statistical fluctuations have a local Ni concentration greater than the mean value characteristic of the alloy. It is possible that also in NiCuSi samples this tendency of clustering for Ni atoms is present.

Previously [21, 22] it has been pointed out that the onset of ferromagnetic order in NiCu is preceded at lower Ni concentration by a type of behaviour similar with mictomagnetism. For all NiCuSi samples having $T_c \geqq 14$ K and $M_s \geqq 0.0052\mu_B$ as determined in the $Ni_{0.60}Cu_{0.32}Si_{0.07}$ alloy, the magnetization curves are typical for a ferromagnet, as seen in Fig. 1.

Further studies are in progress in order to analyse the magnetic behaviour of ternary systems in the composition range where the electron concentration of the diamagnetic matrix is greater than $n_e = 0.6$.

References

[1] S. A. Ahers, M. J. C. Martin, and W. Sucksmith, Proc. Roy. Soc. A **248**, 145 (1958).
[2] A. J. P. Meyer and C. Wolff, C.R. Acad. Sci. (France) **246**, 576 (1958).
[3] N. F. Mott, Proc. Phys. Soc. **47**, 571 (1935).
[4] C. G. Robbins, H. Claus, and P. A. Beck, Phys. Rev. Letters **22**, 1307 (1969).
[5] J. P. Perrier, B. Tissier, and R. Tournier, Phys. Rev. Letters **24**, 313 (1970).
[6] G. M. Stocks, R. W. Williams, and J. S. Faulkner, Phys. Rev. B **4**, 4390 (1971).
[7] J. Inoue and M. Shimizu, J. Phys. Soc. Japan **40**, 1321 (1970).
[8] I. Cosma and E. Burzo, Solid State Commun. **10**, 585 (1972).
[9] E. Burzo, M. Văleanu, and D. Ungur, Solid State Commun. **20**, 991 (1976).
[10] E. Burzo and M. Văleanu, J. Phys. F **12**, 3105 (1982).
[11] M. E. Fisher, Rep. Progr. Phys. **30**, 615 (1967).
[12] L. J. de Jongh and A. R. Miedema, Adv. Phys. **23**, 1 (1974).
[13] J. S. Kouvel and M. E. Fisher, Phys. Rev. **136**, A1626 (1964).
[14] A. Arays, J. appl. Phys. **36**, 1136 (1965).
[15] E. Burzo and I. Lupsa, Solid State Commun. **42**, 369 (1982).
[16] K. J. Standley and K. H. Reich, Proc. Phys. Soc. **68**, 713 (1955).
D. M. S. Bagguley and M. Heath, Proc. Phys. Soc. **90**, 1029 (1967).
[17] I. Lupsa and E. Burzo, J. Magnetism magnetic Mater **37** (1983).
[18] S. Hüfner, G. K. Wertheim, R. L. Cohen, and I. H. Wernick, Phys. Rev. Letters **28**, 488 (1972).
[19] F. M. Ryan, E. W. Pugh, and K. Smoluchowski, J. appl. Phys. **40**, 2269 (1969).
[20] T. J. Hicks, B. D. Rainford, J. C. Kouvel, G. G. Low, and J. B. Comly, Phys. Rev. Letters **22**, 531 (1969).
[21] J. C. Ododo and B. R. Coles, J. Phys. F **7**, 2393 (1977).
[22] C. J. Tranchita and H. Claus, Solid State Commun. **27**, 583 (1978).

(Received September 17, 1982)

phys. stat. sol. (a) 77, 133 (1983)

Subject classification: 11; 1.4; 22.1.2

Sektion Physik der Technischen Universität Dresden[1] *(a),*
Institut für Festkörperphysik und Elektronenmikroskopie
der Akademie der Wissenschaften der DDR, Halle (Saale)[2] *(b), and*
Zentralinstitut für Kernforschung der Akademie der Wissenschaften der DDR, Rossendorf[3] *(c)*

Calculation of Neutron-Induced Defect Clusters in Silicon and Comparison with TEM Investigations

By

T. Gessner (a), M. Pasemann (b), and B. Schmidt (c)

Neutron induced defect clusters in silicon are considered. The calculation is based on a model analogous to the radiation damages by Si^+-ion implantation in Si. The results are compared with high resolution TEM-investigations. The experimental studies in principle confirm these calculations.

Es werden neutroneninduzierte Defekt-Cluster in Silizium untersucht. Die Rechnungen beruhen auf einem Modell, das analog ist zum Strahlungsdamage durch Si^+-Ionenimplantation in Si. Die Ergebnisse werden mit denen von Hochauflösungs-TEM-Untersuchungen verglichen. Die experimentellen Untersuchungen bestätigen im Prinzip diese Berechnungen.

1. Introduction

In many cases the description of neutron interactions with silicon crystal atoms is very important, especially in the field of neutron transmutation doping (NTD) of silicon. As a rule during the production of NTD-Si the FZ p-type silicon crystals are irradiated in a nuclear reactor at a ratio of thermal to fast neutrons of $R = (10$ to $100):1$. The fast neutrons (fluence: $\Phi_s = 10^{15}$ to 10^{17} cm^{-2}) produce radiation damages which are very disordered regions (defect clusters). These defect clusters impair the electrical properties of the silicon crystal. Thermal annealing after irradiation is necessary. The annealing conditions are better adjustable, if the structure and the geometrical dimension of the clusters are known.

In the present paper the cluster dimensions are calculated by means of the fast neutron interactions with silicon crystal atoms and secondary Si–Si interactions. These results are compared with high resolution TEM (transmission electron microscopy) studies.

2. Calculation of the Cluster Dimension

By characterizing the neutron induced defect clusters it is possible to differentiate between primary collision processes (neutrons with Si atoms) and secondary Si collision cascades by the PKA concept [1]. Primary knock-on atoms (PKAs) are silicon atoms which are directly determined by the collision kinetics and the differential cross-section. For elastical interactions [2] the average energy of PKAs $\overline{E}_{Si}$ as a func-

---

[1]) Mommsenstr. 13, DDR-8027 Dresden, GDR.
[2]) Weinberg 2, DDR-4020 Halle (Saale), GDR.
[3]) PSF 19, DDR-8051 Dresden, GDR.

tion of the neutron energy E_n reads

$$\bar{E}_{Si} = E_n\left(1 - \frac{1}{(1+A^2)}(\cos\vartheta + (A^2 - \sin\vartheta)^{1/2})^2\right)\Big/2 = \frac{0.133E_n}{2}. \tag{1}$$

In this equation, A is the silicon mass number and ϑ is the angle between incident and scattered neutrons ($\vartheta = 180°$). For a light-water moderated nuclear reactor the probable neutron energy is $E_n \approx 0.7$ MeV and the average neutron energy is $\bar{E}_n \approx \approx 2$ MeV. Hence, with equation (1) follows $\bar{E}_{Si} = 47$ keV and $\bar{E}_{Si} = 133$ keV, respectively. The energy barrier E_I, from which Si ions originate by ionization interactions, for silicon amounts to $E_I \approx 36$ keV. This value is calculated by applying the Kinchin-Pease equation for the high energy limitation [3]. Therefore, in the secondary collision cascades the PKAs can be concluded to interact as ions with the crystal atoms. At these neutron energies the collision cascades can be described by a model analogous to the radiation damages caused by the Si^+ ion implantation in silicon.

Sigmund and Sanders [4] performed a calculation for Si^+ implantation in Si. The ratio of the standard deviations of the average projected ion range ΔR_{px} and the radiation damage range ΔR_{psx} yields

$$\Delta R_{psx} = 0.845\,\Delta R_{px}\,. \tag{2}$$

In another paper which also Srour et al. [6] used as a basis of their calculation, Sigmund et al. [5] calculated the ratio of $\Delta R_{psx,y}/\Delta R_{px,y}$ along (x) as well as normal (y) to the movement direction of the Si^+ ions. This model can be used at energies up to $E_{Si} \approx 300$ keV.

Based on these calculations, a cylindrical shape is assumed for the neutron induced defect clusters (Fig. 1). It should be pointed out that the length l of the cluster is equal to the longitudinal standard deviation ΔR_{psx} and the diameter D is equal to the transversal standard deviation ΔR_{psy} of the radiation damage distribution resulting from a Si^+ ion implantation in silicon. Table 1 shows the calculated cluster parameters (D, 1) calculated for two characteristic neutron energies. The average projected range R_p of the Si^+ ions is calculated with the LSS theory [7].

Based on this model after irradiating silicon in a nuclear reactor with fast neutrons cylindrical clusters should be generated with an average length of $l = 15$ to 50 nm and a diameter of $D = 8$ to 17 nm. These parameters are a function of the specific neutron energy spectrum of the used nuclear reactor. The careful calculation of the neutron energy spectrum is very complicated or impossible.

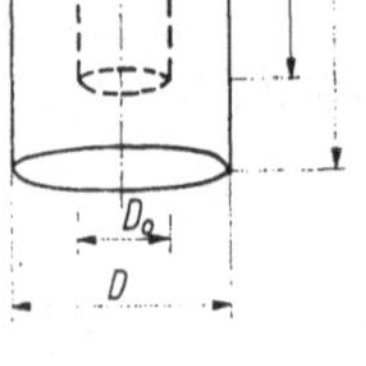

Fig. 1. Cylindrical model of a neutron induced defect cluster with the diameter D and the length l

Table 1

Typical parameters of a neutron induced defect cluster (V_c cluster volume)

E_n (MeV)	E_{Si} (keV)	R_p (nm)	l (nm)	D (nm)	V_c (cm^{-3})
0.7	47	62.2	15.6	8.8	1.13×10^{-18}
2	133	190.4	51.6	16.9	1.16×10^{-17}

Table 2

Summary of calculations and EM investigations of neutron induced defect clusters in silicon crystals

neutron energy (MeV)	method of studies	cluster dimension	ref.
nuclear reactor spectrum	spherical model, carrier mobility measurements	$D \approx 2$ nm	[8, 9]
1	cylindrical cluster model	$D \approx 6$ nm $l \approx 160$ nm	[10, 11]
1 5	spheroidal cluster model	max. axis: 60 nm (1 MeV) 200 nm (5 MeV)	[6]
nuclear reactor spectrum	EM	$D \approx 2.2$ nm	[12]
14	EM	$D \approx 50$ nm	[13]
0.7 2	clcylindrical cluster model	$D \approx 9$ nm $l \approx 16$ nm $D \approx 17$ nm $l \approx 52$ nm	present paper

3. TEM Studies

Earlier papers have characterized neutron induced defect clusters in silicon crystals. First spherical cluster models have been developed, the diameters of which are derived from carrier mobility measurements [8, 9]. Later on electron microscopic (EM) studies have been performed. Table 2 summarizes some papers concerning the calculations or EM investigations of neutron induced defect clusters in silicon crystals. Different cluster darameters are shown to exist in the literature.

Recently high resolution TEM investigations demonstrated the real nature of amorphous cluster regions in silicon, irradiated with heavy ions during HVEM (high-

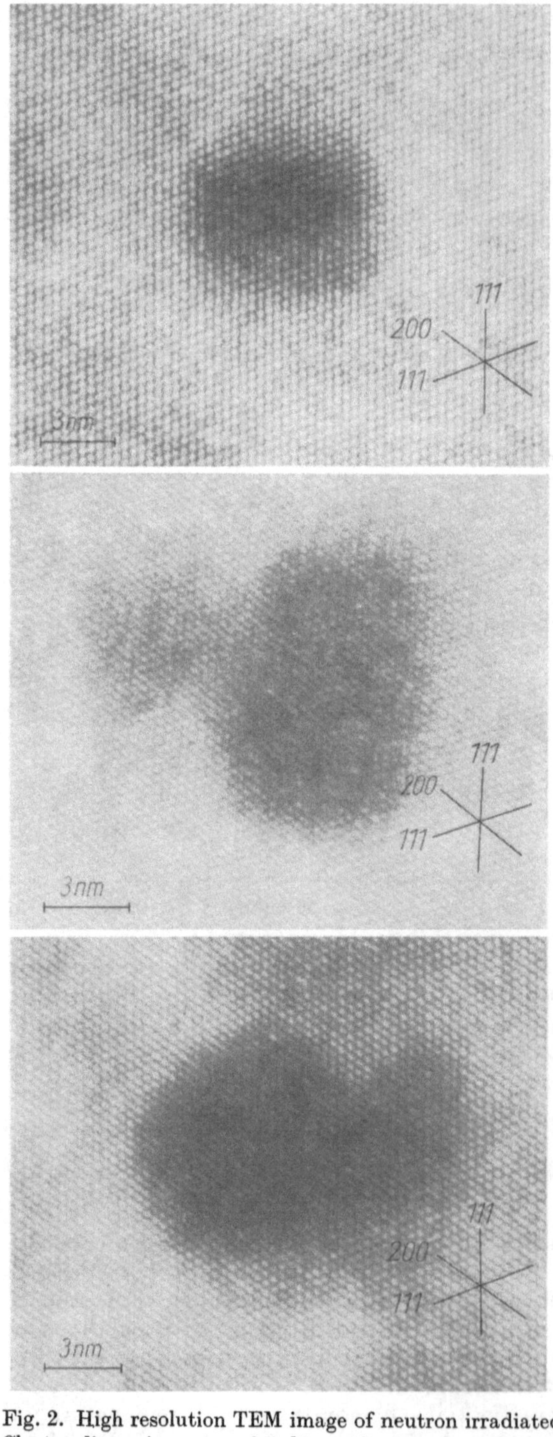

a

b

c

Fig. 2. High resolution TEM image of neutron irradiated ⟨110⟩ silicon in a ≈ 10 nm thick region. Cluster dimension: a) ≈ 6.5, b) ≈ 10, c) ≈ 9 nm voltage electron microscopy) studies [14]. Similar investigations have been carried out on neutron irradiated ⟨110⟩ oriented silicon in the present paper.

In the research reactor of Rossendorf the silicon wafers were irradiated with a fast neutron fluence of $\Phi_s = (5 \text{ to } 6) \times 10^{16}\ \text{cm}^{-2}$. After irradiation and storage at room temperature ($t \approx 30$ days) typical defect regions with variable dimensions were observed. High-resolution crystal lattice images were taken using the three-beam dark-field technique [15]. Under the experimental conditions used one white spot corresponds to two atomic rows separated by 0.136 nm in the ⟨100⟩ direction. Fig. 2 a, b, and c show three defect cluster regions. The dark areas originate from vacancy accumulations. It is obvious that the dark area is almost uniform in contrast, e.g. the defect must have a constant interface with the matrix in its ⟨110⟩ projection. In the projection the defect areas are more or less circular thus supposing a cylindrical shape of the defect cluster. An agglomeration of interstitial atoms was not observed. A diffusion of interstitial atoms to the surface is possible because of the small thickness of the region investigated (≈ 10 nm).

The regions in Fig. 2 b and c exhibit very disordered atoms. This observation is supported by the optical diffractogram in the inset of Fig. 2 c taken from the defected area which clearly shows the amorphous structure besides the reflections of the ⟨110⟩ oriented matrix. The tendency of forming amorphous regions increases with increasing defect dimension. If the cluster dimensions were smaller than 3 nm only vacancy accumulations without an extended atomic disorder were observed.

These experimental results of high resolution TEM studies demonstrate the following conformity with the calculation of neutron induced defect clusters based on a model analogous to the radiation damages caused by the Si^+ ion implantation in Si:

The observed defect regions are not spherical.

The calculated parameters, especially the cylindrical diameter at $E_n = 0.7$ MeV, show a good conformity with the TEM studies.

4. Conclusions

High resolution TEM investigations were carried out to observe for the first time neutron induced defect clusters in silicon after irradiation in a nuclear reactor, which allows a better estimation of the calculated models and the experimental results from the literature. The calculation of these clusters is possible by means of the described model.

Acknowledgements

The authors wish to thank Dr. R. Roß and his coworkers of the central institute of nuclear research Rossendorf for supporting the silicon irradiation in the nuclear reactor and Mr. R. Hillebrand for taking the optical diffractogram.

References

[1] S. Wood, N. J. Doyle, and J. A. Spitznagel, IEEE Trans. Nuclear Sci. 28, 4107 (1981).
[2] P. Sigmund, Appl. Phys. Letters 14, 114 (1969).
[3] G. H. Kinchin and R. S. Pease, Rep. Progr. Phys. 18, 1 (1955).
[4] P. Sigmund and J. B. Sanders, Proc. Internat. Conf. on Application of Ion Beams Semiconductor Technology, Ed. P. Gertin, Grenoble 1967 (p. 215).
[5] P. Sigmund, M. T. Matthies, and D. L. Phillips, Radiat. Eff. 11, 39 (1971).
[6] J. R. Srour, S. Othmer, A. Bahraman, and R. A. Hartmann, IEEE Trans. Nuclear. Sci. 28, 3968 (1981).
[7] J. M. Lindhard, M. Scharf and H. E. Schiott, Kong. Danske Vid. Selsk., mat.-fys. Medd. 33, 14 (1963).

[8] B. R. GOSSICK, J. appl. Phys. **30**, 1214 (1959).
[9] R. S. SWANEPOOL and P. T. WEDEPOHL, Radiat. Eff. **55**, 43 (1981).
[10] R. R. HOLMES, IEEE Trans. Nuclear. Sci. **17**, 137 (1970).
[11] F. G. WIKNER, D. K. NICHOLS, J. NABOR, T. FLANAGAN, H. HORIGE, and V. A. J. LINT, Transient Radiation Effects, Final Report, DA-49-168-AMC-05-X-1 (1967).
[12] J. M. PANKRATZ, J. A. SPARGUEARD, and M. I. RUDEL, J. appl. Phys. **39**, 101 (1968).
[13] M. BERTOLOTTI, in: Radiation Effects in Semiconductors, Plenum Press, New York 1968 (p. 311).
[14] P. WERNER and M. PASEMANN, Ultramicroscopy **7**, 267 (1982).
[15] N. D. ZAKHAROV, M. PASEMANN, and V. N. ROZHANSKI, phys. stat. sol. (a) **71**, 275 (1982).

(Received January 24, 1983)

phys. stat. sol. (a) 77, 139 (1983)

Subject classification: 1.5; 14.3.2; 14.3.3; 22.1.2

Institut für Physik der Werkstoffbearbeitung der Akademie der Wissenschaften der DDR, Berlin[1])

Determination of Semiconductor Parameters and of the Vertical Structure of Devices by Numerical Analysis of Energy-Dependent EBIC Measurements

By

M. KITTLER and K.-W. SCHRÖDER

For a four-layer structure (non-semiconducting layer, upper neutral semiconductor layer, depletion layer, lower neutral semiconductor layer) the EBIC-signal or charge-collection efficiency η, respectively, are calculated as a function of the beam energy E_0 or electron range R, respectively, with the corresponding electrical semiconductor quantities for silicon and the respective geometry values as parameters. Some examples are given for which the measured $\eta(E_0)$-dependences on the basis of the given model curves are fitted by variation of the semiconductor and geometry quantities. The results of a critical analysis of the quantities thus determined are: small diffusion lengths in microscopical volumes can be measured, the vertical structure of semiconductor devices can be investigated, and preparation-related electrically active damaged layers can be detected.

Für eine Vier-Schichtanordnung, bestehend aus nichthalbleitender Deckschicht, oberer neutraler Halbleiterschicht, Sperrschicht und unterer neutraler Halbleiterschicht wird das EBIC-Signal bzw. die Ladungssammlungseffektivität η als Funktion der Strahlenergie E_0 bzw. der Reichweite R mit den zugehörigen elektrischen Halbleitergrößen für Si und den entsprechenden Geometriewerten als Parametern berechnet. Für verschiedene Beispiele werden an gemessene $\eta(E_0)$-Verläufe nach dem gegebenen Modell durch Variation der Halbleiter- und Geometriegrößen Kurven angepaßt. Die auf diese Art bestimmten Größen werden einer kritischen Analyse unterzogen mit dem Ergebnis, daß neben der Messung von kleinen Diffusionslängen L in Mikrobereichen und der Untersuchung der Vertikalstruktur von Halbleiterbauelementen auch schliffbedingte elektrisch aktive Störschichten aufgespürt werden können.

1. Introduction

The EBIC technique is not only a means to investigate the electrical activity of crystal imperfections [1 to 6], it can also be used to measure semiconductor parameters especially recombination properties, and to study the inner structure of devices non-destructively.

Determination of the diffusion length L of minority carriers by means of an electron probe is often made via spatial decay measurements on p–n junctions which are perpendicular to the surface [7 to 10] or measurements near Schottky barriers (SB) [11]. The recombination properties can also be investigated, however, via measurements of the EBIC signal in time, i.e. by determination of the lifetime τ of the minority carriers [12 to 14].

---

[1]) Seestr. 82, DDR-1166 Berlin, GDR.

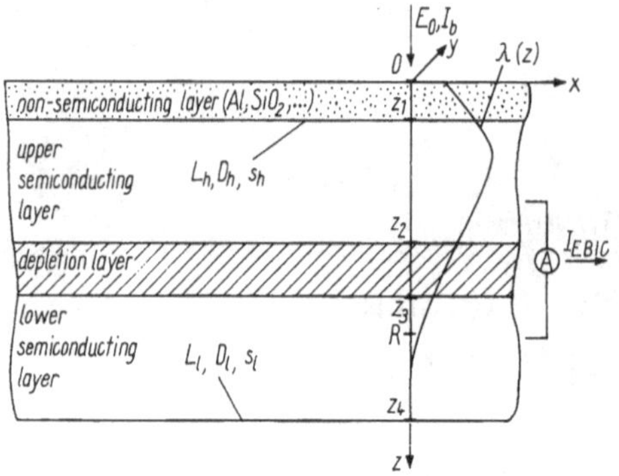

Fig. 1. Schematic representation of sample geometry as used for the model

The nondestructive measurement²) of the inner device geometry (position of p–n junctions, channel lengths, etc.) with EBIC is rendered difficult by resolution limitations due to diffusion. Here the time resolved EBIC technique [15, 16] brings considerable improvement, but at the cost of high experimental effort.

Investigation of small diffusion lengths L is possible on SB by measuring the EBIC signal I_{EBIC} in dependence on the beam energy E_0 [17 to 19]. The EBIC signal results here from the direct bombardment of the SB. The vertical structure of devices with junctions parallel to the surface was also investigated using the dependence of the EBIC signal on beam energy [20, 21].

In the present paper a method will be described which permits both the measurement of L in microscopical volumes on SB or on shallow p–n junctions and the investigation of the vertical structure of devices. This method is based on measurements of the EBIC signal I_{EBIC} in dependence on the beam energy E_0 and of the corresponding beam current I_b. Low injection conditions prevail throughout the paper. Starting from an appropriate model all measured data were processed on a computer. Some examples are given to demonstrate the capability of the method.

2. Model and Theoretical Investigations

Fig. 1 shows a four-layer structure. The non-semiconducting layer $(0, z_1)$, representing, e.g. an Al or SiO_2 layer, is followed by the upper semiconductor layer (z_1, z_2). This field-free semiconductor layer is characterized electrically by the diffusion length L_h, the diffusion coefficient D_h, and the surface recombination velocity s_h at the interface at z_1. The next layer is a depletion layer (z_2, z_3), followed by the lower field-free semiconductor layer (z_3, z_4) of opposite conduction type. The lower layer is characterized by the semiconductor parameters L_l, D_l, and s_l at the interface at z_4. Charge collection of the EBIC signal took place at the surfaces at z_2 and z_3 bounding the depletion layer which are each described by $s \to \infty$.

By appropriate choice of the parameters z_1 to z_4, s_l, etc., this structure can be adapted to various EBIC geometries. For instance, the case $z_1 = z_2$ describes the so-called SB charge collection microscopy (SBCCM) [1], while $z_1 = 0$ describes the case without cover layer, as it is given, e.g. for epitaxial layers. The special case of the lower semiconductor layer being an infinite half-space will be considered in greater detail further below. Now, expressions for the calculation of the EBIC signal will be given. They

²) Often bevelling techniques or other special preparation methods are used to study the inner device geometry. These methods, however, are destructive, and with decreasing device dimensions it becomes difficult to employ them at a certain device.

were derived under the assumption that the layers are infinite in the x and y directions, thus reducing the calculations to the one-dimensional case [22]. Furthermore, the generation of the excess electron–hole pairs was described by the depth–dose function after Everhart and Hoff [23]. The effect of the upper non-semiconducting layer upon the scattering of the primary electrons, which would cause a change in the depth–dose function, was not taken into account. This should introduce only a minor error for Al or SiO_2 layers, because of the mean atomic number Z and density similar to those of Si. A short description of possible derivations of the given expressions is found in the Appendix. For practical purposes the work of Selezneva and Filippov [24] can often be of advantage.

For $z_4 > 1.1R > z_3$ the EBIC signal is made up of contributions from the upper semiconductor, depletion layer, and lower semiconductor. It can then be described by the following expressions:

$$I_{\mathrm{EBIC}} = I_{\mathrm{n\text{-}h}} + I_{\mathrm{f}} + I_{\mathrm{n\text{-}l}}\,, \tag{1}$$

where

$$I_{\mathrm{n\text{-}h}} = \frac{I_{\mathrm{b}}E_0 f}{e_{\mathrm{i}}R} \int\limits_{z_1}^{z_2} \lambda(z/R) \frac{\exp\left((z-z_1)/L_{\mathrm{h}}\right) - \frac{S_{\mathrm{h}}-1}{S_{\mathrm{h}}+1}\exp\left(-(z-z_1)/L_{\mathrm{h}}\right)}{\exp\left((z_2-z_1)/L_{\mathrm{h}}\right) - \frac{S_{\mathrm{h}}-1}{S_{\mathrm{h}}+1}\exp\left(-(z_2-z_1)/L_{\mathrm{h}}\right)}\,\mathrm{d}z\,, \tag{1.1}$$

$$I_{\mathrm{f}} = \frac{I_{\mathrm{b}}E_0 f}{e_{\mathrm{i}}R} \int\limits_{z_2}^{z_3} \lambda(z/R)\,\mathrm{d}z\,, \tag{1.2}$$

$$I_{\mathrm{n\text{-}l}} = \frac{I_{\mathrm{b}}E_0 f}{e_{\mathrm{i}}R} \int\limits_{z_3}^{1.1R} \lambda(z/R) \frac{\exp\left((z_4-z)/L_{\mathrm{l}}\right) - \frac{S_{\mathrm{l}}-1}{S_{\mathrm{l}}+1}\exp\left(-(z_4-z)/L_{\mathrm{l}}\right)}{\exp\left((z_4-z_3)/L_{\mathrm{l}}\right) - \frac{S_{\mathrm{l}}-1}{S_{\mathrm{l}}+1}\exp\left(-(z_4-z_3)/L_{\mathrm{l}}\right)}\,\mathrm{d}z\ ^{3)}\,. \tag{1.3}$$

Here the denotations are I_{EBIC} total EBIC signal; $I_{\mathrm{n\text{-}h}}$ EBIC signal of the upper neutral semiconductor layer; I_{f} EBIC signal of the barrier; $I_{\mathrm{n\text{-}l}}$ EBIC signal of the lower neutral semiconductor layer; I_{b} beam current; E_0 beam energy; f fraction of the beam energy for the generation of electron–hole pairs in the semiconductor, $f_{\mathrm{Si}} \approx 0.9$ for the energy range of interest here [25]; e_{i} average energy for the generation of one pair, $e_{\mathrm{i}} \approx 3.6$ eV for Si [26]; R electron range at beam energy E_0, R (μm) = $= 0.0171E^{1.75}$ (keV) for Si [23]; $\lambda(z/R)$ polynomial to describe the depth–dose function in Si
$\lambda(x) = 0.6 + 6.21x - 12.4x^2 + 5.69x^3$ [23]; L_{h}, L_{l} diffusion lengths of the upper and lower semiconductor layer, respectively; S_{h}, S_{l} reduced surface recombination velocities at the interfaces at z_1 and z_4, respectively: $S_{\mathrm{h}} = s_{\mathrm{h}}L_{\mathrm{h}}D_{\mathrm{h}}^{-1}$, $S_{\mathrm{l}} = s_{\mathrm{l}}L_{\mathrm{l}}D_{\mathrm{h}}^{-1}$, and z_1 to z_4 geometry parameters according to Fig. 1.

Expressions (1) to (1.3) were written for $z_3 < 1.1R < z_4$. For smaller R-values they may be simplified: For $z_2 < 1.1R \leqq z_3$ $I_{\mathrm{h\text{-}l}}$ can be omitted, for $z_1 < 1.1R \leqq z_2$ I_{f} can be omitted as well, and finally, for $0 < 1.1R \leqq z_1$ we have $I_{\mathrm{EBIC}} = 0$.

3) A surface recombination velocities of $s = DL^{-1}$ at z_4 has the effect of an infinite lower half-space upon the recombination taking place. Substituting $s_1 = D_1\mathrm{L}_1^{-1}$, i.e. $S_1 = 1$, into (1.3) gives

$$I_{\mathrm{n\text{-}l}} = \frac{I_{\mathrm{b}}E_0 f}{e_{\mathrm{i}}R} \int\limits_{1.1R}^{z_3} \lambda(z/R)\exp\left[-(z-z_3)/L_{\mathrm{l}}\right]\mathrm{d}z\,. \tag{1.3'}$$

For reasons of lucidity, instead of I_{EBIC} the EBIC signal normalized to the generation conditions, the so-called charge collection efficiency η, is often used. It is

$$\eta = \frac{e_i I_{EBIC}}{I_b E_0 f}. \tag{2}$$

Then the factor $(I_b E_0 f)/(e_i R)$ in (1.1) to (1.3) reduces to $1/R$.

The numerical processing of the given expressions on a desk-top calculator (EMG 666) is a necessary condition for their discussion and for a later evaluation of experimental data. A program was developed, yielding the representation of η as a function of R or E_0, respectively, with L_h, S_h, L_l, S_l, and the geometry values z_1 to z_4 as parameters.

Fig. 2 shows an example of curves for structures with fixed cover-layer thickness z_1, fixed surface recombination S_h, and fixed diffusion length L_h in the upper semiconductor layer, but with different interface depths z_2 and z_3 and different diffusion lengths L_l in the lower semiconductor layer which was assumed as an infinite half-space. An increase in the cover layer thickness z_1 results in a kind of parallel displacement of the groups of curves to the right. Upon decreasing the thickness of the upper semiconductor layer, curve shapes typical for SBCCM as they are shown in Fig. 3 are approached.

As can be seen from Fig. 2 and 3, the calculated $\eta(R)$ and $\eta(E_0)$ curves show a strong dependence on the semiconductor and geometry quantities, characterizing the respective measuring configuration. Fig. 3 will be discussed in more detail, in accordance with our earlier papers [18, 27]. Fig. 3 shows $\eta(E_0)$ for SBCCM in an infinite half-space with $z_1 = z_2 = 0.05$ µm and $z_3 = 0.3$ µm. The L-values for the respective curves are $L = 1$, 5, 10, 20, 30, 40, 50, and 90 µm. It can be seen that η becomes increasingly

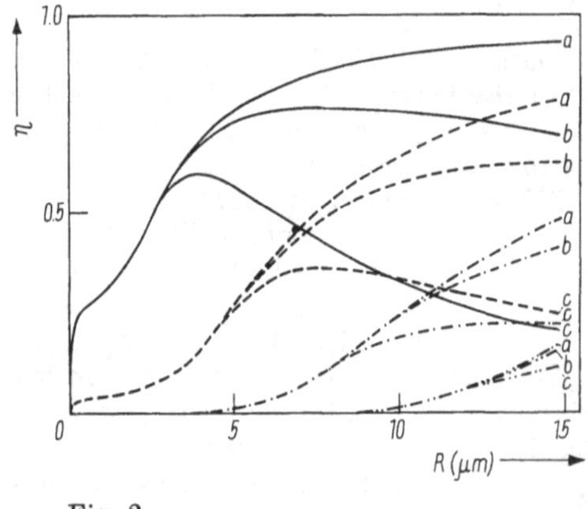

Fig. 2

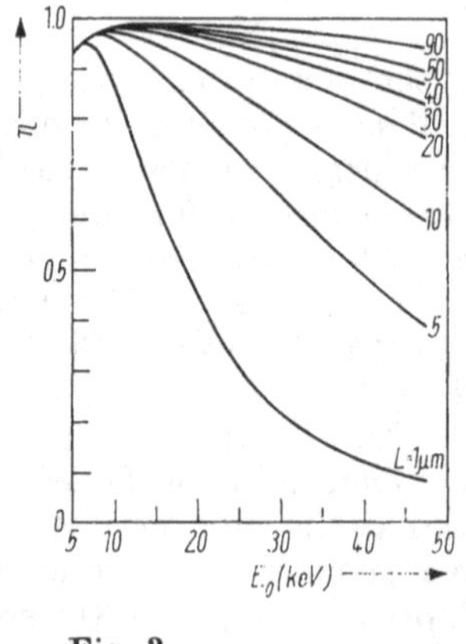

Fig. 3

Fig. 2. Charge collection efficiency $\eta = \eta(R)$ for an infinite half-space with the structure shown in Fig. 1. The parameters are $z_1 = 0.05$ µm, $L_h = 1$ µm, $S_h = 0$, $S_l = 1$. (a) $L_l = 100$ µm, (b) $L_l = 10$ µm, (c) $L_l = 1$ µm, ——— $z_2 = 2$, $z_3 = 2.5$; – – – 4, 4.5; —·—·— 7, 7.5; —··—··— 11 µm, 11.5 µm

Fig. 3. Charge collection efficiency $\eta = \eta(E_0)$ for SBCCM on an infinite half-space with $z_1 = z_2 = 0.05$ µm, $z_3 = 0.3$ µm for variable diffusion lengths L_l

sensitive to changes of L for growing values of the beam energy E_0. For large L-values, however, the sensitivity of η to changes of L becomes increasingly weaker; for $L > > 50$ μm the effect becomes rather small.

3. Experimental

The investigations were carried out with the SEM BS 300 from TESLA, which had to be adapted to our EBIC measurements. The $\eta(E_0)$ curves were measured on SB and on devices with barriers parallel to the surface. E_0 was usually varied in 5 keV steps, smaller intervals were used only in special cases. η was determined from I_{EBIC}, I_b[4]), and E_0 according to equation (2′)

$$\eta = I_{\text{EBIC}}/(250 I_b E_0)\,, \qquad E_0 \text{ in keV}\,. \tag{2'}$$

Equation (2′) is easily derived from (2) by replacing e_i and f by their numerical values for Si. The measured $\eta(E_0)$ relations were fitted by curves according to the above model through variation of the free parameters. If the agreement is acceptable, the corresponding parameter values describe the semiconductor and geometry parameters of the object under study.

3.1 Investigations on Schottky barriers

Results obtained on two samples are reported, which are designated as samples I and II. Sample I is p-type silicon with a resistivity of some Ωcm. After an appropriate cleaning procedure a Schottky contact was prepared by evaporating a thin Al layer. Fig. 4a shows the measuring points $\eta(E_0)$ and the fitted curve. A good fit for high E_0-values was achieved at the expense of the lower E_0-range. The parameters corresponding to the fit are $z_1 = z_2 = 0.14$ μm, $z_3 = 0.9$ μm, and $L_1 = 7.7$ μm, assuming the lower semiconductor as an infinite half-space. This gives a thickness of the Al layer $d_{\text{Al}} \approx 0.14$ μm, a depletion layer width $z_3 - z_2 = w \approx 0.76$ μm, and a diffusion length for electrons in the p-type semiconductor $L \approx 7.7$ μm. The value for L is the quantity of interest in our investigations, but the values for d_{Al} and w give also useful informa-

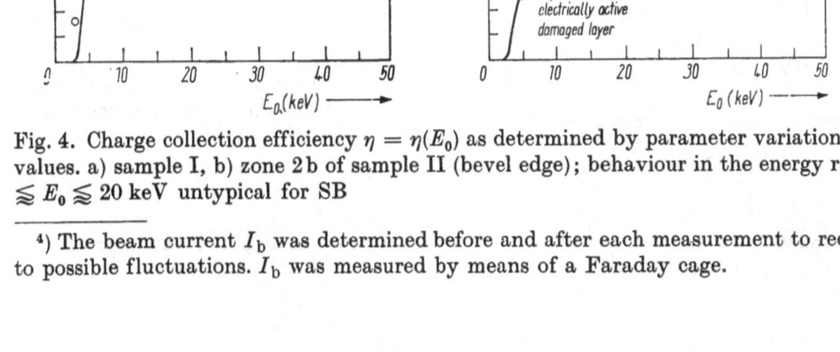

Fig. 4. Charge collection efficiency $\eta = \eta(E_0)$ as determined by parameter variation; (○) measured values. a) sample I, b) zone 2b of sample II (bevel edge); behaviour in the energy range 10 keV $\lessapprox E_0 \lessapprox$ 20 keV untypical for SB

[4]) The beam current I_b was determined before and after each measurement to reduce errors due to possible fluctuations. I_b was measured by means of a Faraday cage.

Fig. 5. EBIC micrograph from the angular bevel of sample II. This sample passed a process involving intrinsic gettering (see Fig. 6 for the geometry)

tion. Sample II consists also of p-type silicon which passed a device fabrication process involving intrinsic gettering [31]. On this sample a small bevel angle ($\alpha \approx 1°$) on a glass plate was prepared on which an Al Schottky contact was produced. Fig. 5 shows an EBIC micrograph obtained from the angular bevel showing three zones of different shades of grey below the device level. Each of the three zones has microdefect contrasts. The following thicknesses of the zones 1 and 2a were roughly estimated from the EBIC micrograph and bevel angle: $d_{zone1} \approx 12$ μm, $d_{zone\ 2a} \approx \approx 7$ μm. Fig. 6 gives a schematic cross section through the measuring arrangement. To obtain information on the efficiency of the getter process, L was determined for the three zones.

Fig. 4b shows the measuring points for $\eta(E_0)$[5]) obtained in zone 2b. They show a behaviour untypical for SBCCM, as there is a saddle between 10 and 20 keV. Indeed, assuming charge collection at a Schottky barrier no satisfying fit was found. Only the assumption of a p–n junction at a depth of about 2 μm made the fit given in Fig. 4.b possible. The lower semiconductor was again assumed to be an infinite half space. The parameters corresponding to the fit are $z_1 = 0.1$ μm, $z_2 = 1.8$ μm, $z_3 = = 2.2$ μm, and $L_l = 3.5$ μm, as well as $L_h = 4.25$ μm, $S_h = 2.5$. This gives a thickness of the Al layer $d_{Al} \approx 0.1$ μm; charge collection takes place at a depth of about 2 μm at a p–n junction with a depletion layer 0.4 μm wide. The diffusion length for electrons is $L \approx 3.5$ μm.

The observed p–n junction is probably due to the formation of an electrically active damaged layer close to the surface, resulting from the angular bevel on a glass plate. On a similar sample the bevel edge was etched after Wright [28] for some minutes prior to the Al evaporation. This gave a SB behaviour as it was observed on sample I.

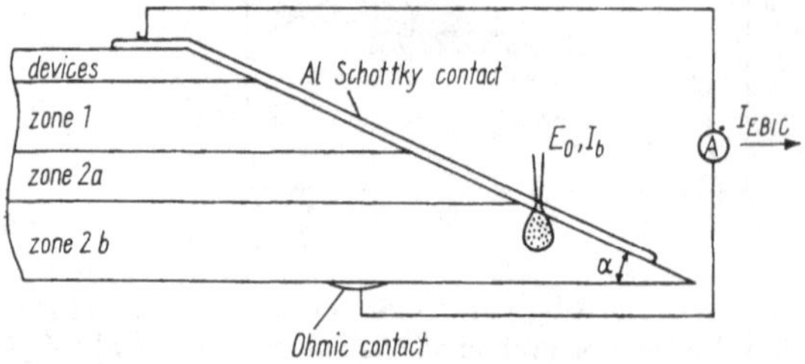

Fig. 6. Schematic cross section through sample II including the measuring configuration

[5]) In the measurement of I_{EBIC} the electron beam was scanned over an area of about 50 × 50 μm² in order to average over a larger area.

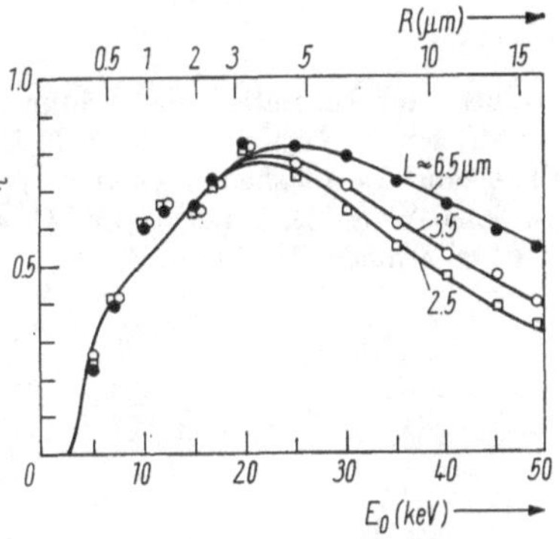

Fig. 7. Measured points (● zone 1, □ zone 2a, ○ zone 2b) and corresponding fits for the three zones of different recombination properties (sample II)

At one sample prepared with a provisional bevelling arrangement, however, a damaged layer (or interface depth according to the model) of as much as 5 μm was found.

Fig. 7 shows both the measuring points[5]) of the zones 1, 2a, and 2b of sample II and the corresponding fits. Accordingly, the following diffusion lengths are found for the respective zones: $L_{\text{zone 1}} \approx 6.5$ μm, $L_{\text{zone 2a}} \approx 2.5$ μm, and $L_{\text{zone 2b}} \approx 3.5$ μm. As can be seen, there is an actual improvement of the electrical characteristics of the material due to the getter process. The L-values determined for the zones 1 and 2a, however, require some critical consideration, since the stacking of zones of different diffusion lengths brings problems which were not taken into account in the underlying model. As the corresponding thicknesses of the layers are, however, not markedly smaller than the ranges R for large E_0-values, and since the diffusion lengths are relatively small, the error in L introduced by the stacking should not be too big. Zone 1 is contained almost completely within the very zone even for the largest E_0-values. This does not hold for zone 2a to the same extent, but the difference in L with respect to zone 2b is relatively small anyway.

3.2 Investigations on devices

Experiments were carried out on different locations of diodes with identical vertical structure, but of different lateral size (sample III: 25 × 25 μm², sample IV: 80 × × 80 μm²). The vertical structure and the three different measurement positions A, B, and C are shown in Fig. 8. The calculations based on the model given above could be reduced to the one-dimensional case, since the layers are of infinite extension in the xy-plane. If curves derived from the one-dimensional model are fitted to the

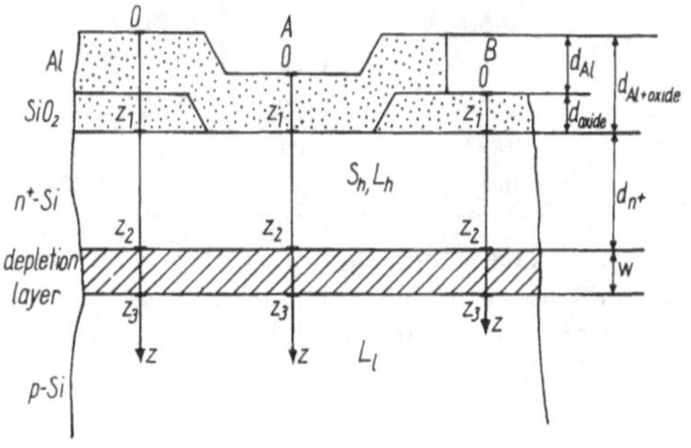

Fig. 8. Schematic representation of the vertical structure of diodes III and IV, also showing the different measurement positions A, B, and C

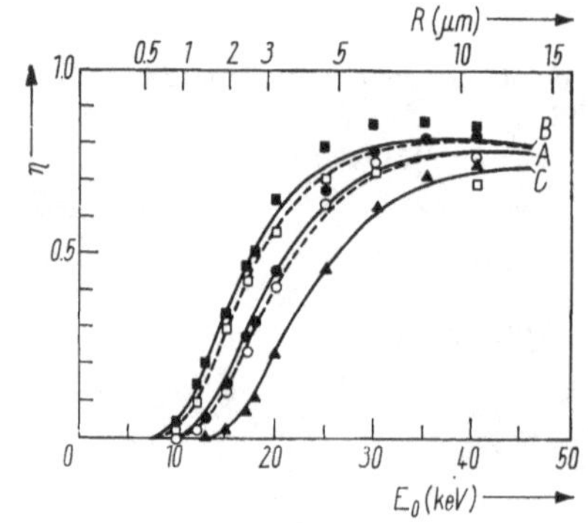

Fig. 9. Measured values and theoretical behaviour of the charge collection efficiency $\eta(E_0)$ on diodes III (full lines and full symbols) and IV (dashed lines and open symbols) at the positions A (●, ○), B (■, □), and C (▲) according to Fig. 8. ●, ■, ▲ diode III (80^2 μm²); ○, □ diode IV (25^2 μm²)

$\eta(E_0)$ measurements obtained on devices which are limited in the xy-plane, errors should be expected. It will be shown, however, that in the case under study nevertheless satisfying results could be obtained.

Fig. 9 shows the $\eta(E_0)$ measuring points and the respective fits for the diodes III and IV at the positions A, B, and C. The different lateral sizes of diodes III and IV give about the same η for small E_0-values; only for large E_0 there are significant deviations. For small E_0 the fit was good, it became worse towards larger E_0-values.

Table 1

Geometry and material parameters at the positions A, B, and C of the diodes III and IV

parameter	position A diode IV	 A diode III	 B diode IV	 B diode III	 C diode III	
	results of fits					
z_1 (μm)	0.95	0.95	0.48	0.50	1.23	
z_2 (μm)	1.95	1.85	1.46	1.40	2.40	
z_3 (μm)	2.60	2.40	2.00	2.00	2.80	
L_h (μm)	0.45	0.45	0.40	0.43	0.37	
L_l (μm)	≈25.0	≈25.0	≈25.0	≈25.0	≈25.0	
S_h	0	0.4	→∞	3	→∞	
	analysis					absolute error Δ
d_{Al} (μm)	0.95	0.95				} 0.21*)
d_{oxide} (μm)			0.48	0.50		
$d_{Al+oxide}$ (μm)					1.23	
d_{n^+} (μm)	1.00	0.90	0.98	0.90	1.17	0.27
w (μm)	0.65	0.55	0.54	0.60	0.40	0.25
L_{n^+} (μm)	0.45	0.45	0.40	0.43	0.37	

The parameters were determined by parameter variation. The thicknesses of the layers are also given (denotations as in Fig. 8); the absolute error could be estimated, since measurements were made at different positions. It shows the good description of the actual sample geometry by the model presented.

*) $\Delta = (d_{Al} + d_{oxide}) - d_{Al+oxide}$

The geometry quantities[6]) z_1, z_2, and z_3 as well as the semiconductor parameters L_h, L_l, and S_h corresponding to the fits are summarized and analysed in Table 1. Besides the quantities d_{Al}, d_{oxide}, $d_{Al+oxide}$, w, and the diffusion length L_{n^+} in the n$^+$-region which are all determined at the positions A, B, and C, their respective errors are given which were determined by comparison.

The value of the substrate diffusion length is probably not very correct due to the bad fit for high E_0-values, and the S_h-values are hardly acceptable due to their large scatter. The geometry values for the vertical structure, however, have an absolute error smaller than 0.3 μm, as was found by comparison of the measured values. The small scatter of the L_{n^+}-values is also quite satisfying. The estimated values for $d_{Al} \approx$ ≈ 1 μm, $d_{oxide} \approx 0.5$ μm, $d_{n^+} \approx 1$ μm, and $w \approx 0.5$ μm correspond to the results of technological test measurements. An additional measurement using the contamination line technique yielded $d_{Al}:d_{oxide} \approx 2:1$ which agrees well with the above data. The value of $L_{n^+} \approx 0.5$ μm is also quite reasonable, since a dopant concentration $N_{dop} \approx$ $\approx 10^{20}$ cm^{-3} in the n$^+$-region results in $L < 1$ μm due to the dominating Auger recombination [29].

4. Evaluation of the Capability of the Method

First let us make some remarks on the measurement of the diffusion length L on directly irradiated SB: Here it is of advantage to extend the measurements to the highest E_0-values attainable on the respective SEM and to take care that the best fit is achieved especially for the high E_0-values. In silicon the measuring range of L lies between some 0.1 μm and some 10 μm ($L \lessapprox 50$ μm). It should be possible to extend this range by improving the measurement accuracy for η, i.e. by improving the measurement accuracy of I_{EBIC} and I_b and by knowing the exact value of E_0. The method gives information on L within a microscopical volume [19]. Irradiating areas of great lateral extension (e.g. with a defocussed electron probe), results, of course, in an increase of the information volume size.

Taking into account the η-values from the lower E_0-range, the analysis allows an estimate of the metal layer thickness, of the depletion layer width, and thus even conclusions as to the dopant concentration, small L supposed. The analysis in this range is also instrumental in the detection of preparation-related effects close to the surface, for instance, of an electrically active damaged layer.

Now some remarks on the investigations of the vertical structure of devices must be made. Regarding the electrical properties of the upper semiconductor layer, our values for L_h were by far more reasonable than those for S_h, contrary to the results of Possin and Kirkpatrick [21][7]). The highest measurable value $L_{h,max}$ is decreased when reducing the thickness $z_2 - z_1$ of the upper semiconductor layer.

Furthermore, this maximum value is also dependent on the recombination properties S_h at the interface at z_1. Our estimates give $L_{h,max} \lessapprox 2(z_2 - z_1)$ for small S_h-values, while for large ones we found $L_{h,max} \lessapprox (z_2 - z_1)$.

The fact that the one-dimensional model yielded a satisfying fit at least for the upper part of the diode structure (despite the limited lateral extension of the diodes) is favoured by the small value of $L_{n^+}|_{0.5\,\mu m} \ll \text{lateral extension}|_{25(80)\,\mu m}$.

Inhomogeneities of the semiconductor quantities neglected in the model (e.g. vertical diffusion length stratification (see Fig. 6), lateral diffusion length differences, devices of small lateral size) require a considerably higher mathematical effort. The

[6]) Assumption of an infinite half-space.

[7]) In [21] $L \to \infty$ is assumed in the analysis which may be possible for the objects under consideration there.

one-dimensional model gives, however, satisfying results also for more realistic cases, if one is aware of the implied neglects and simplifications and is careful in the interpretation of results.

Finally, difficulties should be mentioned which arise, when the upper non-semiconducting layer is thick and has a mean atomic number and density markedly different from silicon. Then the undiscriminating application of the depth–dose function used [23] inevitably leads to erroneous results.

Appendix

Starting from the diffusion equation for the minority carrier concentration $p(\mathbf{r})$

$$\nabla^2 p(\mathbf{r}) - \frac{1}{L^2}\, p(\mathbf{r}) = -\frac{1}{D}\, g(\mathbf{r}) \tag{A1}$$

the expressions used in Section 2 for the EBIC signal

$$I_{\text{EBIC}} = \int \operatorname{grad} p(\mathbf{r})\, \mathrm{d}\boldsymbol{\sigma} \tag{A2}$$

will be derived for a configuration given in Fig. 1.

The surface integral in (A2) is taken over the collecting barriers. Assuming a cylinder-symmetrical Gaussian for the generation function $g(\mathbf{r})$ [30]

$$g(\mathbf{r}) = \frac{g_0}{R}\, \frac{\lambda(z/R)}{2\pi\sigma^2(z, R)} \exp\left(-r^2/(2\sigma^2(z, R))\right), \tag{A3}$$

equation (A1) is suitable written in cylindrical polar co-ordinates

$$\frac{1}{r}\, \frac{\partial}{\partial r}\left(r\, \frac{\partial p}{\partial r}\right) + \frac{\partial^2 p}{\partial z^2} - \frac{1}{L^2}\, p = -\frac{1}{D}\, g(z, r)\,, \tag{A4}$$

if the boundary conditions are independent of r. There are several possibilities in solving equations (A1) or (A4), respectively, differing in the way the Green's function is constructed. While van Roosbroeck [22] uses a representation by Riemann-Stieltjes integrals, and Selezneva and Filippov [24] find solutions by applying the reflection principle, Donolato [30] applies a Hankel's transform to equation (A4) and obtains an ordinary differential equation. In our case it is sufficient to integrate equation (A4) directly with respect to r, i.e. we define

$$\begin{aligned} P(z) &= \int_{-\infty}^{\infty} \int_{-\infty}^{\infty} p(x, y, z)\, \mathrm{d}x\, \mathrm{d}y \\ &= \int_{0}^{2\pi} \int_{0}^{\infty} r p(r, z)\, \mathrm{d}r\, \mathrm{d}\varphi\,. \end{aligned} \tag{A5}$$

Then we get an ordinary differential equation

$$\frac{\mathrm{d}^2 P(z)}{\mathrm{d}z^2} - \frac{1}{L^2}\, P(z) = -\frac{g_0}{DR}\, \lambda(z/R)\,. \tag{A6}$$

The introduction of $P(z)$ is rather evident, since the boundary conditions are independent of r and the collecting barriers are parallel to the surface.

The boundary conditions of (A6) will be

$$\left.\frac{dP(z)}{dz}\right|_{z=z_1} = \frac{s_h}{D_h} P(z)|_{z=z_1}, \tag{A7}$$

$$P(z_2) = 0$$

and

$$P(z_3) = 0$$

$$\left.\frac{dP(z)}{dz}\right|_{z=z_4} = \frac{s_l}{D_l} P(z)|_{z=z_4}, \tag{A8}$$

respectively. Then we have according to (A2)

$$I_{n\text{-}h} = \frac{g_0}{R} \int\limits_{z_1}^{x_h(R)} \lambda(z/R) \frac{\partial}{\partial z'} G_h(z, z')|_{z'=z_2} dz, \tag{A9}$$

$$I_{n\text{-}l} = -\frac{g_0}{R} \int\limits_{z_3}^{x_l(R)} \lambda(z/R) \frac{\partial}{\partial z'} G_l(z, z')|_{z'=z_3} dz, \tag{A10}$$

where $G_h(z, z')$ is the Green's function of (A6) with boundary conditions (A7), while $G_l(z, z')$ is that with respect to (A8). Furthermore, we define

$$x_h(R) = \begin{cases} z_1 & \text{if } 1.1R < z_1, \\ \min(1.1R, z_2) & \text{if } 1.1R \geqq z_1, \end{cases}$$

$$x_l(R) = \begin{cases} z_3 & \text{if } 1.1R < z_3, \\ \min(1.1R, z_4) & \text{if } 1.1R \geqq z_3. \end{cases}$$

Let

$$X_1 = (z_1 + z_2 - z - z')/L_h, \qquad X_2 = (z_2 - z_1 - |z - z'|)/L_h,$$

$$X_3 = (z_3 + z_4 - z - z')/L_l, \qquad X_4 = (z_4 - z_3 - |z - z'|)/L_l,$$

then the Green's functions will be

$$G_h(z, z') = L_h \frac{S_h(\cosh X_1 - \cosh X_2) - (\sinh X_1 + \sinh X_2)}{2\{\cosh[(z_2 - z_1)/L_h] + S_h \sinh[(z_2 - z_1)/L_h]\}}$$

and

$$G_l(z, z') = L_l \frac{S_l(\cosh X_3 - \cosh X_4) + \sinh X_3 - \sinh X_4}{2\{\cosh[(z_4 - z_3)/L_l] + S_l \sinh[(z_4 - z_3)/L_l]\}},$$

respectively. Then the following relations hold:

$$\left.\frac{\partial G_h(z, z')}{\partial z'}\right|_{z'=z_2} = \frac{\exp[(z - z_1)/L_h] - \dfrac{S_h - 1}{S_h + 1} \exp[-(z - z_1)/L_h]}{\exp[(z_2 - z_1)/L_h] - \dfrac{S_h - 1}{S_h + 1} \exp[-(z_2 - z_1)/L_h]}, \tag{A11}$$

$$-\left.\frac{\partial G_1(z, z')}{\partial z'}\right|_{z'=z_3} = \frac{\exp\left[(z_4 - z)/L_1\right] - \frac{S_1 - 1}{S_1 + 1}\exp\left[-(z_4 - z)/L_1\right]}{\exp\left[(z_4 - z_3)/L_1\right] - \frac{S_1 - 1}{S_1 + 1}\exp\left[-(z_4 - z_3)/L_1\right]}, \quad (A12)$$

Substituting the partial derivatives in (A9) and (A10) by (A11) and (A12), respectively, qields the expressions needed for the calculation of the EBIC signal. It should still be mentioned that

$$I_f = \frac{g_0}{R} \int_{z_2}^{x_f(R)} \lambda(z/R)\, dz ,$$

because total charge collection in the depletion layer was assumed, where

$$x_f(R) = \begin{cases} z_2 & \text{if} \quad 1.1R < z_2 , \\ \min(1.1R, z_3) & \text{if} \quad 1.1R \geqq z_2 . \end{cases}$$

Acknowledgements

The authors are indebted to Dr. H. Richter und Dr. C. Becker for support and critical review of the manuscript. Further they wish to thank Mr. W. Seifert for cooperation and stimulating discussions.

References

[1] H. J. Leamy, L. C. Kimerling, and S. D. Ferris, Scanning Electron Microscopy, Vol. I, IITRI, Chicago 1976 (p. 529).
[2] R. B. Marcus, M. Robinson, T. T. Sheng, S. E. Haszko, S. P. Murarka, and L. E. Katz, J. Electrochem. Soc. **124**, 425 (1977).
[3] H. Menninger, H. Raidt, and R. Gleichmann, phys. stat. sol. (a) **58**, 173 (1980).
[4] H. Blumtritt, R. Gleichmann, J. Heydenreich, and H. Johansen, phys. stat. sol. (a) **55**, 611 (1979).
[5] A. Ourmazd and G. R. Booker, phys. stat. sol. (a) **55**, 771 (1979).
[6] M. Kittler and W. Seifert, phys. stat. sol. (a) **66**, 573 (1981).
[7] H. Higuchi and H. Tamura, Japan. J. appl. Phys. **4**, 361 (1965).
[8] F. Berz and H. K. Kuiken, Solid State Electronics **19**, 437 (1976).
[9] C. van Opdorp, Philips Res. Rep. **32**, 192 (1977).
[10] G. Oelgart, J. Fiddicke, and R. Reulke, phys. stat. sol. (a) **66**, 283 (1981).
[11] J. D. Kamm and H. Bernt, Solid State Electronics **21**, 957 (1978).
[12] W. Zimmermann, phys. stat. sol. (a) **12**, 671 (1972).
[13] D. R. Hunter, D. H. Paxman, M. Burgess, and G. R. Booker, Scanning Electron Microscopy: Systems and Applications, Institute of Physics and Physical Society, London 1973 (p. 208).
[14] L. J. Balk, E. Menzel, and E. Kubalek, VIII. Internat. Conf. X-Ray Optics and Microanalysis, Boston 1977.
[15] G. V. Spivak, L. F. Komolova, and G. V. Saparin, Radiotekhnika i Elektronika **21**, 414 (1976).
[16] A. Georges, J.-M. Fournier, J. P. Gonchond, and D. Bois, Scanning Electron Microscopy, Vol. IV, SEM Inc., AMF O'Hare 1980 (p. 69).
[17] C. J. Wu and D. B. Wittry, J. appl. Phys. **49**, 2827 (1978).
[18] M. Kittler, A. Lindert, and W. Seifert, Beiträge Symp. Physikal. Grundlagen zu BE-Technol. d. Mikroelektronik, Frankfurt (Oder) 1979.

[19] J. Y. CHI and H. C. GATOS, J. appl. Phys. **50**, 3433 (1979).
[20] G. V. SPIVAK, G. V. SAPARIN, M. V. BYKOV, N. N. SEDOV, and L. F. KOMOLOVA, Soviet Phys. – Semicond. **3**, 1304 (1970).
[21] G. E. POSSIN and C. G. KIRKPATRICK, Scanning Electron Microscopy, Vol. I, SEM Inc., AMF O'Hare 1979 (p. 245).
[22] W. VAN ROOSBROECK, J. appl. Phys. **26**, 380 (1955).
[23] T. E. EVERHART and P. H. HOFF, J. appl. Phys. **42**, 5837 (1971).
[24] M. A. SELEZNEVA and S. S. FILIPPOV, Preprints 38 and 60, Inst. Prikladnoi Mat. Akad. Nauk SSSR, Moskva 1975.
[25] H. E. BISHOP, Ph. D. Dissertation, Cambridge 1966.
[26] L. REIMER and G. PFEFFERKORN, Rasterelektronenmikroskopie, Chap. 4.7, Springer-Verlag 1977.
[27] M. KITTLER and W. SEIFERT, Crystal Res. Technol. **16**, 157 (1981).
[28] M. WRIGHT, J. Electrochem. Soc. **124**, 757 (1977).
[29] P. A. ILES and S. I. SOCLOF, Proc. 11th Photovoltaic Specialists Conf., IEEE, 1975 (p. 19).
[30] C. DONOLATO, phys. stat. sol. (a) **65**, 649 (1981).
[31] H. RICHTER, in: Proc. XII. Internat. Conf. on Defects in Semicond., Amsterdam 1982, North-Holland Publ. Co., to be published.

(Received January 3, 1983)

phys. stat. sol. (a) 77, 153 (1983)

Subject classification: 14.2

Sektion Physik der Friedrich-Schiller-Universität, Jena[1])

On the Stationary Behaviour of Magnetic Flux Embraced by Asymmetric Two-Junction SQUIDs

By

S. HEINEMANN

The stationary behaviour of an asymmetric two-junction SQUID is described. The qualitatively different shape of the embraced magnetic flux versus applied magnetic flux curves depending on the parameters $\alpha = 2\pi L I_1/\Phi_0$ and $r = I_1/I_2$ is discussed for several constant gate currents I and for the limiting curve between stable and unstable solutions. These results are compared with corresponding critical gate current versus applied flux characteristics.

Es werden die stationären Eigenschaften eines unsymmetrischen Zweikontakt-SQUIDs beschrieben. Für verschiedene konstante Betriebsströme I und für die Grenzkurve zwischen stabilen und instabilen Lösungen wird das qualitativ unterschiedliche Verhalten der Abhängigkeit des inneren vom äußeren Magnetfluß als Funktion der Parameter $\alpha = 2\pi L I_1/\Phi_0$ und $r = I_1/I_2$ diskutiert. Diese Ergebnisse werden mit den entsprechenden Abhängigkeiten des kritischen Betriebsstromes vom äußeren Magnetfluß verglichen.

1. Introduction

Superconducting quantum interference devices (SQUID) consisting of a superconducting ring with two Josephson junctions have been investigated theoretically and experimentally by many authors. Especially, the threshold curves, i.e. the critical current–magnetic flux characteristics [1 to 10], but also maps of potential energy [11 to 13] and dynamical properties (for instance [14, 15]) of such devices have been reported.

There is a substantial interest in the use of the two-junction SQUID as a memory cell [11, 16 to 22], because of its hysteretic behaviour of magnetic flux embraced by the SQUID ring [23]. In some cases the embraced flux is detected by a Josephson junction [19, 20] or by a second SQUID [21, 22] coupled to the memory SQUID. Recently, measurements of the transition between two possible states of embraced magnetic flux have been described [24]. In the present paper, the stationary magnetic flux embraced by asymmetric two-junction SQUIDs as a function of applied magnetic flux and of gate current will be carried out in detail.

2. Basic Equations

In Fig. 1 an electrical equivalent circuit of a two-junction SQUID is shown. A gate current I and a magnetic flux Φ_e can be applied. The flux Φ_e can be injected also directly as $L_e I_e$ by a control current I_e which does not change the results. A sinusoidal dependence of Josephson junction currents on phase differences θ_1 and θ_2, respectively, and circuit inductances as separate lumped inductances L_1 and L_2 are considered for calculation.

The methods used for obtaining the equations and conditions describing a stable state of a two-junction SQUID have been derived adequately before by Schulz-Du Bois and Wolf [6].

[1]) Max-Wien-Platz 1, DDR-6900 Jena, GDR.

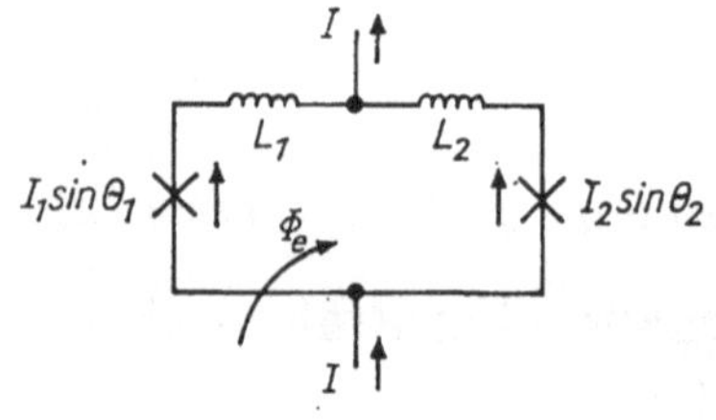

Fig. 1. Electrical equivalent circuit of a two-junction SQUID

The vanishing first derivatives of the potential energy give the equations for the normalized gate current,

$$i = \frac{LI}{\Phi_0} = i_{c\,max} \frac{r \sin \theta_1 + \sin \theta_2}{r+1}, \tag{1}$$

and the flux quantization condition,

$$\frac{\Phi}{\Phi_0} = \frac{\theta_2 - \theta_1 + 2\pi n}{2\pi} = \frac{\Phi_e}{\Phi_0} + \frac{i_{c\,max}}{1+r} \left[pr \sin \theta_1 - (1-p) \sin \theta_2\right], \tag{2}$$

where the following abbreviations are used: the loop inductance $L = L_1 + L_2$, the normalized maximum critical gate current $i_{c\,max} = L(I_1 + I_2)/\Phi_0$, the maximum Josephson current ratio $r = I_1/I_2$, the inductance ratio $p = L_1/L$, and the magnetic flux quantum $\Phi_0 = 2.07 \times 10^{-15}$ Vs. The integer n is the number of the flux quantum state for restricting the phase angles to the range $-\pi < \theta_i < \pi$. The stability conditions due to a minimum of potential energy have been derived from its second derivatives as

$$1 + \frac{2\pi i_{c\,max}}{1+1/r} \cos \theta_1 > 0 \tag{3}$$

and

$$g(\theta_1, \theta_2) = (1+r) \cos \theta_1 + (1 + 1/r) \cos \theta_2 + 2\pi i_{cmax} \cos \theta_1 \cos \theta_2 > 0\,. \tag{4}$$

The condition (4) set equal to zero becomes identical with the solution for critical gate current obtained from the Lagrange multiplier formalism. Therefore, the critical condition $g(\theta_1, \theta_2) = 0$ represents the delimiter between stable and unstable solutions. Condition (3) must be considered for the distinction of a maximum from a minimum of potential energy.

Now, the magnetic flux Φ embraced by the SQUID loop as a function of the applied magnetic flux Φ_e, i.e. the $\Phi - \Phi_e$ characteristic, will be investigated theoretically. This is performed by a method used before [4, 16] for computing the threshold characteristics. The critical condition $g(\theta_1, \theta_2) = 0$ according to (4) can be written as

$$\theta_2 = \arccos \left(\frac{-r \cos \theta_1}{1 + \dfrac{2\pi i_{c\,max}}{1 + 1/r} \cos \theta_1} \right). \tag{5}$$

For an investigation of non-critical stable states, i.e. for constant gate currents i smaller than the critical current i_c ($i < i_c$), (1) can be expressed in the following form:

$$\theta_2 = \arcsin \left\{ (1+r) \frac{i}{i_{cmax}} - r \sin \theta_1 \right\}. \tag{6}$$

Substituting either (5) or (6) into (2) one obtains the applied flux Φ_e and the embraced flux Φ as functions of the phase angle θ_1. Then the dependence of the embraced flux Φ on the applied flux Φ_e for non-critical stable states or the limiting curve between stable and unstable solutions can be calculated.

3. Static Embraced Flux Curves

The $\Phi - \Phi_e$ characteristics have been computed for several parameters $i_{c\max}$, r, and p in the regions of qualitatively different behaviour of the threshold curves given by Fulton et al. [3] and shown in Fig. 2. The parameters $\alpha = 2\pi L I_1/\Phi_0 = 2\pi i_{c\max}/(1 + 1/r)$ and $r = I_1/I_2 \leqq 1$ are the decisive ones for the shape of the static $\Phi - \Phi_e$ curves.

Substituting (1) into (2) one obtains

$$\frac{\Phi_e}{\Phi_0} = \frac{\Phi}{\Phi_0} - pi + \frac{i_{c\max}}{1 + r} \sin \theta_2 \,. \tag{7}$$

Supposing now a constant maximum critical current parameter $i_{c\max} = L I_{c\max}/\Phi_0$ and constant maximum Josephson current ratio $r = I_1/I_2$, equal values of gate current i and of embraced flux Φ give the same phase angles θ_1 and θ_2. Note that phase angles in this case are independent of inductance ratio p, because (1) and $\Phi = (\Phi_0/2\pi)(\theta_2 - \theta_1 + 2\pi n)$ do not contain p. Therefore, a deviation from symmetric inductance ratio $p = 1/2$ leads only to a shift of the embraced flux curve for $i =$ constant along the Φ_e axis,

$$\frac{\Phi_e(p)}{\Phi_0} = \frac{\Phi_e(p = 1/2)}{\Phi_0} + \left(\frac{1}{2} - p\right) i \,. \tag{8}$$

An example of this shift will be given later.

In Fig. 3 a to f the dependence of the embraced flux on the applied flux is plotted. Each figure contains curves for vanishing gate current $i = 0$ (full line), for $i = i_{c\max}/2$ (dotted line), and for $i = -i_{c\max}/2$ (dashed line). The thin line is the limiting curve between stable and unstable solutions due to the critical gate current. For comparison Fig. 4 a and b show the corresponding dependences of the critical gate current on the applied flux (threshold curves). The used parameters are indicated in Fig. 2.

The discussion is started from a symmetric case (Fig. 3a) which lies in region I of Fig. 2. The $\Phi - \Phi_e$ characteristic consists of disconnected curve pieces indicated by the number n of the flux quantum state. Gate currents of different signs but equal absolute values give equal curves. The embraced flux Φ is multivalued for fixed applied flux Φ_e. Increasing values of gate current i lead to a decrease of the length of the flux

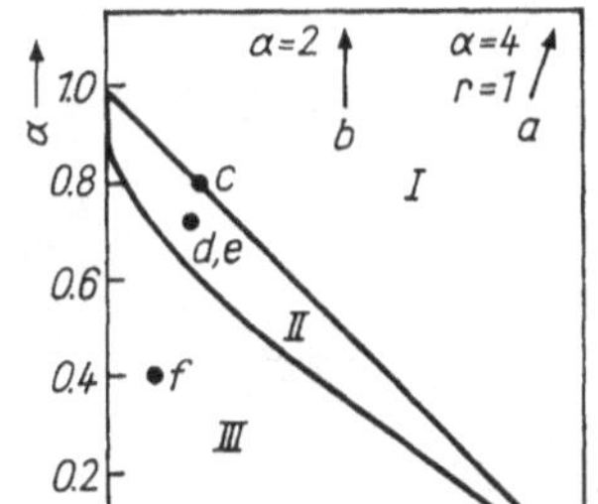

Fig. 2. Diagram of the parameter $\alpha = 2\pi L I_1 \Phi_0$ versus the parameter $r = I_1/I_2$: The boundary lines distinguishing the qualitatively different behaviour of critical current and embraced magnetic flux as a function of the applied magnetic flux [3] are shown. Symbols a to f refer to Fig. 3a to f and to the threshold curves in Fig. 4a and b

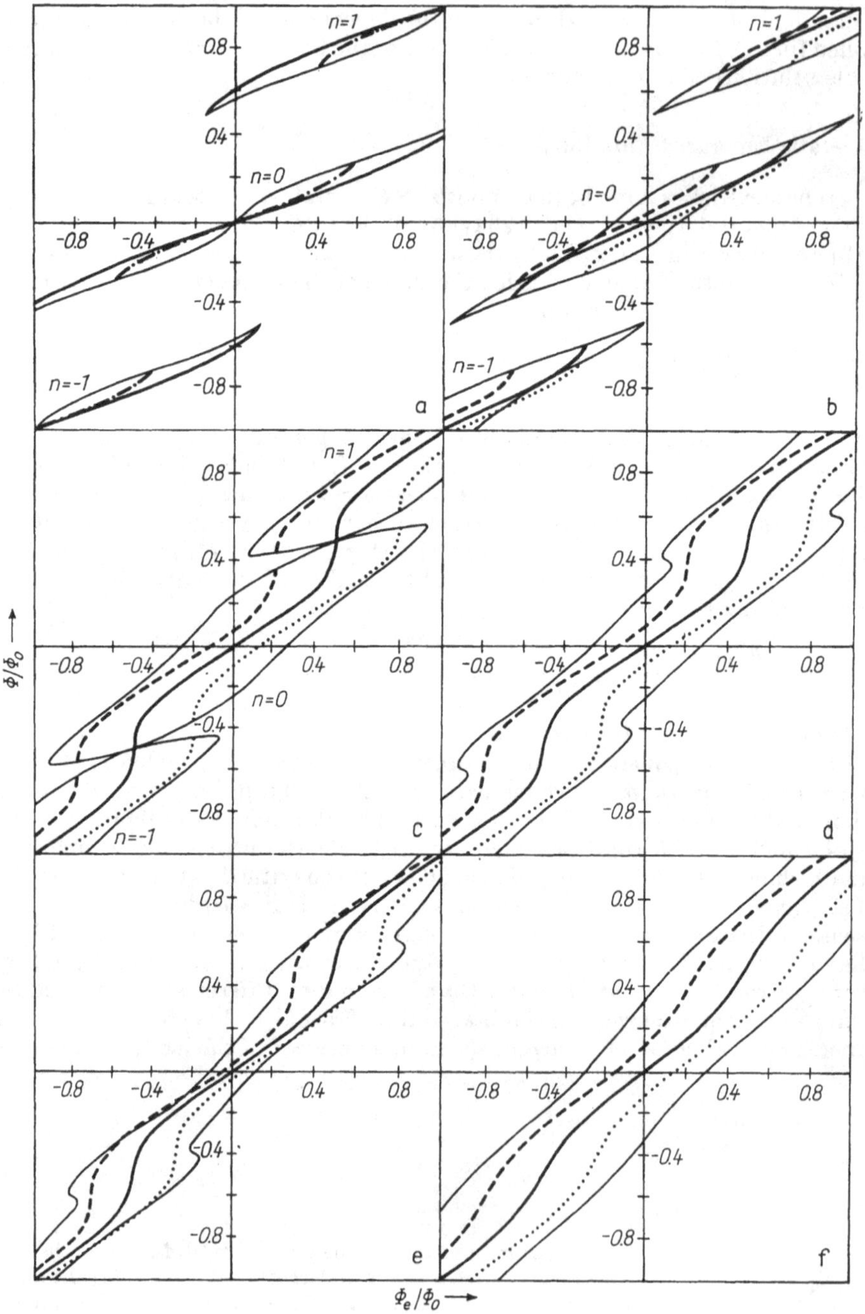

Fig. 3. Embraced magnetic flux versus applied magnetic flux curves of a two-junction SQUID for $i = 0$ (full line), $i = i_{c\,max}/2$ (dotted line), $i = -i_{c\,max}/2$ (dashed line) and the limiting curve (thin line). a) $i_{c\,max} = 1.273$, $r = 1$, $p = 0.5$; b) 0.955, 0.5, 0.5; c) 0.764, 0.2, 0.5; d) 0.751, 0.18, 0.5; e) 0.751, 0.18, 0.7; f) 0.700, 0.1, 0.5

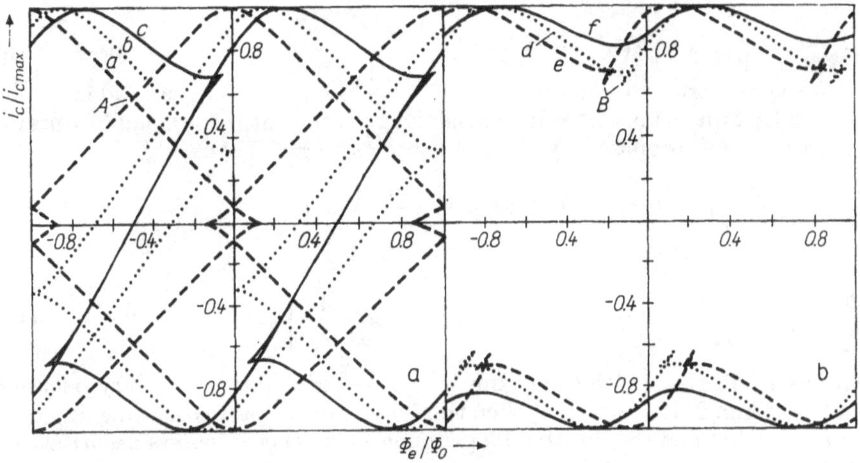

Fig. 4b. Threshold curves of a two-junction SQUID for parameters given in Fig. 3a to f

quantum state relative to the applied flux. For sufficiently large gate currents, there are no overlaps (not shown here) and, therefore, in regions near $\Phi_e/\Phi_0 = 1/2 + m$ (m is an integer) stable, i.e. zero-voltage states do not exist. This occurs if the gate current is larger than the minimum critical gate current (see point A in Fig. 4a). The dependence of minimum critical gate current on SQUID parameters has been investigated by Wunsch et al. [10]. The ends of curve pieces for vanishing and critical gate current in the case $r = 1$ are given by

$$\frac{\Phi_e}{\Phi_0} = \pm \frac{1}{2}(1 + i_{cmax}) + n$$

and

$$\frac{\Phi}{\Phi_0} = \pm \frac{1}{2} + n\,. \tag{9}$$

They are unaffected by the inductance ratio p.

A typical example for region I in Fig. 2 is carried out in Fig. 3b for the case of asymmetric maximum Josephson currents (here $r = 1/2$). Contrary to the symmetric SQUID, now, the curves of positive and negative sign of gate current are shifted against each other. Therefore, the limiting curve is a closed loop where the critical gate current changes its sign in the intersection points with the vanishing gate current curve. Note that a curve of constant gate current i can cross the limiting curve if the critical gate current (see curve b in Fig. 4a) is larger than the gate current i. The minima of a threshold curve (Fig. 4a) near the edges, corresponding to edges of the limiting curve in the flux diagram, can be expressed by

$$\frac{\Phi_e}{\Phi_0} = \pm \frac{1}{2}\left\{1 + \frac{2i_{c\max}}{1 + r}\,[1 - p(1 - r)]\right\} + n\,,$$

$$\frac{\Phi}{\Phi_0} = \pm \frac{1}{2} + n\,,$$

and

$$i/i_{c\max} = \pm(1 - r)/(1 + r)\,. \tag{10}$$

Φ_e and Φ for symmetric inductances ($p = 1/2$) are the same as in (9), independent of r. The ends of flux quantum states for vanishing gate current ($i = 0$) can be obtained by an approximation proposed by Beha [5] supposing $r \leqq 1/2$ as

$$\frac{\Phi_e}{\Phi_0} \approx \pm \frac{1}{2\pi} \left\{ \sqrt{(\alpha + r)^2 - 1} + \arccos \frac{-1}{\alpha + r} \right\} + n$$

and

$$\frac{\Phi}{\Phi_0} \approx \pm \frac{1}{2\pi} \left\{ \frac{r}{\alpha + r} \sqrt{(\alpha + r)^2 - 1} + \arccos \frac{-1}{\alpha + r} \right\} + n\,. \tag{11}$$

These equations are limited by the condition $\alpha \geqq 1 - r$ giving the boundary between regions I and II in Fig. 2. If $\alpha = 1 - r$, then the flux quantum states of vanishing gate current merge (Fig. 3c) and the limiting loops touch each other in points described by $\Phi_e/\Phi_0 = \Phi/\Phi_0 = \pm 1/2 + n$ and $i = 0$ (threshold characteristic c in Fig. 4a). Fig. 3d and e show typical $\Phi - \Phi_e$ curves of region II in Fig. 2. Both flux diagrams differ in the inductance parameter p only. One can see the shift of the constant gate current curves in Fig. 3e ($p = 0.7$) relative to those in Fig. 3d ($p = 0.5$) given by (8). Region II is characterized by merged flux quantum states for gate currents less than a current i_B corresponding to a cusp (point B) in the threshold curve (Fig. 4b). Therefore, the limiting curves due to the critical current form two separate continuous curves for positive and negative gate currents. It is obvious that merged flux quantum states are single-valued relative to the applied flux. Current i_B changes from zero at the boundary between regions I and II to the value $i_B/i_{c\max} = \pm(1 - r)/(1 + r)$ in region III. In region III (Fig. 2) the behaviour of constant gate current curves in the flux diagram (Fig. 3f) is similar to that in region II, but the limiting curves are single-valued and the threshold curves have no cusps (curve f in Fig. 4b). It should be mentioned that equations (10) also describe the minima of the threshold curves in regions II and III, but in region III it is only the absolute minimum of the critical gate current.

4. Conclusion

The three regions of qualitatively different behaviour of an asymmetric two-junction SQUID found by Fulton et al. [3] were verified by an investigation of the dependence of the embraced magnetic flux on the applied magnetic flux. This shows the possibility of different working modes of a two-junction SQUID depending on its asymmetry. These modes, multi-valued or single-valued ones, can be established by a variation of a maximum Josephson current, like I_1, by magnetic flux applied to the junction area as realized in the double SQUID [21, 22]. A single-valued working point can be used for writing an information, but in a multi-valued one, it can be stored in the SQUID.

Acknowledgements

The author likes to thank Dr. W. D. Schmidt and Prof. K. H. Berthel for helpful discussions.

References

[1] A. Th. A. M. de Waele and R. de Bruyn Ouboter, Physica **41**, 225 (1969).
[2] A. Th. A. M. de Waele and R. de Bruyn Ouboter, Physica **42**, 626 (1969).
[3] T. A. Fulton, L. N. Dunkleberger, and R. C. Dynes, Phys. Rev. B **6**, 855 (1972).

[4] WON-TIEN TSANG and T. VAN DUZER, J. appl. Phys. **46**, 4573 (1975).
[5] H. BEHA, Electronics Letters **13**, 216 (1977).
[6] E. O. SCHULZ-DU BOIS and P. WOLF, Appl. Phys. **16**, 317 (1978).
[7] R. L. PETERSON and C. A. HAMILTON, J. appl. Phys. **50**, 8135 (1979).
[8] S. HASUO, H. SUZUKI, T. IMAMURA, and T. YAMAOKA, Japan. J. appl. Phys. **20**, Suppl. 1, 323 (1981).
[9] W. ZAHN, phys. stat. sol. (a) **66**, 649 (1981).
[10] J. WUNSCH, W. JUTZI, and E. CROCOLL, IEEE Trans. Magnetics **18**, 735 (1982).
[11] H. BEHA, IEEE Trans. Magnetics **15**, 424 (1979).
[12] C. D. TESCHE, J. low-Temp. Phys. **44**, 119 (1981).
[13] M. KLEIN and A. MUKHERJEE, Appl. Phys. Letters **40**, 744 (1982).
[14] E. BEN-JACOB and Y. IMRY, J. appl. Phys. **52**, 6806 (1981).
[15] W. D. SCHMIDT and W. KRECH, phys. stat. sol. (a) **68**, 489 (1981).
[16] H. H. ZAPPE, Appl. Phys. Letters **25**, 424 (1974).
[17] P. GUÉRET, Appl. Phys. Letters **25**, 426 (1974).
[18] W. JUTZI, E. CROCOLL, R. HERWIG, H. KRATZ, G. MISCHKE, M. NEUHAUS, J. WUNSCH, A. SCHEIDEL, and H. J. WERMUND, Cryogenics März, 127 (1982).
[19] W. ANACKER, IEEE Trans. Magnetics **5**, 968 (1969).
[20] R. F. BROOM, W. JUTZI, and TH. O. MOHR, IEEE Trans. Magnetics **11**, 755 (1975).
[21] W. D. SCHMIDT and M. MANGOLD, phys. stat. sol. (a) **62**, K175 (1980).
[22] S. HEINEMANN, W. D. SCHMIDT, B. FRISCH, and K. H. BERTHEL, Physica (Utrecht) B, in the press.
[23] A. A. J. MATSINGER, R. DE BRUYN OUBOTER, and H. VAN BEELEN, Physica (Utrecht) **94B**, 91 (1978).
[24] M. NAOR, C. D. TESCHE, and M. B. KETCHEN, Appl. Phys. Letters **41**, 202 (1982).

(Received January 25, 1983)

phys. stat. sol. (a) 77, 161 (1983)

Subject classification: 1.2 and 14.1; 12.1; 21.4

Institute of High Pressure Physics, Academy of Sciences of the USSR, Troitsk[1])

Electroresistance of Cerium at Phase Transitions under Hydrostatic Pressure up to 8.0 GPa

P–T Diagram of the Cerium Phase Transitions

By

L. G. Khvostantsev and N. A. Nikolaev

Direct experimental evidence is given of the existence of β'- and γ'-high pressure phases of cerium. The *P–T* diagram of the phase transitions of cerium is proposed. The behaviour of the cerium resistance in the triple point domains and the dependence of the α-phase resistance on heat treating and its origin are studied.

Экспериментально подтверждено существование β'- и γ'- фаз высокого давления церия. Приводится *P–T* диаграмма фазовых переходов в церии. Изучено поведение электросопротивления церия под давлением в области тройных точек и его зависимость для α-фазы от температурной обработки и от способа ее получения.

1. Introduction

The study of the phase transitions of cerium and the behaviour of its kinetic coefficients under pressure above 3.0 GPa and high tmperatures remained controversial. It was suggested [1] that cerium has the α', β', and γ' high temperature phase sequence under pressure and these phases and the α-phase of cerium have two triple points. This suggestion is supported by the experimental data of the present work.

2. Experiment

The method to measure the kinetic coefficient of cerium under hydrostatic pressure and temperature up to 200 °C in a mixture of methyl and ethyl alcohol was described in [1, 2]. The data of the present work were obtained under pressure at temperature up to 400 °C. The teflon ampoule was filled with petroleum ether hydrostatic pressure up to 5.5 GPa at room temperature [3]. The best way to obtain hydrostatic conditions in the petroleum ether under pressure above 5.5 GPa and high temperature was such as below. The pressure in the ampoule was increased first at room temperature not above 5.0 GPa. After heating the cerium sample above 200 °C at 5.0 GPa the pressure was further increased. The manganin gauge was placed into a groove in the copper cover of the ampoule. The copper cover has thermal contact with a massive die which is held at room temperature (Fig. 1). The groove with the manganin gauge was filled with petroleum ether and closed by teflon diaphragm. Thus the manganin gauge temperature is not above 100 °C when the sample was heated up to 400 °C. The "cold" manganin gauge allows to determine the pressure increment in the ampoule at heating the sample and the liquid surrounding the sample [4].

The hydrostatic pressure was generated in an ampoule of 0.5 cm^3 volume by a "Toroid" high pressure device [5]. The cerium samples with purity of 99.93% had a length of 6 mm and a cross-section of 1×1 mm^2.

[1]) 142092 Troitsk, Moscow region, USSR.

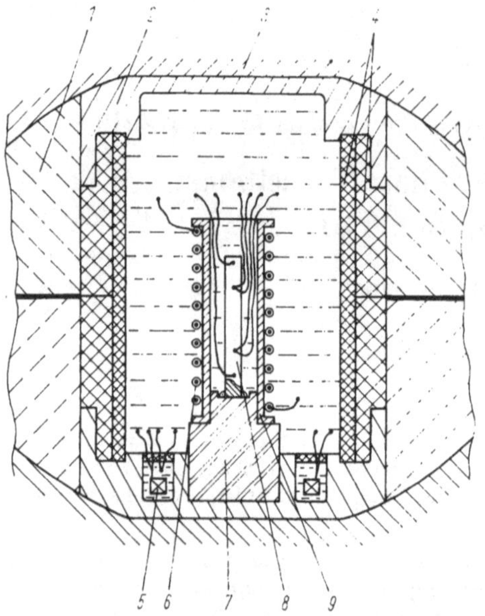

Fig. 1. High pressure–high temperature cell for the kinetic coefficient measurements. 1 catlinite surrounding the ampoule, 2 copper cover of the ampoule, 3 WC die, 4 teflon ampoule, 5 manganin gauge, 6 heater, 7 pyrofillite sample holder, 8 sample with leads, 9 copper cylinder

3. Results and Discussion

The P–T diagram of the phase transitions of cerium is presented in Fig. 2. The dash-dotted line was taken from [6]. The equilibrium line between α- and α′-phases and the line of the reverse β′ → α transition were determined in [1]. The line of the direct α → β′ transition has another location than that of [1].

There is a direct transition from α′- to β′-phase with a marked electroresistance drop by 7% at the isobar 8.0 GPa in a narrow temperature interval (235 to 240 °C) (Fig. 3). The isobaric resistance changes were smaller (≈ 0.5%) during the α′ → β′ transition in a wide temperature interval (225 to 250 °C) at 6.8 GPa because the transition occurs in the vicinity of the triple point of α-, α′-, and β′-phases.

It is impossible to determine exactly the position of the triple point of α-, α′-, and β′-phases and to give a complete P–T phase diagram of cerium because of the lack of

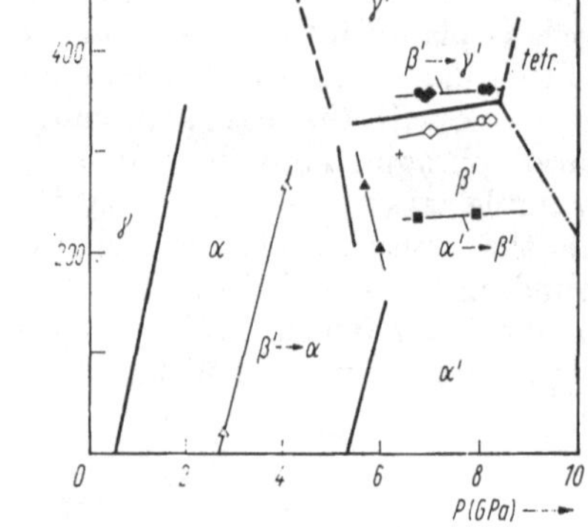

Fig. 2. P–T diagram of the phase transitions of cerium ●, ○ β′ → γ′ transitions, ■ α′ → β′ transition, ▲, △ α ⇄ β′ transitions, + location of the onset of changes in the relative resistance obtained at isobaric heating at 6.3 GPa (see Fig. 6), ◆, ◇ data obtained by differential thermal analysis [8]. The inset shows the hypothetic P–T phase diagram of cerium

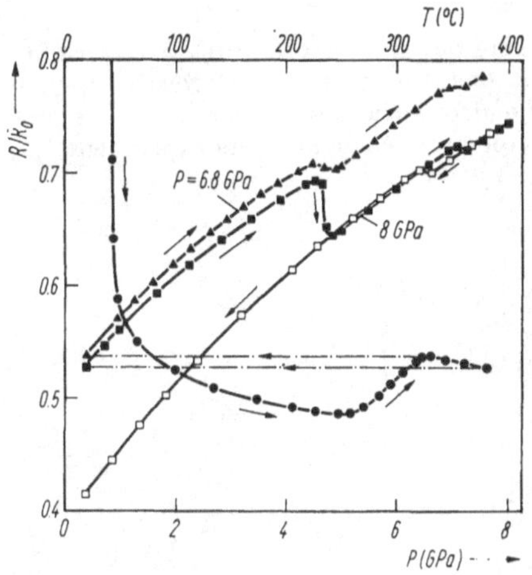

Fig. 3. Pressure and temperature dependences of the relative resistance of cerium. ● isothermal pressure increase, ▲, ■, □ isobaric heating and cooling of the sample, respectively. The arrows show the sequence of the pressure and temperature changes in the experiment

the equilibrium line of the $\alpha' \rightleftarrows \beta'$ transition. There are lines of direct and reverse transitions between α- and α'-phases and between α- and β'-phases. This made it possible to place the equilibrium lines of the $\alpha \rightleftarrows \alpha'$ transition and approximately that of the $\alpha \rightleftarrows \beta'$ transition. There is no reverse $\beta' \to \alpha'$ transition and its line. There is no change of resistance at cooling of the β'-phase up to room temperature (see isobaric curves obtained at 8 GPa in Fig. 3). The triple point has a large domain on the P–T diagram. Therefore, the β'-phase had been obtained at room temperature under nonhydrostatic pressure conditions in the cerium sample surrounded by a solid medium [7]. The β'-phase was obtained also at pressure cycling the sample through the $\alpha \rightleftarrows \alpha'$ transformation [1], although the deterioration of the hydrostatic conditions in the cerium sample itself is small. Decrease of the resistance value after each $\alpha \to \alpha'$ transition during pressure cycling is caused by a formation of small amounts of the β'-phase in a sample (Fig. 5a in [1]). The resistance of the β'-phase is smaller than the resistance of the α'-phase (Fig. 3).

The direct and reverse transitions between β'- and γ'-phases occur with the increase and decrease of temperature at 8 GPa in a narrow temperature interval and they are accompanied by a small ($\approx 0.7\%$) drop and increase of resistance, respectively (Fig. 3). These data conform with data of [8] obtained at 8.4 GPa. The direct $\beta' \to \gamma'$ phase transition occurs in a wide temperature range (340 to 360 °C) at 6.5 GPa with a small drop of the resistance ($< 0.5\%$) because it occurs in the domain of the triple point of α'-, β'-, and γ'-phases.

We have studied also the behaviour of cerium in the domain of the triple point of α-, α'-, and β'-phases and the triple point of α-, β'- and γ'-phases.

The sequence of pressure and temperature changes and the resistance measurements in the domain of the triple point of α-, α'-, and β'-phases are presented in Fig. 4 and 5. The increase ($< 1\%$) of isothermal (265 °C) resistance at 5.7 to 6.0 GPa (Fig. 4) was caused by an incomplete direct $\alpha \to \beta'$ transition [1]. The remaining part of the α-phase transforms to the α'-phase in the domain of the triple point at cooling the sample to room temperature under 6.2 GPa. (All isobars presented are really the pressure intervals, because the pressure changes with temperature). The α'-phase obtained in such a way transforms to the α-phase at pressure decrease from 5.5 to 5.0 GPa at room temperature [1]. The β'-phase is conserved in wide pressure and temperature ranges and transforms to the α-phase at isothermal decrease of pressure at 265 °C and 4.3 GPa. The resistance drops sharply by 7% (Fig. 4).

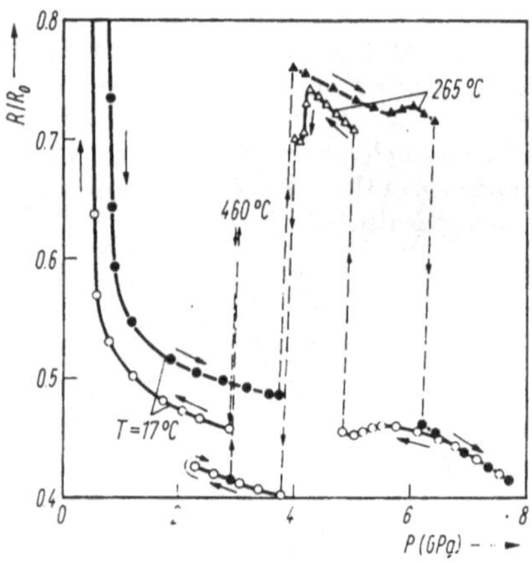

Fig. 4. Pressure dependence of the relative resistance of cerium at increase (●, ▲) and decrease (○, △) of pressure. The arrows show the sequence of the pressure and temperature changes in the experiment

The electroresistance of the α-phase obtained from the β'-phase is much smaller than the resistance of the α-phase obtained from the initial γ-phase. At a pressure of 4.1 GPa and 265 °C the difference of the resistances is 8% and at 3.8 GPa and room temperature 17% (Fig. 4). The difference of the resistance at room temperature is reduced to the half after heating the sample up to 460 °C. The difference of the resistances is conserved with pressure decrease. The resistances of the α-phase obtained from initial the γ-phase and the α-phase obtained from the α'-phase coincides at the same pressure and temperature up to 130 °C (Fig. 3 in [1]). Thus, the α-phase obtained from the β'- phase at any temperature has always a smaller resistance than the α-phase obtained from the initial γ-phase. The result was confirmed also by experiments presented in Fig. 5 and 6.

In the experiment (Fig. 5) the β'-phase was obtained from the α-phase at isothermal pressure increase at 203 °C, i.e. at a temperature well lower than the temperature of the direct α' → β' transition (Fig. 2). This is due to an extended range of the triple point of α-, α'-, and β'-phases over pressure and temperature. The α → β' transition is incomplete and the resistance increase is small. The transformation of the α-phase to the β'-phase becomes more complete at heating the sample up to 260 °C at 6.8 GPa. The resistance of the sample is lower after cooling from 260 to 203 °C. The increase of the resistance during the complete α → β' transition and the dependence of the

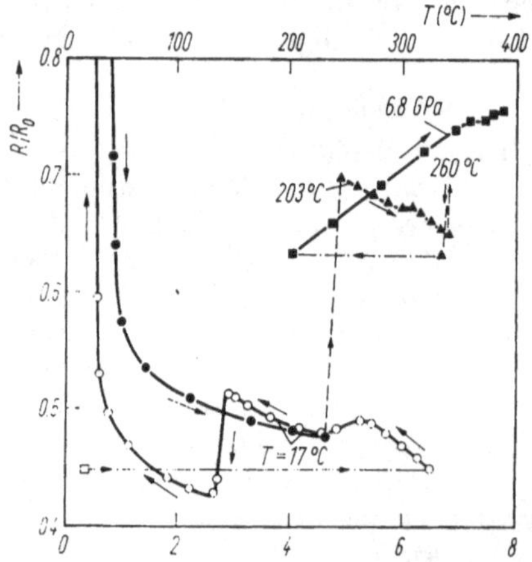

Fig. 5. Pressure and temperature dependences of the relative resistance of cerium. ●, ▲, ○ isothermal increase and decrease of pressure, respectively, ■ isobaric heating of the sample, □ obtained after rapid cooling at 6.8 GPa from 380 to 17 °C. The arrows show the sequence of the pressure and temperature changes in the experiment

resistance of a sample consisting of the β′-phase alone on pressure has to be rather stronger. A more complete transformation of the α-phase to the β′-phase takes place with increase of pressure and temperature in the domain of existence of the β′-phase. Unlike the direct α → β′ transition the reverse β′ → α transition and the drop of the resistance are abrupt.

The small drop of resistance obtained during isobaric heating of the sample at 6.8 GPa (Fig. 5) is related to the β′ → γ′ transition and is the same as that at the isobaric heating at 8 GPa (Fig. 3). A little part of the α-phase remains in the sample after the incomplete α → β′ transition and seems to remain also after the β′ → γ′ transition in the vicinity of the triple point of α-, β′- and γ′-phases (Fig. 2). At cooling the sample to room temperature the α-phase transforms to the α′-phase in the vicinity of the triple point of α-, α′-, and β′-phases. The α′-phase transforms to the α-phase in its turn during a decrease of pressure at room temperature with the appropriate decrease of resistance (Fig. 5). At cooling from 380 °C to room temperature there was a reverse γ′ → β′ transition and the obtained β′-phase transforms to the α-phase at a pressure of 2.9 GPa at room temperature as usually [1]. The cconservation of part of the α-phase at the direct α → β′ transition and in the vicinity of the triple point of α-, β′-, and γ′-phases shows that the direct α → β′ and α → γ′ transitions are "slack" and the triple points of α-, α′-, and β′-phases and α-, β′-, and γ′-phases have a large domain on the *P–T* diagram of cerium. Therefore, the resistance has no anomalies (Fig. 6), neither in the vicinity of the triple point of α-, β′-, and γ′-phases on the 305 °C isotherm nor at increase and decrease of temperature at 6.3 GPa nor in the vicinity of the triple point of α-, α′-, and β′-phases. The resistance curve obtained at decrease of pressure and room temperature shows changes of resistance caused by the α′ → α and β′ → α transitions (Fig. 6). It follows that the α′- and β′-phases were obtained from the α-phase in the vicinity of the triple points.

A small decrease of the resistance at 6.3 GPa (Fig. 6) obtained at cooling the sample is expected to be caused by the reverse transition of part of the γ′-phase to the α-phase.

The presented data show that there are β′-phase and γ′-phase and the triple point of α-, β′-, and γ′-phases. The position of the triple point of α-, β′-, and γ′-phases causes the equilibrium line between α- and γ′-phases to be directed from this triple point to the minimum on the melting curve on the *P–T* diagram (Fig. 2) of cerium, contrary to [1, 6, 8].

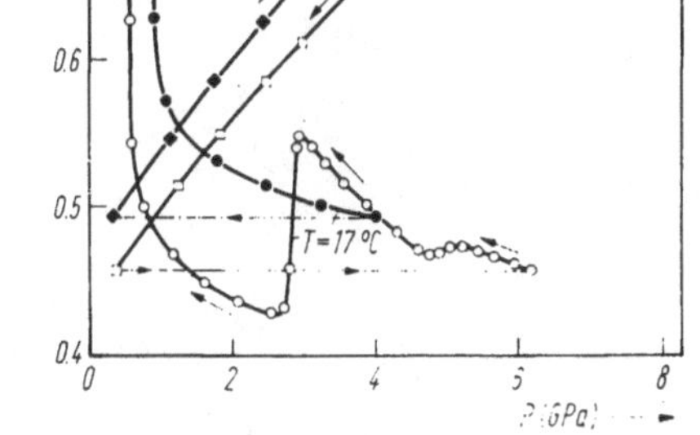

Fig. 6. Pressure and temperature dependences of the relative resistance of cerium. ●, ▲, ○ isothermal pressure increase and decrease, respectively, ◆, ■, □ isobaric heating and cooling of sample, respectively. The arrows show the sequence of the pressure and temperature changes in the experiment

The resistance of cerium decreases at $\gamma \rightarrow \alpha$, $\alpha' \rightarrow \beta'$, and $\beta' \rightarrow \gamma'$ transitions and increases at $\alpha \rightarrow \alpha'$, $\alpha \rightarrow \beta'$, and probably $\alpha \rightarrow \gamma'$ transitions. The thermo-e.m.f. of cerium decreases at the $\gamma \rightarrow \alpha$ transition and increases at the $\alpha \rightarrow \alpha'$ transition considerably [1]. The dependence of the α-phase resistance on heat treatment and the mode of its origin is also observed. The question is now to find a theoretical picture of the change of cerium valency in passing from one high pressure phase to another [9] for the purpose to explain such a complicated behaviour of the kinetic coefficients of cerium under pressure and temperature.

Acknowledgements

We wish to thank Dr. D. I. Khomsky for his interest in the work and A. M. Bukhartsev for useful assistance.

References

[1] L. G. Khvostantsev and N. A. Nikolaev, phys. stat. sol. (a) **68**, 193 (1981).
[2] L. G. Khvostantsev and N. A. Nikolaev, phys. stat. sol. (a) **51**, K57 (1979).
[3] I. D. Barnett and C. D. Bosco, J. appl. Phys. **40**, 3144 (1969).
[4] L. G. Khvostantsev, V. A. Sidorov, L. E. Shelimova, and N. Kh. Abrikosov, phys. stat. sol. (a) **74**, 185 (1982).
[5] L. G. Khvostantsev, L. F. Vereshchagin, and A. P. Novikov, High Temp.-High Press. **9**, 637 (1977).
[6] S. Endo, V. Sasaki, and T. Mitsui, J. Phys. Soc. Japan **42**, 882 (1977).
[7] W. H. Zachariasen and F. H. Ellinger, Acta cryst. **A33**, 155 (1977).
[8] T. E. Antonova, I. T. Belash, and E. G. Ponyatovsky, Fiz. Metallov i Metallovedenie **51**, 131 (1981).
[9] D. I. Khomsky, Uspekhi fiz. Nauk **129**, 443 (1979).

(Received December 1, 1982)

phys. stat. sol. (a) **77**, 167 (1983)

Subject classification: 13.4 and 14.3; 22.2.2

Sektion Physik der Bergakademie Freiberg[1])

Impurity Parameters of Undoped and Zn-Doped p-Type GaP with Two Acceptor Levels Determined by Hall Measurements

By

W. SIEGEL, E. ZIEGLER, and G. KÜHNEL

Hall measurements are performed in the temperature region 60 to 900 K on undoped and Zn-doped p-type GaP crystals grown by the SSD method. The $p(T)$-curves of undoped samples show a small step caused by the presence of a medium deep acceptor with an activation energy (0.18 ± ± 0.02) eV beside the shallow acceptor carbon. Carbon is present in a low concentration in the Zn-doped samples, too. The parameters (activation energy, concentrations) of the two acceptors in both types of samples are determined by an equation cubic in p.

Hall-Messungen werden im Temperaturbereich 60 bis 900 K an undotierten und Zn-dotierten p-GaP-Kristallen durchgeführt, die mittels des SSD-Verfahrens gezüchtet wurden. Die $p(T)$-Kurven der undotierten Proben zeigen eine schwache Struktur etwa bei Raumtemperatur, die auf die Anwesenheit eines mitteltiefen Akzeptors mit einer Aktivierungsenergie (0,18 ± 0,02) eV neben dem flachen Restakzeptor Kohlenstoff zurückzuführen ist. Auch in den Zn-dotierten Proben ist Kohlenstoff in etwa der gleichen Konzentration wie in den undotierten Proben enthalten. Die Störstellenparameter der jeweils zwei Akzeptorarten werden durch Fits mit einer kubischen Gleichung in p bestimmt.

1. Introduction

The analysis of the temperature dependence of the carrier concentration in a semiconductor with the aim to determine the impurity parameters is performed, as a rule, with the standard relation (for p-type material)

$$\frac{p(p + N_\mathrm{D})}{(N_\mathrm{A} - N_\mathrm{D} - p)\, T^{3/2}} = \frac{N'_\mathrm{v}}{g} \exp\left(-\frac{E_\mathrm{A}}{kT}\right), \tag{1}$$

where p is the hole concentration, N_A and N_D are the acceptor, and donor concentration, respectively. E_A is the acceptor activation energy, N'_v $(= 2(2\pi m'_\mathrm{v} k/h^2)^{3/2})$ the reduced density of states of the valence band, and g the degeneracy factor of the acceptor ground level. This relation corresponds to the model of a non-degenerate semiconductor dominated by a single acceptor level partly compensated and without influence of excited states.

Electrical and optical investigations of undoped and Zn-doped p-type GaP crystals grown by the SSD (synthesis, solute diffusion) method indicated that in these samples two acceptor levels exist which both influence the $p(T)$ curves. In this case instead of (1) an equation cubic in p has to be used for the analysis of the Hall data [1].

This paper is concerned with the determination of parameters (activation energy, concentration) for two acceptor species with strong different activation energies in the undoped samples and for two acceptor species with similar activation energies in the Zn-doped samples.

[1]) Silbermannstr. 1, DDR-9200 Freiberg, GDR.

2. Experimental

Polycrystalline GaP ingots were grown by the SSD method [2] in quartz ampoules at temperatures between 900 and 1100 °C.

Rectangular or square single crystalline samples with dimensions of about 2 to 4 mm were prepared from discs cut from these ingots. Ohmic contacts were made welding thin gold wires (50 μm diameter) alloyed with about 5% Zn to the samples by means of a capacitor discharge [3]. These contacts allow measurements in a wide temperatures range (60 to 900 K).

The Hall coefficient R_H and the conductivity σ were measured using the method of van der Pauw and applying a magnetic field of 0.5 T. The hole concentration p follows from R_H by the relation

$$p = \frac{r_H}{eR_H}. \tag{2}$$

The Hall factor r_H was taken to be equal to 1.

3. Results

3.1 Non-intentionally doped p-GaP

Undoped samples in a relatively wide range of the hole concentration at room temperature (2×10^{16} to 1.3×10^{17} cm^{-3}) were available because ingots were grown at different temperatures.

The existence of two acceptor levels in the undoped samples becomes apparent in the lg p–$(1/T)$ curves (Fig. 1) by a weak structure at about room temperature. Such a step in the lg p–$(1/T)$ curves appears only if the two acceptor levels have clearly different activation energies. In this case it is possible to determine the parameters of the shallow acceptor by a fit of the low-temperature part of the $p(T)$ curve with (1) [1]. The error of the fitted parameters depends beside the accuracy of measurements on the temperature range applicable for the fit. Therefore, we have performed the measurements down to about 60 K using a nitrogen cryostat and reducing the pressure over the liquid nitrogen reservoir. The results of the fits for $T \leq 220$ K are listed in Table 1. The average errors of the fitted parameters were estimated to be +15 to −10% for N_A, +30 to −20% for N_D, and ±2% for E_A. The activation energy varies between 42.9 and 48.7 meV depending on the impurity concentration. In [4] it could be shown that the thermal activation energy of impurities determined at low

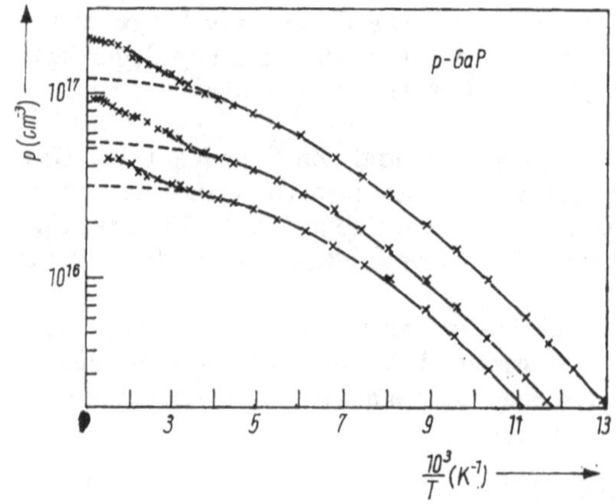

Fig. 1. Hole concentration vs. reciprocal temperature for undoped p-GaP samples

Table 1

Hole concentration p at room temperature and values of the parameters N_A, N_D, E_A, and N'_V/g for undoped p-GaP samples obtained by fitting the low temperature part of the $p(T)$ curves

sample	p (10^{16} cm^{-3})	N_A (10^{16} cm^{-3})	N_D (10^{16} cm^{-3})	E_A (meV)	N'_V/g (10^{14} cm^{-3} K$^{-3/2}$)
21	2.1	3.2	1.1	48.7	5.9
31	3.0	4.5	1.8	47.3	7.3
32	4.1	6.8	2.8	46.7	10.0
33	3.8	5.8	2.0	45.0	6.1
34	5.0	8.3	3.3	44.2	7.1
11	6.0	9.1	3.1	46.2	7.2
12	9.5	12.4	2.7	44.3	8.0
14	9.8	15.0	3.3	43.7	11.0
35	10.6	16.2	4.9	42.9	8.0
36	12.6	17.1	4.1	44.1	9.7

temperatures depends on the concentration of the compensating impurities in the form $E_A = E_{A,0} - \alpha N_D^{1/3}$ (provided the impurity concentrations are not too high).

From the plot E_A versus $N_D^{1/3}$ (Fig. 2) one can conclude that the shallow acceptor is identical with carbon because the regression line yields for $E_{A,0}$ a value (57.5 ± 2) meV which in the accuracy range agrees with the optical value (54 ± 2) meV [5] of the ionization energy of C.

Above room temperature the measured hole concentration deviates clearly from the p-values expected if only the shallow acceptor C would be present (dashed curves in Fig. 2 obtained by extrapolation of the low-temperature fit curve to higher temperatures using (1)). For the fit of the measured $p(T)$ curves in the whole temperature region the cubic equation

$$p^3 + p^2\left[N_D + N'_V T^{3/2}\left(\frac{B_1}{g_1} + \frac{B_2}{g_2}\right)\right] + \\ + pT^{3/2}\left[N'_V N_D\left(\frac{B_1}{g_1} + \frac{B_2}{g_2}\right) - N_V\left(\frac{N_{A1}B_1}{g_1} + \right.\right. \\ \left.\left. + \frac{N_{A2}B_2}{g_2}\right) + N_V'^2 T^{3/2}\frac{B_1 B_2}{g_1 g_2}\right] = \\ = N_V'^2 T^3 \frac{B_1 B_2}{g_1 g_2}(N_{A1} + N_{A2} - N_D); \tag{3}$$

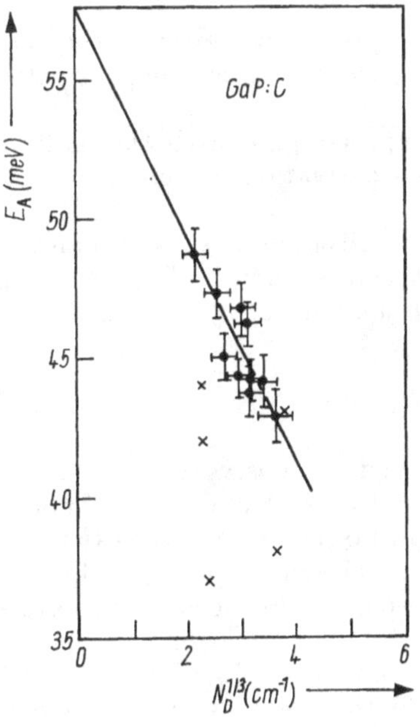

Fig. 2. Dependence of the activation energy of the shallow acceptor on the donor concentration in the undoed p-GaP samples (● this work, × results from C-doped samples [8])

$$B_1 = \exp(-E_{A1}/kT) \quad \text{and} \quad B_2 = \exp(-E_{A2}/kT)\,,$$

is used which can be deduced for the model of a semiconductor dominated by two acceptor levels [1]. N_{A1}, N_{A2} are the concentrations, E_{A1}, E_{A2} the activation energies, and g_1, g_2 the degeneracies of both acceptor levels. As in (1) excited acceptor levels are neglected.

In this case seven free parameters (N_{A1}, N_{A2}, N_D, E_{A1}, E_{A2}, N'_V/g_1, and N'_V/g_2) exist. Their number can be reduced by the reasonable assumption $g_1 = g_2$ and by use of N'_V/g as a fixed parameter for which we have taken the averaged value of the low-temperature fit (8×10^{14} cm^{-3} K$^{-3/2}$). In Table 2 the results of the fits using (3) for the same samples as in Table 1 are listed. In Fig. 1 the good agreement between the experimental points and the fitted $p(T)$ curves (solid lines) can be seen.

Table 2

Values of the concentrations and activation energies of the two acceptor species as well as of the donor concentration in undoped p-GaP obtained by fitting the whole $p(T)$ curves by (3) (index 1 corresponds to the shallow acceptor C)

sample	N_{A1} (10^{16} cm^{-3})	N_{A2} (10^{16} cm^{-3})	E_{A1} (meV)	E_{A2} (eV)	N_D (10^{16} cm^{-3})
21	3.6	1.2	48.0	0.20	1.3
31	4.5	1.3	47.3	0.18	1.8
32	6.9	1.7	46.1	0.17	2.4
33	5.9	1.8	45.2	0.18	1.9
34	8.5	3.1	44.1	0.17	3.4
11	8.7	4.0	46.8	0.16	2.9
12	12.7	4.5	44.6	0.19	2.3
14	17.2	6.4	42.7	0.17	3.6
35	17.5	7.1	43.1	0.16	3.8
36	18.1	6.1	44.7	0.18	3.0

Nearly for all samples the values N_{A1}, N_D, and E_{A1} agree very well with the corresponding values of N_A, N_D, and E_A (Table 1). This good agreement demonstrates the quality of the fits also with (3).

The results of these fits show that in all undoped p-type samples beside the shallow acceptor C a second one exists for which we obtained activation energies between 0.16 and 0.20 eV.

The concentration N_{A2} of the medium deep acceptor in all samples is lower than the concentration N_{A1} of carbon. Moreover, between the concentrations N_{A2} and N_{A1} exists a linear relationship $N_{A2} = cN_{A1}$ (with $c \approx 0.4$) in the concentration range investigated (Fig. 3).

3.2 Zn-doped samples

The hole concentration in the samples from the Zn-doped ingot is noticeably higher than in the undoped samples and varies only between 3×10^{17} cm^{-3} and 6×10^{17} cm^{-3}.

The $p(T)$ curves of the Zn-doped crystals (Fig. 4) show no reference to the existence of two acceptor levels. But, from photoluminescence measurements [6] it was known that beside the dopant Zn the residual acceptor C was present in the Zn-doped samples, too.

In this case of two acceptors with similar activation energies (E_{opt}: Zn — 70 meV [7], C — 54 meV [5]) the $p(T)$ curve is influenced by both acceptors simultaneously

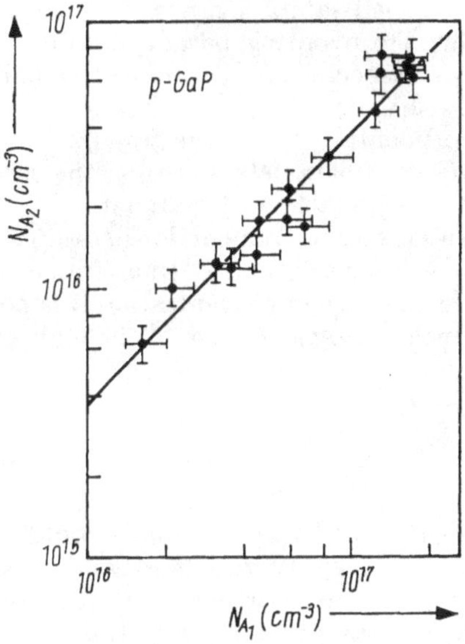

Fig.3

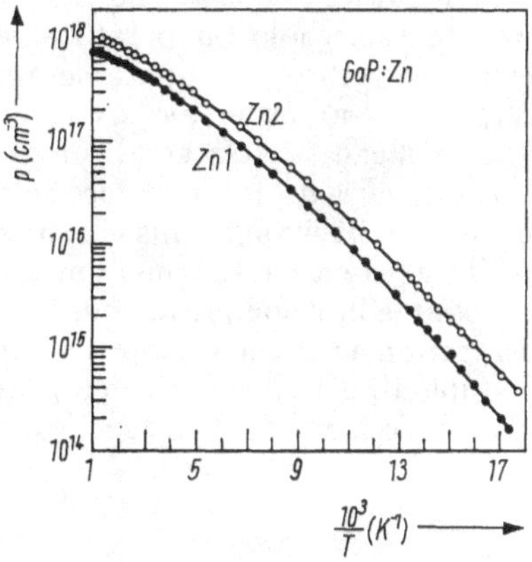

Fig.4

Fig. 3. Concentration N_{A2} of the medium deep acceptor in dependence on the concentration N_{A1} of the shallow acceptor C

Fig. 4. Hole concentration vs. reciprocal temperature for two Zn-doped GaP samples

in a wide temperature region. Therefore, a fit is only possible with (3) [1]. In order to determine parameters for both acceptors measurements of high accuracy in a wide temperature range have to be performed. Measurements down to 77 K were not sufficient to yield reliable values for the parameters. At least measurements down to about 60 K were necessary. The results obtained by fitting two of these $p(T)$ curves by (3) are listed in Table 3.

Table 3

Values of p (at room temperature) and impurity parameters for two Zn-doped GaP samples from fits with (3) (index 1 corresponds to C, index 2 to Zn)

sample	p (10^{17} cm^{-3})	N_{A1} (10^{17} cm^{-3})	N_{A2} (10^{17} cm^{-3})	E_{A1} (meV)	E_{A2} (meV)	N_D (10^{16} cm^{-3})
Zn1	3.4	0.5	7.3	43	59	2.1
Zn4	4.8	1.6	10.6	38	55	5.0

Due to the concentration dependence the activation energies of both acceptors are lower than the optical values given above. But the difference $E_{A2} - E_{A1}$ is nearly the same as that of the optical values which should agree with the thermal activation energies for vanishing impurity concentration [4].

The concentration of the acceptor C is in the range of the values for undoped crystals and the Zn concentration has the expected doping level agreeing with the values determined by chemical analysis.

It is noticeable that the $p(T)$ curves of the Zn-doped samples can be fitted very well in the whole temperature region by (1). But the resulting parameters do not correlate with the real parameters of the impurities. Especially, N_A is much too high, N'_V/g about an order too low, and E_A is an averaged value of E_{A1} and E_{A2}.

This means that a small deviation of the experimental $p(T)$ curve from the calculated one is no criterion for the correctness of the fitted parameters because the used model can be wrong. This can be a source of false interpretations of Hall data.

The existence of the medium deep acceptor species (present in the undoped samples) is possible in principle also in the Zn-doped samples beside Zn and C. But, if its concentration as in the undoped samples is lower than the carbon concentration it is not possible to detect this acceptor level in the Zn-doped samples because of the high Zn concentration.

4. Discussion

The chemical nature of the shallow acceptor in the undoped p-type samples could be found out unambiguously to be C (on P site). In the literature there is hardly information about electrical properties of carbon in GaP up to now. Bortfeld et al. [8] have determined impurity parameters of carbon-doped GaP samples analysing Hall measurements in the temperature region 77 to 300 K. Activation energies between 37 and 44 meV were obtained. The corresponding E_A–N_D points marked in Fig. 2, too, scatter rather strongly and are (except of one) situated clearly below our points, probably, because the measurements were perfomed only at $T \geqq 77$ K. References to the existence of a second acceptor level could not be found because measurements were not at performed temperatures above 300 K.

From the electrical properties of undoped p-type samples published in [9] and [10] it can be concluded that in these samples grown by the SSD method the dominating impurity (in the temperature region 77 to 300 K) is carbon, too. However, in these papers carbon was not identified as dominating acceptor.

In contrast to the shallow acceptor the identification of the medium deep acceptor is difficult.

The value of the activation energy $E_{A2} = (0.18 \pm 0.02)$ eV suggests at first that the medium deep acceptor arises from Si (on P sites) for which by optical measurements a value of 204 meV was obtained [11] (this Si acceptor was up to now not detected by electrical measurements). Moreover, Si was detected in our samples by chemical analysis in a concentration of about 1 to 3 ppm.

However, other facts speak against it: From the literature [12, 13] is known that at the growth temperatures used by us Si is incorporated in GaP nearly completely as donor (on Ga sites). In our undoped samples N_{A2} is either near to or higher than N_D which is the sum of all donors present in a sample. On the other hand, proportionality exists between N_{A2} and N_{A1}. Therefore, one can assume that the concentrations of both acceptors in the melt or their incorporation in the GaP crystal are correlated. Possibly, the medium deep acceptor is a complex centre including C. (For example P vacancies could participate in such a complex because they are expected in SSD–GaP in a high concentration due to the growth conditions.)

It is interesting that with DLTS measurements on GaP-LEDs [15, 16] nearly always an acceptor level (A6) at about 190 meV was found which is mostly the dominating deep level (concentrations up to 10^{15} cm^{-3}). Its correlation with the donor level at 180 meV (D8) is assumed in [15] to be an indication to an amphoteric element such as Ge or Si. But, also the interpretation as a native defect is possible [15].

In (1), (2), and (3) it was not taken in consideration that the Hall factor r_H [17 to 19], the activation energy [4, 20] and the density of states effective mass [21] depend on temperature. This can cause further errors in the fitted parameters but the consideration of these effects is very difficult because of lack of corresponding data for p-GaP. Moreover, it can be estimated that these errors are not so high that they influence the essential results of this paper.

5. Summary

It could be shown that it is possible to determine parameters for two majority impurities present simultaneously in a semiconductor by Hall data analysis not only in the case of a clear difference between the activation energies of both impurity species but also in the case of similar activation energies. A presupposition is that measurements of high accuracy are performed in a wide temperature region.

In undoped p-type GaP samples grown by the SSD method beside the shallow acceptor C a middle deep acceptor with an activation energy of (0.18 ± 0.02) eV could be detected, the nature of which is not clear. C was also detected in a small concentration in the Zn-doped samples by the Hall data analysis.

Acknowledgment

The authors wish to thank the group of Prof. K. Hein (Sektion Metallurgie und Werkstofftechnik der Bergakademie Freiberg) for providing the crystals.

References

[1] E. Ziegler and W. Siegel, Crystal Res. and Technol. **17**, 1015 (1982).
[2] K. Kaneko, M. Ayabe, M. Dosen, K. Morizane, S. Usui, and N. Watanabe, Proc. IEEE **61**, 884 (1973).
[3] W. Siegel, E. Ziegler, and G. Kühnel, Exper. Tech. Phys. **27**, 79 (1979).
[4] J. Monecke, W. Siegel, E. Ziegler, and G. Kühnel, phys. stat. sol. (b) **103**, 269 (1981).
[5] R. Bindemann, R. Schwabe, and T. Hänsel, phys. stat. sol. (b) **87**, 169 (1978).
[6] P. Kirsten, private communication.
[7] V. Berndt, A. A. Kopylov, and A. N. Pikhtin, Fiz. Tekh. Poluprov. **11**, 1977 (1982).
[8] D. P. Bortfeld, B. J. Curtis, and H. Meier, J. appl. Phys. **43**, 1293 (1972).
[9] J. P. Besselere and J. M. Leduc, Mater. Res. Bull. **3**. 797 (1968).
[10] G. Poiblaud and G. Jacob, Mater. Res. Bull. **8**, 845 (1973).
[11] P. J. Dean, C. J. Frosch, and C. H. Henry, J. appl. Phys. **39**, 5631 (1968).
[12] F. A. Trumbore, H. G. Whithe, M. Kowalchik, C. L. Luke, and D. L. Nash, J. Electrochem. Soc. **112**, 1208 (1965).
[13] A. M. Zykov and B. E. Samorukov, Izv. Akad. Nauk SSSR, Ser. neorg. Mater. **11**, 819 (1975).
[14] A. T. Vink, A. J. Bosman, J. A. W. van der Does de Bye, and R. C. Peters, J. Lum. **5**, 57 (1972).
[15] O. Breitenstein, Dissertation A, Leipzig 1980.
[16] W. Schairer, J. Electron. Mater. **8**, 139 (1979).
[17] A. Hackmann, D. Neubert, U. Scherz, and R. Schliff, Phys. Rev. B **24**, 4666 (1981).
[18] J. F. Lin, S. S. Li, L. C. Linares, and K. W. Teng, Solid State Electronics **24**, 827 (1981).
[19] G. Gornik, D. Kranzer, and K. H. Zschauer, private communication.
[20] G. F. Neumark and D. K. Schröder, J. appl. Phys. **52**, 855 (1981).
[21] F. L. Madarasz, J. E. Lang, and P. M. Hemeger, J. appl. Phys. **52**, 4646 (1981).

(Received January 3, 1983)

phys. stat. sol. (a) 77, 175 (1983)

Subject classification: 20.3; 22

Sektion Physik der Humboldt-Universität zu Berlin, Bereich Experimentelle Halbleiterphysik[1] *(a) and Department of Physics, Moscow State University (b)*

Direct Comparison of Photo- and Electroluminescence in $Pb_{1-x}Sn_xTe$ Diode Lasers

By

J. W. TOMM (a), KARIN HERRMANN (a), A. JALYSCHKO (a), A. E. YUNOVICH (b), and V. TEN[2] (b)

A comparison of photoluminescence (PL) and electroluminescence (EL) experiments on one and the same $Pb_{1-x}Sn_xTe$ sample with two principal different measuring methods (determination of the threshold and the peak energy of the stimulated emission) leads to the conclusion that the gain, to have comparable light outputs, is for PL more than one order of magnitude larger than for EL. From the analogous temperature dependences of the PL- and EL-light outputs information is obtained about the ratio of radiative and non-radiative lifetimes of the excited carriers at low temperatures.

Aus dem Vergleich von Photolumineszenz (PL)- und Elektrolumineszenz (EL)-Experimenten an ein und denselben $Pb_{1-x}Sn_xTe$-Laserdioden durch zwei prinzipiell unterschiedliche Meßverfahren (Bestimmung der Schwellanregung, Messung der Lage des Maximums der stimulierten Emission) wird der Schluß gezogen, daß die zur Erreichung vergleichbarer Intensitäten notwendigen Gewinne bei PL mindestens eine Größenordnung höher liegen als bei EL. Der analoge Verlauf der Temperaturabhängigkeiten der Intensitäten der stimulierten Strahlung führt zu Schlüssen bezüglich der Lebensdauern hinsichtlich strahlender und nichtstrahlender Rekombinationsmechanismen bei tiefen Temperaturen.

1. Introduction

Electroluminescence (EL) and photoluminescence (PL) are two well-known methods for exciting stimulated emission in semiconductors. Therefore, it is of principal interest to compare the results of both excitation mechanisms. An attempt to compare the spectra of EL and PL for the same bismuth-doped $Pb_{1-x}Sn_xTe$ ($0.1 \leqq x \leqq 0.22$) in the temperature range $T = 20$ to 200 K was reported earlier [1].

The results presented here were obtained on lead-tin telluride structures of the type n-$Pb_{1-x}Sn_xTe$:Bi/p-$Pb_{1-x}Sn_xTe$/p-PbTe ($0 \leqq x \leqq 0.17$) at 10 to 100 K. As a consequence of the growth temperature for some structures, a junction migration into the PbTe substrate similar to that reported by Eger et al. [2] was observed. PL measurements were carried out for the same diodes. The lasing diodes had cleaved Fabry-Perot resonator geometry with standard dimensions $(500 \times 300 \times 300)\ \mu m^3$. EL measurements were performed in pulse regime in a temperature-stabilized helium cryostat ($\Delta T = 0.01$ K). The PL experiments were carried out on the resonator end faces near the optically active region of the injection laser by a YAG:Nd^{3+} laser ($\hbar\omega = 1.17$ eV) with power densities of about $P = 10^4$ to 10^6 W cm^{-2}.

From both PL and EL experiments the temperature dependence of the peak energy ($\hbar\omega_{max}$), the threshold current density (j_s) or the threshold excitation intensity (I_s), and absolute values of the stimulated emission intensity were measured.

[1]) Hessische Str. 2, DDR-1040 Berlin, GDR.

[2]) On leave of absence from Kalmykzkii University, Elista, USSR.

2. Experiment

The experimental arrangements for EL and PL measurements are shown in Fig. 1. In the place of the Fabry-Perot resonator for EL, in PL experiments works a resonator which is formed by one cleaved resonator face and an effective mirror which is caused by the gradient of the refractive index at the boundary face excited–non-excited medium. The change of the refractive index from the surface to the boundary face was determined to $\Delta\tilde{n} = 0.5$, taking into account that over a distance of some micrometers into the volume more than 10^{17} cm^{-3} carriers are generated by the incident YAG laser beam. The existence of a graduated "resonator" can possibly explain why in PL experiments stable spectral structures, such as modes, were never obtained. It is necessary to emphasize that in both types of experiments the optically active regions cannot principally be identical, but overlap each other.

3. Results and Discussion

3.1 Spectral range of the stimulated peak energy

From Fig. 2 and 3 follows that the peak energy of the stimulated radiation at constant excitation follows in a wide temperature range ($30 \leqq T \leqq 60$ K) the condition

$$\hbar\omega_{\max}(T) = E_g(x, T) + \theta kT \tag{1}$$

($E_g(x, T)$ after Preier [3]). The parameter θ only depends on the excitation mechanism and not on the diode structure. This behaviour is demonstrated for two typical samples, e.g., for PbTe homolaser (Fig. 2) and $Pb_{1-x}Sn_xTe$ heterolaser (Fig. 3). Here for EL the parameter $\theta = 0.5$ is in good agreement with the two-band model [4, 5] for typical values of the gain of about $g \geqq 10^2$ cm^{-1}. For PL θ varies from 1 to 1.3. To explain the greater values of θ for PL in the frame of the same model, values for the gain coefficient of about $g \geqq 10^3$ cm^{-1} are demanded.

Supposing a linear dependence

$$g \sim \frac{1}{L} \ln (R_1R_2)^{-1/2} + \alpha_i \tag{2}$$

for the optically active region in photostimulated surfaces of diode lasers as described above, the demanded large values for g are realistic because of the small length L (some μm) and very small R_2 at the boundary face excited–non-excited medium. The internal absorption losses α_i (maximum of $\alpha_i \approx 10^2$ cm^{-1} according to Antcliffe and Parker [6]) can be neglected due to the high values of $(1/L)$ ln $(R_1R_2)^{-1/2}$.

In the low-temperature range $T < 25$ K, θ strongly increases for both excitation

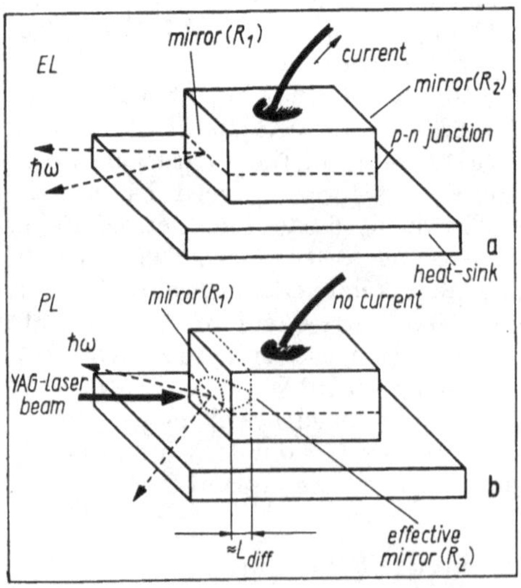

Fig. 1. Experimental arrangement for a) EL and b) PL excitation of laser diodes

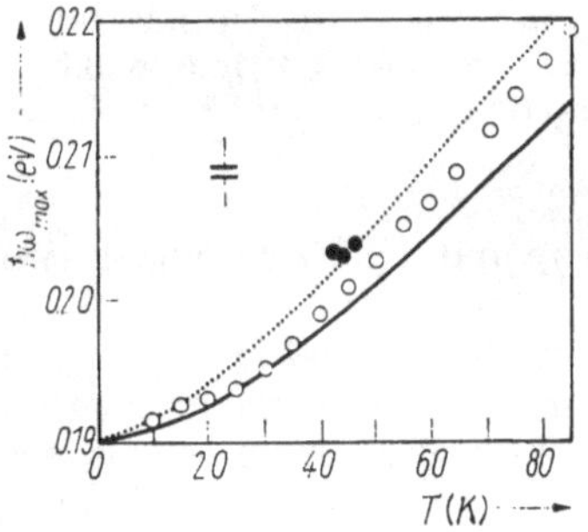

Fig. 2

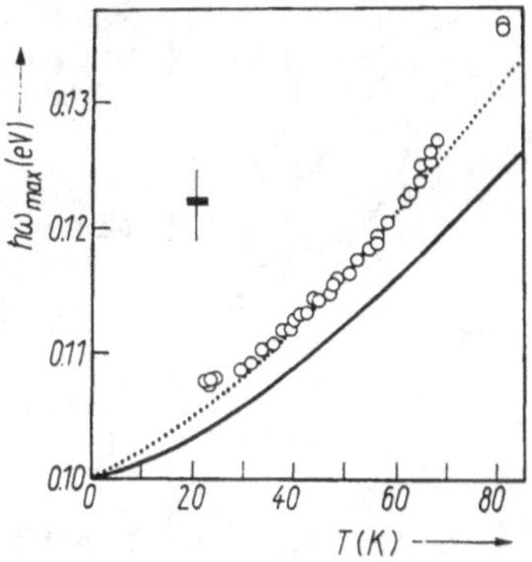

Fig. 3

Fig. 2. Comparison of the temperature dependences of the peak energies ($\hbar\omega_{max}(T)$) for PL and EL with $E_g(T)$ on PbTe homostructures. Excitation energies PL (●): $I = 3 \times 10^5$ W cm^{-2}, EL (○): $j = 5$ kA cm^{-2}; ——— $E_g(x, T)$, ... $E_g(x, T) + kT$

Fig. 3. Comparison of the temperature dependence of the peak energy ($\hbar\omega_{max}(T)$) for PL with $E_g(T)$ on $Pb_{1-x}Sn_xTe$ ($x = 0.165$) heterostructures. Excitation energy $I = 3 \times 10^5$ W cm^{-2}; ——— $E_g(x, T)$, ... $E_g(x, T) + kT$

mechanisms as can be seen from Fig. 2 and 3. We quote that this may be caused by temperature and possibly also a lifetime-dependent increase of the minority carrier concentration at constant injection current or pump intensity. In both cases degenerated minority carrier concentrations may be reached and stimulated emission with $\hbar\omega_{max} > E_g$ is possible. A similar effect of increasing θ-values was found for $T \geqq 60$ K. Earlier discussions that in PL experiments hot carriers cause larger values of θ do not hold, because this effect was found in EL experiments, too. The existence of hot carriers in the active region near the lasing threshold in electrically pumped diode lasers must be excluded. Summarizing we can state that in PL experiments the gain coefficient should be about one order of magnitude larger than in EL experiments if we wish to obtain in both cases comparable luminescence intensities.

3.2 Comparison of the threshold excitations j_s, I_s

The temperature dependences $j_s(T)$ and $I_s(T)$ measured on the same diode laser as shown in Fig. 2 are represented in Fig. 4. To obtain exact values for j_s and I_s at a given temperature, the lasing intensity was determined as a function of the injection current for EL or excitation intensity for PL, respectively. The linear dependence of the lasing intensity on excitation was extrapolated to zero excitation and thus allowed to obtain

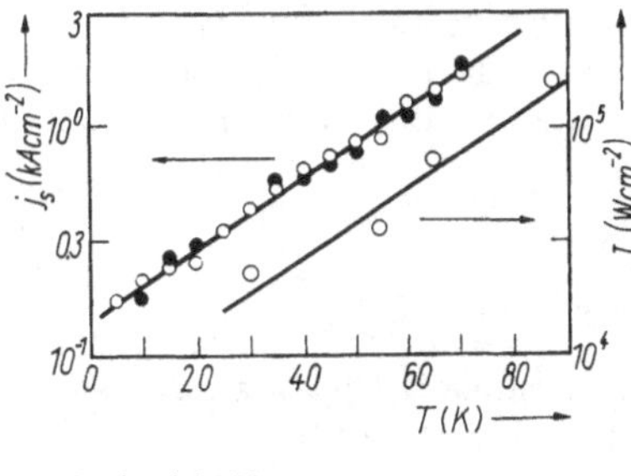

Fig. 4. Comparison of the temperature dependences of the threshold excitation (I_s, j_s) for PL and EL. ○ ● EL for different diodes, ◯ PL on the mirror of the diode

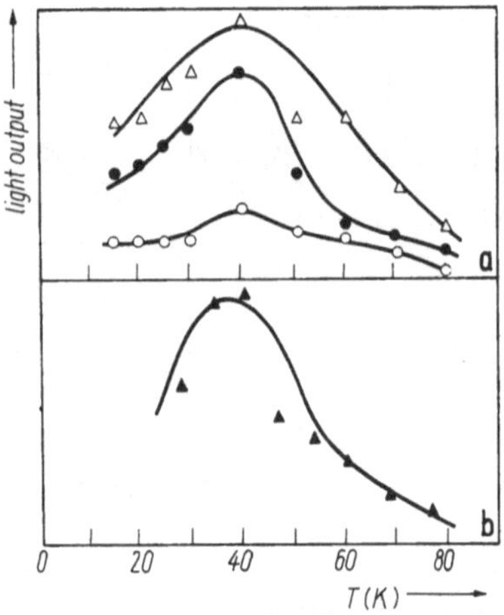

Fig. 5. Comparison of the temperature dependences of the output P for a) EL excitation; △ 11.0 ● 8.3, ○ 5.5 kA cm^{-2}; b) PL excitation; ▲ 3.0×10^5 W cm^{-2}

the exact values for j_s and I_s. From Fig. 4 two informations are extracted:

(i) The parallelity of the curves $j_s(T)$ and $I_s(T)$ allows the conclusion that the recombination mechanisms for both excitation mechanisms are similar.

(ii) From the absolute values of j_s and I_s we can determine the number of excited carriers $\delta n/\delta t$ per time and volume unit, which are necessary to fulfil the threshold condition: For $T = 50$ K, $I_s = 4 \times 10^4$ W/cm^2, and $j_s = 0.88$ kA/cm^2 we find for PL

$$\frac{\delta n}{\delta t} = \frac{2I_s\left(1 - \left(\frac{\bar{n} - 1}{\bar{n} + 1}\right)^2\right)}{\hbar\omega d} = 2 \times 10^{26}\ \text{cm}^{-3}\ \text{s}^{-1} \tag{3}$$

and for EL

$$\frac{\delta n}{\delta t} = \frac{j_s}{ed} = 5.5 \times 10^{24}\ \text{cm}^{-3}\ \text{s}^{-1}\,. \tag{4}$$

The number of photoexcited carriers necessary to fulfil the threshold condition is about one to two orders of magnitude larger than for current-induced carriers. This is in good agreement with our statement from Section 3.1 that at the threshold the gain coefficients for PL and EL differ by more than one order of magnitude.

3.3 Intensity of the stimulated emission

The temperature dependence of the intensity of the stimulated laser emission was determined for constant excitation in the temperature range mentioned above. The result is shown in Fig. 5. Analogous results are obtained by other authors (e.g., Lo [7]). The lasing intensity is proportional to the external quantum efficiency, to the quantum energy of radiation, and to the differences $I - I_s$ and $j - j_s$, respectively,

$$\text{PL:}\quad P \sim \eta_{\text{ext}}(I - I_s)\,\hbar\omega\,, \tag{5}$$

$$\text{EL:}\quad P \sim \eta_{\text{ext}}(j - j_s)\,\hbar\omega\,. \tag{6}$$

For $T > 40$ K the decrease of lasing intensity for PL and EL is easily understood, because excitation differences decrease and the non-radiative recombination mechanisms become more effective [5]. In the low-temperature range $T < 40$ K the decrease of the lasing intensity for both excitation arrangements can only be understood, if

$$\eta_{\text{ext}} = \eta_{\text{int}}\left(1 + \frac{\alpha_i L}{\ln (R_1 R_2)^{-1/2}}\right)^{-1} \tag{7}$$

decreases with decreasing T.

The contribution of $\hbar\omega(T)$ to P can be neglected, because the peak energy shows only a rather weak temperature dependence as shown in Fig. 2 and 3.

4. Conclusions

In Sections 3.1 and 3.2 it was shown that for comparable luminescence intensities the optical gain for PL must be about one order of magnitude larger than for EL. This means that for PL the following equation must be fulfilled:

$$\alpha_i \ll (1/L) \ln (R_1 R_2)^{-1/2} . \tag{8}$$

For EL both terms are of the same order.

It follows from Section 3.3 that for PL and EL the external quantum efficiencies for $T < 40$ K decrease. Due to the very different relations of α_i to $(1/L) \ln (R_1R_2)^{-1/2}$ for current- and photostimulated emission in the same diode the observed decreases of emission intensity for both excitation methods at low temperatures can be explained only, if one assumes a decrease of the internal quantum efficiencies η_i. It is known [5] that for low temperatures the Auger recombination is a non-effective process. Therefore, one possible explanation of the above-described effect consists in an increase of the radiative lifetime or a decrease of the non-radiative lifetime of the minority carriers below 40 K.

References

[1] T. D. Aitikeeva, A. I. Lebedev, A. E. Yunovich, Karin Herrmann, A. W. Jalyschko, and P. Schäfer, phys. stat. sol. (a) **67**, 171 (1981).
[2] D. Eger, A. Zemel, S. Rotter, N. Tamari, M. Oron, and A. Zussman, J. appl. Phys. **52**, 490 (1981).
[3] H. Preier, J. appl. Phys. **50**, 189 (1979).
[4] W. W. Anderson, IEEE J. Quantum Electronics **13**, 8 (1977).
[5] O. Ziep, D. Genzow, M. Mocker, and K. H. Herrmann, phys. stat. sol. (b) **99**, 129 (1980).
[6] G. A. Antcliffe and S. G. Parker, J. appl. Phys. **44**, 4145 (1973).
[7] W. Lo, IEEE J. Quantum Electronics **13**, 8 (1977).

(Received January 31, 1983)

phys. stat. sol. (a) 77, 181 (1983)

Subject classification: 13.4; 22

VEB Zentrum für Forschung und Technologie Mikroelektronik, Dresden[1])

Zum Einfluß einer Temperaturabhängigkeit des Frequenzfaktors auf die Form und Breite von Glow-Peaks

Von

M. Balarin

Der Verlauf eines Glow-Peaks hängt nur von der relativen Temperatur $\vartheta \equiv T/\bar{T}$, bezogen auf die Maximumstemperatur $\bar{T}$, von der dimensionslosen Größe $\bar{y} \equiv k\bar{T}/E$ und von einem Exponenten r ab, der eine zusätzliche Temperaturabhängigkeit des Frequenzfaktors repräsentiert. Damit wird erstmalig für einen variablen Intensitätswert $0 < P < 1$ eine Polynomlösung abgeleitet für die zwei ϑ_i-Temperaturen (Tabellen 1 und 2), bei denen dieses Intensitätsniveau erreicht wird. Die von P abhängigen Polynomkoeffizienten sind in Tabelle 1 und die $\vartheta_i(P_i, r)$ in Tabelle 2 tabelliert. Während bekannt ist, daß aus der Asymmetrie $A \equiv (\vartheta_2 - 1)/(1 - \vartheta_1)$ die Kinetikordnung eines Prozesses gut erkannt werden kann, zeigt sich, daß diese Asymmetrie durch r und auch durch E nur unwesentlich und selbst durch P nur wenig beeinflußt wird. Damit ermöglicht die P-Abhängigkeit der Werte für die Asymmetrie (Tabelle 4) eine unmittelbare Kontrolle darüber, daß für einen Peak eine einheitliche Kinetikordnung vorliegt und ob ein Peak frei ist von Überlagerungen, bzw. bis zu welcher Höhe. Für jedes P kann dann aus der experimentell ermittelten Breite δ die effektive Aktivierungsenergie $E_{eff} = E + rk\bar{T} = \varkappa(P; y, r)\, k\bar{T}^2/\delta$ bestimmt werden; die Koeffizienten $\varkappa$ bzw. $T_\varkappa(P, \bar{y})$ für $r = 0$ sind für die Gesamtbreite sowie für die Nieder- und die Hochtemperaturseite in Tabelle 3 ausgewiesen. Für $P = 0{,}5$ ergeben sie Faktoren für die bekannten Halbwertsbreiten-Methoden. Ein nützlicher Umstand besteht darin, daß über den Parameter r auch der Fall anderer nicht-zeitlinearer Aufheizungsregime bewertet und dabei erhaltene Aktivierungsenergien umgerechnet werden können.

The shape of a glow-peak is caused only by the relative temperature $\vartheta \equiv T/\bar{T}$, related to peak temperature $\bar{T}$, by a dimensionless value $\bar{y} \equiv k\bar{T}/E$ and by an exponent r, representing an additional temperature dependence of the frequency factor. For the first time a polynomial solution is derived for those two temperatures ϑ_i, where intensity $0 < P < 1$ is reached, taken P as a variable. The series coefficients, depending on P, are tabulated in Table 1 and the $\vartheta_i(P_i, r)$ in Table 2. As is well known, from the asymmetry $A \equiv (\vartheta_2 - 1)/(1 - \vartheta_1)$ the kinetic order of a process can be estimated, but this same asymmetry is insensitive to r and even to E, there is only a slight influence by P. This P-dependence enables an immediate control over the uniformity of kinetic order and wether a peak is free from overlapping, or to what intensity. For every P from experimentally obtained width δ an effective activation energy $E_{eff} = E + rk\bar{T} = \varkappa(P; y, r)\, k\bar{T}^2/\delta$ can be determined; the coefficients $\varkappa$ and $T_\varkappa(P, \bar{y})$ for $r = 0$ are given in Table 3, both for the whole width and for the low and the high temperature shoulders. For $P = 0.5$ these are the correct factors for known halfwidth methods. Appropriately, by means of the parameter r cases of other non-time-linear heating runs can be evaluated and activation energies can be transfered.

1. Einleitung

Häufig wird davon ausgegangen, daß die wesentliche Temperaturabhängigkeit für Reaktionsgeschwindigkeiten durch einen exponentiellen Arrhenius-Term $\sim \exp(-E/kT)$ ausgedrückt wird, womit zugleich als Modell- und Bestimmungsparameter die Aktivierungsenergie E eingeführt ist. Aber bereits in der allgemeinen Theorie der absoluten

[1]) Karl-Marx-Str., DDR-8080 Dresden, GDR.

Reaktionsraten [1] wurde damit gerechnet, daß außerdem der vorexponentielle Frequenzfaktor etwas temperaturabhängig sein kann. Mit einer näheren Aufgliederung eines solchen vorexponentiellen Faktors [2, 3], besonders bei der Behandlung von elektronischen Halbleiterproblemen, als ein Produkt aus Ladungsdichte $\times$ Beweglichkeit $\times$ Reaktionsquerschnitt wurden in diesem Faktor zum Teil explizite Temperaturabhängigkeiten erwartet. Diese wurden manchmal als eine Potenzabhängigkeit $\sim T^r$ mit ganz- oder halbzahligem Exponenten r dargestellt; aber auch diese Exponenten sind nicht direkt ableitbare Abhängigkeiten, sondern eher Anpassungen, brauchbar innerhalb beschränkter Temperaturbereiche. Einerseits ist offenkundig, daß eine Variation des Exponenten r von geringerem Einfluß ist, als eine Variation der Aktivierungsenergie E. Mit der steigenden Meßgenauigkeit besteht jedoch der Wunsch, *alle* Reaktionsparameter quantitativer auszuwerten und dazu muß das mögliche Ausmaß des Einflusses eines temperaturabhängigen Frequenzfaktors kritisch bewertet werden. Ein erster Schritt in dieser Richtung [4] zeigte, daß speziell bei der Bestimmung der Aktivierungsenergie mittels einer früher publizierten integralen Methode [5] nach zeitlinearem Aufheizen eine Abweichung $\mathrm{d}E = kT\,\mathrm{d}r$ entstehen würde, falls ein Fehler $\mathrm{d}r$ zugelassen wird.

In der vorliegenden Arbeit soll nun der Einfluß von r auf die Form und Breite von differentiellen Reaktionspeaks oder Glowkurven behandelt werden. Während ein weiterer Reaktionsparameter, die sogenannte Kinetikordnung l, die Symmetrie bzw. Asymmetrie der Glowpeaks bestimmt [6, 7] und vice versa zur Ermittlung von l benutzt werden kann, wird sich hier ergeben, daß r die Symmetrie eines Peaks äußerst wenig verändert und daß — natürlich abhängig von Meßgenauigkeit und Idealität einer Meßkurve — kaum eine reale Chance besteht, phänomenologisch eindeutig festzustellen, ob eine solche Temperaturabhängigkeit vorhanden ist oder gar, welcher Art sie ist. Aus der Breite des Peaks kann eine effektive Aktivierungsenergie ermittelt werden, d. h. genau genommen nur eine Kombination von E und r, nicht aber getrennte Einzelwerte.

In Erweiterung der bekannten Halbwertsmethoden zur Bestimmung der Aktivierungsenergie [8 bis 10] oder der Methode der Wendepunkte [11 bis 13] wird eine allgemeine Lösung für beliebige relative Peakintensitäten P gesucht [14, 15]. Absicht-

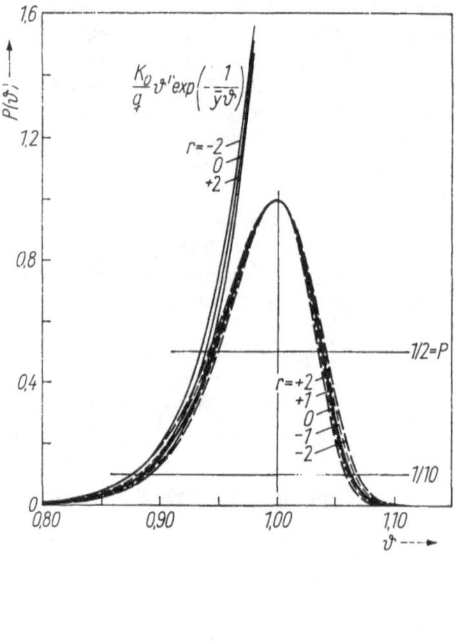

Fig. 1. Temperaturabhängige Glowkurven $P(\vartheta; \hat{y}, r)$ unter dem Einfluß des Scharparameters r für $\hat{y} = 1/25$ nach Gleichung (7). An den Anfangsanstieg sind die zugehörigen Reaktionsraten für $r = -2, 0, +2$ angetragen (dünne Linien); die Differenzen zu den Glowkurven sind bedingt durch den Verbrauch der Reaktionspartner

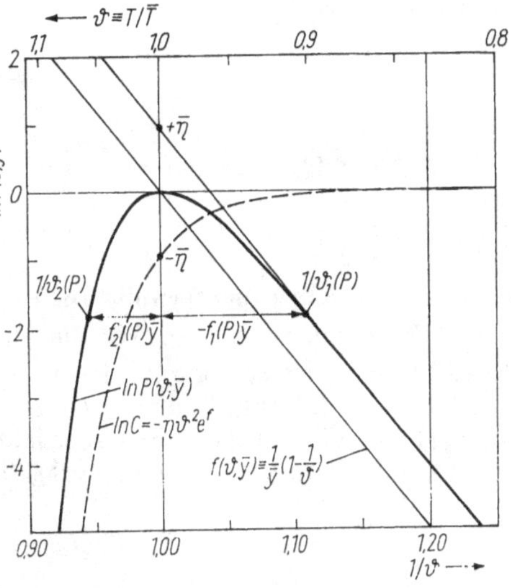

Fig. 2. Die Glowkurve (dicke Linien) in der Form ln P aufgetragen als Funktion der reziproken Temperatur Gleichung (8) hier für $r = 0$ ergibt sich als Summe der Geraden $f(\bar{y}, 1/\vartheta)$, dem Betrag $\bar{\eta}_r \approx \bar{\eta}_0 + 3r\bar{y}^2$ und der mit dem Temperaturanstieg abfallenden Funktion $\eta\vartheta^2 \exp[(1/\bar{y})(1 - 1/\vartheta)]$ für $\bar{y} = 1/25$. Zu jeder Intensität P gehören zwei Lösungen $1/\vartheta_1(P)$ und $1/\vartheta_2(P)$ und damit die jeweiligen Peakbreiten $1/\vartheta_1 - 1/\vartheta_2$, bzw. $1/\vartheta_1 - 1$ und $1 - 1/\vartheta_2$; für den Anfangsanstieg ($P \ll 1$) gilt:

$$\frac{\Delta \ln P}{\Delta(1/\vartheta)} \approx \frac{1 + \ln(1/P)}{1/\vartheta_1 - 1} = \frac{1}{\bar{y}}$$

lich wird dazu neben der Glowkurve $P(T)$ in Fig. 1 auch die transformierte Darstellung als Logarithmus der Intensität in Abhängigkeit von der reziproken Temperatur (Fig. 2) zur Ableitung herangezogen.

2. Normierte Glow-Kurvenschar

Der Reaktionsablauf soll durch folgende kinetische Gleichung beschrieben werden:

$$-\frac{\mathrm{d}C}{\mathrm{d}t} = K(T)\, C^l; \qquad C(t = 0) = C_0 \tag{1}$$

mit $K(T) = K_0(T/T')^{r'}\, \mathrm{e}^{-E/kT}$, wobei C die Konzentration ist, K und K_0 die Reaktionsrate bzw. ihr vorexponentieller Frequenzfaktor, E die Aktivierungsenergie, k die Boltzmann-Konstante, l die Kinetikordnung der Reaktion, r' der Exponent für die Temperaturabhängigkeit des Frequenzfaktors. (T', ein beliebig gewählter Temperaturwert, nur zur Kompensation der Dimension, wird mit Gleichung (3) wieder eliminiert!).

Bei einer konstanten Aufheizgeschwindigkeit[2])

$$q = \frac{\mathrm{d}T}{\mathrm{d}t} = \mathrm{const}$$

ergibt sich aus (1) als *temperaturbezogene Intensität*

$$J(T; E, K_0, l, r) \equiv -\frac{\mathrm{d}C}{\mathrm{d}T} = \frac{K_0 C^l}{q}\, \frac{T^r \bar{T}^{r''}}{T'^{r'}}\, \mathrm{e}^{-E/kT}, \tag{2}$$

wobei r'' der Exponent für eine Temperaturabhängigkeit der Aufheizrate ist.

[2]) Es läßt sich leicht erkennen, daß die gleiche Lösung auch allgemeiner für eine nichtkonstante, stetige Aufheizgeschwindigkeit erhalten wird, die in der Umgebung des Glow-Maximums durch $q(T) = \mathrm{d}T/\mathrm{d}t = \bar{q}(T/\bar{T})^{r''}$ angepaßt werden kann. Der gemeinsame Exponent wird damit: $r = r' - r''$. Bereits hieraus folgt, daß die in der Literatur [16 bis 20] diskutierte hyperbolische Aufheizung (d. h. $r'' = 2$) theoretische Vorteile für die Integration von (2) verspricht.

Aus der *Maximumsbedingung* $\mathrm{d}J/\mathrm{d}T|_{\bar{T}} = 0$ erhält man

$$\frac{K_0 l \bar{C}^{l-1}}{q T'^{r'}} = (1 + r\bar{y}) \frac{E}{k\bar{T}^{2+r'}} \mathrm{e}^{E/k\bar{T}}, \tag{3}$$

wobei $y \equiv kT/E = \bar{y}\vartheta$.

Durch Einsetzen dieser Kombination $K_0/(qT'^{r'})$ nimmt (2) die Form an:

$$J = -\frac{\mathrm{d}C}{\mathrm{d}T} = \frac{c^l}{l\bar{c}^{l-1}} (1 + r\bar{y}) \frac{E}{k\bar{T}^2} \left(\frac{T}{\bar{T}}\right)^r \exp\left\{\frac{E}{k\bar{T}}\left(1 - \frac{\bar{T}}{T}\right)\right\}. \tag{4}$$

Hier zeigt sich deutlich, daß die Ausgangsgleichung und konsequenterweise auch ihre Lösung, nur von der relativen Temperatur $\vartheta \equiv T/\bar{T}$ abhängen. Ebenso tritt die Aktivierungsenergie nur in Relation zu $k\bar{T}$ auf, bzw. $E/k \equiv \bar{T}_0$ ist wie eine für den Prozeß charakteristische Temperatur anzusehen (siehe [5]). Im folgenden substituieren wir daher die letzte Klammer in (4) durch $f(\bar{y}, \vartheta) \equiv (1 - 1/\vartheta)/\bar{y}$. Außerdem beschränken wir uns für die detaillierte Behandlung auf monomolekulare Reaktionen ($l = 1$); der allgemeine Fall $l \neq 1$ läßt sich analog lösen.

Die *Lösung* von (4) wird damit[3])

$$C(\vartheta; \bar{y}, r) = C_0 \exp\left\{-(1 + r\bar{y})\, \vartheta^r \eta_r\, \mathrm{e}^{f(\bar{y}, \vartheta)}\right\} \tag{5}$$

und die *Intensität*

$$J(\vartheta; \bar{y}, r) = \frac{C_0 k}{E} (1 + r\bar{y}) \frac{\vartheta^r}{\bar{y}^2} \exp\left\{f - (1 + r\bar{y})\, \vartheta^r \eta_r\, \mathrm{e}^f\right\}. \tag{6}$$

Durch Normierung auf das eigene Maximum ergibt sich eine *relative Peak-Kurve*

$$P(\vartheta; \bar{y}, r) \equiv \frac{J(\vartheta; \bar{y}, r)}{J(1; \bar{y}, r)} = \vartheta^r \exp\left\{f + (1 + r\bar{y})\, (\bar{\eta}_r - \eta_r \vartheta^r\, \mathrm{e}^f)\right\} \tag{7}$$

$\left(P(\vartheta) \equiv J(T)/J(\bar{T})\right.$ relativer Peakverlauf$\left.\right)$.

Hier sei darauf aufmerksam gemacht, daß jeder beliebige Niveauwert $0 < P_j < 1$ je einmal auf der Niedertemperatur- und auf der Hochtemperaturseite durchlaufen wird. Das heißt, *zu jedem P_j gehören zwei Temperaturen* (vergl. Fig. 1): $T_{j1}(P_j) < \bar{T}$ und $T_{j2}(P_j) > \bar{T}$, bzw. $\vartheta_{j1}(P_j) < 1$, $\vartheta_{j2} > 1$. Im besonderen Fall kann wie bisher in der Literatur am häufigsten betrachtet wurde $P_{j0} = \frac{1}{2}$ sein oder aber $P_{j0} = \frac{1}{10}$ [15] usw.

Offenbar wegen des transzendenten Charakters von Gleichung (6) wurde bisher keine explizite Abhängigkeit $\vartheta_{ji} = \vartheta_{ji}(P_j, r)$ angegeben. Es war dagegen problemlos, P_j als Funktion der relativen Temperatur zu betrachten, allerdings mit dem Nachteil, daß in der Regel nicht auf beiden Temperaturseiten ein exakt gleiches Niveau der Intensität erfaßt wurde und damit keine Möglichkeit bestand, auf jedem Intensitätsniveau die Breite und die Asymmetrie anzugeben.

In dieser Mitteilung wird erstmalig diese explizite Abhängigkeit $\vartheta_{ji} = \vartheta_{ji}(P_{ji}, r)$ mit $i = 1$ und 2 für alle Intensitätsniveaus P_j und für den zusätzlichen Einfluß durch r abgeleitet, um danach für verschiedene P_j die zugehörige *Peakbreite* $(T_{j2} - T_{j1})$ bzw. $(\vartheta_{j2} - \vartheta_{j1})$ und die *Form des Peaks*, ausgedrückt als Verhältnis der Hochtemperatur- zu der Niedertemperaturbreite

$$A(P_j) \equiv \frac{T_{j2} - \bar{T}}{\bar{T} - T_{j1}} = \frac{\vartheta_{j2} - 1}{1 - \vartheta_{j1}},$$

---

3) Aus dem Exponentialintegral ergibt sich [4]

$$\eta_r(y) \equiv \mathrm{e}^{1/y} y^{-2-r} \int_0^y \mathrm{e}^{-1/y'} y'^r \mathrm{d}y' = 1 + \sum_{n=1}^{n'} (-y)^n \frac{\Gamma(n + r + 2)}{\Gamma(n + 2)}; \qquad n' \approx \frac{1}{y} - 2.$$

zu bilden. Für $\ln P(1/\vartheta)$ in Fig. 2 bildet man einen Asymmetrieausdruck unter Berücksichtigung der reziproken Temperaturen:

$$A^*(P_j) = -\frac{f_2(P_j)}{f_1(P_j)} = \frac{1/\vartheta_{j2} - 1}{1 - 1/\vartheta_{j1}} = A\,\frac{\vartheta_{j1}}{\vartheta_{j2}} < A\,.$$

3. Qualitative Abhängigkeit für Peakbreite und Asymmetrie

Zunächst soll eine näherungsweise Beurteilung ausgeführt werden, die *einige charakteristische Korrelationen* gut wiedergibt, ehe im nächsten Abschnitt ein Algorithmus für eine numerisch genaue Lösung angegeben wird, bei der dann allerdings die Übersichtlichkeit eventuell etwas beeinträchtigt wird.

Ausgehend von (7) kann man den Kurvenverlauf für $\ln P$ als Funktion der reziproken Temperatur

$$\ln P\left(\frac{1}{\vartheta}\,;\,\bar{y}, r\right) = (1 + r\bar{y})\,\bar{\eta}_r + f - r\ln\left(\frac{1}{\vartheta}\right) - (1 + r\bar{y})\,\eta_r\vartheta^r\,e^f \tag{8}$$

vorstellen als eine Summation von (vgl. Fig. 2 und zum Anteil (iii) siehe Fig. 3):

(i) einem praktisch konstanten Beitrag

$$(1 + r\bar{y})\,\bar{\eta}_r \approx \bar{\eta}_0 + 3r\bar{y}^2 \approx 1 - 2\bar{y} + (6 + 3r)\,\bar{y}^2\,,$$

(ii) der asymptotischen Geraden $f(\vartheta; \bar{y}) = -1/\vartheta\bar{y} + 1/\bar{y}$, die den bekannten Anfangsanstieg auf der Tieftemperaturseite repräsentiert, d. h., die in direktem Sinne den temperaturbedingten Anstieg des Reaktionsvermögens vertritt,

(iii) einer zu r proportionalen Korrektur $-\,r\ln(1/\vartheta)$, die innerhalb des begrenzten Temperaturbereiches eines beobachtbaren Glowpeaks, d. h. $\vartheta \approx 1$, ebenfalls fast eine Gerade ist entsprechend $\ln(1/\vartheta) \approx 1/\vartheta - 1$, die damit den Anstieg der Hauptgeraden (ii) korrigiert (in Fig. 2 nicht eingetragen!),

(iv) vermindert um den Betrag der stetigen, monotonen Funktion $(1 + r\bar{y})\,\eta_r\vartheta^r\,e^f$, die den Verbrauch der reaktionsfähigen Partner wiedergibt.

Wir wollen jetzt Gleichung (8) nach f auflösen für die beiden verschiedenen Temperaturseiten, wobei Terme der Größenordnung $\bar{y}$ vernachlässigt werden sollen:

a) Auf der Tieftemperaturseite $T_1 < \bar{T}$, $\vartheta_1 < 1$ ist f_1 stark negativ, deshalb $e^{f_1} \to 0$, d. h. die Anfangskonzentration ist noch unvermindert vorhanden; folglich

$$f_1(P; r) = -(1 + r\bar{y})\,\bar{\eta}_r - \ln\left(\frac{1}{P}\right) + r\ln\left(\frac{1}{\vartheta_1}\right)$$

$$\approx -1 + 2\bar{y} - \ln\left(\frac{1}{P}\right) + r\ln\left(\frac{1}{\vartheta_1}\right). \tag{9}$$

b) Vom Peakmaximum bis in die Hochtemperaturseite hinein sei $\vartheta_2 (\gtrapprox) 1$. Es gilt dann

$$\ln P \,(\gtrapprox)\, 1 + f_2 - e^{f_2} + r\ln\vartheta_2 \approx -\frac{f_2^2}{2} - \frac{f_2^3}{6} + r\ln\vartheta_2$$

und damit umgedreht

$$f_2(P; r) \approx \sqrt{\frac{2\ln\left(\frac{1}{P}\right) - 2r\ln\vartheta_2(e^{f_2} - 1)}{1 + \frac{f_2}{3}}} \approx \sqrt{2\ln\left(\frac{1}{P}\right) - 2rf_2\ln\vartheta_2}\,. \tag{10}$$

Um das Maximum bei $\bar{f} = 0$ und $\ln\bar{P} = 0$ verläuft die Peakkurve parabelförmig bis in den Hochtemperaturteil. Auf der Niedertemperaturseite geht die Parabel in eine asymptotische Gerade über. Durch Kombination von (9) und (10) erhält man für die *Breite*

$$\Delta f(P;r) \equiv f_2(P;r) - f_1(P;r) = \frac{E\big(T_2(P) - T_1(P)\big)}{kT_1T_2}$$

$$= 1 + \ln\left(\frac{1}{P}\right) + \sqrt{2\ln\left(\frac{1}{P}\right) - 2r\ln\vartheta_2} - r\ln\left(\frac{1}{\vartheta_1}\right) \qquad (11)$$

und in der Näherung $T_1T_2 \approx \bar{T}^2$ folgt

$$\delta(P;r) \equiv T_2(P;r) - T_1(P;r) =$$

$$= \frac{k\bar{T}^2}{E}\left\{1 + \ln\left(\frac{1}{P}\right) + \sqrt{2\ln\left(\frac{1}{P}\right) - 2r\ln\vartheta_2} - r\ln\left(\frac{1}{\vartheta_1}\right)\right\}. \qquad (12)$$

Ein steigender Wert von r verringert die Peakbreite auf beiden Temperaturseiten geringfügig. In erster Linie wird sie aber durch E und das betrachtete P-Niveau bestimmt. Je größer E ist, umso enger ist der Peak.

Analog erhält man für die *Asymmetrie* der Kurve $\ln P(1/\vartheta)$:

$$A^*(P;r) \equiv -\frac{f_2}{f_1} = \frac{\sqrt{2\ln\left(\frac{1}{P}\right) - 2r\ln\vartheta_2}}{1 + \ln\left(\frac{1}{P}\right) - r\ln\left(\frac{1}{\vartheta_1}\right)}$$

$$\approx \frac{\sqrt{2\ln\left(\frac{1}{P}\right)}}{1 + \ln\left(\frac{1}{P}\right)}\left\{\sqrt{1 - \frac{r\ln\vartheta_2}{\ln\left(\frac{1}{P}\right)}} + \frac{r\ln\left(\frac{1}{\vartheta_1}\right)}{1 + \ln\left(\frac{1}{P}\right)}\right\}. \qquad (13)$$

In dieser qualitativen Abschätzung (13) ist die Asymmetrie als typischer Formfaktor der Glowkurve vollkommen durch P bestimmt; die durch r bewirkte gleichzeitige, geringe Einengung auf beiden Temperaturseiten wirkt sich im Verhältnis für A^* nicht mehr merklich aus.

Im Rahmen dieser qualitativen Kurvendiskussion sei weiter darauf hingewiesen, daß mit Hilfe weniger Werte auf der Tieftemperaturseite die effektive Aktivierungsenergie erhalten werden kann. Üblicherweise ergibt sie sich als Differentialquotient

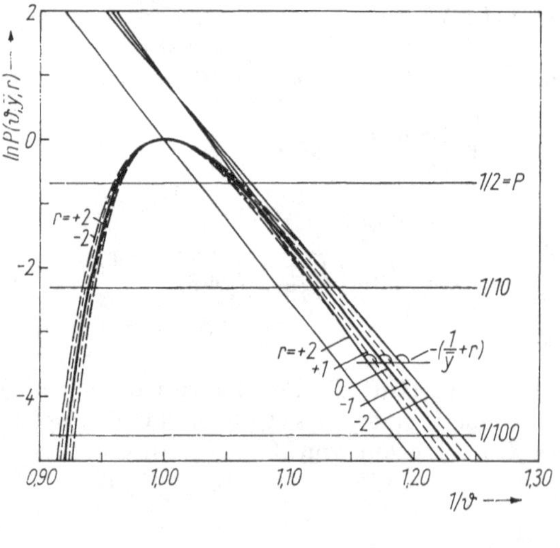

Fig. 3. Einfluß von r nach (8) auf den Verlauf von $\ln P(1/\vartheta; \bar{y}, r)$; alle Asymptoten an den Anfangsanstieg (siehe (15)) schneiden sich auf der Niedertemperaturseite nahe bei $\bar{\vartheta} = 1$ und unterhalb $\bar{\eta}_r \ (\lesseqgtr) \ 1$

von ln P über $(1/kT)$ aus (8)

$$E_{\text{eff}} \equiv -\frac{\mathrm{d}\ln P}{\mathrm{d}\left(\frac{1}{kT}\right)} = E(1 - \mathrm{e}^{f_1} + r\bar{y}) \,. \tag{14}$$

Im Bereich der asymptotischen Tieftemperatur-Geraden strebt e^{f_1} schnell gegen Null, die höchste effektive Steilheit wird erreicht. In der Auswertungspraxis wird jedoch gerade im Anfangsbereich die lokale Ableitung nur unsicher zu erhalten sein wegen Meßpunktschwankungen bei geringer Intensität oder durch noch vorhandene Ausläufer vorausgehender Glow-Peaks. Dann kann man vorteilhafterweise ausnutzen, daß es für die asymptotische Tangente einen weit entfernten und damit gewichtigen Stützpunkt gibt, nämlich den Ordinatenabschnitt $\ln P_0 = \bar{\eta}_0(y) + 3r\bar{y}^2$ für den Abszissenwert $\bar{\vartheta} = 1$ (oder $\bar{f} = 0$ (siehe Fig. 2).

Damit erhält man aus der Messung eines einzigen Tieftemperaturpunktes $P_1(T_1)$ und $\bar{T}$

$$E_{\text{eff}} = -\frac{\Delta \ln P_{\text{asympt}}}{\Delta\left(\frac{1}{kT}\right)} \approx k\,\frac{\bar{\eta}_0 + \ln\left(\frac{1}{P_1}\right)}{\frac{1}{T_1} - \frac{1}{\bar{T}}} \,. \tag{15}$$

Insbesondere für kleine P_1, etwa für $P_1 < \frac{1}{4}$, gibt das für E_{eff} eine gute Bestimmung, ohne jedoch den Beitrag eines möglichen r separieren zu können.

4. Quantitative Polynomlösung

Für eine numerische Auswertung experimenteller Glowkurven $P(T; q, r, E, K_0)$ und zur Ermittlung der einzelnen Aktivierungsparameter ist von den ungekürzten Gleichungen (7) oder (8) auszugehen, die sich in aller Konsequenz aus der kinetischen Differentialgleichung in den Formen (1), (2) oder (4) ergeben hatten. Gesucht ist die funktionelle Abhängigkeit für $\vartheta_i = \vartheta_i(P; r)$, $i = 1, 2$; weiterhin tritt in (7) bzw. (8) noch $\bar{y}$ auf, sowie Größen, die ihrerseits von $\bar{y}$ und ϑ abhängen, wie $f(\bar{y}, \vartheta)$, $\bar{\eta}_r = \bar{\eta}_r(\bar{y})$, $\eta_r = \eta_r(\bar{y}\vartheta)$. Die gesuchte Lösung ist ein Polynom in $\bar{y}$, das relativ schnell konvergiert und in der Regel alternierende Koeffizienten besitzt:

$$\vartheta_i(P; r, \bar{y}) = 1 + a_i\bar{y} + b_i\bar{y}^2 + b_i\bar{y}^3 + d_i\bar{y}^4 + \dots \,. \tag{16}$$

Die Bestimmung der Koeffizienten ist auf zweierlei Wegen möglich [11]. Betrachtet man sie als noch unbestimmte Faktoren, so können (7) oder (8) mit Hilfe von (16) vollständig in Polynome von $\bar{y}^n$ entwickelt und dort die summarischen Koeffizienten für jeden Term $\bar{y}^n$ bestimmt werden. Es muß sich eine rekurrente Beziehung für jeden Koeffizienten aus (16) zu den niederen Koeffizienten von (16) und letzten Endes eine direkte Abhängigkeit vom ersten Koeffizienten a_i ergeben. Für a_i lautet übrigens die Bestimmungsgleichung

$$a_i + 1 + \ln\left(\frac{1}{P}\right) - \mathrm{e}^{a_i} = 0 \,, \tag{17}$$

die im speziellen Fall $P = 0.5$ die zwei Lösungen ergibt:

$$a_1(0.5) = -1.461\,186\,227 \,, \qquad a_2(0.5) = 0.985\,191\,810 \,.$$

Damit ist immer $a_i(P)$ nur eine Funktion von P und stets unbeeinflußt von r. Die nächsten Koeffizienten enthalten dagegen r, z. B.

$$b_i(P; r) = b_{i0} - ra_i; \qquad b_{i0} = a_i^2 + 2 + 2a_ig_i\,,$$
$$c_i(P; r) = c_{i0} + c_{i1}r - r^2a_i; \qquad c_{i0} = a_i^3 + 4a_i - 6 + 7a_ig_i + + 2g_i + 4a_i^2g_i^2 + 4a_ig_i^2 + 2a_i^2g_i^3\,,$$
$$c_{i1} = -5\,(1 + a_ig_i + a_i^2/2)$$

mit $g_i = e^{a_i}/(1 - e^{a_i})$.

Eine rein numerische Bestimmung für einen nächsthöheren Koeffizienten ist auch durch eine Grenzwertbildung möglich, wenn alle niederen Koeffizienten zuvor schon bestimmt sind [6]. Für eine gestaffelte Reihe von P-Werten sind die ersten Koeffizienten in Tabelle 1 angegeben.

Das Polynom (16) stellt eine Entwicklung der Lösung von (7) in der Umgebung des Maximums $\bar{\vartheta} = 1$ dar; für alle physikalisch interessanten Reaktionen mit etwa $\bar{y} = 1/15$ bis $1/35$ konvergiert es ausreichend und sichert auch die Genauigkeit und Konvergenz aller im folgenden noch betrachteten Beziehungen. Entsprechende Wertebeispiele in den Tabellen 2 bis 4 sind immer für ein mittleres $\hat{y} = 1/25$ errechnet worden.

Als ein Beispiel für die Schnelligkeit der Konvergenz ist in den Spalten 3 bis 6 der Tabelle 2 die Größe $a_1 + b_{10}\hat{y} + c_{10}\hat{y}^2 + d_{10}\hat{y}^3$ in unterschiedlichen Näherungen angegeben. Man beachte dabei, daß die sich daraus ergebenden Temperaturen ϑ_1 (auch ϑ_2) im Experiment kaum mit einer solchen Genauigkeit bestimmt werden, auch nicht ihre Differenz $(\vartheta_2 - \vartheta_1)$. Je nach der beabsichtigten Zielstellung wird man sich also auch auf weniger Terme des Polynoms (16) beschränken können.

Mit Hilfe von (16) ergeben sich zu jedem Niveau P

die relative *Peakbreite*

$$\frac{\delta(P; \bar{y}, r)}{\bar{T}} \equiv \vartheta_2 - \vartheta_1 \tag{18}$$

$$\approx (a_2 - a_1)\,\bar{y} + [b_{20} - b_{10} + r(a_1 - a_2)]\,\bar{y}^2 + \dots\,, \tag{18'}$$

das charakteristische *Energie–Breiten-Produkt* (siehe Tabelle 3)

$$\varkappa(P; \bar{y}, r) \equiv \frac{E\delta}{k\bar{T}^2} \tag{19}$$

$$= \frac{a_2 - a_1}{1 + r\bar{y}} + (b_{20} - b_{10})\,\bar{y} + \dots\,, \tag{19'}$$

die *Asymmetrie* (siehe Tabelle 4)

$$A(P; \bar{y}, r) \equiv \frac{\vartheta_2 - 1}{1 - \vartheta_1} \tag{20}$$

$$\approx -\frac{a_2}{a_1}\left\{1 + \bar{y}\left(\frac{b_{20}}{a_2} - \frac{b_{10}}{a_1}\right) + r\bar{y}^2\left(\frac{b_{10}}{a_1} + \frac{b_{20}}{a_2}\right) + \dots\right\}. \tag{20'}$$

In den Tabellen 3 und 4 ist jeweils durch einen Index a, b, c oder d angegeben, bis zu welchem Näherungsterm für $\vartheta_i(P)$ die Berechnung erfolgte. In der vorletzten Spalte von Tabelle 4 ist außerdem das Ergebnis angegeben, wenn in (20′) mit drei Termen abgebrochen wird, nämlich mit der direkten Reihenentwicklung als

$$A_3(P; \bar{y}, r) = -\frac{a_2}{a_1}\left\{1 + \bar{y}\left(\frac{b_{20}}{a_2} - \frac{b_{10}}{a_1}\right) + \bar{y}^2\left(\frac{c_{20}}{a_2} - \frac{c_{10}}{a_1} + \frac{b_{10}b_{20}}{a_1a_2}\right)\right\}. \tag{20''}$$

Von Interesse ist ferner, wie der Kurvenverlauf durch mögliche Variationen von P, E oder r beeinflußt wird; $\bar{T}$ sei fixiert. Dafür sind in den Gleichungen (18′) bis (20′) jeweils auch die Näherungen mit den ersten die Größen $a_i(P)$, $\bar{y}(E)$ und r enthaltenen Termen angegeben. Wir wollen uns dabei für P auf verläßliche experimentelle Werte zwischen $P = 1$ bis 99% beschränken; damit sind auch Singularitäten ausgeschlos-

Tabelle 1

Koeffizienten des Lösungspolynoms $\vartheta_i(P; y, r) = 1 + a_i\bar{y} + (b_{i0} - ra_i)\,\bar{y}^2 + (c_{i0} + c_{i1}r - ra_i^2)\,\bar{y}^3 + d_i\bar{y}^4 \ldots$ für gestaffelte Intensitätswerte P (siehe (16))

	P	$a_1(P)$	$a_2(P)$	b_{10}	b_{20}	c_{10}	c_{20}	c_{11}	c_{21}	d_{10}	d_{20}	
1	0.001	−7.907387	2.325658	64.5209	2.2536	−531.885	0.134	−161.302	−5.634	5769.43	−1.02	1
2	0.005	−6.296474	2.131812	41.6223	1.7072	−280.273	−0.207	−104.056	−4.268	2346.70	−0.33	2
3	0.010	−5.601478	2.033181	33.3350	1.4549	−203.297	−0.316	− 83.337	−3.637	1505.87	−0.26	3
4	0.050	−3.976990	1.748132	17.6645	0.8227	− 82.487	−0.432	− 44.161	−2.057	444.94	0.08	4
5	0.100	−3.264363	1.587134	12.3966	0.5287	− 50.491	−0.396	− 30.992	−1.322	233.35	0.34	5
6	0.200	−2.529759	1.384870	7.9616	0.2231	− 27.695	−0.283	− 19.904	−0.558	107.55	0.61	6
7	0.300	−2.078905	1.235243	5.7275	0.0425	− 17.887	−0.168	− 14.319	−0.106	61.74	0.57	7
8	0.368	−1.841016	1.146040	4.6950	−0.0469	− 13.793	−0.094	− 11.737	0.117	44.64	0.36	8
9	0.400	−1.740933	1.106022	4.2904	−0.0826	− 12.271	−0.060	− 10.726	0.206	38.45	0.29	9
10	0.500	−1.461185	0.985200	3.2525	−0.1738	− 8.601	0.041	− 8.131	0.434	24.66	0.20	10
11	0.600	−1.213741	0.865596	2.4472	−0.2397	− 6.006	0.134	− 6.118	0.599	15.92	0.17	11
12	0.682	−1.023329	0.763895	1.8990	−0.2767	− 4.382	0.203	− 4.747	0.692	10.87	0.0	12
13	0.700	−0.982182	0.740674	1.7886	−0.2827	− 4.071	0.216	− 4.472	0.707	10.09	−0.03	13
14	0.800	−0.751472	0.601245	1.2229	−0.2996	− 2.566	0.281	− 3.057	0.749	5.60	−0.20	14
15	0.900	−0.497019	0.426451	0.7030	−0.2748	− 1.336	0.308	− 1.758	0.687	2.29	−0.28	15
16	0.950	−0.338342	0.304068	0.4338	−0.2270	− 0.770	0.281	− 1.084	0.567	1.25	−0.17	16
17	0.990	−0.145208	0.138505	0.1628	−0.1225	− 0.264	0.169	− 0.407	0.306	−0.03	−0.20	17
18	0.995	−0.101825	0.098482	0.1105	−0.0904	− 0.175	0.128	− 0.276	0.226	0.07	0.0	18
19	0.999	−0.045070	0.044403	0.0468	−0.0427	− 0.072	0.062	− 0.117	0.107	−0.61	0.27	19

Tabelle 2

Beispiele für die Geschwindigkeit der Konvergenz bei Berücksichtigung erster Terme des Lösungspolynoms für einen Fall $y = 1/25$; Temperaturlage ϑ_i und -breite $\delta = \vartheta_2 - \vartheta_1$ in Abhängigkeit von P

	P	$a_1(P)$	$a_1 + b_{10}\hat{y}$	$a_1 + b_{10}\hat{y} + c_{10}\hat{y}^2$	$a_1 + b_{10}\hat{y} + c_{10}\hat{y}^2 + d_{10}\hat{y}^3$	$a_2 + b_{20}\hat{y} + c_{20}\hat{y}^2 + d_{20}\hat{y}^3$	$\vartheta_1(P;\hat{y})$	$\vartheta_2(P;\hat{y})$	$\vartheta_2 - \vartheta_1$	$\frac{1}{\vartheta_1} - \frac{1}{\vartheta_2}$	
1	0.001	−7.9074	−5.3265	−6.1776	−5.8083	2.4159	0.7677	1.0966	0.3290	0.3908	1
2	0.005	−6.2965	−4.6316	−5.0800	−4.9298	2.1997	0.8028	1.0880	0.2852	0.3265	2
3	0.010	−5.6015	−4.2681	−4.5934	−4.4970	2.0909	0.8201	1.0836	0.2635	0.2965	3
4	0.050	−3.9770	−3.2704	−3.4024	−3.3739	1.7804	0.8650	1.0712	0.2062	0.2225	4
5	0.100	−3.2644	−2.7685	−2.8493	−2.8343	1.6077	0.8866	1.0643	0.1777	0.1883	5
6	0.200	−2.5298	−2.2113	−2.2556	−2.2487	1.3934	0.9101	1.0557	0.1457	0.1516	6
7	0.300	−2.0789	−1.8498	−1.8784	−1.8745	1.2367	0.9250	1.0495	0.1244	0.1282	7
8	0.368	−1.8410	−1.6532	−1.6753	−1.6724	1.1440	0.9331	1.0458	0.1127	0.1155	8
9	0.400	−1.7409	−1.5693	−1.5889	−1.5865	1.1026	0.9365	1.0441	0.1076	0.1100	9
10	0.500	−1.4612	−1.3311	−1.3448	−1.3433	0.9783	0.9463	1.0391	0.0929	0.0944	10
11	0.600	−1.2137	−1.1159	−1.1255	−1.1244	0.8562	0.9550	1.0342	0.0792	0.0802	11
12	0.682	−1.0233	−0.9474	−0.9544	−0.9537	0.7532	0.9619	1.0301	0.0683	0.0689	12
13	0.700	−0.9822	−0.9106	−0.9172	−0.9165	0.7297	0.9633	1.0292	0.0658	0.0664	13
14	0.800	−0.7515	−0.7026	−0.7067	−0.7063	0.5897	0.9717	1.0236	0.0518	0.0521	14
15	0.900	−0.4970	−0.4689	−0.4710	−0.4709	0.4159	0.9812	1.0166	0.0355	0.0356	15
16	0.950	−0.3383	−0.3210	−0.3222	−0.3221	0.2954	0.9871	1.0118	0.0247	0.0247	16
17	0.990	−0.1452	−0.1387	−0.1391	−0.1391	0.1339	0.9944	1.0054	0.0109	0.0109	17
18	0.995	−0.1018	−0.0974	−0.0977	−0.0977	0.0951	0.9961	1.0038	0.0077	0.0077	18
19	0.999	−0.0451	−0.0432	−0.0433	−0.0434	0.0428	0.9983	1.0017	0.0034	0.0034	19

Tabelle 3

$\varkappa(P; \hat{y}, r)$ nach Gleichung (19), (19′)

	P	$\varkappa_a(P) = a_2 - a_1$	$\varkappa_b(P; \hat{y}) = a_2 - a_1 + (b_2 - b_1)\hat{y}$	$\varkappa_c(P; \hat{y})$	$\varkappa_{dr}(P; \hat{y})$					$T_\varkappa$	$T_{\varkappa_1}$	$T_{\varkappa_2}$	
					$r = -2$	-1	0	$+1$	$+2$				
1	0.001	10.2330	7.7424	8.5936	8.610	8.401	8.224	8.080	7.969	1411.0	1997.9	4803.3	1
2	0.005	8.4283	6.8317	7.2798	7.538	7.321	7.130	6.966	6.829	1627.7	2353.9	5275.4	2
3	0.010	7.6347	6.3595	6.6842	6.992	6.778	6.588	6.422	6.281	1761.5	2580.5	5550.1	3
4	0.050	5.7251	5.0514	5.1827	5.514	5.325	5.154	5.002	4.868	2251.4	3439.5	6518.1	4
5	0.100	4.8515	4.3768	4.4569	4.766	4.596	4.442	4.303	4.180	2612.4	4094.2	7218.2	5
6	0.200	3.9146	3.0651	3.6489	3.918	3.774	3.642	3.523	3.416	3186.2	5160.5	8328.3	6
7	0.300	3.3141	3.0867	3.1151	3.352	3.226	3.111	3.007	2.913	3729.9	6190.8	9383.4	7
8	0.368	2.9871	2.7974	2.8193	3.037	2.922	2.816	2.721	2.635	4120.2	6938.7	10143.5	8
9	0.400	2.8470	2.6720	2.6916	2.900	2.790	2.689	2.597	2.515	4315.3	7314.6	10524.3	9
10	0.500	2.4464	2.3093	2.3232	2.506	2.410	2.322	2.241	2.169	4998.5	8639.0	11861.6	10
11	0.600	2.0793	1.9719	1.9817	2.139	2.056	1.981	1.912	1.849	5858.9	10320.2	13552.9	11
12	0.682	1.7872	1.7002	1.7075	1.844	1.772	1.707	1.647	1.593	6798.8	12168.1	15407.9	12
13	0.700	1.7229	1.6400	1.6469	1.779	1.710	1.646	1.588	1.536	7049.2	12661.7	15902.9	13
14	0.800	1.3527	1.2918	1.2964	1.401	1.346	1.296	1.250	1.209	8954.1	16430.0	19678.8	14
15	0.900	0.9235	0.8844	0.8870	0.959	0.921	0.887	0.855	0.827	13085.5	24643.9	27899.9	15
16	0.950	0.6424	0.6160	0.6177	0.668	0.642	0.618	0.596	0.576	18790.6	36022.9	39280.5	16
17	0.990	0.2837	0.2723	0.2730	0.295	0.284	0.273	0.263	0.254	42510.0	83413.3	86689.9	17
18	0.995	0.2003	0.1923	0.1928	0.208	0.200	0.193	0.186	0.180	60204.0	118798.8	122061.2	18
19	0.999	0.0895	0.0859	0.0861	0.093	0.090	0.086	0.083	0.080	134680.4	267672.3	271070.6	19

Tabelle 4
$A(P;\hat{y},r)$ nach Gleichung (20), (20′), (20″)

	P	$A_a(P) = -a_2/a_1$	$A_b(P;\hat{y}) = -\frac{a_2+b_2\hat{y}}{a_1+b_1\hat{y}}$	$A_c(P;\hat{y})$	$A_{dr}(P;\hat{y})$			$A_3(P;\hat{y},0)$	
					$r=-2$	0	+2		
1	0.001	0.2941	0.4535	0.3911	0.441	0.416	0.388	0.374	1
2	0.005	0.3386	0.4750	0.4330	0.466	0.446	0.423	0.418	2
3	0.010	0.3630	0.4900	0.4552	0.483	0.465	0.443	0.441	3
4	0.050	0.4396	0.5446	0.5233	0.542	0.528	0.511	0.513	4
5	0.100	0.4962	0.5809	0.5642	0.580	0.567	0.552	0.555	5
6	0.200	0.5474	0.6303	0.6177	0.630	0.620	0.607	0.611	6
7	0.300	0.5942	0.6687	0.6584	0.669	0.660	0.648	0.652	7
8	0.368	0.6225	0.6921	0.6829	0.693	0.684	0.673	0.677	8
9	0.400	0.6353	0.7027	0.6939	0.703	0.695	0.685	0.689	9
10	0.500	0.6742	0.7349	0.7275	0.736	0.728	0.719	0.723	10
11	0.600	0.7132	0.7671	0.7608	0.768	0.761	0.753	0.757	11
12	0.682	0.7465	0.7946	0.7892	0.796	0.790	0.782	0.785	12
13	0.700	0.7541	0.8009	0.7956	0.802	0.796	0.789	0.792	13
14	0.800	0.8001	0.8387	0.8345	0.840	0.835	0.829	0.831	14
15	0.900	0.8580	0.8860	0.8831	0.887	0.883	0.879	0.880	15
16	0.950	0.8987	0.9190	0.9169	0.919	0.917	0.914	0.915	16
17	0.990	0.9538	0.9633	0.9623	0.963	0.962	0.961	0.960	17
18	0.995	0.9672	0.9739	0.9732	0.974	0.973	0.972	0.971	18
19	0.999	0.9852	0.9883	0.9880	0.988	0.987	0.987	0.986	19

sen, die sich physikalisch auszeichnen und die sich für $P = 0$ und für $P = 1$, d. h. $a_i = 0$, ergeben könnten wegen

$$\frac{\mathrm{d}a}{\mathrm{d}p} = \frac{1}{P(1 - \mathrm{e}^{a_i})}, \tag{21}$$

abgeleitet aus (17).

Als relative Variationen ergeben sich:

$$\frac{\mathrm{d}\vartheta_i}{\vartheta_i} = -a_i\bar{y}\left(\frac{\mathrm{d}E}{E} + \bar{y}\,\mathrm{d}r\right) + \frac{\bar{y}}{1 - \mathrm{e}^{a_i}}\,\frac{\mathrm{d}P}{P}, \tag{22}$$

$$\frac{\mathrm{d}\delta}{\delta} = -\left(\frac{\mathrm{d}E}{E} + \bar{y}\,\mathrm{d}r\right) - \frac{1}{(\mathrm{e}^{a_2} - 1)(1 - \mathrm{e}^{a_1})}\,\frac{\mathrm{d}P}{P}, \tag{23}$$

$$\frac{\mathrm{d}\varkappa}{\varkappa} = -\frac{b_{20} + b_{10}}{a_2 - a_1}\,\bar{y}\,\frac{\mathrm{d}E}{E} - \bar{y}\,\mathrm{d}r - \frac{1}{(\mathrm{e}^{a_2} - 1)(1 - \mathrm{e}^{a_1})}\,\frac{\mathrm{d}P}{P}, \tag{24}$$

$$\frac{\mathrm{d}A}{A} = \left(\frac{b_{20}}{a_2} - \frac{b_{10}}{a_1}\right)\bar{y}\,\frac{\mathrm{d}E}{E} + \left(\frac{b_{10}}{a_1} + \frac{b_{20}}{a_2}\right)\bar{y}^2\,\mathrm{d}r - \left(\frac{1}{a_2(\mathrm{e}^{a_2} - 1)} + \frac{1}{a_2(1 - \mathrm{e}^{a_1})}\right)\frac{\mathrm{d}P}{P}. \tag{25}$$

Ausgehend von den Koeffizienten der Gleichungen (21) bis (25) wird noch einmal sichtbar:

Die *absolute Temperaturlage* ϑ_i (immer relativ zu $\bar{T}$) wird, bedingt durch die Kleinheit von $\bar{y}$, nur wenig beeinflußt.

Die relative *Breite* des Peaks verringert sich beim Anwachsen der Aktivierungsenergie; in der gleichen Richtung wirkt ein schrittweiser Anstieg von r.

Die Größe $\varkappa$ wird wieder wegen $\bar{y}$ unempfindlich gegen eine nicht zu große Variation von E; $\varkappa$ ist scheinbar genauso wie δ durch dr zu verschieben, aber der Betrag ist gering (z. B. $-\hat{y} \times 1 \approx -4\%$), so daß in Relation zur üblichen experimentellen Genauigkeit kaum eine Schlußfolgerung über r gesichert werden kann.

Die *Asymmetrie* A ist am allerwenigsten durch Variationen dE oder dr zu beeinflussen; sie ist sogar gegenüber größeren Niveauunterschieden P recht stabil. Die erstgenannten Größen: ϑ_i, aber besonders δ und $\varkappa$, werden dagegen durch P bestimmt. An dieser Stelle sei daran erinnert, daß gerade die Asymmetrie geeignet ist, einen weiteren Aktivierungsparameter, die sogenannte Kinetikordnung, sehr kritisch zu bestimmen [21].

Da die *Aktivierungsenergie* E am besten unter Verwendung von (19) mit Hilfe der experimentellen Peakbreite $\delta(P; r, \bar{y})$ erhalten werden kann, sei nochmals auf (23) hingewiesen. Effektiv ist nur eine Aussage über

$$E_{\text{eff}} = E + k\bar{T}r \tag{26}$$

möglich; es sei denn, ein Autor verfügt anderweitig, zum Beispiel aus einem Modell heraus, über Kenntnisse in bezug auf r. Vergleichen Autoren verschiedene Modelle mit unterschiedlichen r-Werten, oder mit unterschiedlichen Aufheizgesetzen (siehe Fußnote [2])), dann sind wiederum eine Umrechnung und ein Vergleich einwandfrei möglich. Wo keine besondere Veranlassung besteht, wird die gesamte Temperaturabhängigkeit einer Reaktionsrate vollständig der Aktivierungsenergie zugeordnet werden und mit $r = 0$ erübrigt sich jede weitere Diskussion.

Endlich kann für die Ermittlung der Aktivierungsenergie E aus experimentell ablesbaren Werten: $\bar{T}$ und δ bzw. δ_1 oder δ_2 die Gleichung (18) noch vereinfacht und die

Konstanten zusammengefaßt werden zu $\varkappa k \equiv 1/T_\varkappa(P)$

$$E = \frac{\bar{T}^2}{\delta T_\varkappa} \quad \text{beziehungsweise} \quad \frac{\bar{T}^2}{\delta_i T_{\varkappa i}}. \tag{27}$$

Für $\hat{y} = \frac{1}{25}$ und $r = 0$ sind die Temperaturwerte $T_\varkappa$, $T_{\varkappa i}$ in den letzten drei Spalten von Tabelle 2 enthalten; sie können für abweichende r-Werte selbst berechnet werden. Eine Korrektur für in der ersten Auswertung abweichende $\bar{y} \neq \hat{y}$ kann ebenfalls leicht vorgenommen werden, wird im allgemeinen jedoch unerheblich bleiben (siehe [6]). Für eine genaue Bestimmung ist der Fehler der Differenz $\delta(P) = T_2(P) - T_1(P)$ entscheidend, während sich die meist nicht ganz genaue Absolutbestimmung von $\bar{T}$ kaum im Fehler des Ergebnisses niederschlägt. Im allgemeinen ist es wegen der Größe, vor allem aber wegen der erreichbaren Genauigkeit, vorteilhaft, die gesamte Breite δ zu verwenden; in Fällen zur Abgrenzung gegenüber benachbarten Peaks kann es jedoch notwendig werden, auch einzelne Seiten δ_1 oder δ_2 zu benutzen. Gegenüber der bisherigen Praxis der Halbwertsbreiten-Methode besteht jetzt die Möglichkeit, die analoge Auswertung auch auf einem anderen Intensitätsniveau P zur Selbstkontrolle zu wiederholen. Mit niedrigen P-Werten kann damit auch geprüft werden, ob nicht doch eine unaufgelöste Überlagerung mehrerer Peaks vorliegt; diese würde scheinbar zu einer Verbreiterung [22] und zu einer Erniedrigung der E-Werte führen.

Literatur

[1] S. GLASSTONE, K. J. LAIDLER, und H. EYRING, The Theory of Rate Processes, Mc Graw Hill, New York 1941.
[2] P. M. KEATING, Proc. Phys. Soc. 78, 1408 (1961); J. Phys. Chem. Solids **22**, 25 (1961).
[3] R. H. BUBE, Photoconductivity in Solids, Chap. 9.4, John Wiley & Sons, Inc. New York/London 1960
[4] M. BALARIN, phys. stat. sol. (a) **54**, K137 (1979).
[5] M. BALARIN und A. ZETZSCHE, phys. stat. sol. **2**, 1670 (1962); **3**, K387 (1963).
[6] M. BALARIN, J. Thermal Anal. **17**, 319 (1979).
[7] M. BALARIN, Thermochim. Acta **33**, 341 (1979).
[8] R. CHEN, J. appl. Phys. **40**, 570 (1969).
[9] M. BALARIN, phys. stat. sol. (a) **31**, K111 (1975).
[10] D. EDWARDS, Surface Sci. **69**, 301 (1977).
[11] P. L. LAND, J. Phys. Chem. Solids **30**, 1681 (1969).
[12] N. A. LUSHNIKOV, G. A. SHDAN und V. B. SANDOMIRSKII, Fiz. Tekh. Poluprov. **13**, 247 (1979).
[13] M. BALARIN, Fiz. Tekh. Poluprov. **15**, 2216 (1981).
[14] V. N. VERTOPRACHOV and E. G. SALMAN, Thermostimulierte Ströme in anorganischen Stoffen, Nauka, Novosibirsk 1979 (in russisch).
[15] G. RUDLOF, Dissertation, Universität Rostock, 1981.
[16] P. MÜLLER und J. TELTOW, phys. stat. sol. (a) **12**, 471 (1972).
[17] J. U. SULLIVAN, J. Comput. Phys. **9**, 164 (1972).
[18] V. M. GORBATSCHEV, J. Thermal Anal. 17, 187 (1978).
[19] H. ANDERSON, D. HABERLAND und E. WITTE, J. Thermal Anal. 17, 409 (1978).
[20] D. SHAW und H. MOGHADDAM, phys. stat. sol. (a) **65**, 167 (1981).
[21] M. BALARIN, Sitzungsberichte der AdW der DDR Nr. 3/N, (S. 29 bis 46) 1981.
[22] M. BALARIN, Exper. Tech. Phys. **26**, 229 (1978).

(Received September 30, 1982)

phys. stat. sol. (a) 77, 195 (1983)

Subject classification: 1.5 and 3; 22.2

Sektion Physik und Technik elektronischer Bauelemente der Technischen Hochschule Ilmenau[1])

Investigations on the Structure of MOCVD AlN Layers on Silicon

By

U. RENSCH and G. EICHHORN

The AlN layers are deposited from NH_3 and TMA on (111) and (100) Si according to the MOCVD method. The influence of the substrate temperature and of the composition of the gaseous phase on the real structure is investigated in the temperature range from 1073 to 1573 K. At a temperature of 1573 K and a molar ratio of NH_3/TMA $> 400/1$ producing of monocrystalline AlN layers on (111) Si is successful.

Die AlN-Schichten werden nach dem MOCVD-Verfahren aus NH_3 und TMA auf (111) und (100) Si abgeschieden. Es wird der Einfluß der Substrattemperatur und der Gasphasenzusammensetzung im Temperaturbereich von 1073 bis 1573 K auf die Realstruktur untersucht. Bei 1573 K und einem Molverhältnis NH_3/TMA $> 400/1$ gelingt es, auf (111) Si einkristalline AlN-Schichten herzustellen.

1. Introduction

On the basis of the investigations conducted by Manasevit et al. [1] on the AlN preparation using MOCVD, we have developed an experimental variant permitting to deposit AlN layers with a satisfactory homogeneity on different substrates in an induction-heated vertical reactor [2].

For investigating the application possibilities of these AlN layers in microelectronics it is particularly important to find out the influence of experimental conditions on the physical properties and to make the preparation process reproducible.

In the following we will discuss the effects of the deposition temperature, the substrate orientation, and the composition of the gaseous phase on structural perfection. The methods used for analysis were based on reflection electron microscopy, X-ray diffraction, and scanning electron microscopy.

In general, reactions as, for example, that of trimethylaluminium (TMA) with NH_3 are determined essentially by the effects resulting from chemical kinetics. However, the course of reaction will not be examined in this paper.

In order to certain that the experimental results can be compared with each other, such factors as, for example, the distance between gas inlet and substrate, the total flow rate, the absolute concentration of NH_3, and the cooling rate have been kept constant. As substrate we used (111) and (100) silicon wafers with a diameter of 36 mm which have been treated according to [3].

2. Results

2.1 Temperature of the substrate

An exact investigation of the influence of the substrate temperature T_S on the structural perfection of the AlN layers performed by means of reflection electron microscopy revealed the following characteristic features:

[1]) Weimarer Str. 32, DDR-6300 Ilmenau, GDR.

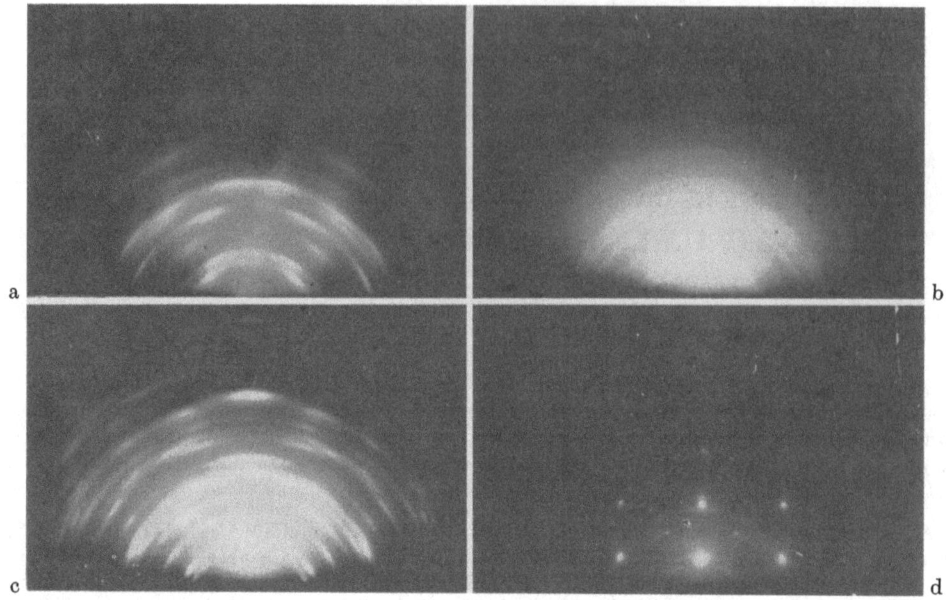

Fig. 1. RHEED patterns from AlN layers on (111)-Si a) $T_S = 1073$ K, ⟨11$\bar{2}$1⟩ fibre axis; b) $T_S =$ $= 1273$ K, textureless; c) $T_S > 1473$ K, ⟨0001⟩ fibre axis; d) $T_S = 1573$ K, electron beam azimuth in [$\bar{1}2\bar{1}0$] direction

At a temperature of $T_S = 1073$ K, AlN layers with a good texture and a ⟨11$\bar{2}$1⟩ fibre axis are formed (Fig. 1a), with the axis being perpendicular to the surface of the substrate. Between 1073 and 1473 K, the degree of texturing is generally lower than below and above this temperature range, which is in many cases equivalent to being textureless (Fig. 1b). From 1473 K on, the structural quality of the layers improves again considerably. The samples show a good to very good texture at 1473 K. Above this temperature, a change in the preferred orientation occurs. The layers show no longer a distinct ⟨11$\bar{2}$1⟩ fibre axis, but predominantly a ⟨0001⟩ orientation (Fig. 1c). At 1573 K, we succeeded in depositing layers with the highest structural perfection namely single-crystalline AlN on (111) silicon (Fig. 1d). The determination of intergrowth resulted in the correlation (0001) AlN || (111) Si, [$\bar{1}2\bar{1}0$] AlN || [$\bar{1}10$] Si, which is in conformity with literature [1, 4, 5]. The electron microscopic examinations and X-ray analysis of the AlN layers made at different points of the substrate revealed a single-crystalline character with a minimum polycrystalline portion.

2.2 Orientation of the substrate

The investigation of the influence exerted by the orientation of the substrate showed that the deposition on (111) and (100) silicon substrates, taking place under the same conditions, brought no serious differences in texture up to 1473 K (e.g. different sharpness of texture, change in the preferred orientation), which was contrary to expectations. The situation was different at 1573 K in so far as we have succeeded in depositing single-crystalline AlN only on (111) silicon so far.

2.3 Composition of the gaseous phase

The analysis of the influence of the composition of the gaseous phase, i.e. of the molar ratio of ammonia to trimethylaluminium (NH_3/TMA), in the temperature range from 1073 to 1573 K showed that the degree of texturing changes only slightly as a function

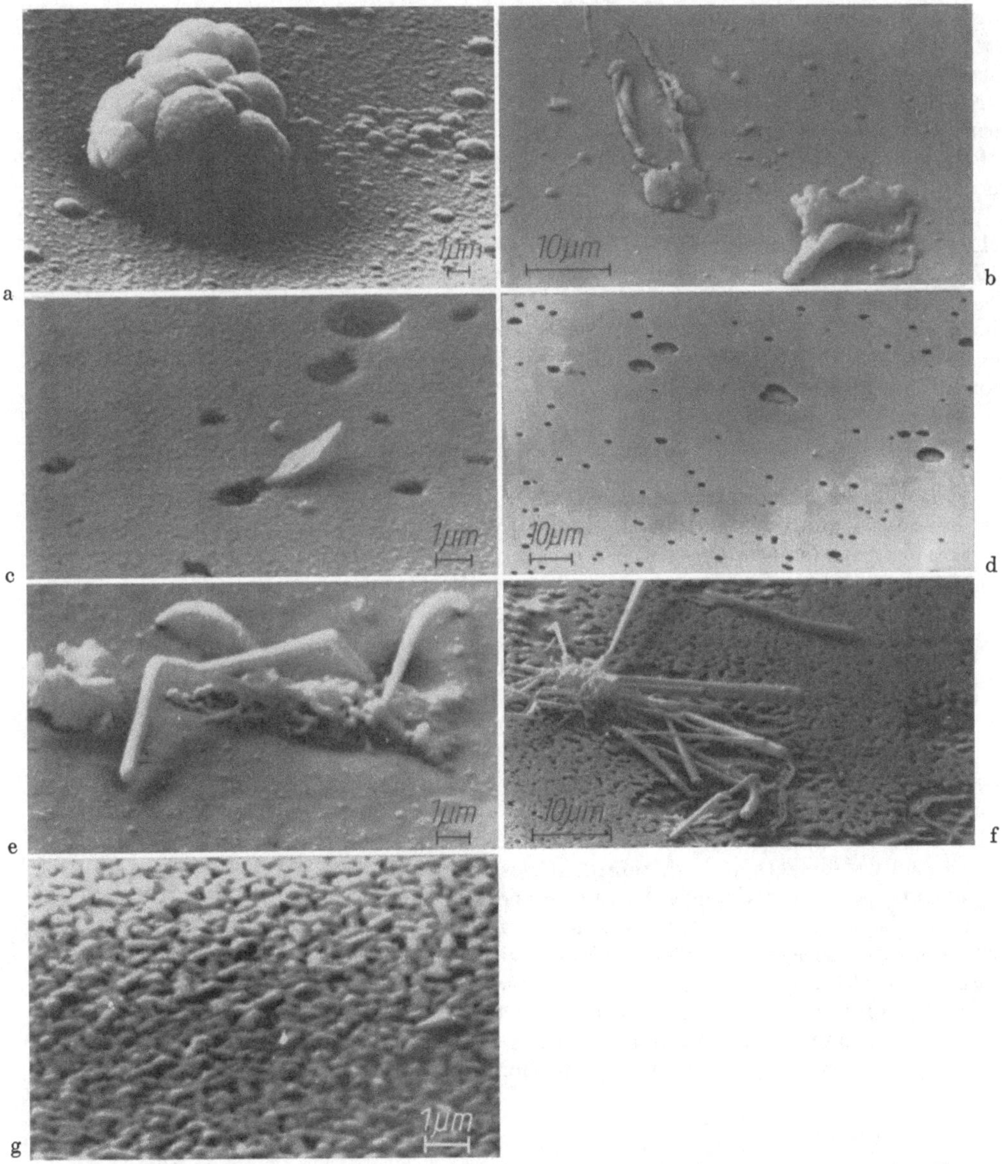

Fig. 2. Scanning electron micrographs of the surface of AlN layers a), b) particles with different shapes; c), d) pinholes; e), f) whiskers; g) surface roughness, film thickness ≈ 1 µm

of the molar ratio. In order to obtain good textures or single-crystalline AlN, the molar ratio should be NH_3/TMA > 400/1 [2], which also applies to other physical and chemical properties.

2.4 Surface morphology

For most applications the quality of the surfaces of the AlN layers plays an important part. Therefore, we examined the surfaces of the layers obtained under various conditions of growth by means of a scanning electron microscope. The layers show various defects which are found in a large number and great variety, especially at high deposition temperatures (1473 to 1573 K). Thus, particles having different shapes and sizes are found on the surface of AlN (Fig. 2 a, b). Furthermore, other disturbances of growth arise in the form of pinholes (Fig. 2 c, d) exerting an unfavourable influence especially on the electrical properties. Their density greatly depends on the perfection of the substrate and is 0 to 350 cm^{-2}. At the present state of the investigations, however, no systematic connection with the production parameters can be discovered.

The surface morphology of relatively thin layers (100 to 200 nm) deposited at 1573 K shows remarkable features. Here we have observed whiskers of different shapes (Figs. 2 e, f) the density of which increases with decreasing TMA content.

The shape of the whiskers suggests the VLS mechanism [6, 7]. The surface roughness increases considerably with growing thickness of the layer (Fig. 2g).

Layers of a thickness of about 1 μm have been polished in a mechano-chemical way for further physical investigations.

3. Discussion

For interpreting the experimental results, it is necessary to have a closer look at the rate of deposition R_D as a function of the substrate temperature T_S as it is represented in Fig. 3. In our experiments we stated the maximum of the rate of deposition between 1173 and 1373 K, with the shape of the curve flattening with increasing molar ratio. At temperatures below 1173 K, R_D decreases quickly. At a substrate temperature of 1073 K, the mobility of the nuclei is still so high that a good texture can develop.

With increasing substrate temperature also R_D goes on increasing, but despite of a higher temperature the quality of the developing textures is not satisfactory any longer, the layers are partially polycrystalline.

At temperatures above 1373 K the rate of deposition decreases because of the preponderance of the homogeneous reaction. This leads to favourable conditions of growth for the nuclei developing on the substrate.

In addition, an improvement is obtained by a more and more decreasing density of the structural defects at high temperatures so that sharply textured or even single-crystalline layers can deposit. In this temperature range, however, one has to expect junction layers that may develop near the phase boundary between AlN and Si, and which create more favourable conditions with respect to the geometry of the lattice. This assumption is confirmed by Auger investigations made by us.

In case of a molar ratio of 400, the rate of deposition is always distinctly greater (20 to 50%) than in case of a molar ratio of 3200 in the temperature range examined. This is reflected in the structural perfection which is lower in case of a molar ratio of 400.

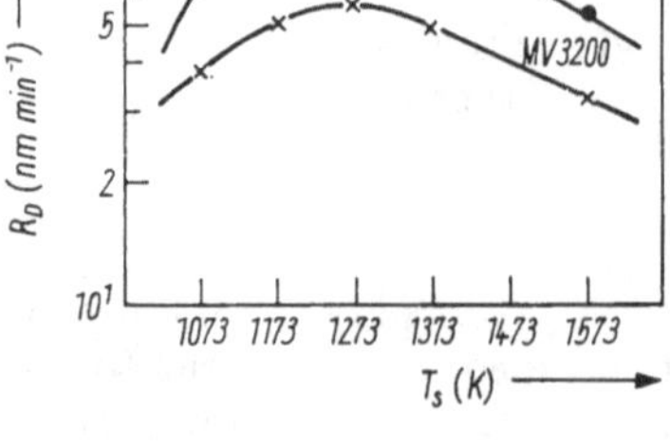

Fig. 3. Rate of deposition R_D as a function of the substrate temperature T_S

Only at temperatures above 1273 K the influence of the orientation of the substrate on the structure of the AlN layers becomes apparent in a distinct manner. Here again there exists a connection with the rate of deposition which is higher by 25% in case of (100) Si than in case of (111) Si at a temperature of 1573 K.

The higher perfection reached on (111) Si, observed also in [8], is certainly favoured by low R_D values. Moreover, the (111) Si surfaces are slip planes. A high mobility of dislocation offers favourable conditions for intergrowth.

4. Summary

The present work investigates the influence of the most important experimental parameters on the real structure of AlN layers. The preparation of the AlN layers is effected by the MOCVD method on the basis of NH_3 and TMA (111) Si and (100) Si served as substrate materials.

The substrate temperature and its influence on the reaction are of decisive importance for the structural perfection of the layers. These show a good texture at a temperature of 1073 K. A further increase of the substrate temperature brings on a deterioration of the structure, but above 1473 K the structure again becomes distinctly better. At a temperature of 1573 K, we succeeded in producing single-crystalline AlN layers.

The deterioration of the quality of the surface occurring in case of increasing substrate temperatures is to be ascribed essentially to the homogeneous reaction.

The course of the reaction and thus the rate of deposition are considered in connection with the perfection and the surface quality of the layers.

The influences exerted by the orientation of the substrate become apparent only above 1273 K. Single-crystalline layers were obtained only on (111) Si and in case of a molar ratio of $NH_3/TMA > 400/1$.

Acknowledgement

We thank all those colleagues of our team who were engaged in this research.

References

[1] H. M. Manasevit, F. M. Erdman, and W. I. Simpson, J. Electrochem. Soc. **118**, 1864 (1971).
[2] G. Eichhorn and U. Rensch, phys. stat. sol. (a) **69**, K3 (1982).
[3] U. Rensch, Dissertation 1983, TH Ilmenau.
[4] M. T. Duffy, C. C. Wang, G. D. o'Clock Jr., S. H. McFarlane, and P. J. Zancucchi, J. electronic Mater. **2**, 359 (1973).
[5] S. Yoshida, S. Misawa, Y. Fujii, S. Takada, H. Hayakawa, S. Gonda, and A. Itoh, J. Vacuum Sci. Technol. **16**, 990 (1979).
[6] T. Ishii, T. Sato, and M. Iwata, Mineralog. J. **6**, 323 (1971).
[7] C. C. Evans, Whiskers, Mills & Boon, Ltd., London 1972.
[8] M. Morita, S. Isogai, N. Mikoshiba, N. Shimizo, and K. Tsubouchi, Japan. J. appl. Phys. **20**, 173 (1981).

(Received February 2, 1983)

phys. stat. sol. (a) 77, 201 (1983)

Subject classification: 1.5; 22.1.2

Sektion Physik/Elektronische Bauelemente der Technischen Hochschule Karl-Marx-Stadt[1]) *(a)*
and Sektion Physik der Friedrich-Schiller-Universität Jena[2]) *(b)*

Characterization of Heteroepitaxial Silicon Layers on Sapphire by Rutherford-Backscattering (RBS) Analyses

By

H. G. SCHNEIDER (a), H. KARGE (b), and B. WEBER (b)

After the discussion of a lattice matching mechanism substrate–deposit by coordination polyhedrons of a process-inherently grown interlayer the results of RBS measurements are described which characterize the heteroepitaxially grown silicon on sapphire relative to their structural perfection. These results are supplementary to corresponding TEM investigations of the interface substrate–deposit. Regions of different degrees of disturbance in certain layer thicknesses over the substrate surface can be determined and quantitatively described.

Nach der Diskussion eines Anpassungsmechanismus Substrat–Deposit über Koordinationspolyeder einer prozeßinhärent gebildeten Zwischenschicht werden die Ergebnisse von RBS-Messungen wiedergegeben, die das hetero-epitaktisch auf Saphir aufgewachsene Silicium hinsichtlich seiner strukturellen Perfektion charakterisieren. Diese Ergebnisse ergänzen entsprechende TEM-Untersuchungen des Interface Deposit–Substrates. Es können Bereiche mit unterschiedlichem Störungsgrad in bestimmten Schichtdicken über der Substratoberfläche nachgewiesen und quantitativ beschrieben werden.

1. Introduction. Preliminary Assumptions on a Lattice Matching Mechanism Silicon–Sapphire

A good understanding of the heteroepitaxy of silicon on sapphire is one of the most important conditions of producing integrated circuits on this insulating substrate material. It is well known that under the usual conditions of chemical vapour deposition (CVD) of silicon on sapphire (temperatures between 950 and 1000 °C, pyrolysis of SiH_4/H_2 or He) an interlayer can be formed by a complicated chemical reaction between Si (or better SiO_x) and Al_2O_3. Such an interlayer formation can be also observed in some other intergrowth cases, e.g. Cu, Ni, NiO, ZnO, and Ti on sapphire, which leads to compounds structurally related to spinel. These interlayers obviously have a lattice matching function between substrate and deposit. This is one of the reasons why the possible interlayer formation between silicon and sapphire is of high technological interest [1 to 10].

The interlayer may consist of an Al–Si–O compound related to the Al silicate Al_2SiO_5. That means a new type of Al–O and additional Si–O bonds arise from this reaction [11]. The thickness of the interlayer (nearly zero up to 60 nm) depends on the specific formation conditions. Stacking faults, microtwins, dislocations, and elastic strains could be observed by usual multibeam TEM lattice images in silicon above this interlayer region but never misfit dislocations in their clear definition and association to the intergrowth (interfacial) region (e.g. [9, 10]), but also locally fault-free silicon and absence of interlayer [12, 13].

[1]) Postfach 964, DDR-9010 Karl-Marx-Stadt, GDR.
[2]) Max-Wien-Platz 1, DDR-6900 Jena, GDR.

If we discuss the problem of epitaxy we always have to take into consideration two decisive crystallographic principles:

1. The more the two intergrowth partners are similar with respect to symmetry, coordination, and type, direction and power of bonding, the better epitaxy between them will succeed.

2. The smaller the lattice misfit of such partners is, the better epitaxy will succeed, too.

But, silicon and sapphire are of very different kinds relative to case 1 and 2.

It seems incompatible with those, that epitaxy experiments under different conditions prove, that it is possible to grow e.g. (100) Si layers on different planes of sapphire, e.g. ($\bar{1}012$), ($11\bar{2}4$), (0001), ($11\bar{2}0$), etc. [11]. But it is also possible to grow (111) and (110) Si layers on such sapphire planes.

It seems reasonable to suppose that bonding between silicon and sapphire occurs by tetrahedrally coordinated oxygen. There are broken bonds at the sapphire surface caused by slight etching and in every case there are traces of external oxygen on the sapphire surface. Therefore, a certain degree of chemical reaction happens at least to such an extent that single [SiO_4] and [AlO_4] tetrahedra or similar units can be established also without detectable interlayer formation in the usual sense. The O–O bond lengths of those tetrahedra and those of the Al_2O_3 units are very similar. The specific properties and behaviour of such tetrahedra, e.g. rotational movements of whole units, displacements of single atoms/ions, i.e. flexibility of bond lengths and angles and substitution of cations (Al–Si), linkage of the coordination tetrahedra of the interlayer with those of the silicon layer, and by the oxygen coordination with the structural basic units of the sapphire promote epitaxial intergrowth between Si and sapphire. In such a way specific residence probabilities of atoms/ions arise which are similar to those of the exact coordination tetrahedra with the specifically necessary orientation which are able to adapt the lattices of silicon and sapphire in a very thin region of some few lattice planes at least.

2. Experimental Procedure of RBS Analyses

The aim of our investigations was the application of Rutherford backscattering (RBS) to analyse the epitaxial silicon layer, especially its region near the interface to the substrate. For this purpose we used a bicrystal system consisting of a 0.95 μm thick (100) silicon layer on ($1\bar{1}02$) sapphire Czochralski grown. The silicon was deposited by a two-step CVD process (SiH_4/H_2 pyrolysis). A high-speed deposition for some seconds was followed by the normal deposition rate of 0.3 μm/min at 1000 °C. In this way we received a very thin interlayer (1 to 2 μm) detected by TEM [15].

The RBS measurements were carried out with a beam of 1.4 MeV He^+ ions. The energy spectrum $Y(E)$ of the backscattered ions is measured at an angle of 170° against the incident beam. For random incidence of He^+ ions the thickness of the epitaxially grown layer can be obtained from the energy width of the silicon part of $Y(E)$ taking into account the tabulated energy loss of He^+ ions in silicon [14]. If the beam of He^+ ions is aligned to the [100] axis of the silicon layer the backscattering yield is reduced by the channeling effect to

$$Y_{[100]} = \chi_{\min}(d)\, Y_{\text{random}}\,.$$

The minimum yield $\chi_{\min}$ increases with growing depth d below the sample surface by scattering of the channeled ions into the random part of the beam (dechanneling). The

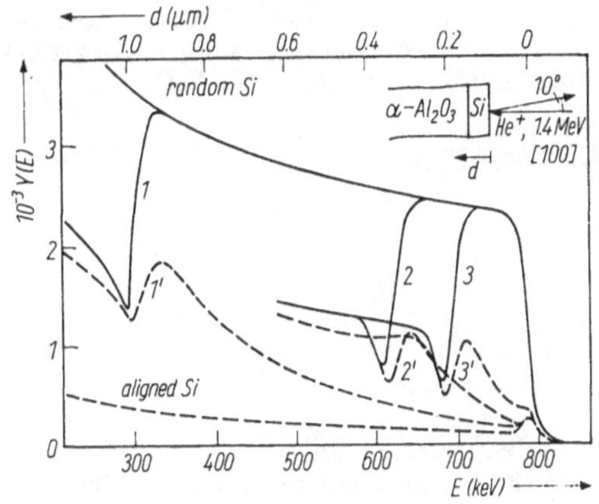

Fig. 1. Energy spectra of backscattered He^+ ions of epitaxial (100) Si layers of various thicknesses on (1$\bar{1}$02) sapphire (α-Al_2O_3)

magnitude

$$A(d) = \frac{\Delta\chi_{\min}(d)}{\Delta d[1 - \chi_{\min}(d)]} - \frac{\Delta\chi_{\min}^0(d)}{\Delta d[1 - \chi_{\min}^0(d)]}$$

($\chi_{\min}^0$ minimum yield of an undistorted Si [100] crystal) is a measure for the concentration of defects [14] according to the model of ion dechanneling by single scattering if only a small defect concentration is to be expected. The quantitative interpretation of the dechanneling function $A(d)$, that means the determination of the depth profile of the defect concentration, presupposes detailed information on the kind of the defects [17]. This condition is not satisfied in this case because of the multiplicity of various defects. Therefore, the epitaxial silicon layer was thinned step by step by anodic oxidation and chemical denudation. The resulting change of $A(d)$ is registered. Thereby a better accuracy of measurement relative to the sapphire–silicon interface region is obtained.

3. Results

Fig. 1 shows some typical energy spectra of backscattered He^+ ions. The diagram also contains the random and aligned spectra of bulk [100] Si (single crystals). Depth scale d (upper abscissa) is related to backscattering by Si atoms. It enables us directly to read off the thickness of the silicon layers. Spectra 1, 2, and 3 are random spectra for $d_{Si} = 180$, 300, and 950 nm. 1′, 2′, and 3′ are the corresponding aligned spectra. It is evident that the aligned yield $Y_{[100]}(E)$ of the epitaxial layer grows more rapidly along with d than in the case of bulk silicon. Thus the epitaxial layers possess a certain

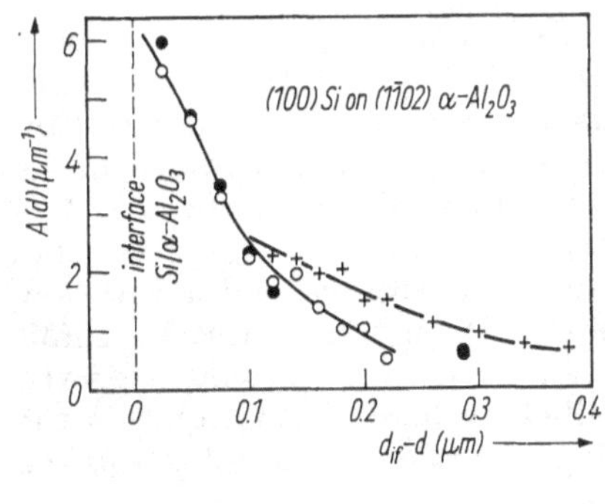

Fig. 2. Defect-induced ascent $A(d)$ of the minimum yield of differently thick (d_{Si}) Si epitaxial layers as a function of the distance (d_{if}–d) from the Si–sapphire interface. ● $d_{Si} = 180$, ○ 300, + 950 nm

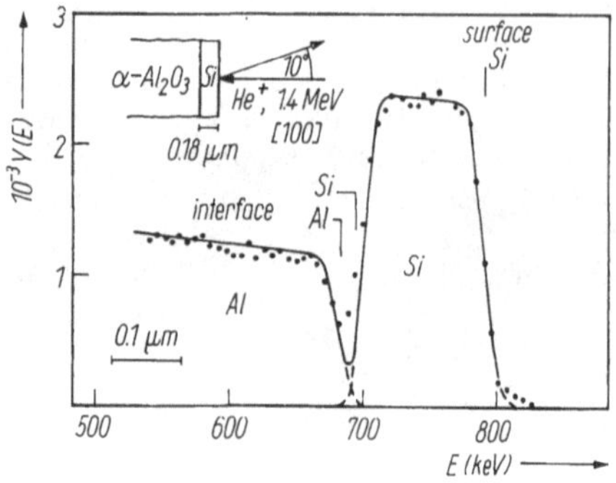

Fig. 3. Comparison of measured (points) and calculated (curve) random spectra of a 0.18 µm thick Si epitaxial layer on sapphire (α-Al_2O_3)

content of lattice defects. Noticeable is the steep rise of the aligned spectra with approach to the interface region. Obviously in the silicon layer the defects which contribute to the dechanneling of the incident He^+ ions attain there the highest amount. In Fig. 2 the sizes $A(d)$ calculated from the measured spectra are plotted versus difference $(d_{if} - d)$, that means versus the distance from the interface (if) Si–sapphire. Directly above the interface in a region of about 100 nm thickness $A(d)$ attains extraordinarily high values which rapidly decrease with increasing $(d_{if} - d)$. Although in all cases the dechanneling as an integral size of the defect concentration continuously decreases up to a distance $(d_{if} - d)$ of about 300 nm, it does not attain the value of the high-quality bulk silicon $(A(d) \equiv 0)$ also up to the sample surface of the initial 0.95 µm thick layer.

Fig. 3 shows the random spectrum measured for a 0.18 µm thick silicon layer on sapphire (points) and the calculated one. For the calculation a sharp boundary between silicon and sapphire was assumed and an energy resolution capability of the detector system of 15 keV was considered. The figure shows a good agreement between the measured and calculated spectra. Insignificant deviations appear only in the region of the silicon–sapphire interface. However, these can be explained by energy loss straggling of the incident He^+ ions in the Si epitaxial layer which was not taken into account in the calculation.

4. Discussion

From the good conformity between the measured and calculated random spectra (Fig. 3) we conclude that the thickness of a possible interlayer between sapphire and silicon must be smaller than 30 nm. This is the depth resolution of the measuring method. Otherwise an interlayer as described in Section 1 which contains Al, Si, and O in corresponding concentrations or concentration profiles had to have a significant effect on the shape of the RBS spectrum.

Lattice matching is, as shown by Fig. 2, not limited to this very thin interlayer. Up to a layer thickness of around 100 nm a lot of lattice defects was integrally identified. The great difference between aligned and random spectra of this layer section (Fig. 1) proves that there cannot be a quasi-amorphous silicon transition layer. On the contrary, the shape of the spectra is typical for crystals containing extended lattice defects with increasing concentration in the direction to the interface substrate–deposit. Such defects have a very wide dechanneling cross-section, and a very low contribution to the direct backscattering [16]. They must be investigated by TEM, especially directly in the interface region [9, 10, 12, 13]. Furthermore it could be proved that the [100] axis of the Si epitaxial layer is tilted by $(6.0 \pm 0.2)°$ towards the (111) plane and with respect to the [$1\bar{1}02$] axis of the substrate. As a result of this the aligned yields in Fig. 1 (curves 1', 2', 3') are nearly equal to the random ones in the substrate part of the spectra. Therefore, the calculation of depth profiles of disordered silicon atoms as demonstrated by Picraux [17] is not suitable in our case. We can only deduce that the upper limit of the relative concentration of disordered silicon atoms in the near interface region is (25 ± 5) %.

The results of our RBS measurements affirm that in the samples in a region of 300 nm thickness above the interface substrate–deposit the defect concentration continuously decreases with increasing distance from the interface. Above this region $A(d)$ reaches a nearly constant value of 0.1 corresponding to a considerably high concentration of about 4% disordered silicon atoms.

Additional measurements of the layer resistance have shown that the conductivity decreases with increase of defect concentration. The layer denudation by anodic oxidation had to be interrupted here because of the disappearing conductivity of the layer at a residual thickness of 180 nm.

Acknowledgements

We gratefully appreciate valuable discussions with Prof. Dr. G. Götz and the technical cooperation of K. Garlipp and U. Barth, Jena.

References

[1] J. Mercier, J. Electrochem. Soc. **117**, 812 (1970).
[2] M. S. Abrahams, C. J. Buiocchi, R. T. Smith, J. F. Corboy, Jr., J. Blanc, and G. W. Cullen, J. appl. Phys. **47**, 5139 (1976).
[3] J. Blanc and M. S. Abrahams, J. appl. Phys. **47**, 5151 (1976).
[4] M. S. Abrahams and C. J. Buiocchi, Appl. Phys. Letters **27**, 325 (1975).
[5] A. G. Cullis and G. R. Booker, Thin Solid Films **31**, 53 (1976).
[6] J. Trilhe, H. Legal, and G. Rolland, J. Crystal Growth **45**, 439 (1978).
[7] T. Hayashi and S. Kurosawa, J. Crystal Growth **45**, 426 (1978).
[8] H. M. Manesevit, J. Electrochem. Soc. **126**, 1785 (1979).
[9] M. S. Abrahams, J. L. Hutchinson, and G. R. Booker, phys. stat. sol. (a) **63**, K3 (1981).
[10] F. A. Ponce and J. Aranovich, Appl. Phys. Letters **38**, 439 (1981).
[11] L. Ickert and H. G. Schneider, Wachstum einkristalliner Schichten, VEB Deutscher Verlag für Grundstoffindustrie, Leipzig 1983.
[12] G. W. Cullen, M. S. Abrahams, W. E. Ham, L. Jastrzebsky, R. T. Smith, M. Blumenfeld, G. Marbeke, and J. Lagowski. Crystal Growth **56**, 281 (1982).
[13] F. A. Ponce, Appl. Phys. Letters **41**, 371 (1982).
[14] W. H. Chu, J. W. Mayer, and M. A. Nicolet, Backscattering Spectrometry, Academic Press, New York 1978.
[15] G. Götz, Ionometrie in: Festkörperanalyse mit Elektronen, Ionen und Röntgenstrahlen, (Ed.) O. Brümmer, J. Heydenreich, K.-H. Krebs, and H. G. Schneider, VEB Deutscher Verlag der Wissenschaften Berlin 1980 (p. 357).
[16] B. Gruska and G. Götz, phys. stat. sol. (a) **67**, 129 (1981).
[17] S. T. Picraux, J. appl. Phys. **44**, 587 (1973).

(Received March 8, 1983)

phys. stat. sol. (a) 77, 207 (1983)

Subject classification: 11; 20.1; 22.8.1

Sektion Physik der Friedrich-Schiller-Universität Jena[1])

Anisotropic Effects in N^+-Implanted $LiNbO_3$

By

SYLVIA JETSCHKE, H. KARGE, and K. HEHL

A model of damage formation in $LiNbO_3$ by N^+ (150 keV, 300 K)-implantation is discussed including the results of RBS, volume expansion, and refractive index investigations of simultaneously implanted, X-, Y- and Z-cuts. For dose values $D_{N^+} \leqq 1 \times 10^{15}\ cm^{-2}$ most of the displaced Nb atoms occupy vacant octahedral lattice sites. Because of the high stability of the crystal structure in the c-direction, a complete amorphization of the implanted $LiNbO_3$ layers is not attained even with $D_{N^+} = 4 \times 10^{16}\ cm^{-2}$.

Anhand der Ergebnisse von RBS-, Volumendilatations- und Reflexionsmessungen an gleichzeitig implantierten X-, Y- und Z-Schnitten wird ein Modell der Defekterzeugung in $LiNbO_3$ durch N^+ (150 keV, 300 K)-Implantation diskutiert. Für Dosiswerte $D_{N^+} \leqq 1 \times 10^{15}\ cm^{-2}$ wird der größte Teil der deplazierten Nb-Atome auf vakanten Oktaederplätzen nachgewiesen. Wegen der hohen Stabilität der Kristallstruktur in Richtung der c-Achse kann eine vollständige Amorphisierung der implantierten $LiNbO_3$-Schichten auch mit $D_{N^+} = 4 \times 10^{16}\ cm^{-2}$ nicht erreicht werden.

1. Introduction

The production of radiation damage in $LiNbO_3$ by ion irradiation is a useful method to change the optical and chemical properties of the material in thin near-surface layers. Several authors [1 to 4] pointed out a remarkable decrease in the refractive index up to saturation values $\Delta n_o/n_o = 0.07 \pm 0.01$ [2, 4] at room temperature implantations. Damage formation and refractive index change are in rough approximation insensitive to ion species and primarily a function of the energy density deposited by nuclear processes [2]. Furthermore, sensibilization of the ion-implanted layers to etching in diluted HF was found [2, 5, 6].

These effects are discussed [4, 6] on the basis of a simple model of damage formation in $LiNbO_3$ (Y-cut) by N^+, O^+, or Ne^+ implantation, deduced from RBS, volume expansion, and etching rate enhancement measurements, as follows:

1. In the low-dose region point defects and simple defect clusters are generated up to a degree of disorder of about 20%. These defects and all property changes anneal out in the temperature range between 500 and 700 K.
2. At doses greater than $10^{15}\ cm^{-2}$ the predamaged layer is converted into a quasi-amorphous region.
3. At further energy deposition drastic volume expansion of the implanted surface regions takes place, connected with a strong increase of the etching rate. The disorder and the refractive index change in such layers are stable at least up to 800 K.

Crystalline $LiNbO_3$ has a pronounced direction dependence of the physical properties [7], especially a high stability of the structural arrangement with respect to the c-axis. Even after strong destruction by ion implantation at stage 3 the anisotropy of the refractive indices ($n_e - n_o$) does not vanish completely in the implanted layer [2, 4].

[1]) Max-Wien-Platz 1, DDR-6900 Jena, GDR.

To clarify this question and to try a localization of the point defects in the low-dose region, we performed a more detailed investigation of both defect formation and refractive index change in N^+-implanted $LiNbO_3$ for the three fundamental cuts (*X*-, *Y*-, and *Z*-cuts).

2. Results

X-, *Y*-, and *Z*-cuts of $LiNbO_3$ single crystals were polished and cleaned chemically including an etching step in diluted HF to obtain a high surface quality of the samples. The Al covered [6] samples were implanted with 150 keV N^+ ions at ion doses D_{N^+} ranging from 2×10^{14} to 4×10^{16} cm^{-2}. Because the ion current density was lower than 1 μA/cm^2, we could be sure that the sample temperature during implantation was lower than 400 K [3]. By simultaneous implantation of the different cuts of the samples identical irradiation conditions were attained for every ion dose.

2.1 Radiation damage

The RBS measurements were carried out with 1.4 MeV He^+ ions. Fig. 1 shows typical energy spectra of the He^+ ions backscattered by the Nb atoms in *Y*-cuts of $LiNbO_3$. The aligned yield is growing slowly with the ion dose up to $D_{N^+} = 1 \times 10^{15}$ cm^{-2}. For higher doses a strong increase of the damage is observed and at $D_{N^+} = 5 \times \times 10^{15}$ cm^{-2} the level of the random yield is reached. This indicates a quasi-amorphous arrangement of the Nb atoms in the implanted layer.

From the RBS spectra the concentration profiles of displaced Nb atoms were calculated with a computer programme [8] based on the multiple scattering theory of ion dechanneling [9]. Because dechanneling is provided by the Li and O atoms as well, we used in our calculations the averaged atomic number 13.6 for the multiple scattering.

The disordered atoms are assumed to be random-like distributed. Therefore, the evaluated values $N'(x)/N_0$ (N_0 total concentration of the Nb atoms) can be interpreted as an effective measure of the radiation damage neglecting the differences in the local arrangement of the disordered atoms seen by the analyzing He^+ beam under different channeling conditions. Fig. 2 shows the depth profiles of the relative concentration $N'(x)/N_0$, which are essentially different for the three crystalline cuts investigated. In the dose region up to 1×10^{15} cm^{-2} the evaluated number of the effective Nb scattering centres is two to three times higher for the *X*-cut in comparison to the *Y*- and *Z*-cuts. For doses of $D_{N^+} = 5 \times 10^{15}$ cm^{-2} the profiles reach the random level in

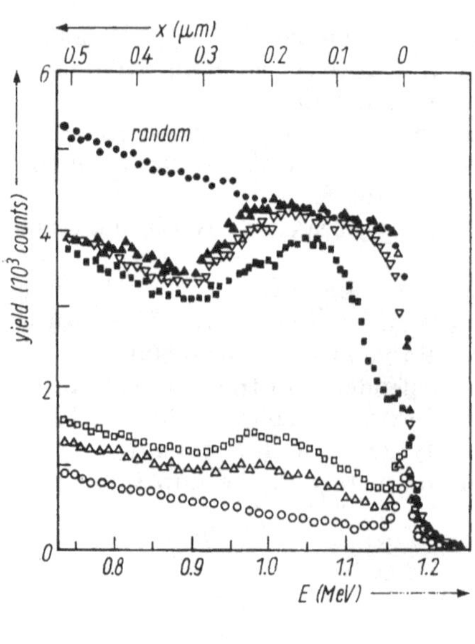

Fig. 1. RBS spectra of N^+ (150 keV, 300 K) implanted *Y*-cut $LiNbO_3$ crystals in dependence on the ion dose D_{N^+} (▲ 2×10^{16}, ▽ 5×10^{15}, ■ $2 \times \times 10^{15}$, □ 10^{15}, △ 5×10^{14} cm^{-2}, ○ unimplanted). In the energy region shown the counts correspond to scattering events on the Nb atoms only

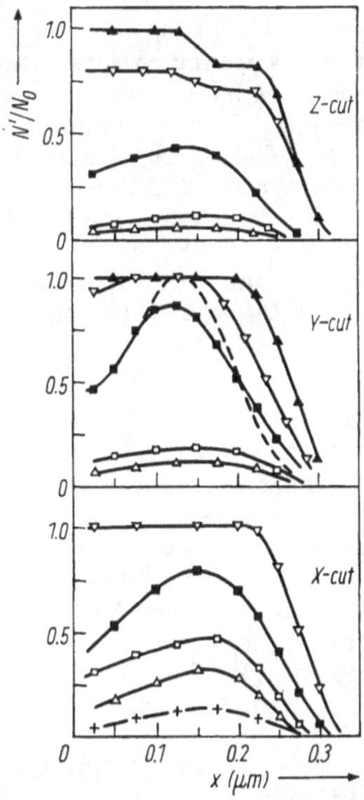

Fig. 2. Damage distributions $N'(x)/N_0$ in N^+ (150 keV, 300 K) implanted $LiNbO_3$ in dependence on the ion dose D_{N^+} for X-, Y-, and Z-cut samples (D_{N^+}: ▲ 2×10^{16}, ▽ 5×10^{15}, ■ 2×10^{15}, □ 10^{15}, △ 5×10^{14}, + 2×10^{14} cm^{-2}). Dashed line: depth distribution of the energy deposited by nuclear processes per unit length [15]

case of the X- and Y-cuts, whereas a considerable channeling effect still exists for the Z-cut. As a summary the averaged values $\overline{N}'/N_0$ versus ion dose D_{N^+} are shown in Fig. 7 for the three different cuts of the $LiNbO_3$ crystal.

2.2 Energy dependence of the dechanneling function

To investigate the defect structure of $LiNbO_3$ after a low-dose implantation ($D_{N^+} = 5 \times 10^{14}$ cm^{-2}) RBS measurements were carried out in dependence on the energy E_0 of the incident He^+ ions ranging from 0.8 to 1.8 MeV. The dechanneling function $P(E_0) = [\chi(E_0) - \chi_1(E_0)]/[1 - \chi_1(E_0)]$ [10] at a depth directly behind the disordered layer ($x = 0.3$ μm) was computed from the measured relative yields χ (implanted layer) and χ_1 (virgin crystalline substrate) of the backscattered He^+ ions. The results for the three fundamental cuts of $LiNbO_3$ are shown in Fig. 3. The values $P(E_0)$ are different for the three cuts in accordance with the different dechanneling efficiencies of the disordered Nb atoms. But for each cut an E_0^{-1}-dependence of the dechanneling function was found. This indicates point-defect-like structures, but no weakly displaced lattice atoms [11]. The measured dechanneling function $P(E_0)$ contains no further information about the localization of the defects themselves.

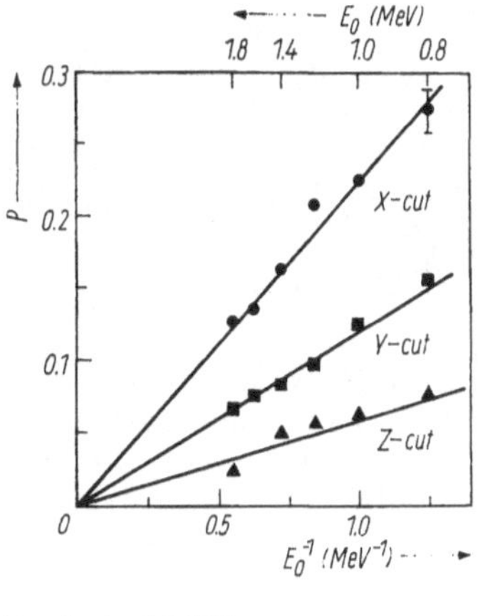

Fig. 3. Dechanneling function $P(E_0)$ vs. E_0^{-1} for X-, Y-, and Z-cut $LiNbO_3$ implanted with N^+ ions (150 keV, 300 K) ($D_{N^+} = 5 \times 10^{14}$ cm^{-2})

2.3 Volume expansion

The volume expansion of the implanted regions relative to the surrounding substrate was studied by means of a Talystep. The measuring conditions have already been described in [6]. The results of the relative volume dilatation $\Delta V/V$ in N^+-implanted $LiNbO_3$ are shown in Fig. 7 in dependence on the ion dose D_{N^+}. The effect in the low-dose region $D_{N^+} \leqq 2 \times 10^{15}$ cm^{-2} is well beyond the detection threshold ($\Delta V/V$ about 1%). The volume expansion becomes measurable at a dose $D_{N^+} > 2 \times 10^{15}$ cm^{-2} and reaches the value of 0.11 ± 0.015 for $D_{N^+} = 4 \times 10^{16}$ cm^{-2}.

Whereas the volume expansion is the same for the X- and Y-cuts within the limits of error, the Z-cut exhibits considerably lower expansions for $D_{N^+} < 2 \times 10^{16}$ cm^{-2}.

2.4 Refractive index

The change of the optical properties was investigated by the measurement of the ratio R/R_0 of the reflectivity of the implanted region and that of the substrate in dependence on the wavelength λ (ranging from 0.4 to 1.1 μm). An experimental arrangement on the basis of a polarization microscope was used for these measurements at normal incidence of light. The beam reflected from the backside of the sample was suppressed by a special geometrical form of the samples. By using polarized light the changes of the ordinary refractive index Δn_o and the extraordinary refractive index Δn_e were separately determined assuming a multilayer system and approximating the experimental data for the optical reflectivity by the calculated ones [12]. The absorption of the $LiNbO_3$ substrate and the implanted layer can be neglected. The index change was supposed to be independent on the wavelength.

Fig. 4 shows the profiles of refractive index change $\Delta n(x) = n_{cryst} - n_{impl}(x)$ in the N^+-implanted X-cut $LiNbO_3$ for both fundamental directions of polarization corresponding to $n_{cryst} = n_o$ and $n_{cryst} = n_e$. The index change increases with dose, but above $D_{N^+} = 5 \times 10^{15}$ cm^{-2} Δn_e remains constant, whereas a further increase of Δn_o takes place. The shape of the index profiles corresponds to the depth distribution of the energy deposited by nuclear processes. The layer thickness increases with ion dose as a consequence of the spread of damage distribution (Fig. 2) and the volume expansion (Fig. 7). In the near-surface region the profiles $\Delta n_e(x)$ may be influenced by outdiffusion of Li atoms during and after implantation, whereas $\Delta n_o(x)$ remains unefected [13].

The profiles $\Delta n_e(x)$ show a smaller layer thickness in comparison to $\Delta n_o(x)$, especially

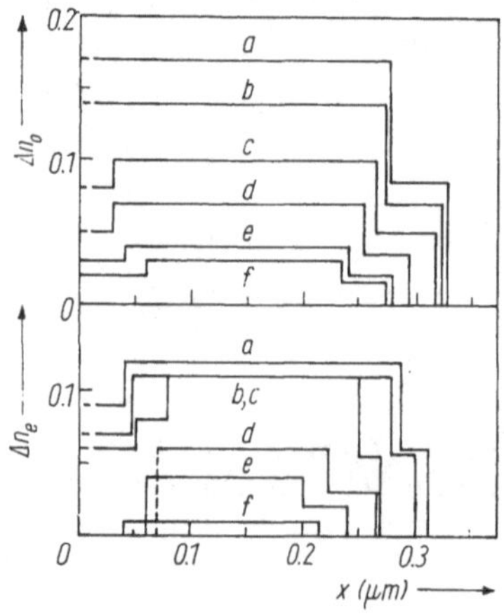

Fig. 4. Depth profiles of refractive index change $\Delta n_o(x)$ and $\Delta n_e(x)$ for N^+ (150 keV, 300 K) implanted X-cut $LiNbO_3$ (D_{N^+}: (a) 4×10^{16}, (b) 2×10^{16}, (c) 5×10^{15}, (d) 2×10^{15}, (e) 5×10^{14}, (f) 2×10^{14} cm^{-2})

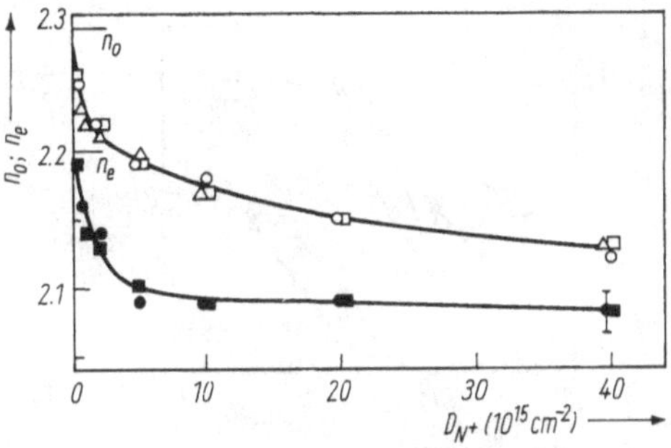

Fig. 5. Refractive indices n_o (empty symbols) and n_e (full symbols) ($\lambda = 632.8$ nm) vs. dose of the implanted N^+ ions, in the depth region between 0.1 and 0.2 μm of the N^+ (150 keV, 300 K) implanted $LiNbO_3$ crystals (*X*- (●, ○), *Y*- (■, □), and *Z*-cut (△)

for the lower dose values. A steeper change of the refractive index Δn_e from the layer to the substrate must be stated because of the remarkably higher interference amplitude in the R/R_0 spectra for n_e. This can be a consequence of different stresses parallel and perpendicular to the *c*-axis.

The index profiles of the simultaneously implanted *X*- and *Y*-cuts are approximately the same for every ion dose. When measuring the reflectivity of the *Z*-cut, the index n_o is effective for all polarization directions of the perpendicularly incident light. The measured Δn_o correspond to that of *X*- and *Y*-cuts.

Fig. 5 shows the refractive indices n_0 and n_e of the depth region between 0.1 and 0.2 μm versus ion dose for the cuts investigated. The decrease of the refractive indices of $LiNbO_3$ by ion implantation is the same for the different cuts, considered within the limits of error $\Delta(\Delta n) = 0.015$. At very high doses ($D_{N^+} = 4 \times 10^{16}$ cm^{-2}) the changes of the indices reach the following values:

$$n_o = 0.165 \pm 0.015\,,$$
$$n_e = 0.120 \pm 0.015\,.$$

The anisotropy ($n_e - n_o$) of -0.09 for the $LiNbO_3$ substrate is reduced in the dose region above 5×10^{15} cm^{-2}, but we observed an anisotropy value $(n_e - n_o) = -0.045$ up to dose values of 4×10^{16} cm^{-2}.

3. Discussion

The radiation damage and volume expansion of N^+-implanted $LiNbO_3$ are summarized in Fig. 7 in dependence on the ion dose. Because the implantation was done 8° off the axis, the primary damage of the $LiNbO_3$ crystals should not depend on the cut of the crystals. This is also indicated by equal refractive index changes in different cuts generated by the same dose values. Therefore, differences of the mean values $\overline{N'}/N_0$ must be discussed at first as differences in the local arrangement of the disordered Nb atoms seen by the analyzing He^+ ion beam along the different channeling directions.

The arrangement of the Nb, Li, and O atoms [14] seen by the analyzing He^+ ion beam falling on the $LiNbO_3$ crystal perpendicularly to the *X*-, *Y*-, and *Z*-cut surfaces is shown in Fig. 6:

(i) *Z-cut*: The channeling of the He^+ ions is mainly provided by Li–Nb rows. Each Li or Nb atom, respectively, occupies a place near the centre of an O octahedron. But there are more octahedral sites along the *z*-direction than Li and Nb atoms, so every third place is vacant.

(ii) *Y-cut*: The rows of different species of atoms are well separated. According to the much higher potential of the Nb rows compared to that of the Li and O rows the channeled He^+ ions are essentially steered by the Nb rows. We have to mention that the vacant octahedral sites will be found near the Nb rows.

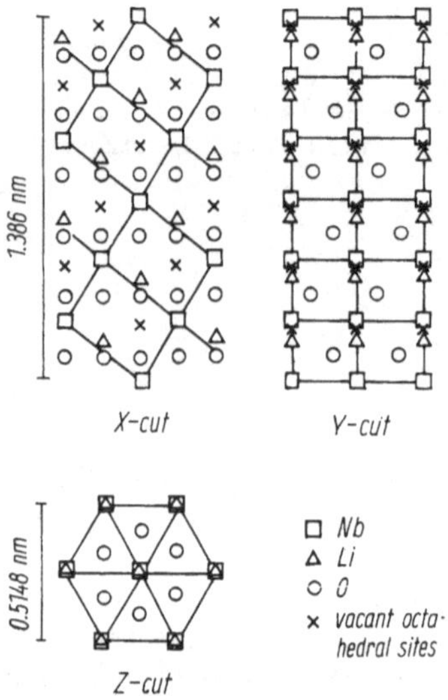

Fig. 6

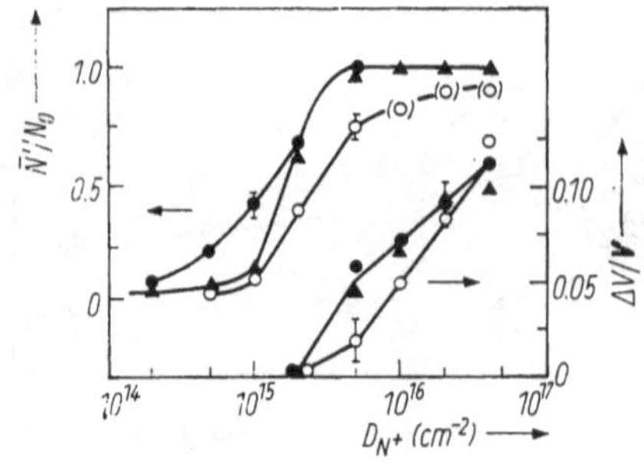

Fig. 7

Fig. 6 Arrangement of the Nb, Li, and O atoms perpendicular to the *X*-, *Y*-, and *Z*-cut surfaces of the $LiNbO_3$ crystals

Fig. 7. Radiation damage (averaged values $\overline{N}'/N_0$) and volume expansion (relative values $\Delta V/V$) in dependence on the ion dose D_{N^+} for *X*- (●), *Y*- (▲), and *Z*-cut (○) samples

(iii) *X-cut*: The situation is nearly the same as in case of the *Y*-cut with the exception that here the rows of vacant octahedral sites are well separated from those of Nb atoms.

Taking into consideration this arrangement of atoms in $LiNbO_3$ crystals, the dose dependence of creation of radiation damage can be subdivided in three distinct regions with different effects on the change of properties in ion-implanted $LiNbO_3$ (Fig. 7).

3.1 Predamage stage ($D_{N^+} \leq 1 \times 10^{15}$ cm^{-2})

For low doses mainly point defects are generated, which yield a rearrangement of Nb atoms on vacant octahedral sites. For the *X*-cut a relatively high RBS yield is achieved because these places are situated far from the rows and near the middle of the channel.

On the contrary, for the *Y*- and *Z*-cuts the rearranged Nb atoms are shielded by Nb atoms on regular lattice sites, and so the RBS yield is smaller.

Comparing the radiation damage of the three cuts for $D_{N^+} = 1 \times 10^{15}$ cm^{-2} we estimated that a contribution of 10% can be attributed to randomly distributed defects, whereas 30% of the value measured for the *X*-cut are caused by Nb atoms occupying vacant octahedral sites. The interpretation of the defects as point defects formed in this stage is supported by the energy dependence of dechanneling. The shift of Nb atoms yields a stress in the lattice but no volume expansion takes place.

The decrease of the refractive indices is already of the order of 30% of the maximum changes, without variation of the anisotropy ($n_e - n_o$). These changes of the refractive indices are mainly caused by the modified bond relations to the O^{2-} ions in $LiNbO_3$.

3.2 Heavy damage stage (1×10^{15} cm^{-2} < D_{N^+} < 5×10^{15} cm^{-2})

In this region heavily damaged clusters are generated which yield a completely random arrangement of the Nb atoms relative to the x- and y-directions. But still there exists a preferential arrangement of the Nb atoms in rows along the z-direction (c-axis) providing a much lower RBS yield in case of the Z-cut.

The volume dilatation, starting at $D_{N^+} > 2 \times 10^{15}$ cm^{-2}, i.e. at a damage level $N'/N_0 > 0.6$, gives an additional contribution to the change of the optical properties due to the decreased density of the $LiNbO_3$. The decrease of the refractive indices reaches 60 or 80% of the maximum changes for n_o and n_e, respectively, with the full preservation of the anisotropy.

3.3 Saturation stage ($D_{N^+} \geqq 5 \times 10^{15}$ cm^{-2})

The damage measured by the displacement of Nb atoms saturates for the X- and Y-cuts in this dose region, and for dose values about 1×10^{16} cm^{-2} the extraordinary refractive index reaches its final value. The damage of the Z-cut shows also a saturation behaviour, but only in the surface region of the implanted layer.

The further increase of the implantation dose yields a considerable volume expansion up to 11%, connected with a further decrease of the refractive index n_o. Although the anisotropy decreases, for the highest doses which can be used avoiding blistering, it has still 50% of the starting value.

4. Conclusions

The results discussed illustrate that it is not possible under our conditions (N^+ implantation at 300 K) to amorphize $LiNbO_3$ crystals even using very high ion doses ($D_{N^+} = 4 \times 10^{16}$ cm^{-2}). The RBS results emphasize that $LiNbO_3$ is more stable in the z-direction than in a direction perpendicular to it, and the ion bombardment produces a shift of parts of the neighbouring rows along the z-direction preserving the channels along the c-axis but completely destroying the channels parallel to the x- and y-direction.

The comparison of the volume dilatation with the radiation damage (Fig. 7) supports the conclusion that the very heavy damage ($N'/N_0 > 0.6$) providing a relaxation of the lattice stress is the reason for the volume expansion. The differences found for the three cuts of $LiNbO_3$ crystals again verify the higher stability against damage of the Z-cut compared to that of the X- and Y-cuts.

Further investigations about the annealing behaviour of the radiation-induced defects in $LiNbO_3$ are to be published. Experiments to amorphize a $LiNbO_3$ crystal by implantation of heavier ions at a temperature of 70 K are planned.

References

[1] D. T. Y. Wei, W. W. Lee, and L. R. Bloom, Appl. Phys. Letters **25**, 329 (1974).
[2] G. L. Destefanis, P. D. Townsend, and J. P. Gailliard, Appl. Phys. Letters **32**, 293 (1978). G. L. Destefanis, J. P. Gailliard, E. L. Ligeon, S. Valette, B. W. Farmery, P. D. Townsend, and A. Perez, J. appl. Phys. **50**, 7898 (1979).
[3] K. Wenzlik, J. Heibei, and E. Voges, phys. stat. sol. (a) **61**, K207 (1980).
[4] H. Karge, G. Götz, U. Jahn, and S. Schmidt, Nuclear Instrum. and Methods **182/183**, 777 (1981).
[5] M. Kawabe, M. Kubota, K. Masuda, and S. Namba, J. Vacuum Sci. Technol. **15**, 1096 (1978).
M. Kawabe, S. Hirata, and S. Namba, IEEE-CAS **26**, 1109 (1979).

[6] H. KARGE, K. HEHL, U. JAHN, and U. KATENKAMP, Radiat. Eff., to be published.
[7] A. RÄUBER, Chemistry and Physics of Lithium Niobate, North-Holland Publ. Co., 1978.
[8] U. FINGER, K. GÄRTNER, H.-D. KOCH, and W. TRIPPENSEE, J. Radioanal. Chem. 28, 49 (1975).
[9] L. MEYER, phys. stat. sol. (b) 44, 253 (1971).
[10] W.-K. CHU, J. W. MAYER, and M.-A. NICOLET, Backscattering Spectrometry, Academic Press, New York 1978.
[11] N. MATSUNAMI, T. GOTO, and N. ITOH, Radiat. Eff. 33, 209 (1977).
[12] K. HEHL and W. WESCH, phys. stat. sol. (a) 58, 181 (1980).
[13] J. G. BERGMAN, A. ASHKIN, A. A. BALLMAN, J. M. DZIEDIC, H. J. LEVINSTEIN, and R. G. SMITH, Appl. Phys. Letters 12, 92 (1968).
[14] S. C. ABRAHAMS, J. M. REDDY, and J. L. BERNSTEIN, J. Phys. Chem. Solids 27, 997 (1966).
[15] D. K. BRICE, Radiat. Eff. 6, 77 (1970).

(Received January 26, 1983)

phys. stat. sol. (a) 77, 215 (1983)

Subject classification: 10.1; 11; 21.1

Institut für Metallphysik und Nukleare Festkörperphysik der Technischen Universität Braunschweig[1])

Velocities of Dislocation Groups in Very Thin Neutron-Irradiated Copper Single Crystals Measured by Slip Line Cinematography

By

H.-H. POTTHOFF

Slip line development on *very* thin flat single crystals of neutron-irradiated Cu (thickness down to only 15 to 20 μm, orientation for single glide, yield region, room temperature) is recorded by high-speed cinematography during tensile deformation. In such very thin crystals glide dislocations on the slip plane must be arranged in a rather simple way. Drops in tensile load occuring during initiation of single slip lines at the Lüders band front indicate that in the beginning of a slip line development dislocation groups traverse the whole glide plane in very short times. Evaluating the data measured for the slip line growth $v_s \geqq 10$ cm/s is found for screw dislocations and $v_e \geqq v_s$ for edge dislocations. For later stages on thin crystals and for all stages on thick crystals ($\geqq$ several 100 μm) slip line development is much slower and slip lines show many cross slip events which then appear to control the mean velocity of the dislocations.

Die Gleitlinienentwicklung auf *sehr* dünnen, flachen Einkristallen aus neutronenbestrahltem Cu (Dicke bis hinab zu nur 15 bis 20 μm, Orientierung für Einfachgleitung, Fließbereich, Zimmertemperatur) wird während der Zugverformung mit Hochfrequenz-Kinematographie aufgezeichnet. Gleitversetzungen müssen in derart dünnen Kristallen recht übersichtlich auf der Gleitebene angeordnet sein. Einbrüche im Lastschrieb, die bei der Entstehung einzelner Gleitlinien an der Lüdersband-Front auftreten, deuten darauf hin, daß die Versetzungsgruppen zu Beginn einer Gleitlinienentwicklung in sehr kurzen Zeiten die ganze Gleitebene überstreichen. Durch Auswertung des gemessenen Gleitlinienwachstums wird $v_s \geqq 10$ cm/s für Schraubenversetzungen und $v_e \geqq v_s$ für Stufenversetzungen gefunden. In späteren Stadien auf dünnen Kristallen und in allen Stadien auf dicken Kristallen ($\geqq$ einige 100 μm) ist das Gleitlinienwachstum erheblich langsamer und die Gleitlinien weisen viele Quergleitstellen auf, welche dann die mittlere Versetzungsgeschwindigkeit bestimmen.

1. Introduction

The velocity of moving dislocations has been measured mainly by means of the following well-known methods:

(i) *TEM in-situ experiments*: The velocity of dislocations directly measured using transmission electron microscopy (TEM) is always found to be slow [1 to 6]. This is not a surprising result, as in the in-situ experiments the tensile stress is increased only up to a value at which the dislocations in the TEM foil just start moving. Nevertheless a lot of information about generation of dislocations, cross slip, and so on can be obtained from such "micro-yield" experiments. On the other hand, the exact values and directions of the tensile stresses are not well known and surface effects may not be negligible.

(ii) *Stress pulses and etch pit investigations*: Using the technique of etch pitting before and after a stress pulse, velocities of dislocations can in principle be measured in a wide range of parameters [7 to 9]. In practice there are many difficulties. So the

[1]) Mendelssohnstr. 3, D-3300 Braunschweig, FRG.

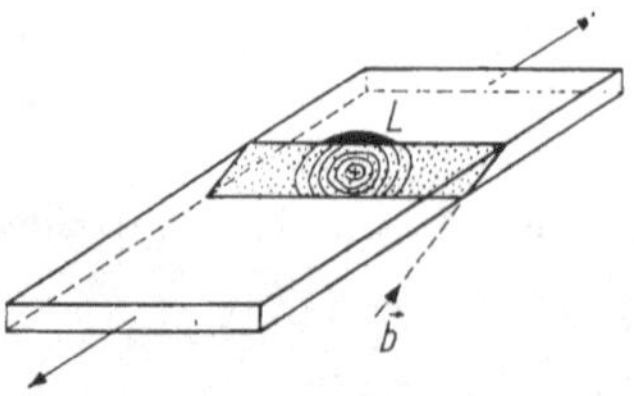

Fig. 1. Schematic drawing of a very thin crystal with slip plane (dotted) and propagating circular dislocation loops. Big arrows indicate direction of tensile stress. The slip line L with wedge-shaped profile at both ends is indicated only at the top surface. About 10^{-3} s after the very first beginning of visible slip L is about 300 μm long and in the centre its height corresponds to about 300 dislocations (situation d) of Section 3.1)

profile of the pulses is not very well known though it is of decisive importance. Furthermore it is not always easy to identify the correct etch pit of a dislocation that has moved, to make sure that the dislocation moves continuously during the whole stress pulse (proportionality between duration of pulses and corresponding etch pit distances), and to check the geometry of the dislocation lines with respect to the surface of the crystal (perpendicular or inclined). The method usually is limited to micro-yield deformations.

(iii) *Studies of slip line development*: If slip lines are generated on the surface of a crystal, their dynamic generation can be used to estimate the velocity of dislocations under realistic conditions of macro-yield deformation (continuous straining, critical shear stress, etc.) [10 to 14]. Again the fundamental problem arises that the geometry of the dislocations (distance from each other, angle relative to the surface, etc.) is not known.

The way we tried to solve this problem is the preparation of *very* thin crystals [15, 16] which are oriented for single glide and show very narrow slip planes (Fig. 1). The specimens, however, are thick enough to exclude problems with surface effects as encountered in TEM [1, 6].

If a dislocation source operating somewhere on a narrow slip plane is emitting loops, the edge-type dislocations will reach the surfaces of the crystal after a very short time and will form a slip line there. The remaining screw-type segments are expected to build up a pile-up which moves sideways to the edges of the slip plane.

This means that the first segment of a slip line on the surface of very thin crystals should be detected directly above the source. If more and more dislocations arrive at the surface, the slip line should increase in height and at the same time in length. The profile of the height along the slip line should be wedge-shaped on both sides of the centre and both tips of the wedges should move to the edges of the crystal during slip line growth. It will be shown below that such a slip line development indeed is the typical one measured experimentally on neutron-irradiated crystals (prepared according to Fig. 1).

Irradiated Cu was chosen for this investigation because of the following reasons:

(i) The obstacles to dislocation motion are well known, i.e. vacancy clusters and Frank loops [2, 17, 18]. Such obstacles will be destroyed by the first dislocations. So the following dislocations are moving in cleared channels. This results in the formation and movement of extended dynamic pile-ups corresponding to prominent slip lines.

(ii) The conditions of deformation remain identical for a considerable strain (10 to 20%, Lüders band propagation with constant load in the yield region). So several comparable measurements can be performed using the same specimen.

2. Preparation and Deformation of Specimens

All preparations were based on bulk cylindrical f.c.c. single crystals (99.999% Cu, grown using Bridgman's technique, oriented for single glide, diameter of 4 mm, length of about 60 mm, irradiated with fast neutrons ($E \geqq 0.1$ MeV) up to a total dose of

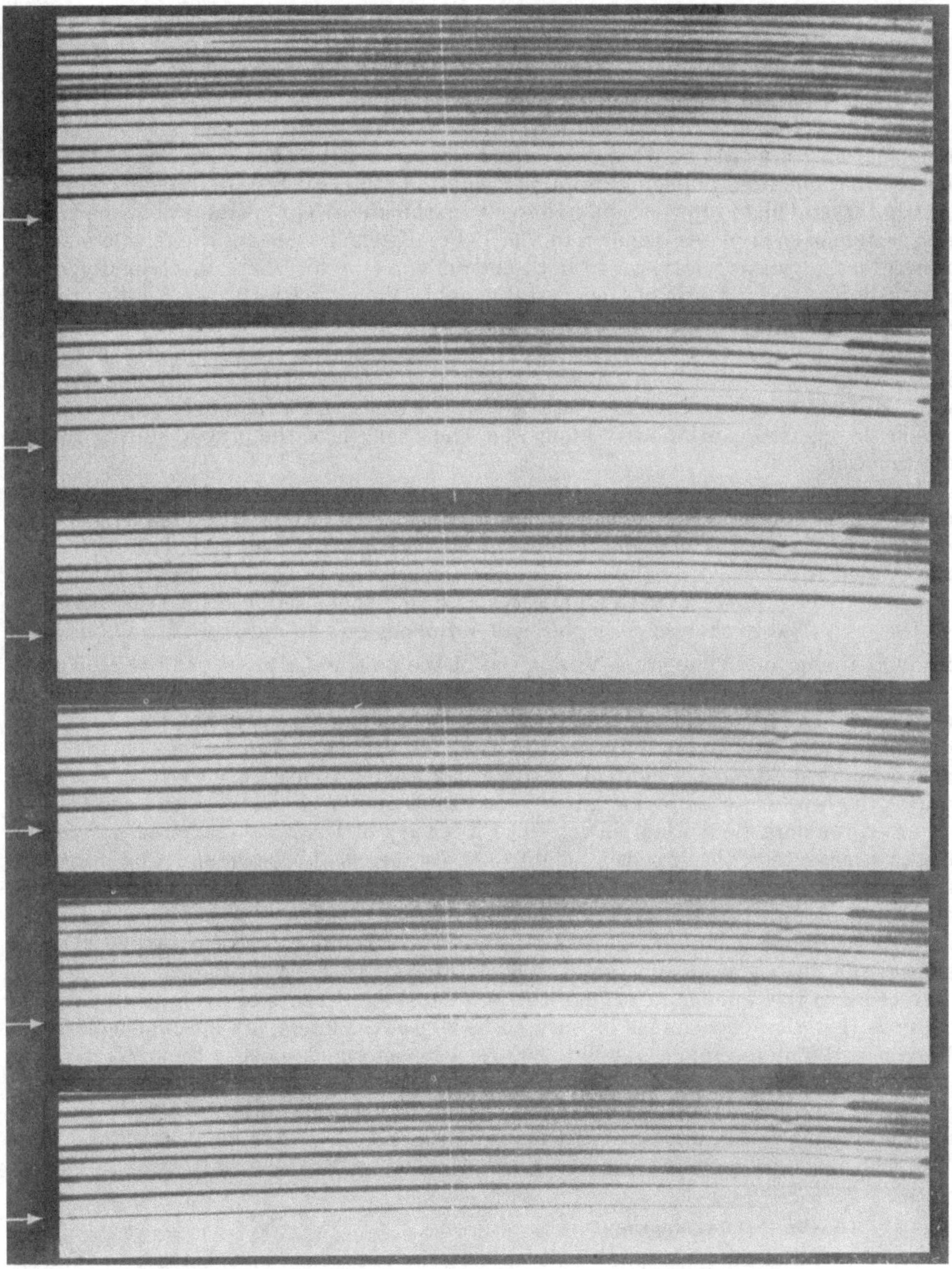

Fig. 2. Typical slip line development in front of the Lüders band. Thickness of the crystal about 20 μm, deformation rate $\dot{l}$ = 1.05 μm/s, crystal length about 20 mm, room temperature, visible slip line length about 400 μm

2×10^{18} neutrons/cm^2). n-Cu is our abbreviation for this material. The crystals were oriented by Laue's technique and mounted on an acid saw to produce slices which were exactly oriented for single glide (angle between Burgers vector and axis of crystal, and angle between Burgers vector and normal of slip plane: $45° \pm 1°$). After acid sawing the crystals had to be thinned down very carefully without any mechanical deformation. In a first step we used electrolytical cloth polishing similar to [19] and in a second one final polishing in an electrolytic bath. This resulted in smooth plane-parallel crystal plates down to only 15 to 20 μm thickness. The geometry of the primary slip system of such plates is shown in Fig. 1. Finally small grips for the tensile machine were glued (cyanoacrylate) on to both ends of the crystal. This had to be done very carefully for the long axis of the crystal should coincide with the tensile direction as exactly as possible.

In the beginning of deformation of such crystals usually several slip lines originated in the neighbourhood of both grips because of grip stresses. Then the slip line development stopped at one end of the crystal and at the other end a Lüders band was formed which propagated continuously along the whole length of the crystal during further deformation.

All measurements were performed using a special tensile machine that can be moved under a light microscope to examine the slip lines on the whole specimen length. This machine is equipped with a small sensitive strain gage ring load cell. The stiffness of the system is governed by the sample. For a crystal of 100 μm thickness it is of the order of 0.15 N/μm. Tests with an additional spring which reduced the stiffness of the system remarkably showed no significant influence on the dynamic development of individual slip lines. The cross-head speed of the machine always was $\dot{l} = 1.05$ μm/s.

We were especially interested in the very first stage of the development of a new slip line which at a time originated in front of the Lüders band. In that very first stage the height of a slip line is only several 10 nm as will be shown below. This is fairly below the resolution of a light microscope. We therefore used a very intensive light source (high-pressure mercury lamp Osram HBO 100 W) focused at the crystal and adjusted for dark field illumination. The intensity of the light scattered at the slip line is a measure of the slip step height [20]. By means of a beam splitting mirror an amount of 5% of the intensity passing the microscope is used to watch the slip line development directly, while the remaining amount of 95% is directed to a Hycam high-speed film camera that can be operated with frame rates of up to 20000/s. A typical slip line development taken at 8000 frames/s is shown in Fig. 2.

In order to give some indication of the statistical significance of the following results we note that three specimens of only 15 to 20 μm thickness, six specimens of 30 to 100 μm, and four specimens of 200 to 600 μm were deformed (most of them 0.7 to 1 mm in width and 10 to 20 mm in length). About 60 slip line developments were recorded successfully on high-speed films and about ten times more could be observed on films with low frame rates (only several 100/s) or by eye.

3. Results

3.1 Slip line development

The evaluation of the high-speed films showed the following typical stages of slip line development (thickness of the crystal 20 μm, deformation at room temperature with a cross-head speed of ≈ 1 μm/s, for a) to d) compare Fig. 2):

a) A short segment of a new slip line originates 5 to 10 μm in front of the foremost slip line of the Lüders band.

b) Its height rapidly increases during the next 10^{-4} s while the tips of the wedge-shaped profile move towards the edges of the crystal.

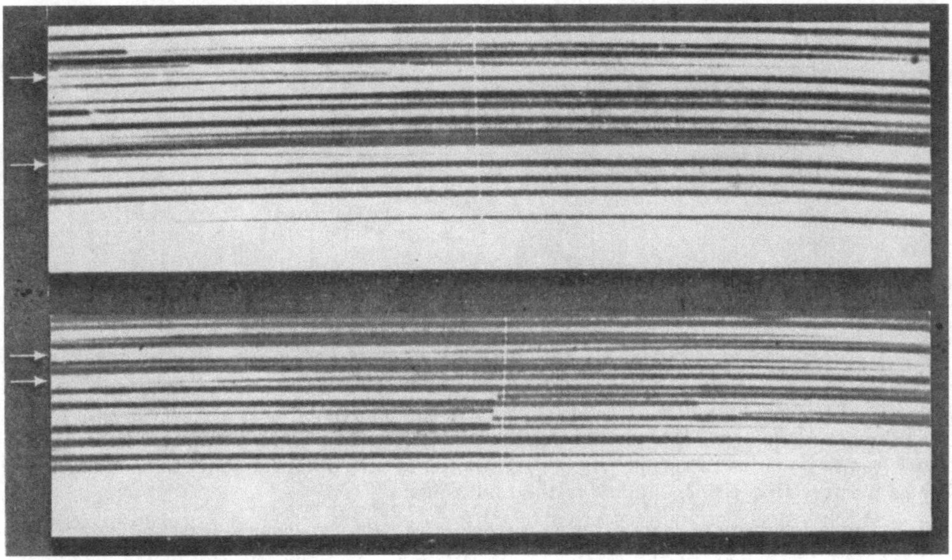

Fig. 3. Slip lines which are not parallel to their neighbours (marked by arrows). Visible slip line length about 400 μm

c) After only several 10^{-4} s the distance between the two tips is about 300 μm and in the centre the height of the slip line is about 50 nm corresponding to about 300 dislocations.

d) The typical situation after about 10^{-3} s is shown in Fig. 1 and typical values are given in its caption. All evaluations of the dislocation velocity discussed below are based on the dynamic behaviour a) to d).

e) For several 10^{-3} s situation d) does not change.

f) After this break the slip line continues growing in a second phase and the two tips of the line move until they reach the edges of the crystal.

g) Up to now no visible cross slip can be found along a slip line.

h) While the development of several new slip lines has already started in front of the old one, this old one gradually increases until it reaches a height h of 200 to 400 nm in a time of the order of 1 s. This growth always is accompanied by visible cross slip.

i) During h) in the spacings between old slip lines sometimes new ones are generated. The growth of such lines is completely different from that described above: It is much slower and more continuous and homogeneous. The final height of these lines is relatively low and often the development of such lines completely stops before the second tip of the line has reached the edges of the crystal. One most surprising observation is that the tips of some of these lines are not parallel to their neighbours (see Fig. 3). This suggests that the dislocations in the tip of the wedge undergo many very short-range double cross slip events.

j) Slip lines which finally lie behind the propagating Lüders band front (20 to 30 lines are in front of them) do not change any more during deformation in the yield region.

k) Each of the slip lines on one surface of a very thin crystal has its corresponding line on the opposite surface (see Fig. 4). This means that slip must have taken place across the whole glide plane. This corresponds to a lengthening of the crystal by about h, i.e. several 100 nm. If the rate of such a lengthening exceeds the external deforma-

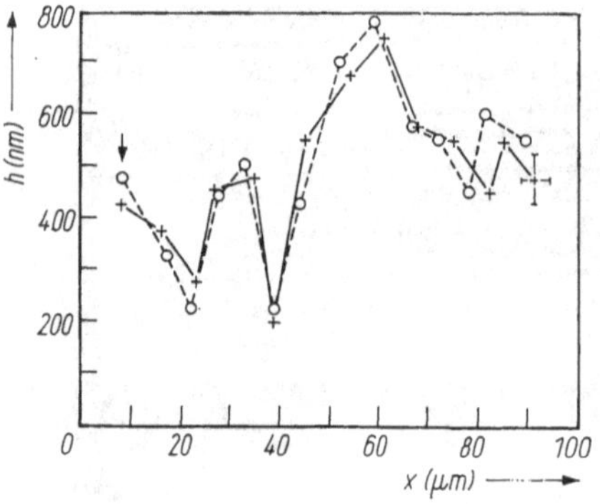

Fig. 4

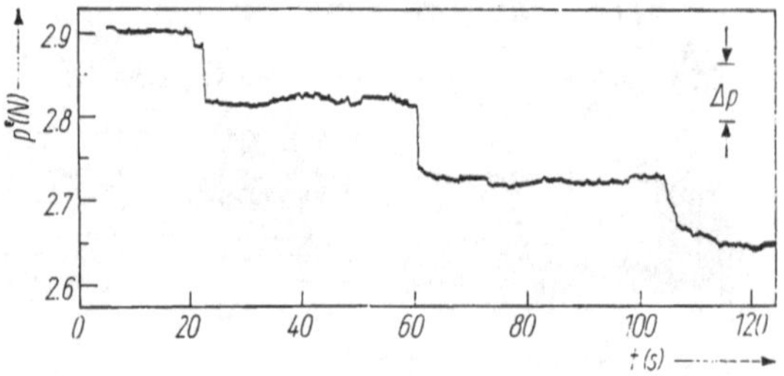

Fig. 5

Fig. 4. Height *h* of slip lines in the front of the Lüders band measured normal to the crystal surface by interference microscopy. *h* is plotted against the coordinate *x* along the tensile axis for both surfaces (top (———) and bottom (– – –)). Thickness of the crystal is about 20 μm. The arrow indicates the very first visible slip line of the Lüders band

Fig. 5. Example for drops of load *p* during a stress relaxation experiment (external deformation rate $\dot{l} = 0$ μm/s, thickness of the crystal about 100 μm). Using the stiffness of the system the expected load drop Δp for a sudden lengthening of the crystal by 400 nm has been calculated and is shown for comparison

tion rate it necessarily will produce a drop in the applied load *p* during the tensile experiment. Experimental *p* versus *t* curves in fact show such load drops (see Fig. 5). The magnitude of the drops indeed corresponds to a lengthening by several 100 nm.

3.2 Velocities of dislocations

In situation d) both ends of the slip line are wedge-shaped and show no visible indications for cross slip (contrary to later stages of development). Thus we assume that all dislocations corresponding to such a slip line have been produced by only one source which may be situated somewhere on the slip plane between the two tips of the wedge-shaped slip line.

3.2.1 $v_{\text{screw}} \approx v_{\text{edge}}$

Assuming first the same magnitudes for edge- and screw-type dislocations the dislocation loops should be circles as indicated in Fig. 1. Based on situation d) which is illustrated in Fig. 1 there are three ways to evaluate dislocation velocities:

1. The front loops of a screw-type group must have moved about 150 μm in only $\leq 10^{-3}$ s. So their mean velocity cannot have been less than about 15 cm/s.

2. Distances between dislocations in pile-ups observed by TEM are not less than about 0.1 μm in n-Cu. Most of the distances are several 0.1 μm [21, 22]. Calculations show as well that the distances of dislocations hardly can be less than about 0.1 μm because of the very strong elastic interactions between the dislocations. That is why we assume a mean distance of $\Delta = 0.2$ to 0.4 μm for the following estimations. The rate of dislocations arriving at the surface $\dot{n} = 300/10^{-3}$ s, that is 3×10^5/s. Using $\Delta = 0.2$ to 0.4 μm the mean dislocation velocity is $v = 6$ to 12 cm/s.

3. Using $\Delta = 0.2$ to 0.4 μm one finds that at the same time only about 75 complete loops can be placed on the slip plane. Therefore, many of the 300 dislocations must have been produced while the development of that slip line was already visible. As we assume that all dislocations (produced till situation d)) originated from only a single source this source must be able to emit about 3×10^5 dislocations/s.

Calculating for example the critical length L of a Frank-Read-type source using the typical yield stress τ_0 of n-Cu ($\tau_0 \approx 23$ MPa) one obtains about $L = 1$ μm. To produce a new loop the "arms" of the source have to glide a mean distance of about $L/2 = 0.5$ μm. As 3×10^5 loops have to be produced in only 1 s the mean velocity of the dislocations cannot be less than $v = \dot{n}L/2 = 15$ cm/s. Similar results can be found for different types of sources.

Thus three different types of estimation yield the same order of magnitude of the mean dislocation velocity $v \approx 10$ cm/s.

3.2.2 $v_{screw} \neq v_{edge}$

Assuming $v_{screw} \gg v_{edge}$ elliptical loops elongated perpendicular to the Burgers vector $\boldsymbol{b}$ must be formed and for the distances between dislocations $\Delta_{screw} \gg \Delta_{edge}$ must be found. This means that the inclinations of the wedges should be much smaller during slip line development than those found in the experiments. Thus, the relation $v_{screw} \gg v_{edge}$ must be wrong.

Assuming the reverse relation $v_{screw} \ll v_{edge}$ elliptical loops elongated parallel to $\boldsymbol{b}$ are formed (see Fig. 1) with $\Delta_{screw} \ll \Delta_{edge}$. Here v_{screw} can be determined as in Section 3.2.1 and the result is the same, $v_{screw} \approx 10$ cm/s. The value of v_{edge} cannot be determined but has to be higher than v_{screw} (according to the assumption $v_{edge} \gg v_{screw}$).

3.3 *Comparison with thick crystals*

All the results mentioned up to now are related to crystals of only 15 to 20 μm thickness. For crystals which are completely similar but about 100 μm in thickness the development of slip lines is very similar. Lines just originate with distances to each other which are a little bit greater and show a trend towards slight enhancement of cross slip and formation of close pairs of slip lines (see Fig. 6).

The arrangement of slip lines on crystals which are about 600 μm thick, however, looks completely different. Lines extending across the whole crystal without visible cross slip and extended wedge-shaped profiles cannot be found anymore even along the most extended lines. From the very first beginning of slip line development numerous points of cross slip can be seen with segments of primary slip lines between them which are only 10 to 100 μm long (see Fig. 6). The mean distance of neighbouring segments is only 2 to 4 μm (much less than in the Lüders front of thin crystals). While segment by segment a "bundle" of slip lines is formed, often even several different segments which are lined up and connected by double cross slip show the same dynamic development in spite of the cross slip between them.

If a "bundle" has reached a certain width, its development stops. Meanwhile about 50 to 100 μm in front of that bundle in the undeformed region one to three other bundles are already growing. In the yield region two different bundles always remain separated by a narrow slice of crystal containing no visible slip lines at all.

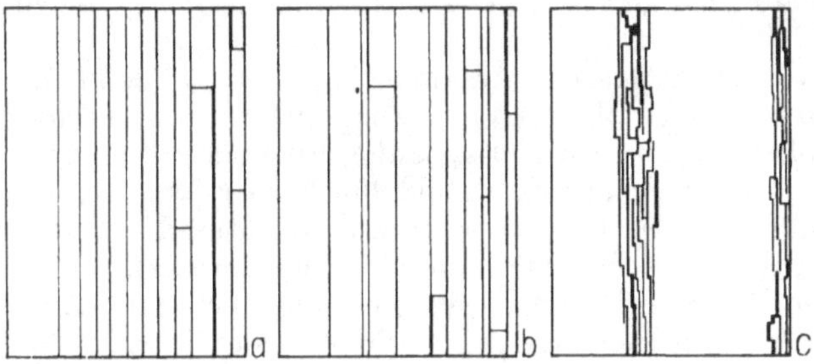

Fig. 6. Arrangement of slip lines on crystals of different thickness: a) 20, about b) 100, and c) 600 μm (schematic)

Very thick crystals (we used cylindrical ones of 4 mm diameter too) show the same arrangement of slip lines as those of 600 μm thickness. The active part of the Lüders band (consisting of growing slip line bundles), however, is more extended (up to about 10 mm).

It should be noted that within the limit of accuracy (which is only 10% to 15% for very thin crystals because of the limited accuracy of determination of the thickness of a sample) no remarkable change of the yield stress with specimen thickness was noticed.

4. Discussion and Conclusion

We found significant differences between slip line development on thin and thick crystals of neutron-irradiated Cu. These differences appear to be due to different amounts of cross slip.

On thin crystals ($\leqq$ 100 μm) no visible cross slip is detected in the very first stages of slip line development. Slip lines are long and wedge-shaped. The velocity of dislocations is found to be high ($\geqq$ 10 cm/s). On thick crystals ($\geqq$ several 100 μm) numerous points of cross slip can be seen even in the very first stages of slip line development. Slip segments are short. No wedges can be detected. The mean velocity of dislocations is found to be low ($\approx$ 0.1 mm/s). In spite of these differences the final heights of the slip lines are similar and the yield stress is the same.

On the one hand, dislocation motion is controlled by irradiation-produced obstacles and accordingly a stronger drag is expected on edges than on screws [23]. Such obstacles, however, will be destroyed by the first dislocations cutting through them [17, 24 to 26]. Thus, "cleared slip channels" are formed providing significantly enhanced mobility for the following dislocations emitted by the source. Because of the obstacle destruction no "inverse pile-ups" [27 to 29] are formed but rather "dynamic pile-ups" [30 to 32] with high dislocation density at their front and possibly very high velocity. On the other hand, by cutting obstacles jogs will be produced along the dislocation lines which are expected to produce more drag on screws than on edges [33, 34]. We find $v_{edge} \geqq v_{screw}$. So cutting of obstacles appears to be not the dominant dragging effect in the first stages of slip line formation but rather dragging of jogs.

During dislocation glide the dislocation length on primary slip planes increases and the probability for cross slip events increases, too. This may be one reason for the enhanced occurrence of cross slip on the extended cross-section of thick crystals. If dislocations have to undergo cross slip events as in later stages of slip line development on thin crystals or in all stages on thick crystals, the moving groups are split by the cross slip events and remain small. Hence the relatively short slip lines should not show wedges as confirmed by the observations. If double cross slip sources will emit new loops, these loops will impede the loops on the old slip plane by forming dipole configurations. These may give rise to local secondary slip which acts as a "strong obstacle" for further motion of primary groups [19, 35 to 39] and may govern the final height of the slip lines.

For very thin n-Cu samples we can summarize: Assuming $v_{screw} \approx v_{edge} \approx 10$ cm/s the very first stages of slip line development observed in the experiments can be understood using a *very* simple model. Assuming $v_{screw} \neq v_{edge}$ the principle of the model remains unchanged and the dislocation velocities should fulfil the relation $v_{edge} \geqq v_{screw} \geqq 10$ cm/s. This relation seems to contradict earlier results measured by Rodloff and Neuhäuser [14]. It agrees with recent measurements by Sinohara et al. [40, 41]. The reason is that the results in [14] refer to the development of "bundles" of slip lines on thick crystals where cross slip reduces the growing rate of slip lines by three to four orders of magnitude. Sinohara et al., however, observed dislocation motion without cross slip in the beginning of deformation as we did in this work.

In a subsequent paper we will additionally account for the bending stresses which are connected with the rotation of the lattice between the slipped and the unslipped region. It will be shown that the influences are not very important for n-Cu, but of decisive importance for alloys like α-brass.

Acknowledgements

The author wishes to thank the Deutsche Forschungsgemeinschaft for financial support and Prof. Ch. Schwink for his interest. He is grateful to Prof. H. Neuhäuser and Mr. K. Bothe for numerous very helpful discussions and comments.

References

[1] T. Imura and H. Saka, Mem. Fac. Engng., Nagoya Univ. **28**, 54 (1974).
[2] E. Johnson and P. B. Hirsch, Phil. Mag. A **43**, 157 (1981).
[3] T. Tabata, H. Fujita, and Y. Nakajima, Acta metall. **28**, 795 (1980).
[4] T. Tabata, N. Ueda, K. Hashimoto, and H. Fujita, High Voltage Electron Microscopy, 1977, Ed. T. Imura and H. Hashimoto, Japan. Soc. Electron Microscopy, Kyoto 1977 (p. 391).
[5] H. Mori and H. Fujita, J. Phys. Soc. Japan **38**, 1342 (1975).
[6] L. P. Kubin and J. L. Martin, Strength of Metals and Alloys, Vol. 3, Ed. P. Haasen, V. Gerold, and G. Kostarz, Pergamon Press, Toronto 1980 (p. 1693).
[7] M. Wada, M. Meshii, and K. Marukawa, phys. stat. sol. (a) **57**, 345 (1980).
[8] M. Kleintges, R. Labusch, H. G. Brion, and P. Haasen, Acta metall. **25**, 1247 (1977).
[9] M. Kleintges and P. Haasen, Scripta metall. **14**, 999 (1980).
[10] R. Becker and P. Haasen, Acta metall. **1**, 325 (1953).
[11] R. B. Pond, Jr., in: Inhomogeneity of Plastic Deformation, Ed. R. E. Reed-Hill, American Society for Metals, Metals Park (Ohio) 1971 (p. 1).
[12] S.-J. Liang and P. D. Pope, J. appl. Phys. **50**, 5726 (1979).
[13] M. J. Makin and J. V. Sharp, phys. stat. sol. **9**, 109 (1965).
[14] R. Rodloff and H. Neuhäuser, phys. stat. sol. (a) **49**, 455 (1978).
[15] R. Medrano and G. Schoeck, phys. stat. sol. (a) **5**, 147 (1971).
[16] J. T. Fourie, Canad. J. Phys. **45**, 777 (1967).
[17] P. B. Hirsch, in: Proc. Conf. Univ. Bristol Vacancies '76, The Metals Society, London 1976 (p. 95).
[18] M. Rühle, F. Häussermann, and M. Rapp, phys. stat. sol. **39**, 609 (1970).
[19] J. S. Ahearn, Jr., J. P. Monaghan, Jr., and J. W. Mitchell, Rev. Sci. Instrum. **41**, 1853 (1970).
[20] H. Neuhäuser and R. Rodloff, Acta metall. **22**, 375 (1974).
[21] J. V. Sharp, Phil. Mag. **16**, 77 (1967).
[22] G. Kroeger, Diploma Work, TU Braunschweig, 1977.
[23] M. Saxlova, Czech. J. Phys. B **19**, 610 (1969).
[24] U. Essmann and A. Seeger, phys. stat. sol. **4**, 177 (1964).
[25] J. E. Foreman and J. V. Sharp, Phil. Mag. **19**, 931 (1969).
[26] J. V. Sharp, Radiat. Eff. **14**, 71 (1972).
[27] A. R. Rosenfield and G. T. Hahn, in: Dislocation Dynamics, Ed. A. R. Rosenfield, G. T. Hahn, A. L. Bement, Jr., and R. I. Jaffee, McGraw-Hill Publ. Co., New York 1968 (p. 255).
[28] S. I. Zaitsev and E. M. Nadgornyi, phys. stat. sol. (a) **9**, 353 (1972).
[29] S. I. Zaitsev and E. M. Nadgornyi, phys. stat. sol. (a) **28**, 49 (1975).
[30] R. J. Arsenault, Phil. Mag. **24**, 295 (1971).
[31] R. J. Arsenault, Scripta metall. **12**, 633 (1978).
[32] R. Rodloff and H. Neuhäuser, phys. stat. sol. (a) **37**, K93 (1976).
[33] A. Seeger, Phil. Mag. **46**, 1194 (1955).

[34] J. P. Hirth and J. Lothe, Theory of Dislocations, McGraw-Hill Publ. Co., New York 1968.
[35] B. J. Hockey and J. W. Mitchell, Phil. Mag. **26**, 409 (1972).
[36] O. Nittono, Japan. J. appl. Phys. **10**, 188 (1971).
[37] P. D. K. Nathanson, P. J. Jackson, and D. R. Spalding, Acta metall. **28**, 823 (1980).
[38] O. L. de Lange, P. J. Jackson, and P. D. K. Nathanson, Acta metall. **28**, 833 (1980).
[39] P. J. Jackson, O. L. de Lange, and P. D. K. Nathanson, in: Dislocation Modelling of Physical Systems, Ed. M. F. Ashby, R. Bullough, C. S. Hartley, and J. P. Hirth, Pergamon Press, Oxford 1981 (p. 337).
[40] K. Sinohara, S. Kitajima, and A. Tanaka, Scripta metall. **15**, 1007 (1981).
[41] K. Sinohara, S. Kitajima, A. Okamoto, and M. Ohta, Trans. Japan. Inst. Metals **23**, 8 (1982).

(Received December 23, 1982)

phys. stat. sol. (a) 77, 225 (1983)

Subject classification: 1.2 and 18.3; 5; 22.8.2

Hochschule für Verkehrswesen „Friedrich List“, Dresden[1]) *(a),*
Zentralinstitut für Kernforschung
der Akademie der Wissenschaften der DDR, Rossendorf (b),
and Forschungsinstitut für Aufbereitung
der Akademie der Wissenschaften der DDR, Freiberg (c)

Structural Transformations and Connected Changes of Magnetic Properties by Mechanical Treatment of $SrFe_{12}O_{19}$

By

U. HEINECKE (a), C. CRUZ[2]) (b), E. WIESER (b), and C. BERNHARDT (c)

Structural transformations caused by mechanical action (vibration mill) in $SrFe_{12}O_{19}$ are investigated using neutron diffraction, X-ray diffraction, and Mössbauer spectroscopy. The occurrence of α-Fe_2O_3-type and spinel-type structures is observed in dependence on milling time. The connection between these structural changes and the magnetic properties (coercivity, saturation magnetization) are discussed. The recovery of the magnetoplumbite structure by annealing after the mechanical treatment is also studied.

An Strontiumferrit, $SrFe_{12}O_{19}$, werden mittels Neutronen- und Röntgenstrahlenbeugung sowie Mößbauerspektroskopie Strukturänderungen untersucht, die durch mechanische Einwirkung (Schwingmühle) erzeugt werden. In Abhängigkeit von der Mahldauer wird das Auftreten von α-Fe_2O_3- und Spinell-Strukturen beobachtet. Der Zusammenhang zwischen diesen Strukturänderungen und den magnetischen Eigenschaften (Koerzitivfeldstärke, Sättigungsmagnetisierung) wird diskutiert. Die Rückbildung der Magnetoplumbit-Struktur beim Tempern wird ebenfalls untersucht.

1. Introduction

The hard magnetic material strontium ferrite has a chemical composition given by the formula $SrO \cdot 6\,(Fe_2O_3)$ and a crystal structure equal to that of magnetoplumbite [1]. The basic magnetic parameters of single crystals are saturation magnetization $I_s = 0.467$ T and anisotropy field $H_A = 1.3$ MAm^{-1} [1]. Coercivity approaches the value of the anisotropy field by dividing the material into smaller and smaller particles. For this purpose different types of mills are used. The diameter of the particles is reduced and lattice defects are formed during the milling process. It is well known that lattice defects reduce coercivity [2 to 4]. But there is a strong influence on saturation magnetization, too. Using a special type of vibration mill I_s has been reduced from 0.47 to 0.08 T [5]. In this paper these materials, i.e. products from different steps of preparation, are investigated in more detail. The results of X-ray diffraction and especially of neutron diffraction and Mössbauer effect measurements are reported. On this basis it is possible to propose a model which explains the mechanism of structural changes due to the mechanical treatment and their connection with the magnetic properties.

---

[1]) PSF 103, DDR-8072 Dresden, GDR.

[2]) Present address: Instituto de Investigaciones Nucleares, Academia de Ciencias de Cuba, Havanna, Cuba.

2. Experimental

In [5] the preparation of the samples is characterized in detail. The powders were made from prefired strontium ferrite with a chemical composition of about $SrO \cdot 5.6\,(Fe_2O_3)$. At first there was a crushing and comminution process.

Parts of this starting material were subjected to an intensive mechanical treatment in a vibrating mill for different times, 1, 2, 4, and 8 h. In order to study annealing effects the powders were tempered in air at 700, 900, and 1000 °C, respectively, for 1 h.

Samples prepared in this way were investigated by Mössbauer spectroscopy, X-ray and neutron diffraction, and by magnetic measurements.

^{57}Fe Mössbauer spectra of all investigated samples were measured at room temperature using a $^{57}Co/Rh$ source. Spectra at lower temperatures down to 80 K were taken from the samples milled for 4 and 8 h, respectively. Using a least-squares computer program the spectra were decomposed in one or two magnetically split sextets of modified Lorentzians and an additional doublet. The corresponding magnetic hyperfine fields of some selected spectra are given in Table 1. As a measure for the electric quadrupole interaction in the magnetically split spectra the value $\Delta S = (v_2 - v_1) - (v_6 - v_5)$ (v_i position of the absorption line i on the velocity scale) is also presented.

Table 1

Observed internal magnetic fields H and values ΔS characterizing the quadrupole splitting in hexagonal strontium ferrite samples after different mechanical and thermal treatments

sample	temperature (K)	H_1 (kA cm^{-1})	ΔS_1 (mm s^{-1})	H_2 (kA cm^{-1})	ΔS_2 (mm s^{-1})
before milling	300	310	−0.8	380	−0.2
1 h milling	300	305	−1.1	380	−0.1
4 h milling	300	307	−0.8	377	−0.2
4 h milling	200	—	—	380	+0.5
4 h milling and annealing at 700 °C	300	313	−0.6	383	+0.3
8 h milling and annealing at 800 °C	300	318	−0.9	385	0.15

Neutron diffraction patterns of all investigated samples were recorded on a conventional two-axis powder diffractometer using a wavelength of $\lambda = 0.118$ nm. The $\lambda/2$ contamination amounted to 0.3%.

X-ray diffraction measurements were carried out using CuKα and CrKα radiations.

3. Results

3.1 Magnetic measurements

The results of the magnetic measurements (saturation magnetization I_s and coercivity $_IH_c$) are shown in Fig. 1 and 2. Of special interest are the I_s-values from the samples with 4 and 8 h milling time before and after annealing. The continual drop of coercivity indicates the increasing influence of lattice defects. Saturation magnetization decreases up to 4 h milling time, too, but a milling time of 8 h leads to an increase again. A temperature treatment of the milled samples causes an increase of saturation magnetization with annealing temperature with one remarkable exception. This concerns the 8 h sample. Here the first annealing step (700 °C for 1 h) decreases I_s.

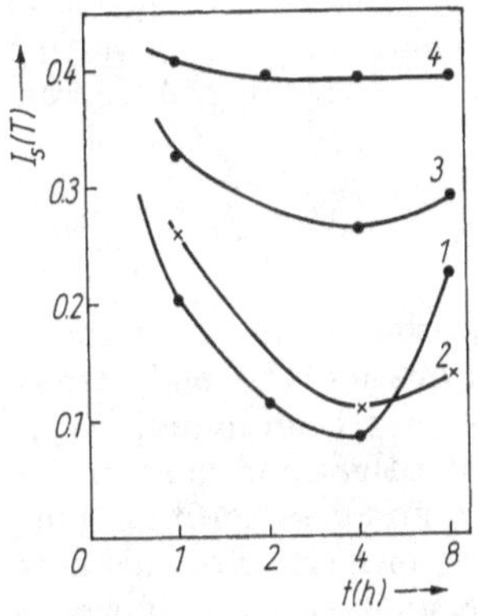

Fig. 1

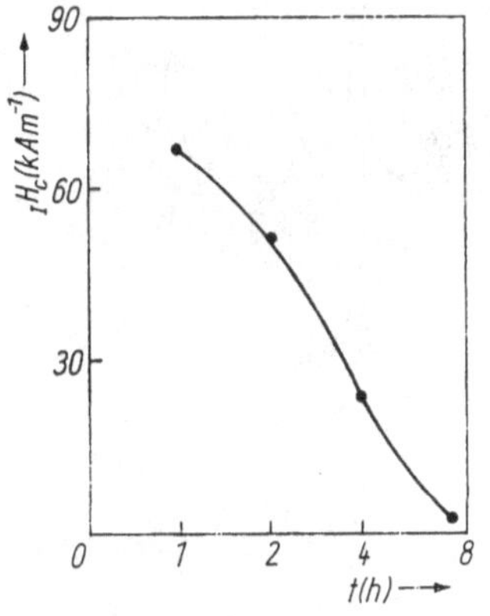

Fig. 2

Fig. 1. Saturation magnetization I_s vs. milling time t for different temperature treatments. (1) without annealing, (2) 700, (3) 900, (4) 1000 °C for 1 h

Fig. 2. Coercivity $_IH_c$ in dependence on milling time t

The $I_s = I_s(T)$ curves are given in [5]. From 1 to 4 h milling time the behaviour is very similar to the $I_s = I_s(T)$ curves of magnetoplumbite. But after 8 h the feature of these curves is changed.

3.2 Diffraction and Mössbauer investigations

The neutron diffraction pattern of the starting material is in good agreement regarding relative line intensities and line positions with the results reported in [6] for $SrFe_{12}O_{19}$. Also the X-ray diffraction measurement confirmed that our starting material was normal single-phase hexagonal strontium ferrite [7].

In the Mössbauer spectrum of this sample two sextets of about equal intensity could be clearly resolved. The subspectrum corresponding to the lower internal magnetic field ($H_1 = 310$ kAcm^{-1}) is assigned to Fe^{3+} ions on the octahedral sites 12k. The sextet with the larger magnetic splitting ($H_2 = 380$ kA cm^{-1}) is assumed to be a superposition of contributions from Fe^{3+} ions on the lattice sites $4f_1$, $4f_2$, and 2a following the interpretation in [8]. This is responsible for the line broadening and the beginning line splitting at the outermost line at positive velocities in this sextet.

3.2.1 *Influence of milling up to 4 h*

The milling process causes structural changes in the strontium ferrite sample. This is reflected by strong changes in the scattering pattern of the neutron diffraction as well as of the X-ray diffraction experiments. As expected the creation of defects and strains leads to increasing line broadening. This effect is especially distinct for (001) reflections pointing to heavy disturbances along the *c*-axis.

The following new qualitative features appear progressively up to 4 h milling time in the scattering pattern: Most strontium ferrite reflections decrease in intensity. Two new reflections appear ($d = 0.185$ and 0.269 nm). The increase in relative intensity of some strontium ferrite reflections is assumed as due to the superposition with reflections of the new developing phase. A comparison with the line positions of α-Fe_2O_3 marked in Fig. 3 shows clearly that a transformation from strontium ferrite into a phase with α-Fe_2O_3 structure takes place. The consideration of the relative line intensities in this new system of reflections confirms this assumption. The lattice

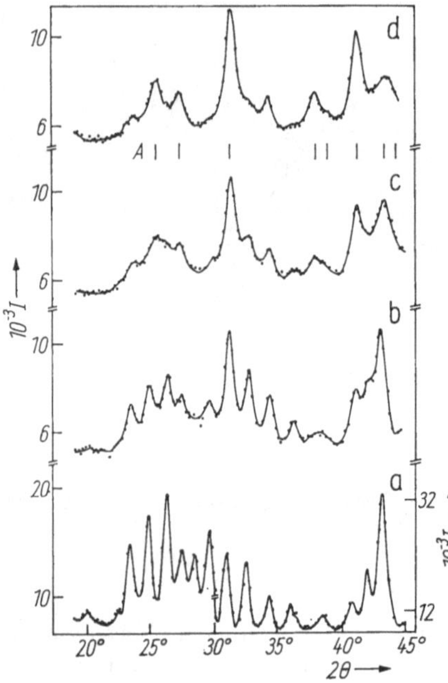

Fig. 3. Neutron diffraction patterns of strontium ferrite samples a) before milling, after b) 1 h, c) 2 h, and d) 4 h milling. *I* intensity, A neutron diffraction lines of α-Fe_2O_3

constant is nearly the same as in pure α-Fe_2O_3. No indications for a second, strontiumrich new phase have been found.

The Mössbauer investigations confirm the formation of α-Fe_2O_3 and give additional information about the properties of this new phase. In the Mössbauer spectra shown in Fig. 4 with increasing milling time the intensity of the strontium ferrite sextets decreases and an additional doublet arises. The disturbance of the strontium ferrite structure leads to a line broadening.

The central doublet is assigned to α-Fe_2O_3. Its isomer shift is characteristic of Fe^{3+}. The quadrupole splitting of 1.0 mm s^{-1} is in good agreement with the value of ≈ 0.92 mm s^{-1} found for α-Fe_2O_3 in the superparamagnetic state [9, 10]. After 4 h milling about 80% of the iron ions contribute to the α-Fe_2O_3 doublet.

Going to lower temperatures the intensity of the doublet decreases. This missing intensity appears now in a magnetically split subspectrum which is superimposed on the outer subspectrum of strontium ferrite.

The corresponding value of the internal magnetic field is in good agreement with values found for small α-Fe_2O_3 particles [10]. As shown in Table 1 the values ΔS characterizing the quadrupole splitting are negative for strontium ferrite. In the 200 K spectrum of the sample after 4 h milling the dominating influence of the α-Fe_2O_3 subspectrum reverses the sign of ΔS_2.

3.2.2 Influence of 8 h milling

The sample after 8 h milling shows a qualitatively different behaviour compared with the samples up to 4 h milling time.

In the diffraction pattern the isolated α-Fe_2O_3 reflections at about 0.220 and 0.270 nm, respectively, are now only very weak as shown by the neutron diffraction result in Fig. 5. The relative line intensities of the other reflections also in no way correspond with a strontium ferrite pattern. But the observed scattering diagrams (X-ray and neutron diffraction) are in accordance with a spinel structure. This is demonstrated by a comparison of Fig. 5 with the line positions of a spinel lattice marked by A in Fig. 5a.

The Mössbauer spectra of this sample show that the spinel phase is magnetically ordered. The spectrum measured at 150 K is typical for a ferrimagnetic oxide with spinel structure containing mainly iron as cations, although octahedral and tetrahedral subspectra are poorly resolved.

A positive ΔS value is characteristic of α-Fe_2O_3 above the Morin transition with the magnetic moments in the plane perpendicular to the *c*-axis. Our measuring temperature of 200 K lies below the Morin transition of bulk α-Fe_2O_3 ($T_M \approx 253$ K). But for

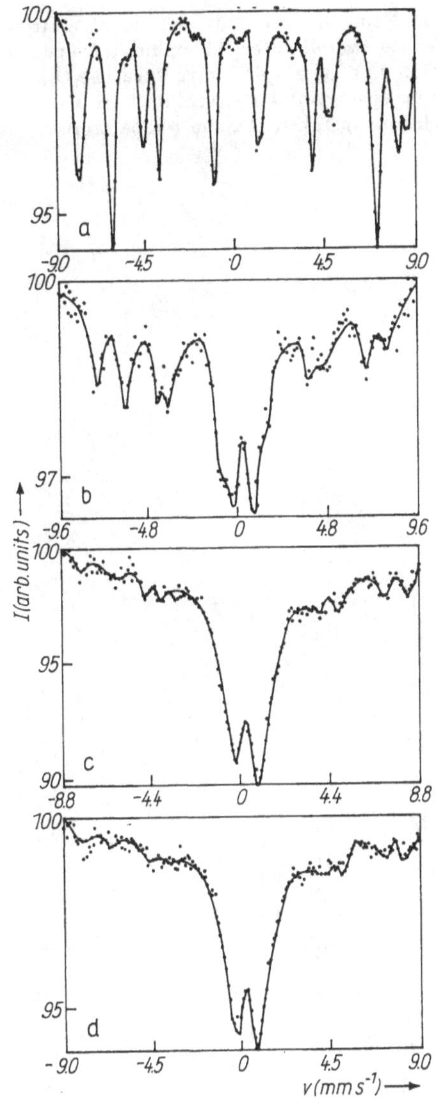

Fig. 4. Mössbauer spectra of strontium ferrite samples taken at room temperature a) before milling, after b) 1 h, c) 2 h, and d) 4 h milling. I intensity in arb. units

small particles T_M decreases [10]. The following consideration shows that in our case the α-Fe_2O_3 particles are indeed very small.

α-Fe_2O_3 is antiferromagnetic with a Néel temperature above room temperature. Therefore, we expect a magnetically split α-Fe_2O_3 spectrum even at 300 K. The occurrence of the doublet shows that α-Fe_2O_3 is formed in very small particles behaving superparamagnetically at room temperature. The gradual appearance of magnetic splitting at lower temperature is connected with the particle-size distribution.

The room temperature spectrum is strongly smoothed. Such a spectrum is characteristic of relaxation effects in connection with superparamagnetism of small particles. The relaxation effect vanishes already at 200 K.

3.2.3 *Annealed samples*

Fig. 6 shows a neutron diffraction pattern and a Mössbauer spectrum of the sample milled 4 h and annealed at 700 °C for 1 h. In the scattering diagram the α-Fe_2O_3 reflections dominate. The strontium ferrite lines are only very weak. The portion of α-Fe_2O_3 remains nearly unchanged by annealing. Also the Mössbauer spectrum corresponds to the superposition of an intensive α-Fe_2O_3 spectrum with a weak spectrum of strontium ferrite. The outer subspectrum is much more intensive than the subspectrum corresponding to the lower internal magnetic field and its ΔS is positive. A further confirmation for a large α-Fe_2O_3 contribution comes from a Mössbauer spectrum measured in an external, longitudinal magnetic field. The relative line intensities of the subspectrum with larger magnetic splitting remain nearly unchanged by the external field as expected for an antiferromagnetic substance.

Superparamagnetic relaxation effects (central doublet) have not been observed in the annealed sample. The α-Fe_2O_3 particles are now obviously larger. The X-ray diffraction pattern after annealing of the 4 h sample at 900 °C for 1 h shows a considerable retransformation from α-Fe_2O_3 into strontium ferrite. But it remains a remarkable part of residual α-Fe_2O_3.

The sample milled 8 h and annealed at 700 °C for 1 h behaves very similar to the

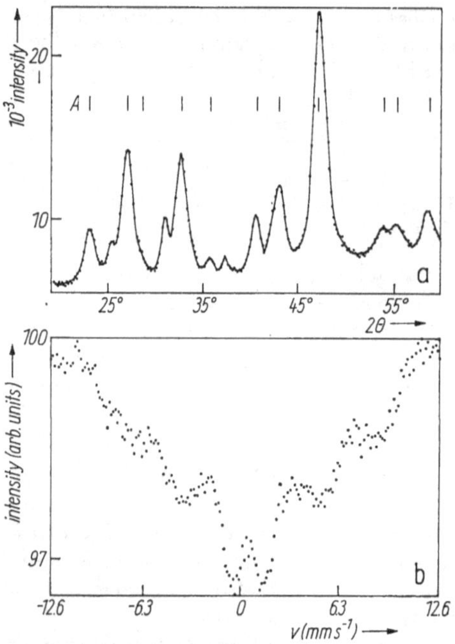

Fig. 5. a) Neutron diffraction pattern of a strontium ferrite sample after 8 h milling; A neutron diffraction line of a spinel structure (Fe_3O_4). b) Mössbauer spectrum of a strontium ferrite sample after 8 h milling taken at room temperature

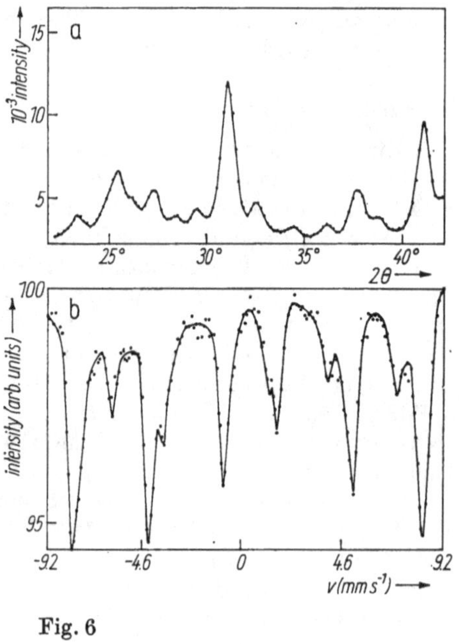

Fig. 6

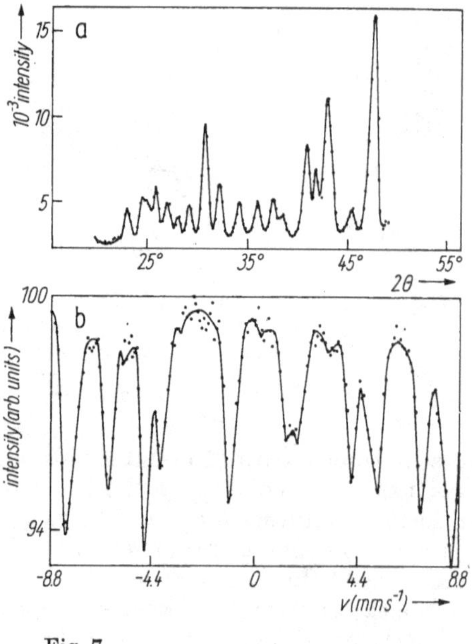

Fig. 7

Fig. 6. a) Neutron diffraction pattern and b) Mössbauer spectrum of the sample milled 4 h and annealed at 700 °C for 1 h

Fig. 7. a) Neutron diffraction pattern and b) Mössbauer spectrum of the sample milled 8 h and annealed at 900 °C for 1 h

4 h sample. The X-ray diffraction pattern shows mainly strong α-Fe_2O_3 reflections. This means a transformation of the spinel structure into an α-Fe_2O_3 structure took place.

A nearly complete retransformation into the strontium ferrite phase is caused by 900 °C annealing for 1 h of the sample milled 8 h. Fig. 7 shows the neutron diffraction pattern and the Mössbauer spectrum of this sample. Both are in good agreement with the results of the starting material in Fig. 3 and 4, respectively. In Fig. 7 weak residual α-Fe_2O_3 reflections are still observed.

4. Discussion

The starting material has magnetoplumbite structure. New peaks appear in the neutron and X-ray diffraction patterns with increasing milling time. This lines correspond to a structure of α-Fe_2O_3 type and finally after 8 h milling time to a spinel structure (e.g. γ-Fe_2O_3).

But no evidence for any other structure, e.g. SrO, is detectable and the results from the annealed samples prove that the original molar ratio SrO/Fe_2O_3 is not changed substantially due to the procedure in the mill.

The sublattice of oxygen and strontium ions is a nearly hexagonal closest-packed structure for magnetoplumbite with a stacking sequence BCACACBABA. Thereby we make no difference between the two kinds of large ions, oxygen and strontium, respectively. We assume that the closest-packed structure of these atoms remains unchanged concerning the planes perpendicular to the c-axis in all stages of mechanical treatment, but the stacking sequence is changed due to shearing operations. It is well known that the planes perpendicular to the c-axis are cleavage surfaces.

The above assumption is supported by some differences between X-ray and neutron diffraction patterns. The half-width of the peaks is much more enlarged and the peaks of magnetoplumbite disappear earlier with increasing milling time in the case of X-ray investigations in comparison with neutron diffraction where the contribution of the oxygen lattice to the scattering intensity is much more pronounced. The shear operations result in a disturbance of the atomic arrangement in direction of the c-axis. This explains the especially strong influence of the mechanical treatment on the (00l) reflections in the neutron diffraction pattern.

It is possible that the special stacking sequence of an α-Fe_2O_3 type structure ABAB... and of the spinel structure ABCABC ... is formed by shearing operations during the milling process. The following discussion shows the corresponding geometrical correlations, but presents no model for the involved shearing operations.

In Fig. 8 a cross-section of the magnetoplumbite structure is given [1]. A stacking sequence of AB ... is developed by two steps as shown below (Fig. 8b). If we continue with these operations we approach also ABC ... including twin formation. In Fig. 8b two planes inside the blocks with spinel structure were shifted one against the other. These planes are marked in Fig. 8a by arrows. This shift may be described by a vector $\boldsymbol{d}' = (n/6)\,(\boldsymbol{e}_{a,1} - \boldsymbol{e}_{a,2})\,a$ with $n = 1, 2$. Other origins and steps are possible, too, but the number of necessary steps is greater. The iron ions from the next neighbourhood of the displaced planes are at first in random interstitial sites. This means that superexchange [1] is disturbed and indeed Mössbauer experiments point to iron ions in a paramagnetic state. Superparamagnetism may arise from the undisturbed regions between these planes. The iron ions diffuse to well defined sites during the annealing process. Here (4 h milling time, 700 °C for 1 h) antiferromagnetism occurs and this is characteristic of α-Fe_2O_3. In this way the conclusions drawn from the Mössbauer spectra in Sections 3.2.1 and 3.2.3 as well as the decrease of the saturation magnetization are explained.

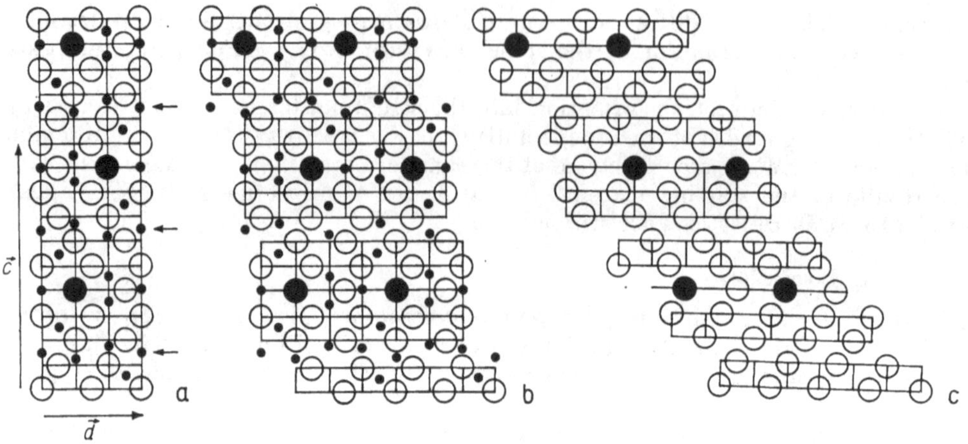

Fig. 8. a) Cross-section of the magnetoplumbite structure [1] D_{6h}^4-P 6/mmc (c-axis vertical). ○ O^{--}, ● Sr^{++}, • Fe^{3+}; $\boldsymbol{d} = \boldsymbol{a}_1 - \boldsymbol{a}_2$; $\boldsymbol{c} \perp \boldsymbol{a}_1, \boldsymbol{a}_2$. b) Transition to stacking sequence ABAB ... by two steps with $\boldsymbol{d}' = (n/6)\,(\boldsymbol{e}_{a_1} - \boldsymbol{e}_{a_2})a$, $n = 1, 2$. c) One possibility to approach the stacking sequence ABC ABC ... by several shearing steps starting from b) (the iron sites are not shown)

A stacking sequence of ABC, characteristic of the spinel structure, needs an essentially greater number of steps (Fig. 8c). This is in correspondence with the enlarged milling time (4 to 8 h).

The iron ions are here in a ferrimagnetic state and two sublattices are detectable in the Mössbauer spectra. The partial transformation from antiferromagnetic to ferrimagnetic ordering explains also the increase of I_s going from the 4 to the 8 h milled sample (see Fig. 1). Annealing of the 8 h milled sample (700 °C for 1 h) leads to an α-Fe_2O_3-type structure. This is not unexpected because already Imai and Senna [11] discuss such a transformation of vibro-milled pure γ-Fe_2O_3 (without Sr^{++}). At temperatures above 400 °C a transformation to α-Fe_2O_3 is favoured against a relaxation of γ-Fe_2O_3. The results of magnetic, Mössbauer, and structural measurements agree with this model. The decrease of I_s from 8 h vibro-milled samples with annealing at 700 °C is explained by the transformation to antiferromagnetic α-Fe_2O_3.

A further increase of annealing temperature leads to a complete recovery of the magnetoplumbite structure and saturation magnetization approaches the single crystal value.

References

[1] J. Smit and H. P. J. Wijn, Ferrites, Philips Technical Library, Eindhoven 1959.
S. Krupicka, Physik der Ferrite, Friedrich Vieweg u. Sohn, Braunschweig 1973.
[2] G. Heimke, Dtsch. Keram. Ges. **39**, 326 (1962).
[3] R. K. Tenzer, J. appl. Phys. **34**, 1267 (1963).
[4] U. Heinecke and H. G. Müller, Wiss. Z. Hochsch. Verkehrsw., Dresden **13**, 219 (1966).
[5] U. Heinecke, C. Bernhardt, and H. Heegn, to be published.
[6] E. F. Bertaut, A. Deschamps, and R. Pauthenet, J. Phys. Radium **20**, 404 (1959).
[7] P. B. Braun, Philips Res. Rep. **12**, 491 (1957).
[8] V. Florescu, D. Barb, M. Morariu, and D. Tarina, Rev. Roumaine Phys. **19**, 249 (1974).
[9] D. Schroeer, Phys. Letters A **27**, 507 (1968).
[10] D. Gündig, H. Bömmel, G. Constabaris, and N. R. H. Lindquist, Phys. Rev. **142**, 327 (1966).
[11] H. Imai and M. Senna, J. appl. Phys. **49**, 4433 (1978).

(Received November 16, 1982)

phys. stat. sol. (a) 77, 233 (1983)

Subject classification: 11 and 14.3.4; 22.1.2

Sektion Physik der Humboldt-Universität zu Berlin, Bereich Atomstoßprozesse der Festkörperphysik[1])

Ion Beam Contacting of Silicon

I. Aluminium in p-Silicon

By

H. KLOSE, A. MERTENS, R. REETZ, and NG. TANG

In ion beam contacting a high-dose ($\geqq 5 \times 10^{16}$ cm^{-2}) and low-energy ion implantation of the contact metal is used to perform a metal–semiconductor contact. For Al-contacts in p-type Si measurements of the contact resistivity and of current–voltage characteristics are made as a function of the implantation parameters, the annealing procedure, and the acceptor density of Si. Due to difficulties to explain the I–U-characteristics on the basis of the common theories of M–S-contacts an attempt is made to describe the characteristics by the properties of the amorphous layer. Contact resistivities below 1×10^{-4} Ω cm^2 are obtained after annealing at 400 °C. The method is characterized by a high reproducibility of the results.

Bei der Ionenstrahlkontaktierung wird eine Hochdosisimplantation ($\geqq 5 \times 10^{16}$ cm^{-2}) von niederenergetischen Ionen des Kontaktmetalls verwendet, um einen Metall–Halbleiterkontakt zu erzeugen. Messungen des spezifischen Kontaktwiderstandes und von Strom–Spannungskennlinien werden an Al-Kontakten in p-Si als Funktion der Implantationsparameter, der Temperbehandlung und der Akzeptorkonzentration des Si durchgeführt. Auf Grund von Schwierigkeiten, die I–U-Kennlinien auf der Basis der bekannten Theorien des M–H-Kontaktes zu erklären, wird versucht, die Kennlinien mittels der Eigenschaften der amorphen Schicht zu beschreiben. Spezifische Kontaktwiderstände kleiner als 1×10^{-4} Ω cm^2 werden nach Temperung bei 400 °C erhalten. Die Methode zeichnet sich durch eine hohe Reproduzierbarkeit der Resultate aus.

1. Introduction

In the present day technology of the very large scale integrated circuits ohmic contacts become more and more important because of the requirement of a low resistance and contact reliability due to reduced device sizes. By ion implantation it is possible to inject ions of any element into a solid state material regardless of solubility limits. Hence there exists the possibility of forming surface alloys which cannot be made by other, more conventional methods [1, 2]. In this field, high-dose ion implantation can be successfully used to produce metallic surface layers [3] or certain metal alloys giving ohmic contacts to semiconductor materials [4].

Ion beam contacting is one of the applications of low-energy high-dose implantation. In this technology implantation of low-energy ions of the contact metal is used to perform a metal—semiconductor contact.

In dependence on the ion dose and energy a high surface metal concentration can be reached so that metallic conduction through the contact occurs. Below the metallic surface a narrow strongly damaged silicon layer exists. The advantages of this ion beam technology consist in the transfer of the metal–semiconductor interface into the bulk of the sample, the independence of surface conditions of the semiconductor due to sputtering, the reduction of the rectifying character of the contacts, a high bond strength, and a good reproducibility of contact properties.

[1]) Invalidenstr. 42, DDR-1040 Berlin, GDR.

Already in 1970 Belevskii et al. [5] have used this technology to prepare ohmic Sn contacts on n-type GaAs. Later, Feuerstein and Kalbitzer [6] have made Al contacts on silicon. Detailed investigations of ion beam contacting are carried out by Reetz [7]. In this paper some essential properties of ion implanted ohmic Al contacts on p-type silicon should be reported and discussed.

2. Experimental

In the investigations ⟨100⟩-orientated B-doped Czochralski-grown silicon wafers with different acceptor densities between 10^{15} and 10^{19} cm^{-3} were used. After thermal oxidation a set of circular windows (area from 10^{-6} to 10^{-3} cm^2) were etched in the 0.3 μm thick oxide layer in order to obtain patterns with defined contact areas.

Implantations were carried out with doses between 1×10^{16} and 5×10^{17} Al^+ cm^{-2} and ion energies ranging from 10 to 25 keV. The target temperature was varied between 80 and 800 K. For the usually used current density ($j \approx 30$ μA cm^{-2}) the temperature rise ΔT of the sample holder was below 50 K for room-temperature implants. Immediately after the implantation an aluminium film was deposited by thermal evaporation in the same vacuum system in order to obtain a sufficient thickness of the metallic layer. This was necessary to realize a good contact between the tungsten probe and the metallic layer (Fig. 1), to avoid difficulties during electrical measurements for different temperature ranges. However, for ion doses $D \geqq 4 \times 10^{17}$ Al^+ cm^{-2} this evaporation step was omitted in many cases, because we did not find any influence of evaporation on the reliability of measurements.

Isothermal annealing of the samples was carried out for 30 min in flowing nitrogen atmosphere in the temperature range from 150 to 500 °C.

In order to characterize the properties of metal–semiconductor contacts formed by implantation current–voltage measurements were used. Fig. 1 shows a cross-section of the sample geometry. The current flows from the tungsten probe on the top contact area through the 200 μm bulk silicon to the conventionally prepared large-area back contact. The total resistance R_t of the sample is determined by

(i) the contact resistance R_c of the top contact which is given by $R_c = \varrho_c/A$, where ϱ_c is the contact resistivity and A the contact area,

(ii) the resistance R_b of bulk material, and

(iii) the contact resistance R_c' of the large-area back contact.

R_c' is assumed to be negligible due to the large area of the back contact. The resistance R_b of the bulk material can be estimated [8]. In the cases of preferentially used material with an acceptor density $N_A \approx 10^{18}$ cm^{-3} R_b is about 8 Ω for a contact area of 10^{-6} cm^2. For contact resistivities lower than 1×10^{-4} Ω cm^2 ϱ_c was determined by the relation

$$\varrho_c = A(R_t - R_b) .$$

The value R_t is directly obtained from current–voltage measurements.

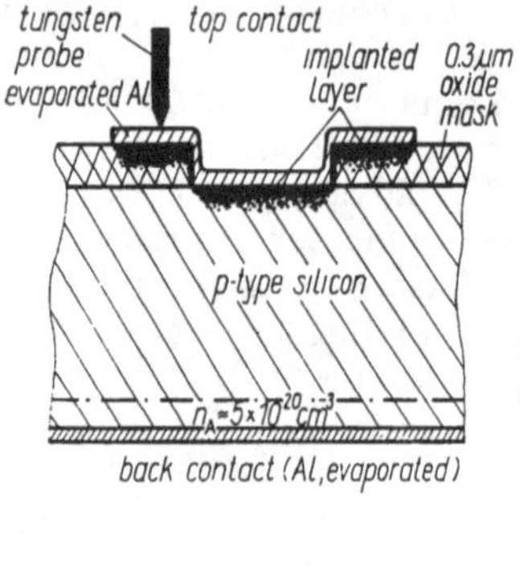

Fig. 1. Schematic view of the sample cross-section. For the top contact the diameter of the circular windows are varied between 10^{-6} and 10^{-3} cm^2. The thickness of silicon wafers is about 200 μm

3. Results

3.1 Generation of the metallic component of the contact

The metallic surface layer of the contact was made by Al-ion implantation. To find out the appropriate implantation energy and dose the depth distributions of implanted aluminium have been calculated for different parameters. For these calculations the influence of (i) the slowly increasing aluminium content on the range and straggling of projectiles, (ii) the swelling of the wafer, and (iii) the sputtering yield were taken into account as in [9]. Fig. 2 shows the calculated distributions of aluminium for ion energies between 7.5 and 20 keV in the dose range from (2 to 5) $\times$ 10^{17} Al^+ cm^{-2}. For comparison furthermore there is shown the aluminium concentration required to form the composition $Si_{0.5}Al_{0.5}$ which has metallic-conducting behaviour. From Fig. 2 it is evident that for nearly all considered parameters of implantation the aluminium concentration is sufficient to form a metallic-conducting surface layer. In order to test this assumption, measurements of the electrical sheet resistivity of implanted layers were made by four-point-probe technique. The measurements for different aluminium doses implanted with a certain energy (15 keV) confirm the metallic-conducting behaviour for dose values $D \geqq 1 \times 10^{17}$ Al^+ cm^{-2} in the 'as-implanted' state and for $D \geqq 5 \times 10^{16}$ Al^+ cm^{-2} after a 30 min annealing at 300 °C.

On the basis of these results the further investigations were concentrated on the properties of contacts got by implantation of 2 $\times$ 10^{17} Al^+ cm^{-2} with 15 keV. Then the thickness of the metallic layer obtained by implantation is about 40 nm and that of the strongly damaged amorphous transition is about 50 nm, as proved by RBS data and SIMS measurements, too.

3.2 Dependence of contact resistivity on implantation temperature and bulk material

Fig. 3 shows the influence of the implantation temperature on the contact resistivity for two different annealing stages ('as-implanted', $T_a = 400$ °C, $T_a = 500$ °C). Low-temperature implantation, characterized by an intermediate amorphous region extending from the metallic surface layer up to an aluminium concentration of about 1×10^{20} cm^{-3} [10], provides obviously substantial lower contact resistivities than

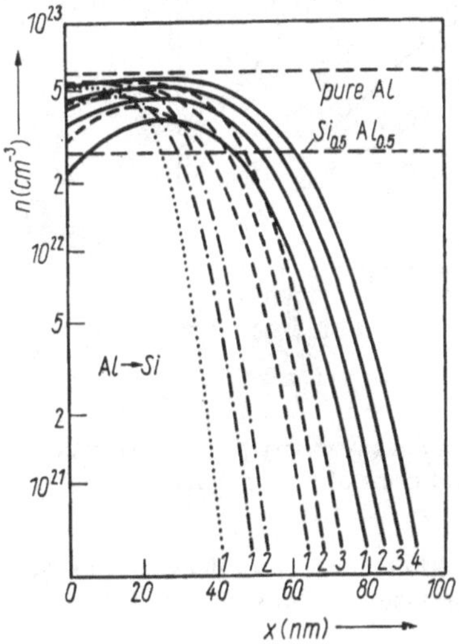

Fig. 2. Calculated distributions of implanted aluminium in silicon for different ion doses: (1) 2 $\times$ 10^{17}, (2) 3 $\times$ 10^{17}, (3) 4 $\times$ 10^{17}, (4) 5 $\times$ 10^{17} cm^{-2} and different energies: ——— 20, — — — 15, —·—·— 10, ·········· 7.5 keV

implantations at 300 or 550 K. Additional annealing of the samples implanted at 77 K leads to a further decrease of contact resistivity, probably due to the insertion of aluminium atoms into the silicon matrix and/or due to the recrystallization of the disordered layer. This is demonstrated by the decrease of contact resistivity (belonging to $T_i = 77$ K) from $5 \times 10^{-2}\ \Omega\ cm^2$ for 'as-implanted' material to values of $1 \times \times 10^{-4}\ \Omega\ cm^2$ received after annealing at 400 °C.

The absolute value of the contact resistivity obtained after a certain annealing procedure depends on the acceptor density of silicon. This is evident from Fig. 4, where the contact resistivity is shown versus the annealing temperature for a standard implantation ($2 \times 10^{17}\ Al^+\ cm^{-2}$, 15 keV) at 77 K with the acceptor density of silicon as parameter. Without annealing ϱ_c differs by about two orders of magnitude. This difference decreases after annealing of the samples. In case of samples with low acceptor density ϱ_c must be calculated from the measured value taking into account the high resistance R_b of bulk silicon.

3.3 Current–voltage characteristics

Fig. 5 shows the current–voltage characteristics of typical contacts obtained by 77 K standard implantations into silicon with an acceptor concentration of $1 \times \times 10^{15}\ cm^{-3}$ and $1 \times 10^{18}\ cm^{-3}$, respectively. The reciprocal temperature represents the parameter of the characteristics. For forward as well as reverse direction the curves show two distinct regions. A strictly ohmic characteristic is evident for the low-voltage region, but for higher voltages the curves indicate a breakdown behaviour. Dependent on the acceptor density, for the reverse direction a low increase of resistivity can be observed in the intermediate region from low to high voltages. With annealing

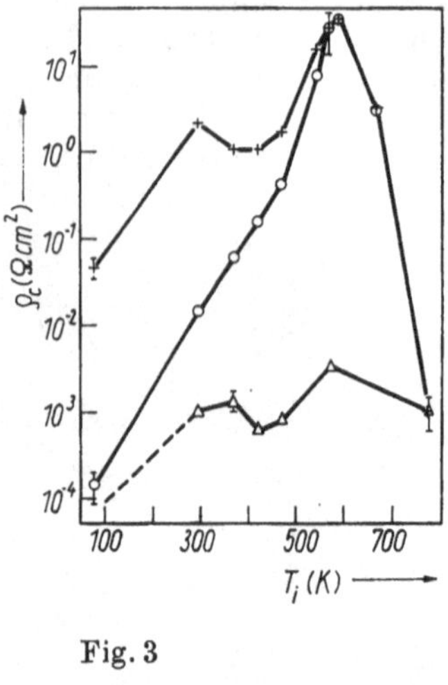

Fig. 3

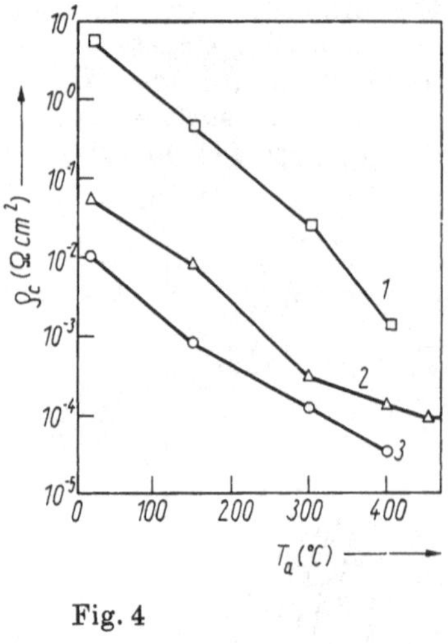

Fig. 4

Fig. 3. Influence of the implantation temperature T_i on the contact resistivity ϱ_c for three different annealing stages (+ as implanted, ○ $T_a = 400$, △ 500 °C). Conditions of ion implantation: 15 keV, $2 \times 10^{17}\ Al^+\ cm^{-2}$ into p-type Si ⟨100⟩ with $N_A \approx 10^{18}\ cm^{-3}$

Fig. 4. Contact resistivity ϱ_c vs. annealing temperature T_a for different acceptor densities of p-type silicon: (1) 10^{15}, (2) 10^{18}, (3) $10^{19}\ cm^{-3}$. Conditions of ion implantation: 15 keV, $2 \times 10^{17}\ Al^+\ cm^{-2}$, $T_i = 77$ K

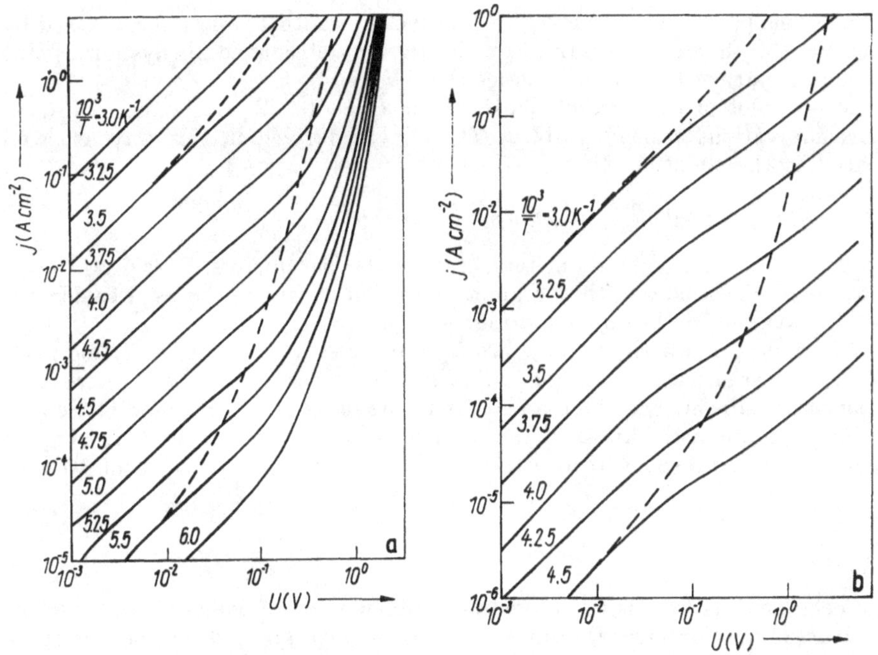

Fig. 5. Typical j–U characteristics for different measuring temperatures T (parameter $10^3/T$). a) p-type Si with $N_A \approx 10^{18}$ cm^{-3}, b) p-type Si with $N_A \approx 10^{15}$ cm^{-3}. ——— reverse, – – – forward direction

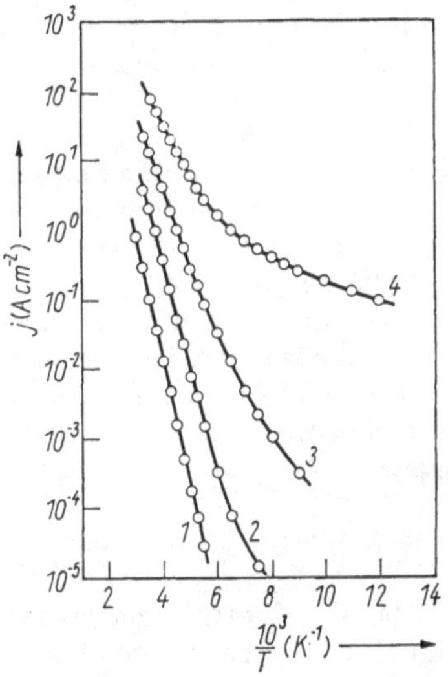

Fig. 6

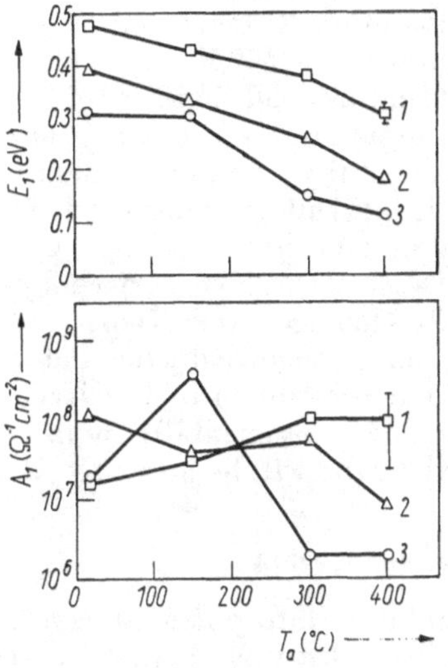

Fig. 7

Fig. 6. Current density j vs. $1000/T$ (measuring temperature) for four different annealing stages: (1) as-implanted, (2) $T_a = 150$, (3) 300, (4) 400 °C. Conditions of ion implantation are the same as those of Fig. 4 and 5a

Fig. 7. Parameters E_1 and A_1 vs. annealing temperature T_a for different acceptor densities of p-type silicon: (1) 10^{15}, (2) 10^{18}, (3) 10^{19} cm^{-3}. Conditions of ion implantation are the same as those of Fig. 4

of the samples this behaviour is reduced. To derive the contact resistivity, defined by dU/dj for $U = 0$ V, the voltage–current deviation was determined always for a point within the ohmic part of the characteristics (10^{-2} V).

Fig. 6 shows a plot of the current densities for $U = 10^{-2}$ V versus the reciprocal temperature for various annealing steps. The curves derived in this way are well approximated by the equation

$$j = (A_1 \exp(-E_1/kT) + A_2 \exp(-E_2/kT)) \cdot U$$

with $E_1 > E_2$ and $A_1 \gg A_2$. At room temperature the parameters E_1 and A_1 determine the value of ϱ_c. Besides ϱ_c the parameters E_1 and A_1 have been used to characterize the properties of ion beam contacting.

In Fig. 7 it is shown how the parameters E_1 and A_1 vary with the annealing temperature. E_1 decreases with increasing annealing temperature and with increasing acceptor density of silicon. The values of A_1 are in the range from 10^6 to $10^8\ \Omega^{-1}\ cm^{-2}$.

Taking into account a thickness of the disturbed layer of about 10^{-5} cm (Fig. 2) these values of A_1 correspond to resistivities of the disturbed layer from 10^{-1} to $10^{-3}\ \Omega$ cm.

4. Discussion

The ohmic behaviour as well as the pronounced dependence of contact resistivity on the temperature of measurement for samples not or weakly annealed represent properties quite different from those of conventionally prepared metal–semiconductor contacts which are characterized by thermionic emission of carriers and/or by tunnelling and thermally stimulated tunnelling through the basis, respectively. On the other hand, there is some similarity to properties of amorphous layers [11, 12]. For high-dose implantation of Al ions into silicon an amorphous layer is built at implantation temperatures below 400 K [10]. Taking into account a possible target heating during implantation this condition is only fulfilled for the 77 K implantations.

But if the properties of contacts are essentially controlled by an amorphous layer, the experimentally found dependence of contact resistivity on the acceptor density of silicon is difficult to understand. The acceptor density is not so important for the resistivity of amorphous silicon but more important for that of the intermediate region between amorphous surface layer and crystalline bulk material [12].

Recently Henniger and Keiper [13] presented a theoretical description of ion-implanted metal–semiconductor contacts with involved amorphous layer. The numerically calculated set of current–voltage characteristics by these authors are similar to those obtained experimentally by us. A detailed comparison of the experimental and theoretical results will be presented in a forthcoming paper.

5. Conclusion

Ion implantation into p-type silicon has been proved to be a successful method for preparing low-resistivity ohmic metal–semiconductor contacts. The resistivity of contacts obtained by this method is lower than the contact resistivity usually obtained by conventional preparation methods, if due to implantation an amorphous layer is formed below the metallic surface layer. After annealing at 400 °C contact resistivities below $1 \times 10^{-4}\ \Omega\ cm^{-2}$ can be obtained.

Further advantages of ion beam contacting are the independence of semiconductor surface conditions, a transfer of the metal–semiconductor interface into the bulk material, the production of ohmic contacts without annealing procedures in special cases of ion implantation technology, a high bond strength, and the very high reproducibility of contact properties. It seems possible to use this physical technology for contacting buried layers, or for the production of microcontact structures for high power chips, too.

References

[1] N. G. Einspruch, VLSI Electronics, Academic Press, New York 1981.
[2] W. A. Grant and I. S. Williams, Sci. Progr. (Oxford) **63**, 27 (1976).
[3] H. Kräutle and S. Kalbitzer, Ion Implantation in Semiconductor and Other Materials, Ed. B. L. Crowder, Plenum Press, New York 1973 (p. 585).
[4] A. Mertens and J. Reckin, Acta phys. Acad. Sci. Hungar. **51**, 371 (1981).
[5] V. P. Belevskii, V. N. Ivanov, and V. M. Yashnik, Prib. i Tekh. Eksper. **6**, 203 (1970).
[6] A. Feuerstein and S. Kalbitzer: Appl. Phys. Letters **22**, 19 (1973).
[7] R. Reetz, Thesis, Humboldt-Universität zu Berlin, 1981.
[8] R. Holm, Electrical Contacts, Springer-Verlag, Berlin-Heidelberg 1967; J. C. Irvin, Bell System techn. J. **41**, 387 (1962).
[9] H. Kräutle, Nucl. Instrum. and Methods **134**, 167 (1976).
[10] F. F. Morehead and B. L. Crowder, Radiat. Eff. **6**, 27 (1970).
[11] M. Morgan and P. A. Walley, Phil. Mag. **23**, 661 (1971).
[12] G. Müller and S. Kalbitzer, Phil. Mag. **B41**, 307 (1980).
[13] U. Henninger and R. Keiper, phys. stat. sol. (a) **74**, 497 (1982).

(Received March 18, 1983)

phys. stat. sol. (a) 77, 241 (1983)

Subject classification: 2; 10.2; 22.1.2

Laboratoire de Spectroscopie II, Equipe de Recherche associée au C.N.R.S. n° 257, Université des Sciences et Techniqués du Languedoc, Montpellier[1])

Departure of Hydrogen from a-Si:H

A Spectroscopic Study

By

C. ANCE and N. M. RAVINDRA

The evolution of hydrogen when samples of hydrogenated amorphous silicon are subjected to heating is investigated. The study is carried out for samples prepared by glow discharge. The present analysis which is essentially spectroscopic in nature leads to the assumption that two kinds of defects exist in a-Si:H, optically active and optically inactive ones. In the second part of this analysis, various spectroscopic parameters and their temperature dependences in the light of Penn and Wemple-Didomenico models are discussed.

Es wird die Entwicklung von Wasserstoff untersucht für den Fall, daß hydrogenisiertes amorphes Silizium einer Wärmeeinwirkung ausgesetzt wird. Die Untersuchung wird für Proben durchgeführt, die mittels Glimmentladung hergestellt werden. Die im wesentlichen spektroskopische Analyse führt zur Annahme, daß zwei Arten von Defekten in a-Si:H existieren: optisch aktive und optisch inaktive. Im zweiten Teil der Analyse, werden die verschiedenen spektroskopischen Parameter und ihre Temperaturabhängigkeiten mit den Modellen von Penn und Wemple-Didomenico diskutiert.

1. Introduction

Studies concerning the properties of hydrogenated amorphous silicon films at high temperatures or after annealing have been of considerable interest. Such studies yield a quantitative and qualitative evaluation of the departure of hydrogen and provide useful information about the properties of a-Si:H and the role played by hydrogen.

Regarding the aspects of annealing, most of the authors [1 to 4] believe that annealing occurs in two stages: the first around 350 °C and the second at about 600 °C. These two temperatures are essentially correlated to the departure of hydrogen. In a recent paper, Deneuville et al. [5] have pointed out that the hydrogen concentration depends on the defects in the Si matrix, and thus the hydrogen sites are controlled by defects.

Recently [6], we proposed a model to determine the hydrogen concentration $C[H]$ from a knowledge of two spectroscopic parameters deduced from optical measurements. This model was also tested for the case of a-Si:H films produced by glow discharge. In this paper, we report our studies concerning the variation of hydrogen concentration especially with reference to the loss of hydrogen. Considering a two-defect-type model, we offer a possible explanation for the departure of hydrogen in a-Si:H films produced by glow discharge.

---

[1]) Place Eugène Bataillon, 34060 Montpellier Cedex, France.

2. Model for Calculating the Hydrogen Concentration

An analysis of the model used here has been published recently [6]. Here, we recall only the main result. Essentially, the hydrogen content in a-Si:H films can be evaluated from the following relation:

$$C[\mathrm{H}] = \frac{1}{3}\frac{n_\mathrm{v}}{n_\mathrm{s}}\left|4 - \sqrt{\frac{E_\mathrm{d}}{2.8}}\right|, \tag{1}$$

where $C[\mathrm{H}]$ is the hydrogen concentration, n_v and n_s are the valence electron densities in amorphous and crystalline silicon, respectively, and E_d is the macroscopic oscillator strength also referred to as the dispersion energy [7]. The value of E_d is obtained from optical moments. Using the values of n_s and density for crystalline silicon ($n_\mathrm{s} \approx 2 \times 10^{23}$ cm^3, density $= 2.33$ g/cm^3), it has been shown earlier [6] that

$$n_\mathrm{v} = 0.0143\left(\frac{E_\mathrm{d}^2}{\varepsilon(0) - 1}\right) \times 10^{23}\ \mathrm{cm}^{-3}, \tag{2}$$

where $\varepsilon(0)$ is the static dielectric constant. The above relations (1) and (2) were obtained taking a value of four for the coordination number and assuming that at $C[\mathrm{H}] = 0$ the network is similar to that of crystalline silicon. In Tables 1 and 2 we report our results concerning the above properties of a-Si:H films produced by glow discharge deposited at temperatures of $T_\mathrm{s} = 350$ and 280 °C, respectively. These data correspond to high rates of deposition (> 150 Å/min). As can be seen in the tables, the calculated values (based on equations (1) and (2)) are in good agreement with the experimental data [8]. For a better understanding of the discussion, we now

Table 1

Values of optical parameters and H content C [H] for GD a-Si:H films prepared at 350 °C

T (K)	95	295	373	473	573	673	723
$n(0)$	3.536	3.561	3.584	3.644	3.688		
		3.622				3.777	
		3.696					3.884
E_d (eV)	39.2	39.5	39.5	40.3	41		
		40.5				42.6	
		40.8					43.6
E_m (eV)	3.41	3.38	3.33	3.30	3.25		
		3.34				3.21	
		3.22					3.10
C [H]	0.08	0.077	0.077	0.063	0.055		
		0.064				0.0325	
		0.057					0.017
E_g (eV)	3.80	3.77	3.72	3.67	3.62		
		3.72				3.57	
		3.59					3.45
$\hbar\omega_\mathrm{p}$ (eV)	16.35	16.34	16.24	16.32	16.53		
		16.45				16.53	
		16.21					16.43
E_0 (eV)	1.71	1.67	1.64	1.59	1.55		
		1.62				1.43	
		1.56					1.33

Table 2

Values of optical parameters and H content $C\,[H]$ for GD a-Si:H films prepared at 280 °C

T (K)	95	295	373	473	573	623	673	723
n (0)	3.287	3.327	3.349	3.385	3.422	3.443		
		3.364					3.491	
		3.452						3.632
E_d (eV)	33.7	34.2	34.6	34.9	35.4	36		
		34.6					36.4	
		35.8						37.9
E_m (eV)	3.44	3.40	3.39	3.34	3.31	3.32		
		3.35					3.25	
		3.28						3.11
C [H]	0.145	0.14	0.135	0.135	0.124	0.118		
		0.134					0.112	
		0.12						0.09
E_g (eV)	3.83	3.78	3.77	3.70	3.67	3.66		
		3.73					3.6	
		3.65						3.45
$\hbar\omega_p$ (eV)	15.23	15.23	15.30	15.20	15.25	15.31		
		15.22					15.30	
		15.32						15.30
E_0 (eV)	1.76	1.70	1.68	1.60	1.54	1.51		
		1.67					1.42	
		1.58						1.26

recall the method used for optical measurements at high temperature [9]. Optical measurements (like transmittance and reflectivity) are made at each temperature and are followed by measurements at room temperature. Eventually, two kinds of parametric values are deduced. After all, during such measurements, the sample is subjected to a certain amount of annealing depending on the time duration of each measurement. In Table 1, under the column referred to as $T = 295$ K, the values of the parameter after the sample is subjected to annealing is noted. The temperature dependence is also reported (up to 723 K). These results are also plotted in Fig. 1. Thus, one can see the variation of hydrogen concentration with temperature T. In this figure, the solid curve represents the variation in hydrogen concentration C[H]

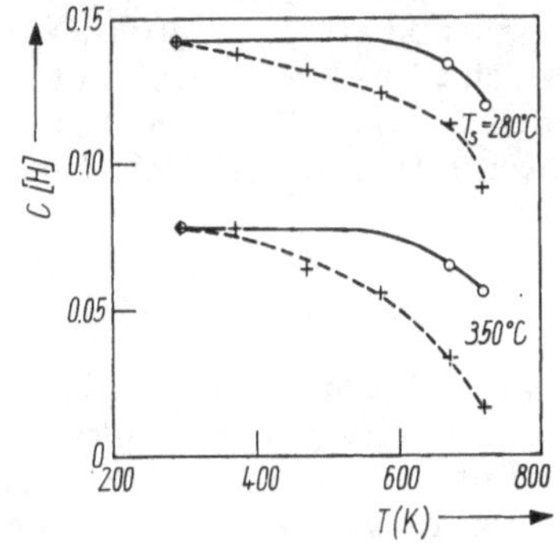

Fig. 1. Variation of hydrogen content with temperature; ——— after annealing, -------- at each temperature

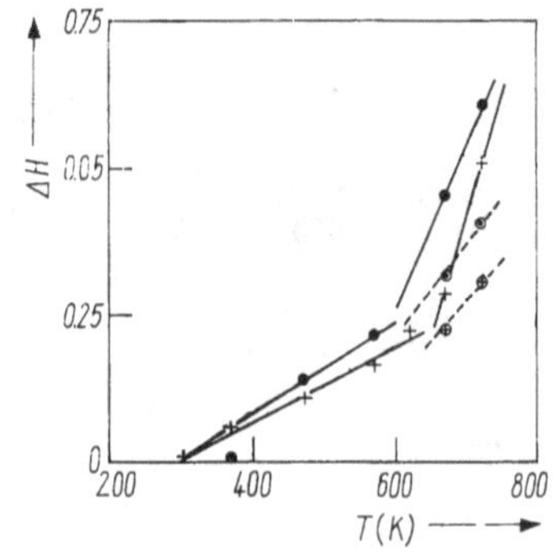

Fig. 2. Variation of ΔH with temperature; ——— refers to hydrogen content in virgin films, – – – refers to post-annealed values of hydrogen content, • $T_s = 350$, + 280 °C

after the samples have been subjected to annealing. As can be seen in the figure, C[H] remains constant until the temperature reaches a value around 620 K. Beyond 620 K, the hydrogen concentration decreases. Such a variation is in good accord with the results reported by Deneuville et al. [10] for hydrogenated amorphous silicon films prepared by RF diode sputtering at high deposition rates ($\geq$ 105 Å/min).

It is to be noted (see Tables 1 and 2) here that the results of C[H] as a function of measurement temperature are reproducible/reversible till a temperature of 573 K. That is, on cooling the system, the value of C[H] at various temperatures is reproduced (including C[H] at ambient, 295 K). However, on heating the sample to 673 K, there is a decrease in the value of C[H] (when evaluated at 295 K). Thus, during this process of heating to 673 K, there has been a loss of hydrogen. On further heating, the loss of hydrogen increases (see Tables 1 and 2). Such a behaviour is seen for both the samples under consideration prepared at temperatures of 350 and 280 °C. The solid curve essentially refers to measurements at ambient (295 K) but plotted at a temperature where such an irreversibility (or loss of hydrogen) has been seen. In Fig. 1, the dashed lines represent the hydrogen concentration C'[H] at various temperatures. As can be seen in the figure, the hydrogen concentration decreases with increasing temperature.

We now define a quantity $\Delta H = C[\mathrm{H}] - C'[\mathrm{H}]$. A plot of ΔH as a function of temperature is presented in Fig. 2. ΔH is seen to increase with increasing temperature. As a first approximation, such a result could be interpreted as a break of weakly bonded hydrogen to a Si site corresponding to a low activation energy defect. In Fig. 2, the solid line represents the difference of hydrogen concentrations in the virgin film (pre-annealed, $T_m = 295$ K) and the hydrogen concentrations in the film evaluated at various temperatures. In the same figure, the dashed line represents such a difference between the post-annealed film and the evaluated hydrogen concentration C'[H] at the corresponding temperature (when irreversibility is seen). As can be seen in Fig. 2, ΔH varies linearly with temperature, but with a change in slope around 620 K for films prepared at $T_s = 350$ °C and at 650 K for films prepared at $T_s = 280$ °C. We refer to this temperature as T_B — the temperature at which there is a "break" in the linear variation of ΔH with T. A look at the results of the two samples under consideration indicates that with increase in hydrogen concentration in the virgin film, ΔH values decrease.

3. Evolution of Hydrogen

Earlier, many authors have reported that hydrogen starts leaving the sample of hydrogenated amorphous silicon at temperature of 620 K. Thus, our results in Tables 1 and 2 and Fig. 1 and 2 are in complete agreement with such a conclusion. Evidently, the results obtained by our model calculation agree well with the scheme developed elsewhere [9] concerning an explanation for the reversibility and the irreversibility phenomena in terms of the departure of hydrogen. At this stage, it would be worthwhile to note that such a scheme would be of immense help in explaining the variation of the dispersion energy E_d and $\varepsilon(0)$. Having made these remarks, we now attempt to explain the possible reasons for the increase in ΔH from room temperature to T_B without any departure of hydrogen and accompanied by a reversible behaviour — that is a reproducibility in the values of the optical parameters when the sample is cooled to room temperature.

The most basic hypothesis involved in the present discussion is that it was earlier observed experimentally [9] that when the samples of a-Si:H films were heated to temperature of 673 K and above, the optical properties were not reproducible when measured at 295 K. But at lower temperatures ($T < 673$ K), the optical properties (values of refractive index and optical gap) were reproducible. Thus, we are led to conclude that such a reversibility and irreversibility is essentially due to the departure of hydrogen.

Earlier, Oguz et al. [11] and Zellama et al. [4] have reported that a-Si:H contains some hydrogen which is infrared inactive. Deneuville et al. [10] find that the total hydrogen content H_0 in a-Si:H films prepared by RF sputtering is greater than the sum of H_1 and H_2 where H_1 and H_2 represent the monohydride and dihydride phase, respectively. Thus, one may conclude that another weakly bonded phase exists. In addition to this, it has also been observed that the difference $H_0 - (H_1 + H_2)$ increases with temperature. We find that this behaviour is very similar to our calculations being presented here. In the light of these results, we may stipulate that in a-Si:H films there exists some hydrogen in the optically inactive sites and some in the optically active sites. In our future discussions, we will refer to them as HOIS and HOAS, respectively. Just as an example, for the sample deposited at 350 °C ($H_0 = 0.08$ in the virgin film), HOIS is about 2% at T_B and this value reaches 4% at 723 K. In fact, to be more precise, HOIS $\geqq 4\%$ at 723 K. However, when the sample is removed to room temperature from 723 K, the hydrogen concentration is about 5.7% in the film. This could be attributed to the fact that hydrogen can occupy different sites (OAS or OIS) depending on the temperature. Further, hydrogen can swing from one site to another. The experimental results of Deneuville et al. [10] as also the results being discussed here seem to indicate that with increase in temperature, hydrogen leaves the optically active site to move over to an optically inactive site. However, until the temperature is less than T_B, hydrogen can return to its original position to become optically active again. When the temperature reaches a value $\geqq T_B$, we suggest that the OIS are saturated and the hydrogen free from OAS can leave the sample. The curves in Fig. 2 suggest that at each temperature there exists an OIS value and with increase in temperature, this value increases. Thus the behaviour of OIS may be identical with that of ΔH.

Thus, from the above study we may conclude that hydrogen does not leave the sample directly but via the OIS which must all be occupied for the departure of hydrogen from the sample. Further, to have the departure of all the hydrogen in the sample, the sample should be heated to high temperatures when all the OIS defects are removed. Unfortunately, we do not have results beyond 723 K. It is however encouraging to note that such an assumption is in good accord with the results of the two temperature stages developed earlier.

As a concluding remark, we may therefore say that the hydrogen leaves the film from OIS when all these sites are saturated. This happens at temperature $\geqq T_B$. Apart from this, the removal of all the OIS leads to a complete departure of hydrogen from the sample.

We now look at samples deposited at low deposition rates and for which the hydrogen concentration is about 25%. Deneuville et al. have earlier reported [10] that the hydrogen leaves the sample as soon as the temperature is increased. Further, ΔH has been seen to have a low value for these samples [10]. We are of the opinion that in this case there exists a little amount of OIS and because of the fact that these sites are quickly saturated, hydrogen leaves the sample at far lower temperatures.

This scheme proposed in the present study agrees very well with the results of post hydrogenation effects in a-Si:H films. Indeed, for a-Si:H films prepared by RF sputtering with hydrogen concentration of 22%, post-hydrogenation does not change the concentration of hydrogen [10, 12]. This can be explained in terms of a saturation of OIS in the sample. In this case, the H concentration remains the same as in the virgin film. Contrary to this, if the OIS are not saturated, the post-hydrogenation will affect the properties of the film. Thus, the amount of hydrogen that a sample can "digest" depends on the availability of OIS and is governed by the total number of OIS. We are of the opinion that our above explanation can be related to a possible flexible network like in non-crystalline solids.

4. Temperature Dependence of Penn Gap

To illustrate the effect of departure of hydrogen on optical properties, we now look at the variation of the Penn gap with temperature. We recall that the Penn gap is a macroscopic optical parameter which gives an account of all the optical transitions in semiconductor materials and it is customary to compare the calculated value of E_g with the energy corresponding to the main peak of $\varepsilon_2(E)$ spectra.

In a recent communication [13] we have reported our work concerning the Penn gap for hydrogenated amorphous silicon films prepared by glow discharge. We have also dealt with detailed calculations which have been shown to be in good agreement with results obtained by Kramers-Kronig analysis [14]. In Fig. 3, we present the variation of E_g with temperature calculated using a Penn-like model [13] and compared with the variation of absorption threshold (rather than fundamental gap for amorphous materials) E_0. One can see that both the variations are very similar. In addition, the difference $\Delta = E_g - E_0$ has roughly the same value irrespective of the temperature ($\Delta \approx 2.1$ eV).

Thus, we can see that if we assume the plasma energy to be roughly a constant with temperature (this is true), the dielectric constant $\varepsilon(0)$ is related directly with the absorption threshold E_0 because $E_g = E_0 + \Delta$. So, we can assume that a simple relation can relate $n(0)$ to E_0.

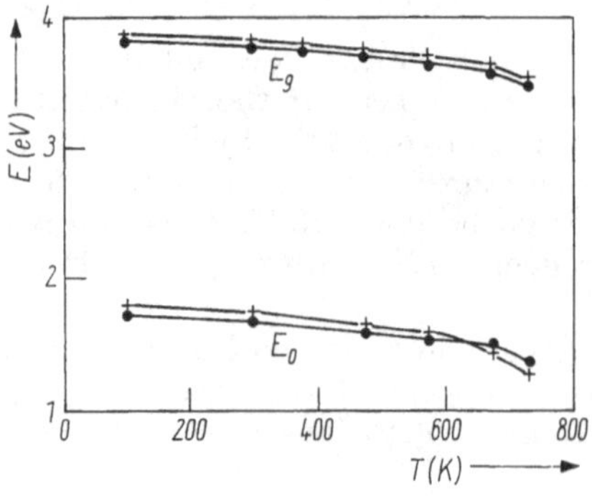

Fig. 3. Variation of E_g (calculated) and E_0 (experimental) with temperature. ● $T_s = 350$, + 280 °C

Table 3
Values of $\Delta E_g/\Delta T$ and $\Delta E_0/\Delta T$ at different temperatures, calculated from the results mentioned in Tables 1 and 2

T (K)	195	334	423	523	623	698
$\Delta E_g/\Delta T$ *)	−1.6	−6.4	−5	−5	− 5	−24
(10^4 eV K^{-1}) **)	−2.4	−1	−5	−4.3	− 5.7	−31
$\Delta E_0/\Delta T$ *)	−2	−3.8	−4.5	−4	−12	−20
(10^4 eV K^{-1}) **)	−3	−2.6	−8	−6	−12	−32

*) Refers to $T_s = 350$ °C.
**) Refers to $T_s = 280$ °C.

In Table 3, we report values of $\Delta E_g/\Delta T$ and $\Delta E_0/\Delta T$ at different temperatures. As can be seen in the table, for $\Delta E_g/\Delta T$, three characteristic values exist; the first before and around ambient temperature, with a slope $\approx -2 \times 10^{-4}$ like the value for crystalline silicon -1.9×10^{-4} [15], the second between room temperature and T_B where the slope is $\approx -5 \times 10^{-4}$, and the third beyond T_B where the slope changes abruptly and reaches a value around -30×10^{-4}. $\Delta E_0/\Delta T$ behaves similarly.

Thus the abrupt change in the slope $\Delta E_g/\Delta T$ and $\Delta E_0/\Delta T$ appears around T_B value (620 K) which is a value near the Debye temperature for silicon ($\theta_D = 647$ [16] at 300 K).

This result may appear to be surprising, yet by definition θ_D is related to the maximum vibrational frequency of atoms in the materials ($\gamma = k\theta_D/h$) and thus θ_D is related to the maximum displacement of atoms. So, at this temperature, H atoms are displaced far enough from their first position to overlap with one another to become a molecule which goes away from the sample.

In this scheme, we assume that H atoms are situated like a three-center bond (Si–H–Si) called WBH by Zellama et al. [4] where each of the hydrogen sites is near enough to become a molecule according to our preceding scheme.

5. Conclusion

In the present study, we show that our model permits an evaluation of loss of hydrogen in a-Si:H silicon films prepared by glow discharge. The results are seen to be in good agreement with the work reported earlier by various workers, for example, with the work of Deneuville et al. [5, 10] as well as Zellama et al. [4]. We have also attempted to explain as to why some authors find hydrogen optically inactive by the existence of two kinds of defects [OIS] (optically inactive site) and [OAS] (optically active site). We assume the available hydrogen to occupy these two sites and swing between them.

Besides, we show the similarity in the variation of the Penn gap and the absorption threshold with temperature, before and beyond T_B (temperature value where departure of hydrogen begins). In the passing, we note that it could be surprising to have a T_B value close to the Debye temperature θ_D. We show that the slope $\Delta E_g/\Delta T$ and $\Delta E_0/\Delta T$ decreases abruptly at T_B.

In the second part of this communication, we show that the Penn gap can be related directly to the absorption threshold E_0 by a simple relation like

$$E_g = E_0 + K \quad \text{with} \quad K \approx 2.1 \text{ eV} .$$

Assuming that the variation of $n(0)$ with temperature must be related directly with the variation of the absorption threshold [17], (by virtue of a simple linear relation between Penn gap and the threshold), the studies of threshold can also lead to an evaluation of spectroscopic properties. Last but not least, we wish to point out some interesting facts of our study concerning the Wemple-Didomenico model.

Concerning the product $E_d E_m$, we note that with temperature it is roughly a constant like $(\hbar\omega_p)$ in the Penn model. With regard to E_m, we can assume from results on $\varepsilon(0)$ and E_d that E_m scales linearly with energy of the absorption threshold E_0 like $E_m = K' + E_0$ with $K' \approx 1.7$. In this context, the reader may refer to the work of Wemple-Didomenico [7, 18] regarding the definitions of spectroscopic terms (like E_d or E_m).

So E_m and E_g have the same linear variation with energy of absorption threshold.

As for as E_d is concerned, we see that E_d increases with temperature. This behaviour is consistent with the decreasing hydrogen content in the sample with removal of defects [19, 20] (density effects).

After having analysed the hydrogen evolution based on a variation of spectroscopic parameters with temperature, we must say that our study is evidently macroscopic and found upon a calculated model which might draw criticism concerning the hypothesis about coordination number and valency of anions [21].

Acknowledgement

One of the authors (NMR) gratefully acknowledges the financial assistance of CIES (France).

References

[1] M. H. BRODSKY, M. H. FRISCH, and S. F. ZIEGLER, Appl. Phys. Letters **30**, 561 (1977).
[2] J. H. MCMILLAN and E. M. PEKISON, Thin Solid Films **63**, 189 (1979).
[3] S. OGUZ and M. A. PAESLER, Phys. Rev. B **22**, 6213 (1980).
[4] L. ZELLAMA, P. GERMAIN, S. SQUELARD, N. BOURDON, J. FONTENILLE, and R. DANIELOU, Phys. Rev. B **23**, 6648 (1981).
[5] A. DENEUVILLE, A. MINI, and J. C. BRUYÈRE, J. Phys. C **14**, 4531 (1981).
[6] C. ANCE, J. P. FERRATON, J. M. BERGER, and F. DE CHELLE, phys. stat. sol. (b) **113**, 105 (1982).
[7] S. H. WEMPLE, J. chem. Phys. **67**, 2151 (1977).
[8] M. MILLEVILLE, W. FUCHS, F. J. DEMOND, H. MANNSPERGER, G. MÜLLER, and S. KALBERTZER, Appl. Phys. Letters **34**, 173 (1979).
[9] J. M. BERGER, J. P. FERRATON, B. YOUS, and A. DONNADIEU, Thin Solid Films **86**, 337 (1981).
[10] A. DENEUVILLE, J. C. BRUYÈRE, A. MINI, H. KALHIL, R. DANIELOU, and E. LIGEON, J. Phys. C **14**, 2279 (1981).
[11] S. OGUZ, R. W. COLLINS, M. A. PAESLER, and W. PAUL, J. non-crystall. Solids **35**, 836, 231 (1980).
[12] F. DE CHELLE, J. M. BERGER, A. DENEUVILLE, J. C. BRUYÈRE, S. P. COULIBALY, J. P. FERRATON, and A. DONNADIEU, submitted to J. non-crystall. Solids.
[13] N. M. RAVINDRA, C. ANCE, J. P. FERRATON, A. DONNADIEU, and S. ROBIN, phys. stat. sol. (b) **115**, 347 (1983).
[14] G. WEISER, D. EWALD, and M. MILLEVILLE, J. non-crystall. Solids **35/36**, 447 (1980).
[15] PETER Y. YU and M. CARDONA, Phys. Rev. B **2**, 3139 (1970).
[16] H. WELKER and H. WEISS, Solid State Phys. **3**, (1956).
[17] N.M. RAVINDRA, C. ANCE, J. P.FERRATOW, J. M. BERGER, F. DE CHELLE, and S. ROBIN, Infrared Physics (in the press).
[18] S. H. WEMPLE and M. DIDOMENICO, Phys. Rev. B **3**, 1338 (1971).
[19] J. P. DE NEUFVILLE, T. D. MOUSTAKAS, and A. F. RUPPERT, J. non-crystall. Solids **35**, 481 (1980).
[20] S. H. WEMPLE, Phys. Rev. B **7**, 3767 (1973).
[21] A. G. REVESZ and S. H. WEMPLE, phys. stat. sol. (a) **72**, 721 (1982).

(Received February 10, 198 3

phys. stat. sol. (a) 77, 249 (1983)

Subject classification: 1.3; 22

University of Antwerp[1]) *(a) and*
Laboratory of Inorganic Chemistry, Materials Science Center, University of Groningen[2]) *(b)*

Clustering and "Phase Ordering" Transitions in Molybdenum (III) Sulfide, $Mo_{2.065}S_3$

By

R. DEBLIECK (a), G. A. WIEGERS (b), K. D. BRONSEMA (b), D. VAN DYCK (a), G. VAN TENDELOO (a), J. VAN LANDUYT (a), and S. AMELINCKX (a)

In the average structure of $Mo_{2.065}S_3$ there are two crystallographically independent zig-zag chains of molybdenum atoms (Mo–Mo = 2.86 Å). X-ray and electron diffraction studies show that the zig-zag chains are distorted in such a way that "four-atom" clusters of Mo are formed; the displacements of the Mo atoms from the average structure are mainly along the chain axes and amount to 0.16 Å. The deformed structure can be described by two independent $\boldsymbol{k}$ vectors which are incommensurate above ≈ 150 K. Important changes in the diffraction patterns are noticed at ≈ 390, 180, and ≈ 110 K. A refinement of the structure at 110 K ($\boldsymbol{k}_1 = [0, \frac{1}{2}, \frac{1}{4}]$, $\boldsymbol{k}_2 = [\frac{1}{2}, \frac{1}{2}, 0]$ shows that $\boldsymbol{k}_1$ is associated with the distortions in one of the two types of zig-zag chains, the distortions in the other zig-zag chain being responsible for the $\boldsymbol{k}_2$ satellites. The behaviour of the satellites can be interpreted in terms of the creation of correlation between the order in adjacent chains, i.e. due to "phase ordering".

In der mittleren Struktur von $Mo_{2,065}S_3$ existieren zwei kristallographisch unabhängige Zick-Zack-Ketten von Molybdänatomen (Mo–Mo = 2,86 Å). Röntgen- und Elektronenbeugungsuntersuchungen zeigen, daß die Zick-Zack-Ketten derart gestört sind, daß „Vier-Atom"-Cluster von Mo gebildet werden. Die Verlagerungen der Mo-Atome von der mittleren Struktur sind hauptsächlich in Richtung der Kettenachsen und betragen 0,16 Å. Die verzerrte Struktur läßt sich durch zwei unabhängige $\boldsymbol{k}$-Vektoren beschreiben, die oberhalb ≈ 150 K inkommensurabel sind. Wesentliche Änderung in den Beugungsdiagrammen werden bei ≈ 390, 180 und ≈ 110 K bemerkt. Eine Verfeinerung der Struktur bei 110 K ($\boldsymbol{k}_1 = [0, \frac{1}{2}, \frac{1}{4}]$, $\boldsymbol{k}_2 = [\frac{1}{2}, \frac{1}{2}, 0]$ zeigt, daß $\boldsymbol{k}_1$ mit den Verzerrungen in einer der beiden Typen der Zick-Zack-Ketten verknüpft ist; die Verzerrungen in der anderen Zick-Zack-Kette sind verantwortlich für die $\boldsymbol{k}_2$-Satelliten. Das Verhalten der Satelliten kann unter der Bedingung von Korrelation zwischen der Ordnung in benachbarten Ketten, d. h. infolge von „Phasenordnungsprozessen", erklärt werden.

1. Introduction

$Mo_{2.065}S_3$ has a basic structure with the molybdenum atoms in 2/3 of the octahedral interstices of a chh type close packing of sulfur atoms (Fig. 1). The Mo atoms are displaced from their ideal positions forming zig-zag chains in the direction of the monoclinic b-axis. There are two crystallographically independent types of zig-zag chains in the monoclinic unit cell of dimensions $a = 6.092$, $b = 3.208$, $c = 8.6335$ Å, $\beta = 102.43°$; the space group being P 2_1/m with all the atoms at the special positions: $\pm (x \frac{1}{4} z)$. The plane of the zig-zag chain of type 1 is approximately parallel to $(\bar{1}01)$ while the plane of zig-zag chain 2 is almost parallel to (001): the plane of the deformed hexagonal layers of sulphur. This basic structure was deduced from powder data by Jellinek [1] and refined also from powder data by De Jonge et al. [2]; at that time it

---

[1]) R.U.C.A., Groenenborgerlaan 171, B-2020 Antwerp, Belgium.

[2]) Nijenborgh 16, 9747 AG Groningen, The Netherlands.

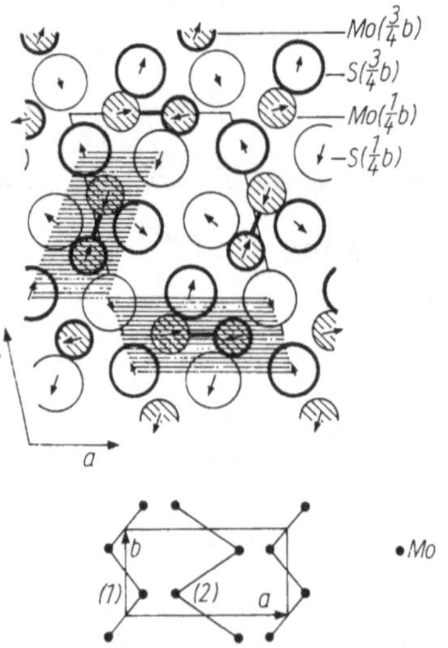

Fig. 1. Structure of $Mo_{2.065}S_3$ according to de Jonge et al. [2]

was concluded from some Weissenberg photographs that the true structure had to be more complex. Low-temperature powder photographs showed a transition to triclinic symmetry at 180 K, the deviation from monoclinic symmetry being small. Physical measurements [2] (resistivity, Hall effect, thermoelectric power, and magnetic susceptibility) showed that the compound is a metal; significant changes were noticed at low temperature indicating phase transitions. It was also found by De Jonge et al. [2] that the compound contains a small excess of Mo compared to the stoichiometry Mo_2S_3. This was confirmed by a recent study by Suzuki et al. [3] and by Krabbes et al. [4]; the latter found a region of existence of $Mo_{2+\delta}S_3$ at 900 °C with $0.008 < \delta < 0.036$.

Single crystals of dimensions suitable for physical measurements and X-ray studies were grown by Hemmel and coworkers [5]. The latter explained the behaviour of the magnetic and electrical transport properties (as measured on single crystals) by abrupt changes in the number of mobile carriers, due to charge density wave type phase transitions. The physical properties were also measured by Rastogi and Ray [6].

In the present paper the modulated (super) structure of $Mo_{2.065}S_3$ as deduced by X-ray and electron diffraction is reported. The low-temperature commensurate (super) structure was determined from diffractometer intensity data.

2. Weissenberg Analysis and Electron Diffraction Studies

The crystals were grown by vapour transport (without added foreign agent); crystals from the same batch were used for the measurements of the electrical transport properties [5]. The crystals occur in the shape of needles and platelets elongated in the direction of the monoclinic *b*-axis. Weissenberg photographs of crystals aligned along *b* showed that most of the crystals contain orientation variants with a common *b*-axis but with *a*- (and *c*-) axes related as given in Fig. 2; the angle between *a* and *a'* is 102.4°, the same value as the monoclinic angle β. We consider this relationship to be due to the congruency of the two types of zig-zag chains in the cell (shaded areas in Fig. 1). Crystals with and without orientation variants showed the same superreflections at room temperature but the temperature behaviour was found to be different.

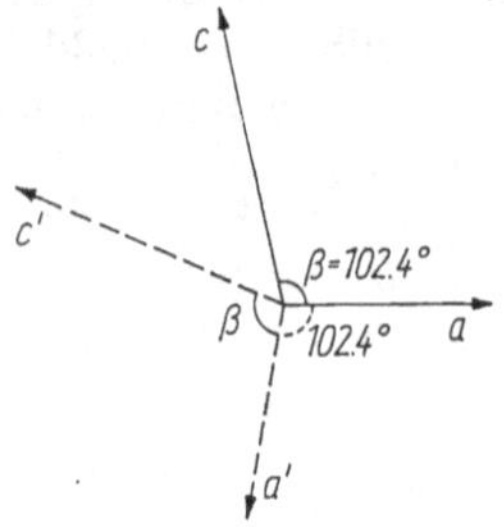

Fig. 2. Relation between *a*, *c* and *a'*, *c'* in multidomain crystals

Superreflections were observed on the rotation photographs indicating a doubling of the monoclinic axis; the extra reflections were very weak for the first interlayer line, those of the second interlayers were stronger and also appeared to be incommensurate with b^*. Weissenberg photographs of the second interlayer showed that there are two types of reflections, those commensurate with $b^*(\frac{1}{2}b^*)$ but incommensurate with a^* and c^* of the subcell and those incommensurate with b^* and commensurate with a^* and $c^*(\frac{1}{2}a^*, c^*)$, respectively. These satellite reflections can be described as being due to two independent distortion vectors $\boldsymbol{k}_1$ and $\boldsymbol{k}_2$ with $\boldsymbol{k}_1 = [p, \frac{1}{2}, q]$ and $\boldsymbol{k}_2 = [\frac{1}{2}, \frac{1}{2} - v, 0]$, where at 300 K $p = -0.041$, $q = 0.214$, $v = 0.059$. These superreflections were very clearly observed in the electron diffraction study (Fig. 3).

X-ray Weissenberg photographs (CuKα radiation) were taken in the temperature range 110 to 390 K. The behaviour of $\boldsymbol{k}_1$ and $\boldsymbol{k}_2$ with temperature is different for monodomain and multidomain (with orientation variants) crystals.

Monodomain crystals:

$$\boldsymbol{k}_1: [0, \tfrac{1}{2}, 0] \xrightarrow{\approx 110\ \mathrm{K}} [0, \tfrac{1}{2}, \tfrac{1}{4}] \xrightarrow{\approx 200\ \mathrm{K}} [p, \tfrac{1}{2}, q] \xrightarrow{390\ \mathrm{K}} \text{disapp.}$$

$$\boldsymbol{k}_2: [\tfrac{1}{2}, \tfrac{1}{2}, 0] \xrightarrow{\approx 150\ \mathrm{K}} [\tfrac{1}{2}, \tfrac{1}{2}, 0.153] \xrightarrow{180\ \mathrm{K}} [\tfrac{1}{2}, \tfrac{1}{2} - v, 0];$$

Multidomain crystals:

$$\boldsymbol{k}_1: \text{short range order } [0, \tfrac{1}{2}, 0] \xrightarrow{\approx 110\ \mathrm{K}} [p, \tfrac{1}{2}, q] \xrightarrow{390\ \mathrm{K}} \text{disapp.}$$

$$\boldsymbol{k}_2: [\tfrac{1}{2}, \tfrac{1}{2}, 0] \xrightarrow{180\ \mathrm{K}} [\tfrac{1}{2}, \tfrac{1}{2} - v, 0]\,.$$

The relevant diffraction features in the reciprocal lattice at $\approx \frac{1}{2}b^*$ are presented in Fig. 4. $\boldsymbol{k}_1 = [p, \frac{1}{2}, q]$ is strongly temperature dependent; for a monodomain crystal $p = 0$, $q = \frac{1}{4}$ at about 200 K. Above that temperature the vector $\boldsymbol{u}$ connecting the two satellites rotates and decreases in length. At 268 K the angle φ between $\boldsymbol{u}$ and $\boldsymbol{c}^*$ is 8° ($\boldsymbol{u}$ almost perpendicular to the a^*, b^* plane), $|\boldsymbol{u}| = 0.466\,|c^*|$ which corresponds to $p = -0.023$, $q = 0.233$. At 300 K the angle $\varphi = 16°$, $|\boldsymbol{u}| = 0.43\,|c^*|$ which corresponds to $p = -0.041$, $q = 0.214$. At still higher temperatures the angle φ increases to about 31° at 330 K. At that temperature the intensities of the $\boldsymbol{k}_1$ satellites decrease in intensity, the satellites disappear at about 390 K. Electron diffraction patterns of the [100] zone taken at high temperature showed streaks due to planes of diffuse intensity perpendicular to b^* and situated midway between the rows of Bragg spots due to the basic structure (Fig. 3c); at lower temperature the $\boldsymbol{k}_1$ satellites become visible. It proves that above 390 K the correlations between distortions of neighbouring chains are of short range. The behaviour of $\boldsymbol{k}_1$ with temperature from room temperature to 390 K was the same for mono- and polydomain crystals; upon cooling from room temperature monodomain crystals showed a lock-in into the commensurate state $\boldsymbol{k}_1 = [0, \frac{1}{2}, \frac{1}{4}]$ which was not observed for a multidomain crystal. Here at about 110 K the $\boldsymbol{k}_1[0, \frac{1}{2}, 0]$-satellites appear as diffuse streaks interconnecting the $[0, \frac{1}{2}, \frac{1}{4}]$ satellites.

The behaviour of $\boldsymbol{k}_2 = [\frac{1}{2}, \frac{1}{2} - v, 0]$ above 180 K was the same for crystals with and without orientation variants. At 390 K, the temperature where the $\boldsymbol{k}_1$ satellites disappear, the $\boldsymbol{k}_2$ satellites are still visible. Electron diffraction showed the $\boldsymbol{k}_2$ distortion to be present up to very high temperatures.

At ≈ 180 K, v becomes 0 (for mono- as well as for multidomain crystals) while the basic structure changes from monoclinic to triclinic ($\alpha = 89.75$, $\gamma = 90.3$). This is accompanied by a twinning as could clearly be seen on a Weissenberg photograph of the hll subcell reflections. The twins are related by a mirror plane perpendicular to the original (monoclinic) b-axis. This phase transition showed a hysteresis of about 18 K (176 K on cooling, 194 K on heating) as measured with a low-temperature powder photograph taken with a Guinier-Simon camera (Enraf Nonius). Crystals with orientation variants did not show the intermediate $\boldsymbol{k}_1 = [\frac{1}{2}, \frac{1}{2}, 0.153]$ as found for a monodomain crystal. At the lowest temperature which could be reached (≈ 100 K), the distortions in a monodomain crystal are described by $\boldsymbol{k}_1 = [0, \frac{1}{2}, 0]$ and $\boldsymbol{k}_2 = [\frac{1}{2}, \frac{1}{2}, 0]$. By quick cooling to 100 K of a monodomain crystal a situation with $\boldsymbol{k}_1 = [0, \frac{1}{2}, \frac{1}{4}]$, $\boldsymbol{k}_2 = [\frac{1}{2}, \frac{1}{2}, 0]$ was obtained.

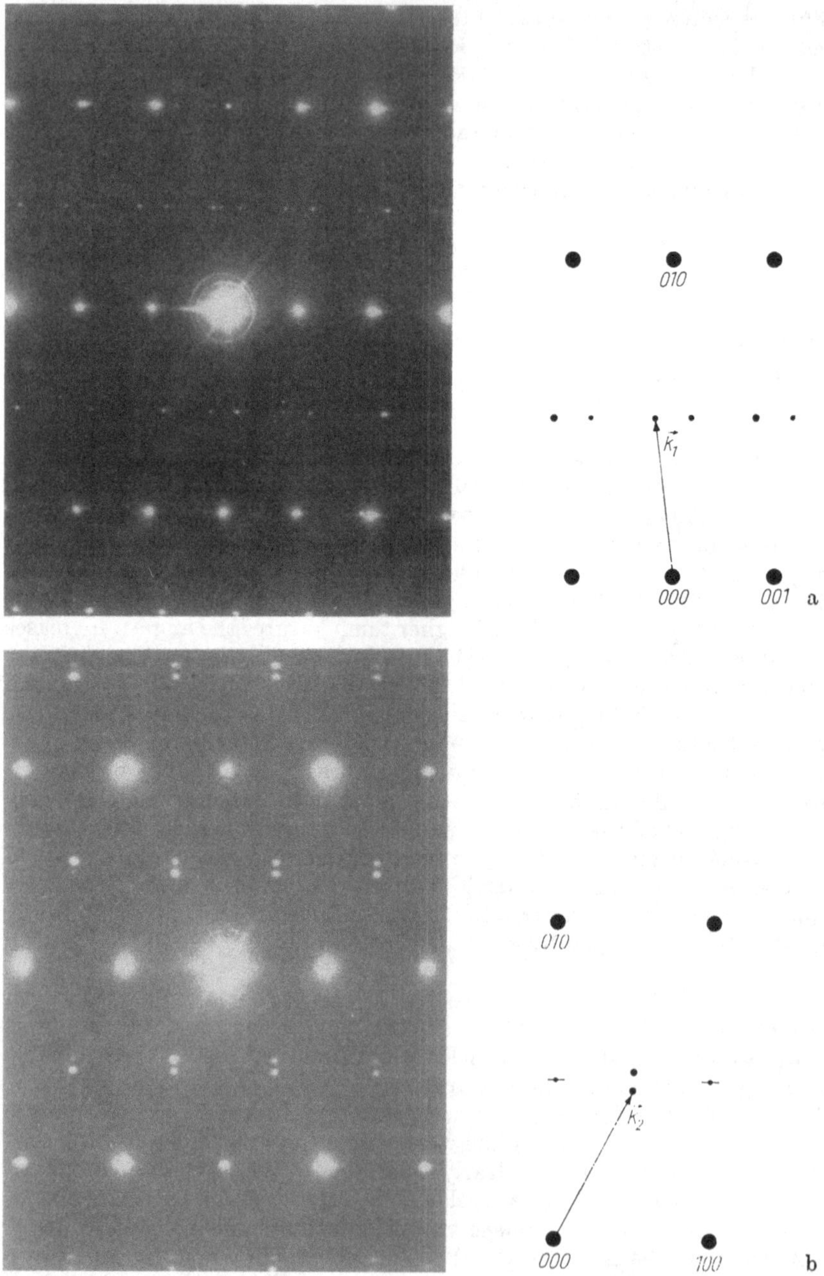

Fig. 3

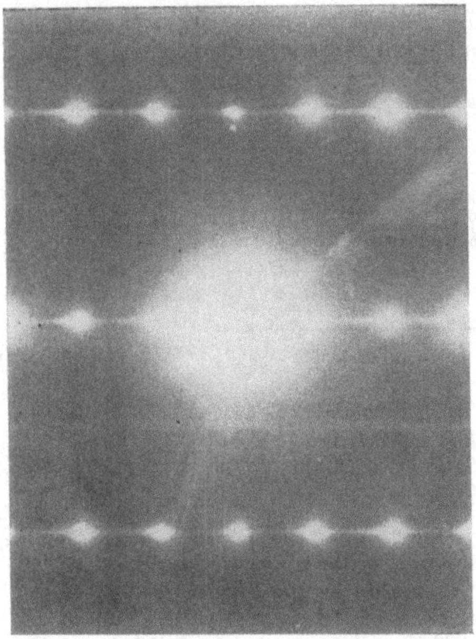

c

Fig. 3. Electron diffraction patterns. Two relevant sections of reciprocal space a) [100]-zone and b) [001]-zone reveal the two types of satellite spots characterized by $\boldsymbol{k}_1$ and $\boldsymbol{k}_2$ on the explanatory and indexed drawings. These sections can be compared with the reconstructed reciprocal lattice in Fig. 4 (room temperature). These patterns result upon cooling from above room temperature where, as shown in c) for the [100]-section diffuse streaks are observed midway the [010]* direction. From tilting experiments it can be concluded that these streaks are due to the intersection of sheets of diffuse intensity with Ewald sphere

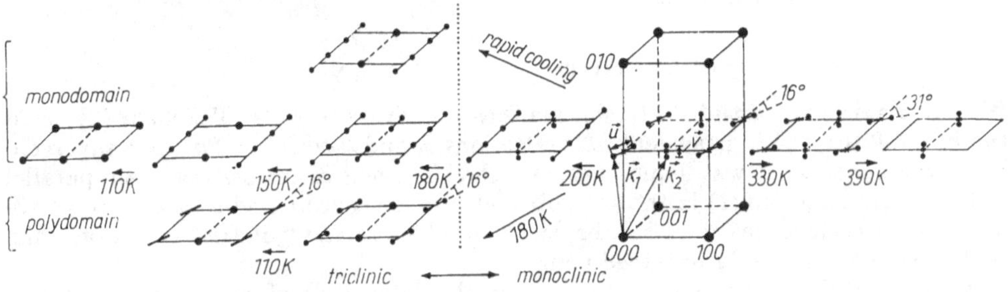

Fig. 4. Relevant diffraction features at phase transitions

3. Data Collection, Subcell Refinement

Intensities of the satellites due to $\boldsymbol{k}_1 = [0, \frac{1}{2}, \frac{1}{4}]$, $\boldsymbol{k}_2 = [\frac{1}{2}, \frac{1}{2}, 0]$ were measured at 100 K on a CAD 4 (Nonius) diffractometer. Reflections due to $\boldsymbol{k}_1$ were measured in an a, $2b$, $4c$ supercell (k and l odd), those due to $\boldsymbol{k}_2$ were measured in a $2a$, $2b$, c supercell (h and k odd). Also a number of subcell reflections were measured. The crystal (monodomain) was a needle of 0.15 mm diameter with a length of about 1 mm. Monochromatic MoKα radiation was used; the measurements were performed with the θ–2θ technique. No correction for absorption was applied. It was found that the intensities of the satellites increase with the index k; which already indicates that the structure will be deformation modulated. Reflections were measured up to $\sin\theta/\lambda = 0.6$ Å^{-1} for the satellites and up to $\sin\theta/\lambda = 0.43$ Å^{-1} for the subcell reflections. The reflections showed monoclinic symmetry, i.e. $I_{hkl} = I_{\bar{h}\bar{k}l'}$, notwithstanding the triclinic symmetry of the subcell. We ascribe this to the twinning introduced into the crystal during the transition at 180 K. The measured data were reduced to 647 independent reflections due to $\boldsymbol{k}_1$, 224 reflections due to $\boldsymbol{k}_2$ and 140 subcell reflections. Refinements were carried out using the XRAY system [7].

A refinement of the subcell reflections showed the coordinates from the powder refinement [2] to be correct; the index R_F was 0.05, individual isotropic temperature factors were used. The coordinates of the atoms from this refinement are compared with those from the powder refinement in Table 1.

Table 1

Atomic coordinates of the atoms in the monoclinic subcell of $Mo_{2.065}S_3$ from a powder refinement (at the left) and the single crystal refinement (right)

	x	z	x	z
Mo(1)	0.310	0.009	0.3101(6)	0.0092(4)
Mo(2)	0.108	0.632	0.1091(6)	0.6312(4)
S(1)	0.505	0.803	0.512(2)	0.804(1)
S(2)	0.962	0.161	0.970(2)	0.156(1)
S(3)	0.729	0.516	0.727(2)	0.514(1)

(i) Single crystal X-ray study at 110 K; true symmetry is triclinic, reflections hkl and $h\bar{k}l$ coincide because of twinning. Cell constants of the monoclinic average unit cell: $a = 6.079$, $b = 3.200$, $c = 8.616$ Å, $\beta = 102.33$ °, space group $P2_1/m$: atoms at $\pm(x\ \frac{1}{4}\ z)$.
(ii) Standard deviations given in parentheses in the last decimal place.
(iii) Powder refinement of data taken at 300 K; cell constants $a = 6.092$, $b = 3.208$, $c = 8.633$ Å, $\beta = 102.43°$.

4. The Superstructure Described by $\boldsymbol{k}_2 = [\frac{1}{2}, \frac{1}{2}, 0]$

We calculated the partial Patterson synthesis as introduced by Takeuchi [8] using the $\boldsymbol{k}_2$ satellites (224 independent hkl reflections, h and k odd) in a $2a$, $2b$, c supercell. From this synthesis it was found that only the Mo atoms of the zig-zag chain parallel to the (001) plane (Mo (1) in the subcell) and sulphur coordinated to these atoms are involved. It became clear that the displacement of Mo is mainly in the b direction while that of sulphur is mainly in the ac-plane.

Because of the monoclinic symmetry of the intensities of the superreflections, a structure-factor calculation and refinement had to be performed in a monoclinic space group of a $2a$, $2b$, c unit cell. We choose the space group Cm because the superreflections are only present for h and k odd. The atomic positions $\pm$ (x, $\frac{1}{4}$, z) in $P2_1/m$ were transformed into positions in Cm: (000, $\frac{1}{2}\frac{1}{2}0$) + ($\frac{x}{2}$, y, z) ($y \approx \frac{1}{4}$), ($\frac{1}{2} - \frac{x}{2}$, 0, $\bar{z}$), and ($\frac{1}{2} - \frac{x}{2}$, $\frac{1}{2}$, $\bar{z}$). These positions are indicated by a, b, and c, respectively; the latter two positions are special positions at the mirror planes. It can easily be shown that an atom on a position of type a contributes to the structure factor of the superreflections when y deviates from the value $\frac{1}{4}$; atoms of types b and c contribute when their x' (= $\frac{1}{2} - \frac{x}{2}$) and/or z' (= $\bar{z}$) coordinates differ. For the refinement the origin was chosen at Mo (1b) the space group being non-centrosymmetric. 224 superreflections together with 140 subcell reflections (with indices transformed into a $2a$, $2b$, c cell) were included; the subcell reflections were taken into account because of the scaling factor. At first, the y coordinate of Mo (1a) was refined, with $y = 0.28$ as a starting value. The intensities of the subcell reflections are not affected while those of the superreflections, apart from the dependence of the scattering factor with θ, only depend upon $(k\delta_y)^2$, δ_y being the deviation from 0.25. The y coordinate refined to a value 0.27. R_F was 0.099 for all reflections. The coordinates of the other atoms were then included in the refinement starting with those of S(1), being sulfur coordinated to four Mo atoms of the (001)-zig-zag chain. The results are given in Table 2 as differences

Table 2
Displacements of the atoms Δx, Δy, Δz found from the $\boldsymbol{k}_2 = [\frac{1}{2}, \frac{1}{2}, 0]$ refinement

	$\Delta x\ (= x_b - x_c)$	$\Delta y\ (= y - \frac{1}{4})$	$\Delta z\ (= z_b - z_c)$
Mo(1a)	—	+0.0240(4)*)	—
Mo(1b)–Mo(1c)	−0.0059(5)	—	−0.0001(3)
Mo(2a)	—	−0.0003(4)	—
Mo(2b)–Mo(2c)	−0.002(1)	—	+0.002(2)
S(1a)	—	+0.004(2)	—
S(1b)–S(1c)	−0.014(3)	—	−0.011(5)
S(2a)	—	−0.006(2)	—
S(2b)–S(2c)	+0.008(3)	—	+0.008(5)
S(3a)	—	−0.001(2)	—
S(3b)–S(3c)	−0.003(3)	—	−0.004(5)

*) Standard deviations given in parentheses are in the last decimal place.

$\Delta y = y_a - \frac{1}{4}$, $\Delta x = x_b - x_c$, and $\Delta z = z_b - z_c$, b at $y = 0$, c at $y = \frac{1}{2}$. A final R_F value of 0.057 was obtained. The temperature factor was kept at the value obtained from the subcell refinement. It is seen (Table 2) that the displacement of Mo (1a): $\Delta y = 0.0240$ (s.d. 0.0005) corresponding to a displacement of $0.0240 \times 6.60 = 0.158$ Å is the largest. Other significant displacements are those of Mo(1b)–Mo(1c) ($\Delta a = 0.0059 \times 12.28 = 0.072$ Å, $\Delta c \approx 0$) and those of S(1). We consider the displacement found for Mo(2) and S(3) as not significant. The displacements are shown in Fig. 5. It is clearly seen that the displacements of S(1) and S(2), being mainly parallel with the ac-plane are a consequence of the (much larger) displacements of Mo(1), keeping the Mo–S bond length invariant.

5. The Superstructure Described by $\boldsymbol{k}_1 = [0, \frac{1}{2}, \frac{1}{4}]$

The partial Patterson synthesis calculated with the $\boldsymbol{k}_1$ satellites (647 independent reflections hkl, h and l odd) in an a, $2b$, $4c$ supercell showed that only the Mo atoms (Mo(2)) of the zig-zag chain parallel to the ($\bar{1}01$) plane and sulphur atoms bonded to these Mo atoms (S(3)) are involved in the displacements. In this case the space group Am was chosen. The atomic coordinates $\pm\ (x, \frac{1}{4}, z)$ in $P2_1/m$ were transformed into coordinates in space group Am of the a, $2b$, $4c$ supercell. In this case each coordinate transforms into positions in Am: (000, $0\frac{1}{2}\frac{1}{2}$) + (a) $x, y, \frac{z}{4}$) $y \approx 0.25$); (b) $\bar{x}, 0, \frac{\bar{z}}{4}$;(c) $x, \frac{1}{2}, \frac{\bar{z}}{4}$; (d) $x, y, \frac{z}{4} + \frac{1}{4}$($y \approx 0.25$); (e) $\bar{x}, 0, \frac{\bar{z}}{4} + \frac{1}{4}$; (f) $\bar{x}, \frac{1}{2}, \frac{\bar{z}}{4} + \frac{1}{4}$. Atoms

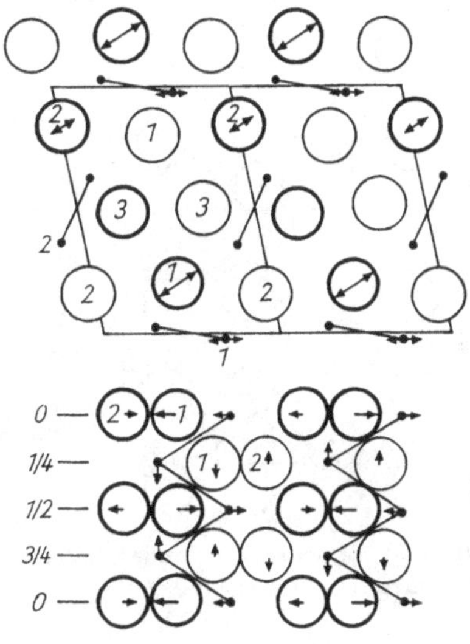

Fig. 5. Displacements of the atoms in the $\boldsymbol{k}_2 = [\frac{1}{2}\ \frac{1}{2}\ 0]$ refinement. Atoms indicated by bold- line circles are at the mirror planes at $y = 0$ and 1/2. The numbers refer to the coordinates given in Table 2, see text. Displacements are enlarged by a factor of 100 and for Δy of Mo by a factor of 50

a, b, c are at a distance c apart from atoms d, e, f, respectively. The refinement was started with the y coordinates of Mo(2a) and Mo(2d) ($\Delta y = 0.025$). In successive runs the coordinates of all atoms were refined with a final R_F of 0.063. The displacements of the atoms are given in Table 3, in the same way as presented in Table 2 for the $\boldsymbol{k}_2$

Table 3
Displacements of the atoms Δx, Δy, Δz found from $\boldsymbol{k}_1 = [0, \frac{1}{2}, \frac{1}{4}]$

	Δx $(= x_b - x_c)$	Δy $(= y - \frac{1}{4})$	Δz $(= z_b - z_c)$
Mo(1a)	—	+0.0004(4)*)	—
Mo(1b)–Mo(1c)	−0.001(2)	—	+0.0016(3)
Mo(1d)	—	−0.0046(4)	—
Mo(1e)–Mo(1f)	−0.002(2)	—	−0.0011(3)
Mo(2a)	—	+0.0272(4)	—
Mo(2b)–Mo(2c)	+0.009(2)	—	+0.0016(3)
Mo(2d)	—	+0.0297(4)	—
Mo(2e)–Mo(2f)	+0.014(2)	—	+0.0023(3)
S(1a)	—	+0.006(1)	—
S(1b)–S(1c)	−0.009(5)	—	+0.000(1)
S(1d)	—	−0.005(1)	—
S(1e)–S(1f)	−0.008(5)	—	+0.001(1)
S(2a)	—	−0.001(1)	—
S(2b)–S(2c)	−0.008(5)	—	−0.001(1)
S(2d)	—	−0.005(1)	—
S(2e)–S(2f)	+0.004(5)	—	−0.001(1)
S(3a)	—	+0.003(1)	—
S(3b)–S(3c)	+0.000(5)	—	−0.008(1)
S(3d)	—	+0.003(1)	—
S(3e)–S(3f)	+0.001(5)	—	+0.008(1)

*) Standard deviations given in parentheses are in the last decimal place.

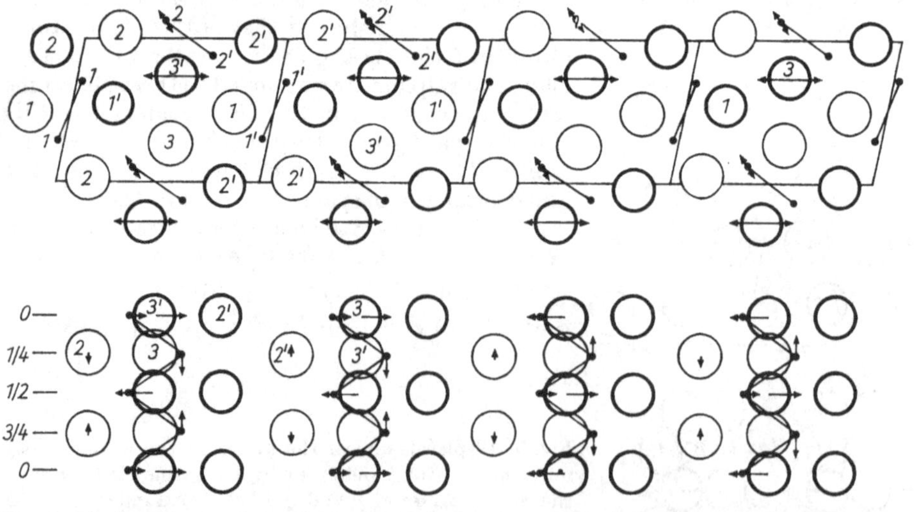

Fig. 6. Displacements of the atoms found from the $\boldsymbol{k}_1 = [0\ \frac{1}{2}\ \frac{1}{4}]$ refinement. Numbers refer to Table 3, see text. Displacements indicated as in Fig. 5. The primed numbers are labelled d, e, f in Table 3 the unprimed a, b, c

refinement. The largest displacements are those in the b direction of Mo(2a) and Mo(2d); the displacements are in phase and about equal in magnitude to that of Mo(1a) in the $\boldsymbol{k}_2$ refinement. The largest (and significant) displacements of sulphur atoms are those of S(3) (bonded to four Mo atoms of zig-zag chain ($\bar{1}01$)). The displacements of S(1a), S(1d), and S(2d) are larger than twice the estimated standard deviation (e.s.d.), these displacements cannot be explained (like those of S(3)) as being due to a relaxation of the sulphur lattice to the large displacement of the Mo(2) atoms. It may indicate that the errors are considerably larger than the e.s.d. In this respect the displacement of Mo(1d), 0.03 Å, is probably not significant. The displacements are shown in Fig. 6. It is seen that atoms at a distance c apart, are in phase with one of the neighbours and in antiphase with the other. The Mo atoms in the ($\bar{1}01$) zig-zag chain form triangular clusters, like those of the Mo atoms of the (001) zig-zag chain in the $\boldsymbol{k}_2$ refinement. The relation with the true structure (in triclinic symmetry) will be discussed below.

6. Relationship Between the Results in Monoclinic Symmetry and the Real (Triclinic) Structure

As has been described above, the symmetry of the average structure changes from monoclinic to triclinic at 180 K, the dimensions of the triclinic cell being very close to those of the monoclinic cell. This change is accompanied by a twinning so that reflections hkl and $h\bar{k}l$ almost coincide. The measured intensities of the subcell reflections as well as those of the superreflections are therefore the average of F^2_{hkl} and $F^2_{h\bar{k}l}$ or

$$\overline{F^2_t}(hkl) = \tfrac{1}{2}\,\{F^2_t(hkl) + F^2_t(h\bar{k}l)\}$$

(t indicating the triclinic case).

We take as an example the distortion in a zig-zag chain as indicated in Fig. 7 a and b; the displacement from the positions in the undistorted structure is δ_y with directions as indicated. We assume a center of symmetry present at the origin (+). In the case of Fig. 7a we find for the superreflections, δ_y small and the form factors f_{Mo} taken as unity,

$$F_t(hkl) = 4\pi k\,|\delta_y|\,\sin 2\pi(hx - \tfrac{1}{8}\,k + lz)\,,$$

$$F_t(h\bar{k}l) = -4\pi\,|k|\,|\delta_y|\,\cos 2\pi(hx - \tfrac{1}{8}\,k + lz)\,,$$

$$\overline{F^2_t}(hkl) = \tfrac{1}{2}\,\{F^2_t(hkl) + F^2_t(h\bar{k}l)\} = 8\pi^2k^2\delta_y^2\,.$$

In the case of Fig. 7b we find

$$\overline{F^2_t}(hkl) = 32\pi^2k^2\delta_y^2\,.$$

For the distortion in monoclinic symmetry (Fig. 7c) one finds

$$F^2_m(hkl) = 16\pi^2k^2\delta_y^2\,.$$

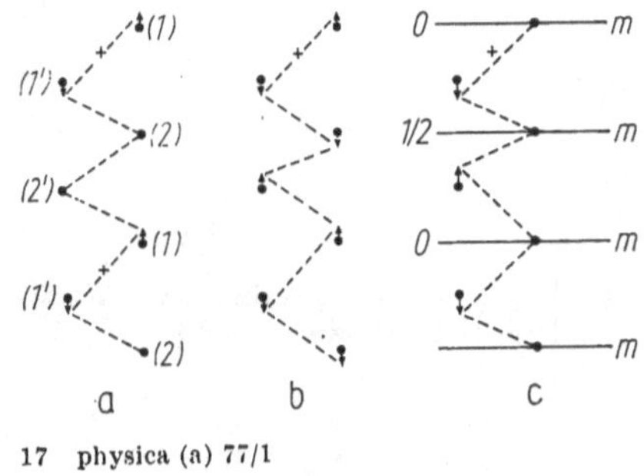

Fig. 7. Displacements of Mo in the zig-zag chain; positions in the average structure are indicated by •; + is the origin: a) triclinic distortion with only atom (1) and (1') involved, centre of symmetry at +; b) as a) but all atoms involved; c) equivalent distortion in monoclinic symmetry (see text)

It is clear therefore that the result obtained from a refinement in monoclinic symmetry (Fig. 7c) is equivalent to the models of 7 a and b in triclinic symmetry including twinning.

We now consider a more complicated type of distortion in triclinic symmetry given in Fig. 8a; the displacements of atom (1) are $\delta_x(1)$, $-\delta_y(1)$, $\delta_z(1)$, those of atom (2) are $-\delta_x(2)$, $\delta_y(2)$, $-\delta_z(2)$, δ positive in all cases. We assume a centre of symmetry at the position denoted by $+$. We then find

$$\overline{F_t^2}(hkl) = 2(p^2 + q^2) + 4pq(\beta^2 - \alpha^2)\,,$$

where

$$p = 2\pi k[\delta_y(1) + \delta_y(2)]\,,$$
$$q = 2\pi[h(\delta_x(1) + \delta_x(2)) + l(\delta_z(1) + \delta_z(2))]\,,$$
$$\beta = \sin 2\pi(hx - \tfrac{1}{8}k + lz)\,,$$
$$\alpha = \cos 2\pi(hx - \tfrac{1}{8}k + lz)\,,$$

x, $\frac{1}{8}$, z being the coordinates in the undistorted structure, the origin is denoted by $+$. An expression of the same form is found for the monoclinic distortion of Fig. 8b

$$F_m^2 = 4(p^2 + q^2) + 8pq(\beta^2 - \alpha^2)$$

with

$$p = 2\pi k\delta_y\,, \qquad q = 2\pi(h\delta_x + l\delta_z)\,.$$

In the case of Fig. 8c the displacements of atom (1) are $-\delta_x(1)$, $-\delta_y(1)$, $-\delta_z(1)$, those of (2) are $\delta_x(2)$, $\delta_y(2)$, $\delta_z(2)$ (δ positive), one finds, assuming a centre of symmetry,

$$\overline{F_t^2}(hkl) = 2(p^2 + q^2) - 4pq(\beta^2 - \alpha^2) \quad \text{(Fig. 8c)}$$

$F_m^2(hkl) = 4(p^2 + q^2) - 8pq(\beta^2 - \alpha^2)$ for the equivalent monoclinic distortion of Fig. 8d.

The displacements in triclinic symmetry can therefore be related to the results obtained from the monoclinic refinements: the displacements δ_y follow from the displacements of the atoms at general positions in the monoclinic space group, the displacements δ_x and δ_z follow from the displacements of the atoms at the mirror planes at $y = 0$ and $y = 1/2$. It is clear that the triclinic displacements are not unique; as the extreme cases we have (i) only displacements of atom (1) or (2), (ii) atoms (1) and (2) have displacements of equal magnitude. Even more possibilities are allowed when we consider the centre of symmetry not to be present. The distortion in the zig-zag chains of Mo atoms from $\boldsymbol{k}_1$ and $\boldsymbol{k}_2$ refinements is of the form given in Fig. 9a; $\delta_y \approx 0.026$, which corresponds to about 0.16 Å; the displacement of the atoms at $y = 0$ and $y = \frac{1}{2}$ is much smaller, about 0.03 Å or less. When all metal atoms are assumed to have the same magnitude of displacement, and a centre of symmetry being present, one finds the result in triclinic symmetry being as given in Fig. 9d with

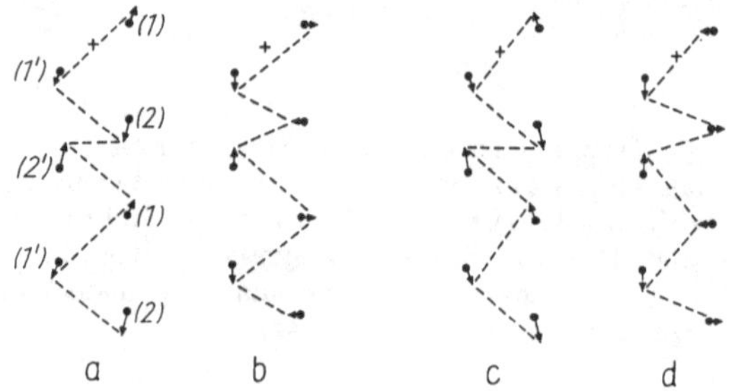

Fig. 8. As Fig. 7, but also displacements in the ac-plane allowed, see text

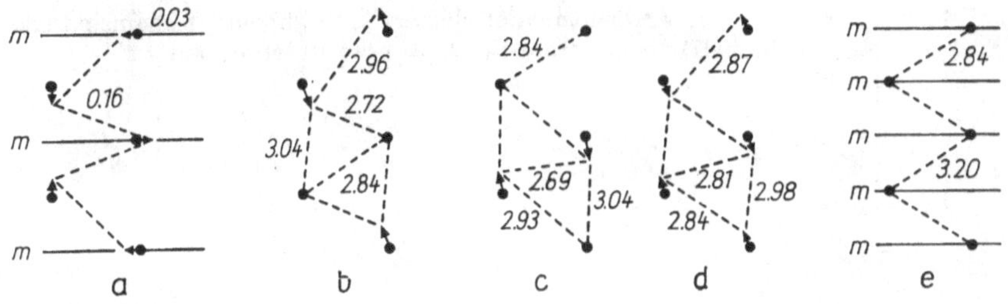

Fig. 9. a) Displacements in the zig-zag chain from the monoclinic refinements. b) Equivalent distortion in triclinic symmetry assuming atoms (1) and (1′) to be displaced. c) as b) but only (2) and (2′) displaced. d) as b) but all atoms displaced with equal magnitude. e) The average chain in monoclinic symmetry. All distances are in Å

distances as indicated. When we consider the case that atom (1) or (2) only (and the atom related by a centre of symmetry) are involved we obtain the results shown in Fig. 9 a and b. We prefer the model of Fig. 9d because all atoms (equivalent in the average structure) are involved in the displacements; it is also seen that the differences between the distances in a cluster are smaller than in case of Fig. 9b and c.

7. Discussion of the Commensurate Structure

It is shown that the symmetric zig-zag chains of Mo atoms, present in the average structure of $Mo_{2.065}S_3$, are in fact strongly coupled four-atom clusters of Mo. One can compare these clusters with four-atom clusters in other d-metal compounds: $FeMo_2S_4$ and $CoMo_2S_4$ [9], $ReSe_2$ [10], Re_2P_5 [11], Re_6P_{13} [12]. Especially for the compounds $CoMo_2S_4$ and $ReSe_2$ the clusters can similarly be considered to be due to distortions in a zig-zag chain as it is the case for $Mo_{2.065}S_3$; the geometry and important distances in the clusters are presented in Fig. 10. The cluster in $ReSe_2$ is centrosymmetric, the deviation from centrosymmetry is small in the cluster of $CoMo_2S_4$; in both clusters the distance between the central atoms is the shortest. The distances of the outer atoms to an atom of a neighbouring cluster: 2.960 Å in $CoMo_2S_4$, 3.077 Å in $ReSe_2$ and 2.87 Å in $Mo_{2.065}S_3$ (model Fig. 9) are also considered as a metal–metal bond like the other (shorter) distances. Each metal atom is therefore involved in the formation of three metal–metal bonds. Since Mo^{3+} has three 4d ($4d^3$) electrons, these electrons are involved in the metal to metal bonding. They are in an almost filled bond, the number of charge carriers (p-type) below 180 K amounts to about 0.03 hole/Mo [5, 6].

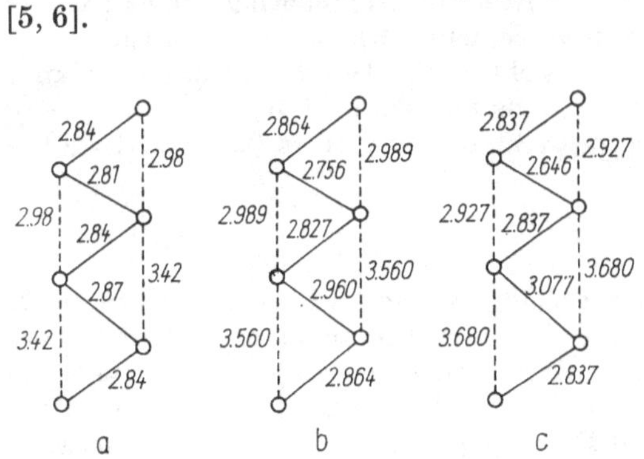

Fig. 10. Distances between atoms in clusters of a) $Mo_{2.065}S_3$, b) $CoMo_2S_4$, c) $ReSe_2$. All distances in Å

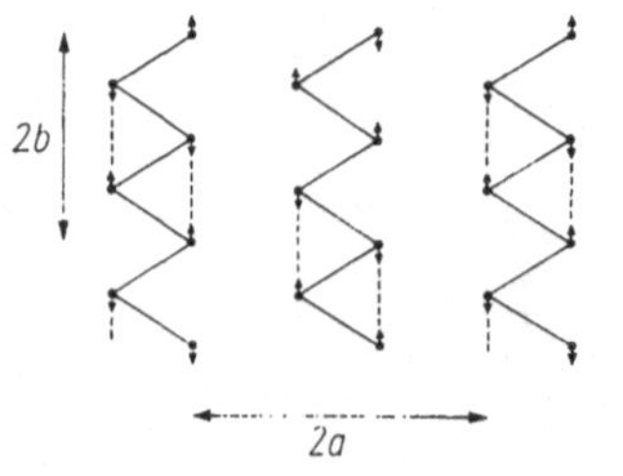

Fig. 11. Arrangement of clusters in neighbouring chains parallel to (001); result from the $\boldsymbol{k}_2 = [\frac{1}{2}\ \frac{1}{2}\ 0]$ refinement

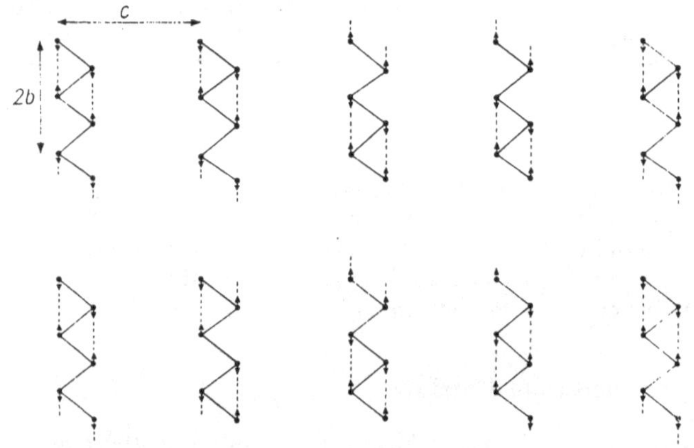

Fig. 12. The two possible cluster arrangements of the chains parallel $(\bar{1}01)$; results in agreement with the $\boldsymbol{k}_1 = [0\ \frac{1}{2}\ \frac{1}{4}]$ refinement

In Nb_2Se_3 and Ta_2Se_3, isotypic with $Mo_{2.065}S_3$, the zig-zag chains are of two kinds; the metal–metal distances in chains parallel to the (001) plane are comparable to those in the pure metal, in the other chains (parallel to the plane $(\bar{1}01)$), the distances are considerably longer [13]. Mo in $Mo_{2.065}S_3$ has one more d electron than Nb and Ta in the corresponding selenides; it explains why the metal–metal bonding in Nb_2Se_3 and Ta_2Se_3 is less strong than in $Mo_{2.065}S_3$. There are no indications for a superstructure (clustering).

In α-ZrI_2 [14], isostructural β-$MoTe_2$ [15, 16, 17] (both with a distorted $Cd(OH)_2$ type structure), and β-ZrI_2 [18] zig-zag chains of metal atoms are present, but whereas α-ZrI_2($4d^2$ for Zr^{2+}) is a semiconductor, β-$MoTe_2$ is a metal.

In $Mo_{2.065}S_3$ the clusters in neighbouring chains parallel to the plane (001) are in a "close-packed" arrangement (Fig. 11). For clusters in neighbouring chains parallel to the plane $(\bar{1}01)$, two arrangements are in agreement with the result of the monoclinic refinement in Fig. 12. Neighbouring chains of the $(\bar{1}01)$ type have a larger distance (and therefore weaker interaction) than the chains parallel to (001); it explains why long-range order in the $(\bar{1}01)$ cluster disappears at a temperature (390 K) where the order in the (001) chains is still present.

8. Phase Transitions

The transition at 180 K is associated with a change in the (001) chains only; the $\boldsymbol{k}_2$ vectors which below 180 K are $\boldsymbol{k}_2 = [\frac{1}{2}, \frac{1}{2}, 0]$ for a multidomain crystal and $\boldsymbol{k}_2 = [\frac{1}{2}, \frac{1}{2}, 0.153]$ for a monodomain crystal, change into $\boldsymbol{k}_2 = [\frac{1}{2}, \frac{1}{2} - v, 0]$, $v \approx 0.059$ above 180 K. At the same temperature the twinning also disappears. High resolution electron microscopy suggests a continuous (CDW type) modulation but another possibility, viz. the formation of twin-related blocks separated by periodic boundaries with an average spacing of $\approx 17b$, (≈ 55 Å), cannot be excluded. In both models some of the Mo atoms are not involved in cluster formation; it explains the increase in the number of charge carriers (0.03 → 0.09 holes/Mo) during the transition at 180 K.

For the distortion $\boldsymbol{k}_2 = [\frac{1}{2}, \frac{1}{2}, 0.153]$ observed in a monodomain crystal below 180 K, clusters in a (001) plane are the same or almost the same as for $\boldsymbol{k}_2 = [\frac{1}{2}, \frac{1}{2}, 0]$ but there is a phase difference between the distortions in successive (001) planes.

The origin of the peculiar behaviour of $\boldsymbol{k}_1$ with temperature above about 200 K remains unexplained. The distortion in each chain has a repeat distance $2b$ ($\boldsymbol{k}_1 = [p, \frac{1}{2}, q]$; while the vector $\boldsymbol{u}$ changes, the intensities of the satellites hardly change. This means that the distortions in a chain are almost the same as found from the $\boldsymbol{k}_2 = [0, \frac{1}{2}, \frac{1}{4}]$ refinement. The transition $\boldsymbol{k}_1 = [0, \frac{1}{2}, \frac{1}{4}]$ to $\boldsymbol{k}_1 = [0, \frac{1}{2}, 0]$ as found for a monodomain crystal, means that the distortions in neighbouring chains become the same. We observed in Weissenberg photographs that the sum of the intensities of $hk(4l+1)$ and $hk(4l-1)$ ($l = 1, 2, \ldots, k$ odd) reflections in an a, $2b$, $4c$ supercell agrees with the intensity of a hkl reflection in the a, $2b$, c supercell (corresponding with the $\boldsymbol{k}_1 = [0, \frac{1}{2}, 0]$ case). For a multidomain crystal we observed diffuse streaks of intensity between the satellites of the $[p, \frac{1}{2}, q]$ distortion; this is due to short-range order of the type $\boldsymbol{k}_1 = [0, \frac{1}{2}, 0]$. The incommensurate distortion will be the subject of a further study.

Acknowledgements

We wish to thank Dr. J. L. de Boer for help during the CAD-4 measurements and Drs. T. Cornelissens for preliminary electron microscopy investigations. We also wish to thank Drs. C. F. van Bruggen and Prof. Dr. F. Jellinek for valuable discussions. One of us, Drs. R. Deblieck wishes to thank the Faculty of Science of the University of Groningen for financial support during his stay in Groningen.

References

[1] F. Jellinek, Nature (London) **192**, 1065 (1961).
[2] R. de Jonge, T. J. A. Popma, G. A. Wiegers, and F. Jellinek, J. Solid State Chem. **2**, 188 (1970).
[3] Y. Suzuki, T. Uchida, M. Wakahira, and M. Taniguchi, Mater. Res. Bull. **16**, 1085 (1981).
[4] G. Krabbes, H. Upperman, and J. Henke, Z. anorg. allg. Chem. **470**, 7 (1980).
[5] R. Hemmel, H. van der Heide, C. F. van Bruggen, C. Haas, and G. A. Wiegers, Proc. 2nd European Conf. Solid State Chemistry, Veldhoven (The Netherlands) 1982.
[6] A. K. Rastogi and R. K. Ray, Bull. Mat. Sci. **3**, 341 (1981).
[7] J. M. Stewart, P. A. Machin, C. Dickinson, H. Ammon, L. Heck, and H. Flack, The X-Ray 76 system. Techn. Rep. TR-446. Computer Science Center, University of Maryland, College Park, Maryland.
[8] Y. Takeuchi, Z. Krist. **135**, 120 (1972).
[9] J. Guillevic, J. Y. Marouille, and D. Grandjean, Acta cryst. **B30**, 111 (1974).
[10] N. W. Alcock and A. Kjekshus, Acta chem. Scand. **19**, 79 (1965).
[11] R. Rühl and W. Jeitschko, Inorg. Chem. **21**, 1886 (1982).
[12] R. Guerin, M. Potel, and M. Sergent, Mater. Res. Bull. **14**, 133B (1974).
[13] F. Kadijk, R. Huisman, and F. Jellinek, Acta cryst. **B24**, 1102 (1968).
[14] D. H. Guthrie and J. D. Corbett, J. Solid State Chem. **37**, 256 (1981).
[15] R. Brown, Acta cryst. **20**, 268 (1966).
[16] R. Clarck, E. Marzeglia, and M. P. Hughes, Phil. Mag. **38**, 121 (1981).
[17] C. Manolikas, J. van Landuyt, and S. Amelinckx, phys. stat. sol. (a) **53**, 327 (1979).
[18] J. D. Corbett and D. H. Guthrie, Inorg. Chem. **21**, 1747 (1982).

(Received February 10, 1983)

phys. stat. sol. (a) **77**, 263 (1983)

Subject classification: 1.3 and 1.4; 21.1; 21.6

Wihuri Physical Laboratory and Department of Physical Sciences, University of Turku[1])

A Study of Non-Periodic Antiphase Boundaries in Long-Period Ordered $Au_3Zn(H)$

I. Antiphase Boundary Planes

By

J. TEUHO and K. MÄKINEN

The behavior of thermal antiphase boundaries, in the one-dimensional long period superstructure $Au_3Zn(H)$ is studied by using both, electron microscopy and electron diffraction. It is found that within non-stoichiometric alloys antiphase boundaries are parallel to the planes of the type $(201)_{f.c.c.}$, $(401)_{f.c.c.}$, $(403)_{f.c.c.}$, or $(405)_{f.c.c.}$, if the periodic direction is taken as $[001]_{f.c.c.}$. This corresponds well to theoretical preferential orientations. In stoichiometric $Au_3Zn(H)$, the boundaries of the ordered domains has most frequently an almost random orientation.

Das Verhalten von thermischen Antiphasengrenzen in der eindimensionalen langperiodischen Superstruktur $Au_3Zn(H)$ wird mittels Elektronenmikroskopie und Elektronenbeugung untersucht. Es wird gefunden, daß bei nichtstöchiometrischer Zusammensetzung die Antiphasengrenzen parallel zu Ebenen von dem Typ $(201)_{k.f.z.}$, $(401)_{k.f.z.}$, $(403)_{k.f.z.}$ oder $(405)_{k.f.z.}$ sind, wenn die periodische Richtung zu $[001]_{k.f.z.}$ gewählt wird. Dies korrespondiert gut mit den theoretischen Vorzugsorientierungen. In der stöchiometrischen $Au_3Zn(H)$-Legierung haben die Grenzen von geordneten Bereichen meistens eine nahezu statistische Orientierung.

1. Introduction

The Au_3Zn alloy transforms from the disordered f.c.c. structure to the long-period ordered phase $Au_3Zn(H)$ below 420 °C. This superlattice is stable at stoichiometric compositions only in the temperature range 420 to 270 °C, and then it transforms to the room temperature modification $Au_3Zn(R_1)$ or $Au_3Zn(R_2)$ [1 to 4].

An antiphase boundary may be formed by an operation in which part of a perfectly ordered crystal is given a vector displacement $\boldsymbol{R}$ with respect to the remainder. As also in the case of stacking faults, $\boldsymbol{R}$ may lie in or out of the plane of the antiphase boundary. These two possibilities are called type 1 and type 2 antiphase boundaries. The antiphase boundary type 1 is conservative, and type 2 is non-conservative. The latter fault can be considered as resulting from the removal or the insertion of one or more atom layers. If the atomic proportion in the inserted or removed slice differs from the stoichiometric one, it causes a change in composition of the crystal [5].

The ordered $Au_3Zn(H)$ structure contains periodic antiphase boundaries with an out-of-step period $M = 2$ [6]. In this paper, however, we deal mainly with the non-periodical antiphase boundaries, formed in connection with the growth process of the periodic domains. Antiphase boundaries of this kind are not often observed in LPS alloys. In the present paper we apply the model proposed for non-conservative antiphase boundaries by Snykers et al. [5] and van Tendeloo et al. [7, 8]. In a subsequent paper, dissociated antiphase boundaries in $Au_3Zn(H)$ will be treated in greater detail.

---

[1]) SF-20500 Turku 50, Finland.

2. Theoretical Consideration of the Domain Structure

The ordered phase $Au_3Zn(H)$ is a tetragonal long-period superstructure produced from four fundamental unit cells of the disordered f.c.c. structure. When ordering takes place, the close-packed planes (111) (all indices correspond to the f.c.c. unit cell) adopt the arrangement shown in Fig. 1a. The unit cell (Fig. 1b) is constructed by stacking close-packed layers in the order ABCA ... in such way that all the nearest neighbours of zinc atoms are gold atoms. If the number of nearest zinc neighbours for gold atom in the close-packed plane is less than two, the structure has greater energy, but such a structure can nevertheless be formed during the ordering process on the antiphase boundary, especially in the non-stoichiometric alloy.

There exist three possible orientation variants. If the different variants on the boundary are facing each other, a twin-like domain structure is formed [9].

3. Preferential Orientation for Non-Conservative Antiphase Boundaries

In order to investigate the different theoretical possibilities for the creation of antiphase boundaries, it is useful to examine the removal (or insertion) of different atom planes. Now we consider the situation in the close-packed plane (111). The removal of the close-packed rows *A* ($[10\bar{1}]$) and *B* ($[01\bar{1}]$) and the second closest-packed rows *D* ($[\bar{2}11]$) and *E* ($[\bar{1}2\bar{1}]$) (Fig. 2) preserves unchanged the stoichiometric relation of the gold and zinc atoms.

In the *C* ($[\bar{1}10]$) and *F* ($[\bar{1}\bar{1}2]$) directions there exist both rows of atoms comprising only gold atoms and also rows comprising gold and zinc atoms in the ratio 1:1. The removal of the rows consisting of gold and zinc creates an excess of gold in the structure. The stacking order remains unchanged in the rearrangement process that follows the removal of Au–Zn rows in the *F* direction, but the ratio of the number of nearest neighbours of the same kind to the number of the nearest neighbours of a different kind becomes more unfavorable in the vicinity of the boundary. The situation is, however, better than in the case *C*. The rearrangement, corresponding to the removal of the Au–Zn row *F* and having an antiphase vector $\boldsymbol{R} = \frac{1}{2}[\bar{1}01]$, creates a non-conservative antiphase boundary in the plane (201) or ($\bar{1}10$).

4. Experimental Method

The specimens were made of pure gold (99.98%), pure zinc (99.99%), and pure copper (99.99%). Induction melting was carried out in an argon atmosphere. The samples were homogenized in a vacuum at 550 or 620 °C for several days and then quenched

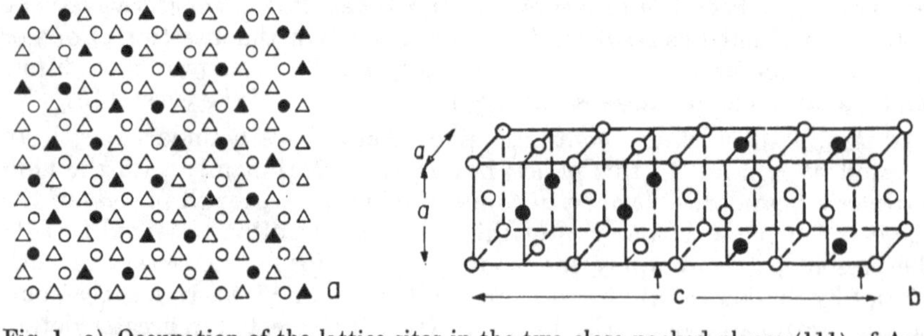

Fig. 1. a) Occupation of the lattice sites in the two close-packed planes (111) of $Au_3Zn(H)$ (upper one marked with triangles and lower one with circles) and b) the long-period unit cell of this structure. Open symbols represent Au atoms (○), closed symbols Zn atoms (●)

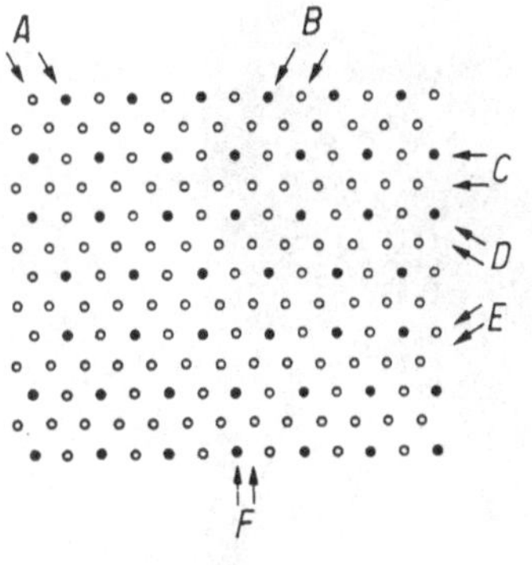

Fig. 2. Derivation of possible types of antiphase boundaries in non-stoichiometric Au_3Zn within the close-packed plane

into water. In order to stabilize the $Au_3Zn(H)$ phase to room temperature, the compositions were slightly non-stoichiometric or an addition of copper was made [4]. The samples were annealed for 6 h at 290 °C, again quenched into water and spark-cut into discs 3 mm in diameter. Electrolytic thinning was performed by using an LiCl/methanol solution [10]. The electron microscopic work was performed with a JEM-6A electron microscope, operating at 100 kV.

5. Results of the Experiments and Discussion

In the present study it was possible to distinguish thermal antiphase boundaries of two different types: i) boundaries which in the formation process of the ordered structure take a nearly random orientation, ii) boundaries which lie most frequently in the preferential planes caused by a deviation from the stoichiometric composition.

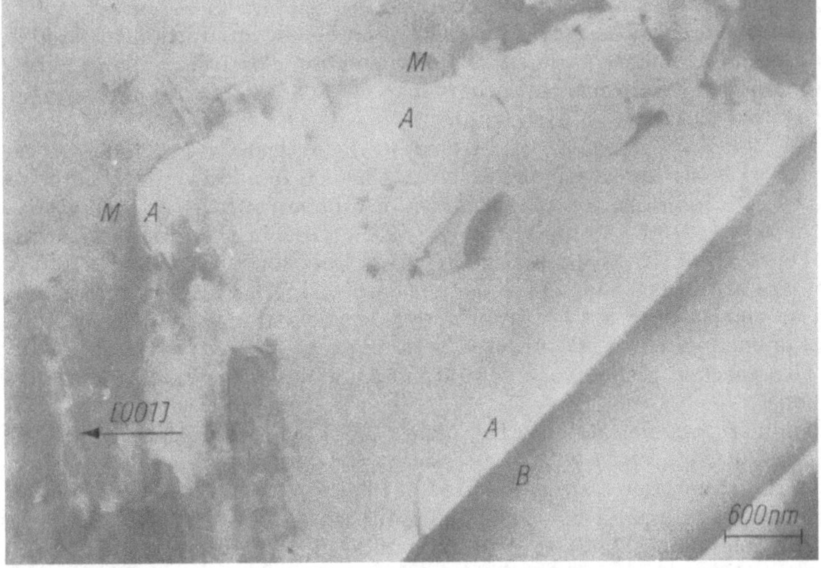

Fig. 3. Bright-field image of the antiphase domain containing a thermal antiphase boundary A/M and a change in the direction of periodicity on the boundary A/B. The specimen $(Au_{68.3}Cu_{8.7})Zn_{22}$ was annealed for 6 h at 290 °C and quenched into water

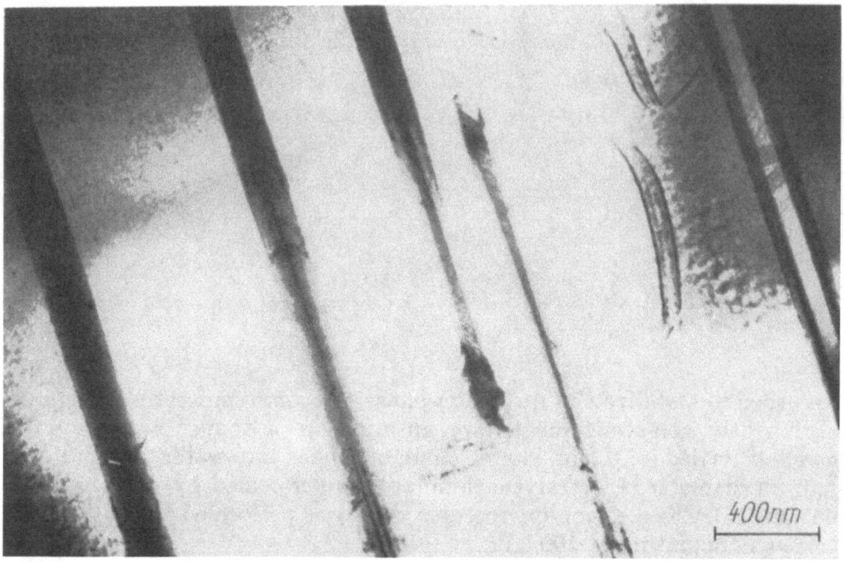

Fig. 4. Bright-field image taken from the same specimen as in Fig. 3

Fig. 3 and 4 are photographs of the stoichiometric alloy $(Au, Cu)_3Zn$ in which 8.7 at% gold is substituted for copper. After heat treatment (6 h at 290 °C; quenched into water) the structure corresponded to the high-temperature phase $Au_3Zn(H)$, according to the diffraction patterns. Fig. 3 shows a thermal antiphase boundary A/M mainly of type i), and the interface of the different orientation variants A/B [9]. In the domain A and the matrix M the direction of the periodicity is the same. The curvature typical of thermal antiphase boundaries is shown in Fig. 4.

A typical non-stoichiometric $Au_3Zn(H)$ structure produces, in addition to periodic domains, also thermal antiphase boundaries, corresponding closely to type ii). These antiphase boundaries are sometimes partially planar and are found in different crystallographic planes near (201) and (011) (Fig. 6).

Some boundary planes differed a little from the theoretical data. Therefore, the better agreement with the observations can be attained, if in addition to the above-mentioned model, the insertion (or the removal) of atomic rows is performed as in Fig. 5. Thus the inserted (or eliminated) atoms are located in the groups of atoms directed to $[0\bar{1}1]$ or $[\bar{1}01]$. In three dimensions these lines lie in the planes (100) or (110). In this case antiphase boundaries can be formed in the planes (401), (403), and (405). These planes differ a little from above-mentioned ones ((201) and $(\bar{1}10)$) but the change of energy, which is connected with an excess of gold, is about equally great. In the close-packed plane every gold atom has at least one zinc atom as the nearest neighbour.

Antiphase boundaries, which lie near the planes (001) (e.g. on the planes (01 19)) are observed often in connection with the dissociated antiphase boundaries [11]. In the plane (111), the atom groups are eliminated in this case along the $[\bar{1}10]$ direction.

In Fig. 6 the twin-like domains lie on the planes $(0\bar{1}1)$ and the antiphase boundaries A, B, C, and D on the planes (043), (201), $(\bar{4}03)$, and (401), respectively.

The non-conservative antiphase boundary can be continued as a conservative one, or it can end up in a twin-like domain T (Fig. 6). The non-conservative boundary can also end up in a dislocation, or in the tip of a twin-like domain.

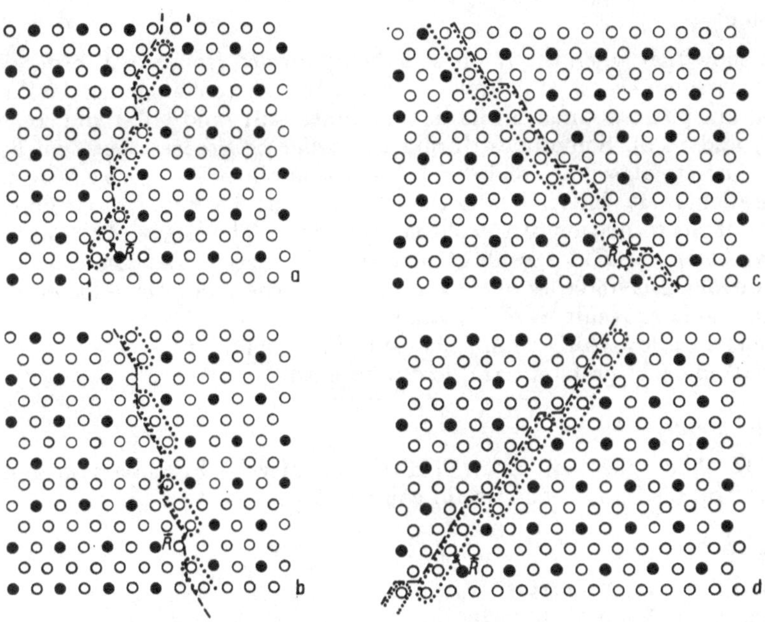

Fig. 5. Theoretical models for non-conservative antiphase boundaries in $Au_3Zn(H)$ structure shown in the plane (111). The inserted atoms are marked with the closed dotted line. The boundaries lie on an average in the planes a) (401), b) (403), c) (405), and d) (401)

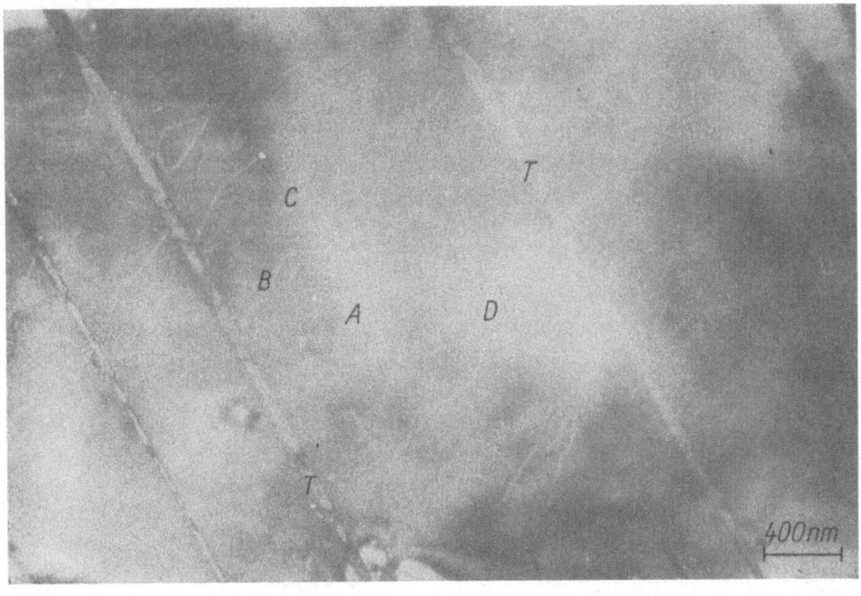

Fig. 6. Thermal antiphase boundaries in $Au_{78}Zn_{22}$ specimen. Heat treatment was the same as in Fig. 3. Antiphase boundaries *A*, *B*, *C*, and *D* lie on the planes (043), (201), ($\bar{4}03$), and (401), respectively

6. Conclusions

Experimental data show that when the *H* phase is being formed from the disordered f.c.c. structure containing an excess of gold with respect to the stoichiometric Au_3Zn, most of the original antiphase boundaries lie in the planes (401) and (403) and some in the planes (201) and (405). Sometimes during the ordering process they can be changed from their theoretical orientation in such a way that the new one may respect local demands more adequately. This indicates that the ordering, beginning at random, is not the only cause of the formation of antiphase boundaries. Thus some of the antiphase boundaries can be produced as a result of deviations from stoichiometry, causing the appearance of antiphase boundaries in certain planes. Earlier, van Tendeloo et al. [7, 8] have observed a similar result in Ni_3Mo alloys.

In the stoichiometric alloy $(Au, Cu)_3Zn$, the antiphase boundary seemed to be curved, showing that no exact preferential orientation exists.

Acknowledgements

The authors wish to thank the late Prof. V. Hovi and Messrs. P. Paalassalo and P. Airo for many fruitful discussions and useful comments.

References

[1] H. IWASAKI, J. Phys. Soc. Japan **14**, 1456 (1959).
[2] H. IWASAKI, J. Phys. Soc. Japan **17**, 1620 (1962).
[3] H. IWASAKI, M. HIRABAYASHI, K. FUJIWARA, D. WATANABE, and S. OGAWA, J. Phys. Soc. Japan **15**, 1771 (1960).
[4] K. SCHUBERT, B. KIEFER, M. WILKENS, and R. HAUFLER, Z. Metallk. **46**, 692 (1955).
[5] M. SNYKERS, P. DELAVIGNETTE, and S. AMELINCKX, phys. stat. sol. (b) **48**, K1 (1971).
[6] J. TEUHO and K. MÄKINEN, phys. stat. sol. (a) **70**, K1 (1982).
[7] G. VAN TENDELOO, P. DELAVIGNETTE, R. GEVERS, and S. AMELINCKX, phys. stat. sol. (a) **18**, 85 (1973).
[8] G. VAN TENDELOO, P. DELAVIGNETTE, J. VAN LANDUYT, and S. AMELINCKX, phys. stat. sol. (a) **26**, 299 (1974).
[9] J. TEUHO and K. MÄKINEN, J. Phys. F **12**, 2165 (1982).
[10] J. TEUHO, V. LEHTIMÄKI, and V. HOVI, J. Phys. E **11**, 510 (1978).
[11] J. TEUHO, to be published.

(Received January 31, 1983)

phys. stat. sol. (a) 77, 269 (1983)

Subject classification: 11; 1.4; 10.1; 21.1

Department of Metallurgy and Materials Technology, University of Surrey, Guildford[1])

TEM Images of Faceted Bubbles and Voids

By

B. COCHRANE and P. J. GOODHEW

A TEM image simulation package is written which is used to investigate the strain field contrast of faceted voids and bubbles. The strain field is modelled either as spherically symmetric or using a Green's function technique and the results of the two methods are compared with each other and with real images. The system demonstrates that a gas pressure of between 0.5 and 0.75 GPa must be present in overpressurised bubbles for significant strain contrast to be observed. It is shown that there is not a linear relationship between the gas pressure and the extent of the strain field contrast and that defocussing the microscope can reduce, or even eliminate, the strain field contrast. Image reversal between voids and overpressurised bubbles is also demonstrated.

Ein Computerprogramm zur Simulation von TEM-Abbildungen wird entwickelt und auf die Untersuchung von fazettierten Hohlräumen und Blasen angewandt. Das Spannungsfeld wird entweder als kugelsymmetrisch angenommen oder nach Greenschen Funktionen entwickelt und die Ergebnisse der beiden Methoden werden miteinander und mit realen Abbildungen verglichen. Es zeigt sich, daß ein Gasdruck zwischen 0,5 und 0,75 GPa notwendig ist, um in Blasen mit Überdruck signifikanten Spannungsfeldkontrast zu erzeugen. Der Gasdruck und die Stärke des Spannungsfeldkontrastes stehen in keinem linearen Zusammenhang und Defokussierung des Mikroskops kann letzteren reduzieren oder sogar auslöschen. Es wird außerdem demonstriert, daß zwischen Hohlräumen und Blasen mit Überdruck eine Umkehrabbildung existiert.

1. Introduction

Radiation produced voids and bubbles are frequently imaged in the electron microscope, and a quantitative assessment of their size and shape can be made using well-understood imaging conditions [1]. However it has not yet proved possible to estimate the gas pressure inside such a cavity from its image in the electron microscope, despite the great significance of being able to assess whether a cavity is at equilibrium with its contained gas, as is often assumed, or is under or over-pressurised [2]. In this paper we therefore address ourselves to the following problems:

a) Can we estimate approximately what pressure of gas is contained in a bubble from the extent of the surrounding strain field contrast?

b) Why is the strain field not always evident even when the diffraction conditions are set to bring it into contrast?

c) If it were experimentally possible to image defects using extremely large $\boldsymbol{g}$ vectors would the extent of the strain field contrast be correspondingly large?

In order to answer these questions a flexible transmission electron microscope image simulation program has been written. In contrast to other simulation attempts this system allows for the simultaneous presence of thickness contrast, diffraction contrast due to the deformation of the host lattice and defocussing effects. The package can handle realistic pore shapes and various strain field configurations. It incorporates the

[1]) Guildford, Surrey GU2 5XH, Great Britain.

microscope aberrations and allows for variation of all the experimental parameters such as defect depth, size and shape, foil thickness and orientation and operating diffraction conditions.

2. Previous Work

In 1963 Ashby and Brown [3] applied Howie and Whelan's two-beam dynamical theory to the case of spherically symmetric strain fields due to spherical inclusions. McIntyre and Brown [4] considered the strain contrast from centres of dilatation whose strength is similar to that of oxids in aluminium. Van Landuyt et al. [5] calculated the contrast from small voids using the two-beam and column approximations and assuming that the voids could be modelled as discs with no surrounding strain field. Ingram [6] synthesised the work of McIntyre and Brown with that of van Landuyt et al. and calculated the contrast of small spherical voids surrounded by spherical strain fields. Rühle [7] calculated the contrast from spherical pores taking account of the defocussing conditions in the microscope and this work was later extended by Rühle and Wilkens [8] to include many beams in the computations. Harrach and Foreman [1] considered faceted cavities when they examined the effects of defocussing on strain free pore images but returned to spherical cavities with spherical strain fields when they included deformation of the surrounding lattice [9]. The limitation of all these approaches is that the strain field of a real, often faceted, cavity is unlikely to be well approximated by a spherically symmetric solution. This assumption has not been made by Sass et al. [10] who incorporated the work of Mura [11] with the dynamical theory to calculate the contrast to be expected from cuboidal inclusions. Faivre [12] also carried out image calculations of these defects. These groups have all made the unfortunately necessary assumption that the elastic constants of the inclusion are the same as those of the surrounding matrix material. The justification assumed by those groups who ignore the strain field is that the defect is in equilibrium with the surrounding matrix material and there will thus be no strain field. This is clearly not the case for helium bubbles grown at low temperatures which can contain a very large overpressure [2]. Lidiard and Nelson [13] also showed that there will be a residual strain field even around an equilibrium bubble.

It is clear from this brief review that although many groups have separately incorporated all the necessary contrast effects none have included them all and hence the conditions under which each major contrast mechanism dominates the image are unclear. The program described here therefore attempts to take account of all the main contributions to the contrast, for a realistic cavity geometry and presents the computed contrast as a simulated two-dimensional image rather than as a set of intensity profiles. The system is based on that used by Thölén [14] to calculate dislocation contrast.

3. Computational Methods

The two-beam and column approximations are used throughout this work and the material is assumed to be isotropic. We implicitly ignore refraction occuring at the pore surfaces as a consequence of using the column approximation. The model is sketched in Fig. 1 and we shall first give the method of including defocus before moving to the details of the in-focus calculation.

The method of modelling the wave propagation from the in-focus to the defocussed image plane is given by Rühle and Wilkens [8] and consists of three steps. The Fourier transform, $\tilde{\psi}$, of the complex wave, ψ, at the in-focus plane is first computed. This is then multiplied by a transfer function, T, which incorporates the microscope aber-

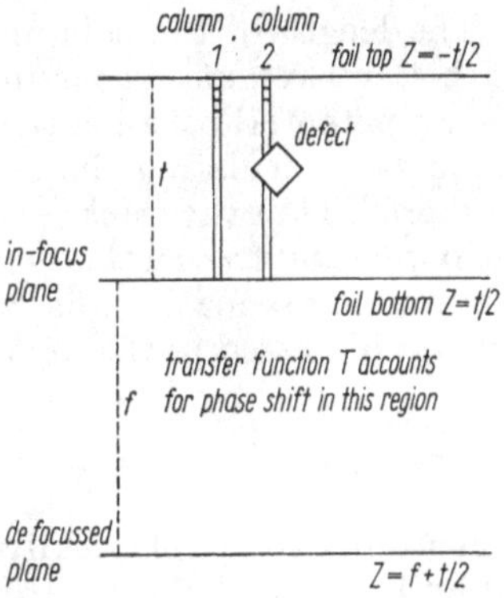

Fig. 1. Model showing the defect geometry

rations and coherence effects. The inverse Fourier transform of $\tilde{\psi}T$ is then taken from which the intensity at the defocussed plane is easily found. Rühle and Wilkens impose a condition of rotational symmetry on their pore in order to obtain an analytical solution for the transforms. This is an unnecessary restriction when fast Fourier transform routines are available.

The Howie and Whelan equations are given by Thölén in the form

$$\frac{d\varphi_0}{dz} = -\frac{\pi\xi_g\varphi_0}{\xi_0^1} + \pi\left(i - \frac{\xi_g}{\xi_g'}\right)\varphi_g\,, \tag{1}$$

$$\frac{d\varphi_g}{dz} = \pi\left(i - \frac{\xi_g}{\xi_g'}\right)\varphi_0 + \left\{-\frac{\pi\xi_g}{\xi_0^1} + 2\pi i(w + \delta_g\beta^1)\right\}\varphi_g\,, \tag{2}$$

where the symbols have their usual meanings and the local tilting of the lattice planes due to the defect is given by the term β'. This function is calculated from $R(z)$, the displacement function at any point within the lattice. Thölén gives the relationship between the wave at the top and bottom of a foil of thickness t in the form

$$\begin{bmatrix}\varphi_0(z+t)\\ \varphi_g(z+t)\end{bmatrix} = \begin{bmatrix}a_{11} & a_{12}\\ a_{21} & a_{22}\end{bmatrix}\begin{bmatrix}\varphi_0(z)\\ \varphi_g(z)\end{bmatrix} = A\begin{bmatrix}\varphi_0(z)\\ \varphi_g(z)\end{bmatrix}, \tag{3}$$

where A is termed the scattering matrix and depends on the local value of w, the Bragg deviation parameter, which depends on the displacement function $R(z)$. As many of the A matrices will be almost identical Thölén first calculates a library of matrices corresponding to a range of w values. He assigns a small range of w, Δw, such that all values within $\pm\Delta w$ of w will have the same scattering matrices. If the displacement is such that the effective w values is outside the range contained within the library then new A matrices are computed.

Referring to Fig. 1, in column 1 the displacement is calculated to be for example R_1 at the top of the first slice. At the bottom of this slice the displacement is again calculated and found to be R_2. In order to find $\boldsymbol{g}\cdot d(R(z))/dz$ we then take

$$\frac{\boldsymbol{g}\cdot d(R(z))}{dz} = \frac{\boldsymbol{g}\cdot(R_2 - R_1)}{\Delta z}. \tag{4}$$

This value is used to calculate w_{eff} at the top of the first slice which is put into the equations to calculate the scattering matrix A_1 in the first slice. The process is then repeated until the lower foil surface is reached. Column 2 passes through the pore and the approach is similar to that for column 1 until the cavity is reached. The A_i matrices are calculated until the column reaches the cavity and then the beam amplitude is assumed unchanged until it leaves the cavity. The length of the column inside the pore is used to calculate the phase change which the wave undergoes in passage through the cavity due to the difference in mean inner potential between the interior of the cavity and the matrix material [7]. The process of calculating the A_i matrices is then resumed until the bottom of the foil is encountered. Thus the thickness and diffraction contrast are both incorporated in this technique and the method of defocussing the image has already been stated. Two separate approximations have been used to model the lattice deformation $R(z)$ and these are discussed in the next two sub-sections.

3.1 Spherical strain field

Computations have been performed assuming that the defect strain field is spherically symmetric. The displacement field used was that suggested by Brown and Mazey [15] for inert gas bubbles in solids and is given in the form

$$u = \varepsilon r_c^3/\varrho^2 \tag{5}$$

with

$$\varepsilon = \frac{-2\gamma + Pr_c}{4\mu r_c}.$$

In these formulae μ is the shear modulus, γ the surface energy, P the pressure in the bubble, r_c the bubble radius, and ϱ the distance from the point under consideration to the centre of the bubble. Thus in order to use this method a radius equivalent to the hole size must be assigned. This point will be discussed later.

3.2 Non-spherical strain field

In this case the pore faces are modelled as sheets of force acting along the face normals. The displacement at each point within the foil is then calculated using the following method. The displacement, U_i, in direction i at a general point $\boldsymbol{r}$, due to a force, F_j, centred at a point $\boldsymbol{r}'$ and in direction j is given by

$$U_i(\boldsymbol{r}) = G_{ij}(\boldsymbol{r} - \boldsymbol{r}')\, F_j(\boldsymbol{r}')\,. \tag{6}$$

For an isotropic body Bullough and Nelson [16] give the perfect lattice Green's function $G_{ij}(\boldsymbol{r})$ in the form

$$G_{ij}(\boldsymbol{r}) = \big(2(1-\nu)\,\delta_{ij}\,|\boldsymbol{r} - \boldsymbol{r}'|_{,pp} - |\boldsymbol{r} - \boldsymbol{r}'|_{,ij}\big)/16\pi\mu(1-\nu)\,, \tag{7}$$

where the comma denotes partial differentiation i.e. $r_{,pp} = \nabla^2 r$ and the rest of the symbols have the usual meanings. After some algebra (7) can be expressed in the form

$$G_{ij}(R) = \frac{1}{16\pi\mu(1-\nu)}\left(\frac{(3-4\,\nu)\,\delta_{ij}}{R} + \frac{x_i x_j}{R^3}\right), \tag{8}$$

where $R = |\boldsymbol{r} - \boldsymbol{r}'|$ and $x_i = (x_k - x_k')$, $i, k = 1, 2, 3$. Three methods have been used to calculate the lattice displacement using (8). The first method merely involved a set of point forces acting at the origin along the ⟨100⟩. This model is only realistic if the pore is very small such that the forces on each face are effectively acting along the axes. A better model is to consider each face as consisting of a finite number of elements, each of which has a force acting normally through its centre. At each lattice point the displacement in each direction must be computed due to each centre of force on each face. These components are then summed to give the total displacements in each direction from which the resultant displacement can be found. This system has been implemented using an FPS120B array processor with up to 100 elements per

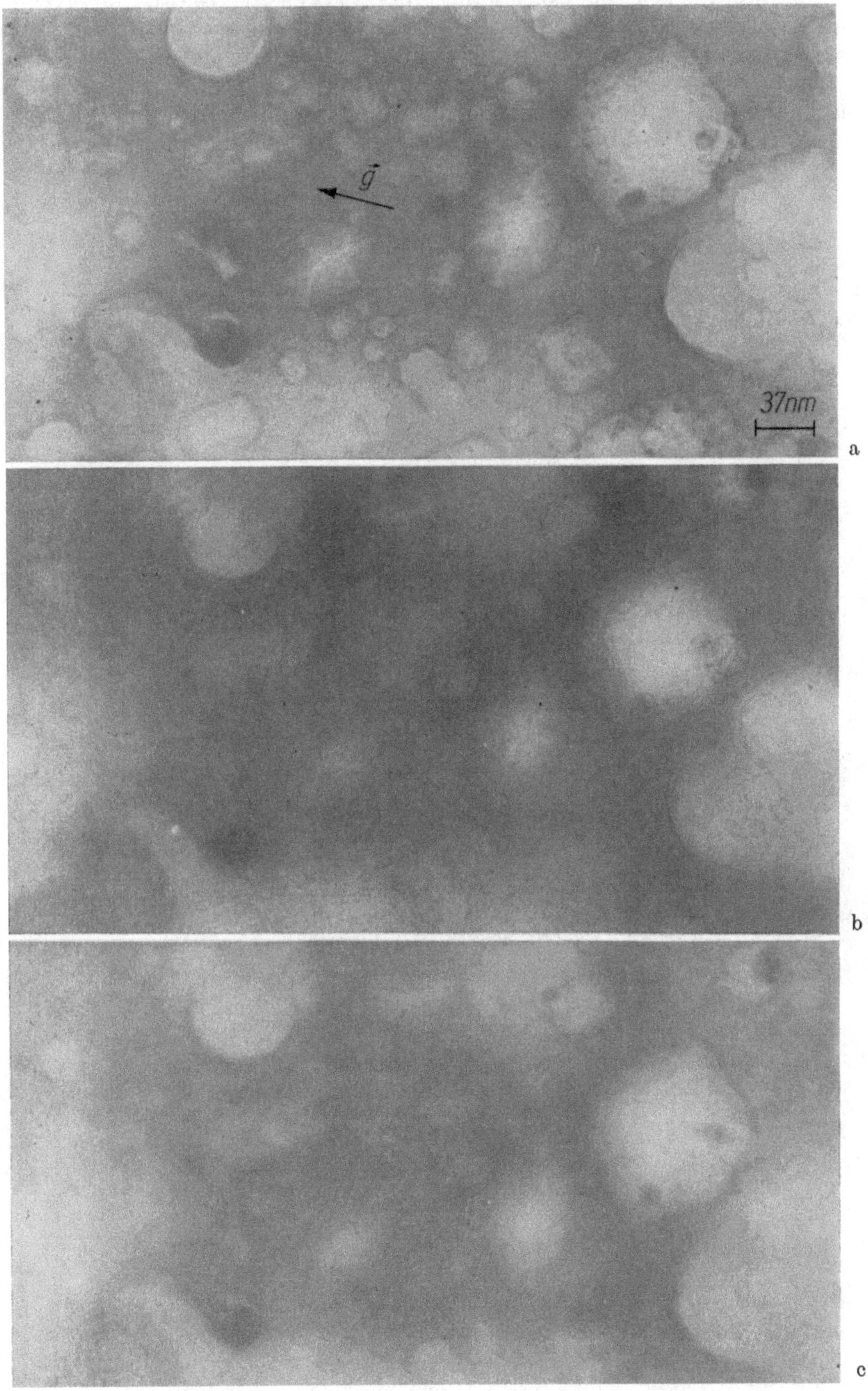

Fig. 2. Helium bubbles in vanadium. $\boldsymbol{g} = 110$; a) $\Delta f = +0.6\ \mu m$, b) $-0.6\ \mu m$, c) 0

face. Unfortunately this is a time consuming process [17] as it typically involves 10^8 passes through the calculation routines. Thus an analytical solution has also been found for the displacement field (Appendix 1). This has not yet been implemented on the computer, however, because of its complexity.

The results presented in the next section were therefore obtained using the "finite element" method described above and are compared with those obtained using the spherical approximation.

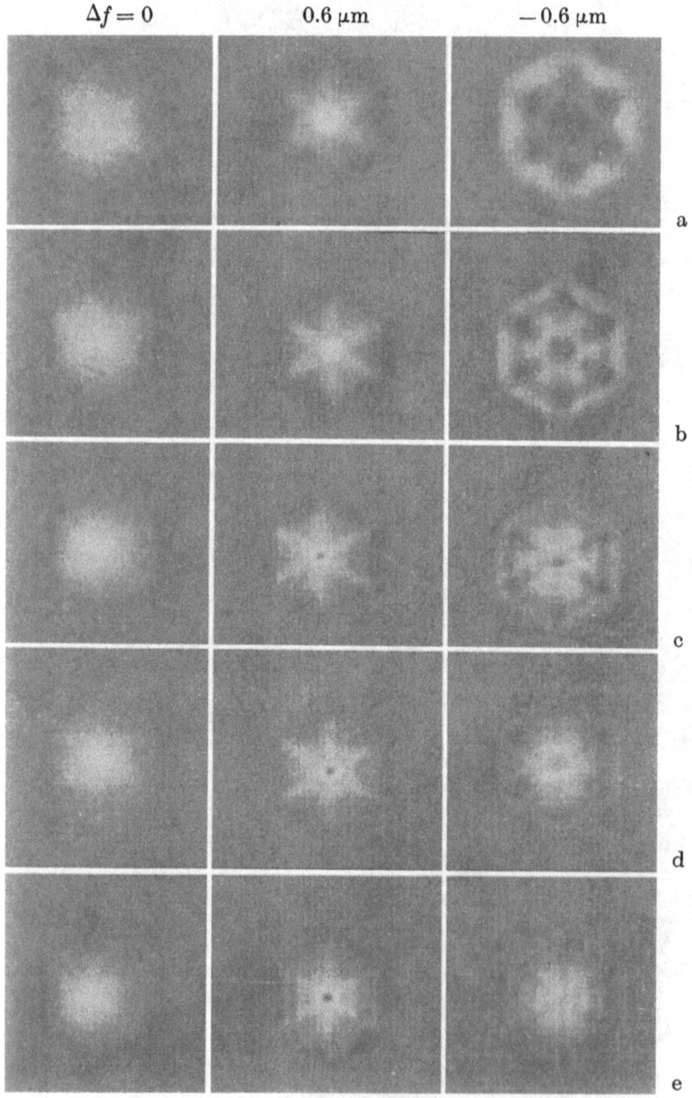

Fig. 3. Through focal series with varying pore size a) cube side $= 0.5\xi_g$, b) $0.75\xi_g$, c) $1.0\xi_g$, d) $1.25\xi_g$, e) $1.5\xi_g$. The defect is at the centre of the foil of thickness $4\xi_g$ and the deviation parameter is 1.0 $\Delta f = 0$, $+0.6$ μm, -0.6 μm

4. Results and Discussion

Fig. 2 is an experimental through-focal series of helium bubbles in vanadium after annealing for 8 h at 1150 °C. The magnification on the prints is 270 K. This series was used for checking the simulated images when varying the defocussing, defect depth, size, and shape with the strain field not in contrast. Fig. 3 shows a computed through-focal series of cubic holes at the centre of a foil of $4\xi_g$ in thickness. A large value of w is used and the cube side is varied from $0.5\xi_g$ to $1.5\xi_g$. It is interesting to note that this series suggests that internal contrast lines become clear when the hole is of side $> 0.75\xi_g$. Analysis of the bubbles shown in Fig. 2 shows this to be in good agreement with actual TEM images. The other parameters, such as foil thickness, foil orientation and operating diffraction conditions have also been varied and the computed images agree well with those obtained experimentally.

Fig. 4 is a series of in-focus images using a spherical approximation for the strain field with a range of pressures contained within the bubble. The details of pressure values and operating conditions are given on the figure. This series suggests that the strain field becomes visible when the gas pressure is between 0.5 and 0.75 GPa in niobium. A cubic bubble of side 20 nm is approximately equivalent using a volume criterion to a spherical bubble of radius 12 nm. Greenwood et al. [18] estimate that dislocation loops will be punched from a bubble when the gas pressure contained within it is of the order of $\mu b/r$. Thus the maximum pressure that can be contained by the 12 nm spherical bubble is about 0.8 GPa and if this approximation can be applied to faceted pores it is clear that the useful range of the pressure variable in image simula-

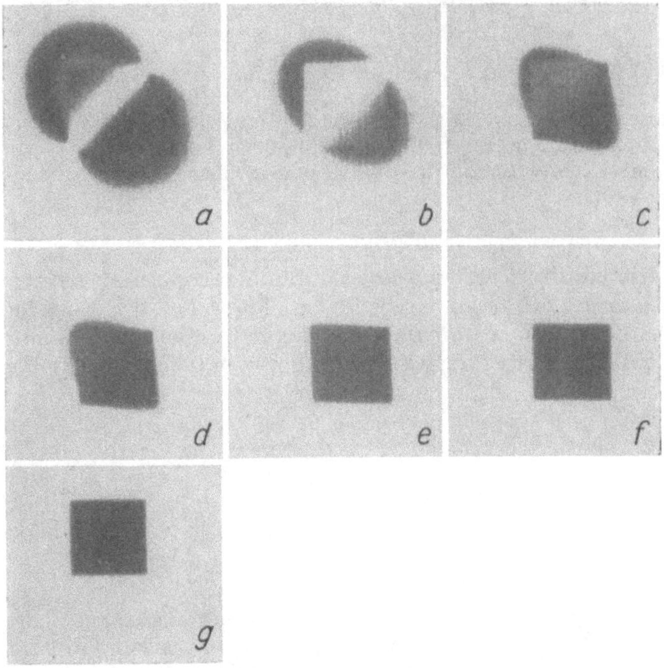

Fig. 4. Cubic pore, modelled with a spherical surrounding strain field, containing various gas pressures: a) 5.0, b) 2.5, c) 1.0, d) 0.75, e) 0.5, f) 0.25, g) 0.10 GPa. The pore is at the centre of a foil of $4\xi_g$ in thickness, $w = 0.0$ and the pore sides are 20 nm. All the images are in focus, $\boldsymbol{g} = 220$ and the electron beam direction is $\langle 100 \rangle$

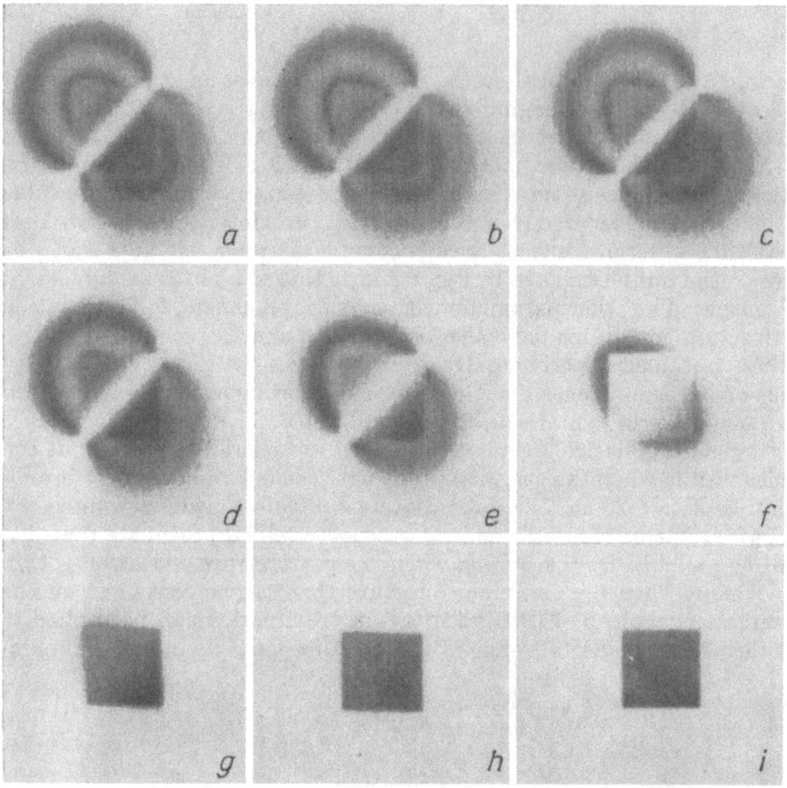

Fig. 5. Cubic pore containing various gas pressures with the strain fields calculated using a Green's function method: a) 10, b) 7.5, c) 5.0, d) 2.5, e) 1.0, f) 0.75, g) 0.5, h) 0.25, i) 0.1 GPa. The pore is at the centre of a foil of thickness $4\xi_g$, $w = 0.0$ and the pore sides are 20 nm. All the images are in focus, $\boldsymbol{g} = 220$ and the beam is $\langle 100 \rangle$

tions of this type is very restricted. Fig. 5 shows a series of bubbles containing a range of pressures of gas under the same diffraction conditions as Fig. 4 but this time the Green's function method has been used to calculate the lattice displacement. Using this model the strain field again appears at a gas pressure between 0.5 and 0.75 GPa.

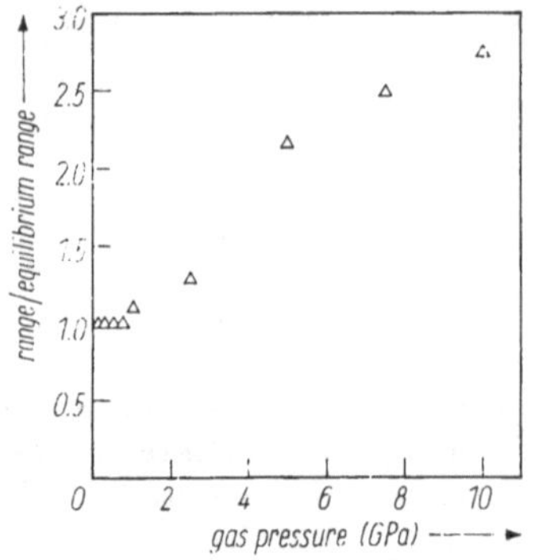

Fig. 6. Graph showing how the ratio of the maximum range of the strain field to the equilibrium range varies with gas pressure in the pore. Both the equilibrium range and the range when gas pressure is evident are measured along the $\langle 220 \rangle$ from Fig. 5

It is clear from a comparison of Fig. 4 and 5 that there is good qualitative agreement between the calculation techniques although the strain contrast is of greater extent in Fig. 5. A possible reason for this is that the perfect lattice Green's function has been used which takes no account of the missing material within the pore. Meissner [19] has given an estimate of the error caused by this approximation which is qualitatively consistent with these results. It is interesting to note that a white fringe appears within the black lobes at high pressures using both strain calculation techniques. This fringe can be seen experimentally in images of very high pressure helium platelets taken by Evans et al. [20]. Profiles of the line of maximum extent of the strain field contrast were plotted from which it is clear that the extent of the strain field is not linearly related to the gas pressure within the pore. Fig. 6 is a graph of maximum range of the strain field against gas pressure.

The effect of defocussing the microscope is displayed as Fig. 7. In the figure the bubble of Fig. 5f is shown at w values of 0.5, 1.0, and 3.0 and with 0, 0.6, and 0.8 μm of underfocus. This series thus shows the image variation with w and it is clear that defocussing can reduce or even eliminate the strain field contrast. Fig. 8 is an experimental image of helium bubbles in niobium/1% zirconium after annealing for 8 h at 1250 °C. The magnification is 600 K. The similarity between these and the simulated images of Fig. 7 is clear. The effect of increasing the modulus of the $\boldsymbol{g}$ vector is shown in Fig. 9. It is evident from this figure that although the extent of the strain contrast is increased with increasing pressure it still remains very limited, even for large $\boldsymbol{g}$. Using a very simple model of spherical pores with spherical strain fields Fig. 10 was produced. Image reversal between voids and overpressurised bubbles can be seen in

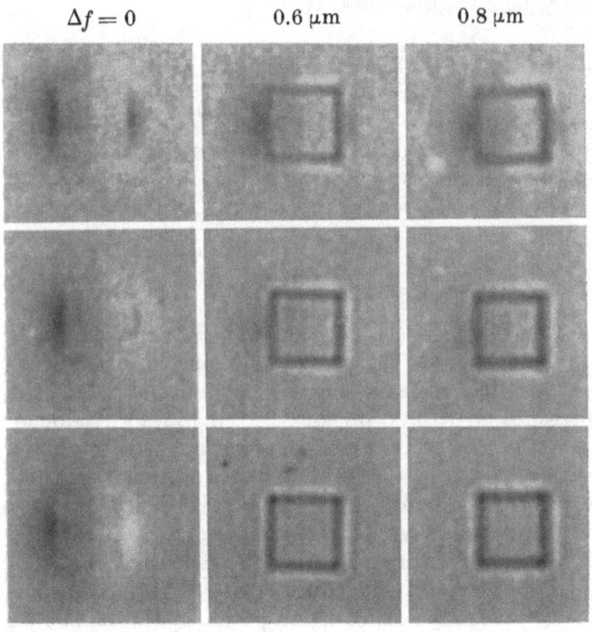

Fig. 7. Effect on the strain field image of varying the defocus at different w values: a) $w = 0.5$ b) 1.0, c) 3.0. The hole is at the centre of a foil of $4\xi_g$ in thickness, the cube sides are 20 nm and $\boldsymbol{g} = \langle 200 \rangle$

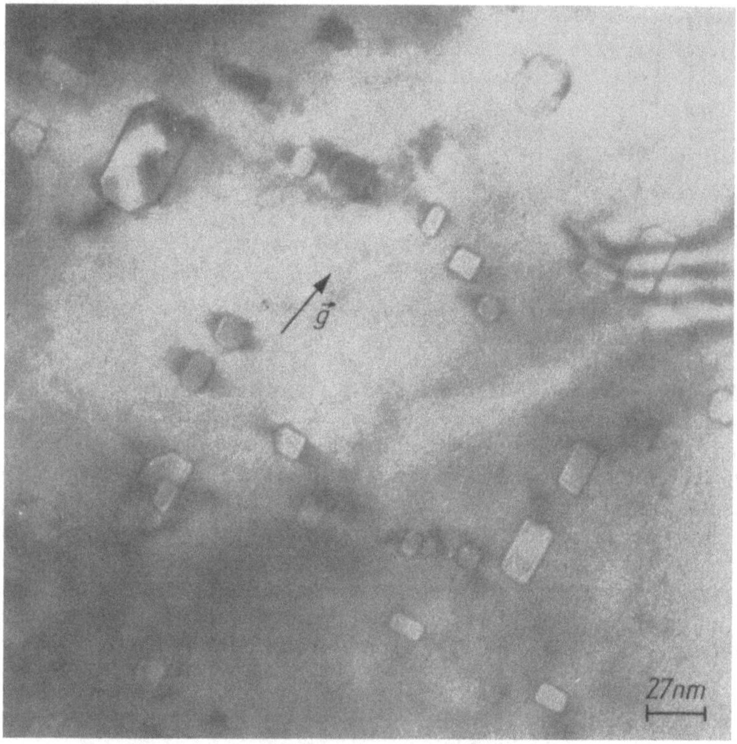

Fig. 8. Helium bubbles in niobium/1% zirconium. $\boldsymbol{g} = 013$

this figure such that if the defect depths are known to lie within certain ranges it should be possible to distinguish between voids and overpressurised bubbles in a manner analogous to that used to distinguish between vacancy and interstitial type clusters.

5. Conclusions

The questions posed in the introduction have largely been answered by this work. It has been shown that although we can not estimate the pressure of gas contained within a bubble from the extent of the strain field we can suggest a pressure at which the strain field becomes visible. This pressure is of the order of 0.5 to 0.75 GPa. Defocussing the microscope, a condition often used to enhance the pore shape, has been shown to reduce, or even remove, the strain field contrast. We have also demonstrated that increasing the $\boldsymbol{g}$ vector to a very large value still only results in a strain field of limited extent and that image reversal occurs between voids and overpressurised bubbles.

The main approximations of this work, which are assuming the material to be isotropic and using only two beams in the wave equations, could both be removed if computing time were not a restriction. If the analytical solution given in the Appendix can be programmed it should make this possible. This work has also shown that until the Green's function method can be made faster the spherical approximation can be used to assess the conditions under which significant strain contrast images can be expected.

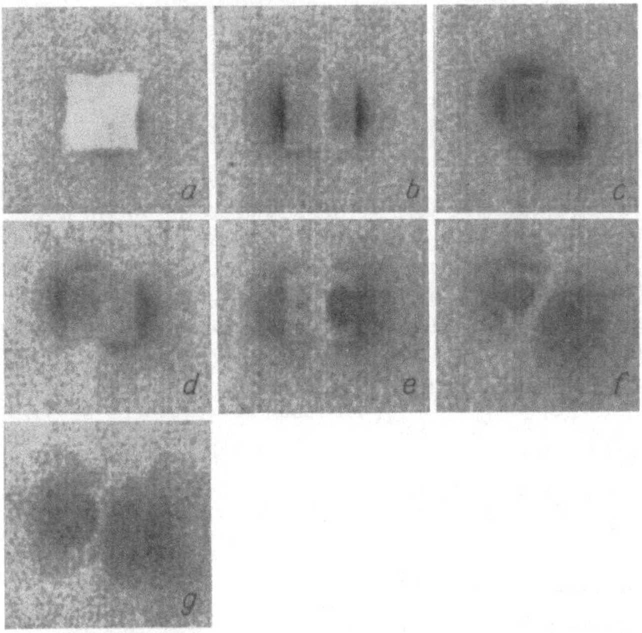

Fig. 9. Effect of varying the operating reflection a) 110, b) 200, c) 220, d) 310, e) 400, f) 420, g) 510. The hole is at the centre of a foil of $4\xi_g$ in thickness, $w = 0.0$ and the gas pressure is 0.5 GPa

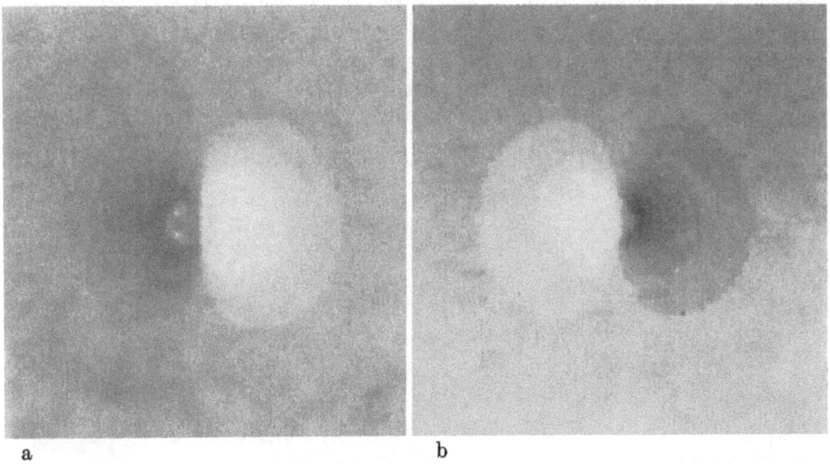

Fig. 10. Image contrast reversal between overpressurised bubbles and voids. The strain field and the pore are both assumed spherical. a) Gas pressure = 1.1×10^{10} GPa, b) 1.1×10^{7} GPa. The equilibrium pressure is 1×10^{8} GPa. The beam is ⟨111⟩, the pore diameter is 5 nm with $w = -0.1$ and the pore depth is $0.5\xi_g$. The diffraction vector is ⟨011⟩ and the images are in focus

Acknowledgements

The authors would like to thank Dr. R. Bullough for valuable advice and Dr. P. Kay for assistance with computing. One of us (B.C.) would like to thank the Northern Irea nd Department of Education for the provision of a research studentship.

Appendix 1

From (6) and (8) we have

$$u_i(r) = \frac{1}{16\pi\mu(1-\nu)}\left[\frac{(3-4\nu)\,\delta_{ij}}{R} + \frac{x_i x_j}{R^3}\right] F_j(r') \,. \tag{A1}$$

In order to arrive at an analytical solution for the displacement at each point within the surrounding matrix we first calculate the displacement due to the forces acting on pairs of opposite faces and then sum these displacements. The procedure is illustrated for one pair of faces, say the faces normal to the X-axis; call these faces A^+ and A^-. We assign general points, $\pm a$, x_2', x_3' to be on the positive and negative of these faces and then let these points vary over the faces by double integration. Letting PL represent the term $F_j(\boldsymbol{r}')/16\pi\mu(1-\nu)$ we then have

$$U_i^{A^\pm}(x_1x_2x_3) = PL\int\limits_{A^+}\left[\frac{(3-4\nu)\,\delta_{ij}}{R^{A^+}} + \frac{X_i^{A^+}(x_1-a)}{(R^{A^+})^3}\right] dx_2'\,dx_3' \,, \tag{A2}$$

$$U_i^{A^\pm}(x_1x_2x_3) = -PL\int\limits_{A^-}\left[\frac{(3-4\nu)}{R^{A^-}} + \frac{X_i^{A^-}(x_1+a)}{(R^{A^-})^3}\right] dx_2'\,dx_3' \,, \tag{A3}$$

where U_i is the displacement in direction i.

$$R^{A\pm} = [(x_1 \pm a)^2 + (x_2 - x_2')^2 + (x_3 - x_3')^2]^{1/2}$$

and

$$X_i^{A\pm} = (x_1 \pm a)\,,\ (x_2 - x_2'),\ (x_3 - x_3')\,.$$

Clearly equations (8) and (A1) imply that the displacement in direction i, due to the stress on face A^+, is obtained by calculating the Green's function times the force acting at a general point on the A^+ face and then integrating this over the area of the face. The other components of the displacement due to the stresses on faces B^+, B^- and C^+, C^- are obtained using (8) and (A1). The required displacement field is given by

$$U_i(x_1, x_2, x_3) = P^*L[(I_i^1(a) - I_i^1(-a)) + (I_i^2(b) - I_i^2(-b)) + \\ + (I_i^3(c) - I_i^3(-c))]\,,$$

where

$$I_i^1(a) = \int_{-b}^{+b} dx_2' \int_{-c}^{+c}\left[\frac{(3-4\nu)\,\delta_{i1}}{R_1} + \frac{X_{i1}(x_1-a)}{R_1^3}\right] dx_3'$$

with $R_1 = ((x_1-a)^2 + (x_2-x_2')^2 + (x_3-x_3')^2)^{1/2}$ and $X_{i1} = (x_1-a), (x_2-x_2'), (x_3-x_3')$.

$$I_i^2(b) = \int_{-a}^{+a} dx_1' \int_{-c}^{+c}\left[\frac{(3-4\nu)\,\delta_{i2}}{R_2} + \frac{X_{i2}(x_2-b)}{R_2^3}\right] dx_3'$$

with $R_2 = ((x_1-x_1')^2 + (x_2-b)^2 + (x_3-x_3')^2)^{1/2}$ and $X_{i2} = (x_1-x_1'), (x_2-b), (x_3-x_3')$.

$$I_i^3(c) = \int_{-a}^{+a} dx_1' \int_{-b}^{+b}\left[\frac{(3-4\nu)\,\delta_{i3}}{R} + \frac{X_{i3}(x_3-c)}{R_3^3}\right] dx_2'$$

with $R_3 = ((x_1 - x_1')^2 + (x_2 - x_2')^2 + (x_3 - c)^2)^{1/2}$ and $X_{i3} = (x_1 - x_1'), (x_2 - x_2'), (x_3 - c)$.

Using the fact that $(\partial/\partial x)(1/R) = -X_i/R^3$ the integrals may be rewritten:

$$I_i'(a) = \left[(3 - 4\nu)\,\delta_{i1} - (x_1 - a)\frac{\partial}{\partial x_i}\right] \int_{-b}^{+b} dx_2' \int_{-c}^{+c} \frac{dx_3'}{R_1},$$

$$I_i^2(b) = \left[(3 - 4\nu)\,\delta_{i2} - (x_2 - b)\frac{\partial}{\partial x_i}\right] \int_{-a}^{+a} dx_1' \int_{-c}^{+c} \frac{dx_3'}{R_2},$$

$$I_i^3(c) = \left[(3 - 4\nu)\,\delta_{i3} - (x_3 - c)\frac{\partial}{\partial x_i}\right] \int_{-a}^{+a} dx_1' \int_{-b}^{+b} \frac{dx_2'}{R_3}.$$

These may in turn be expressed as

$$I_i'(a) = \left[(3 - 4\nu)\,\delta_{i1} - (x_1 - a)\frac{\partial}{\partial x_i}\right] J'(x_1x_2x_3abc)\,,$$

$$I_i^2(b) = \left[(3 - 4\nu)\,\delta_{i2} - (x_2 - b)\frac{\partial}{\partial x_i}\right] J^2(x_1x_2x_3abc)\,,$$

$$I_i^3(c) = \left[(3 - 4\nu)\,\delta_{i3} - (x_3 - c)\frac{\partial}{\partial x_i}\right] J^3(x_1x_2x_3abc)\,.$$

It is then clear that

$$J^2 = J'(x_2x_3x_1bac)\,,$$
$$J^3 = J'(x_3x_1x_2cab)\,.$$

Thus the only integral required is $J'(x_1x_2x_3abc)$ which has the form

$$J'(x_1x_2x_3abc) = \int_{-b}^{+b} dx_2' \int_{-c}^{+c} \frac{dx_3'}{[(x_1 - a)^2 + (x_2 - x_2')^2 + (x_3 - x_3')^2]^{1/2}}\,.$$

This integration is very lengthy but can be performed analytically with the result that (A1) can be used to calculate the displacement field [21].

References

[1] H. V. Harrach and A. J. E. Foreman, Radiat. Eff. **46**, 7 (1980).
[2] P. J. Goodhew and S. K. Tyler, Proc. Roy. Soc. **A377**, 151 (1981).
[3] M. F. Ashby and L. M. Brown, Phil. Mag. **8**, 1083 (1963).
[4] K. G. McIntyre and L. M. Brown, J. Physique C3, **27**, 179 (1966).
[5] J. van Landuyt, R. Gevers, and S. Amelinckx, phys. stat. sol. **7**, 519 (1964).
[6] J. Ingram, J. appl. Phys. **40**, 5030 (1969).
[7] M. Rühle, U.S.A.E.C. Symp. on Radiation Induced Voids in Metals Albany (USA) 1972.
[8] M. Rühle and M. Wilkens, Crystal Lattice Defects **6**, 129 (1975).
[9] D. K. Saldin and H. V. Harrach, Proc. EMAG Conf. Brighton (U.K.), September 1979.
[10] S. L. Sass, T. Mura, and J. B. Cohen, Phil. Mag. **16**, 679 (1967).
[11] T. Mura, Proc. Roy. Soc. **A280**, 528 (1964).

[12] G. Faivre, phys. stat. sol. **35**, 249 (1969).
[13] A. B. Lidiard and R. S. Nelson, Phil. Mag. **17**, 425 (1968).
[14] A. R. Thölén, Phil. Mag. **22**, 175 (1970).
[15] L. M. Brown and D. J. Mazey, Phil. Mag. **10**, 1081 (1964).
[16] R. Bullough and R. S. Nelson, Phys. Tech. **5**, 29 (1974).
[17] K. Ahmed, N. Curran, P. Kay, B. Cochrane, and A. Makiewiez, Proc. Conf. on Progress in Vector and Array Processors, Bristol (U.K.), September 1982.
[18] G. W. Greenwood, A. J. E. Foreman, and D. Rimmer, J. nuclear Mater. (Amsterdam) **4**, 305 (1959).
[19] N. Meissner, Ph. D. Thesis, Surrey University, 1973.
[20] J. H. Evans, A. van Veen, and L. M. Caspers, Nature (London) **291**, 310 (1981).
[21] B. Cochrane, Ph. D. Thesis, Surrey University, 1982.

(Received January 27, 1983)

phys. stat. sol. (a) 77, 283 (1983)

Subject classification: 10.1 and 12.1; 21

Department of Materials Science, Universidad Simon Bolivar, Caracas[1]) *(a) and Department of Materials, Queen Mary College, London*[2]) *(b)*

Temperature and Strain Rate Dependence of the Cyclic Flow Stress of Molybdenum Single Crystals

By

M. ANGLADA (a) and F. GUIU (b)

Measurements are made of the temperature and strain rate dependence of the flow stress of molybdenum single crystals cyclically deformed to saturation in tension–compression in the interval of temperature between 513 and 298 K. The results obtained are compared with the predictions of a recent theory by Seeger for the formation of kink pairs in screw dislocations in pure b.c.c. metals. Rather good agreement is obtained between the experimental results and the theory from which a value of 1.25 eV is obtained for the enthalpy of formation of a pair of kinks.

Die Abhängigkeit der Fließspannung von Temperatur und Dehnungsgeschwindigkeit von zyklisch bis zur Sättigung zug- und kompressionsverformten Molybdäneinkristallen wird im Temperaturbereich von 513 bis 298 K gemessen. Die erhaltenen Ergebnisse werden mit Vorhersagen einer neueren Theorie von Seeger für die Bildung von Knotenpaaren in Schraubenversetzungen in reinen k.r.z.-Metallen verglichen. Ziemlich gute Übereinstimmung zwischen den experimentellen Ergebnissen und der Theorie wird erhalten, woraus ein Wert von 1,25 eV für die Bildungsenthalpie eines Knotenpaares bestimmt wird.

1. Introduction

It is now widely accepted that the strong temperature dependence of the flow stress of pure b.c.c. metals is due to an intrinsic property of the screw dislocations in the b.c.c. lattice. In the past, several models [1 to 3] have been developed based on a suggestion by Hirsch [4] that the screw dislocations are dissociated in several planes. These models could explain qualitatively the strong variation of the flow stress but it is now believed that the dissociation of the screw dislocation should be so small that some concepts used in those models can no longer be considered adequate [5].

In a recent paper, Seeger [5] has developed a simple model for the flow stress of high-purity b.c.c. metals based on the properties of kinks. Seeger proposes that depending on the kink separation the activation energy for the formation of a kink pair should be treated either by a line tension approximation (short distances and high stresses) or by an elastic interaction between kinks (large distances and small stresses).

The aim of the present study is to compare the theory of Seeger with the results obtained for the temperature and strain rate dependence of the flow stress of molybdenum single crystals cyclically deformed to saturation in tension–compression. The method followed here for the determination of the activation parameters is the same as that described in a recent publication [6]. This has the important advantage, over the traditional method based on unidirectional deformation, that the measurements of the activation parameters are carried out under a more stable dislocation substructure.

---

[1]) Apartado 80659, Caracas 1080, Venezuela.

[2]) Mile End Road, London, E1 4NS, Great Britain.

2. Experimental Procedure

The molybdenum crystals used in this investigation were grown by the electron beam-zone melting technique in a vacuum of 10^{-6} Torr, as described in detail elsewhere [7], and had a total interstitial impurity content of about 10 ppm by mass. Cylindrical specimens with $\langle 110 \rangle$ axial orientation were prepared from these crystals by spark-machining and electropolishing. The finished specimens had a gauge length 10 mm long and 4 mm in diameter with shoulders of 5 mm at their ends. The crystals were cyclically deformed in tension–compression at a fixed total strain amplitude of 7.5×10^{-3} and at a total strain rate of $5 \times 10^{-3}\ s^{-1}$ until saturation was reached. Then the temperature was lowered by steps of 20 K whilst maintaining the specimen under zero load. When each desired temperature had been reached, the specimen was deformed for one cycle at the same strain rate and plastic strain amplitude in order to obtain the variation of the flow stress with temperature. The specimen was also deformed for two or three cycles at different frequencies to obtain the strain rate sensitivity of the flow stress at each temperature studied. This procedure, which is illustrated in Fig. 1, was repeated for several temperatures of the interval studied (298 to 513 K). After the lowest temperature ($T = 298$ K) had been attained, the temperature was increased to the initial value ($T = 513$ K) and the specimen was cyclically deformed again to find that the cyclic flow stress reverted, within 4% to its original value. This seems to show that the dislocation substructure did not practically change during the cycles carried out at lower temperatures.

3. Results and Discussion

The flow stress measured at the highest strain of 7.5×10^{-3} has been resolved on the highest stressed $\{112\}$ planes which appear to be the best defined slip planes in tension in the temperature range investigated, according to previous observations [7]. It should be emphasized, however, that the choice of glide plane does not affect the conclusions of this investigation.

The variation of the cyclic flow stress (measured in the tensile cycle) with temperature has been plotted in Fig. 2 where it can be noticed that the rate of change of the flow stress with T depends on T. This is better illustrated in Fig. 3 where the negative of the slope of the τ–T curve has been plotted as a function of T. Following Zwiesele and Diehl [8] we have also plotted in the same graph the parameter $(1/kT)\,(\partial\tau/\partial \ln \dot{\varepsilon})_T$ against T and it can be seen that it has a dependence on T similar to that of $(-\partial\tau/\partial T)_{\dot{\varepsilon}}$. The above parameters are strongly dependent on T in the temperature range between 490 and 365 K, while at lower temperatures they vary only slowly with T.

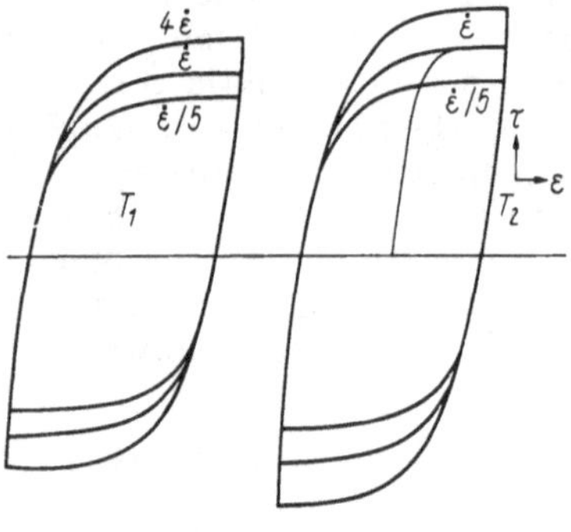

Fig. 1. Hysteresis loops obtained at two different temperatures and different strain rates to illustrate the procedure followed to measure the activation parameters

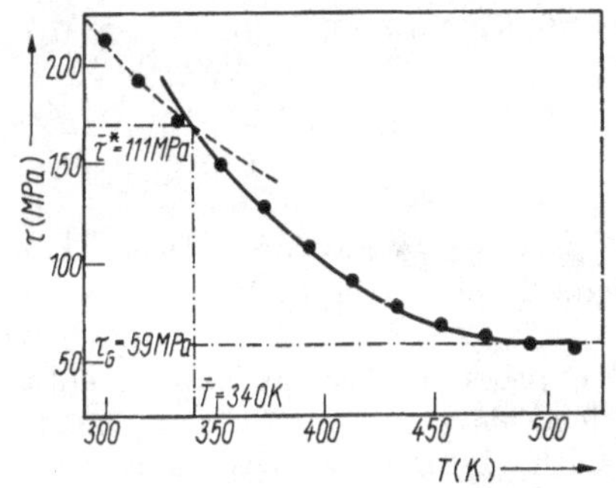

Fig. 2

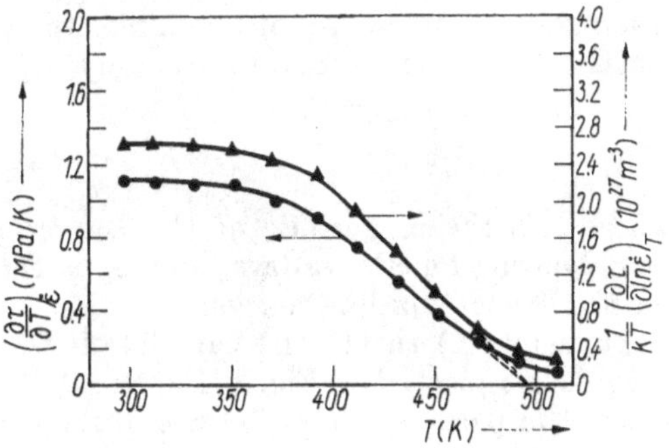

Fig. 3

Fig. 2. Temperature dependence of the tensile cyclic shear stress. The broken lines are the theoretical curves for both the high and low stress range of Seeger's [5] model obtained with $\tau_p = 600$ MPa and $T_k = 495$ K. $A = 6.72 \times 10^{-2}$ MPa$^{1/2}$/K

Fig. 3. Temperature dependence of the experimental parameters, $(-\partial\tau/\partial T)_{\dot\varepsilon}$ and $(1/kT)\,(\partial\tau/\partial \ln \dot\varepsilon)_T$ referred to the tensile (cyclic) shear stress of Mo single crystals

The strain rate sensitivity, $\lambda = (\partial\tau/\partial \ln \dot\varepsilon)_T$, obtained at different temperatures has been plotted against the shear stress in Fig. 4. At very small stresses, λ may be approximated by a linear function of τ. This means that in this stress interval the strain rate can be described by a power law of the effective stress [9], $\dot\varepsilon \sim \tau^{*m^*}$, with a value of $m^* \approx 2.5$ which is less than half that obtained by other workers [10, 11] from direct measurement of dislocation velocities at 300 K.

The theory of Seeger [5] mentioned in Section 1 applies to the temperature interval $T_k/2 < T < T_k$, where T_k is the temperature at which there is a sharp bend in the plot of the flow stress against T. This interval of temperature coincides partly with that studied in the present work, so that we shall compare the present results with the theory of Seeger.

It is assumed, as usual, that the strain rate, $\dot\varepsilon$, is given by an Arrhenius type equation of the form

$$\dot\varepsilon = \dot\varepsilon_0 \exp\left[-(\Delta H/kT)\right], \tag{1}$$

where ΔH is the activation enthalpy for double kink nucleation, $\dot\varepsilon_0$ is treated here as a constant, and k and T have the usual meaning. According to Seeger [5] the activation enthalpy at low stresses is given by

$$\Delta H = 2H_k - 2\alpha\tau^{*1/2}, \tag{2}$$

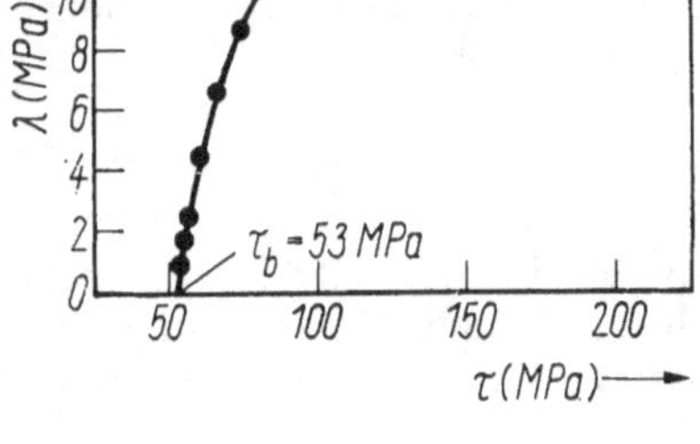

Fig. 4. Variation of the strain rate sensitivity with the tensile cyclic shear stress

where $2H_k$ is the energy of two kinks and τ^* is the effective stress acting on the dislocation line; α is a constant given by

$$\alpha = a\left(\frac{abS_0}{2}\right)^{1/2}, \tag{3}$$

where b is the magnitude of the Burgers vector, a is the separation between two neighbouring Peierls valleys, and S_0 is the pre-logarithmic factor in the expression for the dislocation line tension.

By using (2) in (1) one can obtain theoretical expressions for the experimental parameters plotted in Fig. 3. Firstly, it can be noticed that, assuming that $\dot{\varepsilon}_0$ is constant, the parameter $(1/kT)\,(\partial\tau/\partial \ln \dot{\varepsilon})_T$ coincides with the inverse of the activation volume, i.e.

$$V^{-1} = (b\,\Delta a)^{-1} = -\left(\frac{\mathrm{d}\Delta H}{\mathrm{d}\tau}\right)_T,$$

where Δa is the activation area. In Seeger's theory this parameter is given by

$$\frac{1}{kT}\left(\frac{\partial\tau}{\partial \ln \varepsilon}\right)_T = \frac{1}{b\,\Delta a} = -\frac{A}{\alpha}\,(T_k - T)\,. \tag{4}$$

Similarly, for $(-\partial\tau/\partial T)_{\dot{\varepsilon}}$ the theory gives

$$\left(-\frac{\partial\tau}{\partial T}\right)_{\dot{\varepsilon}} = -2A^2(T_k - T)\,. \tag{5}$$

In these expressions A is a constant given by

$$A = \frac{k}{2\alpha}\ln\left(\frac{\dot{\varepsilon}_0}{\dot{\varepsilon}}\right) \tag{6}$$

and T_k is the "knee" temperature, which is exactly defined as

$$T_k = \frac{2H_k}{k \ln(\dot{\varepsilon}_0/\dot{\varepsilon})}\,. \tag{7}$$

Equations (4) and (5) show that in the region of low stresses where (2) applies the plots of $(b\,\Delta a)^{-1}$ and $(-\partial\tau/\partial T)_{\dot{\varepsilon}}$ against temperature should be straight lines with slopes equal to $-A/\alpha$ and $-2A^2$ respectively and both should have also the same intercept $(T = T_k)$ with the T-axis. From Fig. 3 it can be seen that this is the case, and the value found for T_k is $T_k = 495$ K.

From the slope of the straight lines we obtain

$$A = 6.72 \times 10^{-2}\ \mathrm{MPa}^{1/2}\ \mathrm{K}^{-1}$$

and

$$\alpha = 3.02 \times 10^{-24}\ \mathrm{N}^{1/2}\ \mathrm{m}^2\,.$$

From the parameters plotted in Fig. 3 one can also obtain the factor $\ln(\dot{\varepsilon}_0/\dot{\varepsilon})$ by means of the relationship,

$$\ln(\dot{\varepsilon}_0/\dot{\varepsilon}) = \frac{(-\partial\tau/\partial T)_{\dot{\varepsilon}}}{(1/T)\,(\partial\tau/\partial \ln \dot{\varepsilon})_T} \tag{8}$$

which gives a value of $\ln(\dot{\varepsilon}_0/\dot{\varepsilon}) = 29.5 \pm 1.0$ between 450 and 298 K. At higher temperatures there are some deviations from this value, although this is not significant because (2) does not apply to temperatures above T_k. At very low stresses the possibility of strain fluctuations against the applied stress has to be taken into account so that we shall only consider the interval of temperatures below T_k. The energy of two kinks, $2H_k$, can be determined from (7) and by using the values obtained for T_k and $\ln(\dot{\varepsilon}_0/\dot{\varepsilon})$ one finds $2H_k = 1.25 \pm 0.05$ eV.

Up to this point, the comparison between theory and experiment has been carried out by assuming that the variation of the applied shear stress, τ, is equal to that of the effective stress, τ^*. Now it becomes necessary to know the internal stress, τ_G, if the absolute values of the effective stress ($\tau^* = \tau - \tau_G$) are to be compared with those given by the theory. τ_G can be determined by extrapolation, to $\lambda = 0$, of the linear relationship between λ and τ at low stresses [9]. It is found in this way that $\tau_G = 53$ MPa, which coincides with the value of τ reached by extrapolation of the τ–T curve to the athermal range.

Alternatively, one can consider τ_G as a parameter to be fitted to the experimental results. This is the procedure followed by Zwiesele and Diehl [8] to fit their results on high purity α-iron. This, so called, "fictitious" value of the internal stress [5] is just the total stress at $T = T_k$, that is, at the temperature where τ^* should be zero according to (1) and (2), and from Fig. 2 we find $\tau_G = 59$ MPa which is only 10% higher than that obtained by the alternative method. By using this "fictitious" value for τ_G and the values of T_k and A determined above, the model of Seeger can be very well fitted to the experimental results in the interval between 495 and 350 K as can be seen in Fig. 2, and also in Fig. 5 and 6 which show the stress dependence of the activation area and activation energy respectively. It is important to recall that no such good agreement was obtained in a recent investigation of the temperature and strain rate dependence of the cyclic flow stress of Nb single crystals [6], the reason being that in the case of Nb the value of τ_G obtained from the extrapolation of λ was very different from the "fictitious" value which should have been used to fit the theory to the results. In the present case, however, the difference between the two

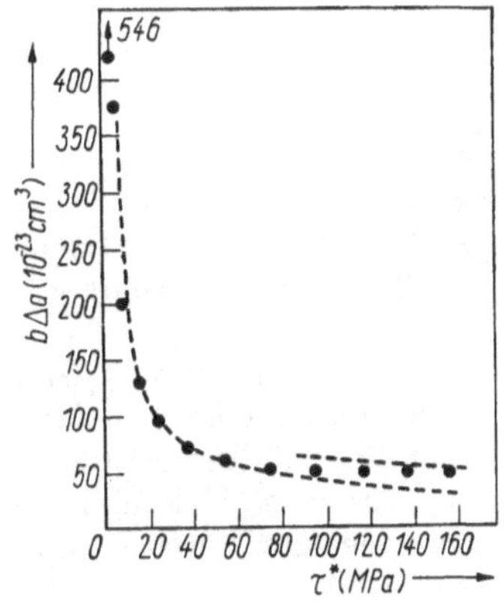

Fig. 5

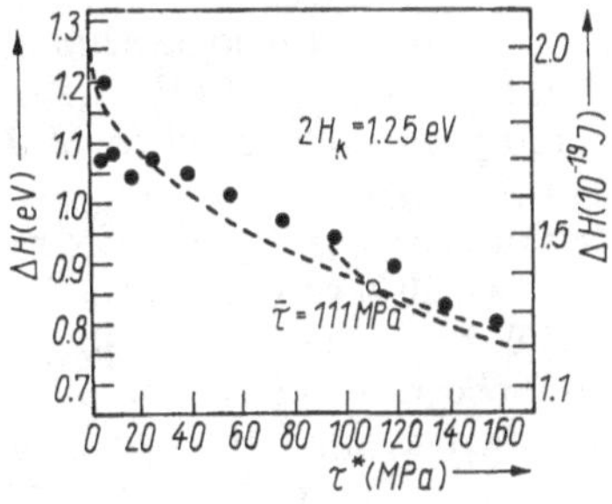

Fig. 6

Fig. 5. Activation volume, $b\,\Delta a$, plotted against the effective shear stress which has been obtained using the value $\tau_G = 53$ MPa for the internal stress. The dashed lines are the theoretical curves for both the high and low stress range of Seeger's [5] model. $T_k = 495$ K, $A = 6.72 \times 10^{-2}$ MPa$^{1/2}$/K, $\alpha = 3.02 \times 10^{-24}$ N$^{1/2}$ m^2, $2H_k = 1.25$ eV, $\tau_p = 600$ MPa

Fig. 6. Activation enthalpy, $\Delta H = -(kT/\lambda)\,(\partial\tau^*/\partial T)_{\dot{\varepsilon}}$ plotted against the effective shear stress calculated using the value $\tau_G = 53$ MPa, for the internal stress. The dashed lines are the theoretical curves for both the high and low stress range of Seeger's [5] model. $2H_k = 1.25$ eV values of τ_G is of 10% only. The reason for finding different values of τ_G is not clear. It could be that by defining the internal stress as that stress at which λ becomes zero, an effective stress τ^* is obtained which includes some contribution arising from energy terms not taken into account in the theory and which do not appear therefore in (2). This contribution could come, for example, from the energy of interaction between kinks and solute atoms, and it is perhaps significant that the agreement between the experimental results and the theory seems to be better when the measurements are carried out with crystals of higher purity, such as is the case of the high purity Nb used by Ackermann [12], and of the molybdenum of the present investigation which has a much lower concentration of interstitials than the Nb used in our previous study.

The value of α determined from the experiment can be compared with that calculated from (3) by using the lattice parameter and the elastic constants which are included in the definition of S_0 $(S_0 \equiv (Gb^2/4\pi)\,(1+\nu)/(1-\nu))$. With $b = 2.73$ Å, $G = 13.40 \times 10^{10}$ N/m^2 and $\nu = 0.305$ we obtain $\alpha = 1.86 \times 10^{-24}$ N$^{1/2}$ m^2 which is about half the experimental value. A similar discrepancy can be detected in the results for Nb [6] and for α-iron [8]. An increase of about 30% in the value of the Poisson ratio used in the calculation would only produce a small change, of about 10%, in the value of α obtained, so that it seems that the discrepancy would not disappear by using anisotropic elasticity.

At high stresses Seeger [5] treats the activation energy for double kink nucleation by using a line tension approximation for the kink–kink interaction and obtains the following expression for the activation enthalpy:

$$\Delta H = 2H_k \left[1 - \left(\frac{\tau^*}{2\sqrt{3}\,\tau_p}\right)\left(1 + \ln\left(\frac{3^{3/2}4\tau_p}{\tau^*}\right)\right)\right] \tag{9}$$

where τ_p is the Peierls stress.

The activation area is given by:

$$b\,\Delta a = \left(\frac{2H_k}{2\sqrt{3}\,\tau_p}\right)\ln\left(\frac{3^{3/2}4\tau_p}{\tau^*}\right) \tag{10}$$

so that it depends only on the logarithm of the effective stress. A similar dependence may be found for $(-\partial\tau/\partial T)_{\dot\varepsilon}$. Both these results are in good qualitative agreement with the observed very slow variation of $(b\,\Delta a)^{-1}$ and $(-\partial\tau/\partial T)_{\dot\varepsilon}$ with T at temperatures below 350 K.

Finally, the temperature dependence of the flow stress and the variation of ΔH, and $b\,\Delta a$ with τ^* which results from the line tension approximation have been plotted in Fig. 2, 5, and 6, respectively, by using the value of τ_p which best fits the results at high stresses and the value of $2H_k$ obtained from the fit at low stresses. Although the interval of τ studied at high stresses is not very wide, the agreement is reasonably good. The value of τ_p obtained by this procedure is $\tau_p = 600$ MPa.

It can be noticed that the stress, $\bar\tau^*$, and the temperature, $\bar T$, at which both approximations given the same value of ΔH are $\tau^* \approx 111$ MPa and $\bar T \approx 340$ K. According to Seeger [5] a good test for the theory is that the extrapolation to $\tau^* = \bar\tau^*$ of the theoretical activation volumes calculated in both approximations should obey the following relationship:

$$1.59 < \frac{\bar V_1}{\bar V_2} < 1.69\,, \tag{11}$$

where $\bar{V}_1$ and $\bar{V}_2$ are the activation volumes at $\tau^* = \bar{\tau}^*$ obtained by using the approximations for high and low stresses, respectively. The value obtained for $\bar{V}_1/\bar{V}_2$ in the present case is 1.65 which emphasizes the good agreement between the experimental results and Seeger's theory for the flow stress of *pure* b.c.c. metals.

Acknowledgements

The authors would like to thank Dr. J. A. Planell who generously supplied the Mo single crystals. One of the authors, Dr. M. Anglada, is indebted to Conicit of Venezuela for providing financial support.

References

[1] M. S. Duesbery and P. B. Hirsch, Dislocation Dynamics, Ed. A. R. Rosenfield, G. T. Hahn, A. L. Bement, Jr., and R. I. Jaffee, McGraw-Hill, New York 1968 (p. 57).
[2] M. S. Duesbery, Phil. Mag. **19**, 501 (1969).
[3] B. Escaig, phys. stat. sol. **28**, 463 (1968).
[4] P. B. Hirsch, 5th Internat. Conf. Crystallography, Cambridge 1960 (oral communication).
[5] A. Seeger, Z. Metallk. **72**, 369 (1981).
[6] M. Anglada and F. Guiu, Phil. Mag. **46**, 881 (1982).
[7] F. Guiu and P. L. Pratt, phys. stat. sol. **15**, 539 (1966).
[8] S. Zwiesele and J. Diehl, Proc. 5th Internat. Conf. Strength of Metals and Alloys, Vol. 1, Pergamon Press, Oxford 1979 (p. 59).
[9] M. Anglada and F. Guiu, Scripta metall. **13**, 103 (1979).
[10] G. C. Das, Ph. D. Thesis, University of London, 1967.
[11] H. L. Preckel, A. Lawley, and H. Conrad, Acta metall. **16**, 337 (1968).
[12] F. Ackermann, Thesis, University of Stuttgart, 1981.

(Received January 27, 1983)

phys. stat. sol. (a) 77, 291 (1983)

Subject classification: 1.1 and 1.4

Materials Science Department, C.E.N./S.C.K., Mol[1])

Accurate Characterization of Subgrain Boundaries in a TEM [2])

By

E. K. POLYCHRONIADIS[3]) and P. DELAVIGNETTE

A method is presented allowing a high precision in the determination of all the elements characterizing a subgrain boundary. These elements are: the rotation axis and its small angle, the boundary plane, the directions of the dislocation lines, the equidistances of the different dislocation families. These informations allow the determination of the length of Burgers vectors of the boundary dislocations. The precision of the method is controlled by Burgers vectors determinations obtained from their contrast analysis in transmission electron microscopy.

Une méthode de détermination à haute précision de tous les éléments qui caractérisent un joint à faible désorientation est présentée. Ces éléments sont: l'axe de rotation ainsi que son angle (petit), le plan du joint, les directions des lignes de dislocations, les équidistances des différentes familles de dislocations. Ces informations conduisent à la détermination de la longueur des vecteurs de Burgers des dislocations du joint. La précision de la méthode a été contrôlée par la détermination des vecteurs de Burgers des dislocations à l'aide de l'analyse de leur contraste au microscope électronique en transmission.

1. Introduction

Different publications appeared about accurate methods for the determination of crystallographic orientations in transmission electron microscopy. These have been reviewed by Ball [1]. Two other publications will also be mentioned concerning more in particular the problem of the characterization of a bicrystal. Young et al. [2] present a method based on the shift and the rotation of Kikuchi lines. A rotation of Kikuchi lines does not allow a high precision determination, therefore Bonnet and Durand [3] present a high precision method based on angular relations between pairs of directions from each crystal element. This method is completely independent of an experimental coordinate system, but it needs complex analytical expressions.

The method presented here has a slightly different purpose: the determination of *all* geometrical elements of interest for a complete characterization of a subgrain boundary will be presented, based on a minimum of information. From the diffraction pattern the following crystallographic data are determined: the small angle rotation operation that brings one crystal element in coincidence with the other crystal element. From the image mode, other geometrical elements are accurately determined. This is applied to the determination of the boundary plane and the dislocation lines constituting the subgrain boundary. Also the distance between two dislocations is calculated. These determinations are performed in an instrumental spherical coordinate system (α, δ) for which α expresses the angle between the tilt axis of a side entry

---

[1]) Boeretang 200, B-2400 Mol, Belgium.

[2]) Work performed on the auspices of the association CEN–IRE–ULB.

[3]) University of Thessaloniki, Physics Department, Thessaloniki (Greece).

goniometer stage and a particular direction and δ expresses the tilt angle indicated by the goniometer stage [4]. Finally, these orientations are readily transformed into crystallographic orientations using the adequate transformation matrix.

It is also the purpose of this paper, using contrast experiments, to determine the Burgers vectors of the dislocations constituting the subgrain boundaries, therefore experimentally establishing the accuracy of the method.

2. Principle of the Method

The small-angle rotation axis characterizing a subgrain boundary may be determined by the different positions of the Kikuchi lines observed on two selected area diffraction patterns taken on each side of the subgrain boundary. (The vector characterized by the two different positions of the Kikuchi lines on these two diffraction patterns will here be called the shift vector.) Two such shift vectors for two different specimen orientations are sufficient for that determination. The rotation angle is determined from the lengths of these shift vectors.

The boundary plane and dislocation line orientations are determined from their projected images in two different specimen orientations (or projection planes). Effective distances between dislocations are determined from their apparent distances in the projected images.

These determinations may also be performed on a stereographic projection.

3. Analytical Expression

Crystal orientations are expressed indifferently in polar (α, δ) or orthonormal (x, y, z) instrumental coordinate systems as precedingly presented [4] (a plane is given by its normal).

If $\boldsymbol{x}_{i1}$, $\boldsymbol{x}_{i2}$, and $\boldsymbol{x}_{i3}$ are three normalized vectors forming an orthonormal coordinate system of a crystal expressed in the instrumental system, the following expression transforms instrumental coordinates into crystal coordinates, i.e. three coefficients proportional to the Miller indices for the cubic system,

$$\boldsymbol{x}_c = \boldsymbol{T}\boldsymbol{x}_i \,,$$

where

$$T = [x_{ij}] \qquad i; j = 1, 2, 3 \,.$$

3.1 Rotation axis

If α_i, δ_i $(i = 1, 2)$ are two directions of shift vectors of Kikuchi lines (called elsewhere (α_{ki}, δ_i) as precedingly defined, since they both represent a direction perpendicular to the rotation axis characterizing the subgrain boundary, this axis is given by the common perpendicular

$$\tan\delta = \frac{\cos\alpha_1 \sin\alpha_2 \cos\delta_2 - \sin\alpha_1 \cos\alpha_2 \cos\delta_1}{\sin\alpha_1 \cos\alpha_2 \sin\delta_1 - \cos\alpha_1 \sin\alpha_2 \sin\delta_2},$$

$$\tan\alpha = \frac{\dfrac{\cos\delta_2}{\tan\alpha_1} - \dfrac{\cos\delta_1}{\tan\alpha_2}}{\sin(\delta_2 - \delta_1)\sin\delta} = \frac{\dfrac{\sin\delta_1}{\tan\alpha_2} - \dfrac{\sin\delta_2}{\tan\alpha_1}}{\sin(\delta_2 - \delta_1)\cos\delta} \tag{1a}$$

or

$$x = \sin\alpha_1 \sin\alpha_2 \sin(\delta_2 - \delta_1) \,,$$

$$y = \sin\alpha_1 \cos\alpha_2 \sin\delta_1 - \cos\alpha_1 \sin\alpha_2 \sin\delta_2 \,,$$

$$z = \cos\alpha_1 \sin\alpha_2 \cos\delta_2 - \sin\alpha_1 \cos\alpha_2 \cos\delta_1 \,. \tag{1b}$$

3.2 Rotation angle

If D_i is the amplitude of the shift vector of the Kikuchi lines, this corresponds to a rotation angle, ϱ_i, given by the following expression derived from the Bragg's law:

$$\sin\frac{\varrho_i}{2}=\frac{D_i}{2L}, \tag{2}$$

where L is the effective camera length. This rotation is a projection of the real rotation characterizing the subgrain boundary. The angle, φ_i, between the two directions (α_i, δ_i) and (α, δ) is given by the following expression:

$$\cos\varphi_i = \cos\alpha\cos\alpha_i + \sin\alpha\sin\alpha_i\cos(\delta-\delta_i) \tag{3a}$$

or

$$\cos\varphi_i = xx_i + yy_i + zz_i\,. \tag{3b}$$

Here the projections of the rotation axis are $(\alpha_{ki}+90°, \delta_i)$ $i = 1, 2$ and the rotation axis is $(\alpha_\varrho, \alpha_\varrho)$. From the projected rotation angle, ϱ_i, it is possible to deduce the rotation angle characterizing the subgrain boundary, ϱ, using the following expression:

$$\tan\frac{\varrho}{2}=\frac{\tan\frac{\varrho_i}{2}}{\cos\varphi_i}\,. \tag{4}$$

3.3 Dislocation lines

If (α_{di}, δ_i) is the orientation of the projection of a dislocation line (projected onto the plane defined by the tilt angle δ_i), the real orientation of the dislocation line is perpendicular to the direction $(\alpha_{di}+90°, \delta_i)$. From two of these determinations it is possible to deduce the real direction of the dislocation line applying (1) to the directions $(\alpha_{di}+90°, \delta_i)$, $i = 1, 2$.

3.4 Boundary plane

The boundary plane is determined by adjusting the electron beam parallel to it, when possible. A general method is also proposed from the determination of two directions parallel to that plane by applying (1). The two directions may be the dislocation line and the intersection between the boundary plane and one of the foil surfaces. This latter direction is determined applying the same procedure as for the dislocation lines.

3.5 Equidistance between dislocations

The subgrain boundaries contain equidistant dislocation sets; their real equidistance is determined from the apparent distance in a projected image. This real distance is calculated by the intermediate of the determination of the intersection line, $(\alpha_{qi}, \delta_{qi})$, between the boundary plane and the projection plane, $(90°, \delta_i + 90°)$, applying (1). Then the angles between this direction and (i) the dislocation line (α_d, δ_d), called φ_{di}, (ii) the projections of the dislocation line $(\alpha_{di}, \delta_{di})$ on each projection plane, called φ_{qi}, are determined using (3). Finally the real equidistance between the dislocations lines is given from its projected images by applying the following relation:

$$d = d_i\,\frac{\sin\varphi_{di}}{\sin\varphi_{qi}}\,. \tag{5}$$

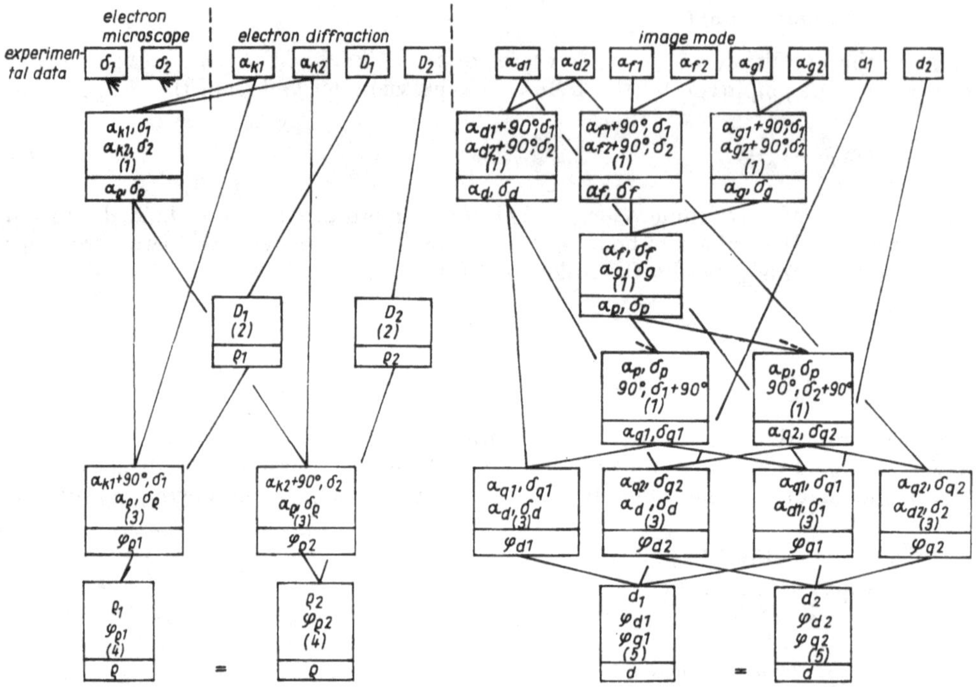

Fig. 1. General diagram illustrating the determination of different elements constituting a subgrain boundary. The different symbols have the following meaning: δ_i tilt angles as measured on the goniometer stage, defining the projection planes of coordinates (90°, $\delta_i + 90°$); α_{ki} direction of the shift vector in each projection plane; D_i length of the shift vector in each projection plane; α_{di} direction of the dislocation lines in each projection plane; α_{fi} direction of the intersection line of the foil plane and the boundary plane in each projection plane; α_{gi} a second direction parallel to the boundary plane (may be chosen parallel to the dislocation lines) in each projection plane; d_i equidistance between dislocations as measured in each projection plane; $(\alpha_\varrho, \delta_\varrho)$ direction of the rotation axis characterizing the subgrain boundary; ϱ_i rotation angle component deduced from the shift vector; ϱ rotation angle characterizing the subgrain boundary; $\varphi_{\varrho i}$ angle between the rotation axis and each of its projections; (α_d, δ_d) direction of the dislocation lines; (α_f, δ_f) and (α_g, δ_g) orientations of two directions parallel to the subgrain boundary plane; (α_p, δ_p) direction of the subgrain boundary plane given by its normal; $(\alpha_{qi}, \delta_{qi})$ direction of the intersection line between the boundary plane and each of its projection planes; φ_{di} angle between (α_d, δ_d) and $(\alpha_{qi}, \delta_{di})$; φ_{qi} angle between $(\alpha_{di}, \delta_{di})$ and $(\alpha_{qi}, \delta_{qi})$; d equidistance between dislocations

Since the total procedure needs the repeated application of a limited number of expressions, its complexity is only apparent as it appears synthetically summarized in Fig. 1. This total procedure has been programmed on a pocket calculator. Its total execution needs less than one minute.

3.6 Burgers vector of the dislocations

The length of the component of the Burgers vector of the dislocations participating in the rotation is given by the well-known expression

$$b = 2d \sin \frac{\varrho}{2}. \tag{6}$$

3.7 Character of the subgrain boundary

The parallel (tilt) and perpendicular (twist) components of the rotation on the subgrain boundary plane are given by

$$\varrho_{\parallel} = \varrho \sin \varphi , \qquad \varrho_{\perp} = \varrho \cos \varphi , \tag{7}$$

where φ is the angle between the rotation axis and the perpendicular to the boundary plane, obtained by applying (3).

4. Field of Application

A subgrain boundary is characterized by sets of parallel and equidistant dislocations giving rise to a twist (rotation around an axis perpendicular to the boundary plane) in the case of a crossed grid of screw dislocations, to a tilt (rotation along an axis parallel to the boundary plane) in the case of a single set of edge dislocations (with a Burgers vector perpendicular to the boundary plane), or to a combination of these two components [5]. All these dislocations induce in the crystal a strain field limited to a layer around the boundary plane of a thickness of the order of the mean distance between the dislocations. Other dislocations or sets of dislocations may also be present; these will not give rise to a relative rotation of the crystal elements separated by the boundary plane, they give rise to longer-range strains. Therefore, the method developed here does not allow the determination of this latter component of the Burgers vector, as illustrated in the second example in Section 6.

5. Specimen Preparation

Tungsten foils have been annealed for 20 d at 1200 °C. They have been electrolytically thinned in 2% NaOH in water at 1.7 V and a current density of 0.6 A/mm^2. The tungsten prepared under these conditions presents grain sizes of the order of 1 to 5 μm in diameter, often containing only few subgrain boundaries. Other thermal treatments have also been performed, but for shorter annealing times the subgrain boundaries present more complex dislocation structures that have not always been analysed in detail.

Subgrain boundaries have also been observed in the DO_{23} structure of Al_3Zr. A more detailed account will be presented elsewhere.

The observations are performed on a goniometer stage of a JEOL 200 CX electron microscope using the single tilt holder.

6. Experimental Results

6.1 First example

A subgrain boundary in annealed tungsten is presented in Fig. 2 in three different orientations. The same dislocations are in contrast for three different reflecting conditions. The orientations of the rotation axis, the dislocation lines, and the boundary plane are determined from the experimental determination of the different projections presented in Fig. 2. The tilt angle is determined with a precision of the order of 10′ since the single tilt holder of a goniometer stage is used. The other orientations are determined on the diffraction pattern or on the micrographs with a precision of the order of 30′ or better. The determination of the shift vector characterizing the projection of the rotation vector on a plane parallel to the reciprocal section shown on the diffraction patterns is obtained by carefully superimposing the Kikuchi maps present on the two diffraction patterns taken on each side of the grain boundary. The shift vector is then given by the double image of the transmitted beam as shown on the three insets of Fig. 2 for the three different orientations.

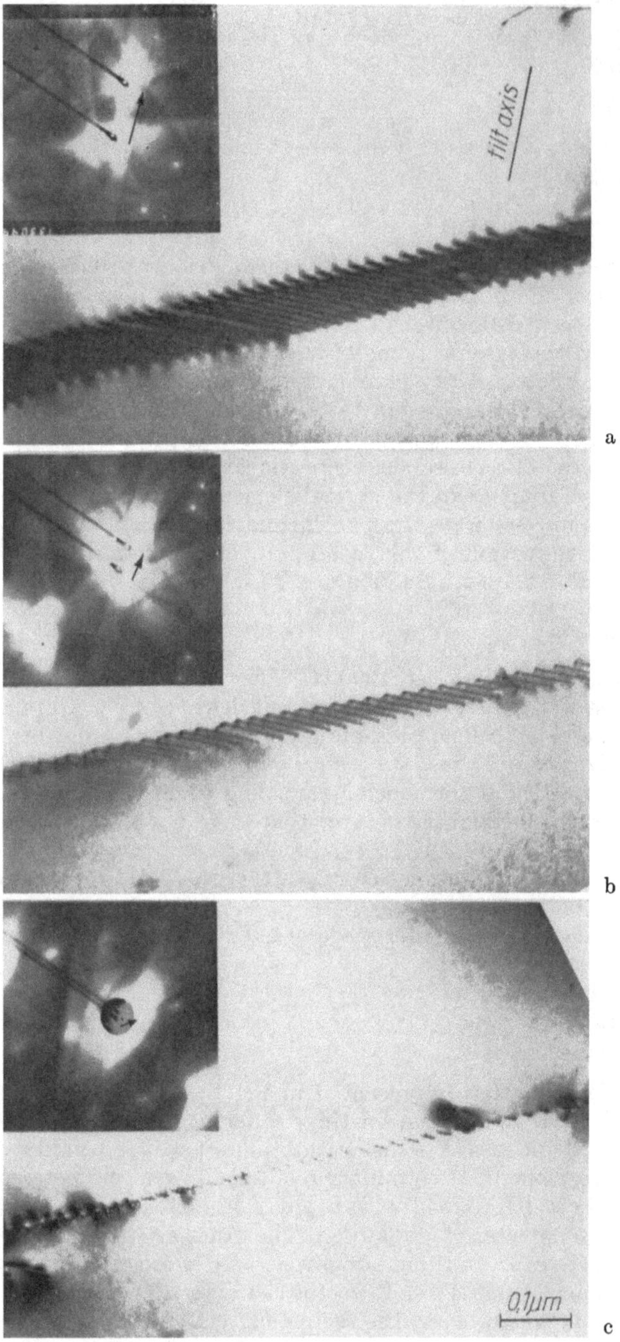

Fig. 2. A subgrain boundary in tungsten in three different orientations. The arrows in the three insets represent the shift vectors. The data that are obtained from them are given in Table 1

Table 1

Data of the subgrain boundary of Fig. 2 in three different orientations called *a*, *b*, and *c* as in Fig. 2

	a	*b*	*c*
δ_i (°)	−29.17	6.17	26.67
α_{ki} (°)	− 5.2	168.9	123.2
D_i (mm)	15	7	1.6
α_{di} (°)	−90.9	88	77
α_{fi} (°)	−68	110.5	113.2
d_i (nm)	10.1	10.1	15.1

The experimental data for the subgrain boundary of Fig. 2, obtained from three different orientations a, b, and c, are summarized in Table 1 (D_i are given in mm for an effective camera length of $L = 1410$ mm). From the procedure presented in Section 3 and Fig. 1, the following results are obtained:

	α	δ	
rotation axis	−94.4°	−60.4°	
rotation angle			42.5′
dislocation line	−90.8°	−60.0°	
normal to the boundary plane	24.5°	31.8°	
distance between dislocations			24.7 nm

The angle between the boundary plane and the rotation axis is 3.1°, therefore the subgrain boundary is virtually a pure tilt boundary. This information may also be obtained by use of the stereographic projection, Fig. 3, with a more limited precision.

Finally the instrumental coordinates of the three crystallographic coordinate axes considered to coincide for each crystal element, are determined. These, expressed in an orthonormal system, are the column vectors of the rotation matrix, Table 2, that transforms instrumental coordinates into crystallographic coordinates, Table 3.

From the calculated orientations of the rotation axis and the boundary plane it is deduced that the subgrain boundary must be formed of edge dislocations whose line orientation is along [0$\bar{1}$1] and with a Burgers vector $\boldsymbol{b}$ = [100]. Moreover from the calculated mean distance between these dislocations and the rotation angle, it is deduced that the length of the Burgers vector, applying (6), is 0.305 nm. These determinations have been experimentally confirmed; in particular, the Burgers vector has been determined: two different typical contrasts are given in Fig. 4. Its Burgers

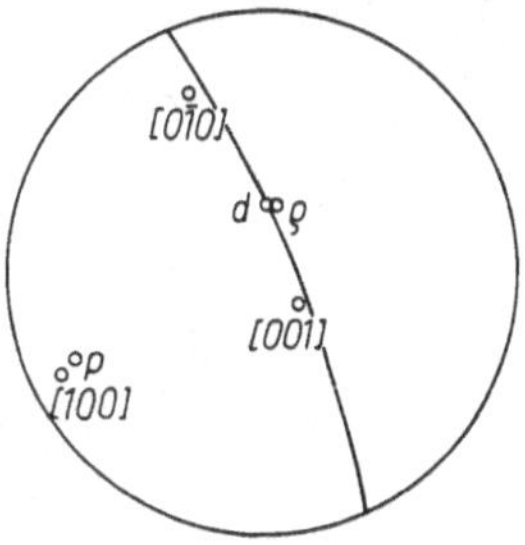

Fig. 3. Stereographic projection of the different elements characterizing the subgrain boundary of Fig. 2. ϱ rotation axis, *d* direction of the dislocation lines, *p* direction perpendicular to the boundary plane, also represented by its trace on the projection

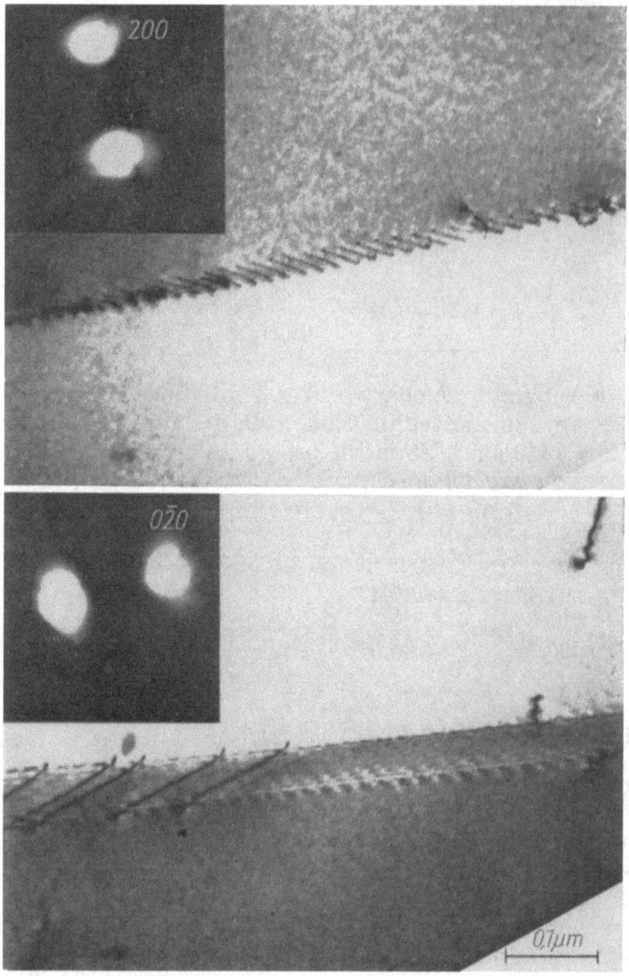

Fig. 4. The subgrain boundary of the first example, Fig. 2, in two different diffraction conditions; a) dislocations in contrast; b) dislocations out of contrast for $0\bar{2}0$

Table 2

Transformation matrix from instrumental coordinates to crystallographic coordinates for Fig. 2

	[100]	[010]	[001]
α	15.2°	110.6°	103.8°
δ	19°	−17.8°	77.5°
matrix	0.905	−0.352	−0.239
	0.403	0.891	0.210
	0.139	−0.286	0.948

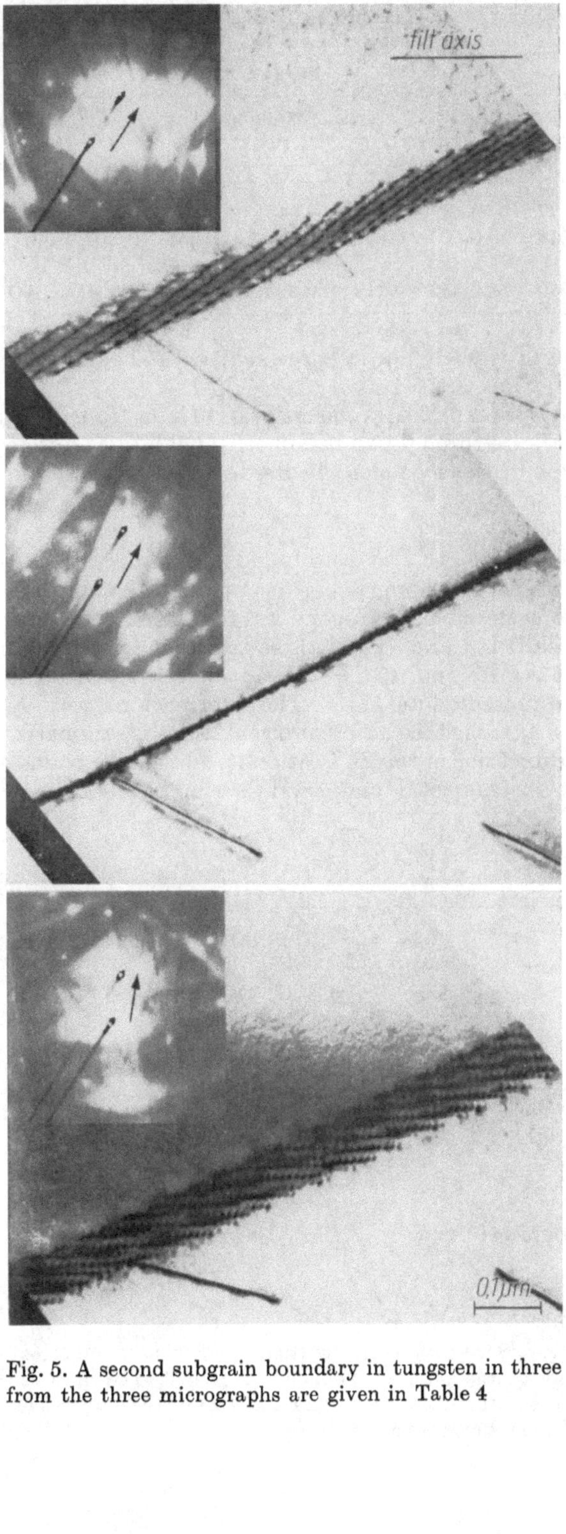

a
b
c

Fig. 5. A second subgrain boundary in tungsten in three different orientations. The data obtained from the three micrographs are given in Table 4

Table 3
Calculated characteristic elements of the subgrain boundary of Fig. 2

	instrumental system					crystallographic system			
	α	δ	orthonormal			orthonormal			[uvw] (hkl)
	(°)	(°)	x	y	z	x	y	z	
rotation axis	−94.4	−60.4	−0.077	−0.492	0.867	−0.148	−0.660	0.737	$\approx 0\bar{1}1$
dislocation line	−90.8	−60.0	−0.014	−0.500	0.866	−0.094	−0.688	0.719	$\approx 0\bar{1}1$
boundary	24.5	31.8	0.910	0.352	0.219	0.996	−0.069	0.064	≈ 100

vector is $\boldsymbol{b} = [100]$, therefore its length is 0.316 nm in agreement with the previous determination.

The deviations of (i) the rotation axis from $[0\bar{1}1]$, (ii) the dislocation lines from $[0\bar{1}1]$, and (iii) the boundary plane from (100) are throught to be effective deviations due to the thin foil state and residual strains in the specimen. These deviations are indeed greater than the experimental precision.

6.2 Second example

A more complex subgrain boundary observed in tungsten after a moderate annealing treatment is presented in Fig. 5. From electron microscope images, dislocations are observed in two different directions called A and B and three different families are observed along the A direction, called A_1, A_2, and A_3. The dislocation structure of the subgrain boundary is schematically represented on Fig. 6. The information given by electron microscope diffractions and images in three different orientations is summarized in Table 4. The rotation axis, the directions of the dislocation lines, and the boundary plane have been calculated in instrumental and crystallographic coordinate

Table 4
Data of the subgrain boundary of Fig. 5

	a	b	c
δ_i (°)	−26.83	−2.67	23.67
α_{ki} (°)	−122	−115.4	77.7
D_i *) (mm)	25.3	23.8	21.6
$\alpha_{di(A)}$ (°)	−146.2	−156.7	3.8
α_{fi} (°)	−157.5	−154.2	24.8
$\alpha_{di(B)}$ (°)	—	−150.2	69.2
$d_{i(A)}$ (nm)	21	—	18.6
$d_{i(B)}$ (nm)	—	—	10.5

*) For an effective camera length of 1410 mm.

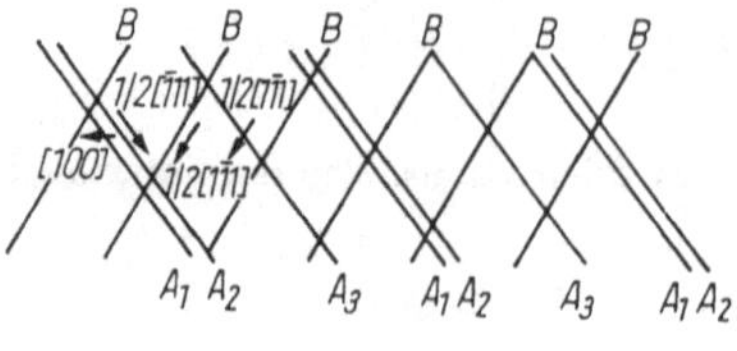

Fig. 6. Schematic representation of the dislocation structure of the subgrain boundary of Fig. 5. The lines represent the dislocations and the arrows the projections of their Burgers vectors

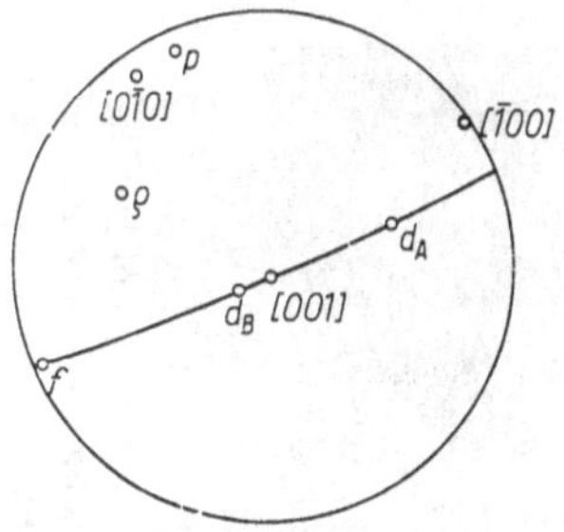

Fig. 7. Stereographic projection of the subgrain boundary of the second example, Fig. 5. The rotation axis ϱ is neither on the boundary plane nor perpendicular to it, the boundary is therefore of a mixed type

systems and are summarized in Table 5, and schematically represented on a stereographic projection, Fig. 7. The rotation angle and the distances between each set of dislocations are also given.

These results show that the subgrain boundary is of a mixed type and may therefore be subdivided in a tilt component and a twist component according to (7). Moreover, the tilt component is parallel to neither dislocation lines A or B, it will therefore be further decomposed in two tilt components which are parallel: the first to the dislocation lines A and the second to the dislocation lines B. These three values, as well as the equidistance between dislocation lines are given in Table 5.

A confirmation of these determinations has only been possible after a complete determination of the Burgers vectors of all four groups of dislocations. From a complete dislocation contrast analysis, the following Burgers vectors have been determined:

$$A1 : \boldsymbol{b} = [100]\,, \quad A2 : \boldsymbol{b} = \tfrac{1}{2}[\bar{1}11]\,, \quad A3 : \boldsymbol{b} = \tfrac{1}{2}[1\bar{1}1]\,, \quad B : \boldsymbol{b} = [1\bar{1}1];$$

six different contrasts are presented on Fig. 8. Since the directions of the dislocation lines are

$$A : \approx [\bar{1}01] \quad \text{and} \quad B : \approx [101]\,,$$

they are all of mixed character and will be decomposed in their edge and screw components for the analysis. Moreover, from all screw components the dislocations

Table 5

Calculated characteristic elements of the subgrain boundary of Fig. 5

	instrumental					crystallographic			
	α	δ	x	y	z	x	y	z	$[uvw]$ (hkl)
rotation axis	−33.7	−47.2	0.832	−0.377	0.407	0.428	−0.837	0.335	$\approx 1\bar{2}1$
dislocation line A	−140.6	−61	0.773	0.308	−0.555	0.804	−0.171	−0.569	$\approx \bar{1}01$
dislocation line B	78.3	80.4	0.203	0.163	0.965	0.254	−0.080	0.963	≈ 101
boundary plane	−64.7	−5	0.427	−0.901	0.079	−0.210	−0.979	−0.011	≈ 010

Rotation angle $\varrho = 62.6'$: $\varrho_{\|A} = 32.0'$, $\varrho_{\|B} = 41.2'$, $\varrho_{\perp} = 45.6'$;
equidistance between groups of three dislocations $A = 32$ nm;
equidistance between dislocations B: 14.5 nm.

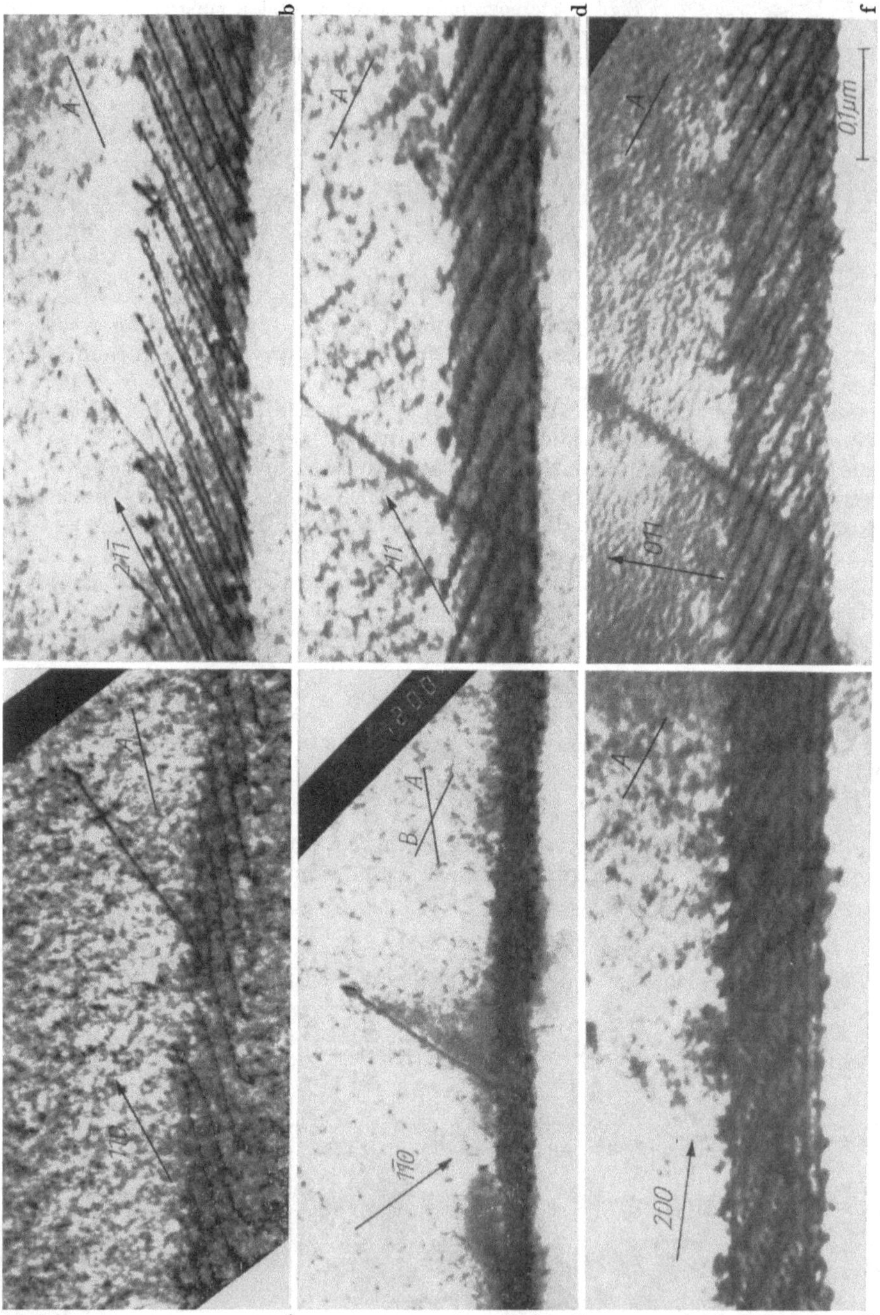

Fig. 8. The subgrain boundary of Fig. 5 in six different diffraction conditions. The two different orientations of the dislocation families are indicated by the lines A and B. All dislocations of the subgrain boundary are visible in c) and e); dislocations A_1 are out of contrast in f); dislocations A_2 are out of contrast in a) and d); dislocations A_3 and B are out of contrast in a), b), and f)

forming a square network of two families of dislocations with the same equidistance will only be considered. Finally from their edge components only their components which are perpendicular to the boundary plane, (010), will be considered for their participation in the tilt components of the rotation axis. The Burgers vectors are indicated in Fig. 6, projected on the boundary plane. They are decomposed in the following components:

$$\begin{array}{llll} & & \text{edge} \quad \text{screw} & \text{edge} \\ A_1: & [100] \to & \tfrac{1}{2}[10\bar{1}] & + \tfrac{1}{2}[101]\,, \\ A_2: & \tfrac{1}{2}[1\bar{1}\bar{1}] \to & \tfrac{1}{2}[0\bar{1}0] + \tfrac{1}{2}[10\bar{1}]\,, & \\ A_3: & \tfrac{1}{2}[1\bar{1}1] \to & \tfrac{1}{2}[0\bar{1}0] & + \tfrac{1}{2}[101]\,, \\ B: & \tfrac{1}{2}[1\bar{1}1] \to & \tfrac{1}{2}[0\bar{1}0] + \tfrac{1}{2}[101]\,. & \end{array}$$

The two edge components perpendicular to the boundary plane of dislocations A_2 and A_3 each at calculated equidistances of 32 nm correspond to a single set with an equidistance of 16 nm therefore giving rise, applying (6), to a rotation angle of $\varrho_{\|A} = 33.9'$ in good agreement with the calculated angle of 32.0′. The edge component of the dislocations B at a calculated equidistance of 14.5 nm gives rise to a rotation of $\varrho_{\|B} = 37.5'$ in good agreement with the calculated angle of 41.2′. Finally, two screw components A_1 and A_2 each at calculated equidistance of 32 nm, corresponding to a single set at an equidistance of 16 nm, associated with the screw components of the B dislocations at approximately the same equidistance (14.5 nm) in a perpendicular direction give rise to a twist rotation angle of 51.3′ in good agreement with the calculated angle of 45.6′.

This complex boundary has also edge components of the dislocations A_1 and A_3 parallel to the boundary plane which will therefore not give rise to a rotation. These give rise to long-range displacement fields, while all the components considered above give rise to displacement fields ranging to distances of the order of their mean mutual equidistances. Table 6 summarizes these results.

This subgrain boundary has also been analysed in other regions for other boundary orientations as illustrated in Fig. 9, regions marked 1, 2, and 3. Their dislocation content varies with the boundary orientation, but in this example the rotation relationship appeared to be unchanged:

	rotation axis		rotation angle
	α	δ	
position 1	−33.7°	−47.2°	62.6′
position 2	−32.2°	−44.1°	67
position 3	−32.3°	−44.5°	60.8′

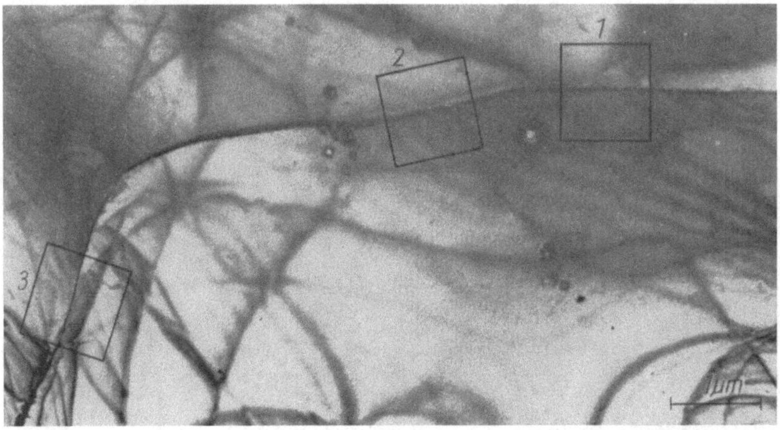

Fig. 9. A general view of the subgrain boundary of the second example. The data for this example have been taken from the region 1, the rotation axis has also been determined for the regions marked 2 and 3

Table 6

Detailed analysis of the dislocation content of the subgrain boundary illustrated in Fig. 5

direction of dislocations	Burgers vector	Burgers vector components giving		
		rise to tilt	rise to twist	no rotation
A_1	$[100] = 0.3165$ nm	—	$\bar{b}_s$ along $[\bar{1}01] = 0.224$ nm	$\bar{b}_c$ along [101]
A_2	$\frac{1}{2}[\bar{1}11] = 0.274$ nm	$\bar{b}_e$ along $[010] = 0.158$ nm	$\bar{b}_s$ along $[\bar{1}01] = 0.224$ nm	—
A_3	$\frac{1}{2}[1\bar{1}1] = 0.274$ nm	$\bar{b}_e$ along $[010] = 0.158$ nm	—	$\bar{b}_e$ along [101]
B	$\frac{1}{2}[1\bar{1}1] = 0.274$ nm	$\bar{b}_e$ along $[010] = 0.158$ nm	$\bar{b}_s$ along $[101] = 0.224$ nm	—

A_2 edge A_3 edge: $d = 16$ nm, $b = 0.158$ nm $\rightarrow \varrho_{\parallel A} = 33.9'$;
B edge: $d = 14.5$ nm, $b = 0.158$ nm $\rightarrow \varrho_{\parallel B} = 37.5'$;
A_1 screw, A_2 screw, B screw: $d = 15$ nm, $b = 0.224$ nm $\rightarrow \varrho_{\perp} = 51.3'$.

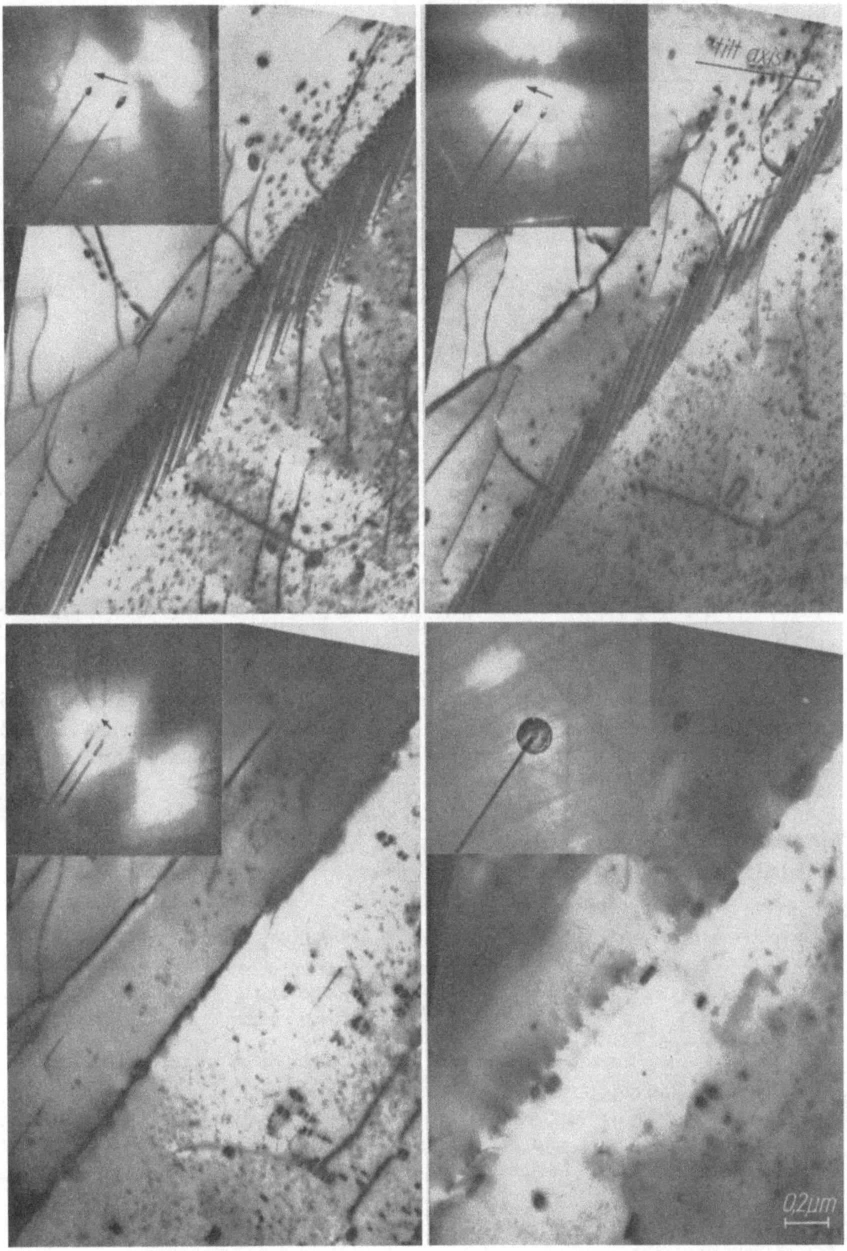

a b c d

Fig. 10. A subgrain boundary in Al_3Zr in four different orientations. The data obtained from them are given in Table 7

6.3 Third example

A last example will be considered in a different material: Al_3Zr. Fig. 10 shows dislocations associated in pairs. The information about that subgrain boundary has been obtained for four different orientations, which are summarized in Table 7. The dif-

Table 7

Data of the subgrain boundary of Fig. 10

	a	b	c	d
δ_i (°)	−18.5	−3	16.5	29.67
α_{ki} (°)	−5.2	−7.5	157.7	51.5
D_i *) (mm)	10.8	7.7	2.8	1.4
α_{di} (°)	−96.6	−99.3	65.6	150
α_{fi} (°)	−117.5	−116.4	65.6	61.5
d_i (nm)	25.6	22.4	—	86.3

*) For an effective camera length of 1410 mm.

ferent calculated orientations, angles, and distances are given in Table 8 and schematically represented in Fig. 11. The subgrain boundary is clearly a tilt boundary. From the rotation angle and the equidistance between dislocations a Burgers vector component perpendicular to the boundary plane of 0.92 nm is deduced. The structure of Al_3Zr is b.c.t. with $a = 0.4013$ nm; $c = 1.732$ nm, it is an ordered structure based on a slightly deformed f.c.c. matrix. It is therefore concluded that the rotation is due to Burgers vector components of $\frac{1}{2}$ [001] per pair of dislocations. The effective total Burgers vector per pair is $\frac{1}{2}$ [111]. The dislocations are dissociated in two partials,

$$\tfrac{1}{2}[111] \rightarrow \tfrac{1}{4}[221] + \tfrac{1}{4}[001] .$$

A detailed account of the behaviour and dissociation of the dislocations in this ordered structure will be treated elsewhere. The subgrain boundary analysis in Al_3Zr allowed a clear confirmation of an unexpected dissociation of dislocations in this structure.

Table 8

Calculated characteristic elements of the subgrain boundary of Fig. 10

	instrumental system					crystallographic system			
	α (°)	δ (°)	orthonormal x	y	z	orthonormal x	y	z	[uvw] (hkl)
rotation axis	−93.7	−64	−0.064	−0.437	0.897	0.979	0.200	−0.007	4.9 1 0 ≈ 100
dislocation line	−94.7	−63	−0.082	−0.452	0.888	0.984	0.177	0.002	5.5 1 0 ≈ 100
boundary	156	16.7	−0.913	0.390	0.117	0	0	1	0 0 1

Rotation angle: 38′;
equidistance between groups of dislocations: 83.1 nm.

7. Accuracy of the Method

Effective determinations of rotation axes in different subgrain boundaries based on three or more different projections have regularly given rotation axis orientation spreads of the order of 0.2°. This spread increases of course when the two reference projections differ by a small angle. Appropriate differences between these angles are of the order of 20°. Dislocation line orientation determinations gave generally spreads

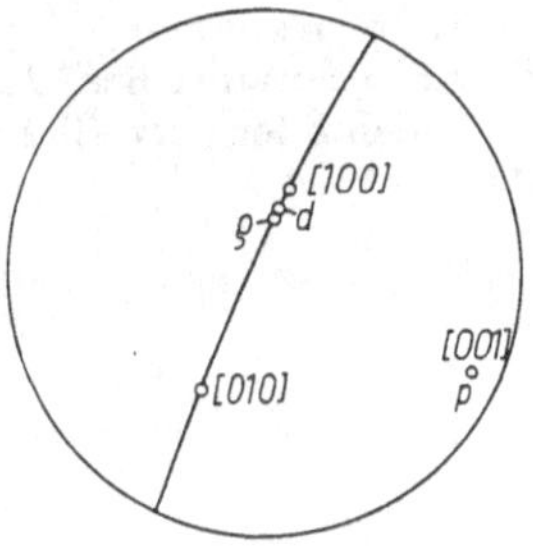

Fig. 11. Stereographic projection of the subgrain boundary of Fig. 10. The subgrain boundary is a pure tilt boundary

of the order of 1°. The rotation angle was generally obtained with a spread of the order of 1 to 2% for angles of the order of 1°. The equidistance between dislocations showed a spread of the results of the order of 5%.

From the calculated values it appears that if two projections are chosen with an angular difference of the order of 40°, and if the two projected directions and the tilt angles are determined with a precision of the order of 0.1°, then this direction is always determined with a precision of the order of 0.2° unless the angle between this direction and the tilt axis is greater than 80°. An alternative determination applicable in this case is treated in Section 8.

From this it is concluded that this method allows a precision of the orientation determination of the order of 0.2° as soon as this is physically significant, i.e. when the crystal is not subjected to a general bending (disappearance of the Kikuchi lines) or when the dislocation lines and boundary planes are effectively clear straight lines or clear planar boundaries. Therefore, it is thought that the precision of the method is generally better than allowed from physical circumstances like bent foils, due to internal strains, surface effects, non-planar boundaries, etc.

8. Orientation Determination of a Plane Parallel to the Tilt Axis

When a plane is approximately parallel to the tilt axis (or a direction is approximately perpendicular to that axis), its precise orientation may be determined on the basis of a different approach. This is the case particularly for the determination of the orientation of a stacking fault from its apparent width in two different projections or also for a segment such as a dislocation segment (perpendicular to the tilt axis) from two projected images.

Its expression will be formulated in the following using the same coordinate system as presented in Section 3. The apparent widths, d_1 and d_2, are measured from projections at tilt angles δ_1 and δ_2 ($-90° < \delta_i < +90°$) and considering the foil horizontal for $\delta = 0$ (the precision of this data is important for the thickness determination, but is not significant for the orientation determination). From Fig. 12, considering

$d_i = A_iC_i$, the apparent width of the fault or the dislocation length;

$\delta_i = \widehat{C_iA_iD_i}$, the tilt angles;

$\delta = \widehat{D_iA_iB_i}$, the inclination of the fault plane or the dislocation line;

$L = A_iB_i$, its width or length;

$E = B_iD_i$, the thickness of the foil,

since

$$A_iC_i = A_iB_i \cos \widehat{D_iA_iB_i} \quad \text{or} \quad d_i = L \cos (\delta + \delta_i) ,$$

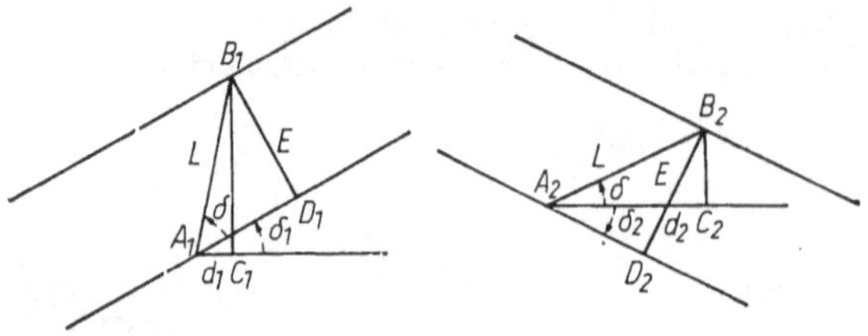

Fig. 12. Schematic representation of a stacking fault or a dislocation line, L, observed in two orientations for illustrating the notations used

therefore,

$$\frac{d_2}{d_1} = \frac{\cos(\delta + \delta_2)}{\cos(\delta + \delta_1)} = \frac{\cos(\delta + \delta_1)\cos(\delta_2 - \delta_1) - \sin(\delta + \delta_1)\sin(\delta_2 - \delta_1)}{\cos(\delta + \delta_1)},$$

from which

$$\tan(\delta + \delta_1) = \frac{\cos(\delta_2 - \delta_1) - \dfrac{d_2}{d_1}}{\sin(\delta_2 - \delta_1)}$$

and calling

$$\delta + \delta_1 = \theta\,,$$

the three following unknowns: defect inclination, δ, length of the defect, L, and thickness of the foil, E, are determined from the experimental data δ_1, δ_2, d_1, and d_2 on the basis of the following expressions:

$$\delta = \theta - \delta_1\,, \qquad L = \frac{d_1}{\cos(\delta + \delta_1)}\,, \qquad E = L \sin \delta\,.$$

9. Conclusion

The method presented here for the determination of orientations is based on a simple formulation and makes use of precise expressions without any approximation. It is applied to the determinations of all the elements characterizing a subgrain boundary, but it is directly applicable to any other direction determination in a microscope equipped with a goniometer stage.

A precision in the direction determination of the order of 0.2° and in the rotation angle of a subgrain boundary of the order of 0.5′ have been routinely achieved.

References

[1] C. J. BALL, Phil. Mag. A **44**, 1307 (1981).

[2] C. T. YOUNG, J. H. STEELE, JR., and J. L. LYTTON, Metallurg. Trans. **4**, 2081 (1973).

[3] R. BONNET and F. DURAND, phys. stat. sol. (a) **27**, 543 (1975).

[4] TH. KARAKOSTAS, G. NOUET, G. L. BLERIS, S. HAGÈGE, and P. DELAVIGNETTE, phys. stat. sol. (a) **50**, 703 (1978).

[5] J. T. FOURIE, P. J. JACKSON, D. KUHLMANN-WILSDORF, D. A. RIGNEY, J. H. VAN DER MERWE, and H. G. F. WILSDORF, Scripta metall. **16**, 157 (1982).

(Received February 10, 1983)

phys. stat. sol. (a) 77, 309 (1983)

Subject classification: 1.4; 22.8

Laboratoire d'Etude et de Synthèse des Microstructures, associée au C.N.R.S., E.S.P.C.I., Paris[1])

Aspect aléatoire de la formation des composés par frittage réactif

Par

C. LACOUR, A. DUBON et P. M. GRANGEON

Après avoir étudié statistiquement la taille et la répartition des vides dans une poudre comprimée on analyse ici la répartition et l'évolution des espèces dans des mélanges binaire, ternaire, quaternaire. L'ensemble de cette étude théorique montre l'extrême importance du diamètre des particules et de leur différence quand il y a plusieurs espèces, non seulement sur la vitesse de frittage, ce qui est bien connu, mais également sur la vitesse de formation des composés définis et sur leur homogénéité à un instant donné. On montre, d'autre part, qu'il est hasardeux de penser avoir intégralement formé un composé défini à partir d'un mélange de poudres et cela d'autant plus que les espèces sont nombreuses, les diamètres des particules gros et différents et les concentrations différentes. Il apparaît d'autre part qu'il existe un rapport idéal des diamètres des particules dans le cas des mélanges, qui conduit à une hétérogénéité minimum; ce rapport idéal étant fonction de la concentration respective des éléments.

In a previous paper the distribution and the evolution of the interparticular spaces of a compact powder, before and during sintering, are studied. Here the evolution of the chemical species in the binary, ternary, and quaternary mixture of powders is analyzed. This theoretical work shows the important effect of the diameter of the particles not only on the rate of sintering, as it is well known, but also on the rate of formation of the defined compounds and on the homogeneity level at a given time. Besides it is shown how difficult it is to form completely a defined compound especially from numerous species with large and various particle diameters and various concentrations. Furthermore it is pointed out that an ideal ratio of the diameters of the particles leads to a minimum heterogeneity. This ratio depends on the respective concentration of the elements.

1. Introduction

Les connaissances actuelles sur l'état initial et l'évolution macroscopique au cours de la diffusion des mélanges pulvérulents binaires, ternaires ou d'ordres supérieurs, présentent des lacunes importantes qui tiennent en partie à des investigations expérimentales difficiles mais également à l'absence d'hypothèses théoriques sur l'aspect et l'évolution macroscopiques des poudres comprimées.

Nous présentons dans ce travail une approche conceptuelle de ces problèmes en nous appuyant sur un travail récent [1] dans lequel nous avons montré que les groupements de particules et des vides interparticulaires dans un agglomérat pulvérulent n'ont pas des dimensions quelconques mais obéissent en fait à une répartition qui peut être définie et qui correspond à une distribution aléatoire des particules dans un espace donné. Dans l'étude antérieure citée [1] nous nous étions penchés sur la description des vides après compression et son évolution au cours du frittage. Ici, nous étudierons la distribution des particules pour différents types de mélanges ainsi que leur évolution que l'on supposera dirigée par la diffusion à l'état solide.

---

[1]) 10, rue Vauquelin, 75231 Paris Cedex 05, France.

2. Etat initial

Les poudres destinées à l'élaboration de composés définis, par frittage, peuvent être obtenues par de nombreux procédés.

Ce qui est recherché dans tous les cas est le meilleur mélange possible c'est à dire l'homogénéité chimique la plus élevée qui permet de former le composé défini dans les meilleures conditions de temps et de température.

Une première remarque est à faire concernant le mélange des poudres. Intuitivement, on peut penser qu'un mélange prolongé permet de tendre vers l'homogénéité chimique. En fait, la probabilité d'obtenir un mélange homogène de deux ou n constituants en quantités égales est aussi faible que celle de tirer alternativement jusqu' à la dernière, une boule de chacune des couleurs également présentes dans une urne. Un mélange est en fait comme nous le verrons, soumis aux lois du hasard et tend au mieux, vers une distribution aléatoire des constituants dans un espace donné. Dans

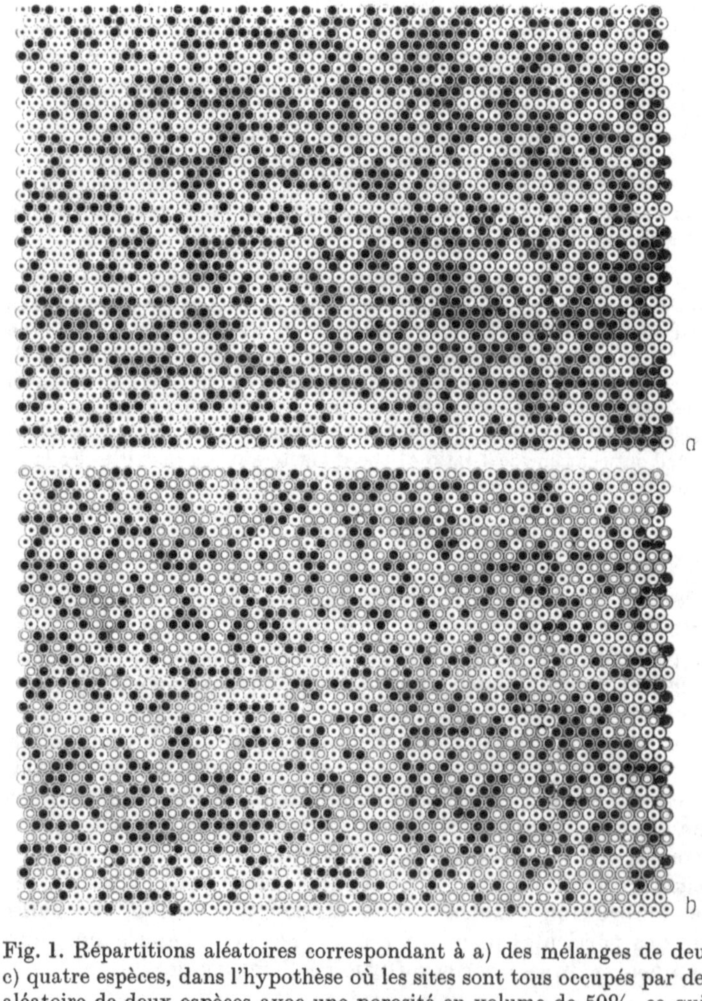

Fig. 1. Répartitions aléatoires correspondant à a) des mélanges de deux espèces, b) trois espèces, c) quatre espèces, dans l'hypothèse où les sites sont tous occupés par des particules. d) Répartition aléatoire de deux espèces avec unc porosité en volume de 50%, ce qui correspond dans le plan à 33% de sites inoccupés. Cette répartition correspond à un état initial expérimental courant

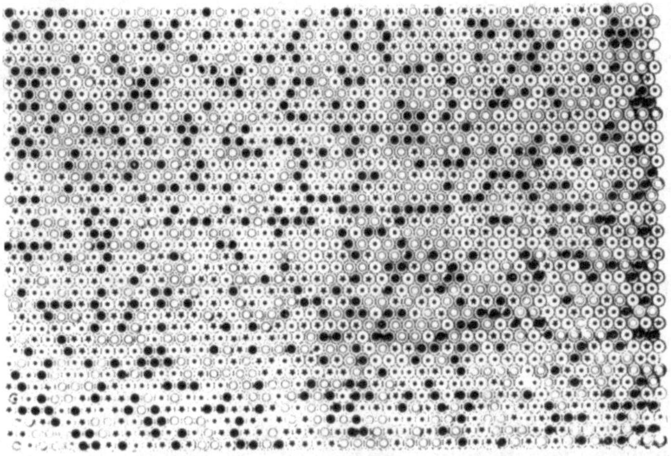

c

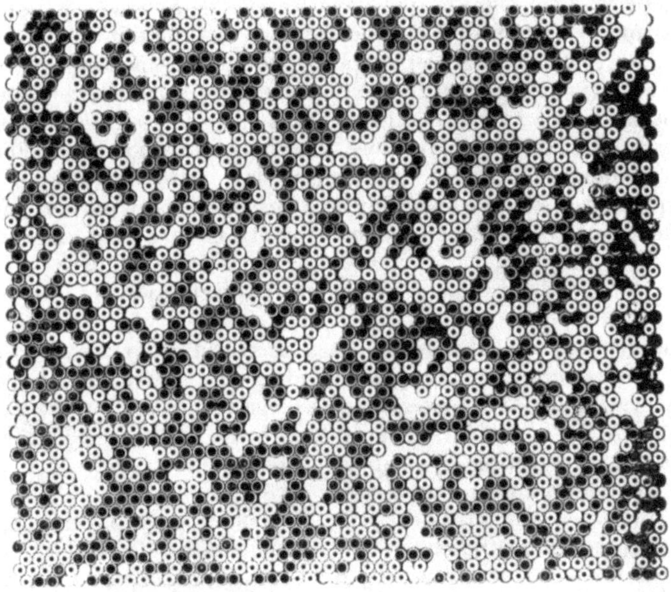

d

le cas des suspensions très dispersées, la distribution des particules est mal connue mais on peut penser que l'homogénéité absolue n'est pas davantage acquise.

Ceci étant, dans les mélanges réels, on peut toujours considérer qu'il existe une relative homogénéité chimique qui peut être définie par le plus petit volume de matière qui conduit, après analyse, à une valeur égale à la composition moyenne (plus ou moins un écart ΔC défini à l'avance) quel que soit l'endroit où ce volume est prélevé. Nous avons attribué au diamètre de ce volume de matière, l'expression: critère d'hétérogénéité (D_h).

Nous avons montré [1] dans le cas où une seule espèce chimique est présente, que les particules sont distribuées aléatoirement dans l'espace qui leur est imparti. Il est naturel de faire l'hypothèse que l'augmentation du nombre d'espèces ne modifie pas la concentration des sites occupés par rapport aux sites disponibles, ni leur loi de distribution propre; pour autant bien entendu, que leurs morphologies ne soient pas trop différentes.

2.1 Etude dans un plan de dimensions finies de l'état initial d'un mélange pulvérulent en fonction du nombre des composants

Les distributions des Fig. 1, obtenues à partir de générations de nombres aléatoires uniformément répartis, montrent les répartitions des particules correspondant au mélange de deux composants (Fig. 1a), de trois composants (Fig. 1b) et quatre composants (Fig. 1c).

Dans chaque cas, on se place dans l'hypothèse d'une porosité initiale de 26% qui correspond à un espace entièrement occupé par un empilement h.c. de particules sphériques et dans le cas d'un mélange binaire dans l'hypothèse d'une porosité en volume de 50% (Fig. 1d), qui correspond, dans le plan à environ 33% de sites inoccupés.

Ces distributions montrent nettement l'existence de clusters et appellent immédiatement deux remarques:

Il n'y a manifestement pas d'homogénéité à courte distance.

On retrouve, dans le mélange, des clusters (amas accidentels de particules d'une même espèce) qui pourraient être interprétés dans les mélanges expérimentaux comme des amas initiaux non dissociés. Ceci signifie que dans les études antérieures dans lesquelles cette interprétation avait été effectivement faite, on pouvait attribuer à tort à un mauvais mélange, ce qui revenait en fait à un groupement naturel dû au hasard. Cependant, l'individualisation parfaite des particules lors du mélange peut ne pas être réalisée et dans ce cas, ce ne seront plus les particules élémentaires qu'il faudra distribuer mais leurs agglomérats mal dissociés. Nous avons dans son principe couvert cette hypothèse en étudiant l'influence du diamètre des particules.

Les valeurs caractéristiques des distributions aléatoires des Fig. 1 sont portées sur le Tableau 1. Dans chaque cas les concentrations des deux, trois ou quatre espèces de particules sont égales, le symbole $n_c(\varphi_1)$ correspond au nombre de clusters de l'espèce φ_1. Le symbole $C_m(\varphi_1)$ est la dimension en nombre de particules, du cluster moyen de l'espèce φ_1 etc.

Pour une distribution donnée la faible différence entre les valeurs caractéristiques des différentes espèces est remarquable et traduit une identité de distributions.

Dans le cas de deux espèces et pour une porosité initiale égale à 26% (dans ce cas tous les sites sont occupés (Fig. 1a) les valeurs caractéristiques ne sont pas indiquées car on se situe au voisinage d'une transition [2] étudiée par les spécialistes de la percolation et qui conduit à un cluster de dimension infinie.

La présentation des distributions complètes des tailles de clusters n'est pas faite ici car une seule distribution par cas a été étudiée, ce qui est insuffisant pour être significatif. Par contre, dans notre étude antérieure [1], référence constante au présent travail, on a montré que la distribution des clusters obéissait à une relation empirique. La distribution experimentale moyenne et les points calculés sont représentés sur la Fig. 2.

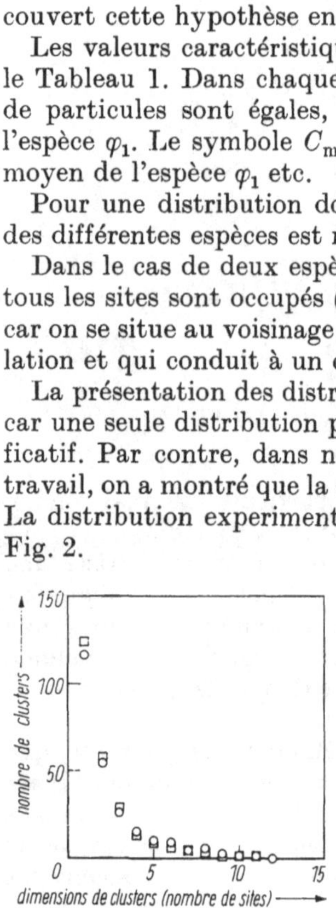

Fig. 2. Moyennes de distributions aléatoires (○) et expérimentales (□) montrant l'identité de distribution des vides interparticulaires entre un modèle théorique et une poudre comprimée

Tableau 1

Valeurs caractéristiques des distributions aléatoires (% de composé formé n_c, C_m) pour deux, trois et quatre espèces; dans un réseau plan de 2500 sites

	pourcentage de sites occupés par les particules	pourcentage de composé formé pour une distance de diffusion de deux particules	$n_c(\varphi_1)$	$C_m(\varphi_1)$	$n_c(\varphi_2)$	$C_m(\varphi_2)$	$n_c(\varphi_3)$	$C_m(\varphi_3)$	$n_c(\varphi_4)$	$C_m(\varphi_4)$	n_c moyen	C_m moyen
deux espèces	67	72	172	5,43	165	5,67					168,5	5,55
(φ_1, φ_2)	100	79										
trois espèces	67	37	255	2,44	249	2,50	238	2,62			247	2,52
$(\varphi_1, \varphi_2, \varphi_3)$	100	54	163	3,88	138	4,60	168	3,77			156,3	4,08
quatre espèces	67	22	245	1,91	251	1,87	243	1,89	248	1,89	246,8	1,89
$\begin{pmatrix}\varphi_1, \varphi_2\\ \varphi_3, \varphi_4\end{pmatrix}$	100	40	200	2,38	185	2,56	187	2,53	182	2,60	188,5	2,52

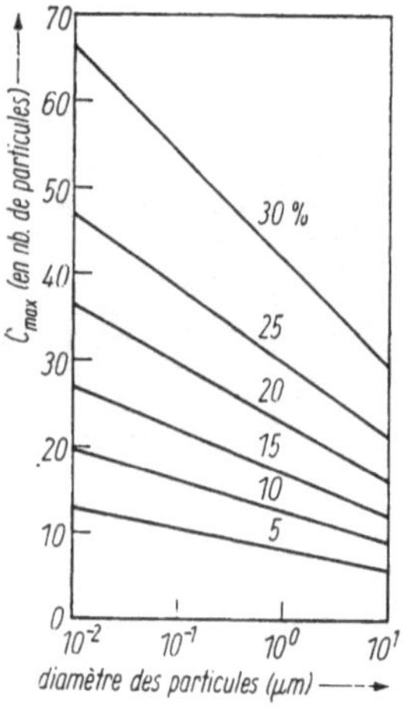

Fig. 3

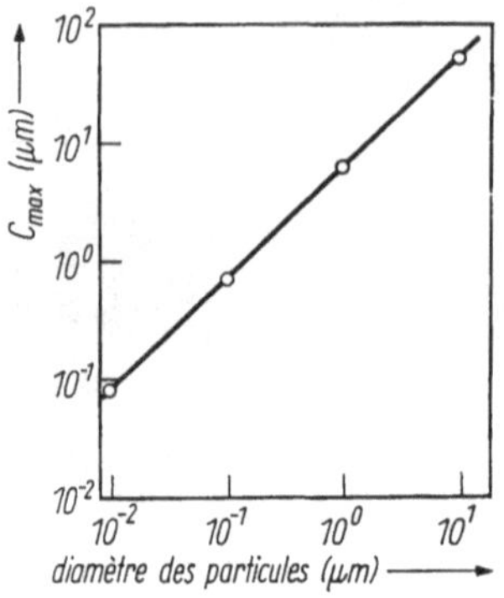

Fig. 4

Fig. 3. Variations de la dimension (en nombre de particules) du plus grand cluster probable en fonction du diamètre des particules et de leur concentration, pour un plan donné

Fig. 4. Variation de la dimension (en unité métrique) du plus grand cluster probable en fonction du diamètre des particules pour une concentration (30%) et un plan donnés

On a montré de la même façon que l'on pouvait calculer le nombre total n_t, les dimensions moyenne C_{moy} et maximale C_{max} des clusters en fonction de la concentration et de la dimension du réseau et plus généralement des distributions des espèces dans une représentation plane d'un mélange. En particulier, on peut représenter pour un plan de surface donnée les variations de la dimension probable du cluster maximum, par exemple, exprimée en nombre de particules, en fonction du diamètre des particules et pour différentes concentrations (Fig. 3). Ces courbes montrent très naturellement que la dimension des clusters croît avec la concentration des particules et décroît quand le diamètre des particules croît.

Mais si l'on considère la dimension des clusters, exprimée non plus en nombre de particules mais en dimension linéaire, on remarque (Fig. 4), toujours pour une surface donnée et pour une concentration donnée, que la dimension du cluster maximum croît avec le diamètre des particules suivant la loi:

$$\frac{C_{max}(\Phi_2)}{C_{max}(\Phi_1)} = \left[\frac{\Phi_2}{\Phi_1}\right]^{k_1}$$

avec $\Phi_{2,1}$ diamètres des particules $\Phi_2 > \Phi_1$, k_1 pente de la droite de la Fig. 4.

Par ailleurs, si l'on considère une quantité donnée de matière, il est évident que le nombre de contacts entre particules croîtra en raison inverse du cube du diamètre des particules ce qui contribuera à une augmentation de la vitesse de frittage d'autant plus grande que les particules seront plus petites.

Ces deux remarques justifient un nouvel intérêt pratique de l'emploi des particules ultrafines dans l'élaboration des matériaux par frittage, indépendamment des difficultés rencontrées dans leur mise en forme [3, 4].

Nous verrons après que l'abaissement du diamètre des particules permet également une augmentation très importante de la vitesse de formation d'un composé par diffusion à partir d'un mélange de plusieurs constituants.

2.2 Etat initial d'un mélange de taille réelle

Dans ce qui précède, pour permettre la visualisation de l'état initial d'un mélange de particules en contact, nous avons raisonné dans un plan et sur un nombre limité de particules. Nous réutiliserons par la suite cette représentation simplifiée pour étudier l'évolution des compacts au cours du frittage réactif. Mais cette représentation ne permet pas de décrire les répartitions dans des échantillons de taille réelle.

On va donc procéder différemment pour évaluer leur loi de distribution.

Cette loi de distribution va être assimilée aux variations de la probabilité d'existence, à un instant donné, de concentrations (n_a, n_b, n_c) d'espèces a, b, c, par exemple, dans un élément de volume v donné.

Dans le cas d'une seule espèce, le terme n n'exprime que la concentration de l'espèce considérée par rapport au volume dans lequel il se trouve. Quand il y aura deux espèces ou plus, l'ensemble des termes: n_a, n_b ... exprimera en plus la composition.

2.2.1 Concentration. Critère d'hétérogénéité de concentration

Dans le cas d'une seule espèce représentée par N particules occupant un volume V, la probabilité qu'il y ait n particules dans un des éléments de volume v au nombre de $n_v = V/v$ s'exprime par la relation (1),

$$P(n, N) = \frac{N!(n_v - 1)^{N-n}}{(N-n)!\, n!\, n_v^N} \tag{1}$$

avec N nombre total de particules dans le volume V, n nombre donné de particules dans un volume v donné, n_v nombre de volumes élémentaires v dans le volume V.

Cette relation peut être calculée numériquement en utilisant l'approximation de Stirling et en passant par les logarithmes.

Il vient après élimination des termes d'ordre inférieur,

$$\ln P(n, N) = n\left(\ln \frac{N}{nn_v} + 1\right) - \frac{N}{n_v}\left(1 + \frac{1}{2n_v}\right) + n\left(\frac{1}{n_v} - \frac{n}{2N}\right) - \\ - \frac{1}{2}\ln\,(2\pi n)\ . \tag{1'}$$

Si l'on introduit le paramètre n^* qui correspond à la concentration moyenne de particules dans l'élément de volume v et dans le cas particulier $P(n^*, N)$, il vient

$$P(n^*, N) \approx \frac{e^{-n^*/2N}}{2\pi n^*} \approx \frac{1}{2\pi n^*} \tag{1''}$$

avec $n^* \ll N$. Le calcul des probabilités de concentrations n de part et d'autre de n^* conduit à une courbe représentative de la loi de Laplace-Gauss ou à une loi de Poisson quand n^* est petit.

Par conséquent, ces probabilités peuvent également s'exprimer par la relation classique

$$P(n^*, N) = \frac{1}{\sigma\sqrt{2\pi}}\ . \tag{2}$$

En égalant (1'') et (2), il vient

$$\sigma \approx \sqrt{n^*} \tag{3}$$

avec σ écart-type de la distribution, N et n_v étant connus par définition, $n^* = N/n_v$ est donc connu et par conséquent l'écart-type.

Dans le cas d'une seule espèce, le problème de la distribution et de sa concentration dans un espace donné est donc résolu. Les réseaux de courbes de la Fig. 5 établis

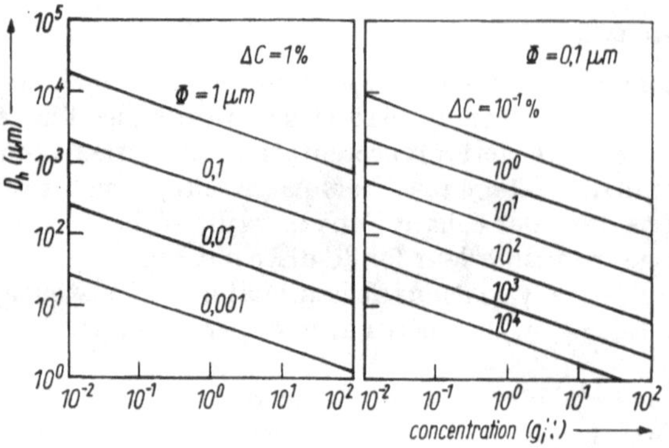

Fig. 5. Evolution du critère d'hétérogénéité de *concentration* D_h en fonction de la concentration globale, du diamètre des particules et ceci pour des écarts de concentration ΔC, compris entre 10^{-1}% et 10^4%

à titre d'exemple, montrent dans le cas précis de particules d'oxyde de Mg en suspension dans un litre d'eau, les variations du critère d'hétérogénéité de concentration D_h (concentration) en fonction de la concentration, du diamètre des particules et ceci pour différents écarts de concentration ΔC défini a priori.

A partir de ces courbes, examinons trois exemples:

Exemple 1: Pour des particules de 10 Å et une concentration de 100 g/l ± 1%, le critère d'hétérogénéité de concentration est de l'ordre de 1 µm. Autrement dit des volumes v de 1 µm de diamètre auront des concentrations équivalentes à 1% près.

Exemple 2: Pour des particules de 1 µm et une concentration de 100 g/l ± 1%, le critère d'hétérogénéité de concentration est de l'ordre de 0,7 mm. Ou, pour comparer avec l'exemple précédent: il sera nécessaire que les éléments de volume v aient un diamètre mille fois plus grand pour que les concentrations dans chacune d'elles soient égales à 1% près.

Exemple 3: Pour des particules de 1 µm et une concentration de 0,01 g/l ± 1%, le critère d'hétérogénéité de concentration est de l'ordre de 20 mm. Ou, si l'on préfère, le rapport des concentrations dans des volumes v de 100 µm de diamètre pourra atteindre 100.

La concentration de 0,01 g/l apparemment très faible, correspond en fait à la présence de 100 ppm d'un élément (impureté volontaire ou involontaire) dans une solution de base de concentration: 100 g/l.

2.2.2 Composition. Critère d'hétérogénéité de composition

Dans un mélange, il n'y a plus une mais plusieurs espèces chimiques. Le problème qui se pose donc est de savoir quelle va être la répartition des concentrations, donc de la composition dans les éléments de volume v.

Pour deux constituants, la probabilité $P(n_a, n_b, N_a, N_b)$ qu'il y ait, dans un volume v: n_a particules de l'espèce a, n_b particules de l'espèce b quand, dans le volume V, il y a $(N_a + N_b)$ particules a et b, est le produit des probabilités $P(n_a, N_a)\, P(n_b, N_b)$ soit

$$P(n_a, n_b, N_a, N_b) = \frac{N_a!\, N_b!\, (N_a - n_a^*)^{(N_a - n_a)}\, (N_b - n_b^*)^{(N_b - n_b)}\, n_a^{*n_a} n_b^{*n_b}}{(N_a - n_a)!\, (N_b - n_b)!\, n_a! n_b! N_a^{N_a} N_b^{N_b}} \tag{4}$$

avec N_a, N_b nombres totals des particules des espèces a, b dans le volume V, n_a, n_b nombres donnés des particules a, b dans un volume v donné.

Dans cette relation, à chaque probabilité est associé une composition mais également une concentration. Nous devrions donc pour étudier le seul paramètre composition faire la somme des probabilités associées à une composition n_a/n_b donnée, telle que

$$\sum_{(-n_{a\,\mathrm{lim}},\,-n_{b\,\mathrm{lim}})}^{(+n_{a\,\mathrm{lim}},\,+n_{b\,\mathrm{lim}})} P\left(\frac{n_a}{n_b}\right). \tag{5}$$

La distribution des probabilités $P(n_a/n_b)$ est discrète et son pas moyen est d'autant plus petit que le nombre moyen de particules par élément de volume est plus grand. Si le nombre moyen de particules est élevé et si la composition considérée est voisine de n^*, l'expression (5) comprend un très grand nombre de termes. Son calcul est très long et n'a pas été entrepris car ces distributions ne correspondent plus à des lois gaussiennes.

Toutefois plus on s'écarte de la composition moyenne C^*, plus le nombre d'éléments de volume ayant cette composition éloignée de C^* diminue. Aux limites, il n'y a plus que deux éléments de volume dont les compositions respectives sont symétriques par rapport à C^*.

On est en mesure de calculer ces compositions limites C_{lim} probables en fonction du nombre, du diamètre des particules et de la concentration des éléments en présence.

Pour cela, on définit la probabilité limite P_{lim} associée à C_{lim}. Il vient

$$P_{\mathrm{lim}} \geqq \frac{2}{n_v} = \frac{2v}{V} \tag{6}$$

en supposant des masses volumiques égales pour les éléments a et b, la composition dans un volume v peut être exprimée par le rapport des volumes de matière, soit

$$\frac{v_a}{v_b} = \frac{n_a^* \pm t_a\sigma_a}{n_b^* \mp t_b\sigma_b}\,\frac{\Phi_a^3}{\Phi_b^3} = C\ , \tag{7}$$

t_a, t_b variables réduites des distributions des espèces a, b, Φ_a, Φ_b diamètres des particules a, b.

La probabilité associée à une telle composition s'écrit

$$P\left(\frac{v_a}{v_b}\right) = \frac{1}{2\pi\sigma_a\sigma_b} \exp\left(-\frac{t_a^2 + t_b^2}{2}\right) \tag{8}$$

et pour l'une des composition limite:

$$P_{\mathrm{lim}} = \frac{1}{2\pi\sigma_a\sigma_b} \exp\left(-\frac{t_{a\,\mathrm{lim}}^2 + t_{b\,\mathrm{lim}}^2}{2}\right) = \frac{2v}{V} \tag{9}$$

soit

$$t_{a\,\mathrm{lim}}^2 + t_{b\,\mathrm{lim}}^2 = -2\ln\left[\frac{4\pi v\sigma_a\sigma_b}{V}\right] = k\ . \tag{9'}$$

La relation entre t_a et t_b étant établie il reste à trouver le maximum de l'expression (7).

Explicitons tout d'abord cette relation en substituant à n_a^* et n_b^* les relations telles que (3) appliquées à σ_a et σ_b et en substituant à σ_a et σ_b leurs expressions ci-dessous en fonction des paramètres élémentaires

$$\sigma_a = \left(\frac{6(1 - P_0)\, vC_a}{\pi\Phi_a^3}\right)^{1/2} \tag{10}$$

et

$$\sigma_b = \left(\frac{6(1 - P_0)\, vC_b}{\pi\Phi_b^3}\right)^{1/2} \tag{10'}$$

avec $(1 - P_0)$ concentration de matière dans le volume V (100 cm^3), C_a, C_b concentrations des éléments a, b.

L'étude de la relation ainsi obtenue fait apparaître un minimum pour la valeur limite de t_a, suivante:

$$t_{a\,\lim} = \frac{-k\sigma_a + \sigma_b[k(\sigma_a^2 + \sigma_b^2 - k)]^{1/2}}{\sigma_a^2 + \sigma_b^2} \tag{11}$$

et l'expression de $C_{\lim}$ devient finalement

$$C_{\lim} = \frac{(\Phi_a^3 C_a)^{1/2}\left[q\left(\frac{C_a v}{\Phi_a^3}\right)^{1/2} + t_{a\,\lim}\right]}{(\Phi_b^3 C_b)^{1/2}\left[q\left(\frac{C_b v}{\Phi_b^3}\right)^{1/2} - (k - t_{a\,\lim}^2)^{1/2}\right]}\,. \tag{12}$$

Nous avons calculé la composition limite $C_{\lim} > C^*$ et le rapport $\left((C_{\lim} - C^*)/C^*\right)_{\max}$ correspondant, en fonction du rapport des concentrations C_a/C_b et du rapport des diamètres des particules Φ_a/Φ_b. Ces résultats sont représentés sur les réseaux de courbes de Fig. 6. Ces courbes montrent essentiellement que l'écart relatif de composition par rapport à la composition moyenne augmente quand le rapport des concentrations C_a/C_b croît et cela d'autant plus que le rapport des diamètres des particules s'éloigne de l'unité.

Mais pour mieux faire apparaitre l'hétérogénéité initiale du mélange binaire en fonction des mêmes paramètres: C_a/C_b et Φ_a/Φ_b, on a établi à partir des courbes de la Fig. 6 les variations du critère d'hétérogénéité maximum pour trois écarts de composition donnés 0,1%, 10% et 10³% (Fig. 7).

L'examen de ces courbes appelle les remarques suivantes:

Cette représentation confirme que toutes choses égales par ailleurs, l'hétérogénéité chimique est d'autant plus élevée que le rapport des concentrations est différent de l'unité.

L'hétérogénéité chimique dépend énormément du rapport des diamètres des deux espèces chimiques en présence.

D'une façon générale elle est d'autant plus élevée que le rapport des diamètre est différent de l'unité. Ceci est d'autant plus vrai que le rapport de concentration n'est pas trop grand et que l'écart de composition que l'on s'est fixé n'est lui-même pas trop élevé.

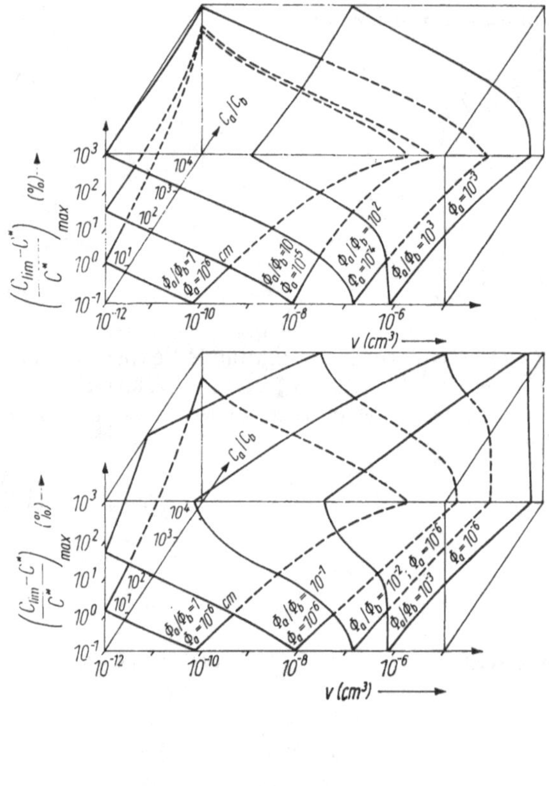

Fig. 6. Variation de l'écart relatif maximum de la composition d'un mélange de deux espèces a et b dans un volume $V = 100$ cm³, en fonction du volume élémentaire v, de la composition C_a/C_b et du rapport Φ_a/Φ_b des diamètres des particules

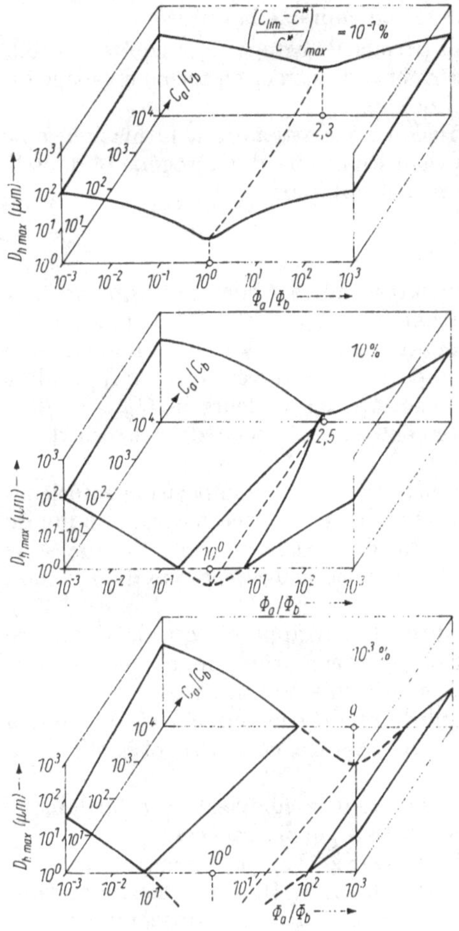

Fig. 7. Variations du critère d'hétérogénéité maximum $D_{h\,max}$ en fonction de la composition, du rapport Φ_a/Φ_b du diamètre des particules a et b et pour trois écarts $\left((C_{lim} - C^*)/C^*\right)_{max}$ fixés à l'avance

Toutefois, l'hétérogénéité chimique minimum n'est pas toujours réalisée quand les diamètres des particules des deux espèces sont égaux. On remarque en effet que pour les rapports de concentration élevés et si l'on accepte de grands écarts de composition le rapport des diamètres conduisant à l'hétérogénéité minimum est voisin de 10. Autrement dit, dans le cas particulier de l'adjonction d'un dopant b de concentration $C_b = 100$ ppm, ce qui correspond sur nos courbes à $C_a/C_b = 10^4$, l'hétérogénéité chimique initiale minimum, avec un écart maximum de composition de 10%, (Fig. 7b) sera réalisée avec des particules de l'espèce b 2,5 fois plus petites que les particules a. Dans ce cas le critère d'hétérogénéité D_h est égal à 1,5 µm.

Il apparaît donc que, contrairement à ce que l'on pouvait penser a priori, les particules de dopant les plus fines possible par rapport à la dimension des particules du composé principal ne conduit pas à l'hétérogénéité minimum. La courbe de la Fig. 7b montre en effet que pour les rapports

$$\Phi_a/\Phi_b = 2{,}5 \quad \text{on a} \quad D_h = 1{,}5\ \mu\text{m}\,,$$

$$\Phi_a/\Phi_b = 10 \quad \text{on a} \quad D_h = 3\ \mu\text{m}\,,$$

$$\Phi_a/\Phi_b = 100 \quad \text{on a} \quad D_h = 14\ \mu\text{m}\,,$$

$$\Phi_a/\Phi_b = 10^3 \quad \text{on a} \quad D_h = 95\ \mu\text{m}\,.$$

Le critère d'hétérogénéité est d'autant plus sensible à la variation du rapport des diamètres Φ_a/Φ_b que l'écart de composition $(C_{lim} - C^*)/C^*$ que l'on s'est fixé est plus élevé.

Autrement dit, plus l'écart acceptable de composition au regard de telle ou telle propriété est élevé plus on pourra admettre des rapports Φ_a/Φ_b grands et ceci quelle que soit la composition.

Cette étude de l'état initial doit permettre aux expérimentateurs d'interpréter au moins qualitativement certains de leurs résultats antérieurs ou leurs projets en fonction de leurs conditions propres concernant leur matériau (diamètres des particules, largeur du domaine monophasé, composition, etc.) ou les moyens d'étude utilisés (microsonde, microanalyse X, analyse ionique, etc.).

3. Evolution macroscopique des mélanges au cours de la diffusion

Après avoir défini le critère d'hétérogénéité qui permet de caractériser un état initial, il parait nécessaire de définir un critère qui permette d'estimer en terme d'espace ou de temps l'évolution probable d'un état initial donné.

La distance macroscopique probable de diffusion pour atteindre la formation complète du composé recherché, correspond à la valeur du critère d'hétérogénéité dans le cas où lui est associé un écart de composition relatif nul. Soit

$$\frac{C_{\text{lim}} - C^*}{C^*} = 0 .$$

Dans le cas simple où $\Phi_a = \Phi_b$, $C_a/C_b = 1$ toutes choses égales par rapport aux conditions précédentes, on a calculé les variations de $(C_{\text{lim}} - C^*)/C^*$ en fonction de v et de Φ (Fig. 8). Il apparait que pour un diamètre donné de particules, il existe une valeur limite de v quand l'écart relatif de composition tend vers 0, ce qui conduit pratiquement à une constance du critère d'hétérogénéité pour les valeurs de $(C_{\text{lim}} - C^*)/C^*$ suffisamment faibles. Ce point apparaît clairement sur le réseau de courbes de la Fig. 9 obtenu à partir des courbes de la Fig. 8.

Sur la Fig. 9, la courbe correspondant aux écarts relatifs de composition inférieurs ou égaux à 10^{-3} est donc la courbe représentative de la distance macroscopique de diffusion en fonction du diamètre des particules. Ces courbes montrent que si le diamètre des particules est multiplié par 100, la distance macroscopique de diffusion est multiplié par 10.

Si l'on raisonne maintenant en temps de diffusion, il apparait que le temps de diffusion nécessaire pour atteindre la formation d'un comprimé donné sera 10 fois plus court quand les particules auront un diamètre 10 fois plus petit.

L'étude de la formation d'un composé de taille réelle peut donc être ramenée à celle d'une zone élémentaire, caractéristique du mélange considéré et égale à la distance macroscopique probable de diffusion.

Nous avons tenté d'étudier, dans une telle zone, dont la composition rappelons-le, est égale à la composition moyenne, l'évolution statistique des espèces.

A partir de distributions identiques à celle de la Fig. 1a, par exemple, dans un mélange binaire, on va considérer que deux particules A et B, qui se touchent vont former au bout d'un certain temps l'équivalent de deux particules du composé final désiré; la distance maximum de diffusion dans ce cas étant égale à deux fois le diamètre des particules. Remarquons que si le mélange était parfaitement homogène et pour une composition telle que $C_a/C_b = 1$, cette très faible distance de diffusion serait suffisante pour former intégralement le composé défini. En fait, il n'en est rien, et seul un certain pourcentage de composé défini sera formé. Pour que le système évolue, il faut imaginer une distance possible de diffusion égale à 3 fois, 4 fois, n fois, le diamètre des particules et ainsi de suite jusqu' à ce que les deux dernières particules A et B aient perdu leur identité chimique. La Fig. 10 représente la quantité Q de

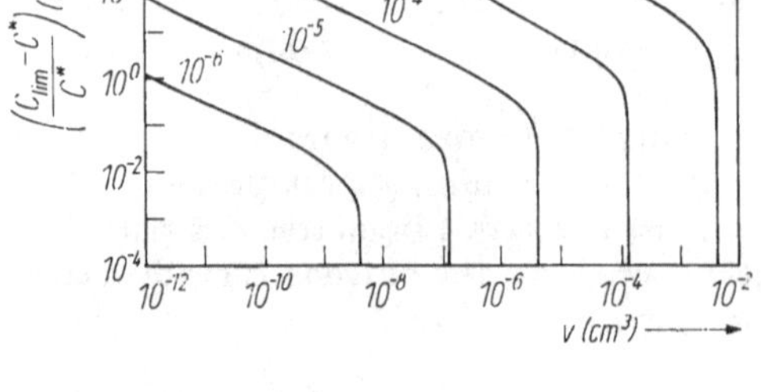

Fig. 8. Dans le cas simple où les concentrations C_a et C_b sont égales ($C_a/C_b = 1$) et où les diamètres Φ_a et Φ_b sont égaux ($\Phi_a = \Phi_b = \Phi$), ce courbes montrent que pour un diamètre de particule donné, le diamètre élémentaire v tend vers une valeur limite, quand l'écart relatif de composition tend vers zéro

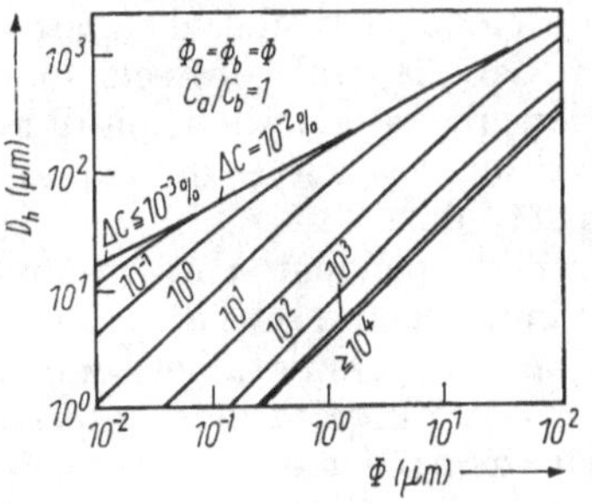

Fig. 9

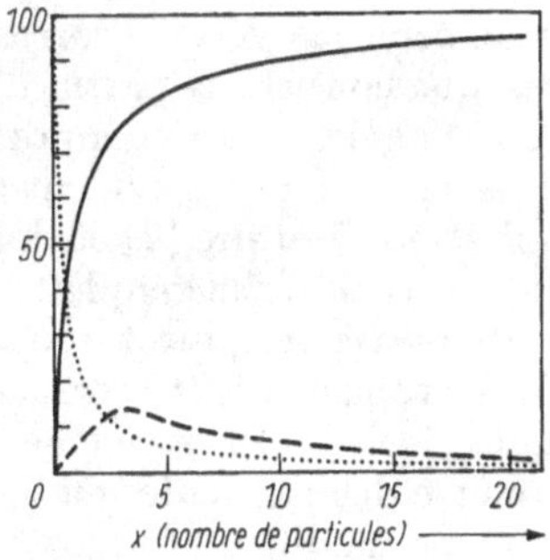

Fig. 10

Fig. 9. Variations du critère d'hétérogénéité de *composition* en fonction du diamètre des particules et de l'écart relatif de composition. La courbe correspondant aux écarts relatifs de composition égaux ou inférieurs à 10^{-3} % correspond à la distance macroscopique de la diffusion la plus probable. $C_a/C_b = 1$, $\Phi_a = \Phi_b = \Phi$

Fig. 10. Dans le cas d'un modèle plan de 2500 sites dans lequel sont distribuées aléatoirement trois espèces différentes en quantités égales on estime par diffusion simulée la formation du composé ternaire Q. ——— composé ternaire, – – – – composés binaires, · · · · · · composés initiaux

composé formé en fonction de la distance de diffusion comptée en nombre de particules dans le cas d'un modèle plan de 2500 sites et pour un système à trois composants. Il apparaît clairement que si la vitesse de formation est très élevée au tout début du traitement, elle devient vite très faible.

Par ailleurs, nous avons fait l'hypothèse de la formation possible de trois composés binaires qui dans les cas réels pourront être partiellement impossibles. Quoi qu'il en soit, l'existence, même partielle, de composés transitoires modifiera les conditions de diffusion du système et par là même, l'allure des courbes de la Fig. 10 dans le sens d'une formation plus lente. Mais ces éventuelles modifications ne changeront pas fondamentalement le principe de l'évolution du système. Par contre, ce qu'il est essentiel de montrer est que ces phases transitoires pourront ne pas être identifiées par analyse. Considérons en effet sur la Fig. 10 l'état du système pour une distance de diffusion $x = 20$ particules. Une analyse globale aux rayons X par exemple ne permettra pas de révéler les 3% de phases transitoires qui sont de surcroit de natures chimiques différentes. Une analyse en microscopie électronique à balayage en émission X ne révèlera pas davantage leur présence à cause de leur petite dimension.

Des analyses beaucoup plus fines mais plus ponctuelles (microsonde, analyse ionique) risquent de conduire aux mêmes résultats négatifs dans la mesure où la probabilité que la zone analysée intègre une phase transitoire peut être très faible.

Seules des études systématiques en microdiffraction électronique ou à la microsonde [5, 6] peuvent permettre de mettre ces phases en évidence et d'en étudier la nature, pour autant, dans le cas de la microdiffraction électronique que ces phases transitoires puissent être visualisées. En tout état de cause, bien qu'une analyse globale conduise à penser que le matériau est intégralement formé, il y aura lieu de suspecter la présence de phases transitoires qui peuvent jouer un rôle extrêmement important sur les propriétés intrinsèques et extrinsèques du matériau.

4. Discussion et conclusions

Nous avons vu que cette étude des mélanges s'appuie sur l'hypothèse d'une distribution aléatoire des espèces chimiques en présence. Cette hypothèse qui s'appuie elle même sur deux séries d'observations expérimentales [1, 5] peut ne pas être ou être incomplètement vérifiée expérimentalement.

Les cas où elle ne sera pas vérifiée correspondent à un mode particulier d'élaboration des mélanges qui consiste à partir d'une solution dans laquelle des sels sont dissous et que l'on congèle. Si au cours de la congélation, il y a cristallisation d'un composé mixte, il se crée un ordre évident dans la répartition des espèces qui élimine la notion de distribution aléatoire [7] et donc ce qui en découle.

Les cas où elle ne sera qu'incomplètement vérifiée correspondent à ceux dans lesquels des amas de particules plutôt que les particules individuelles sont distribuées aléatoirement. Nous avons considéré dans ce cas que les amas se comportainet comme de grosses particules isolées. De la même façon nous avons ignoré l'existence d'un écart-type du diamètre des particules ou des groupements de particules. Mais à notre sens, ces limitations ne modifient pas les conclusions auxquelles nous aboutissons et ne peuvent que renforcer l'hétérogénéité chimique.

Ces réserves étant faites, deux points essentiels ressortent de cette approche théorique:

1. On peut prévoir la microstructure chimique d'un mélange avant et au cours du frittage quand on connaît sa composition moyenne et les diamètres de ses particules. Il apparaît en particulier que l'on est en mesure:

d'établir la distribution la plus probable d'une espèce chimique donnée dans un échantillon macroscopique donné,

après avoir défini un critère d'hétérogénéité, de trouver sa valeur maximum la plus probable en fonction de la composition et du diamètre des particules. On montre à cet égard que pour une composition donnée il existe un rapport idéal entre les diamètres des particules, dans le cas de deux espèces, qui conduit à une hétérogénéité chimique minimum.

Ce point est extrêmement important car il permet non seulement de définir les conditions optimum de formation du matériau mais également, en réduisant le temps de diffusion à sa valeur minimum, de limiter les phénomènes parasites tels que l'effet Kirkendall, les croissances discontinues etc.

Le second point important qui ressort de cette étude, concerne la façon d'aborder le matériau élaboré par frittage.

2. Il apparaît en effet que l'étude de l'évolution macroscopique d'un mélange apporte des informations très importantes sur la microstructure chimique du matériau et que l'étude de son influence sur les propriétés est vraisemblablement déterminante. Nous pensons en effet que pour maîtriser la double relation

$$\text{eléments de base} \leftrightarrow \text{microstructures}\,,$$
$$\text{microstructures} \leftrightarrow \text{propriétés}\,.$$

Il faut associer, aux études microstructurales à l'échelle microscopique, des études macroscopiques susceptibles d'indiquer la fréquence et l'évolution des configurations microstructurales.

Cela revient donc en fait à proposer une approche que l'on peut appeler: aspect macroscopique des microstructures.

Bibliographie

[1] C. Lacour, phys. stat. sol. (b) **96**, 785 (1979).
[2] C. Domd et M. S. Green, Phase Transitions and Critical Phenomena, Academic Press, London/New York 1972.
[3] P. M. Grangeon, Thèse, Paris 1981.
[4] C. Lacour, Science of Sintering **12**, 39 (1980).
[5] C. Lacour et M. Paulus, phys. stat. sol. (a) **27**, 441 (1975).
[6] C. Lacour et M. Paulus, phys. stat. sol. (a) **28**, 71 (1975).
[7] P. M. Grangeon et C. Lacour, Ann. Chim., à paraître.

(Received January 12, 1983)

phys. stat. sol. (a) 77, 323 (1983)
Subject classification: 13.3; 14.3.2; 22.2.1

Institute of Physics (a) and Institute of Electroengineering (b), Electrophysical Research Centre, Slovak Academy of Sciences, Bratislava[1])

GaAs–Anodic Oxide Interface Examination by Deep-Level Transient Current Spectroscopy

By
I. THURZO (a), K. GMUCOVÁ (a), J. ČERVENÁK (b), and T. LALINSKÝ (b)

The deep-level transient current spectroscopy (DLTCS) is applied to study electronic phenomena at GaAs–anodic oxide interfaces. It is found that large-signal excitation pulses lead to a transient response, which analysis may lead to erroneous results as a consequence of the nonlinearity of the system. DLTCS operation with reduced excitation pulses seems to be favourable, since it provides current transient spectra obeying simple theoretical formulae. The data obtained points to a deep acceptor level in the forbidden gap of GaAs, the level being located (0.60 ± 0.03) eV below the conduction band edge. The deep centre is tentatively related to a missing anion in GaAs and it is assumed that the defect is created in the course of anodization.

DLTC-Spektroskopie wird zur Untersuchung der elektronischen Phänomene an der anodischen GaAs–Oxid-Grenzfläche eingesetzt. Es wird gefunden, daß Großsignal-Anregungsimpulse zu einem Übergangsverhalten führen, dessen Analyse fehlerhafte Ergebnisse als Folge der Nichtlinearität des Systems ergeben kann. DLTCS mit verringerten Anregungsimpulsen scheint günstiger zu sein, da sie Stromspektren liefert, die einfache theoretische Formeln befolgen. Die erhaltenen Werte weisen auf ein tiefes Akzeptorniveau in der Bandlücke von GaAs hin, das (0,60 ± 0,03) eV unterhalb der Leitungsbandkante lokalisiert ist. Das tiefe Zentrum wird vorläufig mit einem fehlenden Anion in GaAse verknüpft, wobei der Defekt während der Anodisierung erzeugt wird.

1. Introduction

Despite considerable effort a high-quality gate oxide on GaAs for MIS FET processing has not been found as yet [1]. In connection with GaAs-based MIS diodes it is now definitely established that the main problem resides in strong trapping of charges at the GaAs–oxide interface. The related *C–U* curves show a pronounced hysteresis and in some cases accumulation cannot be achieved. For these reasons even such a reliable method as the conductance technique failed in yielding reliable parameters of the GaAs–oxide interface traps. To characterize the GaAs–oxide interface electrically, some authors suggested to use the phototransient method [2]. Though the latter method enables one to estimate the surface potential for a given bias voltage, the analysis still remains essentially qualitative. Another possibility is to study the interface traps by the deep-level transient spectroscopy (DLTS) introduced by Lang [3]. There are a few attempts to apply DLTS to the GaAs–anodic oxide interfaces [4, 5], however, the results exhibit lack of common features.

Our present work is aimed at the application of the current transient mode of DLTS (DLTCS) to MIS diodes prepared by anodic oxidation of GaAs to obtain complementary data on capture and release of charge carriers in the GaAs–oxide interface

[1]) Dúbravská cesta 9, 842 28 Bratislava, Czechoslovakia.

region. The current-based DLTS was suggested first by Wessels [6] as a tool for investigation of semiconductor junctions and lateron applied to silicon MIS devices [7]. The present results will be confronted with the unified model for Schottky barrier and $A^{III}B^{V}$–insulator interface states proposed by Spicer et al. [8 to 10].

2. Experimental Method

Electrical pulses which reduce the quiescent reverse bias (gate potential) from U_a to U_b are periodically applied to the MIS diode and at a time t_0 after the trailing edge of the excitation pulse the current $j(t_0)$ is sampled by a sample-and-hold circuit. This step is performed repetitively during a thermal scan and in this way the desired DLTCS spectrum is obtained. The current due to emission of charge carriers from a discrete energy level in the gap can be expressed as (n-substrate) [11]

$$j(t_0, T) = \frac{C_0}{C_0 + C_w} \frac{q w_D N_{TT}}{2} e_n \exp(-e_n t_0) , \tag{1}$$

where C_0 and C_w denote the capacitance of the insulator and that of the depletion region, respectively, w_D is the width of the depletion region, N_{TT} the concentration of the traps, and e_n the emission rate for electrons,

$$e_n = \sigma_n v N_c \exp\left[-\frac{E_c - E_t}{kT}\right].$$

Here it is assumed that both N_{TT} and the concentration of shallow donors are spatially uniform and that deep traps within the whole depletion region were excited by a saturating excitation pulse. Should only a part of the depletion region be excited by the filling pulse, then the corresponding Δw_D is to be used in (1) instead of w_D, as will be discussed later. The function $j(t_0, T)$ exhibits a maximum at the temperature T_m, where $e_n(T_m)^{-1} = \tau(T_m) = t_0$. For the case of a temperature-independent capture cross section σ_n one should construct a $\ln [T_m^2 \tau(T_m)]$ versus $1/T$ plot to obtain $\Delta E = E_c - E_t$ as a slope of the Arrhenius plot.

GaAs MOS diodes were prepared by anodization of heavily doped n-GaAs wafers ($N_D \approx 10^{18}$ cm^{-3}) with subsequent deposition of Al dots ($\varnothing = 0.5$ mm) by vacuum evaporation through a mask. The experimental equipment for DLTCS was the same as in our previous work [7], duration of the excitation pulses was always set at 20 ms and their repetition frequency at $f = 40$ Hz.

3. Results and Discussion

A representative DLTCS spectrum of the GaAs MOS diodes is shown in Fig. 1. In the upper part of the figure there is a response to the excitation pulse, which raised the gate potential to $U_b = 0$, the trailing edge of the pulse being depicted schematically by an arrow. The response is composed of essentially two components:

a) transient current due to emission of charges from interface traps,

b) charging current of the geometrical capacitance of the device, which current is slightly increasing with temperature (broken line).

It is obvious that for the latter component the superposition principle is valid, i.e. the charging and discharging currents are equal in magnitude and opposite in direction, see Fig. 1. The emission transient current seems to be composed of two peaks with maxima at T_{m1} and T_{m2} corresponding to 300 and 360 K, respectively. When keeping U_b at zero and varying the reverse quiescent bias voltage U_a, the set of DLTCS spectra, replotted in Fig. 2, was registered. The evolution of the DLTCS

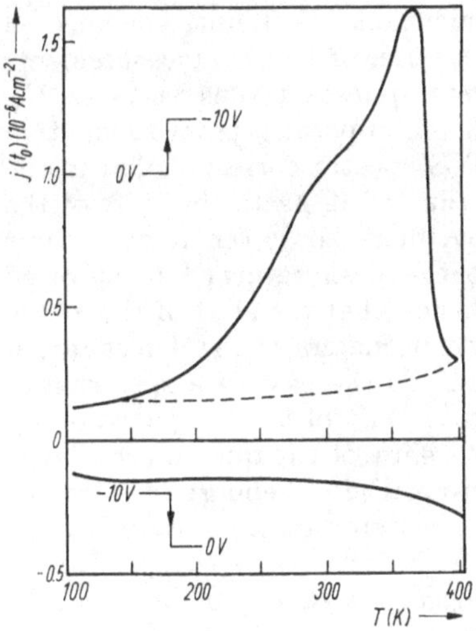

Fig. 1. A representative large-signal deep-level transient current spectrum of a GaAs MOS diode. Charging current of the geometrical capacitance of the device is represented by the dashed line. $t_0 = 5$ ms

spectrum with bias voltage is striking in two aspects. First, the 360 K peak moves to higher temperatures for more negative U_a, second, the 300 K peak disappears at less negative U_a. Following Fig. 2, one would tend to ascribe the 360 K peak to interface states [5] and the 300 K peak to a bulk defect. The latter statement would result from the fact that for more negative gate potentials the depletion region extends more deeply into the bulk of the semiconductor. As will be shown further, the large-signal DLTCS could be misleading when dealing with systems like the GaAs–anodic oxide interfaces. Moreover, we do not know how to start any analysis of the curves from Fig. 2, since neither the shape nor the location of the peaks on the temperature axis are fixed.

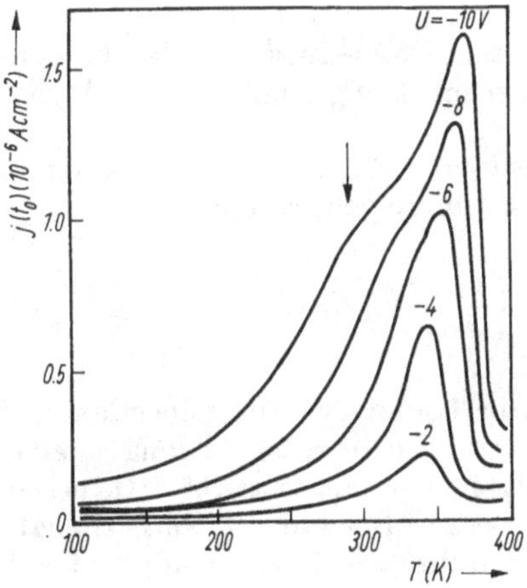

Fig. 2

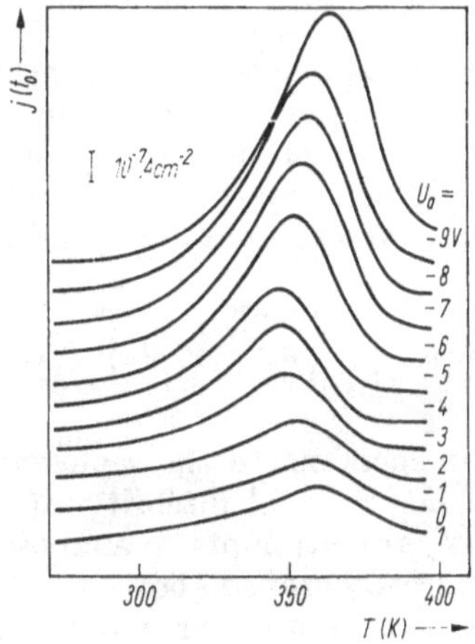

Fig. 3

Fig. 2. Evolution of the large-signal DLTCS spectrum of a GaAs MOS diode with reverse bias voltage. $t_0 = 5$ ms, $U_b = 0$ V

Fig. 3. Small-signal current transient spectra for various bias voltages — space resolved DLTCS. $t_0 = 2$ ms, $\Delta U = -1$ V

To overcome at least to some extent the above-mentioned difficulties connected with the large-signal DLTCS analysis, we performed a series of DLTCS measurements with constant $\Delta U = U_a - U_b = -1$ V and different quiescent bias voltages U_a. The results of such an experiment are shown in Fig. 3. An important point to mention is that the 300 K peak is not contained in the DLTCS spectra corresponding to the "small-signal" case. The spectra are dominated by the 360 K peak, the shift of the latter being evident again. Interestingly, its movement to lower temperatures ceased at $U_a \approx -4$ V and for U_a approaching zero the peak returned to elevated temperatures again. So we have arrived at the conclusion that the shift of the 360 K peak arises from spatial fluctuations of the trapping parameters of the defect rather than from a distribution of the activation energy ΔE. In the case of a spectrum of interface state energies one would expect a *monotonous* shift of the temperature T_m with bias voltage as a consequence of the monotonous shift of the quasi-Fermi level. From this point of view the data from Fig. 3 point to a discrete energy level at the GaAs–anodic oxide interface. A similar conclusion can be found in the work of Kamieniecki et al. [5].

Let us note that the increase of the quantity $j(t_0, T_m)$ with U_a in Fig. 3 can be an indication that the transient phenomena are connected with the excitation of a part of the depletion region in the semiconductor. Since for a given t_0 the function $e_n \exp(-e_n t_0)$ achieves at T_m the same value even for an arbitrary e_n function, the observed increase in the magnitude of the current at T_m may be associated either with Δw_D and/or $N_{TT}(x)$, see (1). Δw_D can be expressed as

$$\Delta w_D = \left[\frac{2\varepsilon_s(U_a' + \Delta U')}{qN_D}\right]^{1/2} - \left[\frac{2\varepsilon_s U_a'}{qN_D}\right]^{1/2} \approx \left[\frac{2\varepsilon_s U_a'}{qN_D}\right]^{1/2} \frac{\Delta U'}{2U_a'^{1/2}},$$

assuming $\Delta U'/U_a' \ll 1$ (U_a' stands for the surface potential). So the revealed dependence of $j(t_0, T_m)$ on U_a is due mainly to the dependence of Δw_D upon $\Delta U'/U_a'^{1/2}$.

If we return now once more to Fig. 2, the large-signal transient distortion, manifested mainly by the apparent peak at nearly 300 K, could be caused by two circumstances:

1) a distortion due to the nonlinearity of the system GaAs–anodic oxide, the non-exponential transient in the time domain experiment being analogous to higher harmonics generation in the frequency domain;

2) spatial fluctuations of the trapping parameters involved in $e_n(x)$ would invalidate equation (1), which equation should be replaced by a more general one

$$j(t_0, T) = \frac{C_0}{C_0 + C_w} q \int_0^{w_D} N_{TT} e_n \exp(-e_n t_0) \left(\frac{w_D - x}{w_D}\right) dx \,, \tag{2}$$

$x = 0$ corresponding to the semiconductor–insulator boundary. Any dependence of e_n on the electric field intensity $E(x)$ is already contained in $e_n(x)$. As long as $e_n(x)$ is unknown, any attempts to analyse the present large-signal case are of little sense.

As was already argued above, the 360 K DLTCS peak isolated in the "small-signal" regime seems to result from emission of charge carriers from a discrete trapping level. To test this hypothesis, we performed DLTCS measurements with t_0 set at different times, the measured curves are replotted at the top of Fig. 4 as solid lines. Let us note that the charging current component was subtracted from the total DLTCS spectrum in each case. The related Arrhenius plot resulted in a straight line, the slope of the line corresponding to an activation energy $\Delta E \approx 0.58$ eV. From the condition $e_n(T_m)^{-1} = t_0$ we obtained an estimate of the capture cross-section $\sigma_n \approx 10^{-16}$ cm^2. ΔE and σ_n

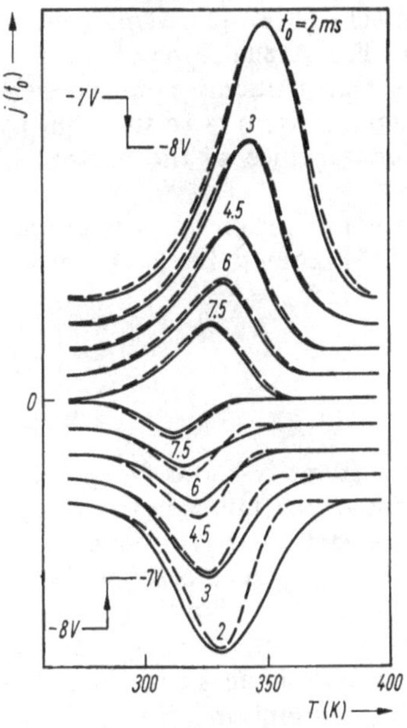

Fig. 4. Emission DLTCS characteristics (top) and refilling DLTCS characteristics (bottom). ——— experiment, - - - - theory

being known, we applied (1) to calculate the emission characteristics. The time dependence of Δw_D was not taken into account. Results of the calculation are presented at the top of Fig. 4 by broken lines. Despite the fact that the calculated curves are slightly broader than the measured ones, the simple theory fits the observed dependence of the magnitude of the current at T_m on t_0 correctly. This finding is an evidence that in the "small-signal" case the emission current decay is *exponential*, as predicted for a discrete energy level. The values of trapping parameters were improved by fitting the experiment according to (1), taking the parameters from the Arrhenius plot as starting values for the computation (Fig. 5).

As complementary data to the emission characteristics, current transient spectra measured during the excitation pulse are recorded at the bottom of Fig. 4. When reducing the reverse bias, refilling of the traps with electrons should take place. It is commonly assumed that the capture rate c_n for electrons from the conduction band is given by the simple relation $c_n = \sigma_n v n_0$, n_0 being the concentration of free carriers. Some authors suggested to determine the capture cross-sections from the trap refilling data [12 to 14]. Normally $c_n \gg e_n$, in our case with $v = 2 \times 10^4\, T^{1/2}\, ms^{-1}$, $n_0 = 10^{24}\, m^{-3}$ the capture time constant $\tau_c = c_n^{-1}$ at $T \approx 360$ K is of the order of 10^{-10} s. This implies that the capture of electrons from the conduction band of GaAs cannot be observed directly in our experiment with the rate window set at a few ms. To maintain the interpretation of the 360 K peak and its refilling counterpart in terms of electron transitions in the depletion region of the semiconductor one should take into account the possibility that there is an intermediate high-resistivity GaAs layer between the heavily doped substrate and the oxide. Then the occupancy of the leve within this layer could be influenced also by the generation rate of minority carriers

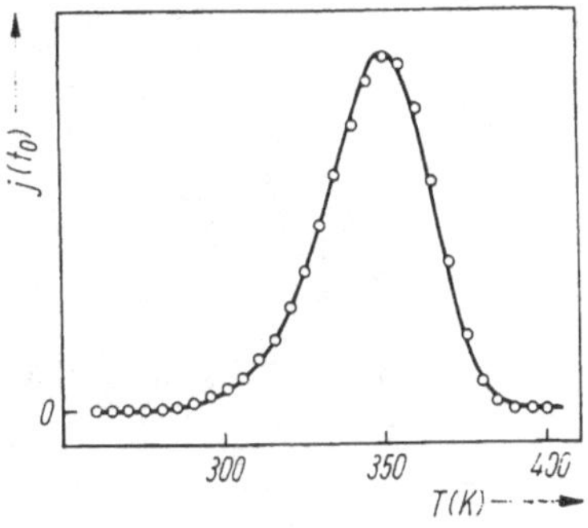

Fig. 5. A comparison between an experimental emission characteristic (solid line) and the best fit (circles). $\Delta E = 0.60$ eV, $\sigma_n = 3.6 \times 10^{-20}\, m^2$, $U_a/U_b = -8\, V/-7\, V$, $t_0 = 2$ ms

this possibility was completely overlooked by Kamieniecki et al. [5]. Constructing an Arrhenius plot for the set of refilling transients from Fig. 4, the slope of the plot changed continuously from 0.45 to 0.85 eV. The refilling transients are non-exponential, as expected for generation–recombination phenomena. Attempts to fit them by a function $\tau^{-1} \exp(-t_0/\tau)$ failed, as is shown by the broken lines at the bottom of Fig. 4 ($\Delta E = 0.67$ eV).

A theory of transient currents due to generation–recombination in a depletion region of a MIS structure was developed by Sah and Fu [15]. They define a characteristic time constant in the form

$$\tau_w = \frac{e_n + e_p}{e_n e_p} \frac{N_D}{N_{TT}}. \tag{3}$$

The rate determining step is the slower one. Now, if we identify tentatively, what we have called emission characteristics, with the generation–recombination transients, there is the problem to decide whether the measured ΔE is confined to the conduction or to the valence band. According to the data on photoemission spectroscopy [8, 9], in n-type GaAs the adsorbed oxygen tends to induce an acceptor-like deep level in the upper part of the band gap. According to Spicer et al. [8 to 10], there is a correlation between the peaks in the density of interface states in MIS devices and the states identified in their fundamental studies. Then under the assumption of $e_p \gg e_n$ the time constant of minority carrier generation $\tau_w \approx e_n^{-1} N_D/N_{TT}$ and the level should be located 0.60 ± 0.03 eV below the conduction band edge. Expression (3) provides an alternative explanation for the shift of the 360 K peak with bias voltage, namely even for spatially uniform emission rates the position of the peak might sensitively reflect spatial fluctuations of both N_D and N_{TT}. At this moment nothing is known about the fate of Te donors at the interface after oxidation. We can only correlate the results of the comparative transient capacitance study performed by Kamieniecki et al. [5], who used n-type substrates with $N_D = 10^{15}$ cm^{-3}, with related results of our previous [11] and present work ($N_D \approx 10^{18}$ cm^{-3}). For a common rate window ($\tau_0 \approx 7.5$ ms), despite the large donor concentration difference and different methods, the discussed DLTS peak was found to be positioned at the same temperature. This finding could speak in favour of the emission mechanism or a space charge relaxation in the oxide layer. However, in the latter case one would expect a discharging transient current even with bias reduced to zero. As is evident from Fig. 1, only the geometrical capacitance will discharge if the bias is reduced to $U_b = 0$. In contrast, the refilling of traps in the depletion region at $U_b = 0$ should not produce external current flow since now $j(t_0, T) \approx U_b\,[\mathrm{d}C/\mathrm{d}t]_{t_0} = 0$ in accordance with experiment [11].

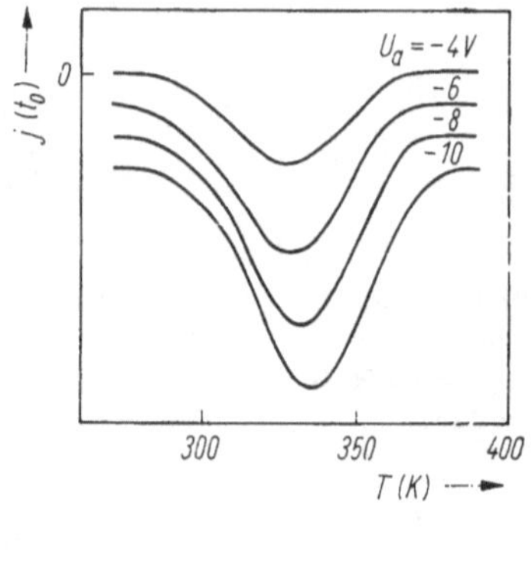

Fig. 6. Small-signal refilling characteristics for various bias voltages. $t_0 = 2$ ms, $\Delta U = 1$ V

The same level is likely responsible for both the emission and refilling DLTCS spectra; it is clear from Fig. 6 that the peak measured on refilling moves with bias voltage in the same manner as the emission peak does (cf. Fig. 3). Moreover, using $\Delta U = 80$ mV ($\approx 3\,kT/q$) we found a complete coincidence of the temperatures of the maxima for both modes of operation. This is in accordance with the simple kinetic equations derived by Miller et al. [16] for the occupancy of a particular level under the assumption of small deviations from equilibrium as well as of validity of the detailed balance. In such situations both the filling and emptying of traps are exponential with a time constant $(c_n + e_p + c_p + e_n)^{-1}$.

Some comments are necessary concerning lacking evidence for the U-shaped continuous distribution of the GaAs–oxide interface states from DLTS, the distribution of which was well established both theoretically and experimentally, see the excellent review by Kreutz [17] and the papers by Spicer et al. [8 to 10]. From this point of view the 360 K peak is expected to be superimposed on a broad DLTS spectrum of fast surface states. Evidently, with the high-resistivity interface layer being present at the interface, one cannot populate the surface states with electrons by means of a saturating pulse and the surface states do not respond to the excitation. There is an independent direct evidence for the existence of such an interface layer with significant deviations from stoichiometry, as was stated by Kreutz [17]. We hope to have demonstrated how to explore the electronic properties of the defects in this layer by DLTS experiments.

Finally, we will discuss the nature of the trapping centre responsible for the appearance of the 360 K peak. There is little doubt that we deal with a defect created during the oxidation, rather than with a deep impurity level. Adopting the unified model [8 to 10], in n-type GaAs the defect should be acceptor-like, e.g. an antisite defect like Ga in an As site. Another possibility would be a point defect identified as an As vacancy or a complex of vacancies. There are a few theoretical predictions [18, 19] that an ideal As vacancy might induce a deep level in the upper half of the band gap of GaAs.

4. Conclusions

The DLTCS spectra of GaAs MOS structures obtained using the "small-signal" regime revealed a single defect level positioned at 0.60 eV from the band edge. Since the position of the corresponding DLTS peak on the temperature scale seems to be independent of the donor concentration in the substrate and the transients behave exponentially with time, it is suggested that we deal with emission rather than generation–recombination phenomena. From the nature of the refilling transients we have concluded that the relaxation phenomena should take place within a high-resistivity interface layer between the conducting substrate and the anodic oxide. It is reasonable to assume oxygen-rich nonstoichiometric GaAs to be present at the interface. Then, it is this high-resistivity region which prevents the MOS devices from accumulation. The correlation of the DLTCS with photoemission spectroscopy and with a few theoretical predictions is suggestive of a defect centre due to the As deficit in the semiconductor at the interface.

References

[1] H. H. Wieder, J. Vacuum Sci. Technol. **17**, 1009 (1980).
[2] F. L. Schuermeyer and H. Hartnagel, J. appl. Phys. **51**, 6279 (1980).
[3] D. V. Lang, J. appl. Phys. **45**, 3023 (1974).
[4] H. Hasegawa and T. Sawada, J. Vacuum Sci. Technol. **16**, 1478 (1979).

[5] E. KAMIENIECKI, T. E. KAZIOR, J. LAGOWSKI, and H. C. GATOS, J. Vacuum Sci. Technol. 17, 1041 (1980).
[6] B. W. WESSELS, J. appl. Phys. 47, 1131 (1976).
[7] I. THURZO and J. JAKUBOVIE, J. Phys. D 14, 1477 (1981).
[8] W. E. SPICER, P. W. CHYE, P. R. SKEATH, C. Y. SU, and I. LINDAU, J. Vacuum Sci. Technol. 16, 1422 (1979).
[9] W. E. SPICER, I. LINDAU, P. R. SKEATH, and C. Y. SU, J. Vacuum Sci. Technol. 17, 1019 (1980).
[10] W. E. SPICER, I. LINDAU, P. R. SKEATH, C. Y. SU, and P. W. CHYE, Phys. Rev. Letters 44 420 (1980).
[11] I. THURZO and T. LALINSKÝ, phys. stat. sol. (a) 71, K83 (1982).
[12] A. ZYLBERSZTEJN, J. appl. Phys., Letters 33, 200 (1978).
[13] J. A. BORSUK and R. M. SWANSON, J. appl. Phys. 52, 6704 (1981).
[14] J. M. NORAS, J. Phys. C 14, 2341 (1981).
[15] C. T. SAH and H. S. FU, phys. stat. sol. (a) 11, 297 (1972).
[16] G. L. MILLER, D. V. LANG, and L. C. KIMERLING, Annu. Rev. Mater. Sci. 7, 377 (1977).
[17] E. W. KREUTZ, phys. stat. sol. (a) 56, 687 (1979).
[18] M. S. DAW and D. L. SMITH, Phys. Rev. B 20, 5150 (1979).
[19] S. DAS SARMA and A. MADHUKAR, Phys. Rev. B 24, 2051 (1981).

(Received February 14, 1983)

phys. stat. sol. (a) 77, 331 (1983)

Subject classification: 14.3; 22.2.1

Institute of Electrical Engineering of the Electro-Physical Research Centre, Slovakian Academy of Sciences, Bratislava[1])

Geometrical Magnetoresistance Effect in Semiconductors with Mixed Conductivity

Determination of the Electron Hall Mobility in Semi-Insulating GaAs

By

J. BETKO and K. MĚŘÍNSKÝ

Low magnetic field geometrical magnetoresistance effect (GMR) in semiconductors with mixed conductivity is analyzed. The relationships between electrical parameters of samples with GMR and Hall configurations are derived. A new method for determination of electron Hall mobility in semi-insulating (SI) III–V materials is proposed using low field GMR measurements. The Hall intrinsic concentration and Fermi level position are calculated. Experimental results on SI-GaAs samples are given.

Der geometrische Magnetowiderstands-Effect (GMR) bei schwachen Magnetfeldern wird in Halbleitern mit gemischter Leitfähigkeit analysiert. Die Beziehungen zwischen den elektrischen Parametern von Proben mit GMR und Hall-Konfiguration wird abgeleitet. Eine neue Methode zur Bestimmung der Elektronen-Hallbeweglichkeit in semiisolierenden (SI) III–V Materialien mit Hilfe von Schwachfeld-GMR-Messungen wird vorgeschlagen. Die Hall-Eigenleitungsträgerdichte und die Lage des Fermi-Niveaus werden berechnet. Experimentelle Ergebnisse von SI-GaAs Proben werden angegeben.

1. Introduction

The geometrical magnetoresistance effect was advantageously utilized for the determination of carrier mobilities in conducting semiconductors [1 to 5]. A method based on the magnetoresistance effect using the influence of sample geometry was proposed by Lakeou et al. [6] for the determination of carrier densities and mobilities also in near-intrinsic semiconductors. The method was demonstrated on Ge and InSb using high magnetic field measurements. The carrier mobilities and concentrations in semi-insulating (SI) GaAs are commonly determined from the resistivity, Hall coefficient, and physical magnetoresistance measurements [7 to 14].

In this paper we analyze the low-field GMR effect in semiconductors with mixed conductivity in order to determine the carrier mobilities and concentrations in SI III–V materials. On the basis of GMR measurements a new method for the determination of electron Hall mobility is proposed which has some advantages with regard to the methods based on Hall measurements. Experimental results on electron Hall mobility, Hall intrinsic concentration, and Fermi level position are obtained with SI GaAs.

2. Set of Equations

The commonly used GMR sample configuration (Corbino disc or a thin rectangular plate with large area contacts) is considered in the following analysis. For the GMR sample the ratio of the electrical resistance R_B in low transverse magnetic field with

[1]) Dúbravská cesta, 842 39 Bratislava, Czechoslovakia.

induction B to the resistance R_0 without magnetic field is given [15] as

$$\left(\frac{R_B}{R_0}\right)_{\text{GMR}} = \frac{\varrho_B}{\varrho_0}\,(1 + \text{tg}^2\,\theta) \tag{1}$$

with

$$\text{tg}\,\theta = \left(\frac{R_{\text{H}B}}{\varrho_B}\right) B\,,$$

where ϱ_B and ϱ_0 are the resistivities with and without magnetic field, resispectively, and $R_{\text{H}B}$ is the Hall coefficient in the magnetic field.

Equation (1) can be rewritten introducing the well-known "physical" quantities S_ϱ, S_H, M_ϱ, and M_H obtained with samples of the conventional Hall configuration (S_ϱ and S_H are the slopes of $1/B^2$ versus $\varrho_0/(\varrho_B - \varrho_0)$ and $1/B^2$ versus $R_{\text{H}0}/(R_{\text{H}0} - R_{\text{H}B})$ plots, respectively, and M_ϱ and M_H are the ordinate intercepts of the same plots). Therefore, using the equations

$$\frac{1}{B^2} = \frac{S_\varrho \varrho_0}{(\varrho_B - \varrho_0)} + M_\varrho\,, \tag{2}$$

$$\frac{1}{B^2} = \frac{S_\text{H} R_{\text{H}0}}{R_{\text{H}0} - R_{\text{H}B}} + M_\text{H} \tag{3}$$

we can obtain from (1) an approximatively linear dependence of the magnetoresistance parameter $\Delta R/R_0B^2$ on B^2,

$$\left(\frac{\Delta R}{R_0 B^2}\right)_{\text{GMR}} = \left(\frac{\Delta R}{R_0 B^2}\right)_{\text{GMR},\,0} - k_{\text{GMR}} B^2\,, \tag{4}$$

where

$$\left(\frac{\Delta R}{R_0 B^2}\right)_{\text{GMR},\,0} = \left(\frac{R_{\text{H}0}}{\varrho_0}\right)^2 + S_\varrho\,, \tag{5}$$

$$k_{\text{GMR}} \approx \left(\frac{R_{\text{H}0}}{\varrho_0}\right)^2 (S_\varrho + 2S_\text{H}) - M_\varrho S_\varrho \tag{6}$$

with

$$\Delta R = R_\text{B} - R_0\,.$$

Equations (5) and (6) were derived using the following two assumptions:

1. $-R_{\text{H}0}/\varrho_0 \gtrapprox \mu_{\text{Hp}}$ (μ_{Hp} denotes the hole Hall mobility), so that the cases with extremely strong dependence of $R_{\text{H}B}$ on B which occurs at $-R_{\text{H}0}/\varrho_0 \to 0$ are eliminated.

2. The magnetic field is low enough to neglect terms of higher order than B^2 with regard to unity.

Substituting expressions (A7) to (A10) for $R_{\text{H}0}/\varrho_0$, S_ϱ, M_ϱ, and S_H (see Appendix) into (5) and (6) we get

$$\left(\frac{\Delta R}{R_0 B^2}\right)_{\text{GMR},\,0} \approx \mu_{\text{Hn}}^2 \frac{b_\text{H} c_\text{H}}{b_\text{H} c_\text{H} + 1}\,(1 + q_{\sigma\text{n}})\,, \tag{7}$$

$$k_{\text{GMR}} \approx \mu_{\text{Hn}}^4 \frac{b_\text{H} c_\text{H}}{(b_\text{H} c_\text{H} + 1)^2}\,[1 + b_\text{H} c_\text{H}\,(q_{\sigma\text{n}} - q_{\sigma\text{n}}^2 + 2q_{\text{Hn}}) + 3q_{\sigma\text{n}} + 2q_{\text{Hn}}]\,, \tag{8}$$

where $b_H = \mu_{Hn}/\mu_{Hp}$, $c_H = n_H/p_H$ (μ_{Hn} and μ_{Hp} are the electron and hole Hall mobilities, and n_H and p_H are the electron and hole Hall concentrations, respectively) and $q_{\sigma n}$ and q_{Hn} are the single-carrier factors defined by (A3) and (A4), respectively. Relations (7) and (8) were derived using additional assumptions $b_H \gg 1$, $q_{\sigma p} \ll b_H q_{\sigma n}$ and $q_{Hp} \ll \ll b_H^2 q_{Hn}$.

From (7) and (8) we can obtain the following single-valued solutions for the electron Hall mobility μ_{Hn} and for parameter $b_H c_H$

$$\mu_{Hn} \approx \left[\left(\frac{\Delta R}{R_0 B^2}\right)_{GMR,0} + \frac{k_{GMR}}{\left(\frac{\Delta R}{R_0 B^2}\right)_{GMR,0}}\right]^{1/2} \left(\frac{1 + q_{\sigma n}}{1 + 3q_{\sigma n} + 2q_{Hn}}\right)^{1/2}, \tag{9}$$

$$b_H c_H \approx \frac{1 + 3q_{\sigma n} + 2q_{Hn}}{(1 + q_{\sigma n})^2\, k_{GMR} \Big/ \left(\frac{\Delta R}{R_0 B^2}\right)^2_{GMR,0} - q_{\sigma n} + q_{\sigma n}^2 - 2q_{Hn}}. \tag{10}$$

Further, from an appropriate relation between μ_{Hn} and b_H [10, 12, 16] we can calculate the value of b_H and, consequently, the Hall intrinsic concentration $n_{Hi} = (n_H p_H)^{1/2}$ and the Fermi level E_F position which are given as

$$n_{Hi} \equiv \frac{n_i}{\sqrt{r_n r_p}} = \frac{1}{e\varrho_0 \mu_{Hn}} \frac{(b_H c_H)^{1/2}}{b_H c_H + 1}\, b_H^{1/2}, \tag{11}$$

$$E_F - E_0 = \frac{k_B T}{2}\left[\ln\left(\frac{b_H c_H N_v'}{N_c^*}\right) - \ln b_H\right] + \frac{k_B T}{2} \ln \frac{r_n}{r_p}, \tag{12}$$

where n_i is the intrinsic concentration, E_0 the middle of the band gap, r_n and r_p the scattering factors for electrons and holes, respectively, N_v' the effective density of valence states and N_c^* the total statistical weight of the conduction band system [17], k_B the Boltzmann constant, T the temperatures, and e the elementary charge.

Here, we introduce three important notes:

1. For undoped SI GaAs or conductive GaAs materials with $c_H \gg 1$ we get from (7)

$$\left(\frac{\left(\frac{\Delta R}{R_0 B^2}\right)_{GMR,0}}{\mu_{Hn}^2}\right)_{c \gg 1} \approx 1 + q_{\sigma n}\,. \tag{13}$$

The nominator on the left side corresponds to the square of the commonly used GMR mobility μ_m in this case. Then, from (13) follows

$$1 + q_{\sigma n} \approx \xi^2\,, \tag{14}$$

where $\xi = \mu_m/\mu_{Hn}$ is the commonly used scattering factor relating GMR and Hall mobilities in N-type conductive materials.

Further, relations (A7), (A8) (see Appendix) and (10) yield for $c_H \gg 1$,

$$(S_\varrho)_{cH \gg 1} \approx q_{\sigma n} \left(\frac{R_{H0}}{\varrho_0}\right)^2, \tag{15}$$

$$\left(\frac{\left(\frac{\Delta R}{R_0 B^2}\right)^2_{GMR,0}}{k_{GMR}}\right)_{cH \gg 1} \approx \frac{(1 + q_{\sigma n})^2}{q_{\sigma n} - q_{\sigma n}^2 + 2q_{Hn}}. \tag{16}$$

Relations (14) to (16) enable one to evaluate the factors $q_{\sigma n}$ and q_{Hn} from Hall and GMR measurements in N-type conductive or undoped semi-insulating GaAs. For these materials the factor $q_{\sigma n}$ is identical with the magnetoresistance factor ξ_H defined by Rode [18].

2. The assumption $-R_{H0}/\varrho_0 \gtrapprox \mu_{Hp}$ which yields $c_H \gtrapprox 2/b_H^2$, is fulfilled at $(\Delta R/R_0B^2)^2_{GMR,0}/k_{GMR} \gtrapprox 0.2$ as can be seen from (7) and (8) for $b_H \gg 1$.

3. Symbols $q_{\sigma n}$, q_{Hn}, $q_{\sigma p}$, and q_{Hp} used in this paper have not the same meaning as those used in our earlier works [9, 14]. The following relationships are valid (asterisk denotes the values used in [9, 14]):

$$q_{\sigma n} = \frac{q^*_{\sigma n}}{r_n^2}, \quad q_{Hn} = \frac{q^*_{Hn}}{r_n^2}, \quad q_{\sigma p} = \frac{q^*_{\sigma p}}{r_p^2}, \quad \text{and} \quad q_{Hp} = \frac{q^*_{Hp}}{r_p^2},$$

3. Experimental Results

The experiments have been performed with two Cr-doped SI-GaAs samples (denoted as M 264 and M 641) and one undoped SI-GaAs sample (denoted as MJ 82). For the measurements of "physical" parameters the Hall sample configuration described in [9] was used. With regard to the "large" width of the samples (> 1 mm) the magneto-concentration effect [19] was neglected.

GMR measurements were made on samples with rectangular plate geometry of dimensions $9 \times 1.8 \times 0.35\ mm^3$ (sample M 264) and $5 \times 5 \times 0.35\ mm^3$ (other samples). The length to width ratio smaller than 0.1 was assumed to be sufficient for GMR sample configuration. Large contacts on opposite areas of the plates have been prepared by rubbing in Ga + In at room temperature or by evaporation and annealing of Au + Ge + Ni [20]. These contacts are suitable for measurements with SI GaAs at temperatures $\lessapprox$ 400 K.

The experimental values of $(\Delta R/R_0B^2)_{GMR}$ at 298 K versus B^2 are plotted in Fig. 1. From these plots the zero field values $(\Delta R/R_0B^2)_{GMR,0}$ and the slopes k_{GMR} were read out and shown in Table 1 together with calculated parameters. For experimental data analysis equations (9) to (12) were used. For both single-carrier parameters $q_{\sigma n}$ and q_{Hn} a value of 0.11 deduced from measurements on undoped SI GaAs materials was assumed. The following relation between b_H and μ_{Hn} was employed [16]:

$$b_H = \left(\frac{r_n}{r_p}\right) K + (\mu_{pL}^{-1} - K\mu_{nL}^{-1}) \frac{\mu_{Hn}}{r_p} \tag{17}$$

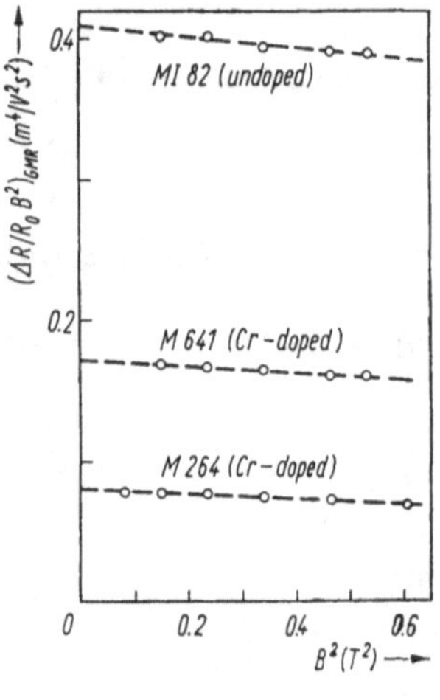

Fig. 1. Dependences of measured parameter $(\Delta R/R_0B^2)_{GMR}$ on B^2 obtained on Cr-doped and undoped Si-GaAs samples at 298 K

Table 1

Experimental values and results obtained on SI GaAs samples with GMR configuration at 298 K

	sample	M 264 Cr-doped	M 641 Cr-doped	MJ 82 undoped
experimental values	ϱ_0 (10^6 Ω m)	7.7	7.8	1.17
	$(\Delta R/R_0 B^2)_{GMR,0}$ ($m^4/V^2 s^2$)	0.080	0.172	0.411
	k_{GMR} ($m^8/V^4 s^4$)	0.018	0.026	0.045
results*)	μ_{Hn} ($m^2/V s$)	0.47	0.48	0.61
	$b_H c_H$	0.49	2.0	150
	b_H	12	13	15
	n_{Hi} (10^{12} m^{-3})	2.9	2.8	2.7
	$E_F - E_0 - 13 \ln (r_n/r_p)$ (meV)	−1	17	71

*) Calculated with $q_{\sigma n} = q_{Hn} = 0.11$, $N'_V = 9.5 \times 10^{24}$ m^{-3}, and $N_c^* = 4.2 \times \times 10^{23}$ m^{-3}.

with $K = 3$, $\mu_{pL} = 0.04$ m^2/Vs, and $\mu_{nL} = 0.80$ m^2/Vs at 298 K (subscript "L" denotes the lattice mobilities). Then

$$b_H = 4 + 18\mu_{Hn} \pm 20\% \tag{18}$$

for $r_n/r_p = 1$ to 1.4 and $r_p = 1$ to 1.2. Relation (18) corresponds (approximately) also to the plots of electron Hall mobility and mobility ratio μ_n/μ_p versus ionized impurity concentration at 300 K due to Walukiewicz et al. [12].

The calculated values of electron Hall mobilities μ_{Hn} are not strongly dependent on single-carrier factors $q_{\sigma n}$ and q_{Hn}. E.g. a variation of 0.01 in $q_{\sigma n}$ or q_{Hn} yields a difference in μ_{Hn} smaller than 1% only. The expected error of b_H in (18) causes an error of ±10% in the determination of n_{Hi} (see (11)) and an error of ± 2.5 meV in the determination of $E_F - E_0 - k_B T/2 \ln (r_n/r_p)$ (see (12)). The values of the Hall intrinsic concentration obtained in Cr-doped samples correspond to an intrinsic concentration value which compares within a factor about $1.5(r_n r_p)^{1/2}$ with the theoretical value n_i deduced from the band gap and effective mass parameters ($n_i \cong 1.9 \times 10^{12}$ m^{-3} at 298 K [17]).

Let us note that in undoped SI GaAs and conductive GaAs (N-type) the ratio $k_{GMR}/(\Delta R/R_0 B^2)^2_{GMR,0}$ is very small and is comparable with $(q_{\sigma n} - q_{\sigma n}^2 + 2q_{Hn})/(1 + q_{\sigma n})^2$

Table 2

Experimental values obtained on a Cr-doped SI GaAs sample (M 264) with Hall and GMR configurations at 400 K and calculated GMR parameters from the Hall measurement data using equations (5) and (6)

	experimental Hall configuration	values GMR configuration	calculated values
$(\Delta R/R_0 B^2)_{GMR,0}$ ($m^4/V^2 s^2$)		0.0377	0.0367
k_{GMR} ($m^8/V^4 s^4$)		0.0047	0.0050 to 0.0065
R_{H0}/ϱ_0 ($m^2/V s$)	−0.0690		
S_ϱ ($m^4/V^2 s^2$)	0.0313		
S_H ($m^4/V^2 s^2$)	0.177		
M_ϱ ($m^4/V^2 s^2$)	−0.10 to −0.15		

(the denominator in (10) is close to zero). The calculated values of $b_H c_H$ and, consequently, n_{Hi} and $E_F - E_0$ are enormously sensitive on factors $q_{\sigma n}$ and q_{Hn} in this case. Good obtained values of the intrinsic concentration may confirm that the chosen values of $q_{\sigma n}$ and q_{Hn} are correct.

For a verification of the basic equations (5) and (6), measurements have been performed on the same sample with Hall and GMR configurations. An agreement between the experimental GMR parameters $((\Delta R/R_0 B^2)_{GMR,0}$ and $k_{GMR})$ and calculated ones from the experimental "physical" parameters $(R_{H0}/\varrho_0, S_\varrho, S_H)$, and $M_\varrho)$ using (5) and (6) is obtained (see Table 2). The confrontation of GMR and Hall measurements was made at higher temperature (400 K) because at room temperature the surface effects complicate the Hall measurements.

4. Conclusion

The analysis of the magnetoresistance of SI-GaAs samples with GMR configuration and its dependence on low magnetic field enables one to determine the electron Hall mobility. The evaluation of single-carrier factors $q_{\sigma n}$ and q_{Hn} is required only, however, these factors do not strongly influence the results. An advantage of the presented method is that the knowledge of the relationship between electron and hole mobilities is not demanded. Further, relative electrical resistance measurements with and without magnetic field are needed only, and the experimental data analysis is simple and yields a single-valued solution. The method can be used universally for both SI and conductive GaAs samples when the assumptions $(\Delta R/R_0 B^2)^2_{GMR,0}/k_{GMR} \gtrapprox 0.2$ and $\mu_{Hn}/\mu_{Hp} \gg 1$ are fulfilled. Due to the lower resistance of GMR samples relative to Hall samples also measurements below room temperature can be done. Measurements on conductive samples or measurements at higher temperatures ($\gtrapprox$ 400 K) require more attention for ohmic contacts preparation. For the calculation of intrinsic concentration and Fermi level position an appropriate relationship between electron and hole mobilities is necessary.

Appendix

In the presence of mixed conduction the Hall coefficient R_{HB} and resistivity ϱ_B may be written [21] as

$$R_{HB} = \frac{R_{HnB}\sigma_{nB}^2 + R_{HpB}\sigma_{pB}^2 + R_{HnB}R_{HpB}\sigma_{nB}^2\sigma_{pB}^2(R_{HnB} + R_{HpB})\, B^2}{(\sigma_{nB} + \sigma_{pB})^2 + \sigma_{nB}^2\sigma_{pB}^2(R_{HnB} + R_{HpB})^2\, B^2}\,, \tag{A1}$$

$$\varrho_B = \frac{\sigma_{nB}(1 + R_{HpB}^2\sigma_{pB}^2 B^2) + \sigma_{pB}(1 + R_{HnB}^2\sigma_{nB}^2 B^2)}{(\sigma_{nB} + \sigma_{pB})^2 + \sigma_{nB}^2\sigma_{pB}^2(R_{HnB} + R_{HpB})^2\, B^2}\,, \tag{A2}$$

where σ_{nB} and σ_{pB} are the single-carrier conductivities and R_{HnB} and R_{HpB} are the single-carrier Hall coefficients for electrons and holes, respectively.

For low magnetic fields we assume a quadratic field dependences of single-carrier quantities,

$$R_{HnB} = \left(-\frac{1}{en_H}\right)(1 - q_{Hn}\mu_{Hn}^2 B^2)\,, \tag{A3}$$

$$\sigma_{nB} = en_H\mu_{Hn}(1 - q_{\sigma n}\mu_{Hn}^2 B^2)\,, \tag{A4}$$

$$R_{HpB} = \left(\frac{1}{ep_H}\right)(1 - q_{Hp}\mu_{Hp}^2 B^2)\,, \tag{A5}$$

$$\sigma_{pB} = ep_H\mu_{Hp}(1 - q_{\sigma p}\mu_{Hp}^2 B^2)\,, \tag{A6}$$

where

$$n_{\rm H} = \frac{n}{r_{\rm n}}, \qquad \mu_{\rm Hn} = r_{\rm n}\mu_{\rm n}, \qquad p_{\rm H} = \frac{p}{r_{\rm p}}, \quad \text{and} \quad \mu_{\rm Hp} = r_{\rm p}\mu_{\rm p}$$

are the Hall concentrations and Hall mobilities of electrons and holes; n, $\mu_{\rm n}$, p, and $\mu_{\rm p}$ are the concentrations and conductivity mobilities of electrons and holes; $r_{\rm n}$ and $r_{\rm p}$ are the scattering Hall factors for electrons and holes; $q_{\sigma{\rm n}}$, $q_{\sigma{\rm p}}$, $q_{\rm Hn}$ and $q_{\rm Hp}$ are factors dependent upon scattering mechanism; and e is the elementary charge.

From (A1) to (A6) with respect to the introduced assumptions we can obtain

$$\frac{R_{\rm H0}}{\varrho_0} = \mu_{\rm Hn} \frac{1 - b_{\rm H}^2 c_{\rm H}}{b_{\rm H}(1 + b_{\rm H} c_{\rm H})}, \tag{A7}$$

$$S_\varrho \approx \mu_{\rm Hn}^2 \left[\frac{c_{\rm H}(1 + b_{\rm H})^2}{b_{\rm H}(1 + b_{\rm H} c_{\rm H})^2} + \frac{b_{\rm H} c_{\rm H} q_{\sigma{\rm n}}}{b_{\rm H} c_{\rm H} + 1}\right], \tag{A8}$$

$$S_{\rm H} \approx \mu_{\rm Hn}^2 \left[\frac{c_{\rm H}(1 + b_{\rm H})^2 (1 - c_{\rm H})}{(b^2 c_{\rm H} - 1)(1 + b_{\rm H} c_{\rm H})^2} + \frac{b_{\rm H}^2 c_{\rm H} q_{\rm Hn}}{b_{\rm H}^2 c_{\rm H} - 1} + \right.$$
$$\left. + \frac{2(b_{\rm H} + 1)\, b_{\rm H} c_{\rm H} q_{\sigma{\rm n}}}{(b_{\rm H}^2 c_{\rm H} - 1)(b_{\rm H} c_{\rm H} + 1)}\right], \tag{A9}$$

$$M_\varrho \approx \mu_{\rm Hn}^2 \left\{\frac{-(1 - c)^2}{(1 + b_{\rm H} c_{\rm H})^2} + \frac{b_{\rm H} c_{\rm H} q_{\sigma{\rm n}}}{b_{\rm H} c_{\rm H} + 1}\left[2 - \frac{b_{\rm H} c_{\rm H} q_{\sigma{\rm n}} \mu_{\rm Hn}^2}{S_\varrho (b_{\rm H} c_{\rm H} + 1)}\right] + \right.$$
$$\left. + \frac{c_{\rm H} \mu_{\rm Hn}^2}{S_\varrho (1 + b_{\rm H} c_{\rm H})^2}\left[(c_{\rm H} - 4 - 2b_{\rm H})\, q_{\sigma{\rm n}} - 2b_{\rm H} q_{\rm Hn} - \frac{2 q_{\rm Hp} c_{\rm H}}{b_{\rm H}^2}\right]\right\}, \tag{A10}$$

where $b_{\rm H} = \mu_{\rm Hn}/\mu_{\rm Hp}$; $c_{\rm H} = n_{\rm H}/p_{\rm H}$; and S_ϱ, $S_{\rm H}$, and M_ϱ are defined by (2) and (3).

References

[1] T. R. Jervis and E. F. Johnson, Solid State Electronics **13**, 181 (1970).
[2] H. Poth, Solid State Electronics **21**, 801 (1978).
[3] J. R. Sites and H. H. Wieder, IEEE Trans. Electron. Devices **27**, 2277 (1980).
[4] P. R. Jay and R. H. Wallis, IEEE Electron. Devices Letters **2**, 265 (1981).
[5] J. Wood, M. J. Howes, and D. V. Morgan, phys. stat. sol. (a) **71**, 207 (1982).
[6] S. Lakeou, S. Cristoloveanu, and A. Chovet, phys. stat. sol. (a) **43**, 213 (1977).
[7] D. C. Look, J. Phys. Chem. Solids **36**, 1311 (1975).
[8] D. C. Look, J. appl. Phys. **48**, 5141 (1977).
[9] J. Betko and K. Měřínský, J. appl. Phys. **50**, 4212 (1979).
[10] D. C. Look, Semi-Insulating III–V Materials, Nottingham 1980, Ed. G. J. Rees, Shiva, Orpington 1980 (p. 183).
[11] K. N. Kang, S. Cristoloveanu and A. Chovet, Semi-Insulating IIII–V Materials, Evian 1982, Ed. S. Makram-Ebeid and B. Tuck, Shiva, Nantwich 1982 (p. 113).
[12] W. Walukiewicz, L. Pawlowicz, J. Lagowski and H. C. Gatos, ibid. (p. 121).
[13] J. J. Winter, H. A. Leupold, R. L. Ross, and A. Ballato, ibid. (p. 134).
[14] Ľ. Hrivnák, J. Betko, and K. Měřínský, ibid. (p. 139).
[15] H. J. Lippmann and F. Kuhrt, Z. Naturf. **13a**, 462 (1958).
[16] Ľ. Hrivnák and J. Betko, phys. stat. sol. (b) **112**, K143 (1982).
[17] J. S. Blakemore, J. appl. Phys. **53**, R123 (1982).
[18] D. L. Rode, phys. stat. sol. (b) **55**, 687 (1973).
[19] S. Cristoloveanu and A. Chovet, J. Phys. C **11**, 3991 (1978).
[20] K. Měřínský, J. Betko, and Ľ. Hrivnák, Elektrotechn. Čas. **32**, 785 (1981).
[21] See for example E. H. Putley, The Hall Effect and Related Phenomena, Chap. 3, Butterworths, London 1960.

(Received February 7, 1983)

phys. stat. sol. (a) 77, 339 (1983)
Subject classificiation: 14.3; 22.8.2

Laboratoire de Recherches sur la Réactivité des Solides, associé au C.N.R.S., Faculté des Sciences Mirande, Dijon[1])

Electrical Conduction Mechanism in Al-, Cr-, and Zn-Substituted Magnetites over the Temperature Range 77 to 1000 K

By
B. GILLOT and F. JEMMALI

From a model of phonon-induced tunneling of electrons from one octahedral site to another, the electrical conductivity of aluminium or chromium slightly substitute and slightly oxidized magnetites $Fe^{3+}[\square_\delta Fe^{2+}_{1-3\delta}Fe^{3+}_{(1-x)+2\delta}M^{3+}_x]O^{2-}_4$ and of zinc–iron ferrites $(Zn^{2+}_x Fe^{2+}_{1-x} Fe^{3+}_2)O^{2-}_4$ is calculated in the cubic phase ($T > T_V$). The theoretical curves $\sigma = f(1/T)$ are in good agreement with the experimental curves, in particular a maximum in the conductivity is obtained above room temperature. In the low temperature monoclinic phases ($T < T_V$), the conductivity is controlled by the diffusion process.

En utilisant un modèle de phonons induits par effet tunnel des électrons d'un site octaédrique à l'autre, la conductivité électrique des magnétites faiblement substituées par l'aluminium ou le chrome et oxydées $Fe^{3+}[\square_\delta Fe^{2+}_{1-3\delta}Fe^{3+}_{(1-x)+2\delta}M^{3+}_x]O^{2-}_4$ ainsi que des magnétites substituées au zinc $(Zn^{2+}_x Fe^{2+}_{1-x} Fe^{3+}_2)O^{2-}_4$ est calculée pour la phase cubique ($T > T_V$) et les courbes théoriques $\sigma = f(1/T)$ s'accordent parfaitement avec les résultats expérimentaux, en particulier elles prévoient un maximum de conductivity pour les températures proches de l'ambiante. Pour la phase monoclinique ($T < T_V$) la conductivité est montrée obéir plutôt à un processus de diffusion.

1. Introduction

In a previous paper [1] it has been shown that the Zener double-exchange model formulated for magnetite [2] and used to describe the electron hopping process may explain satisfactorily the change of conductivity observed near the Verwey transition temperature (T_V) for slightly aluminum or chromium substituted and slightly oxidized magnetites $(Fe^{3+})_A [\square_\delta Fe^{2+}_{1-3\delta}Fe^{3+}_{(1-x)+2\delta}M^{3+}_x]_B O^{2-}_4$ with $0 \leqq x < 0.27$ and $0 \leqq \delta \leqq \leqq 0.040$. For these inverse spinels, when the conduction is largely determined by octahedral sites (B sites), the double-exchange model can account for a sharp decrease of conductivity with x and δ when the temperature drops. These non-stoichiometric spinels exhibit a much diminished conductivity jump at the metal–non-metal transition temperature, with its conductivity above this temperature being lower (cubic phase), and below this temperature being greater (monoclinic phase), than that for stoichiometric magnetite. Moreover, the Verwey transition disappears in zinc-substituted magnetite $(Fe^{3+}_{1-x}Zn^{2+}_x)_A (Fe^{2+}_{1-x}Fe^{3+}_{1+x})_B O^{2-}_4$ $(0 \leqq x \leqq 0.57)$ even for small concentrations of Zn atoms and the electrical conductivity is similar to that of magnetite in the cubic phase [3].

Several models have been proposed to explain the variation of conductivity with temperature above T_V in stoichiometric or non-stoichiometric Fe_3O_4. Haubenreisser [4] has presented a polaron model with the assumption that the ferrimagnetic spin config-

[1]) BP 138, 21004 Dijon Cédex, France.

uration of the Fe ions is the same as that at $T = 0$ K and he has found a maximum in the $\sigma = f(T)$ curves at about 600 K which is caused by the characteristic magnitude of the electron–phonon interaction. On the basis of the Cullen and Callen model [5], the conductivity has also been explain by Lorenz and Ihle [6] by the thermally activated motion of the electrons with a conductivity maximum at 383 K. Srinivasan and Srivastava [7] recently reported that the conductivity in the cubic phase of Zn- and Cu-substituted magnetite arises from phonon-induced tunneling of electrons from one lattice site to another.

In this paper we wish to illustrate how the model of phonon-induced tunneling of electrons can account for the variation of conductivity with temperature from 120 to 700 K in slightly aluminum or chromium substituted and slightly oxidized magnetites and in zinc-substituted magnetites. These finely grained substituted magnetites (crystallite size of about 100 nm) above room temperature undergo a maximum in the conductivity. In the low-temperature monoclinic phase below T_V the conductivity is shown to arise from the diffusion process.

2. Experimental and Samples

The measuring device and the operating conditions have already been reported [8, 9]. The dc conductivity was measured by means of four-probe and two-probe methods on compressed, sintered pellets of about 0.8 cm diameter and 0.2 cm thickness; both techniques gave identical results. For this study the influence of grain size on the electrical conductivity has not been considered but such an influence has been reported in [8]. The two compositions of the selected systems of substituted magnetites have been given in [1, 3].

3. Theoretical Model

As stated earlier, we discuss one conduction mechanism when the electrons which participate in the $Fe^{2+} \leftrightarrows Fe^{3+} + e$ exchange process on octahedral sites are strongly coupled to the lattice and tunnel from one site to the other due to a phonon-induced transfer mechanism which is developed in [7].

Under lattice displacements the Fe^{3+} state on the B site is more stable than the Fe^{2+} state due to the Coulomb interaction and this is illustrated on the simple model of a linear chain of four ions, one Fe^{2+}, one Fe^{3+}, and two oxygen ions, and for maintaining the electrical neutrality of the unit cell, we assign a $-2.5e$ charge to each oxygen ion (Fig. 1). We consider a state Ψ_I in which the Fe^{2+} ions are located at sites $r_{N\pm4n}$, the Fe^{3+} ions at sites $r_{N+2\pm4n}$ and oxygen ions at sites $r_{N+1\pm2n}$ where n can take values from 0 to ∞. In another state Ψ_{II} the sites $r_{N\pm4n}$ are occupied by Fe^{3+} ions and $r_{N+2\pm4n}$ sites by Fe^{2+} ions. Moreover, for a longitudinal optical mode of lattice vibrations and assuming the metal ions to be fixed and the oxygen ions to be moving, Srinivasan and Srivastava [7] calculated the change in lattice energy when the oxygen ions move

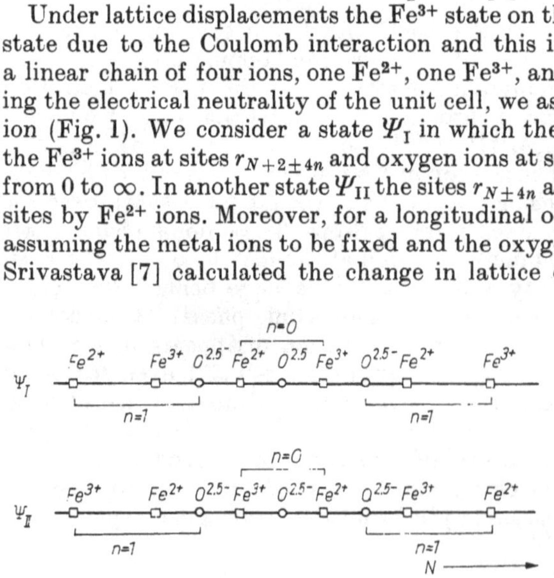

Fig. 1. Scheme of the ground state where the states Ψ_I and Ψ_{II} are degenerate. The Fe^{2+}, Fe^{3+}, and $O^{2.5-}$ ions are located at sites $r^{(Fe^{2+})}_{N\pm4n}$; $r^{(Fe^{3+})}_{N+2\pm4n}$; $r^{(O^{2.5-})}_{N+1\pm2n}$ and $r^{(Fe^{2+})}_{N+2\pm4n}$; $r^{(Fe^{3+})}_{N\pm4n}$, $r^{(O^{2.5-})}_{N+1\pm2n}$; where n can take values from 0 to ∞. This scheme is represented for n, 0, and 1

towards Fe^{3+} compared to the case when they are moving towards Fe^{2+}. Their conclusion is that the change in energy occurs only in the third order in $\delta R/R$ where R is the nearest neighbour distance and this leads to a difference of energy for the displacement of the oxygen ions towards $Fe^{3+}(Fe^{2+})$ of 0.05 eV per cation. Thus, the degeneracy between the Ψ_I and Ψ_{II} states is lifted in the presence of lattice vibrations.

The states Ψ_I and Ψ_{II} are coupled in that the particles which disappear from one appear in the other. Also these states describe a large number of coherent particles, the rate of change of which is exactly the same as for any single particle in them. We may then write the Schrödinger equation of the coupled system in the form

$$\frac{d\Psi_I}{dt} = -\frac{i}{\hbar}(\varepsilon\Psi_I + \gamma\Psi_{II}),$$
$$\frac{d\Psi_{II}}{dt} = -\frac{i}{\hbar}(\varepsilon\Psi_{II} + \gamma\Psi_I), \tag{1}$$

where ε is the energy of a single particle and γ is the coupling coefficient.

We consider the wave function

$$\Psi_I = \sqrt{N_1}\exp(i\Phi_1) \quad \text{and} \quad \Psi_{II} = \sqrt{N_2}\exp(i\Phi_2), \tag{2}$$

where N_i and Φ_i denote the density and the phase, respectively, of the particle in state i (= I, II).

Substituting (2) in (1) and assuming $N_1 = N_2 = N_0$ equal to the density of Fe^{2+} ions, we obtain

$$N_1 = \frac{2}{\hbar}\gamma N_0 \sin(\Phi_2 - \Phi_1),$$
$$N_2 = \frac{2}{\hbar}\gamma N_0 \sin(\Phi_1 - \Phi_2), \tag{3}$$

and $\Phi_2 - \Phi_1 \neq 0$. In the present case $\Phi_1 - \Phi_2 = \pi/2$.

At a finite temperature some of the electrons in the state $\Psi_I(\Psi_{II})$ would be excited to the octahedral vacant sites. As the energy of an electron in Ψ_I is ε, the probability that the Fe^{2+} site in octahedral position is occupied at temperature T in the presence of an external field E is

$$f^{\pm} = \frac{\exp[\beta/2(\varepsilon \pm eEd)]}{\exp[\beta/2(\varepsilon \pm eEd)] + \exp[-\beta/2(\varepsilon \pm eEd)]}, \tag{4}$$

where + (−) sign means that the potential at the Fe^{2+} site is higher (lower) than that of the Fe^{3+} site. In this expression $\beta = 1/kT$ and d is the distance between nearest-neighbour Fe^{2+}–Fe^{3+} ions in the cubic phase.

The current density for $eEd \ll \varepsilon$ is given by

$$J = \frac{2}{\hbar}(f_+ - f_-)\, N_0 ed\gamma \tag{5}$$

and from (4) and (5) we obtain for the dc conductivity

$$\sigma = \frac{J}{E} = \sigma_0 \frac{\beta\varepsilon}{[\exp(\beta\varepsilon/2) + \exp(-\beta\varepsilon/2)]^2} \tag{6}$$

with

$$\sigma_0 = (4N_0\gamma d^2 e^2)/h. \tag{7}$$

In the monoclinic phase, for $T < T_V$ the conductivity is controlled by the diffusion process. The carrier mobility μ is related to the diffusion coefficient D_0 by

$$\mu = \frac{eD_0}{k_B T} = \frac{\sigma}{N_0 e}, \tag{8}$$

where D_0 is given by

$$D_0 = \nu_d d_m^2 \exp(-\beta E_d) . \tag{9}$$

Here E_d is the activation energy, ν_d the vibration frequency of the lattice, and d_m the nearest-neighbour distance. In the monoclinic phase d_m is equal to $d/\sqrt{2}$ where d is the lattice parameter in the cubic phase. From the above equations (8) and (9), the expression for the conductivity is

$$\sigma = \frac{N_0 e^2 \nu_d d_m^2 \exp(-\beta E_d)}{k_B T} . \tag{10}$$

4. Experimental Results

The variation of electrical conductivity over the temperature range 77 to 1000 K for magnetites substituted by trivalent ions $(Fe^{3+})_A (\square_\delta Fe^{2+}_{1-3\delta} Fe^{3+}_{(1-x)+2\delta} M^{3+}_x)_B O^{2-}_4$ was already studied in [1] near the Verwey transition temperature and for zinc-substituted magnetite [3] $(Zn^{2+}_x Fe^{2+}_{1-x} Fe^{3+}_2) O^{2-}_4$ is represented in Fig. 2 to 5 by open circles.

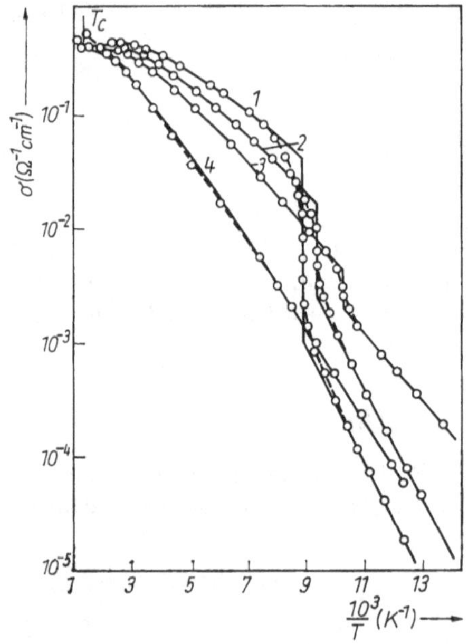

Fig. 2

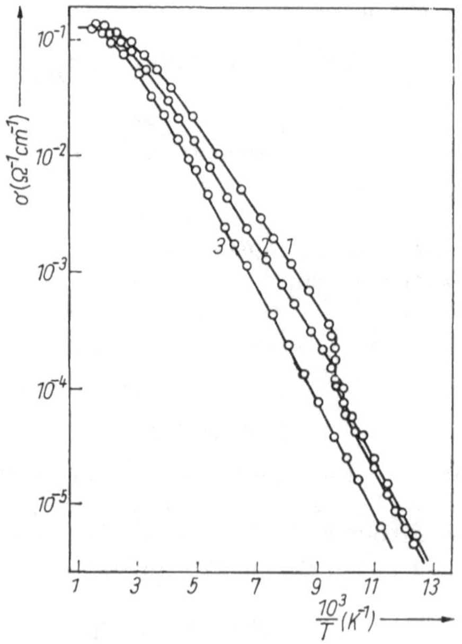

Fig. 3

Fig. 2. Variation of electrical conductivity σ with temperature for polycrystalline samples of pure and slightly oxidized magnetite ($x = 0$) (1) $\delta = 0$, (2) 0.015, (3) 0.025, (4) 0.040. The solid lines represent the theoretical curves

Fig. 3. Variation of electrical conductivity σ with temperature for slightly Cr-substituted and slightly oxidized magnetite (Cr^{3+}, $x = 0.10$). (1) $\delta = 0$, (2) 0.015, (3) 0.025. The solid lines represent the theoretical curves

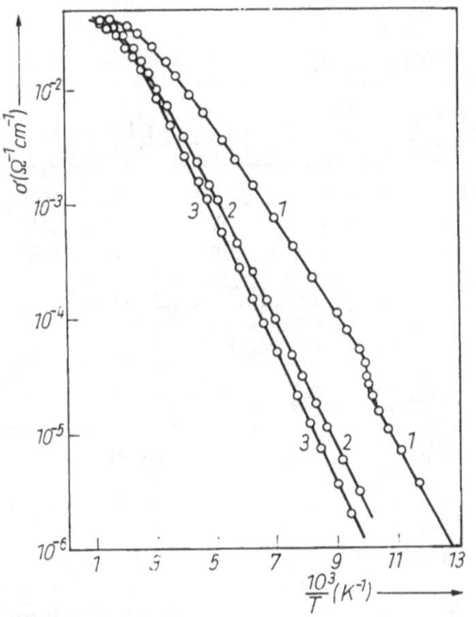

Fig. 4. Variation of electrical conductivity σ with temperature for slightly Al-substituted and slightly oxidized magnetite (Al^{3+}, $x = 0.10$). (1) $\delta = 0$, (2) 0.015, (3) 0.025. The solid lines represent the theoretical curves

The curves of Fig. 2 are relative to pure magnetite ($\delta = 0$, $x = 0$) and slightly oxidized magnetite ($x = 0$, $\delta = 0.015$, $\delta = 0.025$, and $\delta = 0.040$). For compositions $\delta = 0$, $\delta = 0.015$, and $\delta = 0.025$ the spinels exhibit an abrupt change in the electrical conductivity at ≈ 120 K ($x = 0$, $\delta = 0$) and below 120 K ($\delta = 0.015$, $\delta = 0.025$). The conductivity in the cubic phase increases from T_V up to 250 K with a slight maximum at ≈ 300 K. Moreover, for temperatures ≈ 840 K a break of the conductivity–temperature curves (noted T_C in Fig. 2 for magnetite) has been found in the vicinity of the Curie point. The composition $\delta = 0.040$ does not show any phase transition at low temperature nor a maximum in the $\lg \sigma = f(1/T)$ curve.

The curves of Fig. 3 and 4 for slightly Cr- and Al-substituted and slightly oxidized magnetite exhibit also a Verway transition temperature for some compositions. On the other hand, zinc-substituted magnetites (Fig. 5) do not show any phase transition, but exhibit a maximum in the conductivity and a break near the Curie point. It can be seen that as the zinc concentration is decreasing, T_C is moving towards higher temperatures. The values of the transition temperature T_V, of activation energy ε, and E_d above and below T_V as well as the Curie temperature T_C are given in Tables 1 to 3.

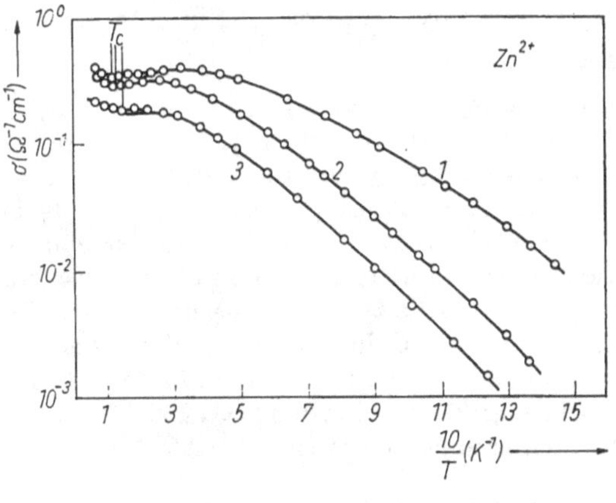

Fig. 5. Variation of electrical conductivity σ with temperature for polycrystalline samples of Zn-substituted magnetite (Zn^{2+}) (1) $x = 0.15$, (2) 0.27, (3) 0.57. The solid lines represent the theoretical curves

Table 1

Physical properties of spinels $(Fe^{3+})_A(Fe^{2+}_{1-3\delta}Fe^{3+}_{2\delta+1}\square_\delta)_BO_4^{2-}$

M^{3+}	x	δ	T_V (K)	ε (eV) $T > T_V$	E_d (eV) $T < T_V$	T_C (K)	σ_0 (Ω^{-1} cm^{-1})	ν_d (Hz)
Fe^{3+}	0	0	119	0.054	0.0109	849	2	7.53×10^{11}
		0.015	109	0.061	0.0105	810	1.91	2.01×10^{12}
		0.025	99	0.069	0.070	735	1.85	7.48×10^{10}
		0.040	—	0.091	—	—	1.76	—

Table 2

Physical properties of spinels $(Fe^{3+})_A(\square_\delta Fe^{2+}_{1-3\delta}Fe^{3+}_{(1-x)+2\delta}M^{3+}_x)_BO_4^{2-}$

M^{3+}	x	δ	T_V (K)	ε (eV) $T > T_V$	E_d (eV) $T < T_V$	σ_0 (Ω^{-1} cm^{-1})	ν_d (Hz)
Cr^{3+}	0.10	0	105	0.090	0.103	0.6	7.98×10^7
		0.015	102	0.098	0.102	0.573	6.56×10^8
		0.025	—	0.110	—	0.555	—
Al^{3+}	0.10	0	99	0.093	0.104	0.185	3.58×10^7
		0.015	—	0.121	—	0.176	
		0.025	—	0.128	—	0.171	

Table 3

Physical properties of spinels $(Fe^{3+}_{1-x}Zn^{2+}_x)_A(Fe^{2+}_{1-x}Fe^{3+}_{1+x})_BO_4^{2-}$

M^{2+}	x	ε (eV)	T_C (K)	σ_0 (Ω^{-1} cm^{-1})
Zn^{2+}	0.15	0.041	800	1.7
	0.27	0.055	753	1.46
	0.57	0.059	671	0.86

5. Comparison with Theoretical Curves

5.1 $T > T_V$

For the compositions previously reported, the values of lg σ as a function of $1/T$ have been calculated from (6) and have been represented by solid lines in Fig. 2 to 5.

5.1.1 *Pure and slightly substituted magnetites*

In Fig. 2, the theoretical curve for pure magnetite is calculated assuming $\sigma_0 = 2\ \Omega^{-1}\ cm^{-1}$. The agreement between theory and experiment is good in the range $140 < T < 850$ K. Above 850 K the results are slightly altered since the lg $\sigma = f(1/T)$ curves are related to magnetic order. The optical phonon frequency ν_L is equal to $\nu_L = \gamma/hE_{def}$ with $\gamma/h = \sigma_0/(4N_0d^2e^2)$ and E_{def} is the lattice deformation energy equal to 1.406×10^{22} J for magnetite. The reduced value of σ_0 in the present case can be attributed to the effects due to the polycrystallinity of the samples [4, 5, 7]. The difference for some samples will be in the density of dislocations and the density of point defects probably mainly due to non-stoichiometry. From the data on the conductivity in a single crystal of magnetite [10] using (6) we obtain $\sigma_0 = 1250\ \Omega^{-1}\ cm^{-1}$; this gives a value of $\nu_L = 3.80 \times 10^{12}$ Hz which is in good agreement with the first vibrational frequency detected in the infrared absorption. Indeed, the value of ν reported for a single crystal of magnetite is 1.80×10^{13} Hz [11, 12].

For the slightly substituted magnetites the agreement between theory and experiment is good (Fig. 2). The values of σ_0 (Table 1) vary linearly with Fe^{2+} concentration N_0 for each composition, N_0 varying with lattice parameter [13 to 15].

5.1.2 Slightly Al- or Cr-substituted and slightly oxidized and Zn-substituted magnetites

For the values of ε and σ_0 reported in Tables 2 and 3, the theoretical values of conductivity were calculated using (6). The agreement with experimental data is satisfactory (Fig. 3 to 5). The different σ_0 are calculated assuming that σ_0 is proportional to the concentration of Fe^{2+} ions which varies with x and δ. For zinc-substituted magnetites this proportionality is also observed by Srinivasan and Srivastava [7].

5.2 $T < T_V$

The theoretical values of σ for compositions for which a Verwey transition temperature is observed have been determined from (10). For pure magnetite the values of the constants taken for the theoretical plot are $N_0 = 1.3512 \times 10^{22}$ cm^{-3}, $d = 0.2968$ nm, $E_d = 0.109$ eV, and $\nu_d' = 7.53 \times 10^{11}$ Hz. The agreement between theory and experiment is good (Fig. 2). For other compositions the agreement with the experimental data is also satisfactory (Fig. 2 to 4). From the lattice parameters determined by X-ray diffraction for different values of x and δ [13 to 15] we can calculated N_0 and d and the values of E_d and ν_d are given in Tables 1 to 3.

Measurement of the Seebeck coefficient, Q, in polycrystalline Fe_3O_4 show that it is negative and constant over the temperature range $T_V < T < 450$ K ($Q = -50$ µV/K). Just near T_V, Q increases rapidly with a reversal of the sign ($Q = +55$ µV/K) and activation energy increase [16 to 18]. For $T < T_V$ the conduction is due to holes as the Seebeck coefficient is positive.

6. Conclusion

For polycrystalline samples $(Fe^{3+})_A [\square_\delta Fe^{2+}_{1-3\delta} Fe^{3+}_{(1-x)+2\delta} M^{3+}_x]_B O^{2-}_4$ with $0 < x < 0.27$, $0 < \delta < 0.040$, and $M^{3+} = Al^{3+}$ or Cr^{3+}, the electrical conductivity is largely influenced by octahedral sites. The mechanism of electron transport based on the tunneling of electrons between Fe^{2+} and Fe^{3+} ions on octahedral sites accounts for a conductivity variation as a function of temperature for $120 < T < 700$ K, in particular through the presence of a maximum in the conductivity at about 300 K. In the temperature range 77 to 120 K the conductivity is shown to arise from the diffusion process. The temperature dependence of electrical conductivity in Zn-substituted magnetites $(Fe^{3+}_{1-x} Zn^{2+}_x)_A (Fe^{2+}_{1-x} Fe^{3+}_{1+x})_B$ $(0 < x < 0.57)$ is similar to that of magnetite in the cubic phase from 77 K to T_C.

References

[1] B. Gillot, J. Phys. Chem. Solids **40**, 261 (1979).
[2] A. Rosencwaig, Canad. J. Phys. **47**, 2309 (1969).
[3] B. Gillot, R. M. Benloucif, and A. Rousset, phys. stat. sol. (a) **65**, 205 (1981).
[4] W. Haubenreisser, phys. stat. sol. **1**, 619 (1961).
[5] J. R. Cullen and E. R. Callen, Phys. Rev. Letters **26**, 236 (1971).
[6] B. Lorenz and D. Ihle, phys. stat. sol. (b) **69**, 451 (1975).
[7] G. Srinivasan and C. M. Srivastava, phys. stat. sol. (b) **103**, 665 (1981).

[8] B. GILLOT, J. F. FERRIOT, and A. ROUSSET, J. Phys. Chem. Solids **37**, 857 (1976).
[9] B. GILLOT, J. Phys. Chem. Solids **38**, 751 (1977).
[10] P. A. MILES, W. B. WESTPHAL, and A. VON HIPPEL, Rev. mod. Phys. **29**, 279 (1957).
[11] R. D. WALDRON, Phys. Rev. **99**, 1727 (1955).
[12] B. GILLOT and F. BOUTON, J. Solid State Chem. **32**, 303 (1980).
[13] B. GILLOT, J. TYRANOWICZ, and A. ROUSSET, Mater. Res. Bull. **10**, 775 (1975).
[14] B. GILLOT, J. F. FERRIOT, G. DUPRE, and A. ROUSSET, Mater. Res. Bull. **11**, 843 (1976).
[15] B. GILLOT, R. M. BENLOUCIF, and A. ROUSSET, Mater. Res. Bull. **16**, 481 (1981).
[16] B. GILLOT, phys. stat. sol. (a) **69**, 719 (1982).
[17] A. J. M. KUIPERS and V. A. M. BRABERS, Phys. Rev. B **14**, 1401 (1976).
[18] C. CONSTANTIN and M. ROSENBERG, Solid State Commun. **9**, 675 (1971).

(Received December 8, 1982)

phys. stat. sol. (a) 77, 347 (1983)

Subject classification: 14.4; 14.3.4; 22.9

Department of Applied Physics, Regional Engineering College, Kurukshetra[1])

Non-Isothermal Depolarization Current Studies in Kapton-H Thermoelectrets

By

R. P. BHARDWAJ[2]), J. K. QUAMARA, K. K. NAGPAUL[2]), and B. L. SHARMA

Thermally stimulated depolarization current characteristics of Kapton-H electrets are studied in the temperature range 303 to 573 K as functions of the polarizing field (133.3 to 500.0 kV cm^{-1}), polarizing temperature (333 to 453 K), polarizing time (2.7×10^3 to 16.2×10^3 s), specimen thickness (2.5×10^6 to 12.5×10^6 nm), heating rate (0.017 to 0.10 K s^{-1}), and storage time (0 to 720 h). The TSDC spectra comprise three maxima in all, namely, β, α, and ρ with their respective locations around 373, 453, and 553 K. These are attributed to dipole-orientation, motion of space charges, and surface charge injection, respectively. A shoulder (β′) emerging around 343 K is suggested to be akin to the main β-peak. The space-charge peak is found to be more sensitive to the forming and storage parameters than to that involving dipoles. The observed dependence of the peak temperature as well as the activation energy for various peaks on polarizing temperature and time is indicative of a continuous distribution of relaxations.

Charakteristiken des thermisch stimulierten Depolarisationsstromes von Kapton-H-Elektreten werden im Temperaturbereich von 303 bis 573 in Abhängigkeit vom Polarisationsfeld (133,3 bis 500,0 kV cm^{-1}), der Polarisationstemperatur (333 bis 453 K), Polarisationszeit ($2{,}7 \times 10^3$ bis $16{,}2 \times \times 10^3$ s), Probendicke ($2{,}5 \times 10^6$ bis $12{,}5 \times 10^6$ nm), Aufheizrate (0,017 bis 0,10 K s^{-1}) und Speicherzeit (0 bis 720 h) untersucht. Die TSDC-Spektren weisen insgesamt drei Maxima auf, β, α und ρ mit ihren entsprechenden Lagen bei etwa 373, 453 und 553 K. Diese werden Dipol-Orientierungen, Wanderung von Raumladungen bzw. Oberflächenladungsinjektion zugeordnet. Von einer Schulter (β′), die bei 343 K auftaucht, wird angenommen, daß sie mit dem Haupt-β-Maximum verknüpft ist. Es wird gefunden, daß das Raumladungsmaximum gegenüber den Bildungs- und Speicherungsparametern empfindlicher ist, als gegenüber denen der beteiligten Dipole. Die beobachtete Abhängigkeit der Maximumtemperatur, sowie der Aktivierungsenergie für verschiedene Maxima von der Polarisationstemperatur und -zeit weist auf eine kontinuierliche Verteilung der Relaxationen hin.

1. Introduction

The aromatic polyimide, poly(4,4′-oxydiphenylene-pyromellitimide), available in thin-film form under the trade name Kapton-H has been traditionally acknowledged as a material of immense practical utility [1] owing to the ability to maintain its excellent physical, chemical, electrical, and mechanical properties over a wide range of temperature and frequency. Its dielectric [2], photoconductive, and photoelectret [3, 4] properties have been the subject of some recent investigations.

In the wake of wide-spread practical applications of polymer electrets a comprehensive knowledge of polarization and depolarization behaviour of Kapton-H films takes on a great deal of significance.

The thermally stimulated discharge current (TSDC) technique [5] has been widely used in order to diagnose and estimate the magnitude of volume electrical polarization.

[1]) Kurukshetra 132119, India.

[2]) Present address: Department of Physics, Kurukshetra University, Kurukshetra 132119, India.

The various maxima appearing in the TSDC spectrum characterize the operative molecular mechanisms through which an electret stores its charge.

Thus, the charge storage and charge decay processes in Kapton can be well understood through the study of thermoelectrets formed by the simultaneous action of heat and electric field to the material, and subsequently monitoring the short-circuit discharge current under a linear thermal program.

In this direction some work has already been reported by Tanaka et al. [6] and the authors themselves [7] wherein the various charging mechanisms operative in Kapton film electrets and the effects of polarizing field, temperature, etc., have been discussed. The present investigation was undertaken in an effort to augment the previous study by extending the ranges of polarizing field and temperature as well as by introducing some additional important parameters like polarizing time, specimen thickness, storage time (electret aging), and depolarization rate (heating rate).

2. Experimental

The material used in the present investigation was procured in the form of thin films of nominal thickness (d) ranging from 2.5×10^6 to 12.5×10^6 nm. The specimens used were 3.0 cm^2 in area and each metallized on both sides by vacuum evaporation of aluminium. The sample was fitted in a specimen holder and was kept in a specially prepared temperature-controlled furnace, shielded against the stray pick-ups and capable of being heated at some arbitrary uniform rate from room temperature to nearly 600 K.

The thermoelectrets were prepared by subjecting the specimen to the desired dc field (E_p) at a certain constant temperature (T_p) for a certain time (t_p) that was supposed to be larger than the relaxation time of the charge carriers. The specimen was then cooled to room temperature within a fixed time (1.8×10^3 s) in the presence of E_p. In order to eliminate the frictional stray surface charges accumulated during the polarization, the electrodes were short-circuited for about 300 s. In order to obtain the thermal variation of depolarization current, the system was short-circuited through an electrometer (Anadigi model E5203) and was heated at a desired linear rate (β) with the help of a resistively heated furnace in conjunction with a precision temperature controller and programmer (Indotherm 457). Further, for the study of storage time (electret aging) effects, the specimen after being polarized was stored in a desiccator containing dry ambient air under non-shorted condition for the predetermined time period (t_s). Throughout the measurements the values of d, t_s, and β were 5.0×10^6 nm, 0 s, and 0.033 K s^{-1}, respectively, unless mentioned otherwise. The activation energies (U) for the different relaxation modes have been calculated using the initial-rise method [8].

3. Results

The TSDC spectra of Kapton-H films in the temperature range 303 to 573 K indicate the presence of three well-resolved prominent maxima in all, situated around 373, 453, and 553 K, identified as β, α, and ρ-peak, respectively. In some cases, notably those corresponding to low values of E_p and T_p, an additional shoulder designated as β' has also been observed around 343 K with an activation energy of approximately 0.1 eV.

The number of peaks actually appearing, their locations, heights, and sharpness are found to be governed by the forming, storage, and depolarization parameters as well as the thickness of the polymeric specimen.

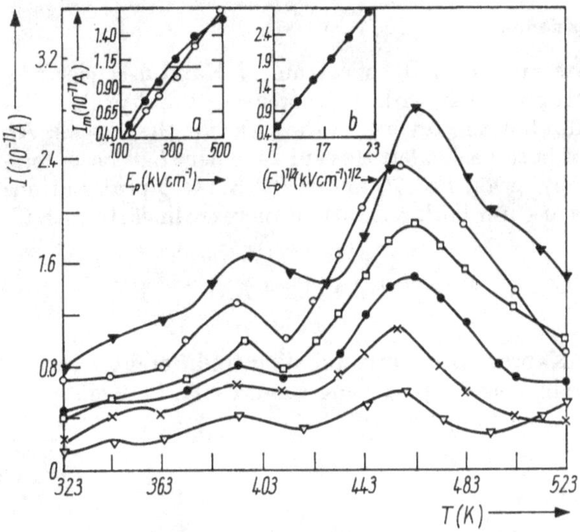

Fig. 1. The TSDC spectra of Kapton-H specimens, polarized with different fields ($T_p = 363$ K, $t_p = 5.4 \times 10^3$ s): ▼ 500.0, ○ 400.0, □ 320.0, ● 250.0, × 200.0, ▽ 133.3 kV cm^{-1}. Inset a) shows peak current as a function of polarizing field. ○ β-peak, ● α-peak. Inset b) shows peak current of α-peak (●) as a function of the square root of the polarizing field

3.1 Field (E_p) dependence

Fig. 1 displays the TSDC characteristics of Kapton samples polarized under identical conditions but different fields ranging from 133.3 to 500 kV cm^{-1}. In each case, two peaks (β and α) are observed with their peak temperatures (T_m) almost independent of E_p, but the peak current (I_m) increasing with the increase in E_p. This increment in case of the α-peak is more pronounced than that in β-peak. I_m for β- and α-peaks shows a linear relationship with E_p and $(E_p)^{1/2}$, respectively (inset of Fig. 1). The values of U and the pre-exponential factor (τ_0) associated with different relaxations alter with E_p, however, the variations are not systematic (Table 1).

Table 1

The activation energy and pre-exponential factor data for the β- and α-relaxation of Kapton-H thermoelectrets

E_p (kV cm^{-1})	T_p (K)	t_p (10^3 s)	β-relaxation		α-relaxation	
			U (eV)	τ_0 (s)	U (eV)	τ_0 (s)
133.3	363	5.4	0.16	22.26	0.34	0.29
200.0	363	5.4	0.18	10.6	0.33	0.38
250.0	363	5.4	0.19	7.74	0.35	0.43
320.0	363	5.4	0.15	32.9	0.34	0.33
400.0	363	5.4	0.24	1.4	0.38	0.09
500.0	363	5.4	0.2	6.04	0.39	0.08
133.3	333	5.4	0.14	33.1	0.32	0.64
133.3	363	5.4	0.16	32.26	0.34	0.29
133.3	393	5.4	0.18	13.6	0.36	0.13
133.3	423	5.4	0.2	6.6	0.38	0.06
133.3	453	5.4	—	—	0.39	0.04
133.3	363	2.7	0.14	39.14	0.33	0.31
133.3	363	5.4	0.16	22.26	0.34	0.29
133.3	363	10.8	0.17	17.6	0.35	0.28
133.3	363	16.2	0.19	9.88	0.38	0.14

3.2 Temperature (T_p) dependence

The effect of polarizing temperature on the TSDC spectrum of Kapton-H electrets has been exemplified in Fig. 2. It is seen that with T_p increasing the β-peak shifts towards the higher-temperature side, but the reverse is the case for the α-peak. An increase in T_p enhances the height of both peaks, but this enhancement has been relatively more pronounced in case of the α-peak. For T_p above 423 K the β-peak flattens and eventually vanishes. The values of U for both relaxations increase linearly with T_p (inset of Fig. 2).

3.3 Time (t_p) dependence

The decay current thermograms of Kapton films corresponding to different values of t_p have been depicted in Fig. 3. The increase in polarizing time results in heightening of the peaks and slight elevation in peak temperatures.

3.4 Thickness (d) dependence

The TSDC spectra have been recorded for Kapton specimens of thickness varying from 2.5×10^6 to 12.5×10^6 nm, polarized and depolarized under identical conditions. The β-peak shows a weak dependence on specimen thickness while the α-peak becomes slightly higher for thinner specimens (figure not shown here). Plotting the total residual polarization involved in the discharge process (i.e. the total charge defined by the area under the TSDC curve) versus specimen thickness gives a nearly straight line (Fig. 4).

3.5 Effect of heating rate (β)

Fig. 5 a, b illustrate the effect of heating rate employed for depolarization on the TSC spectra of specimens polarized identically. For all relaxations both peak intensity and peak temperature increase with increasing β, but maintaining the initial slope almost unaltered.

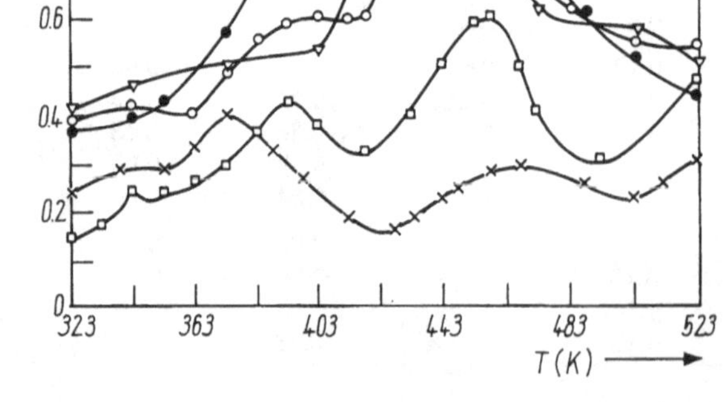

Fig. 2. The TSDC spectra of Kapton-H specimens, polarized at different temperatures ($E_p = 133.3$ kV cm^{-1}, $t_p = 5.4 \times 10^3$ s): × 333, □ 363, ● 393, ○ 423, ▽ 453 K. Inset: The activation energy as a function of polarizing temperature. ○ β-peak, ● α-peak

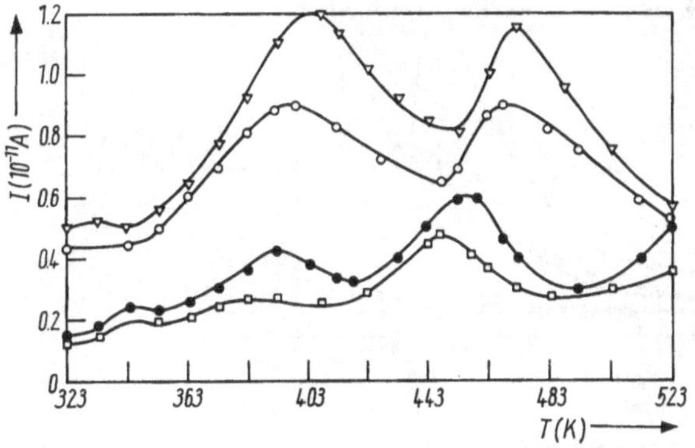

Fig. 3

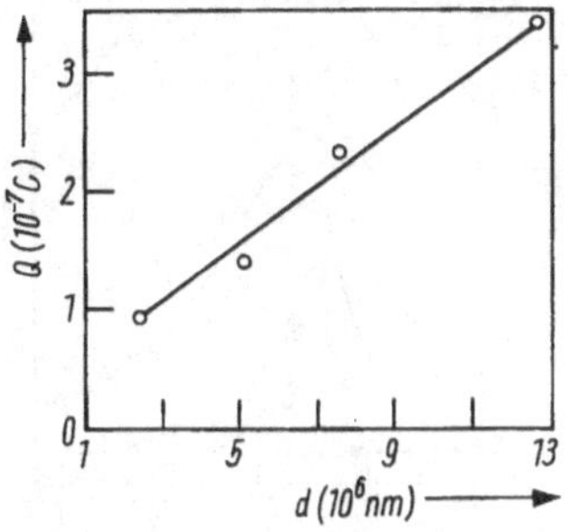

Fig. 4

Fig. 3. The TSDC spectra of Kapton-H specimens, polarized for different times (E_p = = 133.3 kV cm^{-1}, T_p = 363 K): ∇ t_p = 16.2 × 10^3, ○ 10.8 × 10^3, ● 5.4 × 10^{-3}, □ 2.7 × 10^{-3} s

Fig. 4. The total charge released to the external circuit as a function of the thickness of Kapton–H specimens (E_p = 150 kV cm^{-1}, T_p = 353 K, t_p = 5.4 × 10^3 s)

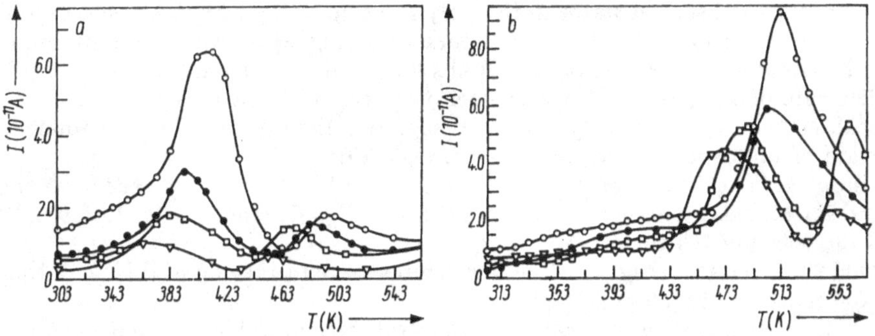

Fig. 5. The TSDC spectra a) comprising β- and α-peaks of Kapton-H specimens (E_p = 280 kV cm^{-1}, T_p = 353 K, t_p = 5.4 × 10^3 s) and b) comprising α- and ρ-peaks of Kapton-H specimens (E_p = = 400 kV cm^{-1}, T_p = 473 K, t_p = 5.4 × 10^3 s), depolarized at different heating rates: ○ β = = 0.100, ● 0.075, □ 0.050, ∇ 0.017 K s^{-1}

3.6 Effect of storage time (t_s)

The influence of storage time (electret aging) on the depolarization behaviour of Kapton film electrets formed and discharged under similar conditions have been depicted in Fig. 6 a, b. An increase in t_s has been observed to suppress the whole TSDC envelope without any noticeable, regular shift in the peak positions. It is interesting to note that the α- and ρ-peaks diminish as a whole in contrast to the β-peak which is affected mainly in the low-temperature region. Further, this obliteration of the residual polarization with time under the α-peak has been stronger than that under the β-peak. In case of long storage time (about 720 h), the discharge currents exhibit a direction reversal in the high-temperature region (above 450 K).

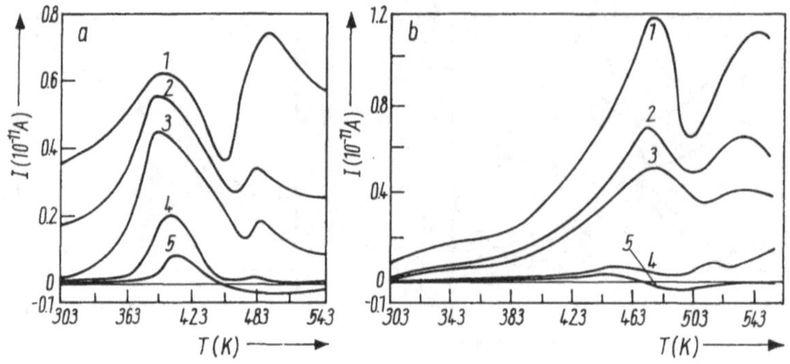

Fig. 6. The TSDC spectra a) comparising β- and α-peaks of Kapton-H electrets ($E_p = 120$ kV cm^{-1}, $T_p = 353$ K, $t_p = 5.4 \times 10^3$ s) and b) comprising α- and ρ-peaks of Kapton-H electrets ($E_p = 300$ kV cm^{-1}, $T_p = 473$ K, $t_p = 5.4 \times 10^3$ s) stored for different periods of time: (1) 0, (2) 24, (3) 120, (4) 360, (5) 720 h

4. Discussion

The first maximum, i.e. the β-peak has been assigned to electric-dipole orientation polarization as a result of the cooperative molecular motion of the residual reactive groups ($>$C=O) and absorbed water in the polymer film [6, 7]. This peak is expected to vanish for sufficiently high forming temperatures (Fig. 2) at which the specimen might lose its absorbed water and the residual side groups also get almost reacted [6]. The dipole origin of the peak has further been verified by the facts that

(i) its peak current varies linearly with the polarizing field (inset of Fig. 1) which is an indication of uniform induced bulk polarization [9];

(ii) in comparison to the other peaks, its magnitude has been less sensitive to the variation in forming and storage parameters [5], viz. E_p, T_p, and t_s (Fig. 1, 2, 6 a, b) as well as the specimen thickness [10];

(iii) during the electret storage it is the low-temperature tail region of this peak that gets affected most [5] (Fig. 6a);

(iv) the activation energy corresponding to this relaxation (Table 1) is in conformity with the theoretically predicted value of $U(\approx 0.2$ eV) for the side-group orientation in a polymer [11].

The next maximum of the thermogram, viz. the α-peak, on which the effects of E_p, t_p, t_s, and d have been relatively more pronounced (Fig. 1, 2, 6 a, b), might be ascribed to space charge polarization. The corresponding apparent activation energy values (Table 1) also support its space charge origin [12].

The unsaturated carbonyl side groups ($>$C=O) present in Kapton can act to a significant extent as traps for charge carriers [3, 13] contributing to their transport and trapping. If the charge is in the form of ions, their transport in the polymer takes place through hopping from one trap site to another [14] while the electrons or holes get immobile in the local traps [10].

The fact that its peak current (I_m) varies non-linearly with E_p but linearly with $(E_p)^{1/2}$ (insets of Fig. 1) reveals that one or more, intrinsic or extrinsic, space charge mechanisms should be jointly operative [13] resulting in an overall non-uniform polarization. These results are in good accord with those reported earlier [6, 7]. Its space charge nature is supported by the electret-aging studies also where this peak is observed to diminish almost as a whole [5] after long storage time (Fig. 6 a, b).

The high-temperature maximum, the so-called ρ-peak, has been found to emerge only for high values of polarizing field and temperature which are known as favourable factors for charge injection into the polymer [15] from the electrodes directly and/or through interfacial layers, i.e. Townsend breakdown [9]. However, in the present case the probability of the latter mechanism could be ruled out because of the use of vacuum-deposited electrodes. The value of activation energy for the same relaxation process (≈ 1.45 eV) is comparable to the energy generally associated with the migration of ions [16]. Thus, the charge injection phenomenon might be held responsible for this peak as the charge carriers injected into the polymer are ionic in nature [16]. The appearance of a prominent peak around 522 K with activation energy 1.15 to 1.4 eV in the TSDC spectra of corona charged Kapton samples [17] provides another sample evidence in favour of the same.

The dependence of the peak temperatures on the forming temperature and time (Fig. 2, 3) is an indication of the existence of a distributed polarization [5, 18]. In case of a dielectric relaxation process the temperature of the maximum is [19]

$$T_m = \left[\beta U \frac{\tau_{T_m}}{k}\right]^{1/2},$$

where τ_{T_m} is the relaxation time at temperature T and k the Boltzmann constant. It follows from the above relation that in case of a single relaxation time (single energy of activation) and a fixed rate of heating, T_m should be fixed, i.e. independent of the polarization temperature and time. At low values of T_p (or t_p) only the fast sub-polarizations could be operative but as T_p (or t_p) is raised, more sub-polarizations (with longer relaxation times) would be activated [5, 20] thereby shifting the maxima towards higher temperatures and enhancing the peak magnitudes. The shift of the peaks with varying heating rates employed during depolarization also affirms the involvement of a distribution in relaxation frequencies [21]. The fact that the values of activation energy for the corresponding decay processes increase with the increase in T_p and t_p (Table 1), and their pre-exponential factors (τ_0) are also not constant (Tables 1) implies that the β- and α-relaxations in the present case are due to a continuous distribution of relaxation times in which both the activation energy and the pre-exponential factor are distributed.

Hence, it should be reasonable to consider the β'-shoulder as a satellite of the main dipole peak (β-peak) and to ascribe it exclusively to the extremely fast reacting dipoles which are expected to be very few in number. Its pertinence to the dipolar origin seems to be justified because its location has been very close to that of the β-peak observed by Tanaka et al. [6] and the corresponding activation energy is comparable with that previously calculated for the β-peak by the authors [7].

No doubt, the approximately linear variation of the total polarization with specimen thickness [Fig. 4] signifies an overall dominance of uniform volume polarization. But at the same time the apparent slight deviation from linearity is significant, as being indicative of the probable presence of some excess charges [22].

Elevation in peak heights and rise in peak temperatures have been the ultimate consequences of the increasing heating rate (Fig. 5 a, b) because the polymer responds less promptly to the higher heating rates and the initial polarization has to be released in a shorter time. Moreover, the afore-said peak shift might be due also to the distributed relaxations in the specimen [21].

The storage truncates the β-peak on the low-temperature side since, during the electret aging these are mostly the faster reacting dipoles that reorient and so discharge first [5].

During the electret storage, an apparent faster erosion in residual polarization under the α-peak than under the β-peak is plausible because

(i) the number of dipoles stored in a polymer electret is generally about 3000 times more than the number of space charges (cf. [5], p. 242), consequently, the loss of a few space charges would be more noticeable than the loss of an almost equal number of dipoles;

(ii) at room temperature the dipoles possess only a nominal mobility whilst, a finite ohmic conduction always exists (cf. [5], pp. 242 and 29) thereby, resulting in the neutralization of space charges.

The reversal of discharge current direction at high temperatures as a result of prolonged electret aging might be attributed to its internal field [23] or the return voltage (cf. [5], p. 242) [10] that builds up within a polymer electret, stored under non-shorted condition.

Acknowledgements

This work was supported by a grant from the Ministry of Education, Government of India. One of the authors (R.P.B.) is thankful to the University Grants Commission, India, for awarding him a Teacher-Fellowship.

References

[1] Technical Information on Kapton Polyimide Film, Du Pont, Bulletin H-1B.
[2] E. SACHER, IEEE Trans. Electr. Insulators **14**, 85 (1979).
[3] J. K. QUAMARA, P. K. C. PILLAI, and B. L. SHARMA, Acta Polymerica **33**, 501 (1982).
[4] J. K. QUAMARA, R. P. BHARDWAJ, P. K. C. PILLAI, and B. L. SHARMA, Acta Polymerica, **34**, No. 6 (1983).
[5] J. VAN TURNHOUT, Thermally Stimulated Discharge of Polymer Electrets, Elsevier Publ. Co., Amsterdam 1975.
[6] T. TANAKA, S. HIRABAYASHI, and K. SHIBAYAMA, J. appl. Phys. **49**, 784 (1978).
[7] J. K. QUAMARA, P. K. C. PILLAI, and B. L. SHARMA, Acta Polymerica **33**, 205 (1982).
[8] C. F. J. GARLICK and A. F. GIBSON, Proc. Phys. Soc. **60**, 574 (1948).
[9] B. GROSS, J. chem. Phys. **17**, 866 (1949).
[10] J. VAN TURNHOUT, Electrets, Chap. 3, Ed. G. M. SESSLER, Springer-Verlag, New York 1980.
[11] W. REDDISH, Trans. Faraday Soc. **46**, 459 (1950).
[12] P. C. MAHENDRU, K. JAIN, V. K. CHOPRA, and PRAVEEN MAHENDRU, J. Phys. D **8**, 305 (1975).
[13] D. K. DAVIES and P. J. LOCK, Electrets – Charge Storage and Transport in Dielectrics, Ed. M. M. PERLMAN, Electrochemical Society, Princeton (N.J.) 1973.
[14] H. SCHER and E. W. MONTROLL, Phys. Rev. B **12**, 2455 (1975).
[15] J. VANDERSCHUEREN and J. GASIOT, Thermally Stimulated Relaxation in Solids, Chap. 4, Ed. P. BRÄUNLICH, Springer-Verlag, Berlin/Heidelberg/New York 1979.
[16] H. GEIGER and K. SCHEEL, Hdb. Phys., Vol. 23, Springer-Verlag, Berlin 1943.
[17] J. K. QUAMARA, P. K. C. PILLAI, and B. L. SHARMA, Acta Polymerica, **34**, 265 (1983).
[18] C. BUCCI, R. FIESCHI, and G. GUIDI, Phys. Rev. **148**, 816 (1966).
[19] T. HINO, K. SUZUKI, and K. YAMASHITA, Japan. J. appl. Phys. **12**, 651 (1973).
[20] H. SOLUNOV and T. VASSILEV, J. Polymer Sci. Polymer. Phys. **12**, 1273 (1974).
[21] J. L. GIL-ZAMBRANO and C. JUHASZ, J. Phys. D **14**, 1661 (1981).
[22] J. VANDERSCHUEREN and J. GASIOT, see [15] (p. 209).
[23] V. M. FRIDKIN and I. S. ZHELUDEV, Photoelectrets and the Electrophotographic Process, Van Nostrand Co., Princeton/New York 1961 (p. 8).

(Received January 27, 1983)

phys. stat. sol. (a) 77, 355 (1983)

Subject classification: 2; 1.5; 21; 21.1

California Institute of Technology Pasadena[1])

Study of Ni–Nb System by Ion Mixing

By

K. T-Y. KUNG, B. X. LIU[2]), and M-A. NICOLET

Amorphous thin films are formed by 300 keV Xe^+ ion mixing of multiple-layered samples in the Ni–Nb (f.c.c.–b.c.c.) system at four different compositions ($Ni_{35}Nb_{65}$, $Ni_{55}Nb_{45}$, $Ni_{65}Nb_{35}$, and $Ni_{80}Nb_{20}$). This confirms the "structural difference rule for amorphous alloy formation" by ion mixing. Some properties of the Ni–Nb amorphous thin films, including their recrystallization temperatures and irradiation stability, are studied. The amorphous alloy formation in this system, as well as the said properties, are compared to those of a similar system, the Ni–Mo (f.c.c.–b.c.c.) system.

Des couches minces amorphes ont été formées par mixage ionique de couches multiples Ni–Nb (f.c.c.–b.c.c.) en utilisant un faisceau Xe^+ à 300 keV. Des alliages Ni–Nb de quatre compositions différentes ($Ni_{35}Nb_{65}$, $Ni_{55}Nb_{45}$, $Ni_{65}Nb_{35}$ et $Ni_{80}Nb_{20}$) ont été étudiés. Les résultats confirment la règle de différence, structurelle pour la formation d'alliages amorphes par mixage ionique. On a étudié quelques propriétés des couches minces amorphes Ni–Nb tel que leur température de recristallisation et leur stabilité sous irradiation. L'article compare leur processus de formation ainsi que les propriétés mentionnées avec les caractéristiques du système similaire Ni–Mo (f.c.c.–b.c.c.).

1. Introduction

Amorphous materials are expected to have novel properties in a number of aspects, and various techniques have been developed to produce amorphous materials in the last two decades; e.g. piston and anvil (liquid quenching), laser zapping (liquid quenching), sputtering and vapor quenching. It is now known that ion mixing (IM) is an alternate method for producing metastable materials. A "structural difference rule for amorphous alloy formation" by IM has been proposed recently, which basically states that an amorphous binary alloy will be formed by IM of a multilayered sample when the two constituent metals are of different crystalline structures [1]. According to this rule, one is able by ion mixing to produce a great variety of amorphous alloys. Furthermore, since the ion irradiation parameters (ion species, energy, temperature, dose) are easy to control, IM offers a good opportunity to study the amorphization processes under differing conditions, in order to achieve an improved understanding of the amorphization mechanism.

In this study, the Ni–Nb system was selected as a representative system of the f.c.c.–b.c.c. combination to investigate the metastable phase formation by IM of multilayered samples. Since Nb ($Z = 41$) is a nearest neighbor to Mo ($Z = 42$) in the periodic table, Nb and Mo atoms have almost equal masses. There should thus be some similarities in IM between the Ni–Nb and the Ni–Mo case. The latter has been studied earlier [1]. It is therefore of interest to compare the IM results obtained for the Ni–Nb system with those observed for the Ni–Mo system.

---

[1]) Pasadena, California 91125, USA.

[2]) Permanent address: Qinghua University, Beijing, The People's Republic of China.

2. Experimental

Ni–Nb multiple-layered thin-film samples were prepared by electron-gun evaporation of high purity Ni (99.9% pure) and Nb (99.99% pure) in an oil-free evaporation system onto about 3000 Å thick SiO_2 layers grown on an organically cleaned n-type (P-doped) Si wafer of ⟨111⟩ orientation. The relative total thicknesses of the Ni and the Nb films were chosen to fix the four desired alloy compositions ($Ni_{35}Nb_{65}$, $Ni_{55}Nb_{45}$, $Ni_{65}Nb_{35}$ and $Ni_{80}Nb_{20}$). The total thicknesses of the Ni–Nb samples, in the range 570 to 640 Å, were designed to equal the $R + \delta R$ (projected range + projected range straggling) of the 300 keV Xe^+ ion employed for the irradiation [2].

The samples were then irradiated with 300 keV Xe^+ ions at room temperature to doses ranging from 3×10^{14} to 1.7×10^{16} Xe^+/cm^2. X-ray diffraction (Read camera) measurement was employed to determine the structures of the mixed layers. Backscattering spectrometry (BS) was used to check that uniform mixing had been achieved and to measure the compositions of the mixed layers. Electrical sheet resistivity (four-point probe) measurements were performed to compare the resistivities of the mixed layers and those of the as-deposited samples.

The amorphous alloys obtained by IM were then post-annealed in a vacuum furnace ($p < 1 \times 10^{-6}$ Torr) to determine their recrystallization temperatures (defined as the temperatures at which the amorphous layers transform into crystalline ones after 1/2 h thermal annealing). The samples were consecutively annealed in 1/2 h periods. The annealing temperature for each period goes from 400 to 900 °C in 50 °C intervals. X-ray diffraction and four-point probe sheet resistivity measurements were performed after each 1/2 h annealing period to detect when the crystallization had occurred.

3. Results and Discussion

3.1 Formation of Ni–Nb amorphous alloys

For all sample configurations, 3×10^{15} Xe^+/cm^2 is the lowest dose required to induce uniform mixing as determined by BS. At that dose, all samples are amorphous, as deduced from halos in the X-ray diffraction patterns. It was confirmed by BS that the samples retained approximately their initial compositions after IM. As shown in Table 1, the sheet resistivities of the ion mixed films are five times those of the as-deposited films. This gives added evidence that the ion mixed films are of amorphous structure. The formation of these amorphous phases confirms that in f.c.c.–b.c.c. binary metallic systems, amorphous phases can be formed by ion mixing.

Table 1
Properties of ion mixed Ni–Nb amorphous alloys

composition	ϱ_s (3×10^{15} Xe^+/cm^2)		T_{a-c} (5×10^{15} Xe^+/cm^2)
	as-deposited	ion-mixed	
$Ni_{35}Nb_{65}$	5.75	24.7 Ω/□	550 to 600 °C
$Ni_{55}Nb_{45}$	4.47	25.1 Ω/□	600 to 650 °C
$Ni_{65}Nb_{35}$	4.94	34.3 Ω/□	500 to 550 °C
$Ni_{80}Nb_{20}$	4.41	28.7 Ω/□	450 to 500 °C

As Nb and Mo have almost identical atomic masses, from the binary collision point of view, they should behave similarly under ion mixing. However, it was found that the Ni–Nb multiple layers are much easier to mix than Ni–Mo. For example, a dose of 3×10^{15} Xe^+/cm^2 induces uniform mixing and amorphizes a $Ni_{35}Nb_{65}$ layer, while a dose of 1.7×10^{16} Xe^+/cm^2 is required to uniformly mix and amorphize a $Ni_{35}Mo_{65}$ layer [1]. This suggests that material properties of the mixed metals, other than their atomic masses, play a role in the ion mixing processes; for example, their melting points (i.e. bonding energies) might have some effect on the mixing processes.

3.2 Thermal and irradiation stability

The amorphized alloys (obtained by irradiating the multilayers to a dose of 5×10^{15} Xe^+/cm^2) were subjected to vacuum annealing at successively increasing temperatures for 1/2 h to find the amorphous-to-crystalline transition temperature T_{a-c}. This temperature is claimed to lie within the 50 K below the annealing temperature at which the first sharp diffraction lines are visible on the X-ray film. The phase transition was also monitored by four-point probe sheet resistivity measurements performed at room temperature after each annealing step. Fig. 1 shows the sheet resistivity versus annealing temperature curve obtained from the $Ni_{55}Nb_{45}$–5×10^{15} Xe^+/cm^2 amorphous alloy. Table 1 shows the transition temperatures T_{a-c} of the four 5×10^{15} Xe^+/cm^2 amorphous alloys. All the Ni–Nb amorphous alloys obtained here have T_{a-c} on the order of 450 to 650 °C. The $Ni_{55}Nb_{45}$ amorphous alloy which has a composition closest to the center of the Ni–Nb phase diagram, has the highest T_{a-c} (600 to 650 °C). The same observation was reported for the Ni–Mo system where the $Ni_{50}Mo_{50}$ amorphous alloy has the highest T_{a-c} (600 to 630 °C) among the three compositions ($Ni_{35}Mo_{65}$, $Ni_{50}Mo_{50}$, and $Ni_{65}Mo_{35}$) investigated [1]. This suggests that for the binary metallic systems there exists an optimal composition for the amorphous alloy formation from the point of view of thermal stability. In the two cases of Ni–Mo and Ni–Nb, the optimal composition is in the middle of the phase diagram.

As reported in the literature, the two metallic glasses, $Ni_{60}Nb_{40}$ obtained by splat quenching [3] and $Ni_{50}Mo_{50}$ obtained by sputter deposition [4], have been subjected to MeV Ni^+ ion irradiation to determine their resistances against radiation damage. The results show that the Ni–Nb glass is more stable than the Ni–Mo glass, though the reasons remain to be explored [4]. We have irradiated the four Ni–Nb multilayered samples up to a dose of 1.7×10^{16} Xe^+/cm^2 to test their irradiation stability. All alloys remained amorphous; that is, an additional dose (1.4×10^{16} Xe^+/cm^2) nearly five times that required for amorphization (3×10^{15} Xe^+/cm^2) does not induce recrystallization. In contrast, the $Ni_{65}Mo_{35}$ amorphous alloy (that requires a minimum dose of 7×10^{15} Xe^+/cm^2 for its formation) dissociates after an additional dose of 1.3×10^{16} Xe^+/cm^2 irradiation [5]. These results are consistent with those obtained from the MeV Ni^+ ion irradiation experiments.

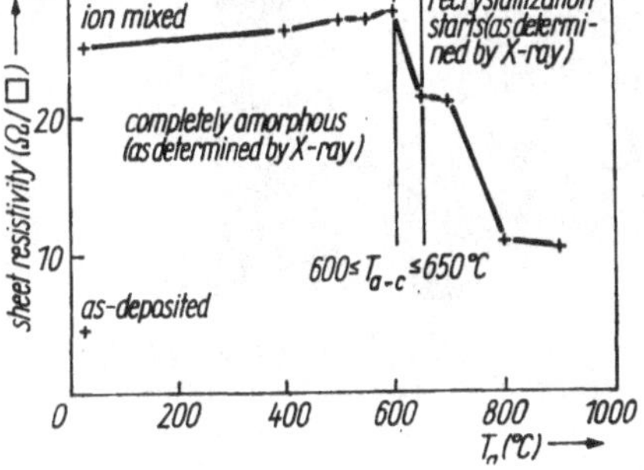

Fig. 1. Sheet resistivity versus annealing temperature T_a for $Ni_{55}Nb_{45}$ – 5×10^{15} Xe^+/cm^2 sample

3.3 A metastable crystalline phase formed in Ni-rich multilayers

In the Ni–Mo system, a metastable crystalline phase (MX) was formed in the Ni-rich sample at the early stage of ion irradiation. The structure of this MX phase was identified to be h.c.p. To check whether this observation is peculiar in this particular system or, it is more or less general for f.c.c.–b.c.c. systems, the $Ni_{65}Nb_{35}$ multilayered samples were irradiated to lower doses ranging from 3×10^{14} to 2×10^{15} Xe^+/cm^2. The similar MX phase was indeed formed after a dose of 5×10^{14} Xe^+/cm^2. Fig. 2 shows the X-ray diffraction photo of Ni–Nb MX phase and Table 2 gives the structural data of MX phase in these two systems. These observations indicate that in these two f.c.c.–b.c.c. systems, before the multilayers goes to complete amorphous, a h.c.p. MX phase forms first which can be considered as a 7/4 Hume-Rothery electronic compound [5]. Though the mechanism involved is not yet clear, we believe that this observation may be helpful in understanding the ion-induced amorphization processes.

Table 2

The metastable electronic compounds formed from the Ni-rich Ni–Mo and Ni–Nb samples by low-dose irradiation

system	overall composition of the multilayer	dose (Xe^+/cm^2)	lattice constants (h.c.p.) (Å)
Ni–Mo	$Ni_{65}Mo_{35}$	3×10^{15}	$a = 3.25$ $c = 5.21$ $c/a \approx 1.60$
Ni–Nb	$Ni_{65}Nb_{35}$	5×10^{14}	$a = 3.28$ $c = 5.22$ $c/a \approx 1.59$

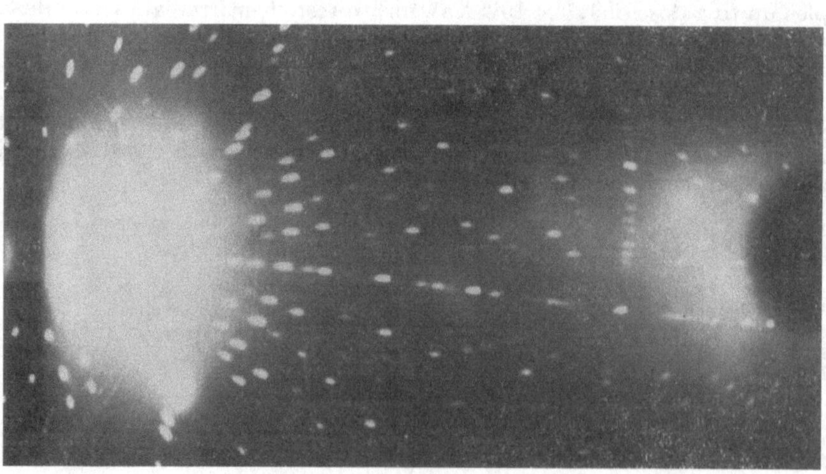

Fig. 2. X-ray diffraction photo of the $Ni_{65}Nb_{35}$ sample irradiated to a dose of 5×10^{14} Xe^+/cm^2 at room temperature. All the sharp diffraction lines are from MX phase of h.c.p. structure. The spots are from Si single crystal underneath

4. Conclusions

1. Amorphous alloys can be formed from multilayered samples in the Ni–Nb (f.c.c.–b.c.c.) system by IM. This conforms with the proposed "structural difference rule for amorphos alloy formation" by IM.

2. The Ni–Nb amorphous alloys so obtained are stable up to temperatures in the range 450 to 650 °C at 1/2 h annealing. In terms of thermal stability there exists an optimal composition for amorphous alloy formation at about equal metal composition.

3. These Ni–Nb amorphous alloys are stable upon further irradiation to an additional dose of 1.4×10^{16} Xe^+/cm^2.

4. The Ni–Mo system also forms amorphous alloys upon IM by Xe^+ ions, but the dose required for amorphizing Ni–Mo multilayers is higher than those for Ni–Nb multilayers. This suggests that some material parameters, not only the masses, affect the IM process. The recrystallization temperatures are in the same general range, and the composition of highest stability is also near the equal composition point. The Ni–Mo amorphous alloys are less radiation resistant than the Ni–Nb amorphous alloys.

Acknowledgements

The authors are grateful to Dr. S. S. Lau (University of California, San Diego) for thoughtful discussion and helpful criticism, and to R. Fernandez (Caltech) for performing the multilayer-sample evaporation. One of the authors (K. T-Y. Kung) wishes to express his gratitude for the support by the Caltech Summer Undergraduate Research Fellowship program (F. Shair, S. P. Krown, and C. Merkel). The ion irradiation part of this study was financially supported in part by the US Department of Energy through an agreement with the National Aeronautics and Space Administration and monitored by the Jet Propulsion Laboratory, California Institute of Technology (D. B. Bickler).

References

[1] B. X. Liu, W. L. Johnson, M-A. Nicolet, and S. S. Lau, Amorphous Film Formation by Ion Mixing in Binary Metal Systems, to be published in: Proc. Internat. Conf. Ion Beam Modification of Materials (IBMM-82), Grenoble (France), September 6 to 10, 1982.

[2] G. Dearnaley, Ion Implantation, North-Holland Publ. Co., New York 1973 (p. 766).

[3] M. D. Rechtin, J. van der Saude, and P. M. Baldo, Scripta metall. **12**, 639 (1978).

[4] J. L. Brimhall, L. A. Charlot, and R. Wang, Scripta metall. **13**, 217 (1979).

[5] B. X. Liu, unpublished results, Caltech 1982.

(Received December 23, 1982)

phys. stat. sol. (a) 77, 361 (1983)

Subject classification: 10.2; 22.5.2

Department of Metallurgy, Mechanics, and Materials Science, Michigan State University, East Lansing[1])

Deformation Structure in LiF Single Crystals

By

K. N. SUBRAMANIAN

The growth of slip bands during deformation and the stress pattern in the deformed LiF single crystals are investigated with polarized light microscopy. A band-like structure consisting of overlapping loops, which might be called deformation bands, is observed in the slip planes usually in a direction perpendicular to the slip direction. These bands are of constant width of approximately 20 μm. It is proposed that these bands are formed by interaction of dislocation loops, originated from randomly distributed sources.

Die Entwicklung von Gleitbänden in deformierten LiF-Einkristallen und die damit verbundenen Spannungsmuster werden mit Polarisationslichtmikroskopie untersucht. Eine Struktur von sich überlappenden Schleifen, die man als „Deformationsbänder" bezeichnen kann, wird in den Gleitebenen beobachtet. Die Mehrzahl der Bänder liegt senkrecht zur Gleitrichtung und ist ca. 20 μm breit. Es wird vermutet, daß sich diese Bänder durch Wechselwirkung zwischen den Versetzungsschleifen zufällig verteilter Quellen bilden.

1. Introduction

The dislocation structure of deformed crystals of rock-salt structure is mostly studied by using etch-pit technique or by observing thin sections made out of deformed crystals under an electron microscope [1 to 12]. Thus one is limited to either the surface or a thin section of bulk and as a result has only a partial picture of the dislocation structure of the bulk specimens. To understand the work hardening behavior of materials, it is necessary to know the three-dimensional dislocation structure in the deformed bulk crystals. Polarized light microscopy of transparent crystals, therefore, appears to be a suitable technique [13 to 17]. Although it cannot reveal fine details of the dislocation structure in deformed crystals, it can provide information about deformation structure of bulk specimens. Lithium fluoride is a transparent crystal and is suitable for polarized light microscopic studies.

Lithium fluoride is an ionic crystal and has the NaCl structure. Extensive reviews on the deformation behavior of NaCl-structured materials are available in the literature [5, 6, 9]. The slip systems in this structure are one of the simplest. The primary slip planes are $\{110\}$ and the secondary slip planes are $\{001\}$. The slip directions are $\langle 1\bar{1}0\rangle$ in both cases. Initiation of secondary slip is difficult due to the high stresses needed for moving dislocations in the secondary slip planes [1]. In NaCl structure no two primary slip planes have the same Burgers vector and hence cross-slip at room temperature is not probable [18 to 21]. Various advantages of NaCl-structured crystals for the study of dislocation structure are also discussed in the literature [9].

[1]) East Lansing, Michigan 48824, USA.

2. Experimental

The LiF single crystals were deformed in compression using Instron testing machine. The growth of the slip bands was studied on the {100} with linearly polarized light in dark field. Detailed observations of the damage in the slip bands were made with monochromatic light (mercury green) in a photoelastic set up and also with a modified petrographic microscope. The stress fields of the dislocations are such that they can be observed with polarized light. In the dark field observations, where the directions of polarization of the polarizer and the analyzer are perpendicular to each other, the stress fields due to dislocations appear as bright lines provided the slip vector has a certain orientation with the direction of polarization.

Thin specimens were prepared from bulk crystals, previously deformed in compression along the [100] direction, by grinding them with emery paper so that the broad surfaces were parallel to one of the slip planes active during the deformation of the bulk crystal. The thin crystals were subsequently polished with a metallographic polishing wheel and then chemically polished with fluoboric acid for at least 1 h. Fluoboric acid polishes at a rate of approximately 20 μm/min, thus at least 1 mm of the crystal was removed from each side. This prolonged polishing ensured the removal of the damage introduced during the grinding and mechanical polishing. Finally the specimens were polished with distilled water to obtain smooth surfaces. The thickness of the specimens after the above treatments was of the order of 0.025 mm. A petrographic microscope was used for studying the damage in the slip plane. Observations were made with linearly and circularly polarized monochromatic light in both bright and dark fields. The microscope stage on which the specimen was mounted was rotated to obtain various relative orientations of the slip vector with respect to the direction of polarization.

3. Results and Discussion

Transmitted p larized light microscopic observations of deformed LiF single crystals are presented in Fig. 1 through 3. These specimens were not chemically etched to reveal dislocations after deformation. Orthogonal slip bands in a slightly deformed LiF crystal can be seen in Fig. 1. In the dark-field observations where axes of polarizer and analyzer are at 90° to each other, the slip bands can be clearly seen only when the axis of polarizer is at 45° to the Burgers vectors (which will be parallel to the slip bands in this case). There was complete extinction when the axis of polarizer was either parallel or perpendicular to the orthogonal slip bands. Since these slip band patterns were present in the plane polarized and as well as in the circularly polarized light observations, they were isochromatics. Isochromatics are due to unequal principal stresses and their intensities are proportional to the difference in the principal stresses. The orthogonal slip bands in the (100) plane of one of the specimens prepared for the (011) observation are shown in Fig. 2. Damage in the slip plane in one of the stress concentrated regions in the slip plane of such a specimen is presented in Fig. 3. The crystal was mounted on the microscope stage and rotated successively while the polarizer and the analyzer were kept fixed to obtain different orientations of the Burgers vector with respect to the direction of polarization. The parallel set of bright lines in these pictures are due to slip in the slip planes which are the conjugate of the slip plane under observation. The conjugate slip bands appear to be spaced more or less evenly. The striking feature of this set of pictures is the band structure along the direction A′A″ in each picture. These bands represent a certain stress pattern and are in the plane of observation. They are different from the parallel set of bright lines due to the intersecting slip. Each of these bands appear to be made up of loops that

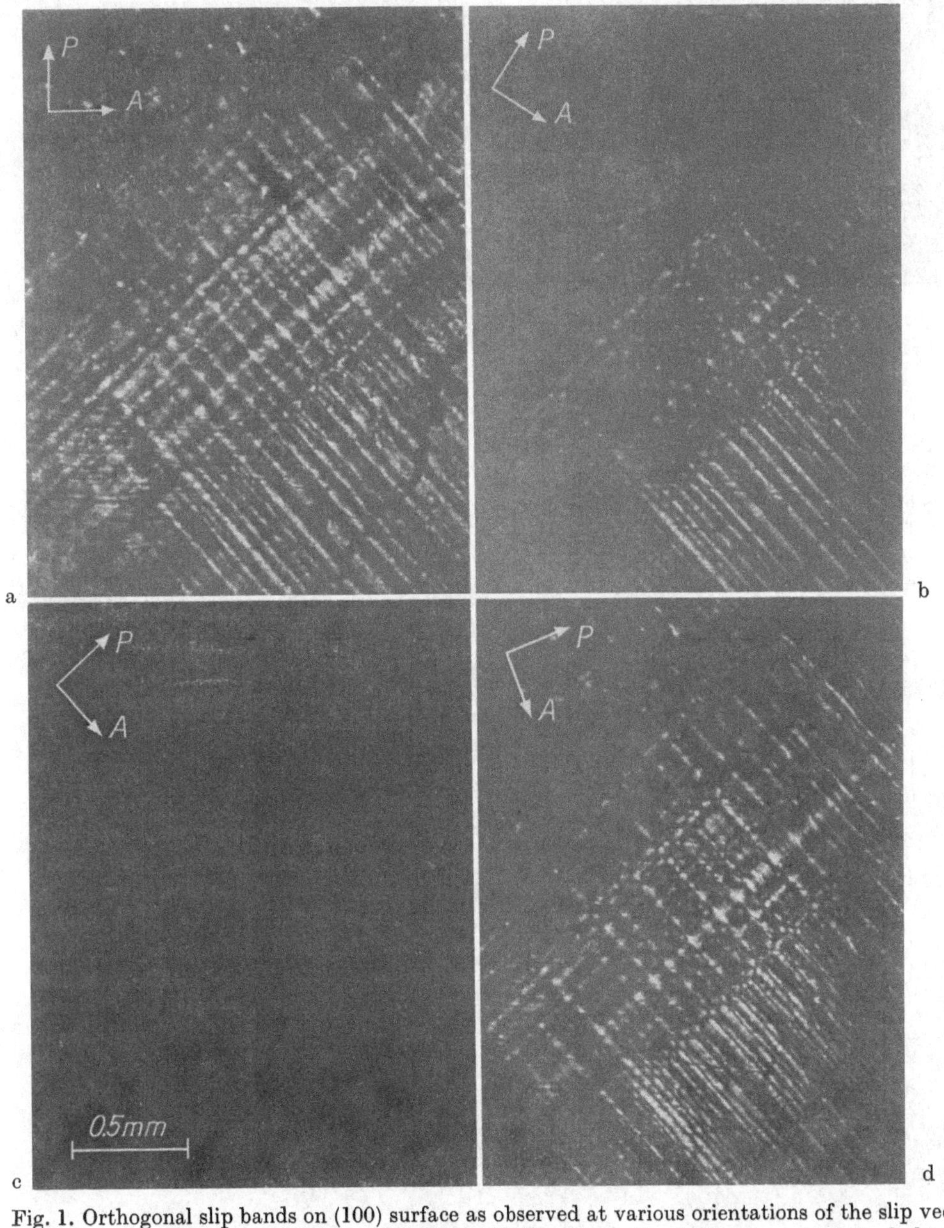

Fig. 1. Orthogonal slip bands on (100) surface as observed at various orientations of the slip vector with the direction of polarization in dark field. Direction of polarization: a) Vertical, b) 30°, c) 50°, d) 70° to the vertical. P and A represent axis of polarizer and analyzer, respectively

have lined up in a direction perpendicular to the slip vector. These loops become prominent only when the orientation of the Burgers vector is in the range of 20° to 30° with respect to the direction of polarization.

The bands observed are approximately of constant width of 20 μm. They are evenly spaced and are usually perpendicular to the slip vector. Similar band structure was

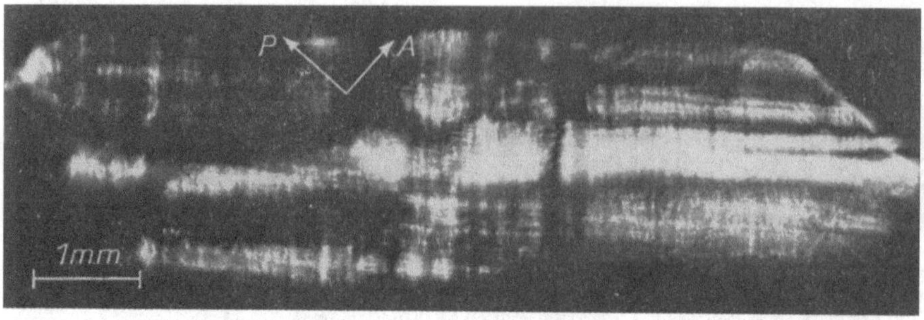

Fig. 2. Orthogonal slip bands on the (100) surface of a specimen for the observation of damage in the (011) slip plane

also observed in the active slip plane of a sodium chloride crystal deformed under compression along [100].

These bands are similar to the ones observed in zinc where they are classified as deformation bands [22]. It is to be discussed later in this section that intersecting slip may create collinear sources which may contribute to the formation of the band structure. However, this mechanism cannot explain the band structure formation in zinc because basal plane is the only active slip plane at a moderate stress level and

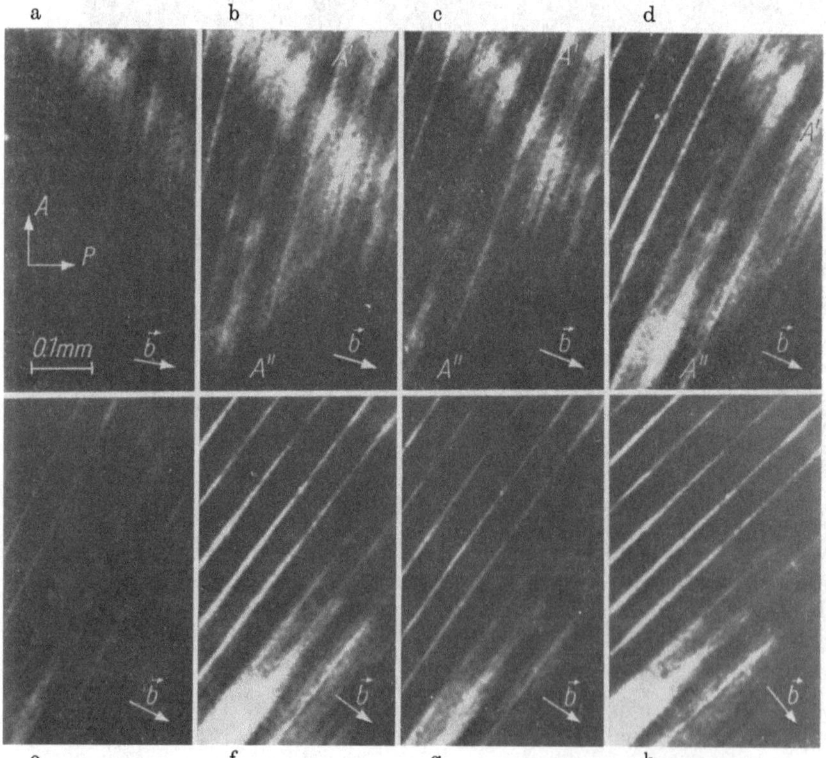

Fig. 3. Polarized light observation of the bands in the slip plane. Bands making a) 10°, b) 15°, c) 20°, d) 25°, e) 30°, f) 35°, g) 40°, h) 45° to the direction of polarization

at room temperature. There should be a general mechanism of band formation regardless of intersecting slip. Extending the idea originated by Frank, Wei has proposed such a mechanism based on the interaction of dislocation loops originating from sources in neighboring atomic planes [22]. A theoretical justification for such a model has been presented recently [23]. The dislocation sources distributed at random are most likely non-coplanar. Some of them may lie in nearby parallel planes. In a set of parallel slip planes in which the dislocation loops are expanding due to the applied stress the loops should have the same Burgers vector. When these expanding loops approach each other the interacting segments can be either edges or screws of the opposite signs, or mixed segments whose components are of the opposite signs. In the case of screw interaction, since cross-slip is difficult in NaCl-structured crystals, the attractive force will cause the two screw components to arrange themselves parallel to one another in a plane perpendicular to the slip planes when the separation is a minimum. In the case of a large number of dislocation loops being generated from two neighboring sources and interacting in their screw components, all the screw components will line up in similarly parallel pairs of opposite signs resulting in the formation of a screw-lock. The early state of the screw-lock formation can be accomplished relatively easily by the movement of a few pairs. However, after a large number of pairs have been locked, any further addition of the pairs becomes difficult because it requires the movement and rearrangement of all the pairs locked in already. The screw-lock becomes a barrier against which the subsequent dislocations may pile up thereby making the source inoperative. It seems that screw-locking is a possible cause of band formation in materials that do not have easy cross-slip. It explains why the loops observed in these bands usually line up in a direction perpendicular to the Burgers vector.

For interaction of two arrays of edge dislocations in nearby slip planes the plus-minus pair formed by the two leading dislocations acts as a barrier against which the

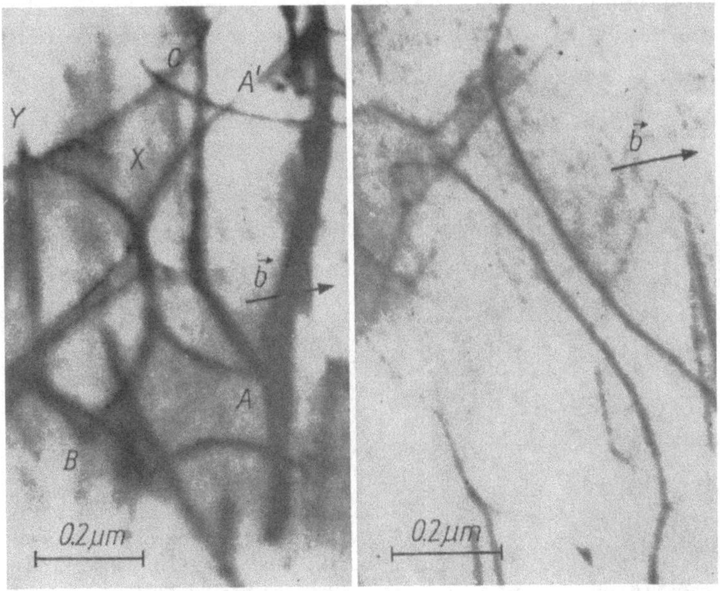

a b

Fig. 4. Transmission electron micrographs of fatigued MgO specimens. a) is from [23] rest of the dislocations pile up. Even if the process repeats itself, it cannot proceed too far since the force field will be against such a formation. It can be shown that the interaction between mixed dislocations may also lead to the formation of screw-lock because it has the lowest strain energy.

Although lithium fluoride is well suited for studies using etch-pit, polarized light, X-ray, and decoration techniques, a direct observation of the dislocations in lithium fluoride using transmission electron microscopy is difficult because of its poor thermal conductivity and low melting point. On the other hand, magnesium oxide, which has the same crystal structure as lithium fluoride and exhibits similar deformation behavior, is well suited for transmission electron microscopy. Transmission electron micrographs of magnesium oxide single crystals deformed in fatigue at room temperature are presented in Fig. 4. Fig. 4a gives a clear picture of the formation of these dipoles. At points X and Y there are edge and screw dipoles, respectively. It seems that these dipoles are formed as a result of the interaction of the loops AA′, BB′, and CC′. Fig. 4b shows the interaction of two dislocations of mixed orientation. These observations are quite consistent with the proposed mechanism.

In a few instances the loops lined up along the line of intersection of the slip plane under observation with various other primary slip planes as shown in Fig. 5 where the bands BB′, are along the line of intersection of a non-conjugate slip plane with the slip plane under observation. These bands were not well-developed, and were of poor contrast compared to the bands formed along the direction perpendicular to the Burgers vector.

This observation suggests another possible mechanism of band formation in materials with intersecting slip. A mechanism of band formation based on intersecting slip has been described in detail in a recent article [23]. According to this mechanism dislocations moving in two intersecting slip planes acquire jogs and produce sources in parallel slip planes. Dislocations produced by such sources will interact to form a band structure along the line of intersection of the slip planes. For non-conjugate slip planes the intersection line is inclined to the slip vector as seen in Fig. 5. From this point on the general mechanism of loop interaction discussed previously, takes

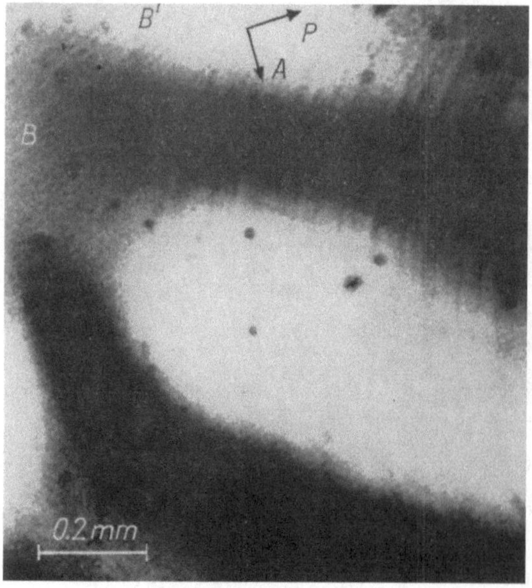

Fig. 5. Band structure observed along non-orthogonal slip plane intersections in a thick specimen deformed in three-point bending. Burgers vector making 20° to the direction of polarization

over. Screw-locks are formed in the case of conjugate slip plane intersection. In the case of non-conjugate slip plane intersections the interacting dislocations are mixed in nature. Since the dislocation loops are aligned before the interaction takes place, they would have to increase their lengths by a large proportion to form screw-locks. So the interacting dislocations may stay as mixed dipoles when the band structure is not fully developed. Infrequent observations of bands in directions other than the one perpendicular to the slip vector, suggest that the mechanism of band formation due to intersecting slip may not be important. Further, such a mechanism cannot account for the bands observed in zinc single crystals deformed by basal slip.

4. Summary

The essential features of the band structure observed in the slip plane of deformed LiF can be summarized as following:

1. The bands were usually perpendicular to the slip vector.
2. There was some evidence of bands developing along the line of intersection between the non-conjugate slip planes and the slip plane under observation.
3. The bands were typically of constant width of approximately 20 μm.
4. They were made up of loops usually interlinked by segments parallel to the Burgers vector.

References

[1] J. J. Gilman, Acta metall. **7**, 608 (1959).
[2] J. J. Gilman, Trans. AIME **209**, 449 (1957).
[3] J. J. Gilman and W. G. Johnston, J. appl. Phys. **27**, 1018 (1956).
[4] J. J. Gilman, W. G. Johnston, and G. W. Sears, J. appl. Phys. **29**, 747 (1958).
[5] J. J. Gilman and W. G. Johnston, Dislocation and Mechanical Properties of Crystals, John Wiley, 1956 (p. 116).
[6] J. J. Gilman and W. G. Johnston, Solid State Phys. **13**, 148 (1962).
[7] J. J. Gilman, J. appl. Phys. **30**, 1584 (1959).
[8] W. G. Johnston and J. J. Gilman, J. appl. Phys. **31**, 632 (1960).
[9] J. Washburn, Electron Microscopy and Strength of Crystals, Interscience Publishers, 1963 (p. 301).
[10] W. Elkington, G. Thomas, and J. Washburn, J. Amer. Ceram. Soc. **46**, 307 (1963).
[11] J. Washburn, G. W. Graves, A. Kelley, and G. K. Williamson, Phil. Mag. **5**, 991 (1960).
[12] J. Washburn, Strengthening Mechanisms in Solids, ASM, Metals Park (Ohio) 1960 (p. 51).
[13] J. F. Nye, Proc. Roy. Soc. **A198**, 190 (1949).
[14] B. K. Kear and P. L. Pratt, Acta metall. **6**, 457 (1958).
[15] W. L. Bond and J. Andrus, Phys. Rev. **101**, 1211 (1956).
[16] R. Bullough, Phys. Rev. **110**, 620 (1958).
[17] S. Mendelson, J. appl. Phys. **32**, 1999 (1961).
[18] A. J. McEvily, Jr. and E. S. Machlin, Fracture, John Wiley, 1959 (p. 450).
[19] A. J. McEvily, Jr. and E. S. Machlin, Trans. MS AIME **221**, 1086 (1961).
[20] T. H. Alden, Trans. MS AIME **227**, 1103 (1963).
[21] T. H. Alden, Appl. Phys. Letters **2**, 107 (1963).
[22] C. T. Wei, Acta metall. **5**, 435 (1957).
[23] K. N. Subramanian, phys. stat. sol. (a) **74**, 683 (1982).

(Received Februry 8, 1983)

phys. stat. sol. (a) 77, 369 (1983)
Subject classification: 1.2 and 14.2; 22.8.1

Institute of Solid State Physics, Latvian State University, Riga[1])

Crystallographic Properties and Superconductivity of Solid Solutions of the $BaBi_xPb_{1-x}O_3$ System

By

L. A. SHEBANOV, V. YA. FRITSBERG †, and A. P. GAEVSKIS

The results of X-ray studies of the crystallographic characteristics of solid solutions of the $BaBi_xPb_{1-x}O_3$ system in the temperature range from −180 to 500 °C are reported. The relation between the appearance of superconductivity and the anomalies of structure parameters observed in the range $x = 0.15$ to 0.25 as well as the ordering of the perovskite lattice is discussed.

Es werden die Ergebnisse der Untersuchung der kristallographischen Parameter von festen Lösungen $BaBi_xPb_{1-x}O_3$ im Temperaturbereich von −180 bis 500 °C mitgeteilt. Der Zusammenhang zwischen den Charakteristiken der Superleitfähigkeit und den experimentell beobachteten Anomalien der Strukturparameter im Konzentrationsbereich $x = 0{,}15$ bis 0,25 sowie dem Ordnungsgrad von Elementen des Perovskitkristallgitters wird diskutiert.

1. Introduction

The discovery of superconductivity in the system $BaBi_xPb_{1-x}O_3$ $(0.05 \leqq x \leqq 0.35)$ in 1974 by Sleight and coworkers [1, 2] has caused intense studies of the properties of ABO_3-type perovskite, forming double and triple oxides, where B positions in the lattice are occupied by the heavy ions of lead and bismuth. These materials are interesting for a number of unique properties:

a) a high transition temperature to the superconductive state ($T_t = 12.5$ K for the compound $BaBi_{0.25}Pb_{0.75}O_3$ [2],

b) a rather high specific resistance (10^{-3} to 10^2 Ωcm near T_t) in the ordinary state (as compared to the known superconductive metals and alloys),

c) resistance to prolonged action of rather high temperatures (900 to 1000 °C).

These properties make the perovskite superconductors rather promising for cryogenic devices of different types. At the same time they are of certain general scientific interest. Really, the $BaBi_xPb_{1-x}O_3$ system is a good example of compound superconductors, not containing any transition elements, with the highest transition temperature known at the present time. A slight distortion from the cubic symmetry below some $T_k \gg T_t$ raises the problem about the character of the relation between the superconductivity and the possible presence of spontaneous polarization in the solid solutions under discussion [3, 4].

The present experimental study of the crystallographic properties of $BaBi_xPb_{1-x}O_3$ solid solutions in the concentration range $0 \leqq x \leqq 0.4$ is an attempt to reveal the effects of concentration and structure on the superconductivity of the compounds.

2. Experimental

The relevant solid solutions are known to have a unit cell of orthorhombic symmetry [5] which is conveniently described in terms of pseudo-monoclinic ($a = c$) distortion of the cubic original cell. Since the distortion is rather small, one may determine the

[1]) 8 Kengaraga str., 226063 Riga, USSR.

crystallographic parameters by the analysis of X-ray diffraction line splitting of the cubic perovskite structure based on the crystal homology [6, 7]. The characteristics of the line splitting (the number of components and their relative intensities) in the case of pseudo-monoclinic distortion are determined by two parameters, $\cos\beta$ and $y = (b/a \sin\beta) - 1$ as well as by their ratio $y/\cos\beta$ [7, 8]. For quantitative calculations in the case of small distortions ($|\beta - 90°| < 1°$) simplified expressions of the quadratic forms given in [6] may be used. In our case the cell parameters were obtained from reflections with $\sum_i h_i^2$ equal to 8 or 12 when the following formulae are valid:

$$\beta = \arccos 3 \frac{\sin\theta_{22\bar{2}} - \sin\theta_{222}}{\sin\theta_{22\bar{2}} + \sin\theta_{222}}, \tag{1}$$

$$y = \frac{3}{2} - \frac{\sin\theta_{22\bar{2}} \sin\theta_{222}}{\sin\theta_{20\bar{2}} \sin\theta_{202}}, \tag{2}$$

$$a = c = \frac{\lambda\sqrt{3 - 2y}}{\sin\beta\sqrt{\sin\theta_{22\bar{2}} \sin\theta_{222}}}, \tag{3}$$

$$b = a(1 + y)\sin\beta\,. \tag{4}$$

The corresponding components of the 222 and 220 lines (CuK_α radiation, Ni β-filter, diffractometer "DRON-2") were separated from the complex profile by the method described in [8]. It has to be noticed that for the examined compounds the approximate equality $\cos\beta \approx -y$ is valid which is supported by the data for $BaBi_{0.25}Pb_{0.75}O_3$ given in [5].

Ceramic samples of 15 mm diameter and 2 mm thick obtained by thermal reaction of $BaCO_3$, Bi_2O_3, and PbO_2 in stoichiometric ratio were examined.

Table 1

Crystallographic characteristics of solid solutions at $T = 20$ °C

concentration x (mol%)	$a = c$ (Å)	b (Å)	β	$\sqrt[3]{V}$ (Å)
0	4.2734	4.2494	90.32°	4.2654
5	4.2754	4.2567	90.25°	4.2691
15	4.2771	4.2629	90.19°	4.2723
20	4.2808	4.2681	90.17°	4.2765
25	4.2841	4.2699	90.19°	4.2796
30	4.2893	4.2720	90.23°	4.2835
40	4.3059	4.2871	90.25°	4.2996

3. Results and Discussion

From our results (Table 1) it follows that within the range $0 \leqq x \leqq 0.4$ a pseudo-monoclinic ($a = c$) distortion of the unit cell is characteristic of the solid solutions of the $BaBi_xPb_{1-x}O_3$ system (the common unit cell symmetry is orthorhombic, point group mm2). The corresponding data at $T = 20$ °C are given in Table 1. A distinct minimum at $x = 0.15$ to 0.25 is characteristic of the monoclinic angle versus concentration. This may be a possible reason for the erroneous conclusion [9] of a tetragonal distortion of $BaBi_{0.3}Pb_{0.7}O_3$ from neutron diffraction study, which seems to be due to the insufficient resolution of that method. Our experiments in the range of concentrations $0 \leqq x \leqq 0.4$ unambiguously point to a small but quite distinct splitting of the 222 diffraction line which is not characteristic of the tetragonal unit cell. It has to be noted that the depth of the minimum of the concentration curve $\beta(x)$ is growing as the temperature is lowered and, consequently, this property is more pronounced near the transition to the superconductive state.

We have studied the thermal behaviour of the unit cell parameters of the $BaBi_xPb_{1-x}O_3$ system in the range from −180 to 500 °C. The qualitative features are the same for all the solid solutions within $0 \leq x \leq 0.4$. The results for $BaBi_{0.05}Pb_{0.95}O_3$ are shown in Fig. 1. The characteristic features for other compounds are the same, the particular points being shifted along the temperature axis. The monoclinic-to-cubic phase transition is strongly diffused and divided into four intervals. E.g., in the case of $BaBi_{0.05}Pb_{0.95}O_3$ they are:

I: $T = -180$ to 200 °C; a small but distinct splitting of the 222 diffraction line due to the monoclinic distortion. The monoclinic angle and line splitting decrease rapidly with the growth of temperature.

II: $T = 200$ to 360 °C; a very small monoclinic distortion weakly dependent on temperature. Components of the 222 line are not resolved.

III: $T = 360$ to 410 °C; X-ray diffraction pattern corresponds to a cubic unit cell; the thermal behaviour of the integrated intensities is critical.

IV: $T > 410$ °C; cubic unit cell, regular thermal behaviour of the integrated intensities.

Transitions I → II ($T_{k_2} \approx 200$ °C) and II → III ($T_{k_1} \approx 360$ °C) display features characteristic of phase transitions of first and second order, respectively (Fig. 1, top). The distortion of the unit cell allows possible polar or antipolar ordering at $T < T_{k_1}$.

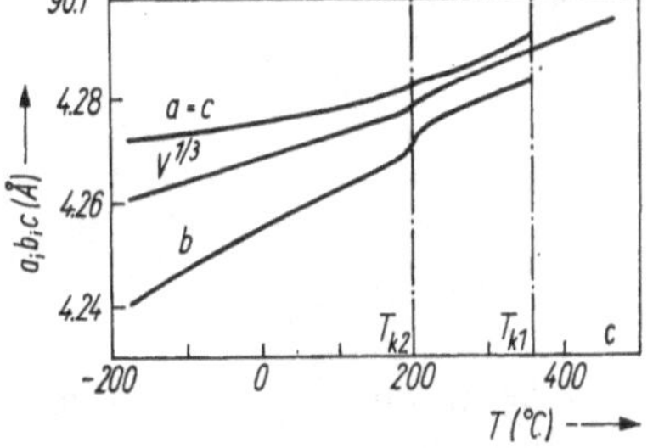

Fig. 1

Fig. 2

Fig. 1. Parameters of the pseudo-monoclinic unit cell $a = c$, b, β, $\sqrt[3]{V} = \sqrt[3]{a^2 b \sin \beta}$, and the linear thermal expansion coefficient α of $BaBi_{0.05}Pb_{0.95}O_3$ as functions of temperature. Results of dilatometric experiments are shown by the dashed line

Fig. 2. The monoclinic angle β (at $T = 20$ °C) and the temperature of monoclinic-to-cubic phase transitions T_{k_1} as functions of concentration in $BaBi_xPb_{1-x}O_3$ solid solutions. The data on T_{k_1} for pure $BaBiO_3$ are given according to [12]

According to the X-ray data the linear thermal expansion coefficient of $BaBi_{0.05} \cdot Pb_{0.95}O_3$ is equal to $\alpha = (12.4 \pm 0.4) \times 10^{-6}\ K^{-1}$ at room temperature. The X-ray data are consistent with the results of dilatometric measurements (Fig. 1, top).

The temperature of the II → III transition (to a cubic unit cell X-ray diffraction pattern) as a function of concentration is shown in Fig. 2. The minimum at concentrations $x = 0.15$ to 0.25 correlates with the character of the monoclinic angle variation versus concentration (Table 1).

The reasons for the T maximum at $x = 0.25$ long before the metal-to-semiconductor transition at $x = 0.35$ are vividly discussed at present [5]. The Hall effect measurements of carrier density n as a function of concentration x also show a distinct maximum near $x = 0.25$. The maximum values of carrier density n are 1.8×10^{21} [10] or $4.5 \times 10^{21}\ cm^{-3}$ according to the data given in [11].

In accordance with the present ideas [2] the metallic conductivity of pure $BaPbO_3$ is due to the overlap of the 2p valence band of oxygen with the empty 6s band of lead. If the bismuth atoms are assumed [2] to be in the state Bi^{4+} providing one electron to the conduction band, the decrease of n at $x = 0.25$ is rather unexpected. It has been suggested [10] that long before a gap appears in the energy spectrum of electrons part of Bi atoms change their state to Bi^{3+} and Bi^{5+}, which are characteristic of the semiconducting compound $Ba_2Bi^{3+}Bi^{5+}O_6$ [12], the total concentration of the carriers thus being decreased. The reasons for this anomaly and its microscopic nature are not discussed in [10].

It should be noted that structural studies up to the present time have not revealed any remarkable anomalies of crystallographic parameters as functions of concentration neither in the region of $x = 0.25$, nor in the region of the metal-to-semiconductor transition. Our experiments show (Table 2) that besides the minimum of the monoclinic angle in the range of concentrations $0.15 \leq x \leq 0.25$ an anomalous behaviour in the ratio of X-ray diffraction intensities of lines with different parity of indices hkl is also observed. Such an effect is obviously related to the change of the ordering of sublattice B of the perovskite structure ABO_3. Really, in the solid solutions examined one has to consider at least three sorts of ions of different radii occupying the position B: Pb^{4+}, Bi^{3+}, Bi^{5+}, and also bismuth ions of surrounding-induced valence +4 in the range of x where superconductivity appears. The ordering of Bi^{3+} and Bi^{5+} in pure $BaBiO_3$ is discussed in [12] where, in contrast to [5], a complete ordering (long-range order) of bismuth ions is obtained. The unit cell volume becomes multiple at complete ordering, and the so-called superstructure X-ray diffraction lines appear

Table 2

The ratio of X-ray diffraction line intensities I_{210}/I_{211} as a function of concentration in $BaBi_xPb_{1-x}O_3$ solid solutions in the cubic phase ($T = 465$ °C)

concentration (mol%)	calculated	experimental
0	0.1001	0.1077 ± 0.0040
5	0.1022	0.1179
15	0.1065	0.1698
20	0.1084	0.1201
25	0.1106	0.1146
30	0.1120	0.1151
40	0.1168	0.1085

along with the relative intensity redistribution of the main lines. The analysis of the data of Table 2 shows that the unusually large values of I_{210}/I_{211} in the range $0.15 \leqq x \leqq 0.25$ are due to the partial ordering of the ions at these concentrations. The calculation shows that in the case of completely random distribution of Pb and Bi ions a weak and monotonous dependence of I_{210}/I_{211} on concentration is expected (Table 2), since the difference in atomic factors of X-ray scattering for Pb and Bi ions is small and the unit cell size of $BaPbO_3$ is close enough to that of $BaBiO_3$.

The observed partial ordering together with the maximum closeness to the cubic symmetry (minimum β) obviously must provide an increase of the coherence length as well as of the electron mobility at the expense of minimizing the effects of the fluctuating random potential. At last, the break-down of the superconductive state must occur at a higher temperature as compared to the case of a disordered lattice. This also explains the variation of $T_c(x)$ in the $BaBi_xPb_{1-x}O_3$ system and the results of Hall experiments [10, 11].

References

[1] A. W. SLEIGHT, U.S.P. 3932 315 (1974).
[2] A. W. SLEIGHT, J. L. GILLSON, and P. E. BIERSTEDT, Solid State Commun. 17, 27 (1975).
[3] E. G. FESENKO, E. G. SHUVAYEVA, and YU. I. GOLTSOV, Kristallografiya 17, 419 (1972).
[4] A. M. GABOVICH, L. V. GAVRILYUK, D. P. MOISEEV, E. A. PASHITSKII, A. F. PRIHOTKO, S. K. UVAROVA, and A. S. SHPIGEL, Ukr. fiz. Zh. 24, 674 (1979).
[5] Y. KHAN, K. NAHM, M. ROSENBERG, and H. WILLNER, phys. stat. sol. (a) 39, 79 (1977).
[6] YU. N. VENEVTSEV, S. P. SOLOVEV, and G. S. ZHDANOV, Zav. Labor. 27, 1112 (1961).
[7] M. F. KUPRYANOV and V. S. FILYPYEV, Kristallografiya 8, 356 (1963).
[8] L. A. SHEBANOV, Uchen. Zap. Latv. Univ. 189, 150 (1974).
[9] D. E. COX and A. W. SLEIGHT, Solid State Commun. 19, 969 (1976).
[10] D. P. MOISEEV, S. K. UVAROVA, and M. B. FENIK, Fiz. tverd. Tela 23, 2347 (1981).
[11] TRUONG D. THANH, A. KOMA, and S. TANAKA, J. appl. Phys. 22, 205 (1980).
[12] D. E. COX and A. W. SLEIGHT, Acta cryst. B35, 1 (1979).

(Received December 23, 1982)

phys. stat. sol. (a) 77, 375 (1983)

Subject classification: 13.4 and 20.3; 22.8.1

Moscow Institute for Rentgenology and Radiology (a)
and Moscow State University[1] *(b)*

Enhancing and Quenching of Intrinsic Luminescence and Characteristic Features of Calcium Tungstate Phosphorescence in the Presence of Lanthanides

By

S. I. GOLOVKOVA (a), A. M. GURVICH (a), A. I. KRAVCHENKO (b), V. V. MIKHAILIN (b), and A. N. VASILEV (b)

It is found that doping of $CaWO_4$ by impurities of the lanthanide series (their concentration in the charge 10^{-6} to 10^{-5} g-atom/mol) causes the increase of $CaWO_4$ intrinsic luminescence up to 40 to 50% with ultraviolet excitation. Further increase of the lanthanide concentration results in quenching of the luminescence. Both effects approach the maxima in the region of the deepest depression in the excitation spectrum of $CaWO_4$ at 12 eV. Lanthanides lead to the increase of $CaWO_4$ phosphorescence. Mechanisms of the phenomena mentioned are discussed.

Обнаружено что примеси ряда лантаноидов при концентрации их в шихте 10^{-6} до 10^{-5} g-atom/mol вызывают усиление собственной люминесценции $CaWO_4$, достигающее при возбуждении в области вакуумного ультрафиолета 40 до 50%. Дальнейшее увеличение концентрации лантаноидов приводит к тушению люминесценции. Оба эффекта достигают максимума в области наиболее глубокого провала в спектре возбуждения $CaWO_4$ — при $E = 12$ eV. Лантаноиды увеличивают фосфоресценцию $CaWO_4$. Рассмотрен механизм описанных явлений.

1. Introduction

Calcium tungstate is known to show intrinsic luminescence, i.e. luminescence not associated with impurities ($\lambda_{max} = 425$ nm). Impurities, however, can considerably affect it: reduce the yield and increase the light sum storage [1, 2]. Previous investigations [3, 4] have shown group V elements, transitional d-elements, and lanthanides to be most effective in this respect. Lanthanides unlike group V elements quench effectively not only luminescence excited by quanta with energy $E > 6.2$ eV but also emission due to direct excitation of luminescent centres in the band with maximum at 5.1 eV. A resonant mechanism of quenching taking place in this case along with the recombination mechanism is responsible for this. Low concentrations (10^{-6} to 10^{-5} g-atom/mol) of the majority of lanthanides lead to the increase of afterglow which causes blackening of optically non-sensitized X-ray film. This can be due to the formation of effectively positively charged electron trapping centres like $Ln_{Ca}^{\cdot}$[2]), and in the case of Ln which tends to transitions to the tetravalent state due to the formation of hole trapping centres [2, 4].

In the present paper the effect of Ln on the intrinsic luminescence of $CaWO_4$ is investigated not only by X-ray and long-wavelength ultraviolet excitation as it was

[1]) 117234 Moscow, USSR.

[2]) Indexes throughout the article relate to the lattice sites with point defects (here impurity defects), points mark effective positive charges, primes effective negative charges.

reported earlier but in the vacuum ultraviolet region as well. Besides, phosphorescence spectra in the presence of Ln were measured. New experiments made the effects observed earlier clearer, revealed new features, and clarified the origin of the effect of rare-earth impurities on $CaWO_4$ luminescence.

2. Experimental Technique

We used powder phosphors prepared by heating at 800 °C of precipitated $CaWO_4$ with $CaCl_2$ as a flux and controlled quantities of lanthanides. For details see [2, 4].

An improved version of monochromator VMR-2 was used to obtain luminescence excitation spectra. Gold (1200 lines/mm) and tungsten (2400 lines/mm) coated gratings served as dispersive elements. Hydrogen and helium discharge tubes provided exciting radiation in the region 5 to 14 eV. Excitation in the range 12 to 30 eV was performed by synchrotron radiation. A filter SS-5 was used to select $CaWO_4$ luminescence.

A one-beam scheme was applied to measure the luminescence yield. At each wavelength of exciting radiation (λ) we measured the emission intensity of both $CaWO_4$ and sodium salicilate which has a constant quantum yield in the range under investigation (5 to 30 eV). To avoid errors due to instability of excitation sources the results of luminescence measurements were normalized to the simultaneously detected total intensity of the exciting radiation. Operation of the experimental arrangement and data processing were on-line with a minicomputer "Elektronika D3-28".

The experimental arrangement used to obtain phosphorescence spectra comprised monochromator MDR-3, photomultiplier FEU-79, direct-current amplifier, and potentiometer KSP-4. Phosphorescence spectrum recording has been carried out with a speed of 200 nm/min, it started about a second after switching off the excitation. The sample was irradiated with X-rays (X-ray generator RUM-7, applied voltage 30 kV, Al filter 0.1 mm) and with the Hg line at 254 nm (BUF-15 with a filter UFS-1).

3. Enhancing and Quenching of $CaWO_4$ Intrinsic Luminescence by Lanthanides

The effect of rare-earth impurities on the luminescence of $CaWO_4$ excited in a wide range of quantum energies was found to vary substantially in different regions of the spectrum. Fig. 1 presents typical luminescence intensity spectra versus praseodymium concentration, Fig. 2 the effect of Yb, Pr, and Tb impurities on the spectra, i.e. the effect of exciting quantum energy on the relative change of radiation intensity $(I_0 - I)/I_0$ due to doping (I_0 is the luminescence intensity of undoped $CaWO_4$, I that of doped samples). From the analysis of recent and earlier data in relation to the excitation spectra (Fig. 2b) we come to the following conclusions:

1. Doping by impurities of the rare-earth series Pr, Sm, Gd, Tb, Ho, Tm, and Yb in concentrations 10^{-6} to 10^{-5} g-atom/mol (the actual concentration may be still less since in the absence of coactivators not the full quantity of Ln impurity in the charge penetrates into $CaWO_4$ lattice) enhances the luminescence. In the vacuum ultraviolet region the intensity gain amounts to 40 to 50%.

2. Further increase of Ln concentration up to 10^{-3} to 10^{-2} g-atom/mol causes quenching of luminescence.

3. Both enhancing at small concentration and quenching at relatively large concentrations show up more definitely in the region of the dip in the $CaWO_4$ excitation

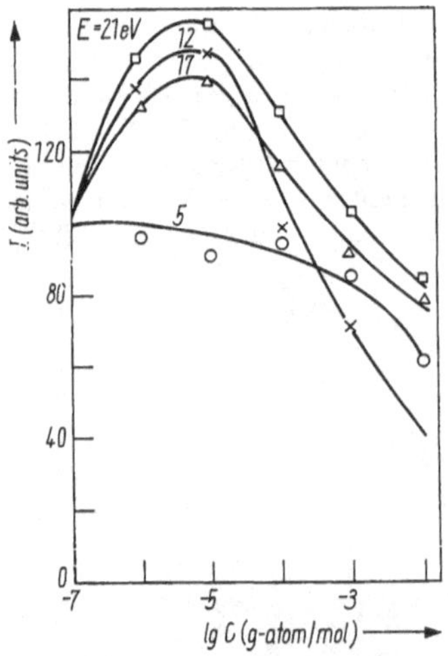

Fig. 1

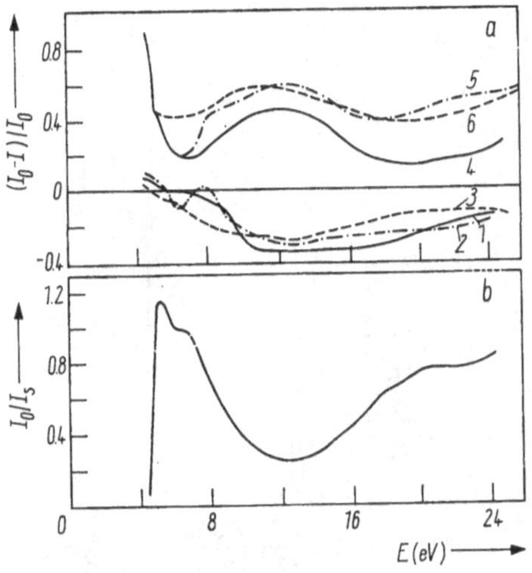

Fig. 2

Fig. 1. $CaWO_4$ luminescence intensity versus Pr concentration, C, in the charge. Energies of exciting photons are marked at the curves

Fig. 2. a) Effect of Yb (1), (4), Pr (2), (5), and Tb (3), (6) on the $CaWO_4$ luminescence; impurity concentrations (in g-atom/mol): (1) 10^{-5}, (2) and (3) 10^{-6}, (4), (5), (6) 10^{-2}. b) Excitation spectrum of undoped $CaWO_4$. I_s means the luminescence intensity of sodium salicilate

spectrum at 12 eV. This is not the case for Ce, which is an exception in the lanthanide series.

4. Ln impurities, their effect on the luminescence being a function of quantum energy of excitation, can distort the shape of the excitation spectra.

4. The Effect of Lanthanides on $CaWO_4$ Phosphorescence

The lanthanide-free $CaWO_4$ phosphorescence spectrum consists of a wide band, its position is determined by the remaining impurities of group V elements As, Sb, Nb, and Ta [3]. Doping with lanthanides does not effect the phosphorescence spectrum in most cases. Tb, Pr, and Eu show however, a different behaviour: as it is seen from Fig. 3 the phosphorescence of $CaWO_4$ phosphors, containing these impurities Tb, Pr, and Eu shows their characteristic series of lines. In spite of the low concentration of Tb, the phosphorescence spectrum of $CaWO_4$: Tb mainly consists of the lines arising from the transitions from the lowest excited level $^5D_4$ occurring possibly due to the sensitizing effect of $CaWO_4$ intrinsic luminescence centres.

It was found earlier that Eu quenches the $CaWO_4$ phosphorescence [4], which causes blackening of a non-sensitized film with the sensitivity band at $\lambda \leqq 500$ nm. But besides the effect mentioned Eu as well as Pr and Tb show intrinsic phosphorescenc in the red spectral region being built-in in the lattice of $CaWO_4$ in a trivalent state. This state is favoured by charge compensation by pentavalent tungsten.

The data presented in Fig. 3 were obtained with X-ray excitation of luminescence. Besides, experiments have been performed using the Hg line at 254 nm as excitation source and isochromatic film RM-6 as recorder (spectral sensititivity range at $\lambda \leqq \leqq 640$ nm). In this case we observed neither $CaWO_4$: Tb and $CaWO_4$: Pr phosphores-

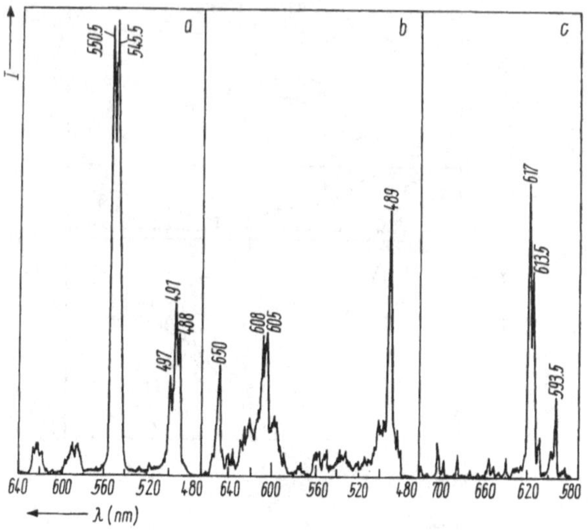

Fig. 3. Doped $CaWO_4$ phosphorescence spectra excited by X-ray radiation; impurity concentration 10^{-4} g-atom/mol. a) Tb, b) Pr, c) Eu. No correction for spectral sensitivity of the experimental arrangement was made. The dominating lines are marked

cence nor afterglow due to other lanthanides. The only exception was Eu; $CaWO_4$:Eu 10^{-4} g-atom/mol) phosphors show a weak phosphorescence with a spectral distribution determined by Eu^{3+} ions.

5. Discussion

Lanthanides built-in on the cation sites of $CaWO_4$ lattice form, as it was already mentioned, electron traps $Ln^{\cdot}_{Ca}$ with their charge compensated by intrinsic defects, W'_W (penta-valent tungsten) or V''_{Ca} (calcium vacancy) [5, 6]. If $Ln^{\cdot}_{Ca}$ concentration increases, their tendency to associate with compensating defects increases as well. It stimulates, on one hand, resonant quenching of intrinsic luminescence arising from W'_W [1 to 3] and, on the other hand, favours the creation of hole traps by lanthanides liable to the transition to the tetravalent state. Such a behaviour is typical for Tb^{3+} and Pr^{3+} which have, respectively, one and two electrons in the 4f shell in excess of the stable configuration. Hole trap levels created by these ions seem to be deeper than levels of $(AO_4)'_{WO_4}$ type centres, where A is a group V element (Fig. 4). This is the origin of the decisive effect of Tb and Pr on the phosphorescence spectrum of Tb- and Pr-doped calcium tungstate. Their emission arises from the recombination of electrons with Tb^{4+} and Pr^{4+} and the transport of the recombination energy to the trivalent ions, Tb^{3+} and Pr^{3+}, created. The extraordinarily strong phosphorogenic effect of Tb and Pr, mentioned earlier [2, 3], is due to the creation of deep hole traps. At the same time the quantum energy 4.9 eV (254 nm line) is not sufficient to ionize directly these impurities in $CaWO_4$ crystals.

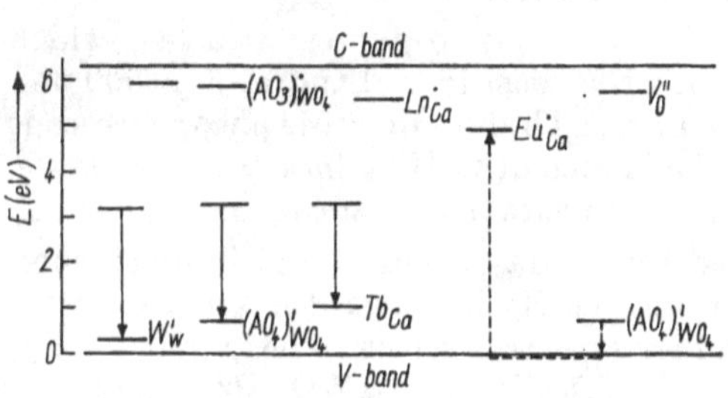

Fig. 4. Simplified scheme of energy levels of $CaWO_4$ doped with lanthanides and group V elements. The solid arrows are radiative transitions, the dashed arrows are charge transfer transitions accompanied by hole capture by the centre $(AO_4)'_{WO_4}$

Eu^{3+}, which lacks one electron for the stable configuration ($4f^7$), forms a deeper electron trap than other Ln^{3+}. The emission characteristic of Eu^{3+} can occur due to the recombination of the trapped electrons with holes ejected from $(AO_4)'_{WO_4}$-type centres. Since almost all electrons captured are trapped by Eu^{3+} traps deeper than that created by $(AO_4)'_{WO_4}$ centres, the probability of electron and hole recombination at $(AO_4)'_{WO_4}$ centres in the presence of Eu drastically decreases and leads to the quenching of phosphorescence which causes blackening of optically non-sensitized films.

Irradiation with the 254 nm line can cause transitions with charge transfer (Fig. 4) resulting in transitions of electrons from the valence band to the levels Eu_{Ca} and trapping of holes by $(AO_4)'_{WO_4}$ centres. Subsequent hole release leads to their recombination with captured electrons accompanied by Eu^{3+} excitation.

Enhancing of $CaWO_4$ intrinsic luminescence in the presence of small concentrations of lanthanides is probably the most interesting effect. We have observed it earlier investigating X-ray luminescence of $CaWO_4$ and some other systems containing Ln [7, 8]. The gain in emission intensity with X-ray excitation did not usually exceed 10 to 15%. But with VUV excitation it amounts to 40 to 50% and even more. Such effect of Ln can be explained on the assumption that even $CaWO_4$ prepared from very pure raw materials possesses quenching centres acting according to the Auger mechanism [7, 9]. Transitions of electrons from these centres to deeper levels formed by Ln prevent quenching.

The strongest enhancing and quenching effect of impurities in the region of the deepest dip in the excitation spectrum can be due to the small penetration depth of ultraviolet radiation in this region in $CaWO_4$. Only a thin surface layer is excited where the concentration of W'_W-type centres is reduced. The competing effects of other centres are more clearly visible on these conditions. Then non-uniform distribution of Ln impurities in the surface layer in the absence of coactivators can also play a role.

It was often stated earlier [10, 11] that every group WO_4^- can serve as an emission centre. This conclusion was made in the absence of data on the quenching and enhancing effect of impurities on the $CaWO_4$ luminescence [12]. The phenomena discussed in the present paper, mainly the enhancing of intrinsic luminescence of $CaWO_4$ by extremely small concentrations (10^{-6} to 10^{-5} g-atom/mol) of lanthanides, give evidence that this is not the case: not all WO_4 groups are emission centres.

Acknowledgements

The authors thank R. V. Katomina, A. A. Mikhalev, M. I. Tombak, and T. A. Sokolova for preparation of the phosphors.

References

[1] A. M. Gurvich, R. V. Katomina, A. I. Kravchenko, V. V. Mikhailin, M. I. Tombak, and A. N. Vasilev, in: Internat. Conf. Defects in Insulating Crystals, Abstracts, Zinātne Publ. House, Riga 1981 (p. 243).

[2] A. M. Gurvich, R. V. Katomina, A. A. Mikhalev, M. I. Tombak, and H. Wagner, Z. phys. Chem. 282, No. 3 (1982).

[3] A. M. Gurvich, V. B. Gutan, B. N. Meleshkin, V. V. Mikhailin, A. A. Mikhalev, and M. I. Tombak, J. Lum. 15, 187 (1977).

[4] A. M. Gurvich, R. V. Katomina, A. A. Mikhalev, and M. I. Tombak, in: Proc. III. Union Conf. Luminescence Detectors and Transformers of X-Ray Radiation, Izd. VNIILyuminoforov, Stavropol 1980 (p. 39) (in Russian).

[5] K. Nassau, J. Phys. Chem. Solids 24, 1511 (1963).

[6] A. M. Morozov, E. G. Reut, A. I. Ryskin, and P. P. Feofilov, Ucheniye zapiski Kazanskogo Univ. Vol., 129/1, No. 5, 1969 (p. 5) (in Russian).

[7] A. M. Gurvich, Izv. Akad. Nauk SSSR, Ser. fiz. **45**, 281 (1981).
[8] T. A. Birman, S. I. Golovkova, A. M. Gurvich, R. V. Katomina, and N. P. Soshchin, see [4] (p. 59).
[9] M. K. Sheinkman, N. E. Korsunskaya, I. V. Markevich, and T. V. Torchinskaya, Izv. Akad. Nauk SSSR, Ser. fiz. **40**, 2290 (1976).
[10] G. Blasse and A. Bril, Philips tech. Rev. **31**, 304 (1970).
[11] E. G. Reut, Izv. Akad. Nauk SSSR, Ser. fiz. **43**, 1186 (1979).
[12] F. A. Kröger, Some Aspects of Luminescence of Solids, Elsevier Publ. Co., Amsterdam/New York/London 1948.

(Received November 10, 1982)

phys. stat. sol. (a) 77, 381 (1983)

Subject classification: 14.3.3; 22.9

Solid State Physics Laboratories,
Department of Physics, University of Saugar, Sagar[1])

Spontaneous Electric Current from Polyvinyl Alcohol Induced by the First Heating Run

By

A. U. SHAREEF, K. K. SARAF, and A. P. SRIVASTAVA

A spontaneous current (SC) is found from metal(1)–polyvinyl alcohol–metal(2) systems. Variation of SC from room temperature to just below the softening temperature of the polymer is recorded. The maximum value of the SC during first heating is found to be of the order of 10^{-6} A and its thermograms exhibit two transitions at 130 and 180 °C (heating rate 6 K/min). Thermograms after first heating become independent of number of heating runs and attain a maximum value of 10^{-8} A in subsequent re-heatings, and display a single transition at about 65 K. Studies on spontaneous current emission (SCE) from first heating are discussed in detail. A temperature dependence of the open circuit voltage is also reported. It is proposed that the SCE is a water activated phenomenon, which is also influenced by the transitional changes of the polymer.

Самопроизвольный (спонтанный) ток получается из металла(I)-поливинил-спирта–металла(2) системы. Вариация, самопроизвольного тока из комнатной температуры, которая чуть меньше, чем смягчённая температура полимера, описана. Максимальное значение самопроизвольного тока при первом нагреве получилось до 10^{-6} A и его термограммы показывают два перехода при температуре 130 и 180°C (скорость нагрева 6 K/min). После первого нагрева термограммы становятся независимыми от количества нагрева и получается максимальное значение 10^{-8} A при последующем повторном нагреве и показывают один переход в приблизительно 65 K градусов. Исследования самопроизвольного тока эмиссии (СТЭ) от первого нагрева подробно обсуждали. Зависимость температуры напряжение холостого хода тоже описали. Предлогаем, что самопроизвольный ток эмиссия (СТЭ) это явление активированной (возбуждённой) воды, которое влиялось переходными изменениями полимера.

1. Introduction

Some polymers sandwiched between short-circuited metal electrodes are found to generate an electric current [1 to 3] which is weak at room temperature but becomes quite appreciable at elevated temperatures. This phenomenon is called spontaneous current emission (SCE). The literature in this field [1 to 4] has suggested that the origin of SCE is of electrochemical nature. It is assumed that some reaction takes place between metallic electrodes and ions present in the insulator, but it has not yet been possible to identify the actual reaction that takes place. The knowledge of the true nature of SCE is necessary because it is always present as a background current in all electrical measurements that involve polymeric dielectrics. At present the metal–polymer–metal system is considered a poor solid state galvanic cell as it cannot deliver any useful power [4], but such cells can be made to deliver higher power at elevated temperatures.

The purpose of the present work was to study in detail the phenomenon of SCE in thin films of polyvinyl alcohol (PVA) with different combinations of vacuum-coated

[1]) 151-A, 5-Civil Lines, Sagar 470001 (M.P.) India.

electrodes. We found that the nature and magnitude of SCE from a non-preheated sample and from a sample which had gone through a heat-and-cool cycle are very different. The maximum value of short-circuit current during the first heating of a virgin specimen is found to be of the order of 1 μA, and its thermogram exhibits two transitions at 130 and 180 °C (heating rate 6 K/min). Thermograms after first heating become independent of number of heating runs and can attain a maximum value of 10^{-2} μA in subsequent reheatings, and display a single transition at about 65 °C.

In this paper studies on SCE from the first heating run of M_1–PVA–M_2 systems are discussed in detail, and the open-circuit voltage across the electrodes has also been recorded as a function of temperature. It is proposed that the SCE from PVA is a sorbed-water-activated phenomenon. In a recent communication the results of the investigations on SCE from second and subsequent reheatings of M_1–PVA–M_2 structures are intended to be reported.

2. Experimental

Commercial grade PVA, supplied by Robert Johnson England, was used in this study. Thin films of thickness 10 μm were grown from aqueous solution of the polymer on clean substrates of various metals by isothermal immersion technique [5]. Aluminium or silver was vacuum evaporated on the upper surface of the film to obtain a sandwich configuration. Electrical contacts for measurement were obtained by sticking thin silver wire by means of a highly conductive dag. Samples were kept in an electrically shielded chamber whose temperature can be varied with the help of a resistive-heating arrangement. The output from samples was fed to an electrometer amplifier (EA-815, Electronic Corporation of India Ltd.) for measurements of short-circuit current and open-circuit voltage.

3. Results

3.1 Short-circuit current

SCE thermograms obtained from Al–PVA–Cu/Zn/Ni/Pb/Al structures, when heated at a uniform heating rate of 4 K/min, from room temperature to just below the softening temperature (200 °C), are shown in Fig. 1. It can be seen that the curves are generally similar in shape, exhibiting two well-resolved maxima centred at 108 and 170 °C (± 10 K). The direction of the current as per accepted convention [4] is found to be negative, that is it flows from the lower (Cu, Zn, Ni, or Pb) to the upper (Al) elec-

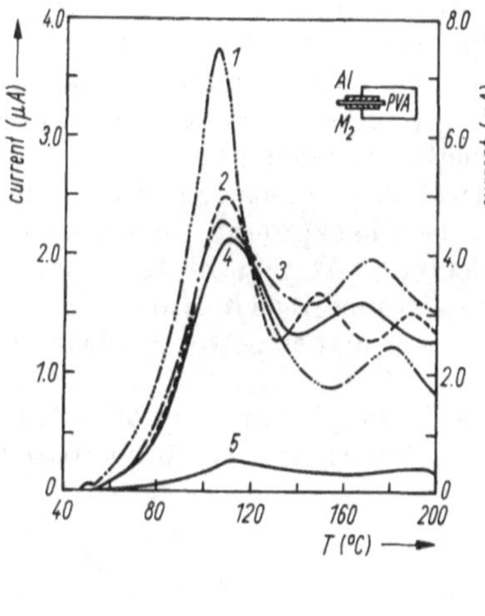

Fig. 1. Short-circuit current during first heating for Al–PVA–M_2 systems. Curves 1, 2, 3, 4 and 5 correspond to lower electrode (M_2) of Cu, Zn, Ni, Pb, and Al, respectively. Heating rate 4 K/min. (Curve 1 refers to right hand scale)

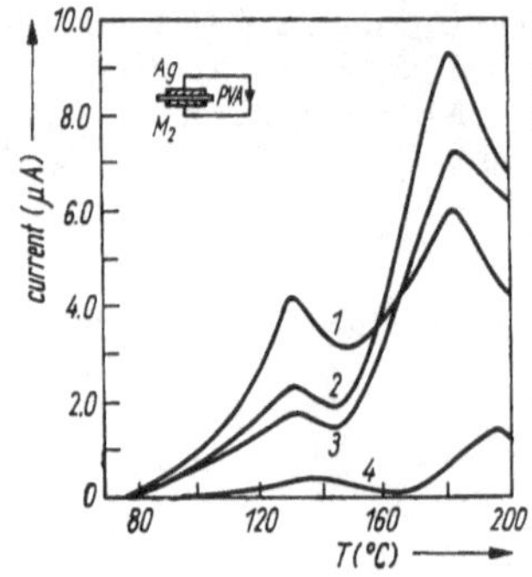

Fig. 2. Short-circuit current during first heating for Ag–PVA–M_2 systems. Curves 1, 2, 3, and 4 correspond to lower electrode (M_2) of Al, Zn, Pb, and Ag, respectively. Heating rate 6 K/min

trode through the external circuit. In Fig. 2 the SCE spectra from Ag–PVA–Zn/Pb/Al/Ag systems are presented obtained by heating the samples at a rate of 6 K/min. The direction of the current in all these cases is found to be positive, that is, from upper (Ag) to lower (Zn, Pb or Al) electrode in external circuit. As the work functions of Cu, Zn, Ni, and Pb are greater than that of Al, and the work functions of Zn, Pb, and Al are smaller than that of Ag, it can be inferred that the current always flows from an electrode of higher work function to one of lower work function. If similar electrodes are used, the current is found to be of a magnitude much lower than that in case of dissimilar electrodes (curve 5 in Fig. 1 and curve 4 in Fig. 2) and its direction differs from sample to sample, being positive in some samples and negative in others.

A comparison of Fig. 1 and 2 shows that high heating rate increases the current magnitude and sharpens the peaks of SCE spectra, and displaces them towards the high-temperature side by about 10 to 20 K. To confirm this conclusion, identical samples of Ag–PVA–Al were heated with different heating rates, viz. 2, 4, and 6 K/min. The results are shown in Fig. 3 (curves 3, 2, and 1).

An occasional peak of varying magnitude is also found at about 65 °C in the SCE spectra of some of the samples used in this investigation (curve 5, Fig. 3). Further, we have examined the effect of water molecules absorbed on the film surface. This was done by exposing the sample to saturated humidity for different lengths of time. A typical result is shown in Fig. 3 (curve 4). The current from a moisture-soaked sample has a very high value at and near room temperature, but in the high temperature region it is always less than the current obtained from a virgin specimen untreated by moisture.

Curve 6 in Fig. 3 shows the nature of SCE during an uncontrolled cooling of the Ag–PVA–Al structure. The current falls very rapidly and becomes negligible even at 100 °C or so. The direction of the current during cooling remains the same as it was

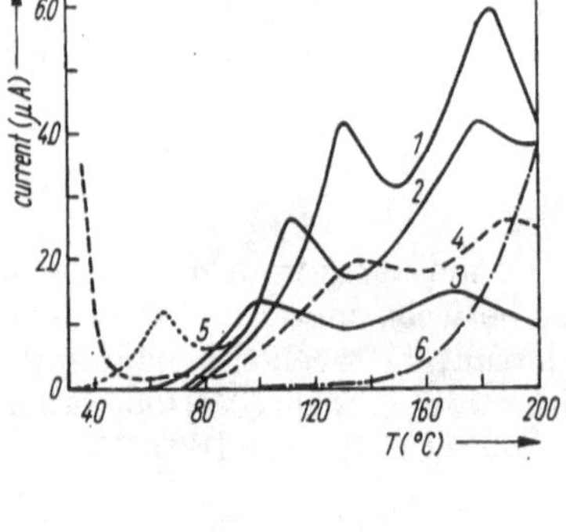

Fig. 3. Short-circuit current during first heating for Ag–PVA–Al system under different conditions. Curves 1, 2, and 3 correspond to heating rates of 6, 4, and 2 K/min, respectively. Curve 4 shows the current spectrum obtained from the sample after keeping it in saturated humidity for 20 h (heating rate 6 K/min). Curves 5 shows an additional peak observed in some samples. Curve 6 shows the current during uncontrolled cooling

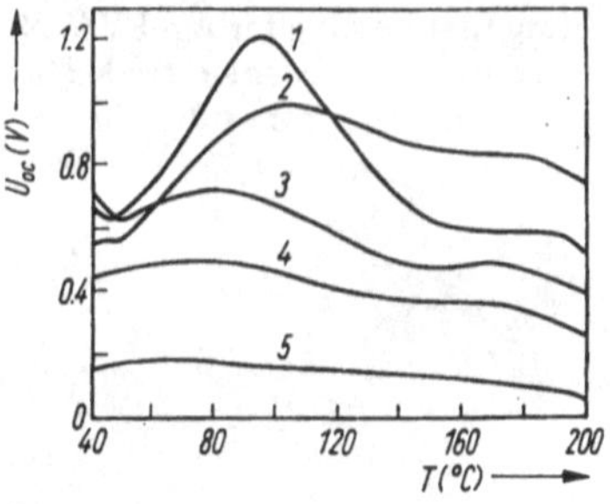

Fig. 4. Open-circuit voltage U_{oc} during first heating for Al–PVA–M_2 systems. Curves (1) to (5) correspond to lower electrode (M_2) of Cu, Ni, Pb, Zn, and Al, respectively. Heating rate 4 K/min

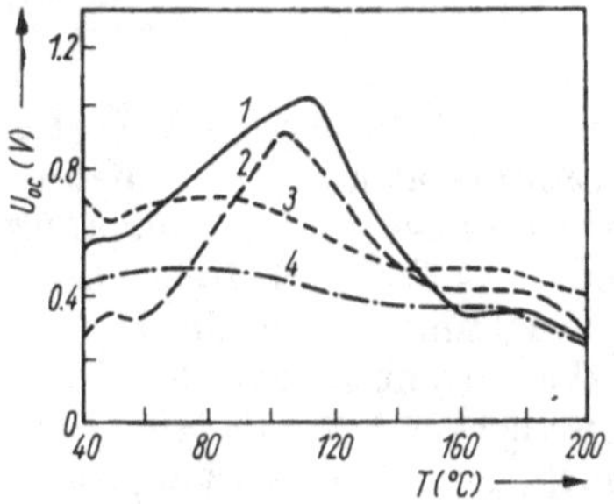

Fig. 5

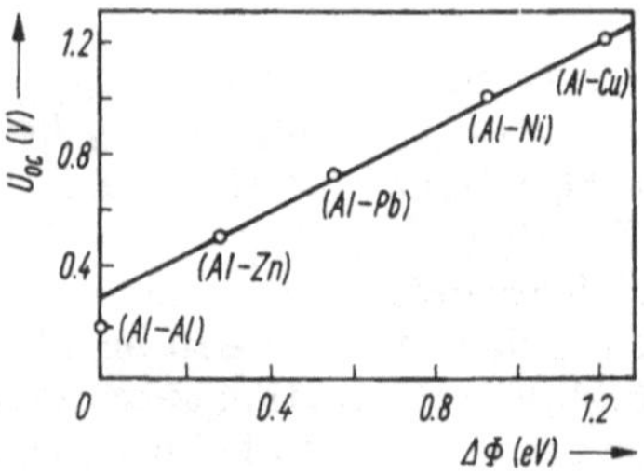

Fig. 6

Fig. 5. Open-circuit voltage during first heating for (1) Ag–PVA–Zn (difference in work functions of electrodes) $\Delta\Phi = 0.94$ eV) (2) Ag–PVA–Pb ($\Delta\Phi = 0.66$ eV) (3) Al–PVA–Pb ($\Delta\Phi = 0.56$ eV) (4) Al–PVA–Zn ($\Delta\Phi = 0.28$ eV)

Fig. 6. Peak U_{oc}, corresponding to U_{oc} vs. temperature curves of Al–PVA–M_2 structures, as a function of work function difference $\Delta\Phi$ of respective electrodes

during the heating of the sample. PVA, thus, belongs to category A of the classification of Sawa et al. [1].

3.2 *Open-circuit voltage (U_{oc})*

Fig. 4 shows the temperature dependence of U_{oc} during the first heating run of a virgin Al–PVA–Cu/Ni/Pb/Zn/Al sample with a heating rate of 4 K/min. U_{oc} first rises with temperature exhibiting a broad peak at about 95 °C, then it tends to become steady at higher temperatures. It is also clear from Fig. 4 that U_{oc} magnitudes as well as sharpness of the transitional peaks are proportional to the work function difference of the electrode metals. In Fig. 5 U_{oc} plots of Ag–PVA–Zn/Pb systems are shown along with those of Al–PVA–Zn/Pb for comparison, which confirms the conclusion made above. The peak U_{oc} for each of the curves of Fig. 4 is plotted against the work function difference of the corresponding electrodes (Fig. 6), and it is seen that their relation is linear.

4. Discussion and Conclusion

A number of workers have described that metal surfaces may acquire a net positive or negative charge due to electron and/or ion transfer between metal and dielectric [6, 7], and it has also been shown [8, 9] that the relationship between the magnitude of the charge transferred and the difference in work function is linear. Various other interpretations have been put forward for the origin of the SCE [1 to 3, 10 to 12].

The spontaneous current emission seems to be an activation-controlled phenomenon, at least in those dielectrics which contain sorbed water. Though the exact nature of the activation centres is not clear, the detail manner in which the water molecules are accomodated in the polymer structure controls the mechanism of the emission process. Grinter and Bowlt [13] have described that the activation centres at the surface of the specimen are activated by adsorbed water and the centres within the bulk are activated by absorbed water. The heating process could alter the quasi-equilibrium position of an activation centre and could thus lead to the movement of charge carriers.

The peaks observed at about 65 and 170 °C ($\pm$ 10 K) can be attributed, respectively, to dissociation of loosely bound (adsorbed) and strongly bound (absorbed) water molecules in PVA. As T_g of the polymer is around 65 °C [14], it becomes more crystalline above this temperature by releasing loosely bound water. Further, it has been reported [15] that the water which is strongly bound with PVA splits off at the maximum rate at 160 °C.

The first heating run empties all the activation centres and destroys some of them. The surface-activation centres reappear by adsorbing moisture from the surroundings when the polymer is cooled below T_g, and they again produce a peak around 65 °C at subsequent reheating. This peak being comparatively weak is often masked by the stronger current peak occurring at 130 °C in the initial heating run.

The sharp peak in SCE (and also in U_{oc}) thermograms at about 130 °C is probably due to structural transitions of PVA. In addition to first- and second-order transitions, discontinuous changes in various other physical properties of PVA have been observed at about this temperature [16]. It has been demonstrated from infrared [17] and NMR data that the molecular motion in PVA crystallites becomes more intense about 130 °C. This increased mobility allows rapid crystallisation of incompletely crystallised samples [18] resulting in a lower density of activation centres.

The SCE thermogram is not repeated in subsequent reheatings because the first heating run liberates the absorbed water molecules and causes an almost complete crystallisation of the polymer. As observed by Nagai et al. [19], PVA can be readily crystallised by heating, and a few minutes of heat treatment is sufficient for maximum crystallisation. Valentine [20], who has reviewed the relation between the sorption of moisture by various kinds of polymers (including PVA) and their crystallinity, concludes that if a polymer can form well-defined crystallites, then these crystallites usually are inaccessible to H_2O vapour.

SCE at room temperature from a moisture-soaked sample has a high magnitude because water in contact with metal can give rise to an electric potential by orientation of its dipolar molecules and by the flow of electrons from or into the electrodes. Further, the presence of dissolved oxygen molecules and hydrogen ions (H^+) makes the interface similar to that between an electronic conductor and an ionic conductor [7]. Long exposure of the sample to moist air causes galvanic corrosion and pitting [21] of electrodes, and as a consequence a decrease in current magnitude is observed in the high-temperature region.

The size and shape of transitional peaks, and their location on the temperature scale depend on the heating rate because of the fact that the polymers change their properties gradually over a certain interval of temperature.

Acknowledgements

The authors are indebted to Prof. R. K. Shrivastava, Head of Physics Department, University of Saugar, for providing the necessary facilities. One of the authors is thankful also to the University Grants Commission, New Delhi, India for the award of a Research Fellowship.

References

[1] G. Sawa, D. C. Lee, and M. Ieda, Japan. J. appl. Phys. **16**, 359 (1977).
[2] S. Radhakrishna and S. Haridoss, Indian J. Phys. **53**, 103 (1979).
[3] P. K. C. Pillai and M. Mollah, J. macromol. Sci., Phys. **B16**, 327 (1979).
[4] M. Ieda, G. Sawa, S. Nakamura, and Y. Nishio, J. appl. Phys. **46**, 2797 (1975).
[5] A. C. Rastogi and K. L. Chopra, Thin Solid Films **18**, 187 (1973).
[6] Y. Murata and S. Kittaka, Japan. J. appl. Phys. **18**, 421 (1979).
[7] J. O'M. Bockris and A. K. N. Reddy, Modern Electrochemistry, Vol. I, Plenum Press, New York 1977.
[8] D. K. Davies, J. Phys. D **2**, 1549 (1969).
[9] A. Chowdry and C. R. Westgate, J. Phys. D **7**, 713 (1974).
[10] S. Radhakrishna and S. Haridoss, J. appl. Phys. **49**, 301 (1978).
[11] F. Gutmann and H. Keyzer, J. chem. Phys. **50**, 550 (1969).
[12] A. R. Tiwari, S. K. Shrivastava, K. K. Saraf, and A. P. Srivastava, Thin Solid Films **70**, 191 (1980).
[13] M. R. Grinter and C. Bowlt, J. Phys. D **9**, L61 (1976).
[14] H. Futaba and J. Furuichi, High Polymer (Japan) **71**, 51 (1954).
[15] H. F. Mark and N. G. Gaylord, Encyclopedia of Polymer Science and Technology, Vol. 14, Interscience Publishers, New York 1971.
[16] K. Ishikawa and K. Miyasaka, Rept. Progr. Polymer Phys. Japan **7**, 93 (1964).
[17] E. Nagai, Polyvinyl Alcohol, Ed. I. Sakurada, The Soc. of Polymer Sci., Japan 1955.
[18] K. R. Tubbs, J. Polymer Sci. **A3**, 4181 (1965).
[19] K. Nagai, S. Mima, S. Kuribayashi, and N. Sagane, Chem. High Polymers (Japan) **12**, 199 (1955).
[20] L. Valentine, J. Polymer Sci. **27**, 313 (1958).
[21] E. A. Gulbransen, McGraw-Hill Encyclopedia of Science and Technology, Ed. D. N. Lapedes, Vol. 3, McGraw-Hill Book Company Inc., NewYork 1971.

(Received January 27, 1983)

phys. stat. sol. (a) **77**, 387 (1983)

Subject classification: 2 and 14.3.4; 22.1.3

V. Kapzukas State University, Vilnius[1])

Features of Hot Carriers in Amorphous Selenium

By

G. Juška and K. Arlauskas

The temperature and field dependences of both, drift mobilities and effective free paths of electrons and holes heated by the electric field are investigated. The non-activated hot hole mobility as well as carrier free paths increase with kinetic energy. The highest possible electric field is limited by the avalanche due to the impact ionization by holes and electrons.

Die Abhängigkeit der Beweglichkeit und effektiven freien Weglänge heißer Ladungsträger von der Temperatur und der Feldstärke wird untersucht. Die unaktivierte Löcherbeweglichkeit und freie Weglänge nimmt mit der kinetischen Energie zu. Es wird gezeigt, daß das elektrische Feld in amorphen Selen durch eine Lawine von Löcher- und Elektronen-Stoßionisation begrenzt ist.

1. Introduction

In disordered structures the heating of charge carriers is considered to be unlikely due to the smallness of a mean free path. However, our experiments on the interband impact ionization in a-Se has proved it to be possible [1, 2]. The presence of impact ionization is based on the following facts: (i) for the quantum efficiency exceeding unity, the pulse of hole drift current increases exponentially with time due to the hole multiplication and the appearance of the pulse tail due to the drift of electrons ionized by holes; (ii) the total hole quantum efficiency β_p as well as the partial contributions of ionized electrons and holes to β_p are specifically dependent on the electric field; (iii) an exponential dependence of quantum efficiency β_p on the thickness of the selenium layer is observed.

Further detailed study of the impact ionization allows to discover certain properties of hot holes and electrons. In this paper the temperature and field dependences of the nonactivated hole mobility as well as the mean free paths of electrons and holes are presented.

2. Experimental and Results

A conventional time-of-flight technique has been used to study carrier mobility and impact ionization. A film of amorphous selenium (3 to 200 μm) is placed between two isolating layers in order to exclude carrier injection from electrodes. The carrier photogeneration on the selenium surface is carried out using strongly absorbed light of a pulsed sub-nanosecond nitrogen laser through semi-transparent metal electrode and isolating layer. A single pulse of drift current through the 50 Ω waveguide feed is registered by the oscilloscope (0 to 5 GHz). The drift mobility is evaluated from measurements of transit time. The quantum efficiencies β_p and β_n are computed from the transit pulse area, when either the anode or the cathode is illuminated. The evaluation of the electric field in a-Se is described in [1].

It was shown in [1] that in the insulator–selenium–insulator system the electric field is proportional to the applied voltage up to a certain value $E = E_m$. A further

[1]) Universiteto 3, 232734 Vilnius, USSR.

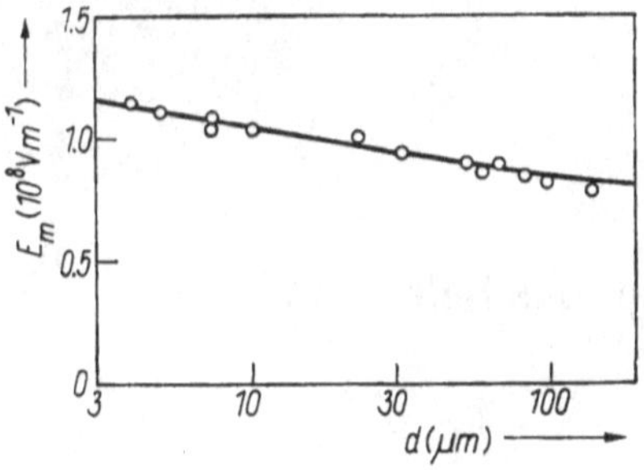

Fig. 1

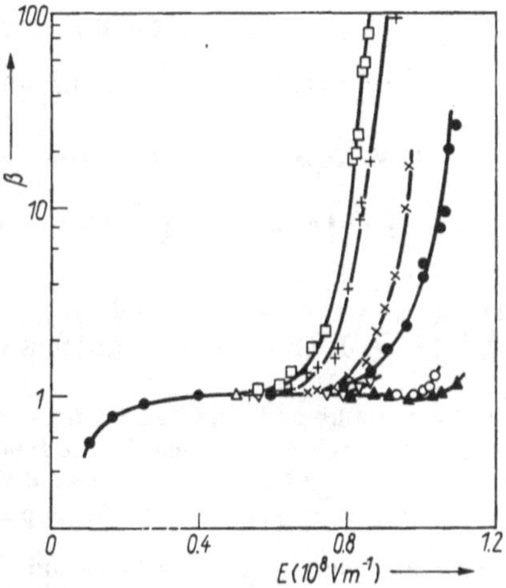

Fig. 2

Fig. 1. Dependence of the highest possible field E_m on thickness. Circles are experimental values, solid line is calculated according to the avalanche condition (5)

Fig. 2. Field dependences of quantum efficiencies for holes (thickness of the selenium layer: ● 4, × 7, + 33, □ 103 μm) and electrons (▲ 4, ○ 5, ▽ 59 μm)

increase in voltage does not change the electric field. The values of the highest electric field E_m versus the thickness d of the selenium layer are shown in Fig. 1. The field dependences of β_p and β_n are presented in Fig. 2.

It was observed experimentally [1] that the hole mobility at $E \approx 10^8$ V/m lost its temperature activated dependence characteristic of disordered structures. The temperature and field dependences of the non-activated hole drift mobility μ_p are shown in Fig. 3 and 4, respectively. The hole mobility μ_p is independent of layer thickness in the range studied. The drift mobility of electrons is temperature activated up to the electric field E_m and therefore $\mu_n(E)$ results are not presented.

3. Discussion

The experiment has been carried out in the regime of small charge drift, i.e. at a constant electric field. Besides, due to the interband ionization the charge transfer occurs over the whole thickness of the selenium layer: a hole transfers charge from the ionization site to the cathode, and an electron to the anode. These circumstances simplify the calculation of the quantum efficiency.

At fields where the impact ionization by electrons is still absent $((\beta_n - 1) \ll 1)$, each photogenerated hole ionizes $[\exp(\gamma_p d) - 1]$ pairs during its drift, where γ_p is the coefficient of impact ionization by holes. In this case, the quantum efficiency is $\beta_p = \exp(\gamma_p d)$.

At higher fields this impact ionization will develop to an avalanche when the number of pairs produced by electrons exceeds the number of photogenerated holes. This avalanche also will develop by the multiplied "dark" carriers independently of their density. Consequently, it would limit the highest possible electric field E_m. Considering the facts mentioned above together with the experimentally obtained relation $\beta_p - 1 \gg \beta_n - 1$ (see Fig. 2), it follows that at $E < E_m$ the inequality $\gamma_n d \ll 1$ must be valid. Hence, in the electric field region where the ionization by electrons occurs ($\beta_n > 1$) the pairs are mainly ionized by holes at the cathode (since here $\gamma_p d \gg$

$\gg 1$), whereas they are uniformly created in case of ionization by electrons (since $\gamma_n d \ll 1$). Summing up the charges, photogenerated as well as ionized by electrons and holes when the anode is illuminated, one obtains

$$\beta_p = \exp(\gamma_p d) + \frac{\gamma_n}{\gamma_p} [\exp(\gamma_p d) - 1]^2 \sum_{i=2}^{\infty} \left\{\gamma_n d \left[\frac{\exp(\gamma_p d) - 1}{\gamma_p d} - 1\right]\right\}^{i-2}$$

$$= \exp(\gamma_p d) + \frac{[\exp(\gamma_p d) - 1]^2 \cdot \gamma_n}{\gamma_p \left[1 - \gamma_n d \left(\frac{\exp(\gamma_p d) - 1}{\gamma_p d} - 1\right)\right]}, \qquad (1)$$

while for the case of photogeneration at the cathode

$$\beta_n = (1 + \gamma_n d) \sum_{i=1}^{\infty} \left[\gamma_n d \left(\frac{\exp(\gamma_p d) - 1}{\gamma_p d}\right)\right]^{i-1} = \frac{1 + \gamma_n d}{1 - \gamma_n d \left(\frac{\exp(\gamma_p d) - 1}{\gamma_p d} - 1\right)}, \qquad (2)$$

where i is the ionization number.

γ_p and γ_n may be determined from (1) and (2) as functions of experimentally measured values of β_p, β_n as

$$\gamma_p = \frac{1}{d} \ln\left(\frac{\beta_p - 1}{\beta_n} + 1\right), \qquad (3)$$

$$\gamma_n^{-1} = \frac{(\beta_p - 1)\, d}{(\beta_n - 1) \ln\left(\frac{\beta_p - 1}{\beta_n} + 1\right)} - 1. \qquad (4)$$

The values of γ_p are determined from the experimental values of β_p at fields, where $\beta_n - 1 \ll 1$, i.e. when the ionization is due to holes only. In such a case, the assumption about the ionization site has no influence on the γ_p-value. The values of γ_n calculated from (4) are slightly underrated as compared with the real values. The independence of the computed values of γ_p and γ_n on the sample thickness together with a strong dependence of β_p and β_n on d affirms the reliability of the data obtained.

When the denominator in (1) and (2) equals zero, β_p and β_n increase infinitely due to the avalanche. This gives the criterion for a maximum possible field E_m,

$$\gamma_n(E)\, d \left(\frac{\exp(\gamma_p(E)\, d) - 1}{\gamma_p(E) d} - 1\right) = 1. \qquad (5)$$

The field dependences of γ_p and γ_n calculated according to (3) and (4) as well as criterion (5) were used for the calculation of E_m. A good agreement between calculated and experimental results (see Fig. 1) points to the fact that E_m is determined by the avalanche.

The coefficient of impact ionization γ is defined by the effective carrier path λ between the events of inelastic scattering. In case of elastic scattering λ exceeds the real distance between the collisions by the factor by which the carrier mean energy exceeds the energy dissipated in a single collision. Assuming an exponential distribution of carrier paths, one obtains

$$\gamma = \frac{\alpha}{\lambda} \exp\left(-\frac{\Delta\varepsilon}{e\lambda E}\right),$$

where α is the ionization probability of a pair, $\Delta\varepsilon$ the interband impact ionization energy. In crystalline semiconductors $\Delta\varepsilon$ is determined from the energy and momentum conservation laws. In disordered structures the momentum of a particle is uncertain due to the small value of the mean free path. Therefore, the value of $\Delta\varepsilon$ is expected to be between the band gap value (≈ 2.3 eV) and half the energy of a quantum which may photogenerate two carriers (≈ 3 eV) [3]. Due to the smallness of λ as compared to the Coulomb radius it follows that α should be defined by the probability of pair separation, i.e. by the Onsager dissociation, similarly as the quantum efficiency [4]. The region of non-photoactive absorption narrows at $E = 10^8$ V/m. For quantum energies of 2.3 eV the dissociation probability of a photogenerated pair approaches unity. Therefore, the uncertainty of the mean free path between the events of inelastic scattering is due to the assumption that $\alpha = 1$ as well as due to the unknown correct shape of the distribution function. Since in disordered structures the mobility increases with carrier energy this distribution function may be flatter than chosen above.

Fig. 3 and 4 show a correlation between the field and temperature dependences of μ and λ, which should indicate its increase with energy. However, two more possibilities of explanation for the $\lambda/\Delta\varepsilon$ temperature dependence exist. First, if we assume that λ is temperature independent, then $\Delta\varepsilon$ must be temperature dependent as $\Delta\varepsilon = \Delta\varepsilon_0 - \eta T$. In Fig. 3 experimental data agree quite well with the solid line which represents the relation $\lambda_0/(\Delta\varepsilon_0 - \eta T)$, where η is the temperature coefficient of the fundamental absorption edge ($\eta = 10^{-3}$ eV/K, $\Delta\varepsilon_0 = 2.3$ eV). The second possibility involves a linear increase of α with temperature. However, it is impossible to explain the experimental dependences of $\lambda/\Delta\varepsilon$ on the field by means of physically acceptable $\Delta\varepsilon(E)$ or $\alpha(E)$ Consequently, λ increases with the electric field and thus with the carrier energy.

Extrapolation of λ to low electric fields allows to evaluate the field E_c, at which the carrier heating starts ($E_c = 3kT/2e\lambda$). At 300 K E_c is 1.8×10^7 and 3.3×10^7 V/m for holes and electrons, respectively. It should be noted that approximately at these values of E the field dependence of drift mobility passes from a power law, typical for a stochastic transport mechanism, to an exponential dependence [1, 5]. In [5] this phenomenon is attributed to the heating of charge carriers.

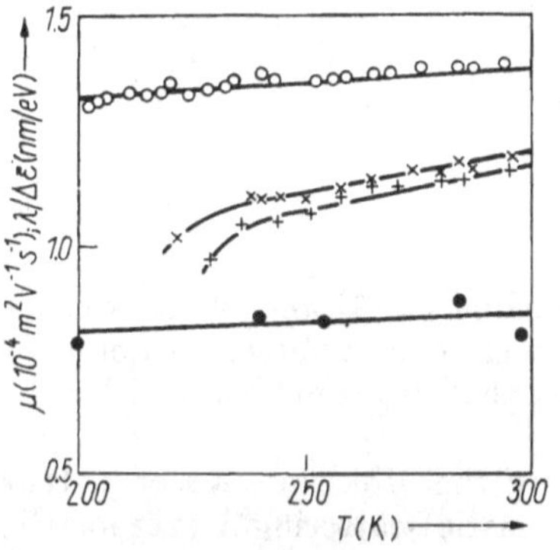

Fig. 3

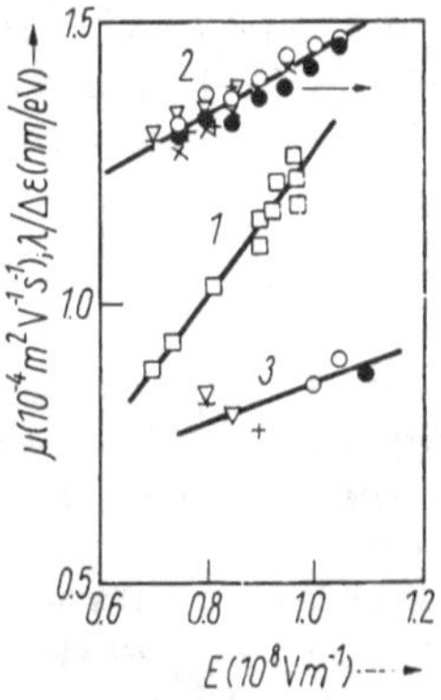

Fig. 4

Fig. 3. Temperature dependences of hole mobility (× $E = 9.5 \times 10^7$ V/m, + $E = 9 \times 10^7$ V/m) and effective free path of holes (○ $E = 9.7 \times 10^7$ V/m) and electrons (● $E = 1.13 \times 10^8$ V/m)

Fig. 4. Field dependences of hole mobility (1) ($d = 31$ µm) and effective free paths of holes (2) and electrons (3). Symbols of thickness are the same as in Fig. 2

Small values of the non-activated hot hole mobility as well as of carrier mean free path agree well with the small thermalization distance of carriers heated by light. This distance has been obtained from the interpretation of quantum efficiency data as Onsager dissociation of photogenerated pairs [4]. The smallness of μ and λ points to a strong interaction between the carriers and the lattice despite the fact that they are heated up to ≈ 0.3 eV.

References

[1] G. Juška and K. Arlauskas, phys. stat. sol. (a) **59**, 389 (1980).
[2] G. Juška and K. Arlauskas, Proc. Internat. Conf. Amorphous Semiconductors-80, Kishinev 1980 (p. 53).
[3] I. A. Akimov, Y. A. Cherkasov, and M. I. Cherkashin, Sensibilizirovannij fotoeffekt, Nauka, Moskva 1980 (p. 72).
[4] D. M. Pai and R. C. Enck, Phys. Rev. B **11**, 5163 (1975).
[5] J. M. Marshall and A. E. Owen, phys. stat. sol. (a) **12**, 181 (1972).

(Received January 10, 1983)

phys. stat. sol. (a) **77**, 393 (1983)

Subject classification: 13.4 and 20.3; 22.8.2

Institute of Physics, Teacher's Academy, Opole[1])

X-Ray Luminescence of YAG:Nd^{3+}

By

A. NIKLAS and W. JELEŃSKI

The X-ray luminescence of the YAG:Nd crystals in the visible and UV range of the spectrum is studied. The position of the $^2F2_{5/2}$ level is determined to 37810 cm^{-1}. The observed spectrum is ascribed to the transition from the afore-mentioned level to the SLJ states of the Nd^{3+} ions in YAG.

Die Röntgenlumineszenz von YAG:Nd-Kristallen wird im ultravioletten und sichtbaren Spektralbereich untersucht. Die Lage des Niveaus $^2F2_{5/2}$ wird zu 37810 cm^{-1} bestimmt, das beobachtete Spektrum wird den Übergängen von diesem Niveau zu Zuständen des Nd^{3+}-Ions zugeschrieben.

1. Introduction

The garnet crystals doped with trivalent rare ions RE^{3+} are of considerable interest because of their good laser properties. The luminescence properties of garnets excited with light are known relatively well, whereas studies devoted to the luminescence of garnets excited with X-rays, reactor radiation, electrons, and vacuum ultraviolet are rather scarce [1 to 5].

The continuation of these studies seems to be very suitable because of the possibility to obtain a laser action in the YAG:RE^{3+} crystals excited with radiation other than visible light. Generation of light in the YAG:Nd crystals excited by electron bombardment has been recently reported in [1]. The generation possibility in YAG:Nd crystals excited with X-rays has been also considered [2]. Contrarily to the intensively investigated luminescence of YAG:Nd crystals, connected with transitions from the $^4F_{5/2}$ and $^4F_{3/2}$ levels to the $^4I_{9/2}$ and $^4I_{11/2}$ levels, the studies of the luminescence accompanying the transitions from the S level [6] are rather rare [3, 4] and the detailed data on the transitions from the S-level to the $^4I_{9/2}$, $^4I_{11/2}$, $^4I_{13/2}$, and $^4I_{15/2}$ levels are still lacking.

In the present communication we wish to present the results of our measurements of the luminescence spectra of YAG:Nd crystals in the interval from 44000 to 9000 cm^{-1} together with a description of the observed transitions.

2. Experimental Procedure

The investigated YAG crystals containing 0.7 at% of Nd were grown by the Czochralski method. The luminescence spectra were recorded at room temperature with the aid of GDM-1000 and SPM-2 monochromators and EMI9684B and FEU-100 photomultipliers. The samples were irradiated with X-rays (Cu, 50 kV, 12 mA).

3. Results and Discussion

In Fig. 1 to 4 the X-ray luminescence (RL) spectra of the investigated crystals in the selected spectral range are presented and the corresponding electron transitions are indicated. The RL spectrum recorded at room temperature is considerably broader

[1]) Ul. Oleska 48, 45-052 Opole, Poland.

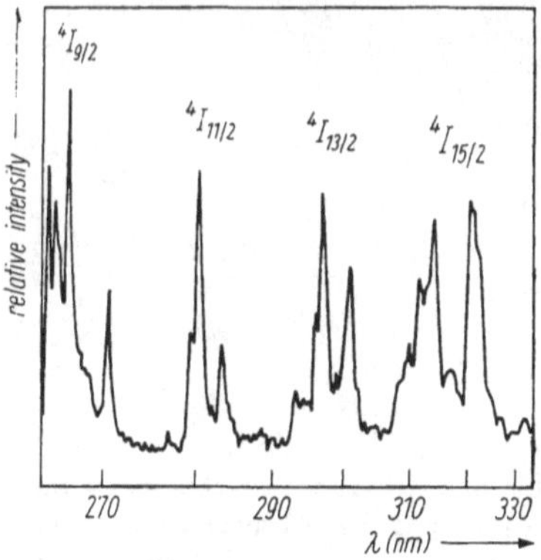

Fig. 1. Ultraviolet emission of YAG:Nd at 300 K under X-irradiation. The transition is $^2F2_{5/2} \rightarrow {}^4I_{9/2}$, $^4I_{11/2}$, $^4I_{13/2}$, $^4I_{15/2}$

that than observed for crystals excited with light. Due to the high energy of X- rays, the probability of direct excitation of the Nd^{3+} ions is negligibly small. In such a case the deciding role is ascribed to the energy transfer from lattice defects to the acitvator ions [2, 7 to 10].

The RL spectra for both pure and Nd-doped YAG show a characteristic emission band with a maximum at 31900 cm^{-1} (320 nm, Fig. 5). The mechanism of this emission ascribed to lattice defects, is still subject of controversy [3, 8]. The majority of authors agree that it is connected with transitions from excited states to the ground state of defects with a possibility of radiationless energy transfer from defects to the RE^{3+} ion. Such radiationless processes of the energy transfer enable the observation

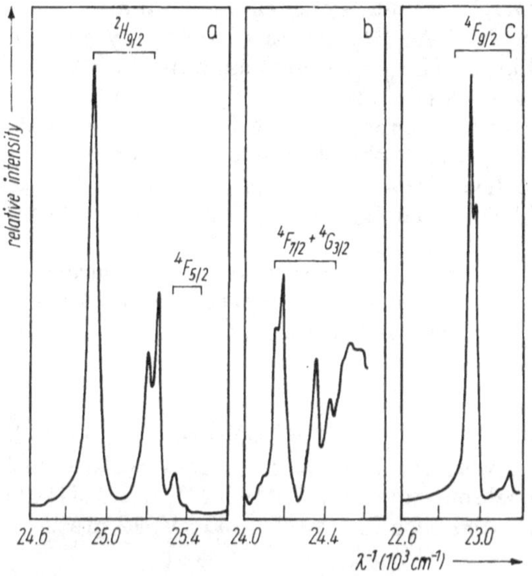

Fig. 2

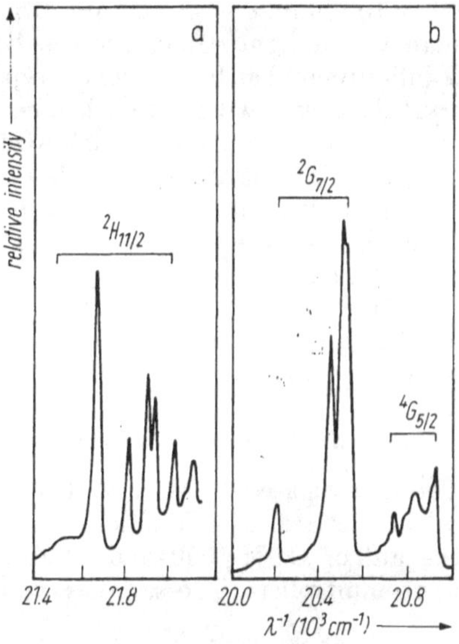

Fig. 3

Fig. 2. X-ray luminescence spectrum of YAG:Nd at 300 K. The transition is $^2F2_{5/2} \rightarrow {}^2H_{9/2}$, $^4F_{5/2}$, $^4F_{7/2} + {}^4G_{3/2}$, $^4F_{9/2}$

Fig. 3. X-ray luminescence spectrum of YAG:Nd at 300 K. The transition is $^2F2_{5/2} \rightarrow {}^2H_{11/2}$, $^4G_{5/2}$, $^2G_{7/2}$

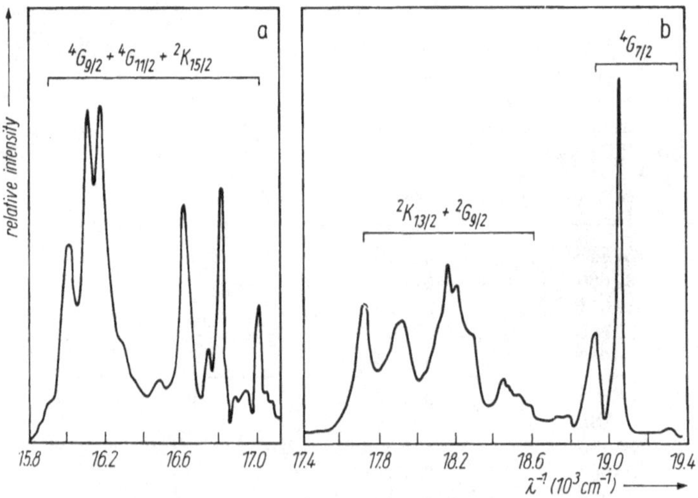

Fig. 4. X-ray luminescence spectrum of YAG:Nd at 300 K. The transition is $^2F2_{5/2} \rightarrow {^4G_{7/2}}$, $^2K_{13/2} + {^2G_{9/2}}$, $^4G_{9/2} + {^4G_{11/2}} + {^2K_{15/2}}$

of the RL in the infrared, visible as well as in the ultraviolet range of the spectrum.

Taking into account the results of investigations of the cathodoluminescence of YAG:RE^{3+} (RE = Ce, Tb, Eu, Gd) crystals [7 to 10] we suppose that the excitation of the Nd^{3+} ions occurs as follow:

1. Under action of the X-rays the "lattice" electrons (e.g. from O^{2-}) are liberated, as a result of which hot electrons and holes are created;

2. due to the interaction of hot electrons with the lattice they are multiplicated and thermalized;

3. fast trapping of the electrons and holes in the vicinity of RE^{3+} ions leads to the creation of bound excitons;

4. the Nd^{3+} ions are excited due to the radiationless energy transfer from recombining bound excitons.

The ultraviolet absorption (at room temperature) of YAG:Nd crystal has been investigated by Krupke [11] who discovered the absorption lines at 39210, 38020, and 33890 cm^{-1}, corresponding to the transitions from the $^4I_{9/2}$ level to the $^2F2_{7/2}$, $^2F2_{5/2}$, and $^2H1_{11/2} + {^2D2_{5/2}}$ levels, respectively.

On the basis of the experimentally obtained RL spectrum in the ultraviolet range (Fig. 1) it has been found that the energy corresponding to the emission observed

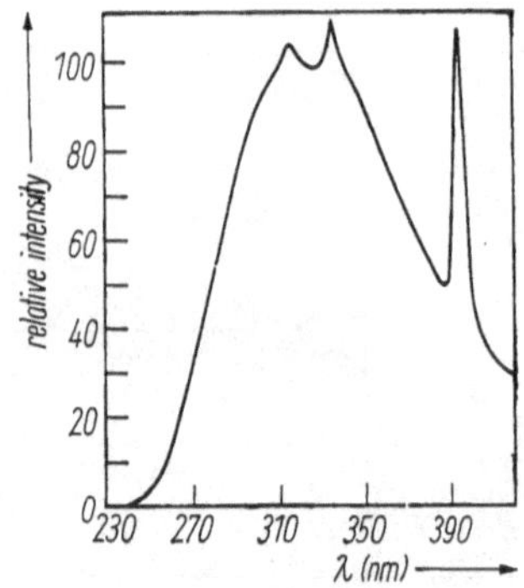

Fig.5. Ultraviolet emission of YAG at 300 K under X-irradiation

Table 1

Transitions between the $^2F2_{5/2}$ level and the SLJ states of the Nd^{3+} ions in YAG, registered in the visible and UV ranges of the spectrum

assignment	energy levels of Nd^{3+} in YAG [12] (cm^{-1})	transition from $^2F2_{5/2}$ to state SLJ Nd^{3+} ions		$\triangle E$ (cm^{-1})	$\triangle$ (E (cm^{-1})
		calc. (cm^{-1})	obs. (cm^{-1})	theor.	exper.
$^4I_{9/2}$	0	37810	37810	857	855
	130	37680	37679		
	199	37611	37613		
	308	37502	37509		
	857	36953	36955		
$^4I_{11/2}$	2002	35808	35803	519	517
	2029	35781	35778		
	2110	35700	35699		
	2147	35663	35676		
	2468	35342	35335		
	2521	35289	35286		
$^4I_{13/2}$	3922	33888		585	574
	3930	33880	33852		
	4038	33772			
	4052	33758	33760		
	4442	33368			
	4507	33303	33278		
$^4I_{15/2}$	5757	32053	32080	977	1025
	5812	31998	31949		
	5933	31877			
	5963	31847	31847		
	6570	31240			
	6585	31225	31153		
	6640	31170	31055		
	6732	31078			
$^4F_{5/2}$	12370	25440	25460	149	110
	12432	25378	25410		
	12519	25291	25350		
$^2H_{9/2}$	12575	25235	25260	265	330
	12607	25203	25212		
	12623	25187	25180		
	12819	24991	24960		
	12840	24970	24930		
$^4F_{7/2}$ +	13363	24447	24425	270	265
+ $^4S_{3/2}$	13433	24377	24360		
	13563	24247			
	13572	24238	24240		
	13596	24214	24190		
	13633	24177	24160		
$^4F_{9/2}$	14626	23184	23180	290	290
	14678	23132	23120		
	13793	23017	23000		
	14819	22991	22960		
	14916	22894	22890		

Table 1 (continued)

assignment	energy levels of Nd^{3+} in YAG [12] (cm^{-1})	transition from $^2F2_{5/2}$ to state SLJ Nd^{3+} ions		$\triangle E$ (cm^{-1})	$\triangle E$ (cm^{-1})
		calc. (cm^{-1})	obs. (cm^{-1})	theor.	exper.
$^2H_{11/2}$	15743	22067	22030	376	360
	15838	21972	21950		
	15870	21940	21910		
	15957	21853	21830		
	16103	21707	21685		
	16119	21691	21670		
$^4G_{5/2}$	16849	20961	20940	198	200
	16992	20818	20840		
	17047	20763	20740		
$^2G_{7/2}$	17241	20569	20536	334	324
	17268	20542	20516		
	17322	20488	20456		
	17575	20235	20212		
$^4G_{7/2}$	18723	19087	19050	263	250
	18822	18988	19000		
	18843	18967	18940		
	18986	18824	18800		
$^2K_{13/2}$ +	19154	18656	18670	894	935
+ $^2G_{7/2}$	19194	18616	18549		
	19294	18516	18490		
	19470	18340	18340		
	19543	18267	18290		
	19596	18214	18269		
	19620	18190	18210		
	19651	18159	18170		
	19814	17996	17950		
	20048	17762	17735		
$^4G_{9/2}$ +	20730	17080	17100	1176	1185
+ $^4G_{11/2}$ +	20773	17037	17060		
+ $^2K_{15/2}$	20790	17020	17040		
	20803	17007	17030		
	20962	16848	16845		
	21029	16781	16810		
	21080	16730	16740		
	21110	16700	16700		
	21159	16651	16665		
	21162	16648			
	21522	16288	16290		
	21593	16217	16220		
	21661	16149	16165		
	21697	16113	16100		
	21767	16043			
	21791	16019	16020		
	21872	15938	15950		
	21906	15904	15915		

equals 4.68 eV (37810 cm^{-1}). In the nearest neighbourhood of this level there is a group of absorption lines $^2F2_{5/2}$ [11].

Among the transitions from the level 37810 cm^{-1} the transitions from the level $^4F_{3/2}$ to the levels $^4I_{9/2}$ and $^4I_{11/2}$ as well as from $^4F_{5/2}$ to the $^4I_{9/2}$ level have been also observed in the infrared range. The positions of these lines in the RL spectrum show a very good agreement with those observed in the photoluminescence of the investigated YAG:Nd crystals.

A detailed analysis of the RL spectrum showed that among the lines connected with the Nd^{3+} ions there are also additional groups of lines, which we ascribe to the Tb^{3+} and Dy^{3+} ions. Some of these lines overlap with those connected with the Nd^{3+} ions thus influencing the accuracy of the determination of the positions of these lines in the RL spectrum. The presence of traces of rare earth ions has been confirmed by the mass spectrometry of the raw material used for growing the investigated YAG:Nd crystals.

The data collected in Table 1 show a good agreement between the experimental data and those calculated theoretically under the assumption that the RL spectrum observed in the ultraviolet and visible wavelength ranges is connected with transitions from one of the levels $^2F2_{5/2}$ to the lower levels of the Nd^{3+} ion in YAG crystals.

References

[1] u.Y K. VORONKO, E. L. NOLLE, V. V. OSIKO, and M. N. TIMOSHECHKIN, Zh. eksper. teor. Fiz., Pisma **13**, 125 (1971).
[2] YU. K. VORONKO, V. V. OSIKO, A. M. PROHOROV, and I. A. SHCHERBAKOV, Trudy FIAN SSSR, **60**, Izd. Nauka Moscow 1972 (p. 3).
[3] N. S. ROOZE and N. A. ANISIMOV, Optika i Spektroskopiya **38**, 627 (1975).
[4] YU. K. VORONKO, B. I. DENKER, V. V. OSIKO, A. M. PROHOROV, and M. N. TIMOSHECHKIN, Dokl. Akad. Nauk SSSR **188**, 1258 (1969).
[5] SH. A. VAHIDOV, E. M. IBRAGIMOVA, B. KAIPOV, G. A. TAVSHUNSKII, and A. A. YUSUPOV, Radiacionnye yavleniya v nekotorykh kristallakh, Izd. FAN Uz. SSR, Tashkent 1977 (p. 59).
[6] G. H. DIEKE and H. M. GROSSWHITE, Appl. Optics **2**, 675 (1963).
[7] D. J. ROBBINS, B. COCKAYNE, B. LENT, C. N. DUCKWORTH, and J. L. GLASPER, Phys. Rev. B **19**, 1254 (1979).
[8] D. J. ROBBINS, B. COCKAYNE, J. L. GLASPER, and B. LENT, J. Electrochem. Soc. **126**, 1213 (1979).
[9] D. J. ROBBINS, B. COCKAYNE, J. L. GLASPER, and B. LENT, J. Electrochem. Soc. **126**, 1221 (1979).
[10] D. J. ROBBINS, B. COCKAYNE, B. LENT, and J. L. GLASPER, J. Electrochem. Soc. **126**, 1556 (1979).
[11] W. F. KRUPKE, IEEE J. Quantum Electronics **7**, 153 (1971).
[12] A. A. KAMINSKII, Lazernye Kristally, Izd. Nauka, Moscow 1975; Laser Crystals, Springer-Verlag, 1981.

(Received March 16, 1983)

phys. stat. sol. (a) 77, 399 (1983)
Subject classification: 14.4.1; 22.9

Department of Physics, Indian Institute of Technology, New Delhi[1])

Dielectric Dispersion in Triglycine Sulphate Single Crystals

By
A. K. Batra, S. C. Mathur, and A. Mansingh[2])

The dielectric dispersion in pure and doped ferroelectric triglycine sulphate (TGS) single crystals is investigated in the frequency range 200 Hz to 100 kHz in both, ferro- and paraelectric phases. The results obtained strongly suggest that the dielectric dispersion and reduction in dielectric loss in doped TGS crystals can be attributed to domain wall motions.

Die dielektrische Dispersion in reinen und dotierten ferroelektrischen Triglyzinsulfat (TGS)-Einkristallen wird im Frequenzbereich von 200 Hz biw 100 kHz sowohl in der ferro- als auch in der paraelektrischen Phase untersucht. Die erhaltenen Ergebnisse weisen stark darauf hin, daß die dielektrische Dispersion und die Verringerung der dielektrischen Verluste in dotierten TGS-Kristallen auf Domänenwandbewegungen zurückgeführt werden können.

1. Introduction

The response of electric polarization to an applied field gives useful information about the mechanism of polarisation in a material. Such studies have been extensively used in polar liquids and polymers to understand the mechanism of relaxation and dipole–dipole interaction. In dielectric solids and amorphous semiconductors, dielectric measurements at different frequencies provide localized states and give a frequency dependent conductivity, while for band conduction the conductivity is almost independent of frequency. In ferroelectrics, just as in any dielectric the mechanism by which polarization responds to varying fields includes electron effects and atomic effects of greater inertia. At low frequencies all mechanisms respond, while at high frequencies only some do so, thus giving rise to dispersion. In ferroelectrics, there is an additional dispersion process associated with the domain walls. A ferroelectric domain is a region of homogeneous polarization in a ferroelectric crystals. In each domain, the polarization is in different direction. The demarcation between two domains is referred to as a domain wall and the wall is thought of as being so thin that it has a much smaller volume than the bulk material of the domains. The electrical response of domain walls has important consequences on the dielectric character of the ferroelectrics. The oscillating response of the wall to an applied ac field may fail at higher frequencies, giving rise to a dispersion due to this. Apart from these mechanisms, sometimes the large conductivity of samples in the presence of barrier capacitance can also give rise to a dielectric dispersion, the so-called the Maxwell-Wagner type dispersion [1, 2].

The dielectric dispersion in TGS can be roughly divided into two regions: (i) microwave region: the dispersion in this region has been extensively studied [3 to 9]. It has been observed both in para- and ferro-electric phases near the transition temperature. Evidence of critical slowing down (collective interaction of dipoles) has also been presented by an increase in relaxation time near the transition temperature [7]. There is still a controversy as to whether the dispersion is due to single relaxation or multiple relaxation times [5]; (ii) audio and radio frequency range: the dispersion in this region has usually been observed only in the ferroelectric phase. Several studies have been

[1]) Hauz Khas, New Delhi 110016, India.
[2]) Delhi University, Delhi—7.

reported [1 to 25] but no conclusive picture has yet emerged. This low-frequency dispersion has been attributed due to domain wall motion [1, 12]. Choe et al. [14] studied the dependence of dielectric loss in unpoled TGS upon sample thickness and suggested that domain wall mation is the source of the dielectric losses. More recently Jonscher and Dube [13] have reported low-frequency dielectric dispersion in TGS. They have shown that dielectric loss and constant follow an empirical relation of the type, $\chi(\omega) \sim \chi''(\omega) \sim \omega^{n-1}$, which is consistent with the empirical relation for most of the solid dielectrics.

Most of the studies of dielectric dispersion have been reported for undoped TGS crystals. It is considered desirable to study the dielectric dispersion in pure and doped TGS crystals in the audio frequency range to shed light on the mechanism of relaxation. Further, this information is of significant importance for many practical applications where the knowledge of the dielectric loss is required.

2. Experimental

Single crystals of TGS were grown from seeds by slow cooling of a saturated solution of glycine along with the dopants and sulphuric acid mixed in stoichiometric ratio at an initial rate of 0.3 K/d to a maximum of 1 K/d under identical conditions so as to balance out the effects of growth conditions. The details of sample preparation and measurements are described elsewhere [21]. The dielectric measurements were made by employing a weak alternating field (0.5 V/cm) at various frequencies (200 Hz to 100 kHz) by a Genrad 1615-AP capacitance measuring assembly. The samples 1 to 2 mm thick and 1 to 2 cm^2 in area were chosen for the measurements. The choice of a relatively large thickness for the sample ensured a low-intensity measurement field at all frequencies used. This practically eliminated the effect of measurement voltage on the dielectric constant (ε') and loss (ε''). Further, this enables one to obtain the frequency dependence of ε' and ε'' as characteristics of the material rather than of samples.

3. Results

The variation of dielectric constant (ε') and loss (ε'') of TGS in the frequency range from 200 Hz to 100 kHz are shown in Fig. 1 a and b respectively. The Cole–Cole plots are shown in Fig. 2. The curves of ε' and ε'' versus frequency at a typical temperature

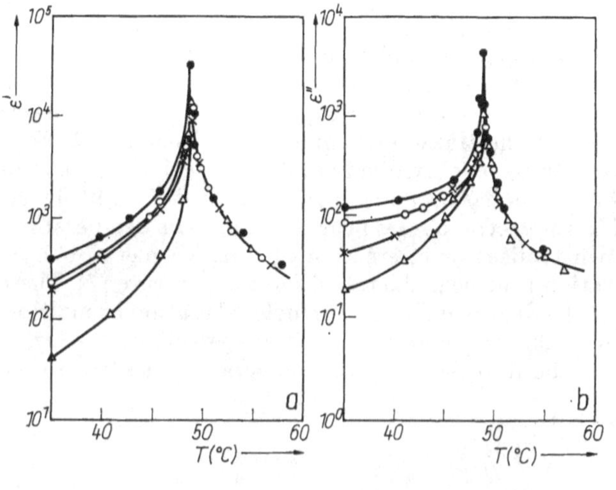

Fig. 1. a) Real part (ε') and b) imaginary part (ε'') of dielectric constant vs. temperature of pure TGS crystal. ○ f = 0.2, ● 1.0, × 10.0, △ 100.0 kHz

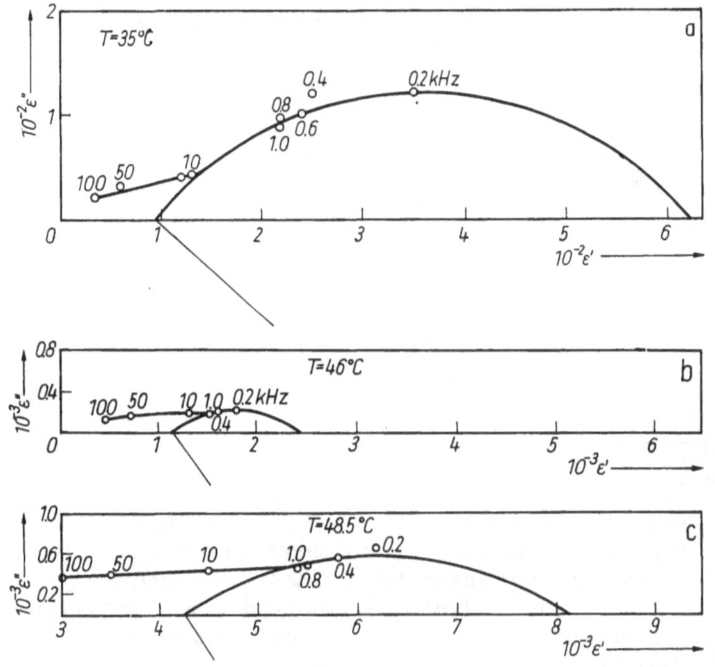

Fig. 2. Cole–Cole plots of pure TGS crystal at different temperatures: a) $T = 35$, b) 46, c) 48.5 °C

Table 1

Relaxation parameters for the low-frequency dispersion region of pure TGS crystals

temperature (°C)	ε_{01}	$\varepsilon_{\alpha 1}$	τ (10^{-3} s)	activation energy, W (eV)
35.0	620	94	0.43	
46.0	1170	240	0.29	0.60
48.5	9200	4250	0.25	

for pure and doped TGS crystals are shown in Fig. 3 and 4. Table 1 summarizes the relaxation parameters for the low-frequency dispersion region for a pure TGS crystal.

4. Discussion

4.1 Low-frequency dielectric dispersion in pure TGS crystal

It may be noted from Fig. 1 a and b that dielectric constant and loss in the paraelectric phase are independent of frequency. Dielectric dispersion is observed only in the ferroelectric phase. Curves of ε' and ε'' versus frequency (Fig. 3) at a typical temperature (46 °C) show that a clear dispersion is obtained in the ferroelectric phase. A broad peak in ε'' around a frequency of 20 kHz is accompained by a monotonous decrease in dielectric constant, this behaviour is indicative of a dispersion. Dielectric constant and loss can be represented on a Cole–Cole plot and are defined by

$$\varepsilon_{LF}^* = \varepsilon_{\alpha 1} + (\varepsilon_{01} - \varepsilon_{\alpha 1}) \frac{1}{(1 + j\omega\tau)^{1-\alpha}},$$

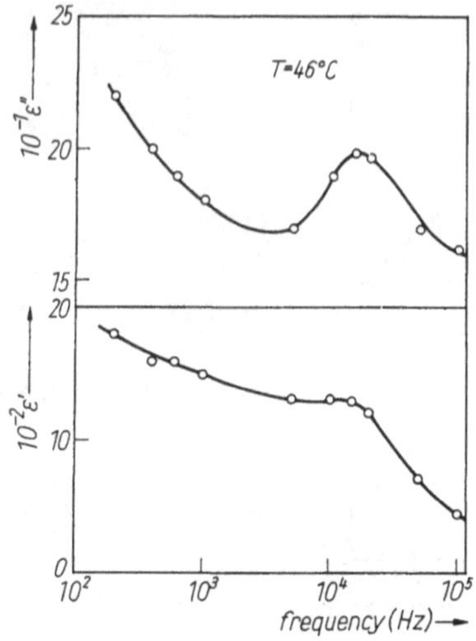

Fig. 3. ε' and ε'' versus frequency of pure TGS crystal at $T = 46$ °C

where ε_{01} and $\varepsilon_{\alpha 1}$ are, respectively, the low- and high-frequency limiting values of dielectric constant for the dispersion region, τ is the relaxation time and α the distribution parameter. The relaxation frequency was found to increase with increasing temperature, but the distribution parameter shows little variation with temperature. In the range of temperature studied dielectric relaxation times seem to be thermally activated and can be expressed by

$$\tau = A \exp\left(\frac{W}{kT}\right),$$

where A is a constant, k the Boltzmann constant, and W the activation energy. The activation energy W is found to be about 0.60 eV, which is slightly greater than the dc activation energy (0.4 eV) and is slightly lower than that reported by Gittetta [1].

The above analysis shows that although the Debye model can be applied to the relaxation observed, it may not be due to dipolar relaxation, since in the present studies, it does not show a critical slowing down process and the relaxation time does not increase as the transition temperature is approached (Table 1). Thus one can rule out

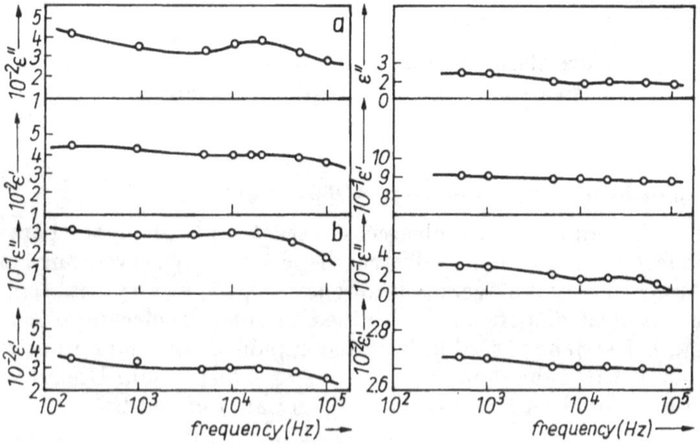

Fig. 4. ε' and ε'' versus frequency of doped crystal at $T = 46$ °C. a) doped with 5% $Fe_2(SO_4)_3$, b) doped with 0.25% p-nitroaniline, c) doped with 30% L-alanine, d) doped with 5% L-alanine

the possibility of dipolar relaxation within the domains. The fact that the dispersion is observed only in the ferroelectric phase and not in the paraelectric phase rules out the possibility of this dispersion to be due to surface capacitance and bulk conductivity of the sample. Studies on the thickness-dependent dielectric constant show that surface capacitance is equally effective both in ferro- and paraelectric phases near the transition temperature [15].

The above discussion completely rules out dipolar relaxation and relaxation due to surface capacitance and probably the low-frequency dispersion is caused by the domain wall motion. The relaxation time decreases with increasing temperature (as one approaches the transition temperature) for domain wall relaxation [25, 17]. This has been reported by Fousek et al. [12] and Gurk [17] and is in agreements with our results. Gurk [17] has suggested a phenomenological model to explain the low-frequency dispersion in ferroelectric Rochelle salt due to domain wall relaxation. To establish qualitatively the results of TGS doped with impurities which do change the domain mobility, configuration and density are discussed in the following section.

4.2 Low-frequency dielectric dispersion in doped TGS crystal

In doped TGS crystals, three types of dopant, viz. amino acid, organic compound having NH_2 group, and inorganic metal ion, have been used. It has been suggested that the mechanisms involved in changing the domain wall configuration, density, and mobility in the three types of dopant are not same. These impurities change the parameters to different degrees [21 to 23]. The dopants studied are L-alanine, p-nitroaniline and Fe^{+++} ions. Fig. 4c and d show the variation of ε' and ε'' with frequency at 46 °C for TGS crystals doped with different amounts of L-alanine. These curves show no significant variation of ε' with frequency. However, there is a variation of ε'' with frequency. The variation in ε'' decreases with increasing concentration of L-alanine. In these crystals the peak in ε'' around 20 kHz is not observed as in the case of pure TGS. It is interesting to note that there is an order of magnitude decrease in the value of ε''. The decrease and absence of dielectric relaxation may be understood in the basis of the domain walls present and domain dynamics. On doping with L-alanine the crystal tends to become a single-domain poled one [22]. An increase in the concentration of L-alanine brings the crystals closer to the single-domain structure. There is a marked decrease in the effective area of the moving domain walls. Since there are still domain walls present even at higher concentration, the loss factor persists. It is also reasonable to believe that a single-domain poled crystal may not be able to contribute to the relaxation process.

In the case of p-nitroaniline-doped TGS crystals, there is a decrease in the values of ε' and ε'' as compared with those of the pure TGS crystal. The decrease in the values of ε' and ε'' increases with increasing concentration of p-nitroaniline. It can be seen from Fig. 4b that there is no significant variation in ε' with frequency. No pronounced maxima in ε'' are observed below 20 kHz. The decrease in ε'' for p-nitroaniline-doped TGS crystals is not as marked as for L-alanine-doped TGS crystals. It has been proposed that nitroanilines may not be producing single-domain crystals [21]. There are domain walls present which give rise to the loss factor. However, the decrease in ε'' as we increase the concentration suggests that there is decrease in the number of domain walls and the walls are becoming immobile. Immobile or pinned domain walls may not be able to contribute to the relaxation process.

It can be seen from Fig. 4a that there is a slight decrease in the value of ε' with frequency for Fe-doped TGS crystals. But a maximum is observed in the variation of ε'' with frequency near 20 kHz. The values of ε'' are smaller than for pure TGS crystal. Most of the inorganic impurities form chelates with TGS and lead to frittering of the domain structure of TGS crystals [23 to 24]. The chelate formation restricts the possible movement of domain walls, which in turn lowers the measured value of ε''. These trapped domain walls might be contributing to the relaxation at some particular frequency. It may be pointed out that the dc conductivity of the Fe-doped TGS crystal is higher [21], while the magnitude of dielectric loss and the peak value, however, are lower than those of undoped TGS crystals. This suggests that the low-frequency dielectric dispersion may not be due to the high conductivity of domain walls as suggested by Gilletta [1]. But it is more likely due to the domain wall motion and relaxation under the applied ac field as has been suggested by Fousek et al. [12, 20].

The qualitative picture formed by changing the domain structure with the help of impurities shows that the low-frequency dispersion is due to domain wall motion and their pinning by impurities decreases the magnitude of dispersion.

5. Conclusion

The present studies lead to the conclusion that the dispersion in the frequency range 200 Hz to 100 kHz is associated with the domain wall motion. In order to establish the exact mechanism of domain wall motion responsible for dispersion, further studies in samples for which domain structure and density are known, have to be extensively carried out. However, it is clearly demonstrated that the magnitude of dielectric dispersion in TGS crystals decreases by doping. The decrease in loss factor (ε'') in doped crystals is due to the reduction in domain walls and pinning of domains. This suggests that the doped crystals may be useful in making transducers with no frequency dispersion.

References

[1] F. Gilletta, phys. stat. sol. (a) **12**, 143 (1972).
[2] K. Kamayoshi, Sci. Rep. Res. Inst. Tôhoku Univ. **A19**, 124 (1967).
[3] H. E. Mueser, Proc. European Meeting Ferroelectricity, Saarbrücken, Eds. Mueser and Petersson, Stuttgart 1970 (p. 91).
[4] R. M. Hill and S. K. Ichiki, Phys. Rev. **128**, 1140 (1962).
[5] G. Luther and H. E. Mueser, Z. Naturf. **24a**, 389 (1969).
[6] L. Lurio and E. Stern, J. appl. Phys. **31**, 1125 (1960).
[7] H.-G. Unruh and H.-J. Wahl, phys. stat. sol. (a) **9**, 119 (1972).
[8] G. Luther, phys. stat. sol. (a) **20**, 227 (1973).
[9] G. Luther, phys. stat. sol. (a) **20**, K25 (1973).
[10] Landoldt/Börnstein, Numerical Data, New Series, Vol. 9, Springer 1969.
[11] P. Lauginie and F. Gilletta, Proc. European Meeting Ferroelectricity, Saarbrücken, Eds. Mueser and Petersson, Stuttgart 1970 (p. 219).
[12] J. Fousek and V. Janousek, phys. stat. sol. **13**, 195 (1966).
[13] A. K. Jonscher and D. C. Dube, Ferroelectrics **17**, 533 (1978).
[14] H. M. Choe, J. H. Judy, and A. Vanderziel, Ferroelectrics **15**, 185 (1977).
[15] Gulshan Rai, Ph. D. Thesis, Delhi University, Delhi 1977.
[16] I. S. Zheludev, Physics of Crystalline Dielectrics, Plenum Press, New York 1971.
[17] P. Gurk, phys. stat. sol. (a) **10**, 407 (1972).
[18] R. C. Miller and A. Sevage, Phys. Rev. **112**, 755 (1958).
[19] H. H. Weider, J. app. Phys. **27**, 413 (1956).
[20] J. Fousek, Czech. J. Phys. **B15**, 412 (1965).
[21] A. K. Batra, Ph. D. Thesis, Indian Institute of Technology, Delhi 1981.
[22] E. T. Keve, Phillips tech. Rev. **35**, 247 (1975).
[23] F. Moravek and J. Novotny, Czech. J. Phys. **B7**, 23, 885 (1973).
[24] J. Stankowska and E. Czosnowska, Acta. phys. Pol. **A43**, 641 (1973).
[25] A. Man Singh and S. R. J. Eswar Prasad, J. app. Phys. **48**, 4307 (1977).

(Received January 3, 1983)

phys. stat. sol. (a) 77, 405 (1983)

Subject classification: 14.4 and 15; 10.2; 22.5.4

Department of Physics, University of Gorakhpur[1])

Electrical Transport Studies of Copper Mercuric Iodide

By

H. B. Lal and O. P. Srivastava

Measurements are given of electrical conductivity (σ) and thermoelectric power (S) of copper mercuric iodide (Cu_2HgI_4) (which becomes a solid superionic conductor above 344 K) are reported both, in the α- and β-phase. In the α-phase it behaves as a Cu^+-ion solid superionic conductor in the temperature range 345 to 410 K. The heat of transport (Q) and activation enthalpy (h_m) for Cu^+ ions in this phase both are found to be equal to 0.11 eV. These results are explainable by free ion as well as lattice gas models. In the β-phase (below 344 K) the electrical conductivity is mainly extrinsic. In relatively pure material, it is due to interstitial Cu^+ ions whose heat of transport and activation enthalpy are evaluated to 0.27 and 0.28 eV, respectively.

Es werden die Meßergebnisse der elektrischen Leitfähigkeit (σ) und der Thermospannung (S) von Kupferquecksilberjodid (Cu_2HgI_4) (das oberhalb 344 K ein fester Superionenleiter wird) sowohl in der α- als auch in der β-Phase mitgeteilt. In der α-Phase verhält es sich als fester Cu^+-Ionen-Superionenleiter im Temperaturbereich 345 bis 410 K. Die Transportwärme (Q) und die Aktivierungsenthalpie (h_m) für Cu^+-Ionen werden in dieser Phase zu 0,11 eV gefunden. Diese Ergebnisse lassen sich erklären sowohl durch freie Ionen als auch mit Gitter-Gasmodellen. In der β-Phase (unterhalb 344 K) ist die elektrische Leitfähigkeit hauptsächlich extrinsisch. In relativ reinem Material wird sie durch Cu^+-Zwischengitterionen hervorgerufen, deren Transportwärme und Aktivierungsenthalpie zu 0,27 bzw. 0,28 eV bestimmt werden.

1. Introduction

In recent years considerable interest has been taken in the development, study, and application of solid electrolytes popularly known as solid superionic conductors (SSIC). Useful solid electrolytes which have been developed and studied during the last fifteen years are mostly silver ion conductors [1 to 6]. Since copper is cheaper compared to silver, it is more meaningful to search for copper-based SSIC for practical applications. Various attempts [7 to 13] have been done in this direction during the last few years and many good copper ion SSIC have been reported. To proceed in this direction, it seems important to understand the electrical transport mechanism in copper salts, particularly in those which are known to undergo a phase transition from low-conductivity β to high-conductivity α-phase. One of such materials, namely copper iodide, has recently been studied by Verma et al. [14]. Copper mercuric iodide (Cu_2HgI_4) is another such material. This material was prepared and studied by Ketelaar as early as in 1934 [15]. It undergoes a phase transition around 343 K from β to α-phase. Above 343 K it exists in the α-phase and is iso-structural with α-Ag_2HgI_4. In this phase the anion arrangement is face-centred cubic [16]. The three cations are uniformly distributed over the 4d sites, those that are occupied by Zn^{2+} ions in zinc-blende or by Ag^+ ions in γ-AgI. Although Cu_2HgI_4 is known to us for the last fifty years, there exist only few [17 to 20] studies regarding electrical transport in this solid. Further, there also exists a wide diversity in reported electrical conductivity data as is evident from Table 1. The reason for this difference in data is not very clear, but

[1]) Gorakhpur 273001, India.

Table 1

Electrical conductivity of Cu_2HgI_4 in α- and β-phases as obtained by various workers

sample phase	sample form	electrode	σ (Ω^{-1} m^{-1})	T (K)	E_a (eV)	Q (eV)	ref.*)
			$\approx 10^{-3}$	340	0.59	—	[18]
α-Cu_2HgI_4	pellet	Cu-foil	5.5×10^{-4}	342	0.46	—	[20]
			2.3×10^{-1}	345	0.11	0.11	[PS]
			3.6×10^{-6}	287	0.39	—	[18]
β-Cu_2HgI_4	pellet	Cu-foil	7.5×10^{-6}	310	0.76	—	[20]
			9.9×10^{-2}	300	0.28	0.27	[PS]

*) PS: Present study.

one of the reasons for the large scatter in data seems to be the experimental conditions (e.g. pelletizing pressure of the pellet, thermal history, sintering and annealing of the pellet, electrode material used for measurement, and temperature). This necessitates to investigate the conditions in which measurement reveals the bulk value of the material and is free from other external influences. Our systematic studies in this direction are presented in this paper.

2. Material Preparation and Experimental Details

Cu_2HgI_4 has been prepared by repeated firing around 500 K of the pelletized stoichiometric mixture of CuI and HgI_2. Both these materials with purity of 99.99% were procured from M/s Bonds India. No weight loss has been observed in firing and the compound is formed by the following solid state reaction:

$$2\,CuI + HgI_2 = Cu_2HgI_4\,. \tag{1}$$

For the measurement of electrical conductivity (σ) the powder samples were pressed into pellets with typical dimensions of 9.6×10^{-5} m^2 area and 1×10^{-3} to 5×10^{-3} m thickness at pressures ranging from 2.0×10^7 to 8.5×10^7 kp m^{-2}. For thermoelectric power (S) measurement the dimensions of the pellets were 9.6×10^{-5} to $11.1 \times$

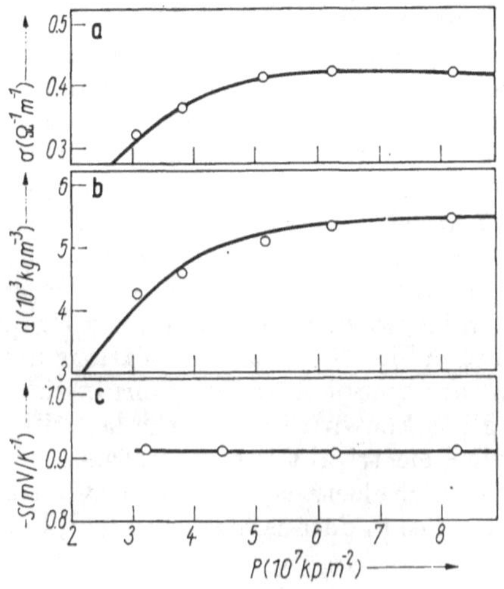

Fig. 1. Plots of a) electrical conductivity σ at $T = 390$ K; b) density d, and c) thermoelectric power S as a function of pelletizing pressure P

$\times 10^{-5}$ m area and thickness 5×10^{-3} to 10×10^{-3} m. Thin copper foils mechanically pressed on the faces of the pellets served as electrodes. The details of both σ and S measurement are discussed elsewhere [21, 22].

3. Results

Fig. 1a shows the plot of electrical conductivity (σ) as a function of pelletizing pressure P at constant temperature using copper foil electrodes. The measurements are done on pressed pellets after the source of pressure has been removed. It is seen from this figure that σ significantly depends upon P. Slightly larger values are always obtained for pellets made at higher pelletizing pressure. However, it is seen that the increase of σ up to a pressure of about 4×10^7 kp m^{-2} is faster, but for $P > 6 \times 10^7$ kp m^{-2} a trend to constancy is observed in these values. Practically the values are independent of P values when it exceeds 6×10^7 kp m^{-2}. The relatively low conductivity observed for pellets prepared at low pressure is likely to be due to the presence of a less conducting air space and grain boundaries. These regions are considerably reduced at higher pelletizing pressure. This is further evidenced by the measurement of the density d of these pellets. The d versus P plot is shown in Fig. 1b. It is very similar to the σ versus P plot. The density also tends to approach the density of the polycrystalline (melted and cooled) sample at higher values of P. Fig. 1c shows the plot of S versus P. S values are nearly constant for pellets made at higher P, however, slightly higher S values are obtained for pellets made at lower pressure. From these observations we conclude that σ, d, and S are nearly constant and independent of P for pellets made at higher pressure. Electrical conductivity of well sintered pellets has also been measured as a function of ac signal frequency f using copper foil electrodes. Fig. 2a shows the plot of σ versus log f at constant temperature. The conductivity seems to be independent of f indicating that in such conducting solid the grain boundary effects are insignificant. From all these observations one can conclude that measurements done on highly pressed and well sintered pellets using good copper electrodes give the bulk values for the material.

It has been pointed out by several authors [5, 23] that electrodes play an important role in the measurement of both σ and S. For a better measurement stringent criteria

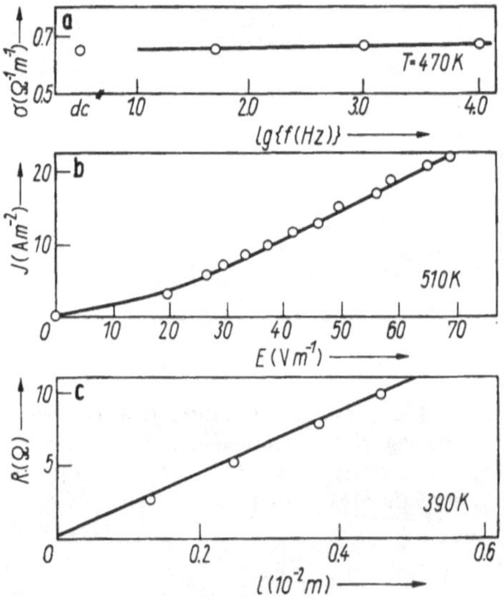

Fig. 2. a) Plot of electrical conductivity σ vs. applied signal frequency f at 470 K. The dc value of σ is shown separately. b) Plot of current density J vs. applied dc electric field E at 510 K. c) Plot of resistance R vs. thickness l for pellets of Cu_2HgI_4 with area $= 9.6 \times \times 10^{-3}$ m^2

Table 2
Electrical conductivity (σ) of Cu_2HgI_4 using different electrode materials at 350 K*)

electrode	electrical conductivity (Ω^{-1} m^{-1})	
	freshly prepared **)	after heating ***)
copper foil	0.14	0.27
silver foil	0.13	0.25
platinum foil	0.13	0.26
silver point	0.12	0.22
aluminium point	0.09	0.19

*) Conductivity value have been corrected for $T = 350$ K from σ vs. T data.
**) Electrode mechanically pressed or painted and measurement done by heating the sample slowly. Heating is done after applying the electrode for the first time.
***) Heating the pellet with electrode up to 500 K and measurement done by cooling the sample slowly.

are an ohmic contact at the electrode–pellet interface and the contact resistance to be minimum. To get better contacts several materials (e.g. copper, silver, and platinum foils, silver and aluminium paints) were tried as electrode materials. The electrical conductivity has not been found to be much dependent upon electrode material (Table 2). Better values of σ are, however, obtained for each case after heating the material (pellets with electrodes on both sides) to a temperature close to its melting point and cooling it back to room temperature. Probably the material softens near its melting point and the electrode–pellet interface contact becomes smoother and better. To look into the nature of contacts dc currents through pellets with copper electrodes have been measured at different applied voltages. A typical plot of J (current density) versus E (applied electric field) is shown in Fig. 2b for low fields. Although at very low fields the J versus E curve is not linear, at moderate field it becomes linear insuring ohmic contact at the electrode–pellet interface. To have an idea of the value

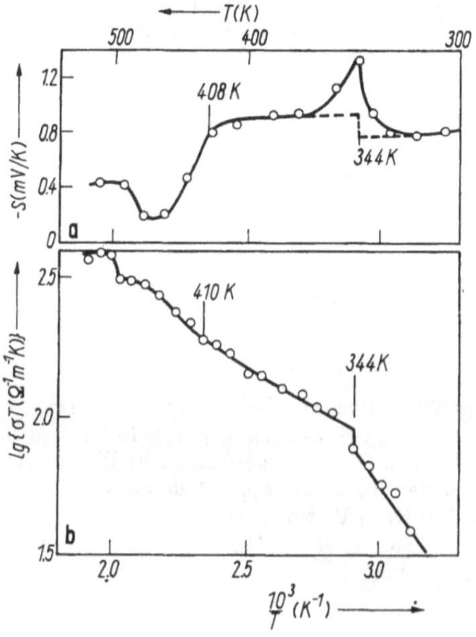

Fig. 3. Plots of a) thermoelectric powers S and b) electrical conductivity σ vs. inverse absolute temperature (T^{-1}) for well sintered pellets of Cu_2HgI_4 using Cu-foil electrodes

of contact resistance at the copper electrode–pellet interfaces, the resistance of a few well sintered pellets having the same area of cross-section but different thicknesses have been measured at fixed temperature. A typical plot is shown in Fig. 2c. The plot is a good straight line cutting the resistance axis on the positive side at $l = 0$. This intercept gives the value of contact resistance as $\approx 0.2\ \Omega$. Lead resistances also become of the order of fractions of Ω on the higher-temperature side. Thus contact as well lead resistances are significant only in the solid superionic phase, where σ exceeds $0.1\ \Omega^{-1}\ m^{-1}$ and the pellet resistance becomes less than $5\ \Omega$.

Electrical conductivity of several well annealed pellets made at $P > 6 \times 10^7$ kp m^{-2} having different dimensions have been measured as functions of temperature. The average values are shown as lg (σT) versus T^{-1} plot in Fig. 3. No hysteresis has been observed and values are nearly the same in heating and cooling cycles. The plot is linear in the temperature ranges 300 to 344 K and 345 to 410 K and can be represented by the relation

$$\sigma T = C \exp\left(-\frac{E_a}{kT}\right), \tag{2}$$

where C is a constant and E_a the activation energy. The values of C and E_a are found to be $9.70 \times 10^5\ \Omega^{-1}\ m^{-1}$ K and 0.28 eV and $2.19 \times 10^3\ \Omega^{-1}\ m^{-1}$ K and 0.11 eV in the temperature ranges 300 to 344 K and 345 to 410 K, respectively. Above 410 K, the conductivity shows an irregular variation with T. Comparing our data with those reported by earlier workers (Table 1) we find that our values of conductivity are larger and the activation energies much lower both above and below the transition temperature (T_p). However, there is fair agreement between the reported values of T_p (66.6 °C by Suchow and Pond [18], 69.0 by Shibota et al. [20], and 71 °C in the present study).

The thermoelectric power (S) has also been measured for a few thick pellets, ($l > 4 \times 10^{-3}$ m). S values have been found negative in the whole temperature range. Thus as per convention [24], positive ions are the dominant charge carriers in this solid over the entire temperature range. The plot of S versus T^{-1} is shown in Fig. 3. It is also linear between 300 and 344 K and between 350 and 408 K and can be expressed by the relation

$$S = -\frac{Q}{eT} + H\,. \tag{3}$$

The values of Q and H are 0.27 eV and 0.03 mV and 0.11 eV and -0.64 mV for the temperature ranges 300 to 344 K and 350 to 408 K, respectively. S also shows an irregular variation with T^{-1} above 408 K.

4. Discussion

In a conventional ionic solid where only one type of charge carriers is dominating in electrical conduction the expression for σ is given by the relation [25]

$$\sigma T = (NN')^{1/2} \exp\left(-\frac{H_f}{2kT}\right) \frac{q^2 \nu f a^2}{k} \exp -\left(\frac{h_m}{kT}\right), \tag{4}$$

where N and N' are the number of ions at normal and interstitial sites per unit volume. H_f is the enthalpy of formation of a defect pair, q the charge on the ion, ν the jump frequency, f the correlation factor which differs from unity if consecutive jumps of the charge carriers are not independent, a the jump distance of charge carriers, and h_m the activation enthalpy for migration. The first exponent gives the number of charge carriers and the second their mobility. Howard and Lidiard [26] have developed an expression for the total thermoelectric power of pure ionic conductors. Taking their general formula, an expression for the total homogeneous thermoelectric power has been obtained by us [21]. This expression can be generalized to any ionic conductor in which only one type of ions is the dominant charge carrier. On this consideration it is easy to show that S of Cu_2HgI_4 (Cu^+ ion conducting) can be given by the expression

$$S = -\frac{Q}{eT} + H\,, \tag{5}$$

where

$$H = \frac{1}{e}\left(\frac{\partial g}{\partial T}\right) + k \ln\left(\frac{n}{N}\right) + S_1\,. \tag{6}$$

Q is the heat of transport for Cu^+ ion, g the work required to bring a Cu^+ ion from a state of rest at infinity into an interstistial position in the crystal at constant temperature and pressure, n and N the number of mobile Cu^+ ions and normal sites per unit volume, respectively, and S_1 the partial entropy of a Cu^+ ion in copper metal.

It is evident from log (σT) versus T^{-1} and S versus T^{-1} plots that Cu_2HgI_4 has a phase transition around 344 K. The high-temperature phase as per convention is α and low-temperature phase β. It is convenient and more appropriate to discuss our result separately in these phases.

(i) α-Cu_2HgI_4: It is well known [16] that α-Cu_2HgI_4 is a cationically disordered solid, where I ions stay in definite lattice positions and Cu^+ and Hg^{2+} ion are uniformly distributed on sites whose number is large compared to the number of these ions. The bigger size and large charge leading to larger Coulomb repulsion of Hg^{2+} ions compared to Cu^+ ions does not favour the movement of Hg^{2+} ions. Thus Cu^+ ions dominate the electrical conduction. These ions may be thought of as existing in a quasi-molten state and all of them can be considered mobile. Thus the concept of thermally generated defects is not valid in this phase and consequently the expression for σ reduces to

$$\sigma T = C \exp\left(-\frac{h_m}{kT}\right), \tag{7}$$

where C is a temperature-independent constant. Also in the expression for H (6), the first and third terms are small and such that the increase of one of them nearly compensates the decrease of the other, n is already shown to be temperature independent. Thus H for this compound in the α-phase can be taken almost constant. Thus log (σT) versus T^{-1} as well as S versus T^{-1} curves should be straight lines. This is experimentally true in the temperature range 345 to 410 K. These linear plots give $h_m = E_a = 0.11$ eV and $Q = 0.11$ eV. Various models, e.g. free ion [27], lattice gas [28], ionic polaron [29], and extended lattice gas [30] have been proposed in the literature for SSIC. They are briefly described by us elsewhere [14]. Our result $E_a = Q$ favours the free ion and lattice gas models. The discrepancy above 410 K in both σ and S variation might be arising due to the movement of Hg^{2+} or electronic current but the region is not important for a detailed investigation. The only remark which seems meaningful is that α-Cu_2HgI_4 behaves as a superionic solid in the temperature range 345 to 410 K.

(ii) β-Cu_2HgI_4: In β-phase, Cu_2HgI_4 is a normal ionic solid in which Cu^+ ions are still the dominant charge carriers. In pure Cu_2HgI_4, defects are thermally generated. Since Frenkel defects are normally more mobile compared to vacancies, we shall consider the mobility of only these defects. The number of such defects n at any temperature can be expressed by the relation

$$n = (NN')^{1/2} \exp\left(-\frac{H_f}{2kT}\right), \tag{8}$$

where N' and N are the number of interstitial and normal sites per unit volume. Since both are of the same order, we can write

$$\frac{k}{e} \ln\left(\frac{n}{N}\right) = -\frac{H_f}{2eT}. \tag{9}$$

Putting this value into (6) and noting that the first and third terms in this equation are small, the expression for S becomes

$$S = -\frac{1}{eT}\left(Q + \frac{H_f}{2}\right) + S_2\,, \tag{10}$$

where S_2 is a small constant. Equations (7) and (10) favour the linear variation of log (σT) and S with T^{-1} for the pure solid with a slope of $H_f/2 + h_m$ and $Q + H_f/2$, respectively. H_f is expected to be of the order of 2 eV. Although the variations of log (σT) and S with T^{-1} are linear from room temperature to 344 K, the value of their slopes are too small to support the above theory of intrinsic conduction. This is probably due to the fact that the temperature is too low for an intrinsic region in this solid and the variation below 344 K represents the extrinsic region. Thus the variation of both σ and S in this region should depend upon impurities. For a relatively pure sample the number of interstitial ions may be taken as constant. In such a situation H becomes nearly constant and exp ($-H_f/2kT$) of the conductivity expression becomes ineffective. Thus the linear variation of log (σT) and S with T^{-1} is expected to yield a slope giving h_m and Q values for the mobile ion in this phase. Thus Q and h_m for Cu^+ ions in β-Cu_2HgI_4 are 0.27 and 0.28 eV, respectively.

Acknowledgement

Financial assistance of CSIR, India, for this work is thankfully acknowledged.

References

[1] W. van Gool, Fast Ion Transport in Solids, North-Holland Publ. Co., Amsterdam 1973.
[2] G. D. Mahan and W. L. Roth, Superionic Conductors, Plenum Press, New York 1976.
[3] S. Geller, Solid Electrolytes, Springer Verlag, 1977.
[4] M. Lazzari and B. Scrosati, J. Power Source **1**, 33 (1977).
[5] K. Shahi, phys. stat. sol. (a) **41**, 11 (1977).
[6] S. Chandra, Superionic Solids: Principle and Application, North-Holland Publ. Co., Amsterdam 1981.
[7] T. Takahashi, J. appl. Electrochem. **3**, 79 (1973).
[8] T. Takahashi, O. Yamamoto, and S. Ikeda, J. Electrochem. Soc. **122**, 83 (1975).
[9] A. F. Sammels, J. Z. Govgovtas, and B. B. Owens, J. Electrochem. Soc. **122**, 83 (1975).
[10] T. Takahashi, N. Wakabayashi, and O. Yamamato, J. Electrochem. Soc. **123**, 129 (1976).
[11] R. H. Dahn, S. Hackwood, R. G. Lineford, and J. M. Pollock, Nature (London) **272**, 522 (1978).
[12] T. Takahashi, O. Yamamoto, and A. Sawai, J. appl. Electrochem. **8**, 161 (1978).
[13] T. Takahashi, O. Yamamoto, S. Yamada, and S. Hayashi, J. Electrochem. Soc. **126**, 1654 (1979).
[14] B. K. Verma, V. Pratap, and H. B. Lal, Japan. J. appl. Phys. **20**, 1665 (1981).
[15] J. A. A. Ketelaar, Z. Krist. A **87**, 436 (1934).
[16] J. A. A. Ketelaar, Trans. Faraday Soc. **34**, 874 (1938).

[17] H. WIEDERSICH and S. GELLER, in: The Chemistry of Extended Defects in Non Metallic Solids, Ed. L. EYRIG and M. O. KEEFEE, North Holland Publ. Co., Amsterdam 1970 (p. 629).
[18] L. SUCHOW and G. R. POND, J. Amer. Chem. Soc. **75**, 5242 (1953).
[19] H. HOSHINO, J. Phys. Soc. Japan **10**, 197 (1953).
[20] S. SHIBOTO, H. HOSHINO, and M. SHIMOJI, J. Chem. Soc. Faraday Trans. **70**, 1409 (1974).
[21] K. SHAHI, H. B. LAL, and S. CHANDRA, Indian. J. pure appl. Phys. **13**, 1 (1975).
[22] B. K. VERMA, O. P. SRIVASTAVA, and H. B. LAL, phys. stat. sol. (a) **64**, 467 (1981).
[23] D. O. RALEIGH, in: Electroanalytical Chemistry, Ed. Q. BARD, Springer Verlag, 1972.
[24] S. M. GIRVIN, J. Solid State Chem. **25**, 65 (1978).
[25] A. B. LIDIARD, in: Hdb. Phys., Vol. 20, Ed. S. FLÜGGE, Springer-Verlag, Amsterdam 1957 (p. 246).
[26] R. E. HOWARD and A. B. LIDIARD, Phil. Mag. **2**, 1462 (1957).
[27] M. J. RICE and W. L. ROTH, J. Solid State Chem. **4**, 294 (1972).
[28] W. J. PARDEE and G. D. MAHAN, J. Solid State Chem. **15**, 310 (1975).
[29] C. P. FLYNN, Point Defect and Diffusion, Oxford University Press, London 1972.
[30] G. D. MAHAN, Phys. Rev. B **14**, 780 (1976).

(Received January 18, 1983)

phys. stat. sol. (a) 77, 413 (1983)

Subject classification: 20.3; 13.4; 22.5.2

Department of Physics, Nagpur University[1])

A Phenomenological Model for Supralinearity and Sensitization in LiF-TLD 100

By

S. V. MOHARIL

The supralinearity and sensitization factors of LiF-TLD 100 are calculated on basis of the model which is most satisfactory in explaining the experimental results. The model attributes the supralinearity and sensitization to the removal of competition of luminescence centres. The predictions of the model are in excellent agreement with the experimental results.

Die Supralinearitäts- und Sensibilisierungsfaktoren von LiF-TLD 100 werden auf der Grundlage eines Modells berechnet, das die experimentellen Ergebnisse am befriedigendsten erklärt. Das Modell führt die Supralinearität und Sensibilisierung auf die Beseitigung von Konkurrenz der Lumineszenzzentren zurück. Die Vorhersagen des Modells befinden sich in ausgezeichneter Übereinstimmung mit den experimentellen Ergebnissen.

1. Introduction

LiF doped with Mg and Ti (LiF-TLD 100) is now extensively used in thermoluminescence dosimetry [1]. The phenomena of "supralinearity and sensitization" in TL of this material have received attention of several workers [2 to 5]. One of the earliest models [6] to explain the supralinearity postulated the presence of deep traps which exhibited early saturation. No deep trap with such properties was found, and hence the model was discarded. Several models have been proposed thereafter; merits and drawbacks of these models have been discussed [7]. No single model can explain all the experimental facts. The model which at present appears most satisfactory attributes the supralinearity to the presence of non-luminescence centres which compete with the luminescence centres. The sensitization is attributed to the removal of this competition by filling the non-luminescent centres during pre-irradiation. The model sounds plausible in that it qualitatively explains most of the experimental facts. No attempt has been made to explain the "quantitative" results, e.g. supralinearity and sensitization factors have not been calculated. In fact, quantitative results have seldomly been given for any model. The Z-centre model of Nink and Kos [8] furnishes these details, but the model fails for other reasons [7].

Here, an attempt is made to calculate supralinearity and sensitization factors on the basis of a model suggesting the presence of non-luminescent centres. It is shown that the calculations are in excellent agreement with the experimental results. It is also shown that the model can bring together various pieces of experimental evidence.

2. Experimental Results

The experimental results which have to be explained by any model are briefly given in the following.

---

[1]) Nagpur 440010, India.

The response of LiF-TLD 100 to γ-rays is linear up to a dose of 10^3 rd; it becomes "supralinear" for higher doses. The maximum supralinearity factor of 3.5 is observed for a dose of 3×10^4 rd [7]. The dosimetry peak (peak V) saturates at about 10^5 rd and then it declines.

If the material is exposed to ionizing radiation, heated beyond peak V, and quickly cooled to room temperature, its response to subsequent exposures is enhanced. This is known as "radiation induced sensitization". The sensitization factor is a function of previous exposure (pre-irradiation dose). There is very little sensitization for doses $< 10^3$ rd. The sensitization factor increases with dose; it is maximum for 10^5 rd (whereas maximum supralinearity is observed at 3×10^4 rd) and then it declines [7]. The sensitized material exhibits no supralinearity.

Supralinearity is reduced when the material is exposed to radiations of high LET [9]. Sensitization is also very much reduced if the pre-exposure is made with such radiations.

3. Model and Calculations

3.1 *Supralinearity*

It is assumed that in the virgin sample empty traps and luminescence centres exist along with a fixed number of non-luminescent centres. Exposure to ionizing radiations fills the traps and an equal number of luminescence centres are produced at a linear rate. During the read-out, the charges are released from the traps. A fraction of them is lost non-radiatively due to capture by non-luminescent centres. (The non-luminescent centres which have captured charges are rendered ineffective for the next exposure until the material is given a suitable annealing treatment.) The remaining charges recombine at the luminescence centres leading to TL emission. The number of filled traps and luminescence centres goes on increasing with dose, but the non-luminescent centres are fixed and soon get exhausted (for a dose of 10^3 rd). For higher doses the fraction of charges destroyed non-radiatively goes on decreasing for this reason. This leads to "supralinearity". At still higher doses ($> 3 \times 10^4$ rd), the number of traps filled does not rise linearly but tends to saturate (broken curve a of Fig. 1). For doses exceeding 10^5 rd, there is a permanent damage to traps and hence the response declines (circles on full curve a of Fig. 1). The damage can be removed only after suitable thermal annealing treatment (400 °C, 1 h).

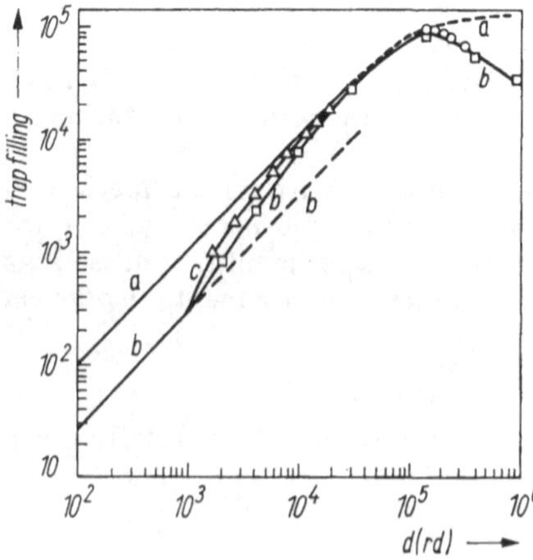

Fig. 1. Dependence of thermoluminescence on dose d. Full curve a shows the trap filling. The dashed curve a shows the limitation due to saturation and circles show the decline in number of filled traps due to damage. At this stage, the difference between curves a and b is too small to be distinguished in the figure; hence numerical values are given in Table 1. Curve b gives actual TL response calculated from (2). The squares represent the experimental points. The dashed curve b shows the apparent line of linearity. Curve c gives the response calculated from (1)

Let n_l be the rate (per unit dose) at which the traps and luminescence centres are filled. Let n_{0v} be the number of non-luminescent centres in the virgin sample. Let α be the fraction of charges going to non-luminescent centres during read-out. Then the TL output T_0 is proportional to $(1 - \alpha)\, n_l d$, where d is the dose. Thus, it varies linearly with dose. However, the number of charges destroyed non-radiatively, viz. $\alpha n_l d$, has the upper limit n_{0v}. For doses exceeding the limiting dose at which this limit is surmounted the TL output will be $n_l d - n_{0v}$.

Thus, we have for the TL output

$$T_0 = \begin{cases} (1-\alpha)\, n_l d & \text{for} \quad d < 10^3\ \text{rd}\,, \\ n_l d - n_{0v} & \text{for} \quad d > 10^3\ \text{rd}\,. \end{cases} \tag{1}$$

For doses exceeding 3×10^4 rd, the filled traps are not $n_l d$ but assumed to be given by curve a of Fig. 1 (also see column 3 of Table 1).

Table 1

Trap filling as a function of dose

dose d (rd)	traps filled ($10^4\ n_l$)			damage (%)	traps combining radiatively ($10^4\ n_l$)
	linear range (full curve a of Fig. 1)	saturation (dashed curve a)	damage (circles on curve a)		
$< 3 \times 10^4$	$n_l d$	—	—	—	$(1-\alpha)\, n_l d$
4×10^4	—	3.84	—	—	3.43
5×10^4	—	4.65	—	—	4.25
6×10^4	—	5.4	—	—	5.01
7×10^4	—	6.1	—	—	5.71
8×10^4	—	6.8	—	—	6.4
10^5	—	8.0	8.0	0.0	7.6
2×10^5	—	9.1	7.8	14.0	7.7
3×10^5	—	10.0	6.7	33.0	6.6
4×10^5	—	11.6	5.66	52.2	5.6
5×10^5	—	12.0	5.04	58.0	5.0
6×10^5	—	12.2	4.5	63.2	4.5
7×10^5	—	12.3	4.1	66.7	4.1
8×10^5	—	12.4	3.8	69.4	3.8
9×10^5	—	12.45	3.5	72.0	3.5
10^6	—	12.5	3.3	73.6	3.3

The number of traps filled rises linearly with dose (as shown by the full curve a of Fig. 1. After 3×10^4 rd traps tend to saturate (broken curve a), as given in column 3. The saturation value attained would have been 12.5×10^4, but there is damage to traps for doses exceeding 10^5 rd (column 5), and for such doses the actual number of filled traps is given by column 4 (circles on curve a) and not by column 3.

The response obtained with these assumptions is shown in curve c of Fig. 1. The model parameters used are $n_{0v} = 10^3 \alpha n_l$ and $\alpha = 0.72$. The maximum supralinearity factor of 3.48 is obtained at 3×10^4 rd in excellent agreement with the experimental results. At lower doses, however, the calculated supralinearity factor (S_p) values are higher than the experimental values. The onset of the supralinearity at 10^3 rd is rather abrupt. This is due to the fact that α is assumed to be constant and the saturation of non-luminescent processes is abrupt. A more realistic assumption will be that

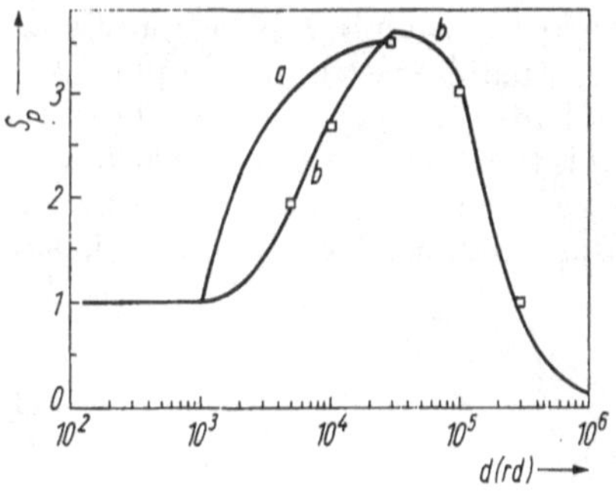

Fig. 2. Variation of supralinearity factor S_p with dose. (a) From (1), (b) from (3). Squares represent experimental data [7]

the probability of the released charge going to luminescence or non-luminescent centres will depend on their relative concentrations. Thus, the fraction of charges destroyed non-radiatively during the read-out is not a constant, but given by a relation of the type

$$\alpha = \alpha_0 \left[1 - \exp\left(-\frac{n_{0v}}{n_1 d}\right)\right], \tag{2}$$

where α_0 is a constant.

Thus, for low doses $(n_1 d \ll n_0)$ $\alpha - \alpha_0$, as $n_1 d$ (number of luminescence centres) becomes comparable to n_{0v}, α goes on decreasing the variation being smooth. The supralinearity factor S_p will be given by

$$S_p = \frac{1-\alpha}{1-\alpha_0}. \tag{3}$$

The response was calculated with (2). For doses exceeding 3×10^4 rd, α was calculated using (2) and the number of traps filled was obtained from curve a of Fig. 1. Curve b of Fig. 1 represents the calculated response. It agrees excellently with the experimental results (points on the curve). The corresponding S_p values are shown in Fig. 2 (curve b). These curves (Fig. 1, curve b and Fig. 2, curve b) were obtained for model parameters $\alpha_0 = 0.76$ and $n_{0v} = 6 \times 10^3 n_1$ which gave best agreement with the experimental results. The maximum S_p-value of 3.59 is obtained for 3×10^4 rd. The curve fitting was also attempted for various experimental results existing in the literature. The best values of α_0 ranged between 0.72 and 0.78 and that of n_{0v} between 3×10^4 and 6×10^4 rd.

3.2 Sensitization

The proposed mechanism of sensitization is twofold. The number of non-luminescent centres in the sensitized material, n_{0s}, is less than n_{0v}. Again, during the post-irradiation thermal annealing, not all the luminescence centres are destroyed.

Thus, if d_p is the pre-irradiation dose, then during pre-irradiation $n_1 d_p$ traps were filled and an equal number of luminescence centres were effective. Out of the $n_1 d_p$ traps, $\alpha n_1 d_p$ were destroyed non-radiatively and $(1 - \alpha)\, n_1 d_p$ contributed to TL. Hence, during post-irradiation thermal anneal $(1 - \alpha)\, n_1 d_p$ luminescence centres were destroyed by recombinations and $\alpha n_1 d_p$ ones are left behind. Hence, during the test $n_1 d_t$ traps are filled, where d_t is the test dose. The luminescence centres in the sensitized material are $n_1 d_t$ (due to test dose) plus $\alpha n_1 d_p$ (left behind). The non-luminescent centres are

$$n_{0s} = n_{0v} - \alpha n_1 d_p .$$

Hence, in the sensitized material the fraction of charges destroyed non-radiatively, α_s, is given by

$$\alpha_s = \alpha_0[1 - \exp(-n_{0s}/n_{ls})] = \alpha_0 \left[1 - \exp\left(-\frac{n_{0v} - n_l d_p \alpha}{n_l d_t + k\alpha n_l d_p}\right)\right]. \quad (4)$$

Here, α itself is a function of pre-irradiation dose d_p and given by (2) when d is replaced by d_p. The sensitization factor, S/S_0, is

$$\frac{S}{S_0} = \frac{1 - \alpha_s}{1 - \alpha_t}, \quad (5)$$

where α_t is the α-factor corresponding to the test dose d_t and can be obtained from (2) by substituting d_t for d.

For a fixed test dose d_t, α_s can be calculated as a function of pre-irradiation dose d_p using (2) and (4). The sensitization factor then can be obtained through (5).

One of the important results regarding the sensitization is that the maximum S/S_0 is observed for 10^5 rd, whereas the maximum supralinearity factor is attained at 3×10^4 rd [7]. To explain this, it is necessary to assume that though the traps saturate after 3×10^4 rd, the luminescence centres continue to grow up to 10^6 rd (the validity of this assumption will be discussed later).

Thus, for $d_p > 3 \times 10^4$ rd the number of luminescence centres left behind will be $n_l d_p$ minus the number of traps destroyed radiatively (which is not equal to $(1 - \alpha) n_l d_p$, but less due to the saturation effect and can be obtained from curve a of Fig. 1). Similarly, n_{0s} at these doses will be n_{0v} minus the number of non-luminescent centres filled during pre-exposure, which is not $\alpha n_l d_p$, but less due to the saturation effect.

The sensitization factor can be calculated as a function of pre-irradiation dose using (2), (4), and (5) and curve a of Fig. 1. The data of curve a is also given in Table 1. The calculations for a test dose 10^2 rd are presented in Fig. 3. It is seen that there is very little sensitization ($S/S_0 \approx 1$) for doses $< 10^3$ rd. It increases with d_p. For $d_p = 3 \times 10^4$ rd, $S/S_0 = 3$. For doses exceeding 3×10^4 rd, S/S_0 would have decreased as the number of non-luminescent centres filled would have gone down. However, the number of luminescence centres left behind increases rapidly owing to the fact that the number of traps filled during pre-exposure is less (saturation effect) than that of luminescence centres (which do not saturate but continue to grow linearly). This results in increasing sensitization. The maximum sensitization factor of 3.85 is attained at 10^5 rd. Pre-irradiation doses in excess of this result in permanent damage to traps, so that the traps filled during the test dose, then are not $n_l d_t$ but some fraction of it. The fraction is governed by the damage which is shown in column 5 of Table 1.

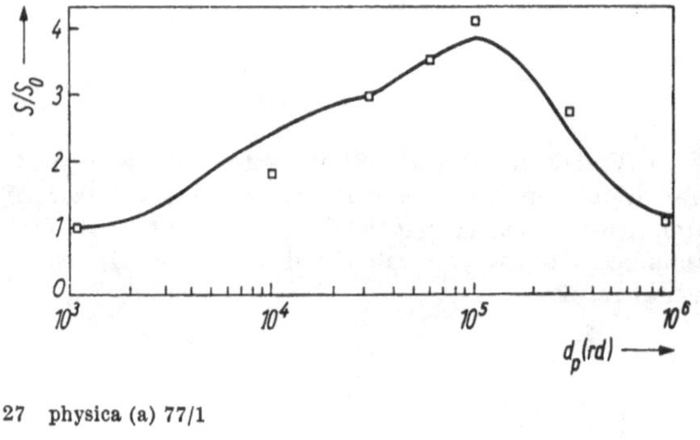

Fig. 3. Dependence of sensitization factor S/S_0 on pre-irriadiation dose d_p. The curve is calculated for test dose of 10^2 rd. Open squares represent experimental points [7, 10]

It is seen that the rise in S/S_0 after 3×10^4 rd is rather abrupt, because it is assumed that all the luminescence centres which are in excess over the filled traps are left behind. Actually, during the post-irradiation thermal anneal some of the luminescence centres will get annealed out even after all traps are emptied. By taking this into account a smoother variation can be obtained. The important fact is that the maximum S/S_0 is different from maximum supralinearity factor and is attained at a different dose. The experimental results also show [10] that a maximum sensitization factor of just above four is attained at 10^5 rd when the area method is used. In the peak-height method the maximum S/S_0 is between 7 and 8, but as pointed out earlier [10], this is due to the narrow width of peak V as peak IV is absent in the sensitized material.

From (5) the sensitization factor can be calculated as a function of test dose, e.g. there will be hardly any sensitization for test doses in excess of 10^4 rd. This is consistent with the experimental facts. Systematic calculations for various lower doses can be presented, but no such experimental results are available at present for comparison.

3.3 Sensitization and post-irradiation thermal annealing treatment

It has been commented [7] that S/S_0 should decrease with increasing post-irradiation thermal annealing temperature T_p as more and more luminescence centres will be annealed out and less will be left behind. However, it is to be noted that the loss of sensitization will not be directly proportional to the number of luminescence centres annealed out, but it will be governed through (4) and (5).

Here we make an attempt to match the calculated variation in sensitization with T_p with the particular experimental results available. The pre-irradiation dose is 10^5 rd and the test dose is 100 rd. Even if it is assumed that the luminescence centres are annealed out as fast as shown in Fig. 4 (curve a), the corresponding loss of sensitization (curve c calculated through (4) and (5)) is gradual till $T_p = 360$ °C. Above this temperature, the filled up non-luminescent centres start recovering (curve b), and hence there is a rapid loss of sensitization. For $T_p = 400$ °C, all the non-luminescent centres n_{0v} have recovered, there is little sensitization left; the sample returns to its original state. This agrees excellently with the experimental data [7] (curve d).

Curves a and b of Fig. 4 are assumed in the above analysis. One can obtain the actual curves from experimental measurements, predict the variation of S/S_0 with T_p through (4) and (5), and then compare with the experimental results. Such measurements are not available at present, but the calculations of this section show that it is possible to explain the variation of S/S_0 with T_p in the present model.

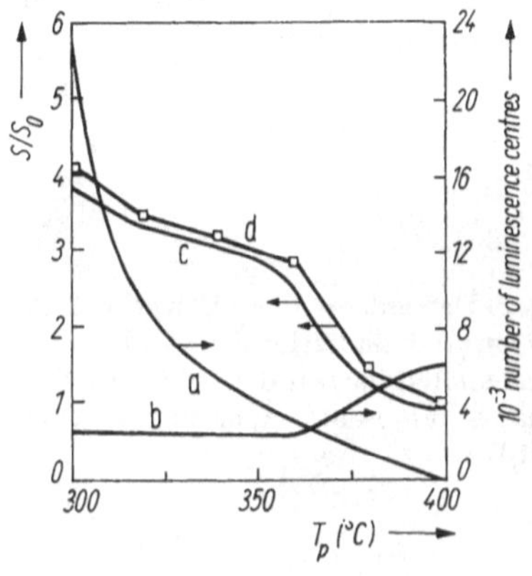

Fig. 4. Variation of sensitization factor S/S_0 with post-irradiation annealing temperature T_p. (a) Number of luminescence centres left behind, (b) number of non-luminescent centres, (c) calculated variation, (d) experimental results

3.4 Supralinearity in the sensitized material

The response of the sensitized material which has been exposed to pre-irradiation doses in excess of 10^4 rd will be almost linear. There will hardly be any supralinearity observed as a large number of luminescence centres are left behind and a maximum number of the non-luminescent centres are filled during the pre-exposure. This agrees with the experimental results. It is also evident that for $d_p < 10^3$ rd very few non-luminescent centres are filled ($n_{0s} \approx n_{0v}$); also the number of luminescent centres left behind is small compared to n_{0s}. Such samples will exhibit supralinearity. With increasing d_p, more and more non-luminescent centres will be rendered ineffective and/or the number of luminescence centres left behind will go on increasing. The supralinearity will also go on decreasing. In fact the quantitative results can be obtained using (2) to (5); but as there are no such experimental results available for comparison, the presentation of these results is deferred.

3.5 Exposure to high LET radiations

A remarkable experimental result is that reduced supralinearity is observed when exposure is to high LET radiations such as α-rays [9]. The TL response is also about ten times lower than that for γ-rays. This can be explained in the present model. Thus the rate at which traps are filled, viz. n_1, is about ten times smaller for α-rays as compared to γ-rays. The onset of the supralinearity will thus be shifted to doses ten times higher (10^4 rd). S_p will start increasing for doses exceeding 10^4 rd. At 10^5 rd, the S_p-factor should be 2.74 if the model parameters of the previous sections are used. The results of Jain [9], however, indicate that the S_p-factor is quite low even for the dose 10^5 rd. To explain this it is necessary to assume that during α-irradiation additional non-luminescent centres are produced. For example, if it is assumed that the exposure to α-radiation produces equal numbers of luminescence and non-luminescent centres, then the S_p-factor of 1.6 is obtained for 10^5 rd. This also implies that there is little sensitization if pre-exposure is made to radiation of high LET. On the other hand, if pre-exposure is made by γ-rays, then there should be sensitization for not too high test doses ($< 10^3$ rd) irrespective of the type of radiation used. All these predictions are qualitatively borne out by the known experimental results. The predictions of the model cannot be quantitatively tested as there are no such experimental results available at present.

4. Physical Possibilities

In the preceding sections the nature of the traps and luminescence centres has not been specified. It will now be shown that the assumptions of the model are at least plausible in view of the properties of the actual lattice defects which can act as traps and recombination centres. The essential properties required to explain the experimental results are

(i) traps and luminescence centres are filled at a linear rate;

(ii) traps saturate for doses $> 3 \times 10^4$ rd, and for doses exceeding 10^5 rd there is permanent damage to traps;

(iii) luminescence centres grow linearly till 10^6 rd;

(iv) there exists a fixed number of non-luminescence centres.

It will be interesting to see which of the various lattice defects fulfil these requirements. It is known that the electron centres such as F- and Mg-related centres (225, 250, 310, and 380 nm bands) grow linearly with dose up to 10^6 rd [11]. These can thus be the luminescence centres of the present model. The traps then are of hole type [12] or "trapped interstitial halogen atoms" [13]. It is known that for heavy doses the clustering of these defects occur, and these clusters anneal out only at high temperature. This can explain the saturation of the traps and the damage which can be removed only by high-temperature (400 °C) anneal. The existence of the non-luminescent centres has been known since long [14]. The defects which can play this role (presumably related with the OH impurity) have been suggested. Several other possibilities can also be considered. Due to the lack of an adequate understanding of TL, it is not possible to specify the details of the physical processes leading to sensitization and supralinearity. All the same, it is seen that the assumptions of the model are consistent with the properties of the known lattice defects.

5. Conclusions

Thus, the present model can successfully explain a large number of experimental facts. It can be transformed to a physically realistic situation. The model remains phenomenological at this stage because of our inadequate understanding of the "mechanism of defect production", photon emitting step during TL, etc. It will be worthwhile to test the predictions of the model.

If the model parameters obtained here are any indication, then it can be surmised that the luminescent processes in LiF-TLD 100 are highly inefficient at low doses in that about 70 to 80% of the filled traps are annealed out non-radiatively. If the non-luminescent centres are got rid of, then about three- to fivefold sensitization would result. At present, this is achieved by filling the non-luminescent centres by pre-exposure. However, for re-using it is necessary to anneal the dosimetry material at 400 °C. The sensitization is also lost during this process. It may prove useful to find another convenient way of eliminating the non-luminescent centres.

Acknowledgements

The financial assistance from the Council of Scientific and Industrial Research, New Delhi is gratefully acknowledged. The author enjoyed many fruitful discussions with Dr. S. P. Kathuria, Dr. A. R. Lakshmanan and Dr. B. Chandra of Bhabha Atomic, Research Centre, Bombay. He also thanks his senior colleague Dr. B. T. Deshmukh for providing constant encouragement.

References

[1] K. G. Vohra, Bull. Rad. Prot. **2**, 37 (1979).
[2] V. K. Jain, phys. stat. sol. (a) **60**, 351 (1980).
[3] V. K. Jain, phys. stat. sol. (a) **66**, 341 (1981).
[4] B. Chandra, A. R. Lakshmanan, and R. C. Bhatt, phys. stat. sol. (a) **66**, 335 (1981).
[5] A. R. Lakshmanan, R. C. Bhatt, and S. J. Supe, J. Phys. D **14**, 1683 (1981).
[6] N. Suntharalingam and J. R. Cameron, Phys. Med. Biol. **14**, 397 (1969).
[7] A. R. Lakshmanan, R. C. Bhatt, and K. G. Vohra, phys. stat. sol. (a) **53**, 617 (1979).
[8] R. Nink and H.-J. Kos, phys. stat. sol. (a) **35**, 121 (1976).
[9] V. K. Jain, Nuclear Instrum. and Methods **180**, 195 (1981).
[10] B. Chandra, A. R. Lakshmanan, and R. C. Bhatt, phys. stat. sol. (a) **57**, 323 (1980).
[11] N. Takeuchi, R. Nink, and H.-J. Kos, J. Mater. Sci. **15**, 2933 (1980).
[12] M. R. Mayhugh, R. W. Christy, and N. M. Johnson, J. appl. Phys. **41**, 2968 (1970).
[13] F. Sagastibelza and J. L. Alvarez-Rivas, J. Phys. C **14**, 1873 (1981).
[14] T. G. Stoebe and S. Watanabe, phys. stat. sol. (a) **29**, 11 (1975).

(Received November 12, 1982)

phys. stat. sol. (a) **77**, 421 (1983)

Subject classification: 16; 22.9

A. F. Ioffe Physico-Technical Institute, Academy of Sciences of the USSR, Leningrad[1])

Absence of 3.39 μm Photoconductivity in the Quasi-One-Dimensional Organic Semiconductor $Cs_2(TCNQ)_3$

By

V. N. Semkin, R. M. Vlasova, and G. Yu. Yashin

In the temperature range 4.2 to 300 K under modulated 3.39 μm laser illumination, the dependencies of photoresponse on modulation frequency, as well as polarization and phase shift are measured in $Cs_2(TCNQ)_3$ single crystals. An interpretation of the results obtained is presented; the photosensibility and photocarrier lifetime values of less than 10^{-26} MΩ^{-1} W^{-1} and 10^{-14} s, respectively, suggest that the standard band pattern for $Cs_2(TCNQ)_3$ is unsuitable.

В интервале температур 4.2 до 300 K под действием модулированного лазерного освещения с $\lambda = 3.39$ μm сняты поляризационные и фазовые зависимости сигнала фотоотклика на монокристаллах $Cs_2(TCNQ)_3$, а также зависимость его от частоты модуляции. Полученные данные обьяснены с точки зрения обычных тепловых явлений. Фоточувствительность монокристаллов $Cs_2(TCNQ)_3$ оказывается ниже 10^{-26} MΩ^{-1} W^{-1}, а время жизни фотоносителей меньше 10^{-14} c, что говорит о неприменимости стандартной зонной картины для описания процессов переноса заряда в этих кристаллах.

1. Introduction

The ground state structure for quasi-one-dimensional organic systems has recently been widely discussed in the literature. One can hardly claim, however, that there is a unified model to describe the experimental evidence now available. This is true even for the simplest TCNQ salts known to date in numerous modifications [1].

It has become common to describe the properties of low- and intermediate-conducting salts on the basis of a band-like model in which the gap exists in the one-particle excitation spectrum. That such a gap may actually exist seems to be suggested by potential lattice periodicty, in the first place, or the Mott-Hubbard or Peierls transitions [2]. On the basis on the observed temperature dependence of the dc dark conductivity and thermopower [3], it is tempting to regard $Cs_2(TCNQ_3)$ as one of such band-gap systems. The $Cs_2(TCNQ)_3$ reflection spectra in the 0.04 to 4 eV frequency region obtained by our group [4, 5] at room temperature and their Kramers-Kronig analysis provide good evidence for an absorption peak ($\hbar\omega = 0.36$ eV or 2900 cm^{-1}) due to the transfer of electrons along the TCNQ stacks. Since the dark conductivity activation energy (0.3 eV) and the intramolecular electronic transition energy are nearly equal, there is good reason to believe that the peak is of interband nature in analogy to a semiconductor. On the basis of this assumption photoconductivity along the b-axis should occur at 0.36 eV illumination. To test this assumption an investigation of single crystals irradiated by 3.39 μm laser light at 4.2 to 300 K was carried out, the results of which are reported below.

---

[1]) 194021 Leningrad, USSR.

2. Experimental Techniques

Grown by recrystallization from $Cs_2(TCNQ)_3$ acetonitrile solution, the crystals had monoclinic symmetry (space group $P2_1/c$, $z = 2$) [6]). TCNQ stacks produced from TCNQ molecular triades were directed along the b-axis. The samples (typical size $5 \times 0.5 \times 0.1$ mm^3) with the most perfect surface were selected for the subsequent study. Silver electrodes made by vacuum evaporation were used as shown in the insets of Fig. 1. Periodic He–Ne gas laser (3.39 μm, 10 mW) beam pulses were obtained by means of an obturator (with a set of disks). To determine the anisotropy of photoresponse the beam was depolarized by a $\lambda/4$ plate and then repolarized using a germanium polarizer. The degree of polarization was close to 0.98, the error in the polarization angle measurements being about 2°. A fast detector response was used to normalize our data.

3. Experimental Results

Taking into account the anisotropy of the reflection coefficient $R(\alpha)$, α being the angle between the plane of polarization and the b-axis, the light power dissipated within a crystal $I(\alpha)$ may be written as

$$I(\alpha) = I_0[1 - R_\perp + (R_\perp - R_\parallel) \cos^2 \alpha] . \qquad (1)$$

Fig. 1 shows the dependence of the photoresponse $Q(\omega, \alpha, T)$ obtained at room temperature with modulation frequencies equal to 0, 0.2×10^2, and 2.5×10^3 s^{-1} on the direction of polarization, normalized to $I(\alpha)$ and divided by $Q(\omega, 0, T)$.

It can be seen that at $\omega = 0$, $Q(0, \alpha, T)$ is practically independent of α, but as ω rises the dependence of Q on α becomes appreciable. In the temperature range 150 to 300 K the dependence of $Q(\omega, \alpha, T)$ on T was obtained for several crystals. Typical results are shown Fig. 2. The strong temperature dependence of Q cannot be ascribed

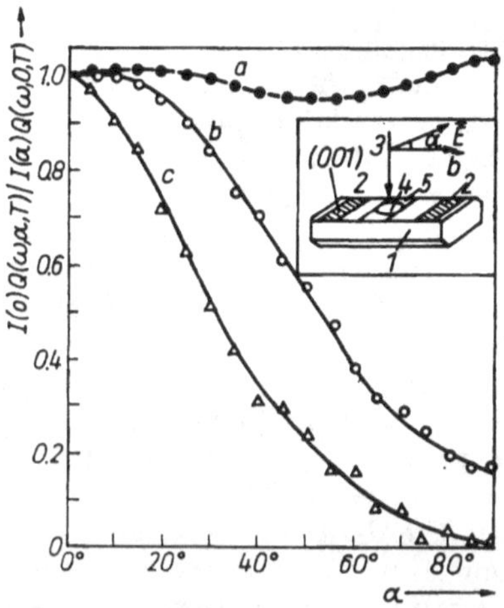

Fig. 1

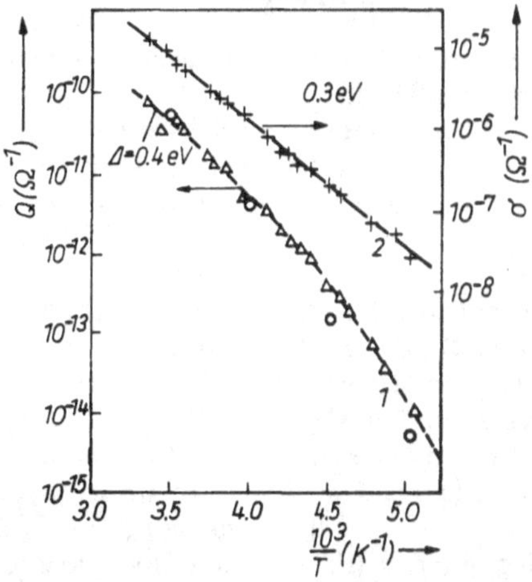

Fig. 2

Fig. 1. Angular dependence of normalized photoresponse: ● $\omega = 0$, ○ 2×10^2, △ 2.5×10^3 Hz. The inset shows 1 the $Cs_2(TCNQ)_3$ crystals, 2 electrodes, 3 laser beam, 4 real light spot, 5 type of the light spot used in the model

Fig. 2. Temperature dependences of the photoresponse Q (crosses) and dark conductivity (triangles); circles calculated from (3)

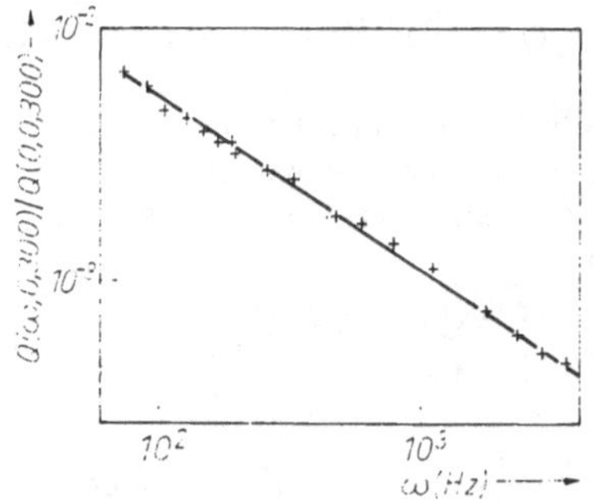

Fig. 3. Frequency dependence of the photoresponse Q, $T = 300$ K, $\alpha = 0$

to the factor $\exp(-\Delta/kT)$ alone; it will be noticed that the temperature dependence is described by a steeper curve than the dark conductivity (curve 1).

In the frequency region from 70 to 3.8×10^3 s^{-1}

$$Q(\omega, 0, 300\ K) \sim Q_0/\omega$$

is valid within reasonable accuracy (see Fig. 3). The photoresponse phase shift was measured at the frequencies $\omega = 10^2$ s^{-1} and 2×10^3 s^{-1} and found to be equal to $\Delta\varphi = -\pi/2$.

4. Interpretation

Despite the fact that the absorption at 3.39 μm is of electronic nature (provided $\boldsymbol{E} \parallel \boldsymbol{b}$), both frequency dependence and phase shift of the response seem to indicate a thermal origin of the observed effects. Moreover, it appears that temperature and polarization dependences of Q may also be interpreted in terms of thermal phenomena.

Assuming that all absorbed light is instantly transformed to heat, one can imagine the following pattern of the processes under consideration. All energy dissipation occurs within the length l (α-dependent). If $\boldsymbol{E} \parallel \boldsymbol{b}$ ($\alpha = 0$), then $l_\parallel \approx 0.1$ μm while if $\boldsymbol{E} \perp \boldsymbol{b}$ ($\alpha = \pi/2$) the penetration length $l_\perp$ approaches 1.5 μm. Therefore, the spatial distribution of the heat sources comes into play, the distribution length being essentially dependent on α. The dissipated energy consists of two parts: the first is constant in time and determines the constant temperature regime of the sample; the second is periodic in time and modulates the temperature of the sample in the vicinity of the heat source, the thickness of the heated layer being

$$\lambda_T \sim \sqrt{\varkappa/\varrho c \omega}\,, \tag{2}$$

where $\varkappa$ is the heat conduction coefficient, ϱ the density, c the specific heat, and ω the modulation frequency. Under these conditions one obtains $\lambda_T \approx 5 \times 10^{-5}\omega^{-1/2}$m ($\omega$ measured in s^{-1}). At high frequencies the value of λ_T is small, and in the $\omega \to \infty$ limit the temperature oscillations occur only in that part of the sample which is transparent to light.

For an experimental geometry as shown in Fig. 1, the main part of the electric current should flow near the sample surface within a layer of $\lambda_j \approx 1$ μm thickness (because of the large conductivity anisotropy, $\sigma_\perp/\sigma_\parallel \approx 10^{-3}$, present in $Cs_2(TCNQ)_3$). At high frequencies the thickness of the current layer heated and the amplitude of its temperature oscillations decrease. This should result in polarization dependences like those shown in Fig. 2. Clearly, the polarization dependence could arise at such frequencies only when λ_T becomes comparable with the greatest penetration length;

the higher ω is (the smaller λ_T) the stronger is this dependence. This is what seems to be suggested by our experiments.

It is possible to consider these effects quantitatively in terms of a model based on an isotropic $\varkappa$ and an experimental arrangement similar to that shown in Fig. 1.

If $\omega > \varkappa/(\varrho c l_\perp^2) \approx 10^3\ \mathrm{s}^{-1}$ and the light modulation is sinusoidal, the model yields

$$Q(\omega, \alpha, T) \sim \frac{I(\alpha)}{l(\alpha)} \frac{\exp(-\Delta/kT)}{\omega \varrho c} \frac{e^{i\omega t}}{[\lambda_T/l(\alpha)]^2 - i} \int\limits_0^\infty \frac{\exp[-x/l(\alpha)]}{1 + (x/\lambda_j)^2}\, dx\,. \qquad (3)$$

The exponential factor $\exp(-\Delta/kT)$ results from the temperature dependence of dark conductivity; $[1 + (x/\lambda_j)^2]^{-1}$ is the distribution of the current density in the X-direction (at a point equidistant from both electrodes). Equation (3) should certainly be considered only as a rough approximation: there are many factors which might cause deviations in the actual experimental situation (such as a slightly different law of current density distribution near the surface or a great effect of surface defects on transport properties, etc.).

In spite of the possible inaccuracies, (3) should provide a reasonable fit for the dependence of our experimental data on ω and α as well as on temperature, provided the dependence of concentration c on temperature is taken into account. An increase of c with increasing T will give a stronger temperature dependence of $Q(\omega, \alpha, T)$ than $\exp(-\Delta/kT)$. In case $c(T)$ undergoes a two-fold increase in the temperature range shown in Fig. 3, (3) should be an adequate representation of our experimental data. While we do not know $c(T)$ experimentally, we feel that such an assumption should be viable as it is known [5] that there are many high-frequency phonon branches in the system.

5. Discussion

The estimates obtained above seem to point to a possibility of interpreting the dependences of the photoresponse $Q(\omega, \alpha, T)$ in terms of thermal effects, the strongest arguments being provided by the dependence of Q on ω (as ω^{-1}) and the photoresponse phase $\Delta\varphi = -\pi/2$. This scheme also accomodates the polarization and temperature dependences of Q not alike those obtained experimentally. As far as thermal effects are concerned, the most interesting and unusual discovery is a strong polarization dependence of Q which is much stronger than the dependence of dissipated energy on α, and, as has been shown above, is the result of the dependence of penetration depth on α.

From the known sensitivity of our experimental arrangement the photosensitivity of $Cs_2(TCNQ)_3$ single crystals must be less than $10^{-26}\ \mathrm{M\Omega^{-1}\,W^{-1}}$. The photocarrier lifetime τ calculated on the basis of the standard band model will be less than 10^{-14} s, much less than in conventional semiconductors. This estimate demonstrates the non-applicability of the usual band picture to such a substance as $Cs_2(TCNQ)_3$ as far as transport phenomena are concerned. The validity of this picture is doubtful irrespective of the nature of the band gap. A possible explanation for the small value of the photoresponse might be that there exist some phonon branches whose frequencies overlap the gap value as determined from the temperature dependence of dark conductivity in $Cs_2(TCNQ)_3$ [5]. If this assumption is true, some mixed electron–phonon states are likely to exist.

The alternative interpretations, on the other hand, may make use of complicated energy spectrum or mobility edge concepts.

Acknowledgements

We would like to thank V. V. Kuzina and Dr. A. I. Sherle for supplying the samples; we are also grateful to Prof. I. F. Shchegolev, Prof. V. N. Bogomolov, and Prof. A. R. Regel for useful discussions.

References

[1] I. F. Shchegolev, phys. stat. sol. (a) **12**, 9 (1972).
[2] L. N. Bulaevsky, Uspekhi Fiz. Nauk **115**, 263 (1975).
[3] W. J. Siemons, P. E. Bierstedt, and R. G. Kepler, J. chem. Phys. **39**, 3523 (1963).
[4] R. M. Vlasova, A. I. Gutman, N. F. Kartenko, L. D. Rosenshtein, L. S. Agroskin, G. V. Papayan, L. P. Rautian, and A. I. Sherle, Fiz. tverd. Tela **17**, 3529 (1975).
[5] M. V. Belousov, A. M. Vainrub, R. M. Vlasova, and V. N. Semkin, Fiz. tverd. Tela **20**, 107 (1978).
[6] C. J. Fritchie and P. Arthur, Acta cryst. **21**, 139 (1966).

(Received January 11, 1983)

See Short Note by J. MAHY et al.

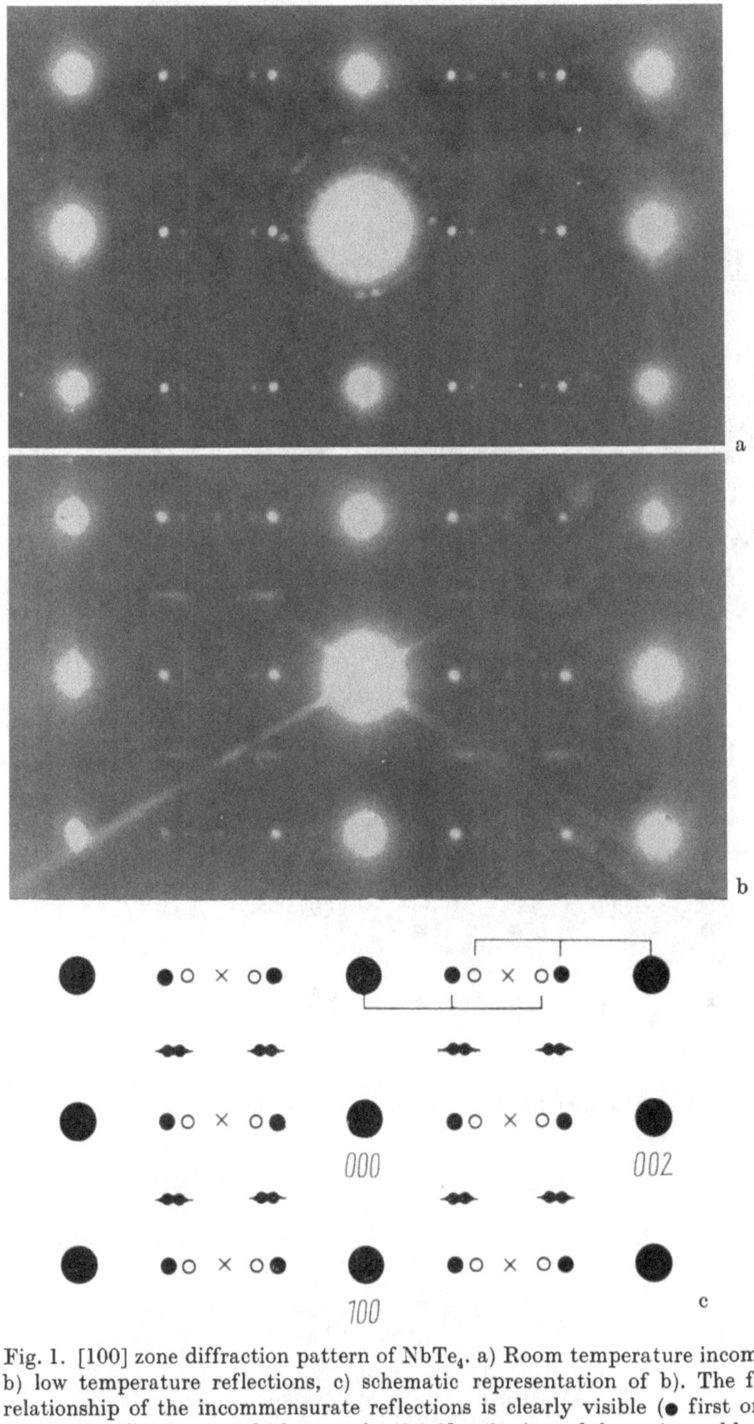

Fig. 1. [100] zone diffraction pattern of $NbTe_4$. a) Room temperature incommensurate reflections, b) low temperature reflections, c) schematic representation of b). The first and second order relationship of the incommensurate reflections is clearly visible (● first order; ○ second order). The weak reflections marked × are the $(0\ 0\ 2l + 1)$ sites of the reciprocal basic lattice, which have become non-extinct due to the modulation of the lattice

See Short Note by J. MANY et al.

a

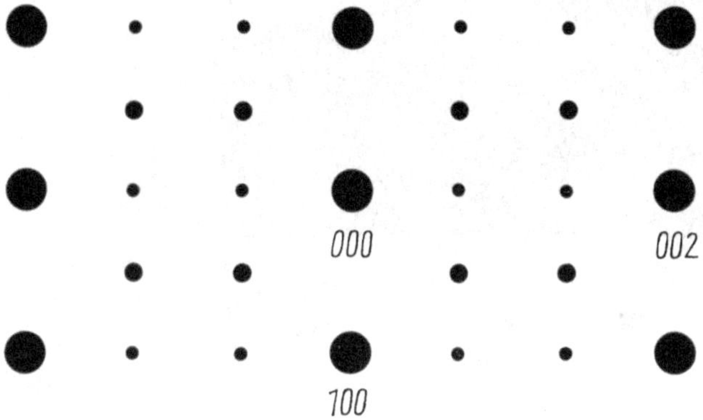

b

Fig. 2. a) [100] zone diffraction pattern of $TaTe_4$, b) schematic representation of a)

See Short Note by K. Abdelmoula et al.

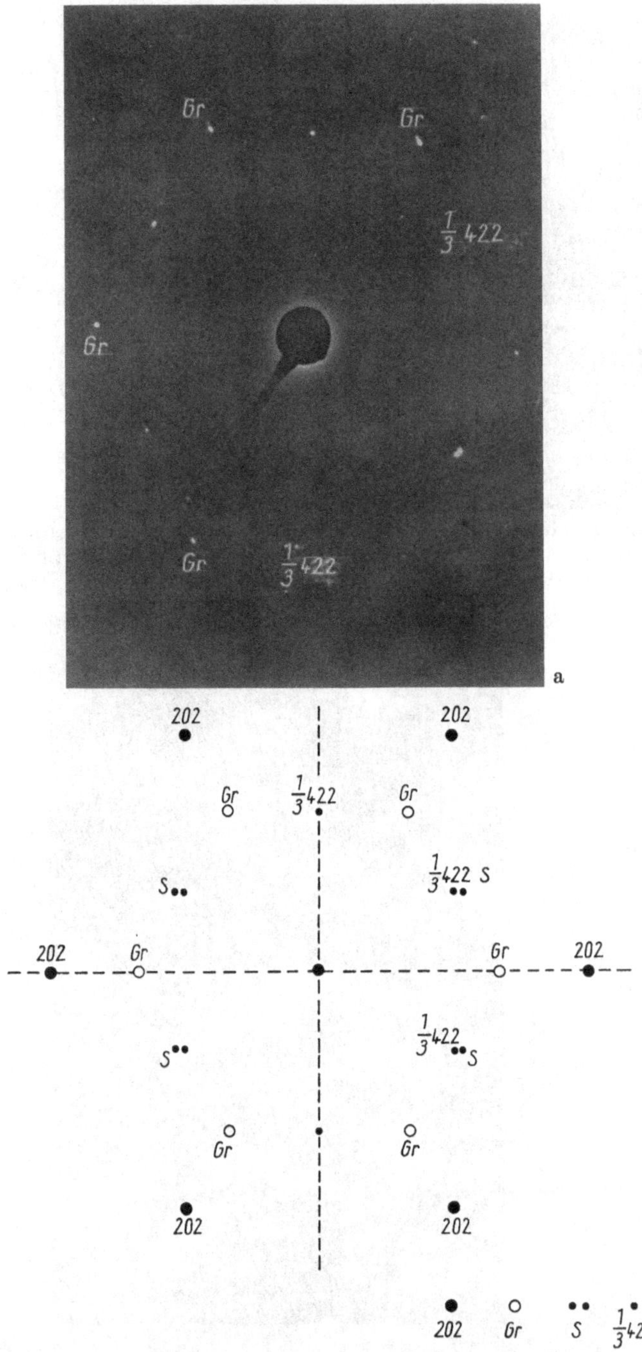

a

b

Fig. 1. a) Diffraction pattern obtained from a gold crystallite. One observes the $\frac{1}{3}$ (422) spots, with satellites. The spots marked Gr are graphite spots. $g(\frac{1}{3}(422)) = 0.400$ Å^{-1}; b) drawing of a diffraction pattern of a gold crystallite with only one direction of reconstruction. One can see that there is no longer a sixfold symmetry, though there remain symmetries over the directions shown by the dashed lines

See Short Note by K. Abdelmoula et al.

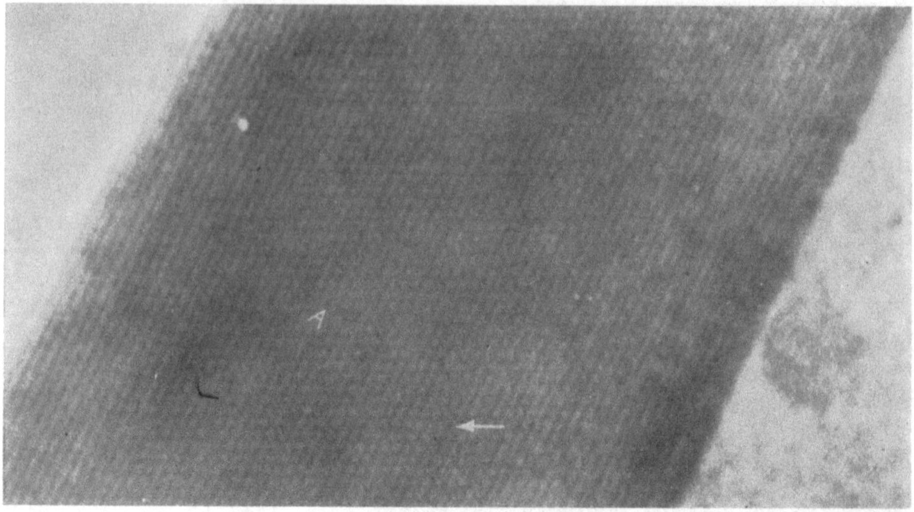

Fig. 2. High resolution image obtained by interfering all diffracted beams inside an aperture corresponding to 0.45 Å^{-1} in gold reciprocal space. The region marked A exhibits three directions of fringes, concurring in white spots. The arrow indicates the fringe directions with the larger interfringe spacing

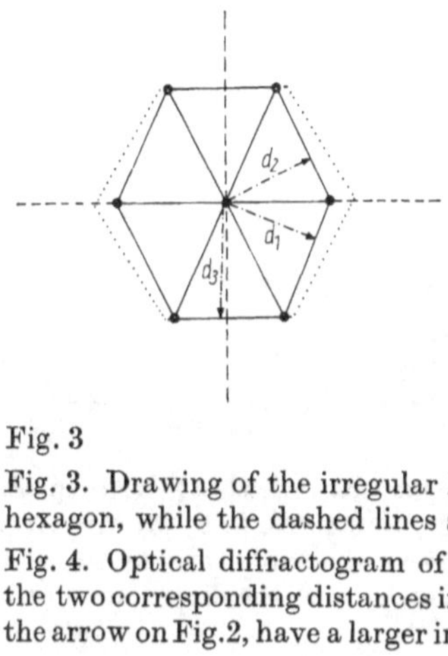

Fig. 3

Fig. 4

Fig. 3. Drawing of the irregular hexagonal array shown on Fig. 2. The dotted lines show a regular hexagon, while the dashed lines show the remaining symmetries. We have $d_1 = d_2 < d_3$

Fig. 4. Optical diffractogram of the region around A in Fig. 2. The distance AB is smaller than the two corresponding distances in the other directions, which confirms that the fringes indicated by the arrow on Fig.2, have a larger interfringe spacing than the fringe systems in the two other directions

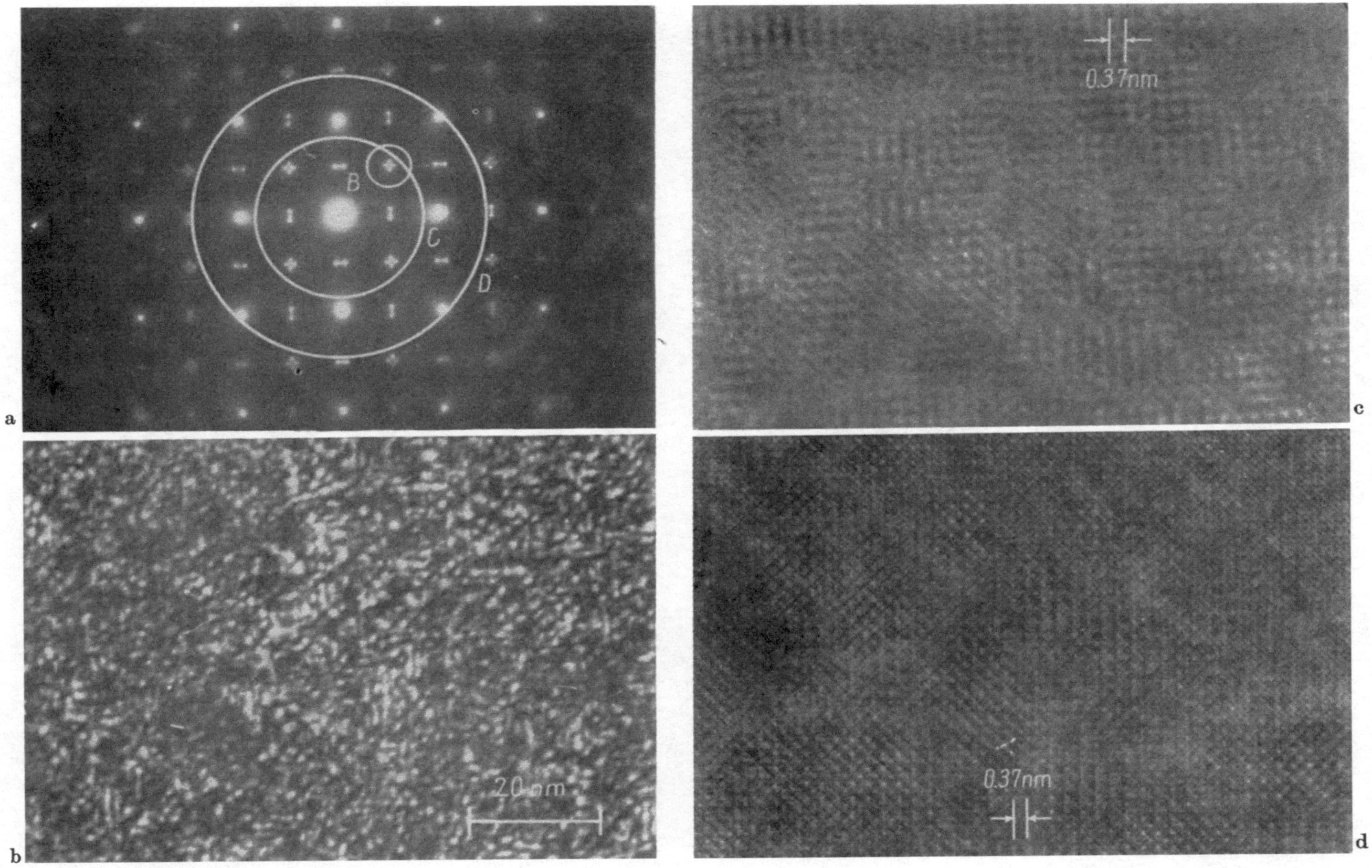

Fig. 1. a) [001]* diffraction pattern of $Cu_{70}Pd_{30}$ briefly annealed at 340°C. The circles B, C, D indicate the size of the objective aperture used to obtain the images reproduced in Fig. 1b to d, respectively; b) dark field image only selecting the four reflections around the $110_{\text{f.c.c.}}$ position (aperture B); c) bright field image excluding all basic f.c.c. reflections (aperture C); d) bright field image (aperture D) imaging the superlattice as well as the basic f.c.c. lattice

Short Notes

phys. stat. sol. (a) 77, K1 (1983)
Subject classification: 1.2 and 1.4; 22

Rijksuniversitair Centrum Antwerpen[1)]

Electron Diffraction Evidence for Superstructures in $TaTe_4$ and $NbTe_4$[2)]

By

J. MAHY, J. VAN LANDUYT, and S. AMELINCKX[3)]

$TaTe_4$ and $NbTe_4$ are isostructural transition polychalcogenides which can be treated as one-dimensional crystals with regard to their electronic structure. The materials crystallize tetragonally in chains consisting out of square anti-prisms of eight Te atoms, containing the cations and stacked along the c-axis, while sharing the square face with each other /1, 2/.

It was reported by means of X-ray investigations /1, 4/ that $TaTe_4$ as well as $NbTe_4$ exhibited a superstructure formed by tripling the c-axis or fiber axis and doubling the a-axis.

Recent electron diffraction experiments on $NbTe_4$, however, showed incommensurate reflections close to $2c_0^*/3$ /3/. Also a low temperature superlattice in this compound was reported by the same author. Because of this apparent contradiction it was considered worthwhile to examine $TaTe_4$ and $NbTe_4$ with electron diffraction techniques, to establish the existence of possible superstructures.

Crystals of $NbTe_4$ and $TaTe_4$ were grown from the elements in evacuated quartz tubes with iodine as transport agent in a temperature gradient. The diffraction patterns of both compounds were observed as a function of temperature from liquid nitrogen temperature (77 K) on up to the temperature at which the specimens undergo an irreversible, probably chemical, transformation (about 700 K).

Although the basic reciprocal lattices are similar for $NbTe_4$ and $TaTe_4$ the observed superlattice reflections have different characteristics. The incom-

---

1) Groenenborgerlaan 171, B-2020 Antwerp, Belgium.

2) Work performed under the auspices of the association R.U.C.A./S.C.K. and with financial support of the I.I.K.W.

3) Also at S.C.K.-C.E.N., Boeretang, B-2400 Mol, Belgium.

mensurability of the room temperature (RT) superlattice reflections in $NbTe_4$ as reported by Boswell et al. /3/ is confirmed as well as the occurrence of a low temperature (LT) superlattice. In the present experiments we also determined the incommensurate component along the c^*-axis to be 0.62 c_o^*, with c_o the subcell lattice parameter, the a^* and b^* components of this reflection being both equal to 0. It should be pointed out that the low temperature reflections appear in doublets, spaced about 0.07 c_o^*, and markedly streaked along c^*, the a^* and b^* components being equal to 1/2.

They remain visible in a broad temperature range, notably in thick specimens, where they can be observed even up to room temperature as circular areas of scattered intensity. Fig. 1a and b (see on the photo pages before the Short Notes part) show a [100] zone diffraction pattern of $NbTe_4$ at RT and LT, respectively.

In apparent contradiction with the results reported in /3/, tilting experiments along the c-axis revealed the presence of lines of diffuse intensity overlapping the interval (c_S^*, $2c_o^*/3$), c_S being the superlattice parameter. These lines are situated in a plane perpendicular to the c^*-axis, and become less intense upon lowering the temperature.

A [100] zone diffraction pattern of $TaTe_4$ is presented in Fig. 2 (see on the photo pages before the Short Notes part). In this compound a superlattice is formed, characterised by a tripling of the c-axis and a doubling of the a- and b-axis of the basic lattice, confirming the X-ray data presented in /1, 4/. No additional reflections occur upon lowering the temperature of the specimen to 77 K.

It should be mentioned that upon heating $NbTe_4$ as well as $TaTe_4$, the RT superspots tend to disappear at about 650 K, whilst the basic reflections persist up to 700 K. Upon cooling from 650 K, however, no significant change of the intensity or sharpness of the RT superspots was observed. X-ray microanalysis failed to reveal the presence of intercalated iodine along the chain, a phenomenon which cannot be excluded, since according to Meerschaut et al. /8/ and Gressier et al. /9/ it was observed in the structurally similar compounds $NbSe_4$ and $TaSe_4$.

$TaTe_4$ and $NbTe_4$ are reported to show diamagnetic behaviour /2, 4/ suggesting strongly that no unpaired d-electrons are present on the pentavalent cations. Since the interchain Te-Te distance approaches the single bond distance it is likely, in order for the bonds to be saturated, that one electron is localised on each cation /1, 2/. In order to account for the observed diamagnetism, the cations should form isolated pairs along the chain, or, more generally, one-dimensional clusters of cations.

Assuming the shift of the cations from their equilibrium positions halfway the more or less rigid anion squares to be periodic, this would cause a deformation modulated structure, made visible by the satellite reflections in the diffraction pattern.

Apart from the correlation effects due to electron-electron interaction, resulting in the clustering of cations along the chain with the formation of a covalent type of chemical bond /6/, another related mechanism generating a deformation modulated structure could be active in the compound. Electron-phonon coupling in a chain, as originally described by Peierls /7/ could also provide a cation shift as a consequence of the particular shape of the Fermi surface, tending to modulate the electron density. Evidence for the occurrence of a Fermi surface driven charge density wave coupled to a periodic lattice deformation (CDW/PLD) should then be found in the analysis of the temperature dependence of physical data (electric and thermo-electronic quantities).

The information derived from these experiments should give us a better insight into the mechanism generating the observed modulations in the crystal structure of $NbTe_4$ and $TaTe_4$.

References

/1/ K. SELTE and A. KJEKSHUS, Acta chem. Scand. 18, 690 (1964).
/2/ F. HULLIGER, in: Structure and Bonding, Springer Verlag, 1968 (p. 124).
/3/ F.W. BOSWELL, A. PRODAN, and J.K. BRANDON, Coll. Internat. C.N.R.S., Paris 1981 (E. 1).
/4/ E. BJERKELUND and A. KJEKSHUS, Z. anorg. allg. Chem. 328, 235 (1964).
/5/ K. SELTE and A. KJEKSHUS, Acta chem. Scand. 19, 258 (1965).
/6/ C. HAAS, Solid State Commun. 26, 709 (1978).
/7/ R.E. PEIERLS, The Quantum Theory of Solids, Clarendon Press, Oxford 1955 (p. 108).

/8/ A. MEERSCHAUT, P. PALVADEAU, and J. ROUXEL, J. Solid State Chem. 20, 21 (1977).

/9/ P. GRESSIER, L. GUEMAS, and A. MEERSCHAUT, Acta cryst. B 38, 2877 (1982).

(Received March 9, 1983)

phys. stat. sol. (a) 77, K5 (1983)
Subject classification: 1.6; 21.6

Department of Physics, Tunis University (a),
C.R.M.C. 2, Luminy, Marseille[1] (b), and
GMET and ERA 545, Toulon University[2] (c)

Reconstructed Surface Structure of (111) Gold Platelets, Observed by High Resolution Techniques

By
K. ABDELMOULA (a), J.J. MÉTOIS (b), and G. NIHOUL (c)

Surface reconstruction of (111) gold surface has been previously observed by Melle and Menzel /1/ using RHEED techniques, then by Yagi et al. /2/ and Heyraud and Métois /3/, using transmission electron microscopy. These studies have shown that the surface net is compressed in the ⟨110⟩ direction, as compared with the bulk lattice.

To get deeper knowledge on this superstructure, we thought that high resolution techniques could be used. The problem is to obtain a structure image corresponding mainly to the upmost plane which forms the surface. As the so-called forbidden reflexions (1/3) (422) and their satellites correspond to diffraction on lines of atoms in the bi-dimensional top layer, we have obtained high resolution images from interference between the diffracted beams on these lines, and have studied them by optical methods.

Experimental conditions a) Gold crystallite preparation In UHV, gold vapour was deposited on in situ cleaved (00.1) graphite surfaces. A detailed description of this technique has been done elsewhere /4/.

We thus obtain crystallites, usually in epitaxy on the graphite: $(111)_{Au} \parallel (00.1)_{Gr}$ with $|11\bar{2}|_{Au} \parallel |01.0|_{Gr}$. The graphite is then peeled off, outside the vacuum chamber. The lateral size of the crystallites is between 100 and 200 Å, and their thickness has been shown to be about one tenth of their size in the ⟨112⟩ direction /3/.

b) High resolution conditions High resolution images were obtained with a JEOL 200 CX, operated at a voltage of 200 kV (corresponding wavelength λ = = 0.0251 Å), using a goniometer stage (rotation angle $\approx 10^{o}$); the beam di-

---

1) CP 913, 13288 Marseille Cédex, France.

2) 83130 la Garde, France.

vergence was 10^{-3} rad, while the spherical aberration coefficient C_s was measured as 1.0 mm. All beams corresponding to reciprocal distance g inferior to 0.45 $Å^{-1}$ were kept to interfere, and form a structure image ($g(1/3)(422) = 0.40$ Å).

To avoid problems with the remaining graphite substrate, we restricted our investigations to gold crystallites on the edge of the graphite flakes, so that, due to the small thickness of the graphite, no fringes belonging to the substrate were observed.

The images were analyzed with a laser beam: a Fourier transform of the image was thus obtained (optical diffractogram).

Experimental results a) Diffraction pattern Fig. 1a (Fig. 1 see on the photo pages before the Short Notes part) shows a diffraction pattern obtained from a gold crystallite. One observes the $(1/3)(\bar{4}22)$ spots with satellites: these can be indexed $(1/3)\langle\bar{4}22\rangle + \alpha\langle 1\bar{1}0\rangle$ where α is a value around 1/21. There are three equivalent directions $\langle 1\bar{1}0\rangle$ so that there are three possible domains. But we often observe only one direction for one crystallite, and, for simplicity, we have restricted our high resolution studies to crystallites with only one domain. In that case the diffraction pattern loses the apparent sixfold symmetry of bulk gold, but keeps some symmetries as shown on Fig. 1b, where we have drawn the diffraction pattern for a crystallite with only one domain.

b) High resolution image Fig. 2 (see on the photo pages before the Short Notes part) shows a structure image formed with beams up to the $(1/3)(\bar{4}22)$ and their satellites. The crystallite exhibits an elongated shape and is situated at the edge of a graphite flake: no graphite fringes are observed. The average lateral dimension is 300 Å, and the thickness is estimated 30 Å. At this resolution, the upmost surface of the crystallite appears not perfectly flat. However, one can find a large region where the crystallite is in a symmetrical position (marked A on Fig. 2): the three directions of fringes are clearly visible, and, at their intersection, they form white dots. The general appearance of the image in this region is an hexagonal array.

The distance between two fringes was carefully measured, in the three possible directions (1/3) [422]. For one direction, shown by the arrow on Fig. 2, the reticular distance is larger than for the two others, so that the array appears as shown on Fig. 3: it is an irregular hexagonal array, but retaining the symmetries which were still appearing on the diffraction pattern.

c) Optical diffractogram Fig. 4 (see on the photo pages before the Short Notes part) shows the optical diffractogram obtained from a laser beam passing through the film corresponding to Fig. 2, in the region marked A. One observes six diffracted beams corresponding to the three fringe directions. Measurement of the distance between each spot and its opposite, confirms the distances measured on the image: the distance between the two spots labelled A, B is 45.5 mm, while the distance between the two other possible couples, is 49 mm. So there is no doubt, that the white spots on the high resolution image shown on Fig. 2 do not form a regular hexagonal array.

Discussion Bulk gold seen along the ⟨111⟩ direction, can be described as a piling of hexagonal bi-dimensional layers (the usual A, B, C packing). The (1/3) ($\bar{4}22$) spots occur as diffracted beams from one (or two) hexagonal layers. They nearly disappear every three slices. If the bi-dimensional top layer is reconstructed, that is to say if the regular hexagonal array is distorted and becomes an irregular hexagonal array, the diffracted beams from the first slice will be different, and will not cancel with the diffracted beams from the bulk layers, so that an image formed with these beams, will give some information on the top layer and the reconstruction on the surface.

The obtention of high resolution images with irregular hexagonal arrays shows that there is a link between these images and the top surface of the crystallite. However, to get real information, and in particular to be able to compare the images to the different models proposed in the literature, one needs to compute the images, taking into account the passage of electrons through the reconstructed surface layer and through the underlying bulk gold.

This is presently being done and will be published later.

References

/1/ H. MELLE and E. MENZEL, Z. Naturf. 339, 282 (1978).

/2/ K. YAGI, K. KOBAYASHI, N. OSAKABE, Y. TANISHIRO, and G. HONJO, Surface Sci. 86, 174 (1979).

/3/ J.C. HEYRAUD and J.J. MÉTOIS, Surface Sci. 100, 519 (1980).

/4/ J.J. MÉTOIS, J.C. HEYRAUD, and Y. TAKEDA, Thin Solid Films 51, 105 (1978).

(Received March 24, 1983)

phys. stat. sol. (a) 77, K9 (1983)
Subject classification: 1.4; 21.1

Rijksuniversitair Centrum Antwerpen[1)]

Real Space Evidence for the Occurrence of Microdomains in "Disordered" Cu_3Pd[2)]

By
G. VAN TENDELOO and S. AMELINCKX

In the past there has been a vast amount of speculation on possible models for short range ordered and partially "disordered" alloys. On the one hand homogeneous models have been proposed based on the use of SRO parameters to describe the correlation in the occupation of pairs (or triplets) of lattice sites, connected by given lattice vectors. Such SRO parameters can be determined experimentally from the intensity distribution in the diffuse X-ray scattering /1 to 3/.

On the other hand, models have been proposed based on the presence of small ordered domains, called "microdomains" inbedded in a disordered matrix /4, 5/.

Finally a state of intermediate order, between SRO and LRO, has been defined in a number of binary systems. The crystal is thought to be built from "atom clusters" satisfying a certain composition condition: the cluster relation /6, 7/. A close relation with the microdomain model has been pointed out /8, 9/.

As early as 1968 /4/ it was claimed that microdomains had been revealed by means of transmission electron microscopy using dark field imaging in a SRO spot of Ni_4Mo. However, such evidence was not judged convincing and it was considered necessary to wait until microscopes with a sufficient instrumental resolution became available, before clear cut evidence could be expected. Present day microscopes allow the resolution of interatomic distances under suitable conditions and it therefore became possible to obtain satisfactory direct evidence.

---

1) Groenenborgerlaan 171, B-2020 Antwerp, Belgium.

2) With financial support of the I.I.K.W.

We have studied several alloys by means of high resolution electron microscopy and electron diffraction; in particular in Cu_3Pd the evidence allows a clear conclusion. In this short note we limit ourselves to this alloy.

Fig. 1a (Fig. 1 see on the photo pages before the Short Notes part) shows the [001] diffraction pattern of a specimen with composition $Cu_{70}Pd_{30}$ quenched from above 800 ^{o}C and briefly annealed at 340 ^{o}C; it exhibits the typical features characteristic of the "short range ordered alloy". The "superlattice" spots of the $L1_2$ type with h + k = even are broken up in crosses, whereas those with h + k = odd have the shape of small segments. The "spots" i.e. the maxima form in fact an incommensurate pattern. It has been shown that the size of the crosses and the length of the segments depends on composition and can be related to the Fermi wave vector $\vec{k}_F$ /10, 11/.

A dark field image taken by only selecting spots belong to a cross (aperture B in Fig. 1a), shows the spotty pattern of Fig. 1b. The domain shape can be recognized; the APB's are preferentially in cube planes.

A bright field high resolution image along the [001] zone, using the direct beam, the first shell of f.c.c. spots, as well as the "superstructure" spots, is reproduced in Fig. 1d. In the thin parts of the specimen the f.c.c. basic structure is revealed; in the thicker parts the superstructure becomes evident. The brightest dots reveal the minority atom columns (i.e. the Pd columns) of the structure. The random fragmentation in domains, containing somewhere from 9 to 25 unit cells, and separated by APB's is revealed.

Since the structures within the domains are all derived from the same basic f.c.c. lattice, they are correlated; only a finite number of relative positions is possible; four in our case.

In the contact region between two domains the configuration of bright dots is again a square one, but the mesh size is smaller and it is rotated over 45° with respect to the one within the domains. It is suggested that this pattern is due to inclined APB's along which columns belonging to two overlapping domains and consisting partly of Pd are revealed.

In order to eliminate redundant information we also produced images, using only the central beam, as well as all "superlattice" reflections, but avoiding all basic reflections (aperture C) in Fig. 1a. Such an image is shown in Fig. 1c, the domain size is now even more clearly visible; also the antiphase relation between domains is quite evident.

The observations described here are in many respects consistent with the "correlative microdomain model" described in /12/ and which explains satisfactorily the geometry of the diffraction pattern. However, whereas it was assumed in this model that relatively well ordered microdomains are dispersed in a disordered f.c.c. matrix, our observations suggest, at least in the case of $Cu_{70}Pd_{30}$, that the crystal is in fact finely fragmented in microdomains containing all conceivable translation variants. Orientation variants are not present in this case, since both basic and ordered structure, belong to the same cubic point group. Work in progress on short range ordered Ni_4Mo leads to similar conclusions.

References

/1/ J.M. COWLEY, J. appl. Phys. 21, 24 (1950); Phys. Rev. 77, 669 (1950); 120, 1648 (1960).

/2/ R. KIKUCHI, Phys. Rev. 81, 988 (1951); J. chem. Phys. 19, 1230 (1951); 60, 1071 (1974).

/3/ P.C. CLAPP, J. Phys. Chem. Solids 30, 2589 (1969).

/4/ E. RUEDL, P. DELAVIGNETTE, and S. AMELINCKX, phys. stat. sol. 28, 305 (1968).

/5/ P.R. OKAMOTO and G. THOMAS, Acta metall. 19, 825 (1971).

/6/ R. DE RIDDER, G. VAN TENDELOO, and S. AMELINCKX, Acta cryst. A32, 216 (1976).

/7/ R. DE RIDDER, G. VAN TENDELOO, D. VAN DYCK, and S. AMELINCKX, phys. stat. sol. (a) 38, 663 (1976).
R. DE RIDDER, D. VAN DYCK, G. VAN TENDELOO, and S. AMELINCKX, phys. stat. sol. (a) 40, 669 (1977).
R. VAN DYCK, R. DE RIDDER, G. VAN TENDELOO, and S. AMELINCKX, phys. stat. sol. (a) 43, 541 (1977).

/8/ G. VAN TENDELOO, Mater. Sci. and Engng. 26, 209 (1976).

/9/ J.P. CHEVALIER and W.M. STOBBS, Acta metall. 27, 285 (1979).
R. DE RIDDER, D. VAN DYCK, G. VAN TENDELOO, J. VAN LANDUYT, and S. AMELINCKX, phys. stat. sol. (a) 39, 383 (1977).

/10/ S. KUBO and K. ADACHI, J. Phys. Soc. Japan 35, 776 (1973).

/11/ K. OHSHIMA, D. WATANABE, and J. HARADA, Acta cryst. A32, 883 (1976).

/12/ S. HASHIMOTO, Acta cryst. A30, 792 (1974).

(Received March 31, 1983)

phys. stat. sol. (a) 77, K13 (1983)

Subject classification: 8; 22.9

Institute of Chemical Physics and Biophysics, Academy of Sciences of the Estonian SSR, Tallinn[1)]

Thermal Conductivity of Pressed Pellets of TCNQ Complexes. Manifestation of Amorphous Structure

By

B. P. BESPALOV and A. M. VAINRUB

The thermal conductivity of TCNQ complex monocrystals has been measured only for the TTF-TCNQ complex because of the small size of the crystals /1/. For a large number of other TCNQ complexes polycrystalline pressed pellets have been investigated /2 to 4/. The aim of this note is to show radical distinctions between thermal conductivity values and temperature dependences of TCNQ complex monocrystal and polycrystalline samples. We report here our results on polycrystalline pellets of high-conductive TTF-TCNQ and semiconducting simple TEA-TCNQ, and complex TEA-$TCNQ_2$ complexes of TCNQ with triethylammonium. The thermal conductivity of TTF-TCNQ pressed pellets is several ten times less than that of monocrystals. We explain the fact as a result of a partial amorphousness, inevitably appearing during the procedure of pellet pressing.

The thermal conductivity was measured using an absolute stationary method in the temperature range T = 50 to 400 K. The measuring cell was the same as in /3/. An accuracy of $\pm 7\%$ for the absolute value was achieved in quartz etalon control measurements. TCNQ complex crystals were grinded in a mortar and pressed at 10^8 Pa in a vacuum (10^{-2} Pa) press. TTF-TCNQ and TEA-TCNQ powders cannot be pressed at room temperature. Strong pellets containing no cracks were obtained by pressing, being heated up to the temperature of 80 $^{\circ}$C. Their density was close (90 to 95%) to the crystal density.

Fig. 1 shows the measured temperature dependences of the thermal conductivity. The curves for TEA-TCNQ and TEA-$TCNQ_2$ become close at temperatures higher than T = 300 K and the thermal conductivities are about K = = 2 mW/cmK. The same value was obtained for a number of TCNQ complexes with molecular cations /4/. The thermal conductivity of TTF-TCNQ is larger,

1) Lenini pst. 10, 200001 Tallinn, USSR.

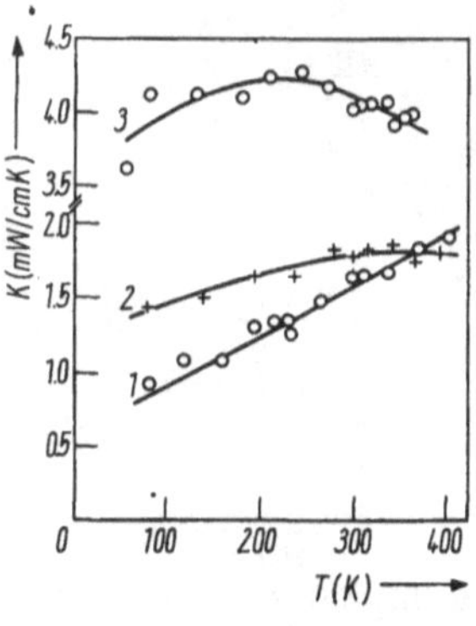

Fig. 1. Thermal conductivity of TCNQ complexes: (1) $TEA-TCNQ_2$, (2) TEA-TCNQ, (3) TTF-TCNQ

$K \approx 4$ mW/cmK. For this high-conductive complex the total thermal conductivity consists of the lattice K_1 and the electronic K_{el} contributions. Since there are no reasons to presume essential differences between K_1 values for TTF-TCNQ and the other complexes, we evaluate $K_1 \approx 2$ mW/cmK and $K_{el} \approx 2$ mW/cmK for TTF-TCNQ pellets. Then utilizing the free-electron Lorentz number $L = 2.44 \times 10^{-5}$ $mW/\Omega K^2$, we get an estimate for the electrical conductivity $\sigma \approx 300\ \Omega^{-1}cm^{-1}$. The value is typical of the metallic conductivity along the crystal b axis of TTF-TCNQ and may be supposed to be less for pellets with random crystal orientation. This discrepancy may be caused by the inexactness of separation of the electronic contribution from the total thermal conductivity.

The obtained lattice thermal conductivity value $K_1 \approx 2$ mW/cmK is 40 times less than $K_1 = 80$ mW/cmK along the TTF-TCNQ monocrystal high-conductive b axes /1/. Using the well-known formula for the phonon gas thermal conductivity $K_1 = (1/3)c_1 v_s l$ we obtain the phonon mean free path for the crystal $l \approx 200$ Å and for the pellet $l \approx 5$ Å close to the intermolecular TCNQ-TCNQ distance, ≈ 3.2 Å. For the lattice contribution to the specific heat $c_1 = 1.3 \times 10^5$ J/mol K /5/ and the sound velocity $v_s = 2.5 \times 10^5$ cm/s /6/ of TTF-TCNQ crystal for T = 60 K were used.

According to /2, 3/ the intergranular thermal resistances of TCNQ complex pellets are small and thus cannot be the reason of low thermal conductivities of pellets. We can also assume that the lattice contributions to the thermal conductivity along the perpendicular directions are of the same order of magnitude as along TCNQ chains. A weak anisotropy of the TTF-TCNQ sound velocities /6/ supports the assumption.

Probably the low thermal conductivity of pellets is due to a large number of crystal lattice defects, i.e. partly amorphous structure, appearing due to grinding and pressing crystals. An amorphous structure was observed in X-ray patterns of TCNQ complex thin films obtained by rubbing crystals on the glass substrate /7/. The conclusion is supported by the short phonon mean path close to the intermolecular distance and a weak temperature dependence of the thermal conductivity typical of amorphous substances and glasses /8/.

We can also note the probable manifestation of the partly amorphous structure in the low-temperature specific heat of TCNQ complexes. The phonon frequency density for the amorphous state does not depend on the phonon frequency and causes a specific heat contribution proportional to temperature T. Such T term together with the usual Debye lattice T^3 term was reported for TTF-TCNQ pressed pellets /9/.

References

/1/ M.B. SALOMON, J.W. BRAY, G. DE PASQUALI, R.A. GRAVEN, G. STUKY, and A. SCHULTZ, Phys. Rev. B 11, 619 (1975).

/2/ A.V. PETRÓV, U.G. NURULLAEV, R.M. VLASOVA, V.V. KUZINA, and A.I. SHERLE, Fiz. tverd. Tela 14, 1838 (1972).

/3/ A.V. PETROV, U.G. NURULLAEV, and R.M. VLASOVA, Inzhen.-fiz. Zh. 25, 3432 (1973).

/4/ R.M. VLASOVA, U.G. NURULLAEV, A.V. PETROV, and M.I. CHERKASKIN, Fiz. tverd. Tela 15, 3432 (1973).

/5/ R.A. GRAVEN, M.B. SALOMON, G. DE PASQUALI, R.M. HERMAN, G. STUCKY, and A. SCHULTZ, Phys. Rev. Letters 32, 769 (1974).

/6/ T. TIEDJE, R.R. HAERING, M.H. JERICHO, W.A. ROGER, and A. SIMPSON, Solid State Commun. 23, 713 (1977).

/7/ R.M. VLASOVA, A.I. GUTMAN, L.D. ROZENSHTEIN, and N.F. KARTENKO, phys. stat. sol. (b) 47, 435 (1971).

/8/ A.S. OKHOTIN, A.S. PUSHKARSKII, and V.V. GORBACHEV, Teplofizicheskie svoistva poluprovodnikov, Atomizdat, Moscow 1972.

/9/ R. VISWANATHAN and D.C. JOHNSTON, J. Phys. Chem. Solids 36, 1093 (1975).

(Received February 23, 1983)

phys. stat. sol. (a) 77, K17 (1983)

Subject classification: 13.3; 22.1.2; 22.6

Kando College for Electrical Engineering, Institute of Computer Technology, Székesfehérvár[1] (a) and
Physical Institute of the Technical University, Budapest (b)

Energy Dependence of the Si-SiO_2 Interface State Cross-Section Measured by the TSC Method

By

A. HÜBLER (a) and D. MARTON (b)

In the past decade several methods have been commonly used for the measurement of interface state densities in MOS structures. First of all, there are the ac capacitance and conductivity measurements and that of quasi-static capacitance by means of which the interface state distributions in the middle of the gap can be determined reliably. On the basis of the conductivity measurements some capture cross-section data have been published, too /1 to 4/. Similar data have also been published as obtained from the photoionization of interface states /5/ and from the DLTS (deep level transient spectroscopy) method as well /6/.

In the early seventies Simmons and co-workers /7, 8/ suggested the use of thermal non-equilibrium methods to investigate the interface states (TSC - thermally stimulated currents - methods). It seems to be very useful to investigate MOS structures, since in principle the D(E) spectrum can be obtained with it as close to the band edges as desired, which is not so when using the C-U and the conductance methods, and it provides information on the capture cross-section of the states, too.

Despite all these advantages the TSC method has had only very limited applications for MOS investigations, and even when used, generally it was not in its original form /9/. Only a few attempts have been made to use the TDRC (thermal dielectric relaxation current) method for the evaluation of capture cross-sections on MOS. A characteristic example is Ning's work /10/, whose results show that σ_n could have been determined only within eight orders of magnitude. There are several experimental σ_n and σ_p values shown in Fig. 1 as they were published by different authors. The collection contains only those

---

1) Vörös Hadsereg útja 45, Székesfehérvár, Hungary.

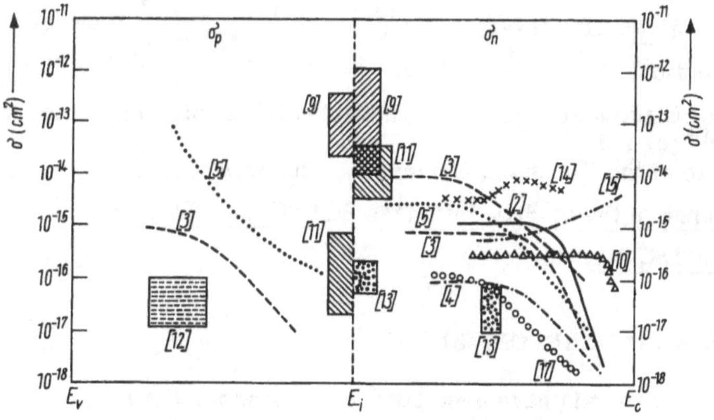

Fig. 1. Capture cross-section data obtained by different authors. In the lower part of the gap ($E < E_i$) the data are for holes (σ_p), in the upper one for electrons (σ_n). The data have been obtained in conductivity measurements except that of Tredwell and Viswanathan (DLTS) /5/ and Yamashita et al. (TSSP) /9/

data, of which energy and cross-section of the states were reported at the same time, or they could be found out unambiguously from the reports. In spite of the trends the collected data show, it is clear that the complete solution of this problem can be obtained only after further measurements.

In this note we want to complete these data with some thermal experimental results, extending our measurements over a wide range of the gap.

To evaluate the interface state density from the TSC measurements, one has to know by all means ν, the so-called attempt-to-escape frequency being closely connected with the cross-section of states. Simmons and Taylor /8/ have shown that lg ν is to be evaluated as

$$\lg \nu = \frac{T_1 \lg \beta_1 - T_2 \lg \beta_2}{T_1 - T_2} + \text{const},$$

where T_1 and T_2 are the temperature of the TSC current maximum of measurements with two different temperature ramps β_1 and β_2, respectively. Owing to the very small currents to be detected and the shape of the TSC curves, T_1 and T_2 can be evaluated only with an error of about 0.5 K. (In our cases TSC currents were in the range of $I = 10^{-13}$ to 10^{-11} A, and curves were not pointed enough.) Using typical data the error of lg ν in this case can be still within ± 2 orders of magnitude. As errors made when reading the

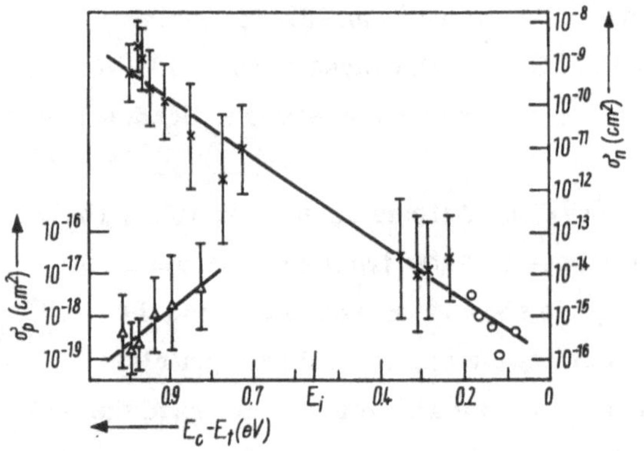

Fig. 2. σ_n versus $E_c - E_t$ and σ_p versus $E_c - E_t$ curves obtained in TSC measurements. The mean value and the standard deviation of 12 TDRC measurements are indicated in the figure. The IDRC results are the mean values of two measurement series; × n - TDRC, o n - IDRC, Δ p - TDRC

peak maximum can be regarded as random, the value of lg ν is determined by the mean value of repetitive measurements. In this case the error is proportional to $N^{-1/2}$ when using N pairs of curves in the calculation.

Assuming that the state density spectrum does not change rapidly with energy, the carriers, coming from the same states in measurements with two different β, are to be identified even beyond the current maximum. E.g. according to /7/ the temperature dependence of the generation current is given by

$$J_g(T) = \frac{Aq\beta(E_c - E_v)}{T} D(E) ,$$

where A and $(E_c - E_v)$ are the area of the sample and the gap energy, respectively. If we can identify ranges on the two current versus temperature curves, where D(E) is monotonous (e.g. the sections preceding the maximum or the ones following it), we know that in these ranges the quantity $J T \beta^{-1}$ is the same only when the same states are discharged. Using the corresponding T values of our two measurements evaluated with a successive method, we can determine the cross-section of states at a given energy level.

Experiments were carried out on capacitor samples based on 2 Ωcm n-Si and 8 to 10 Ωcm p-Si, by means of a complex measuring system described in a previous work /18/. The samples were oxidized in an HCl containing atmosphere (4 vol% HCl) at $T_{ox} = 1100\ ^{o}C$. The thickness of the oxide was 110 nm, Al dots of $A = 1.15x10^{-2}\ cm^2$ area produced by electron beam evaporation were used as metal electrodes.

We repeated the TSC measurement 10 to 12 times on each samples, the $\sigma \sim E_c - E_t$ relation evaluated from these results (from 45 to 66 pairs of curves) can be seen in Fig. 2. The figure shows the mean value and the standard deviation of σ. Since with our equipment we could not get close enough to the majority carrier band edge by the TSC method, we used the IDRC (isothermal dielectric relaxation current) method as well. These data were determined as mean value of two measurement series. On the basis of the results presented on the figure, the cross-section of the states can be evaluated as $\sigma = \sigma_o \exp(m\Delta E)$, where $\sigma_{on} = 8.67x10^{-17}\ cm^2$, $m_n = 15.64\ eV^{-1}$, and $\Delta E = E_c - E$ for n-type material, $\sigma_{op} = 4.90x10^{-20}\ cm^2$, $m_p = 16.99\ eV^{-1}$, and $\Delta E = E - E_v$ for p-type, respectively.

Note that the above dependence could be the reason of the temperature dependence of capture cross-sections. However, in this case one has to find some correlation between the β and the measured cross-sections. The lack of it correlates to the results of Katsube and Kakimoto /19/ about the weak temperature dependence of cross-sections.

References

/1/ E.H. NICOLLIAN and A. GOETZBERGER, Bell Syst. Tech. J. 46, 1055 (1967).

/2/ H. DEULING, E. KLAUSMANN, and A. GOETZBERGER, Solid State Electronics 15, 559 (1972).

/3/ A. GOETZBERGER, E. KLAUSMANN, and M.J. SCHULTZ, CRC Crit. Rev. Solid State Sci. 6, 1 (1976).

/4/ M. MORITA, K. TSUBOUCHI, and N. MIKOSHIBA, Appl. Phys. Letters 33, 745 (1978).

/5/ T.J. TREDWELL and CR. VISWANATHAN, Appl. Phys. Letters 36, 462 (1980).

/6/ W.E. DAHLKE and D.W. GREWE, Solid State Electronics 22, 893 (1979).

/7/ J.G. SIMMONS and H.A. MAR, Phys. Rev. B 8, 3865 (1973).

/8/ J.G. SIMMONS and G.W. TAYLOR, Solid State Electronics 17, 125 (1974).

/9/ K. YAMASHITA, M. IWAMATO, and T. HINO, J. appl. Phys. 49, 2866 (1978).

/10/ H.H. NING, J. appl. Phys. 49, 5997 (1978).

/11/ J.A. COOPER and R.J. SCHWARTZ, Solid State Electronics 17, 641 (1974).

/12/ G. DECLERCK, R. VAN OVERSTRATEN, and G. BROUX, Solid State Electronics 16, 1451 (1973).

/13/ G. KAPLAN, Solid State Electronics 24, 89 (1981).

/14/ N.M. JONGSON, Appl. Phys. Letters 34, 803 (1979).

/15/ K.L. WANG, IEEE Trans. Electron Devices 26, 819 (1979).

/16/ R.J. KRIEGLER, T.F. DÉVÉNYI, K.D. CHIK, and J. SHAPPIR, J. appl. Phys. 50, 398 (1979).

/17/ T.C. POON and H.C. CARD, J. appl. Phys. 51, 5880 (1980).

/18/ I. KRAFCSIK and D. MARTON, Solid State Electronics 22, 751 (1979).

/19/ T. KATSUBE and K. KAKIMOTO, J. appl. Phys. 52, 3504 (1981).

(Received March 23, 1983)

phys. stat. sol. (a) 77, K23 (1983)

Subject classification: 14.2; 1.3; 21.6

Institute of Solid State Physics,
Academy of Sciences of the USSR, Chernogolovka[1)]

Superconductivity of Pd-Au-H Solid Solutions

By

V.E. ANTONOV, T.E. ANTONOVA, I.T. BELASH,
E.G. PONYATOVSKII, and V.I. RASHUPKIN

It was revealed in /1/ that hydrogen implantation into alloys of palladium with noble metals (copper, silver, and gold) can convert them to superconductors whose superconducting transition temperature, T_k, reaches much higher values than that of Pd-H solutions. The discovery of this effect has stimulated to a considerable extent the extensive studies of the last few years on the superconducting properties of hydrogen solutions in the palladium alloys. A great number of works has been conducted in order to analyse what lies at the bottom of the effect (see review /2/).

Further experiments have shown, however, that at least in the case of the Pd-Cu-H and Pd-Ag-H solutions with f.c.c. metal sublattice the existence of superconductivity with anomalously high T_k values is not an equilibrium property of massive, homogeneous samples /3/. The technique for hydrogen compression to high pressures enabled us to saturate the $Pd_{60}Cu_{40}$ and $Pd_{80}Ag_{20}$ alloys (having nearly optimum compositions for achieving the highest T_k values on hydrogen implantation /1/) with hydrogen up to concentrations also approximating the optimum ones according to /1/. It turned out that at H-to-metal atomic ratio $n \lessapprox 1$ the $T_k(n)$ dependence for the $Pd_{80}Ag_{20}$-H solutions is close to that for the Pd-H solutions, and in the $Pd_{60}Cu_{40}$-H solutions superconductivity is absent at $n \lessapprox 0.6$ and $T \gtrapprox 2$ K.

The present note is devoted to the investigation of hydrogen solutions in the Pd-Au alloys containing 3, 9, 22.5, 50, and 75 at% Au. The ingots were melted from Pd (99.99%) and Au (99.999%) in an induction furnace in vacuum. After a 10 h homogenization in vacuum at 1000 °C and water-quenching these ingots were rolled into strips 0.15 mm thick, then subjected to stress-relief annealing in vacuum at 1000 °C for 5 min and again quenched in water. The

1) 142432 Chernogolovka, Moscow District, USSR.

specimens in the form of discs of 5 mm diameter were cut out of these strips.

Hydrogenation of Pd-Au samples was conducted by exposure under hydrogen pressures up to 8 GPa. Hydrogen compression was performed by the method suggested in /4/ (a detailed description of the method is given in /5/). The pressure and the temperature were measured accurate to ± 0.3 GPa and ± 10 ^{o}C, respectively. After the hydrogenation was completed, the high-pressure chamber was rapidly cooled down to ≈ -120 ^{o}C, then the pressure was lowered and the samples were taken out of the chamber and placed for storage into liquid nitrogen to prevent hydrogen losses. At atmospheric pressure a pronounced release of hydrogen from specimens began at $T \gtrsim -50$ ^{o}C. The hydrogen concentration was measured to an accuracy of 5%; the technique is described in details in /6/. The T_k values were determined by the induction method at $T \geqq 2$ K. An X-ray study was carried out by a phototechnique at $T = -190$ ^{o}C using a DRON-2.0 diffractometer with FeK_{α} radiation.

Under normal conditions Pd and Au form continuous disordered solid solutions with the f.c.c. lattice /7/. According to /8/, the gold alloying of palladium results in a rapid decrease in the critical temperature of the separation of the Pd-H solutions into the two isomorphous phases, γ_1 and γ_2, poor and rich in hydrogen, and at $T \gtrsim 25$ ^{o}C hydrogen should form continuous interstitial solutions (based on the f.c.c. metal sublattice) with the Pd-Au alloys containing $\gtrsim 17$ at% Au. Our recent experiments, however, have shown that at high pressure some new phase transitions followed by a diffusional redistribution of atoms of the metallic matrix may take place in the Pd-Me-H systems (dissolution into the poor and rich in palladium phases in the Pd-Ni-H /9/ and Pd-Pt-H /10/ systems; ordering in the $Pd_{60}Cu_{40}$-H system /6/), the temperature, T^*, above which these transitions occur with a detectable rate, being approximately equal for all the systems in hand running as high as ≈ 250 ^{o}C. Besides, gold was found to form a hydride with rather complicated (probably, orthorhombic) structure at $P_{H_2} \gtrsim 2.8$ GPa and $T \gtrsim 300$ ^{o}C /11/.

To examine whether analogous phenomena occur in the Pd-Au-H system, we have studied two series of samples. The samples of the first series were exposed at $P_{H_2} = 6.5$ GPa and $T = 200$ $^{o}C < T^*$ for 24 h. The second ones were exposed initially at 350 ^{o}C for 24 h and then at 200 ^{o}C for another 24 h. The X-ray study has shown the samples of both series to be single-phase and to have near values of the parameters, a, of their f.c.c. metal sublattices, see

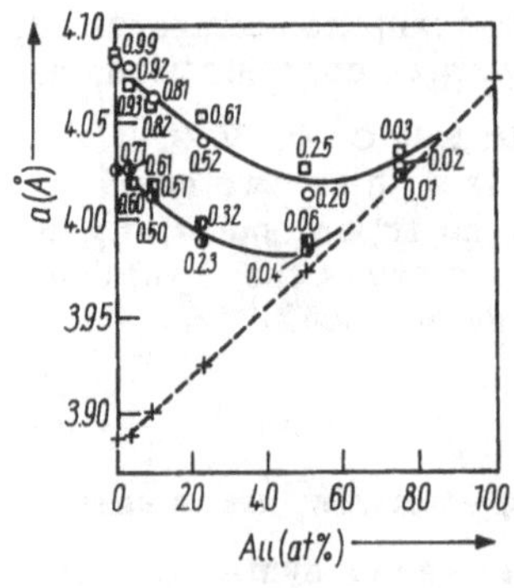

Fig. 1. Parameters a of the f.c.c. metal lattice at -190 °C and atmospheric pressure for the initial Pd-Au alloys (+) and for the Pd-Au-H samples exposed at P_{H_2} = 6.5 GPa: (o) 200 °C for 24 h; (□) 350 °C for 24 h and then 200 °C for 24 h; (◐, ■) the same samples after partial hydrogen release at normal conditions

Fig. 1. The hydrogen concentrations of these samples proved to be also the same within the experimental error (the n-values are listed in Fig. 1). As is seen from Fig. 1, the samples of both series exhibited close values of a and n as well after partial losses of hydrogen through several days exposure at normal conditions (a noticeable release of hydrogen from the samples ceased in several hours; the diffraction pattern showed relatively narrow lines that evidenced a homogeneous hydrogen distribution over the samples volume, though its concentration of the samples was not thermodynamically equilibrium for normal conditions, compare with /8/). These results allowed one to conclude that no irreversible phase transitions (analogous to those observed in some other Pd-Me-H system /6, 9, 10/ and Au-H system /11/) took place at P_{H_2} = 6.5 GPa and T ≦ 350 °C in the Pd-Au-H system under study.

One more fact corroborating this conclusion is worth presenting. It has been shown in /12/ that at n ≲ 0.7 the dependences $\Delta V_o(n)$ of the increment of the unit-cell volume are very close for all the investigated hydrogen solutions in the f.c.c. disordered palladium-based alloys. The dependence $\Delta V_o(n)$ obtained in /12/ is plotted in Fig. 2 by a broken line. To extend the dependence

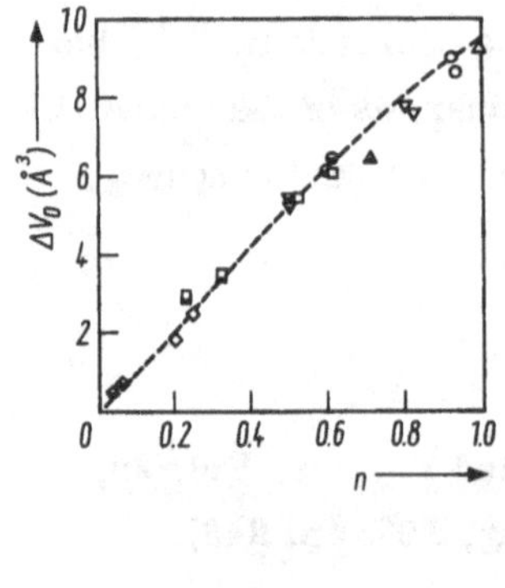

Fig. 2. Increase in the volume of the unit cells, ΔV_o, versus hydrogen concentration, n, of the Pd-Au alloys containing △ 0, o 3, ▽ 9, □ 22.5, ◇ 50 at% Au. Half-blackened symbols stand for the same samples but after partial release of hydrogen under normal conditions. The broken line - see the text

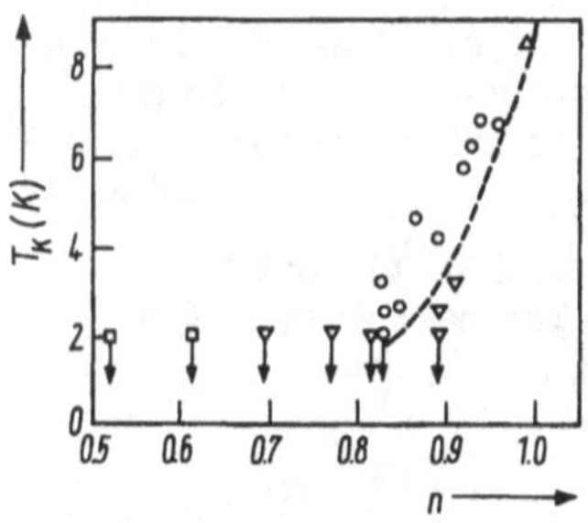

Fig. 3. Superconducting transition temperature, T_k, as a function of hydrogen concentration, n, of the Pd-Au alloys with △ 0, ○ 3, ▽ 9, □ 22.5 at% Au. Symbols with arrows stand for the samples possessing no superconductivity at $T \geqq 2$ K. The broken line shows the $T_k(n)$ dependence for the Pd-H solutions /2/

for $n \gtrsim 0.6$ we used the data of /13/ for the Pd-H γ_2 solutions. As is seen from Fig. 2, the values $\Delta V_0(n)$ for the samples of both series synthesized in the present work agree well with this dependence. Thus, one can ascertain that only ordinary γ-solutions on the base of the starting disordered Pd-Au alloys are formed in the Pd-Au-H system at high pressure and $T \leqq 350$ °C.

The samples of hydrogen solutions in alloys containing $\geqq 22.5$ at% Au obtained in the present work were revealed to possess no superconductivity at $T \geqq 2$ K. To get a more comprehensive information on the $T_k(n)$ dependences for the $Pd_{97}Au_3$-H and $Pd_{91}Au_9$-H solutions we have prepared additional samples by exposure to $P_{H_2} \leqq 8$ GPa and $200 \leqq T \leqq 350$ °C for 24 h. The width of the steps observed on the temperature dependences of the signal of disbalance of an ac bridge in the ranges of sample transition to the superconducting state did not exceed 0.15 K which was indicative of a homogeneous hydrogen distribution over the samples volume. The T_k values were estimated from the step midpoint locations. The results of measurements are depicted in Fig. 3. As is seen from the figure, the $T_k(n)$ dependences for the $Pd_{97}Au_3$-H and $Pd_{91}Au_9$-H solutions are close to that for the Pd-H solutions.

Thus, the gold alloying of palladium, as well as copper and silver alloying /3/, does not result in a distinct increase in T_k of the Pd-H solutions in the case of the sample hydrogenation under conditions close to thermodynamical equilibrium ones. As to the anomalously high T_k values observed in /1/, their appearance seems to be attributable to the specific properties of the metastable state of the thin (≈ 1500 Å) hydrogen-bearing layer obtained during hydrogen implantation at a low temperature.

References

/1/ B. STRITZKER, Z. Phys. 268, 261 (1974).

/2/ B. STRITZKER and H. WÜHL, in: Topics in Applied Physics, Vol. 29, Ed. G. ALEFELD and J. VÖLKL, Springer-Verlag, 1978 (p. 243).

/3/ V.E. ANTONOV, I.T. BELASH, E.G. PONYATOVSKII, and V.I. RASHUPKIN, Zh. eksper. teor. Fiz., Pisma 31, 451 (1980).

/4/ I.T. BELASH and E.G. PONYATOVSKII, Patent No. 741105 (USSR).

/5/ E.G. PONYATOVSKII, V.E. ANTONOV, and I.T. BELASH, Proc. 8th AIRAPT/19th EHPRG Conf., 17 to 22 August 1981, Vol. 1, Uppsala (Sweden) (p. 80).

/6/ V.F. DEGTYAREVA, V.E. ANTONOV, I.T. BELASH, and E.G. PONYATOVSKII, phys. stat. sol. (a) 66, 77 (1981).

/7/ M. HANSEN and K. ANDERKO, Constitution of Binary Alloys, McGraw-Hill Publ. Co., Inc., New York/Toronto/London 1958.

/8/ M. MAELAND and T.B. FLANAGAN, J. phys. Chem. 69, 3575 (1965).

/9/ V.E. ANTONOV, T.E. ANTONOVA, I.T. BELASH, E.G. PONYATOVSKII, V.I. RASHUPKIN, and V.G. THIESSEN, phys. stat. sol. (a) 77, 71 (1983).

/10/ V.E. ANTONOV, T.E. ANTONOVA, I.T. BELASH, E.G. PONYATOVSKII, and V.I. RASHUPKIN, phys. stat. sol. (a) 77, No. 2 (1983).

/11/ V.E. ANTONOV, T.E. ANTONOVA, I.T. BELASH, A.E. GORODETSKII, and E.G. PONYATOVSKII, Dokl. Akad. Nauk SSSR 266, 376 (1982).

/12/ B. BARANOWSKI, S. MAYCHRZAK, and T.B. FLANAGAN, J. Phys. F 1, 258 (1971).

/13/ J.E. SCHIRBER and B. MOROSIN, Phys. Rev. B 12, 117 (1975).

(Received February 10, 1983)

phys. stat. sol. (a) 77, K29 (1983)

Subject classification: 20.1 and 20.3; 11; 22.6

Departamento de Física del Estado Sólido, Facultad de Ciencias, Universidad del País Vasco, Lejona[1] (a) and
Departamento de Física del Estado Sólido, Facultad de Ciencias Físicas, Universidad Complutense, Madrid (b)

Cathodoluminescence and Optical Absorption in 20 kV Electron Irradiated MgO Single Crystals

By

A. REMÓN (a), J. PIQUERAS (a), J. LLOPIS (b), and
C. BALLESTEROS (b)

The cathodoluminescence (CL) emission from MgO single crystals observed in the scanning electron microscope (SEM) changes during the irradiation with the electron beam of the microscope /1/. The variation of the CL intensity depends on the wavelength and is also a function of the irradiation conditions, mainly of the beam current. With high currents the emission increases from the beginning of the irradiation while in the case of lower currents the increase is preceded by an initial decrease. In the present work the optical absorption of crystals irradiated in the SEM has been measured in order to investigate whether the CL variations are related to absorption variations. If this is the case, the results from previous works concerning absorption of defects or impurities in MgO could be used to interpret the CL changes.

The MgO single crystals used were grown by W. & C. Spicer with a purity of 99.9%. Samples of about 5x5x1 mm^3 were cleaved from larger crystals, along (100) faces. The samples were polished in phosphoric acid at 60 ^oC during 5 min and the absorption spectrum between 200 and 800 nm was recorded with a Beckman UV 5260 spectrophotometer. The crystals were subsequently irradiated in a scanning electron microscope Cambridge S4-10 at 20 kV with a current of 10^{-7} A during 80 min and the CL spectra were recorded before and after the irradiation. The optical absorption after the irradiation was also measured. Some crystals received an additional irradiation during 3 h and the CL and absorption spectra were recorded again.

The absorption of the irradiated crystals was also measured after accumulative annealings of 10 min at fifteen temperatures between 90 and 950 ^oC.

1) Lejona (Vizcaya), PB Apartado 644, Bilbao, Spain.

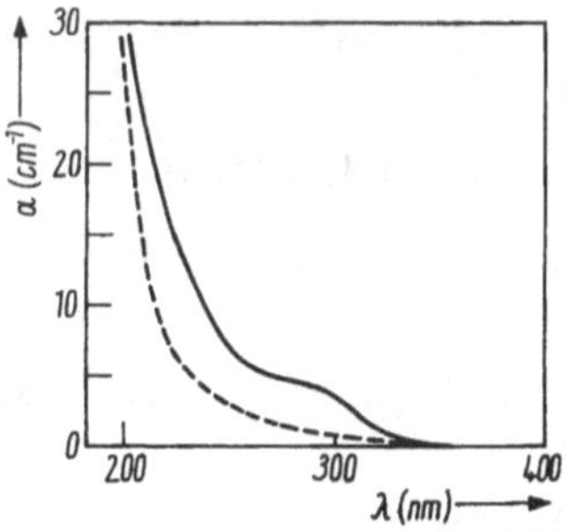

Fig. 1. Optical absorption of an irradiated crystal (upper curve). The dashed curve represents the absorption before the irradiation

The first irradiation produced a decrease of the CL emission and the second an increase above the initial level. The irradiation produces a small absorption band at about 285 nm and a similar increase of the absorption at wavelengths lower than this value. The spectrum does not change above 325 nm. Fig. 1 shows the absorption spectrum of an irradiated crystal. In the crystals which received two irradiations the effect of the second irradiation is additive but not in a proportional way. The first irradiation produces a higher increase of optical absorption than the second, although the latter corresponds to longer time, which indicates that the effect tends to saturate. Both irradiations produce variations of the optical absorption with the same signs and variations of the CL with opposite signs, which shows that the changes of CL observed during the irradiation cannot be associated to changes of optical absorption in the wavelength range considered. The 285 nm band is due to the presence of Fe^{3+} /2/ generated from Fe^{2+} during the irradiation. According to the present results the Fe^{3+} concentration does not determine the CL intensity, which agrees with previous results on Fe doped MgO /3/.

Annenkov et al. /4/ attribute the F-type center density increase which they observe during bombardment with 100 kV electrons and high excitation density to the creation of anion vacancies. In the scanning electron microscope high excitation densities are used which could produce the observed CL variations if new vacancies were formed as in the case described in /4/. The present results do not show, however, in the irradiated crystals the absorption band of the F and F^+ centers at 250 nm but a general increase of the absorption at wavelengths below the 285 nm band. The increase can be associated to impurities, for instance iron has been reported to contribute to the absorption in this region, or may be an artifact of increased scattering. The absorption in the irradiated crystals changes with the above-mentioned thermal treatments. The variations are small but qualitatively reproducible for the different crystals. The absorption shows a maximum after the 350 °C annealing, then

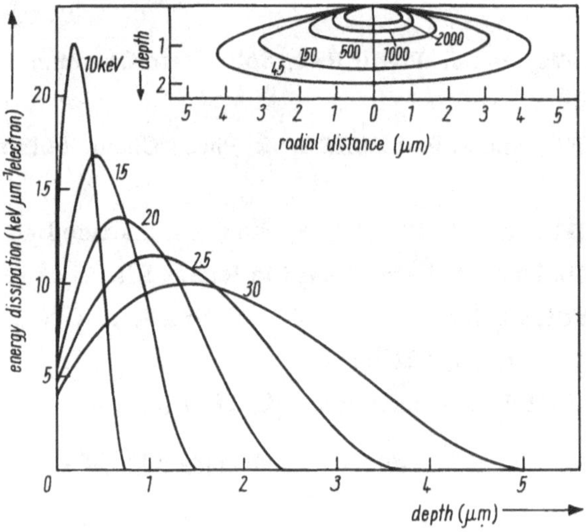

Fig. 2. Energy dissipation curve of electrons in MgO

decreases with increasing annealing temperature, and has a minimum value after the 500 ^{o}C annealing. The thermal treatments did not produce any specific variation of the absorption at 250 nm indicative of a possible presence of F centers. In order to check whether the absorption below 285 nm was a surface effect which could mask other radiation effects the irradiated samples were polished for short times in phosphoric acid and the optical absorption was then measured. The absorption in the region below 285 nm decreases slowly with the polishing time, and seems to be generated in the irradiated volume rather than in a thinner surface layer. The volume affected by the irradiation corresponds to a rather thin region as can be observed in Fig. 2 which shows the curves of electron energy loss in MgO obtained from the Gruen function given by Everhart and Hoff /5/.

It can be deduced from these results that the mechanism proposed in /4/ for defect formation does not apply to the experimental conditions of the electron microscope. The present results show also that the CL variation produced by electron irradiation in the SEM is not associated to a detectable corresponding variation of the optical absorption.

References

/1/ C. BALLESTEROS, J. LLOPIS, and J. PIQUERAS, Solid State Commun. 43, 739 (1982).

/2/ R.W. SOSHEA, A.J. DEKKER, and J.P. STURTZ, J. Phys. Chem. Solids 5, 23 (1958).

/3/ C. BALLESTEROS, J. LLOPIS, and J. PIQUERAS, 4th Conf. Lattice Defects in Ionic Crystals, Dublin 1982, to be published in Radiat Eff.

/4/ YU.M. ANNENKOV, A.P. SURZHIKOV, A.D. PROGREBNYAK, and V.P. SURZHIKOV, Zh. tekh. Fiz. 50, 222 (1980).

/5/ T.E. EVERHART and P.H. HOFF, J. appl. Phys. 42, 5837 (1971).

(Received February 23, 1983)

phys. stat. sol. (a) 77, K33 (1983)

Subject classification: 11 and 14.3; 13.4; 22.2.1; 22.8

V.D. Kuznetsov Siberian Physico-Technical Institute, Tomsk[1] (a) and A.F. Ioffe Physico-Technical Institute, Academy of Sciences of the USSR, Leningrad[2] (b)

Electrical Properties of Electron-Irradiated GaAs and $ZnSiAs_2$ Solid Solutions

By

V.N. BRUDNYI (a), A.I. POTAPOV (a), YU.V. RUD (b), and M. SERGINOV (b)

The solid solution of GaAs and $ZnSiAs_2$ is a possible material for light emitting source fabrication. But up to now these solutions are obtained as p-type samples only with high free hole concentration ($p \approx 1\times10^{19}$ cm^{-3}) and low resistivity ($\varrho \approx 10^{-3}$ to 10^{-1} Ωcm) at room temperature (RT) /1/. It is assumed that the "hole" conduction in these semiconducting compounds is modified by the grown-in defects which are introduced during crystal growth because of lack of perfect stoichiometry. On the basis of a thermodynamic analysis the anti-site defects such as $[IV_{II}]$ and $[IV_V]$ were predicted to be omnipresent in all II-IV-V_2 ternary compounds. For example, the p-type conductivity of most of the Zn-II-V_2 ternary compounds is assigned to the presence of acceptor-type defects such as $[Si_{As}]$ /2/. Therefore, the successful application of GaAs-$ZnSiAs_2$ compounds for making devices necessitates a better understanding of the lattice defects.

The present investigation is undertaken to determine whether electron bombardment induced point defects have any influence on the electrical properties of GaAs-$ZnSiAs_2$ solid solutions.

The samples for investigation ($p = 1$ to 2×10^{19} cm^{-3}, $\varrho \approx 3$ to 4×10^{-2} Ωcm at RT) are obtained by melting of Zn, Si, and As_2 elements at 1100 °C in the stoichiometric ratio with subsequent homogenization at 1000 °C during 20 h. The electron ($\approx$ 2 MeV) irradiation up to cumulative fluence 1×10^{19} cm^{-2} is carried out at a dose rate of 5×10^{13} electrons $cm^{-2}s^{-1}$ at RT using a linear accelerator.

---

1) 634050 Tomsk, USSR.

2) 194021 Leningrad, USSR.

The effect of ≈ 2 MeV electron irradiation on the resistivity of 1 (2GaAs) - 1 $ZnSiAs_2$ and 2 (2GaAs) - 3 $ZnSiAs_2$ samples is shown in Fig. 1. $\varrho(D)$ in both samples increases with fluence achieving saturation $\approx 10^5$ Ωcm at RT and $\approx 10^{12}$ Ωcm at 77 K. This is attributed to free hole trapping by deep donor traps introduced by bombardment. The free hole removal rate $(p_o - p_D)/D$ is estimated to ≈ 30 cm^{-1} at RT, here, p_o, p_D are the free hole concentration before (o) and after (D) electron irradiation. At prolonged electron bombardment ($D \gtrsim 3x10^{18}$ cm^{-2}) the resistivity in the investigated samples goes through a maximum and then decreases.

In Fig. 2 the temperature variation of ϱ for several bombardment steps is shown. The $\varrho(T)$ dependence near RT in the investigated samples can be expressed by

$$\varrho(T) \sim \exp(E/kT) \ . \qquad (1)$$

Here E is the activation energy of this conduction mechanism. The E value increases upon irradiation from ≈ 0 in unirradiated samples to ≈ 0.24 eV in heavy electron irradiated samples ($D \approx 3x10^{18}$ cm^{-2}). An increase of E due to irradiation could be attributed to a shift of the Fermi level towards midgap.

Below ≈ 130 K lg $\varrho(T)$ versus $(kT)^{-1}$ in heavy electron irradiated samples shows a deviation from linearity. It is shown that the relation

$$\varrho(T) \sim \exp\left(\frac{T_o}{T}\right)^{1/4} , \qquad (2)$$

where $T_o \approx (10^7$ to $10^8)$K is in agreement with the experimental data for the temperature range below ≈ 130 K. Fig. 2 shows the results plotted in this way. The value of T_o for high electron irradiated samples is about the same as for amorphous semiconductors /3/. Therefore we suggest that at low temperature the mechanism of conduction in heavy electron bombarded samples is by variable-range hopping via radiation induced localized states in the forbidden gap. This is the reason of the resistivity decreases at electron fluences higher than $3x10^{18}$ cm^{-2} in Fig. 1.

The change of resistivity resulting from ≈ 2 MeV electron irradiation of p-type conductivity GaAs-$ZnSiAs_2$ samples recovers in three well-defined stages which are centered near 110 to 180, 220 to 300, and 400 to 500 °C with the main annealing stage near 220 to 300 °C. Fig. 3 shows that annealing occurs in the direction of restoring the resistivity, i.e. donor-type defects

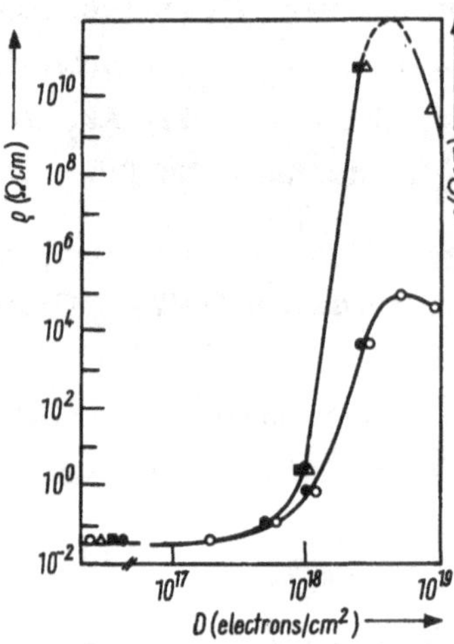

Fig.1

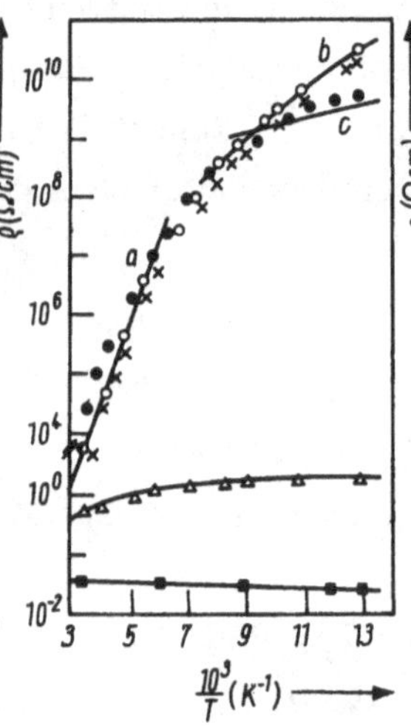

Fig.2

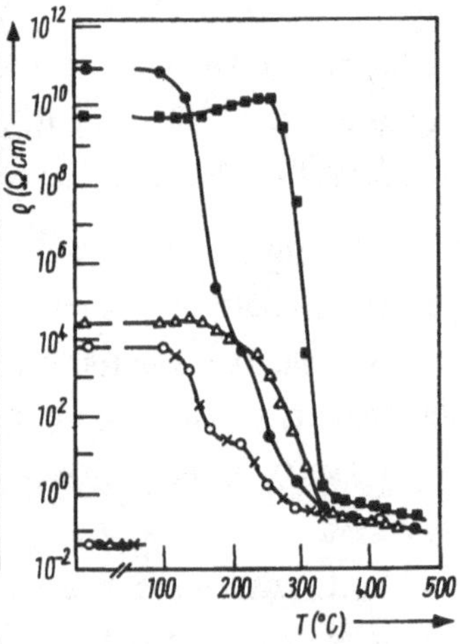

Fig.3

Fig. 1. Change of resistivity with 2 MeV electron bombardment in 1(2 GaAs) - -1 $ZnSiAs_2$ (o, △) and 2(GaAs) - 3 $ZnSiAs_2$ (●, ■): (o, ●) RT, (△, ■) 77 K

Fig. 2. Temperature variations of resistivity in electron irradiated samples of 1 (2 GaAs) - 1 $ZnSiAs_2$ (■, △, o, ●) and 2 (2 GaAs) - 3 $ZnSiAs_2$ (×): (■) before irradiation (△, o, ●, ×) after irradiation up to: $1x10^{18}$ (△), $3x10^{18}$ (o ×), $1x10^{19}$ cm^{-2} (●). (a) $\varrho(T) \sim \exp(0.24\ eV/kT)$; (b) $\varrho(T) \sim \exp(T_0/T)^{1/4}$; (c) $\varrho(T) \sim \exp(T_0/T)^{1/4}$

Fig. 3. Isochronal (600 s) annealing of ≈ 2 MeV electron irradiated 1 (2 GaAs) - -1 $ZnSiAs_2$ (o, ●, △, ■) and 2 (2 GaAs) - 3 $ZnSiAs_2$ (×) samples: (o, ●, ×) $D = 3x10^{18}$ cm^{-2}; (△, ■) $D = 1x10^{19}$ cm^{-2}. (o, △, ×) RT, (●, ■) 77 K

are annealed in the temperature range 110 to 500 °C. The interval of thermal stability of radiation induced defects in the investigated samples is close to that of electron bombardment induced defects in GaAs /4/ and $ZnSiAs_2$ /5 to 7/ compounds.

The results presented here demonstrate that in solid solutions of GaAs and $ZnSiAs_2$ the electrical properties change readily at electron irradiation near RT as in the early investigated binary GaAs /4/ and ternary $ZnSiAs_2$ /5 to 7/ semiconducting compounds. The fact that the electron bombardment leads to a decrease of the free hole concentration suggests that donor-type defects are involved during electron irradiation. The electron bombardment is useful for isolating region fabrication in solid solutions of GaAs and $ZnSiAs_2$. Semi-isolating samples ($\approx 10^5$ Ωcm at RT, $\approx 10^{12}$ Ωcm at 77 K) of 1(2 GaAs) - 1 $ZnSiAs_2$ and 2 (2 GaAs) - 3 $ZnSiAs_2$ were produced after electron irradiation at 300 K.

References

/1/ N. A. GORUNOVA and YU.V. VALOV (Ed.), Semiconducting II-IV-V_2 Compounds, Izd. Soviet Radio, Moscow 1974.

/2/ V.S. GRIGOREVA, V.D. PROCHUKHAN, YU.V. RUD, and A. A. YAKOVENKO, Doping of Semiconductors, Nauka, Moscow 1975 (p. 8) (in Russian).

/3/ E. A. DAVIS and N. F. MOTT, Phil. Mag. 22, 903 (1970).

/4/ V.N. BRUDNYI, M. A. KRIVOV, A.I. POTAPOV, and V.I. SHAKHOVTSOV, Fiz. Tekh. Poluprov. 16, 40 (1982).

/5/ V.N. BRUDNYI, M. A. KRIVOV, A.I. POTAPOV, V.D. PROCHUKHAN, and YU.V. RUD, Fiz. Tekh. Poluprov. 12, 1109 (1978).

/6/ V.N. BRUDNYI, M. A. KRIVOV, A.I. POTAPOV, V.D. PROCHUKHAN, and YU.V. RUD, Radiation Physics of Semiconductors and Related Materials 1979, Tbilisi State University Press, Tbilisi 1980 (p. 680).

/7/ V.N. BRUDNYI, M. A. KRIVOV, A. MAMEDOV, A.I. POTAPOV, V.D. PROCHUKHAN, and YU.V. RUD, phys. stat. sol. (a) 60, K57 (1980).

(Received February 28, 1983)

phys. stat. sol. (a) 77, K37 (1983)

Subject classification: 1.4 and 14.4.1; 22.8.1

Institute of Physics, Rostov (Don) State University[1)]

Ferroelectric Phase Transition in Heteroepitaxial Films of (Ba, Sr)TiO_3

By

V.M. MUKHORTOV, YU.I. GOLOVKO, V.A. ALESHIN,
E.V. SVIRIDOV, VI.M. MUKHORTOV, V.P. DUDKEVICH, and
E.G. FESENKO

$Ba_xSr_{1-x}TiO_3$ films, where x = 0.5 to 1.0, were deposited on (100) MgO crystal cuts by rf sputtering of polycrystalline targets of corresponding composition in oxygen atmosphere /1/ by the technique described in /2/. The film composition was determined by the X-ray fluorescent analysis technique with relative error ≈ 3% (modernized X-ray spectrometers DRUS-3 and KRUS-5). Stoichiometry disturbances were not observed. The fact of heteroepitaxial growth and orientation relations between the films and substrates and also the unit cell parameters were established by X-ray diffraction (diffractometer DRON-3 (CuK_α)). For further studies of the phase transition specimens were selected with complete parallel orientation (crystallographic directions [100], [010], and [001] of MgO and $Ba_xSr_{1-x}TiO_3$ are parallel) and good crystal quality that made possible to determine the film unit cell parameters with a sufficiently small error (±0.0003 Å for the parameter c and ±0.0006 Å for the parameter a) by the reflections 005 and 034. All the films (more than 100 films) at room temperature were tetragonal with the crystallographic axis c perpendicular to the substrate surface and with the parameters of the unit cell a smaller, and c larger than in bulk specimens of the same composition. For instance, for $Ba_{0.7}Sr_{0.3}TiO_3$ films a = 3.955 Å, c = 3.997 Å, c/a = 1.010 ± ± 0.001 and for the bulk specimens a = 3.970 Å, c = 3.982 Å, c/a = 1.003 /3/. The temperatures of ferroelectric phase transitions, which were, for the first time, observed by us from the temperature dependences of the unit cell parameters, appeared elevated by 40 to 60 °C. For illustrating this peculiarity and some others of the phase transition, in Fig. 1 are shown the temperature dependences of unit cell parameters and the physical widths $\Delta 2\Theta$ of the 005 reflection of $Ba_{0.7}Sr_{0.3}TiO_3$ film with thickness ≈ 2 μm.

---

1) prospekt Stachki 194, Rostov (Don), USSR.

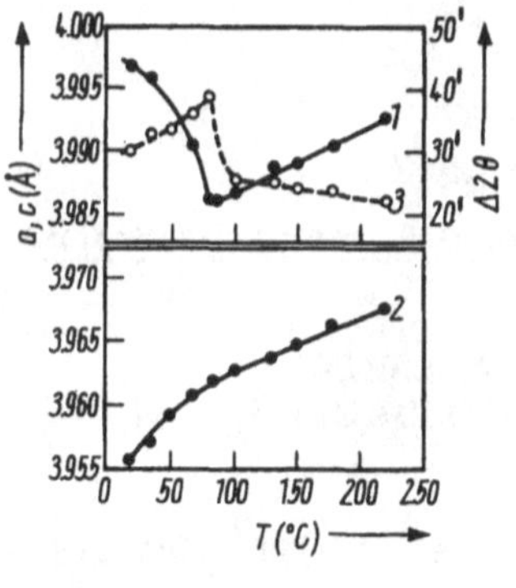

Fig. 1. Temperature dependences of unit cell parameters c (curve 1), a (curve 2) and widths $\Delta 2\Theta$ of 005 reflection (curve 3) of $Ba_{0.7}Sr_{0.3}TiO_3$ film 2.0 μm thick

It may be seen that at temperature ≈ 80 °C a phase transition occurred (instead of 35 °C for bulk specimens /3/). The break on the a(T) and c(T) curves at T ≈ 80 °C and also the maximum on $\Delta 2\Theta$(T) curve at the same temperature (the broadening of reflections in $BaTiO_3$ crystals in the vicinity of the ferroelectric phase transition temperature was, for the first time, observed by Kanzig /4/) were, in particular, indicative of phase transition. However, as distinct from the bulk specimens, the transition was not into the cubic phase, but into the tetragonal paraelectric phase (the patterns of domain etching were not traced if etching was carried on at T > 80 °C). The effective coefficients of linear expansion of films in the paraphase determined by the slope of the a(T) and c(T) curves appeared equal to $\alpha_a = (9.7 \pm 3.2) \times 10^{-6}$ K^{-1} and $\alpha_c = (11.3 \pm 1.8) \times 10^{-6}$ K^{-1}. Using formulae of /5/ where the model of rigid coupling of the film with the substrate was considered and several other simplifying suppositions were made, it may be shown that

$$\alpha_a = \alpha_c \ , \tag{1}$$

$$\alpha_c = \alpha + \frac{2S_{12}}{S_{11} + S_{12}} (\alpha_o - \alpha) \ , \tag{2}$$

$$\sigma_1 = \sigma_2 = \frac{(\alpha_o - \alpha)\Delta T}{S_{11} + S_{12}} \ , \tag{3}$$

where α_o and α linear expansion coefficients of substrate and films, respectively; S_{12} and S_{11} elastic compliances of film; $\sigma_1 = \sigma_2$ mechanical stresses at flat biaxial compression (plus) and extension (minus) appearing when cooling from the temperature of condensate deposition T_s to the temperature $T > T_k$ at the expense of the difference between α_o and α; $\Delta T = T_s - T$.

If, when calculating by the formulae (1) to (3), we take the values of film physical constants equal to the corresponding physical constants of $BaTiO_3$ crystals /6/ (i.e.

$\alpha = 9.8 \times 10^{-6}\,K^{-1}$, $S_{11} = 8.33 \times 10^{-13}\,cm^2/dyn$, and $S_{12} = -2.68 \times 10^{-13}\,cm^2/dyn$) and use the value $\alpha_0 = 11.0 \times 10^{-6}\,K^{-1}$ (found experimentally), then $\alpha_a = 11.0 \times 10^{-6}\,K^{-1}$, $\alpha_c = 8.6 \times 10^{-6}\,K^{-1}$, and $\sigma_1 = \sigma_2 = 1700\,kg/cm^2$ (at $\Delta T = 800\,^oC$). The calculated values of effective linear expansion coefficients fairly well agree with the experimental data. The value of mechanical stresses at flat equal-axis compression, $\approx 1700\,kg/cm^2$, according to the estimation carried out in the framework of the common thermodynamical formalism using the expansion coefficients of the thermodynamical potential of $BaTiO_3$ crystals /7/ is sufficient for elevating the Curie temperature by 35 to 40 ^oC. However, when calculating the deformations caused by thermal stresses they appear smaller than observed. The tetragonality c/a, related with these deformations, directly higher than the temperature of the ferroelectric phase transition,. as calculations reveal, should be equal to 1.002 (for $BaTiO_3$), whereas, in reality, this value is near to 1.006 to 1.007 and feebly depends on the barium content. Apparently, the tetragonal paraelectric phase should not be regarded as an elastically deformed cubic phase, characteristic of the bulk specimens. In our opinion the tetragonality may be induced by an anisotropic distribution of point defects (interstitial atoms and vacancies) in the process of heteroepitaxial film growth. Point defects are peculiar to the films of complex oxides obtained by vacuum methods /8/. In the case of isotropic distribution in the particles of fine-grained $BaTiO_3$ powders synthesized at low temperature, these defects cause considerable changes of unit cell parameters /9/.

The problem about the nature of structural alterations at phase transitions in heteroepitaxial ferroelectric films has not been previously considered. It is hardly possible to expect a quick and unambiguous solution of this problem. The experimental data presented here only open the period of accumulating experimental material.

References

/1/ VL.M. MUKHORTOV, V.P. DUDKEVICH, V.S. BONDARENKO, V.S. MIKHALEVSKII, and E.G. FESENKO, Izv. Akad. Nauk SSSR, Ser. inorg. Mater. 11, 2010 (1975).

/2/ V.M. MUKHORTOV, YU.I. GOLOVKO, VL.M. MUKHORTOV, V.P. DUDKEVICH, and E.G. FESENKO, Zh. eksper. teor. Fiz., Pisma 5, 1175 (1979).

/3/ L. A. SHABANOV, in: Phase Transitions in Ferroelectrics with Perovskite Structure, Latvian State Univ., Riga 1974 (p. 150)(in Russian).

/4/ W. KANZIG, Helv. Phys. Acta 24, 175 (1951).

/5/ V. P. DUDKEVICH, YU.I. GOLOVKO, V.V. KULESHOV, A.M. MARGOLIN, L. KALISH, and E.G. FESENKO, in: Semiconductors-Ferroelectrics, Rostov State University, Rostov-on-Don 1978 (p. 124)(in Russian).

/6/ F. YONA and D. SHIRANE, Ferroelectric Crystals, Mir, Moscow 1965.

/7/ L. P. KHOLODENKO, Thermodynamic Theory of Ferroelectrics of Barium Titanate Type, Zinate, Riga 1971 (in Russian).

/8/ V. P. DUDKEVICH and E.G. FESENKO, Physics of Ferroelectric Films, Rostov State University, Rostov-on-Don 1979 (in Russian).

/9/ Yu. I. GOLOVKO, M.G. RADCHENKO, R.V. KOLESOVA, V.P. DUDKEVICH, and E.G. FESENKO, Kristallografiya 25, 203 (1980).

(Received February 10, 1983)

phys. stat. sol. (a) 77, K41 (1983)

Subject classification/ 14.3.4 and 20.1; 22.1.1

S.V. Starodubtsev Physico-Technical Institute, Academy of Sciences of the Uzbek SSR, Tashkent[1)]

On the Minimum Coefficient of Light Absorption by Free Carriers under the Conditions of Exclusion in "Pure" Semiconductors with Non-Injecting Contacts

By

D.A. ARONOV and V.V. RUBINOV

As it was shown by us in /1/, the maximum decrease in the coefficient α of light absorption by free carriers, and consequently in the light intensity being absorbed, corresponds to the conditions of complete exclusion in "pure" semiconductors /2/. In the case of n-type semiconductors with concentration of residual donors N_D and that of intrinsic carriers n_i, the absorption coefficient is /1/

$$\alpha \approx \sigma_n N_D \ll \sigma_n n_i \, , \tag{1}$$

where σ_n is the cross section of photon absorption by electrons (in Ge, at light wavelength $\lambda \approx 10$ µm, $\sigma_n \approx 4\times10^{-17}$, $\sigma_p \approx 1.4\times10^{-15}$ cm^2 /3/). It follows from (1) that the value of α is proportional to the impurity concentration N_D and that it is possible to decrease the light absorption coefficient to as small values as desired if the material purity is made higher and higher by removing impurities. This note is intended to show that there is such a small (minimum) value of the coefficient α that with carrier exclusion it is impossible to attain a smaller value by reducing the concentration of impurities, and to define the optimum values of semiconductor, light, and exclusion parameters corresponding to the minimum value of α.

Indeed, according to /1/, equation (1) corresponds to voltages

$$U \gtrsim U_{ce} \equiv \frac{(b+1)n_i^2}{2N_D(n_o+p_o)} \left(1+\frac{2b}{b+1}\frac{N_D}{p_o}\right) \frac{L^2}{\mu_n \tau_o} \approx \frac{(b+1)n_i L^2}{4\mu_n N_D \tau_o} \tag{2}$$

applied to a semiconductor sample. It is evident that the higher the material purity and the longer the sample, i.e. the higher the concentration of intrinsic

---

1) 700060 Tashkent, USSR.

carriers n_i as compared with the impurity concentration N_D, the greater are the applied voltages. On the other hand, the value of current flowing under these conditions which is determinable from /1/

$$I_{ce} = 2 q \mu_n N_D \frac{S}{L} U_{ce} = \frac{q(b+1) n_i^2 L S}{(n_o + p_o) \tau_o} \left(1 + \frac{2b}{b+1} \frac{N_D}{p_o}\right) \approx \frac{q n_i (b+1)}{2 \tau_o} LS \quad (3)$$

is practically independent of the concentration N_D and proportional to the volume LS of the semiconductor from which all intrinsic carriers are removed by exclusion. Since, according to (2) and (3), the power being released while the exclusion current is flowing

$$W_{ce} \equiv I_{ce} U_{ce} \approx \frac{(b+1)^2 q n_i^2 L^3 S}{8 \mu_n \tau_o^2 N_D} \quad (4)$$

(i.e. the current heating of the crystal lattice increases causing higher concentrations of free carriers), the light absorption coefficient rises due to this effect. Consequently, there is such minimum value of $\alpha = \alpha_{min}$ that it is impossible to attain a smaller one by lowering infinitely the level of impurity concentration in semiconductors. In this connection, below we consider the allowable values of voltage and current corresponding to complete carrier exclusion but nevertheless not causing current heating of the crystal exceeding that resulting from light absorption.

It follows from (4) that the power being released in the case of a square-aperture semiconductor sample (the electrode length l is equal to the sample length L, and the cross section area S = ld = Ld where d is the sample thickness) is dependent on the sample geometry d, L (especially on L), carrier lifetime (τ_o), and material purity (N_D). For Ge at T = 300 K, the values of coefficients in (4) are b = 2, $n_i = 2.5 \times 10^{13}$ cm^{-3}, $\mu_n = 3750$ cm^2/Vs, the typical values of lifetime τ_o = 100 to 500 µs, and the presently attainable values of $N_D = 10^{13}$ to 10^9 cm^{-3}. At worst, if the aperture is large (L = 1 cm), lifetime is short (τ_o = 100 µs), impurity concentration is moderately low ($N_D = 10^{11}$ cm^{-3}), and d = 1 mm, it follows from (4) that $W_{ce} \approx 3$ W. A decrease in L to one half results in 16 times lower W_{ce}, and a five-fold increase in τ_o reduced it by 25 times.

According to the above analysis, the full power being released with illumination and exclusion at $l = L$ is

$$W = W_{ce} + W^{e}_{abs} \approx \frac{(b+1)^2 q n_i^2 L^4 d}{8 \mu_n N_D \tau_o^2} + W_o \sigma_n N_D d , \qquad (5)$$

where W_o is the power of luminous radiation incident on the sample. Since W_{ce} rises and W^{e}_{abs} drops with decreasing concentration N_D, there is an optimum value of N_D corresponding to minimum values of W and α:

$$N_D = N_D^{opt} \equiv \frac{\sqrt{q}\,(b+1) n_i L^2}{2 \tau_o \sqrt{2 \mu_n \sigma_n W_D}} , \qquad (6)$$

$$W_{min} \approx \frac{n_i}{\tau_o} (b+1) L^2 d \sqrt{\frac{q \sigma_n W_o}{2 \mu_n}} , \qquad (7)$$

$$\alpha_{min} \approx \frac{n_i}{2 \tau_o} (b+1) L^2 \sqrt{q \sigma_n / 2 \mu_n W_D} , \qquad (8)$$

i.e. under the optimum conditions, N_D^{opt}, W_{min}, and α_{min} are determined by the semiconductor parameters (n_i, τ_o, b, μ_n, σ_n), its geometry (L, d), and the power of incident radiation (W_o): the value of W_{min} is directly proportional and that of N_D^{opt} is inversely proportional to the square root of the incident radiation power. In the case of Ge at the above values of τ_o, typical values of L = 0.5 to 1 cm, and d = 1 mm, and with W_o varying in a wide range (from 0.1 to 100 kW), equation (6) gives $N_D^{opt} = 2\times10^{11}$ to $3\times10^{12}\,cm^{-3}$ which result is quite acceptable from the standpoint of complete exclusion realization and "pure" germanium production process; the power W_{min} being released in this case does not exceed 1 W whereas without exclusion it varies from 0.1 W at W_o = 100 to 120 W at W_o = 100 kW, and the light absorption coefficient $\alpha_{min} = 2\times10^{-6}$ to $10^{-3}\,cm^{-1}$ which value is essentially lower than that without exclusion when $\alpha_{min} = 1.2\times10^{-2}\,cm^{-1}$ /4/. It follows from these results that it is expedient to use the effect of complete exclusion in semiconductors for a considerable reduction of light absorption by free carriers not at any radiation power but only at rather great values of $W_o \geqq 1$ kW.

It is also important that the range of working voltage variation proves to be quite acceptable at practically all reasonable values of semiconductor parameters and incident light power. Indeed, according to (2) and (6), we obtain the following equation for the voltage corresponding to complete carrier exclusion under the optimum conditions:

$$U_{ce}^{opt} \approx \sqrt{\sigma_n W_o / 2 q \mu_n} \ . \tag{9}$$

The voltage U_{ce}^{opt} calculated from this equation with W_o varying from 0.1 to 100 kW ranges within 2 to 60 V while the values of I_{ce} = 0.3 to 6 mA according to (3).

On the basis of the above analysis we conclude that the ranges of allowable working voltages and currents are quite acceptable and allow the realization of complete exclusion in "pure" germanium (which essentially reduced the light absorption by free carriers), and at the same time the current heating of the semiconductor combined with residual light absorption in these ranges of voltages and currents is lower by many times than that caused by light absorption without exclusion. The latter conclusion means that the use of the exclusion effect under the optimum conditions ensures temperature stability of semiconductors even under the action of very powerful (W_o = 1 to 100 kW) long-wave radiation.

References

/1/ D.A. ARONOV and V.V. RUBINOV, phys. stat. sol. (a) 73, K1 (1982).

/2/ D.A. ARONOV, R. MAMATKULOV, and V.V. RUBINOV, phys. stat. sol. (a) 68, 61 (1981).

/3/ H.Y. FAN, in: Semiconductors and Semimetals, Vol. 3, Ed. R.K. WILLARDSON and A.C. BEER, Academic Press, New York/London 1967.

/4/ D.A. ARONOV and V.V. RUBINOV, Dokl. Akad. Nauk Uzbek SSR No. 10, 37 (1982).

(Received January 31, 1983)

phys. stat. sol. (a) 77, K45 (1983)

Subject classification: 17; 21

Department of Industrial Chemistry, Ibaraki University, Hitachi[1)]

Interaction of Mechanically Rubbed Aluminum with Polar Organic Vapors Found by Measurements of Optically Stimulated Exoelectron Emission under the Influence of Applied Electric Potential

By

Y. MOMOSE, M. OHKUBO, and J. GUNJI

The optically stimulated exoelectron emission (OSEE) from abraded aluminum corresponds to a photoelectric effect controlled by the adsorption of oxygen and water vapor /1, 2/. The interaction with organic compounds also caused a remarkable change in the emission intensity /3/; that is, when the emission from ground aluminum powder was measured using a Geiger counter, it was strongly influenced by the kind of organic vapor in the Geiger-counter gas. A previous investigation /4/ demonstrated that when the potential of rubbed aluminum specimens (applied voltage, AV) was negatively varied by applying voltage between the specimen and the grounded grid of a Geiger counter /5/ with Ar-C_2H_5OH counter gas, an emission peak was observed in the OSEE intensity (EI) versus AV curves. The purpose of this work was to examine the EI versus AV curves for rubbed aluminum specimens when organic vapors (CH_3OH, C_2H_5OH, n-C_3H_7OH, $(CH_3)_2CO$, and CH_3CN) were contained in the counter gases.

Commercial aluminum sheets (purity > 99.5%) degreased with benzene and then annealed in vacuum for 1 h at 300 °C were used. The flat metal sheets were rubbed under vacuum (0.67 Pa) (specimen I) or ambient air (16 °C, about 50% relative humidity) (specimen II) for 5 min with the same material wound around an iron bar using a magnetic stirrer at a speed of 200 rpm. After the rubbing specimen II was annealed in air for 1 h at 300 °C and subsequently exposed to dry air for 20 h (specimen III). The measurements of the OSEE from these specimens were perfo.med at a certain value of the stimulating-light intensity (I_p) from a small bulb (5.9 V, 340 mA; wavelength > 295 nm) in the same way as describeu previously /4/. The counter gas was composed of Ar (11200 Pa) and the organic vapor (2700 Pa) (Table 1). The AV was increased and

1) Hitachi, Ibaraki 316, Japan.

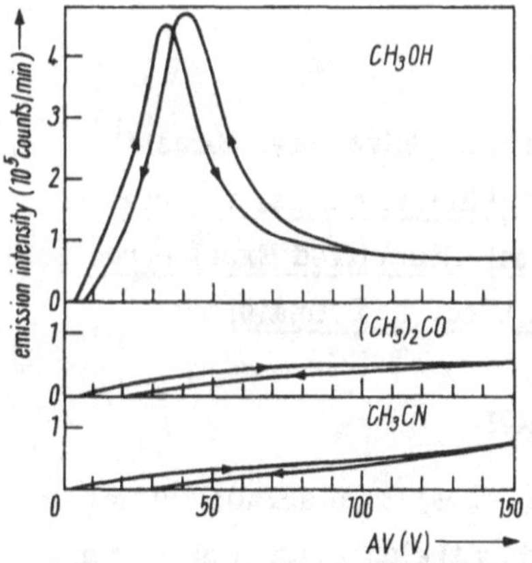

Fig. 1. OSEE intensity versus applied voltage (AV) curves for specimen I. Arrows on the curves (→ and ←) indicate that AV was increased and subsequently decreased, respectively

subsequently decreased at a speed of 5 V/min in the 0 to 150 V region.

Fig. 1 shows the EI versus AV curves. When CH_3OH is contained in the counter gas, a peak is clearly observed. Other alcohols gave similar peaks. We refer the values of EI and AV at these maxima as EI_{max} and AV_{max}, respectively. In the case of $(CH_3)_2CO$ and CH_3CN shown in Fig. 1, there is no peak and the emission is much weaker than for CH_3OH. The results for specimens I and II are summarized in Table 1. The appearance of the peak can be explained in terms of the interaction modes between the alcohols (ROH) and the surface hydroxyl group (AlOH) bound to thin aluminum oxide layer on the metal surface. As described in a previous paper /4/, an emission peak was also observed in the EI versus I_p curves. Therefore, the accumulation of electrons is considered to occur at the surface layer by the action of both AV and I_p, causing alteration in the orientation of an electric dipole formed by the adsorbed alcohol molecules. Below AV_{max} a dipole like ROH_2^+-O^-Al is formed by the hydrogen bond between the oxygen atom of ROH and the hydrogen atom of AlOH, lowering the work function. When the AV is increased, the number of emitted electrons increases together with the accumulation of electrons at the surface layer, and at the same time the emitted electrons are directed toward the grid and thus efficiently collected by the anode of the counter. Therefore the emission intensity increases. Above AV_{max} the interaction mode of RO^--H_2^+OAl becomes predominant, because when the metal is more negatively charged with increasing AV, the electron density on the oxygen atom of AlOH at the surface increases. This interaction mode increases the work function. For $(CH_3)_2CO$ and CH_3CN, the alteration in the orientation of the dipole is more difficult than for the alcohols, because the acidity constant (pK_a) as a conjugate acid decreases in the order: RCH_2CN (pK_a = 25) > $RCOCH_2R$ (19 to 20) > RCH_2OH (18) > CH_3OH (16) /6/. The large values of EI_{max} for the alcohols are associated with the fact that the acidity

Table 1

Characteristics of OSEE intensity (EI) versus applied voltage (AV) curves[a)]

specimen	organic vapor of counter gas	EI_{max} [b)] (10^5 cpm)	AV_{max} [b)] (V)	EI at AV = 150 V (10^5 cpm)
I	CH_3OH	4.50	34	0.75
		4.75	40	
	C_2H_5OH	2.30	8	1.00
		2.55	19	
	n-C_3H_7OH[c)]	3.25	7	1.35
		3.65	8	
	$(CH_3)_2CO$	no peak	no peak	0.55
	CH_3CN	no peak	no peak	0.76
II	CH_3OH	3.45	50	0.90
		4.95	61	
	C_2H_5OH	2.35	9	0.65
		3.70	29	
	CH_3CN	no peak	no peak	0.15

a) These values were obtained from the curves giving the medium value of AV_{max} during the increase in AV with the alcohols and EI at AV = 150 V with $(CH_3)_2CO$ and CH_3CN of two to seven measurements made under the same conditions.

b) The upper and lower values for each alcohol were obtained during the increase in AV and during the decrease in AV, respectively.

c) This counter gas was composed of Ar (11900 Pa) and n-C_3H_7OH (2700 Pa).

constants of the conjugate acids (ROH_2^+) of the alcohols ($pK_{ROH_2^+} = -2$) are much greater than those for $RCNH^+$ (-10) and R_2COH^+ (-7) /6/. It can also be seen from Table 1 that the value of AV_{max} for the alcohols decreases in the order: $CH_3OH > C_2H_5OH \geqq n\text{-}C_3H_7OH$. This finding may be associated with a difference in the rate at which the interaction of the alcohols with the surface takes place.

For specimen III there was no peak during the increase and subsequent decrease in AV in the EI versus AV curve measured using $Ar\text{-}C_2H_5OH$ counter gas and the emission was rather weak ($< 2 \times 10^3$ cpm).

The authors wish to thank Prof. Y. Tamai of Tohôku University for his interest and encouragement.

References

/1/ J.A. RAMSEY, Surface Sci. 8, 313 (1967).

/2/ E. LINKE and K. MEYER, Surface Sci. 20, 304 (1970).

/3/ Y. MOMOSE, Y. IGUCHI, S. ISHII, and K. KOMATSUZAKI, J. phys. Chem. 80, 1329 (1976).

/4/ Y. MOMOSE, T. ISHII, and T. NAMEKAWA, J. phys. Chem. 84, 2906 (1980); Proc. 6th Internat. Symp. Exoelectron Emission and Applications, Ahrenshoop (GDR) 1979 (p. 86).

/5/ Y. MOMOSE and T. NAMEKAWA, J. phys. Chem. 82, 1509 (1978).

/6/ N.L. ALLINGER, M.P. CAVA, D.C. DE JONGH, C.R. JOHNSON, N.A. LEBEL, and C.L. STEVENS, Organic Chemistry, Worth Publishers, New York 1971.

(Received December 23, 1982)

phys. stat. sol. (a) 77, K49 (1983)

Subject classification: 1.3; 10.1; 12.1; 21; 21.1

School of Engineering, Okayama University (a)
and Okayama University of Science[1] (b)

Effects of Grain Boundary on Age Hardening and Reversion in Al-Zn Alloys

By

M. OHTA (a), T. KANADANI (b), A. SAKAKIBARA (a), H. YAMADA (a), and M. YAMADA (a)

Aging and reversion in Al-Zn alloys were studied by means of Vickers microhardness test with careful attention to the positions of impressions. Specimens, 30x5x0.4 mm^3, were prepared by hot-forging and cold-rolling the ingots of alloys. The sizes of grains in each specimen were larger than 3 mm in diameter. Dependence of hardness on the crystalline orientation was not observed.

Fig. 1a shows an isochronal annealing curve in H_v (Vickers hardness number) at the load of 50 g, where the open circles show the means of measured values and the vertical segments show the standard deviations. In this case the hardness was determined without paying attention to the positions of impressions. Compared with the small standard deviations above 160 ^oC where reversion almost finished, H_v's scattered considerably during and before the reversion. This large scattering was considered to be caused by the heterogeneity of hardness in the aged state or on the halfway of reversion. Then careful attention was payed to the positions of impressions and three curves of reversion were obtained according to the following three positions, respectively: (i) about 200 μm or more from the grain boundary (G.B.), (ii) about 70 μm from the G.B., and (iii) just on the G.B.. The results are shown in Fig. 1b. It is clearly seen that the large scattering of H_v's in Fig. 1a is mainly due to the difference in hardness at different positions.

As the dependence of the H_v's on the positions in the grain was observed in the aged state, the aging process was investigated for each position. Fig. 2a to c show isothermal aging curves of the Al-12 wt% Zn alloy at 20 ^oC after quenching from various temperatures. When quenched from temperatures higher than 450 ^oC, the hardness numbers after sufficient aging time become the same everywhere, although they are different from each other in the early stage of

1) Okayama, Japan.

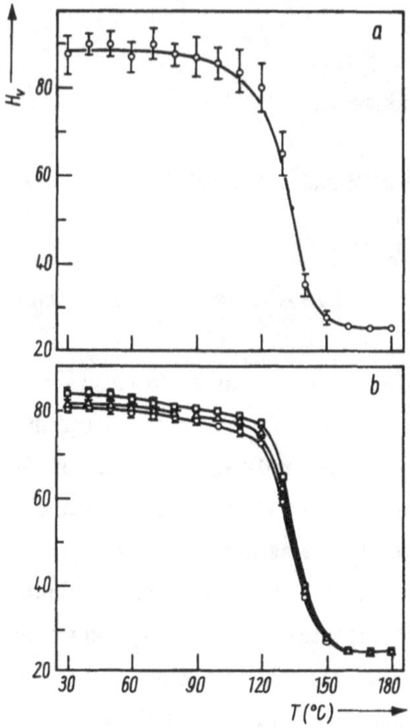

Fig. 1. Isochronal annealing curves in H_v of the Al-10 wt% Zn alloy aged for 1000 min at 30 °C after quenching from 350 °C. Annealing time at each temperature was 10 min; a) random positioning of impressions, b) variation with the positions of impressions, □ 200 μm or more from the G.B., △ about 70 μm from the G.B., ○ just on the G.B.

aging according to the distances from the G.B. On the other hand, when quenched from 350 °C, the H_v's are different from beginning to end, and the final hardness near the G.B. seems to be suppressed. These results suggest that the aging behavior was affected by the difference in vacancy concentration which varied with the quenching temperature and distance from the G.B.

Isochronal annealing curves of the Al-12 wt% Zn specimen aged after quenching from 350 °C were distinctly separated according to the positions of impressions as in Fig. 1b. Fig. 3 shows the isochronal annealing curves of the specimen aged after quenching from 450 °C. Annealing curves begin to separate from each other and decrease separately. This may be due to the presence of difference in the size distribution of G.P. zones in the aged state which was not detectable in hardness test, or may be caused by the rate of reversion varying with the distance from the G.B.

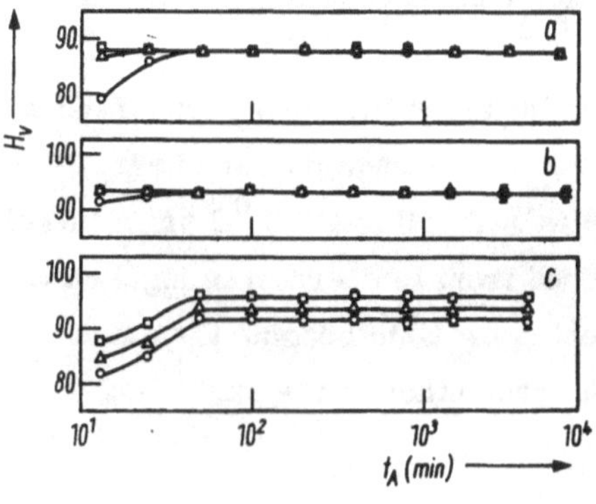

Fig. 2. Variation of isothermal aging curves of the Al-12 wt% Zn alloy at 20 °C with the positions of impressions and the quenching temperatures; □ 200 μm or more from the G.B., △ about 70 μm from the G.B., ○ just on the G.B. a) T_Q = 550 °C, b) T_Q = 450 °C, c) T_Q = 350 °C

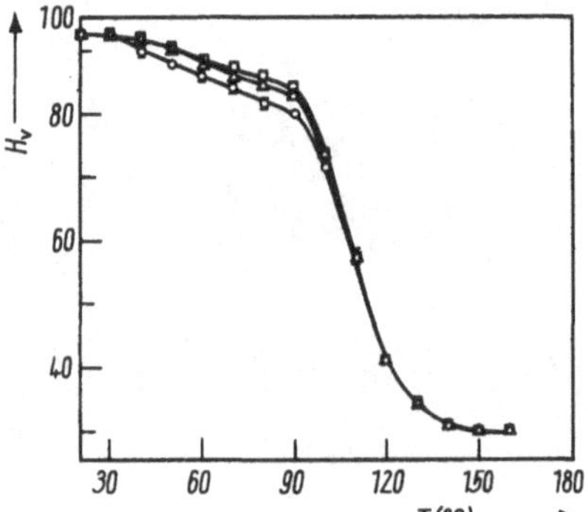

Fig.3

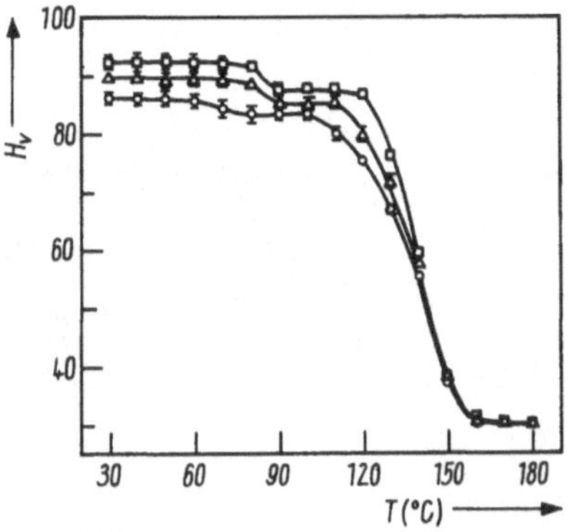

Fig.4

Fig. 3. Variation of isochronal annealing curves of the Al-12 wt% Zn alloy with the positions of impressions. The specimen was aged for 1000 min at 20 °C after quenching from 450 °C before the annealing; □ 200 μm or more from the G.B., Δ about 70 μm from the G.P., o just on the G.B.

Fig. 4. Isochronal annealing curves of the Al-12 wt% Zn alloy aged for 1000 min at 30 °C after two-step quench: 350 °C → 130 °C, 30 s → 0 °C; □ 200 μm or more from the G.B., Δ about 70 μm from the G.B., o just on the G.B.

Heterogeneity of the G.P. zone growth due to vacancy concentration heterogeneity was previously suggested by Vigier and Merlin to interpret the aging behavior after a two-step quench /1/. Fig. 4 shows isochronal annealing curves of Al-12 wt% Zn alloy which was quenched from 350 to 130 °C, held there for 30 s, quenched again into iced water and then aged at 30 °C. Three separate curves are obtained for different positions of impressions. This may be the effect of the heterogeneity of vacancy concentration caused by the same reason as the case of direct quenching shown in Fig. 1b and 2c. On the other hand, two stages of reversion are observed clearly in all curves. These two steps are considered to be corresponding to the reversion of the G.P. zones formed in the regions of low and high solute concentration of the fluctuation /2/. And it is considered that the heterogeneity of the G.P. zone growth caused by the fluctuation of the solute concentration is detected everywhere in the grain.

References

/1/ G. VIGIER and J. MERLIN, phys. stat. sol. (a) 58, K199 (1980).

/2/ M. OHTA, M. YAMADA, T. KANADANI, and A. SAKAKIBARA, phys. stat. sol. (a) 52, K21 (1979).

(Received February 17, 1983)

phys. stat. sol. (a) 77, K53 (1983)
Subject classification: 13.4 and 16; 22.2.2

Semiconductor Physics and Technology Research Laboratory,
University of Sofia[1)]

A Model for Photoconductivity of Fe-Doped GaP

By

A.S. POPOV, D.B. KUSHEV, and SH. CHADRAABAL

In our previous work /1/ measurements of photoconductivity (PC) spectra of Fe-doped GaP were reported. The existence of several deep levels within the forbidden gap of GaP was demonstrated. Some of these levels are connected with the presence of iron ions in the samples.

Three types of samples were investigated:

i) n-GaP:Fe of low Fe concentration ($\varrho \approx 10^2$ Ωcm),

ii) p-GaP:Fe of higher Fe concentration ($\varrho \approx 10^4$ Ωcm),

iii) n-GaP undoped ($\varrho \approx 1$ Ωcm).

The crystals were prepared by floating zone melting and were doped with iron during the crystallization process. The crystals were grown under identical stoichiometric conditions at the melting point 1467 $^{\circ}$C and a phosphorus vapour pressure of 33 atm.

The spectra were taken at 77 and 300 K and were treated following the method proposed by Ridley /2/ and used in /3/ and /4/. The electron transition energies are listed in Table 1. These ionization energies are determined from the intersection of $(\sigma_\nu h\nu)^{2/3}$ verus $h\nu$ lines with the energy axis. The spectra are shown in Fig. 1a to c.

Firstly we shall comment on the transitions observed in PC spectra of both Fe-doped and undoped samples - 1.6 eV and 1.8 eV. The two transitions may be associated with the contribution of gallium vacancies in different charge states. The high concentration of equilibrium defects (vacancies) is provided by a small deviation from the stoichiometric conditions of crystallization, which can well take place during the technological process. As it is shown in /5/, the neutral vacancy in the gallium sublattice V^o_{Ga} forms a level at about 0.3 eV above the valence band edge. It has an electron configuration t_2^3, spin S = 3/2, L = 0 and can be recorded by the ESR method. It is this level that may be involved in the

1) bul. Anton Ivanov 5, 1126 Sofia, Bulgaria.

Table 1

Summary of photoionization threshold energies obtained from Fig. 1a to c

samples	T(K)	photoionization energies (eV)						
p-GaP:Fe	77		0.70	0.99	1.39	1.59		
4/2	300	0.29		1.04	1.48	1.59	1.16	
n-GaP:Fe	77				1.37	1.60		1.85
7/7	300			1.06	1.34	1.60	1.20	1.81
undoped GaP	77					1.60		1.78
F2	300					1.59		1.72

1.8 eV transition, observed in n-type material. It is not registered in p-type material since the larger content of iron results in larger disorder of the lattice.

Since the background concentration of the free carriers is about 10^{16} cm^{-3}, it is possible to observe the second charge state of the gallium vacancy V_{Ga}^- with electron configuration t_2^4. This level is at 0.67 to 0.7 eV and may participate in the 1.6 eV electron transition.

In the PC spectra of Fe-doped GaP crystals several transitions, connected with various charge states of iron atoms, are observed. It is known that the iron atom occupies the Ga site in the lattice and can exist in neutral $3d^5$ or Fe^{3+}, or Fe^0 state, or in the state with a trapped electron ($3d^6$ or Fe^{2+} or Fe^-), or with two trapped electrons (Fe^{1+} or $3d^7$ or Fe^{2-}). The Fe^- state has a ground ^5E and an excited $^5T_2$ term, which are often observed, whereas the Fe^{2-} state is registered only by ESR /6/. If the state is neutral Fe^0, the electron is in the valence band.

The Fe^- state forms a level at 0.84 to 0.87 eV above the valence band and was registered by several methods /7/. Now we observe a transition at 1.4 eV from this level to the conduction band for both n- and p-type GaP:Fe.

When an electron is trapped on the neutral Fe^0 state an excited $^5T_2(Fe^-)$ state with electron configuration $e^2t_2^4$ may be formed. A deep acceptor level at

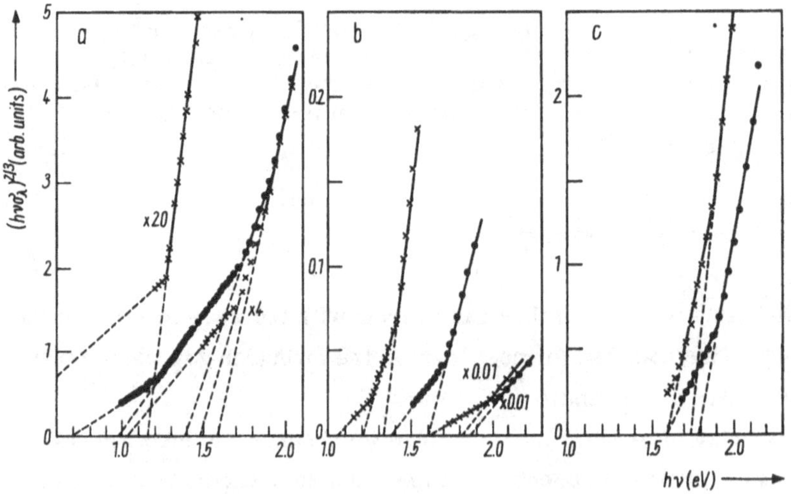

Fig. 1. Spectral dependences of the photoconductivity at 77 (●) and 300 K (×) plotted in terms of the Ridley model for samples: a) p-GaP:Fe 4/2, b) n-GaP:Fe 7/7, and c) undoped n-GaP F2

1.2 eV above the valence band corresponds to this state. The energy 1 to 1.06 eV observed in PC spectra of n- and p-type crystals may be due to electron transitions from this level to the conduction band. The transition from the $^5T_2$ level of the Fe^- center to the conduction band should not be observed in the spectrum of the ideal crystal because it is forbidden by selection rules. However, we assume that this transition can occur in a real crystal because of disorder. The electron can drop from the excited $^5T_2$ to the ground ^5E state which is known as intracenter transition. It is well known that the energy of this intracenter transition in GaP:Fe is 0.41 eV. We suppose that the level 0.87 eV observed in our absorption measurements corresponds to the ^5E level of the Fe^- center. However, the electron transition from the Fe^- center to the neutral gallium vacancy is converted to the second charge state according to the scheme

$$Fe^-(^5T_2) \xrightarrow{e} V^o_{Ga} = Fe^o + V^-_{Ga} .$$

This transition was reported /8/ for GaP:Fe crystals in photo-ESR measurements. In addition to this case iron atoms and gallium vacancies can form complexes if they are close neighbours. Such a model was proposed in /9/. According to this model a considerable delocalization of the d-electron wave

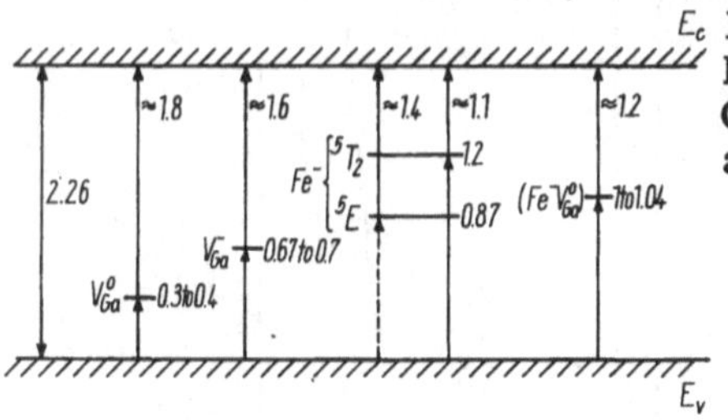

Fig. 2. Proposed diagram for deep energy level position and electron transitions in GaP:Fe. Transitions observed in PC (——) and absorption (- - - -) spectra

functions in $3d^6$ state occurs and they can overlap with the wave function of the neutral vacancy. This complex forms a level in the forbidden gap which must be between the states of Fe^- and V^0_{Ga}.

In the PC spectra of n- and p-type samples we observe an electron transition about 1.2 eV, which is absent in undoped samples. It corresponds to a level 1 to 1.04 eV above the valence band. We suppose that this level arises from the complex mentioned above. A similar level has been observed /10/ in PC of n-i-n structures of GaP:Fe, but its origin was not commented.

Finally we will mention the weak transition at about 0.3 eV in the PC spectra of p-type GaP:Fe. A similar ionization energy has been observed in the absorption spectra of n-GaP:Fe /11/ and has been explained with the existence of a doubly charged donor, but it is not the case here. The transition from the valence band to the level of the neutral gallium vacancy in p-type material may be responsible for this energy.

As a result of the discussion we suggest a diagram for the electron transitions in PC spectra of GaP:Fe (Fig. 2).

References

/1/ A.S. POPOV, D.B. KUSHEV, and SH. CHADRAABAL, Proc. III Bulgarian-Greek Symp., Varna 1981, to be published.

/2/ B.K. RIDLEY, J. Phys. C 13, 2015 (1980).

/3/ H.R. SZAWELSKA and J.W. ALLEN, J. Phys. C 12, 3359 (1979).

/4/ S. FUNG, R.J. NICOLAS, and R.A. STRADLING, J. Phys. C 12, 5145 (1979).

/5/ N.P. ILIN and V.F. MASTEROV, Fiz. Tekh. Poluprov. 10, 836 (1976).

/6/ U. KAUFMANN and J. SCHNEIDER, Solid State Commun. 21, 1073 (1977).

/7/ L.A. DEMBEREL, A.S. POPOV, and D.B. KUSHEV, phys. stat. sol. (a) 52, 653 (1979).

/8/ B.F. MASTEROV, B.E. SAMORUKOV, and V.K. SOBOLEVSKII, Fiz. Tekh. Poluprov. 12, 2346 (1978).

/9/ N.P. ILIN and V.F. MASTEROV, Fiz. Tekh. Poluprov. 14, 375 (1980).

/10/ A.V. LYUBCHENKO, V.G. MAKARENKO, I.E. MARONCHUK, E.A. MERISOVA, and E.A. SALKOV, Ukr. fiz. Zh. 24, 1506 (1979).

/11/ A.A. KOPILOV, A.N. PIKHTIN, and A.S. POPOV, Fiz. Tekh. Poluprov. 15, 770 (1980).

(Received March 16, 1983)

phys. stat. sol. (a) 77, K59 (1983)
Subject classification: 9; 22.1.2

Semiconductor Research and Development Centre, Bucharest[1)]

On the Diffusion Profile of Boron in Silicon at High Concentrations

By
F. GĂISEANU

If there is a permanent contact between the doping source and the semiconductor surface, the boron diffusion profile in silicon, at concentrations higher than the intrinsic carrier concentration, is not correctly described by the complementary error function. Starting from the assumption that the diffusion coefficient varies with the first power of the impurity concentration, the profile has been approximated with polynomials which agrees with experimental data concerning arsenic diffusion in silicon /1/, but not with boron diffusion in silicon /2/. It has been also shown that the curve given by the empirical relation

$$\frac{C}{C_o} = 1 - \left(\frac{x}{x_j}\right)^{2/3} \tag{1}$$

(where x_j is the junction depth), satisfactorily fits the experimental data of boron diffusion in silicon. These data are represented as C/C_o versus x/x_1 for $C/C_o \geqq 0.01$, where C_o is the surface concentration and x_1 is the depth for which $C/C_o = 0.01$ /2/. Nevertheless, relation (1) is not a solution of the diffusion equation.

In this note an analysis of the boron diffusion profile in silicon at high surface concentration ($C_o > 10^{20}$ cm^{-3}) is presented, starting from the following expression of the diffusion coefficient /3/:

$$D = D_o\left(\frac{C}{C_o}\right)^m , \tag{2}$$

where m is a positive number ($m > 0$) which must be determined.

Relation (2) partially expresses an explicit dependence of the diffusion coefficient on the concentration because D_o depends on it, as well. During the diffusion process, after the limit solubility at the semiconductor surface has been attained, D_o may be considered a constant. Relation (2) is a favourable form for numerical integration of the diffusion equation with a diffusion coefficient depending on concentration

---

1) St. Erou Iancu Nicolae 32B, R-72996 Bucharest, Romania.

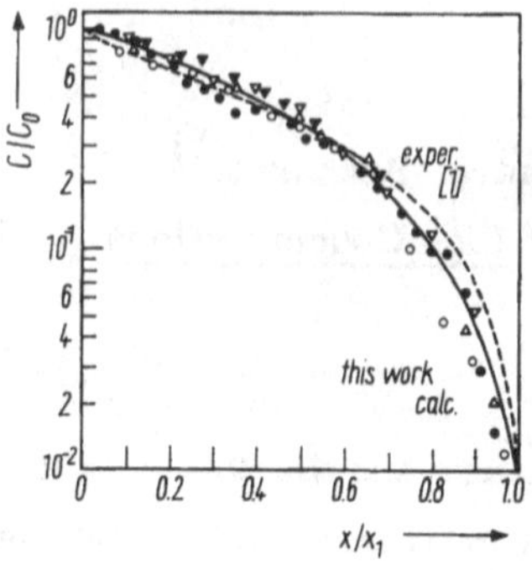

Fig. 1. Theoretical curve C/C_o versus x/x_1 for $D = D_o(C/C_o)^{1/2}$ (full line) and the empirical curve (dashed line) compared with the experimental data on B diffusion in Si for $C > 10^{20}\ cm^{-3}$. Data of the present work: bulk resistivity of silicon sample was 0.5 Ωcm, $x_j = 1.07\ \mu m$, $C_o = 2.5 \times 10^{20}\ cm^{-3}$
▽ T = 1050 °C, t = 60 min /2/;
●○T = 1000 °C, t = 20 min /5/;
△ T = 1150 °C, t = 35 min /6/;
▼ T = 1100 °C, t = 20 min, this work

$$\frac{\partial C}{\partial t} = \frac{\partial}{\partial x}\left(D(C)\frac{\partial C}{\partial x}\right) . \tag{3}$$

In Fig. 1 the curve C/C_o versus x/x_1 (full line) is plotted; it has been numerically calculated from (3) /4/, for m = 1/2. This representation shows an evident advantage consisting in the fact that it does not depend on D_o, thus allowing a comparison with the experimental data of the boron diffusion profile at high concentration obtained after various diffusion processes /5/, at various temperatures /2, 5, 6/.

In the present note boron diffusion was performed at 1100 °C for 20 min. The diffusion profile was obtained by successive etching of silicon layers and by measuring the sheet resistivity after each step using the four-point method. Data processing was made by means of Evans and Donovan's procedure /7/.

In Fig. 1 the empirical curve (dashed line) given by relation (1) is also plotted. The good agreement between the theoretical calculations and the experimental data over the investigated concentration range suggests that the form (2), showing the dependence of diffusion coefficient on concentration with m = 1/2, is a good representation of the boron diffusion coefficient for boron diffusion in silicon at high concentrations. The relation

$$D = D_o\left(\frac{C}{C_o}\right)^{1/2} \tag{4}$$

allows the estimation of the boron diffusion profile in silicon if the value D_o, corresponding to the surface concentration, is known.

References

/1/ R.B. FAIR, J. appl. Phys. 43, 1278 (1972).

/2/ R.B. FAIR, J. Electrochem. Soc. 122, 800 (1975).

/3/ R. L. WEISBERG and J. BLANC, Phys. Rev. 131, 1548 (1963).
/4/ W. KASS and M. O'KEEFFE, J. appl. Phys. 37, 2377 (1966).
/5/ P. NEGRINI, A. RAVAGLIA, and S. SOLMI, J. Electrochem. Soc. 125, 609 (1978).
/6/ G. L. VICK and K. M. WHITTLE, J. Electrochem. Soc. 116, 1142 (1969).
/7/ R. A. EVANS and R. P. DONOVAN, Solid State Electronics 10, 155 (1968).

(Received January 6, 1983)

phys. stat. sol. (a) 77, K63 (1983)
Subject classification: 14.3 and 15; 22.8.1

Department of Physics, Osmania University, Hyderabad[1)]

A Study of the Semiconducting Properties of Mixed Mn-Mg Ferrites

By
P.V. REDDY and T.S. RAO

A comprehensive understanding of the conduction mechanism in a mixed ferrite system can be obtained by fitting the experimental data such as electrical conductivity and thermoelectric power to a theoretical model. Besides this, a theoretical model provides a method of evaluating certain important parameters of a semiconductor such as the energy gap, lattice frequencies, activation energies involved in the lattice deformation of a semiconductor. As such it is thought that it would be of interest to arrive at a theoretical model for the case of the manganese-magnesium ferrite system using experimental data of electrical conductivity and thermo-electric power and the results of such a study are presented in this communication.

The ferrite samples used in the present investigation have been prepared by the conventional double sintering process using manganese carbonate, magnesium carbonate, and ferric oxide (all 99% pure), the final sintering temperature being 1330 $^{\circ}$C. The details of method of preparation are given in an earlier publication /1/. The well-known two-probe method /2/ has been used to study the electrical conduction in these ferrites over a temperature range 300 to 650 K to an accuracy of $\pm 2\%$. The Seebeck voltage for all the samples of this ferrite system has also been studied from room temperature to about 500 K in order to find the type of conduction. As the ferrites under study are very good thermal conductors, a pointed probe is used to maintain a temperature gradient between the upper and the lower surfaces of the ferrite semiconductor. A Keithley electrometer (model 610C) has been used for measuring the thermal e.m.f. developed during the process.

The Seebeck coefficients (Q) have been computed from the observed values of Seebeck voltage for five mixed Mn-Mg ferrites, having the compositional formula $Mn_{1-x}Mg_xFe_2O_4$, the values of x being 0, 0.25, 0.50, 0.75, and 0.90, respectively. The five specimens are denoted as Mn I, Mn II, Mn III,

1) Hyderabad 500007, India.

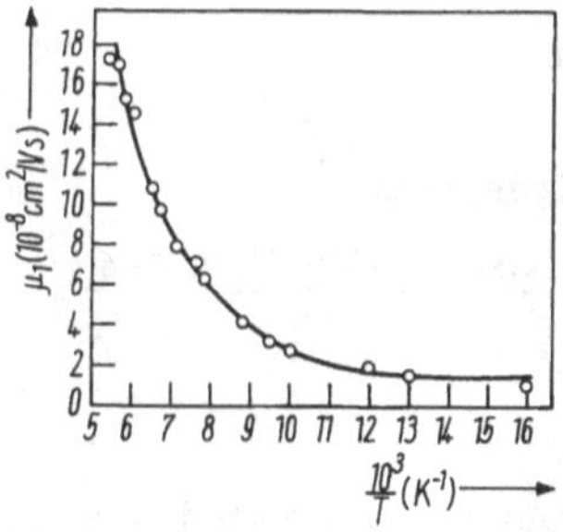

Fig. 1. Plot of mobility of charge carriers versus 10^3/T for $MnFe_2O_4$

Mn IV, and Mn V, respectively. For the specimens Mn I, Mn II, and Mn V, the Seebeck coefficients are found to be negative indicating that they are n-type semiconductors, while the specimens Mn III and Mn IV are found to be p-type semiconductors as their Seebeck coefficients are positive. Chemical analysis /2/ of the specimens using an atomic absorption spectrophotometer (model Varian Techtron A 60) has been carried out to corroborate the results obtained from the thermoelectric power studies. From this study it has been found that n-type ferrites have cations slightly in excess of what is indicated by the ideal formula, while there is a marked deficiency of cations in the case of p-type mixed ferrites.

The concentration of charge carriers (both electrons and holes) has been computed from the values of the Seebeck voltage at different temperatures. Further, assuming that the number of electrons contributing to the conduction to be equal to the number of Fe^{2+} ions in an n-type ferrite, the electron mobility (μ_1) has been calculated for the specimens Mn I, Mn II, and Mn V using the relation,

$$\sigma = n e \mu_1 .$$

Similarly assuming that the number of holes contributing to the conductivity to be equal to the number of Mn^{3+} ions in a p-type semiconductor, the hole mobilities (μ_2) have been calculated in the case of Mn III and Mn IV. The mobilities (μ_1 and μ_2) thus obtained along with those obtained in the case of cobalt and nickel ferrite systems /3, 4/ have been included in Table 1.

Fig. 1 to 3 show plots of μ_1 and μ_2 versus 10^3/T in the case of the Mn-Mg ferrite system. It can be observed from the figures that the mobilities of electrons and holes in the case of all the ferrites under study increase with increase of temperature. This behaviour is similar to that exhibited by many semiconducting materials /5 to 8/. The ordinary band theory of solids cannot be applied to explain the conduction mechanism of these ferrites, as it would result in high values of mobilities and also in a continuous decrease of mobility with increase of temperature both of which contradict the experimental ob-

T a b l e 1

Comparison of parameters required to describe the semiconducting properties of different mixed ferrites

No.	parameter	$Mn_{1-x}Mg_xFe_2O_4$ (present study)	$Co_{3-x}Fe_xO_4$ /1/	$NiFe_2O_4$ /2/
1	q_1 (eV)	0.21 to 0.05	0.25 to 0.175	0.11 ± 0.02
2	q_2 (eV)	0.27 to 0.14	0.51 to 0.475	0.28 ± 0.02
3	ν_1 (s^{-1})	9.25×10^8	6.5×10^{12}	4×10^{10}
4	ν_2 (s^{-1})	2.20×10^{10}	1.0×10^{14}	5×10^{11}
5	α (eV)	0.15	0.025	0.02
6	β (eV)	0.16	0.15	0.03
7	2Δ (eV)	0.46	0.55	1.15
8	μ_1 (cm^2/Vs)	1.00×10^{-8} to 1.50×10^{-6}	0.7 to 1.3×10^{-4}	$(2.5 \pm 0.5) \times 10^{-5}$
9	μ_2 (cm^2/Vs)	$1.00 \times 4.10 \times 10^{-6}$	0.6 to 2.1×10^{-8}	6.5×10^{-8}

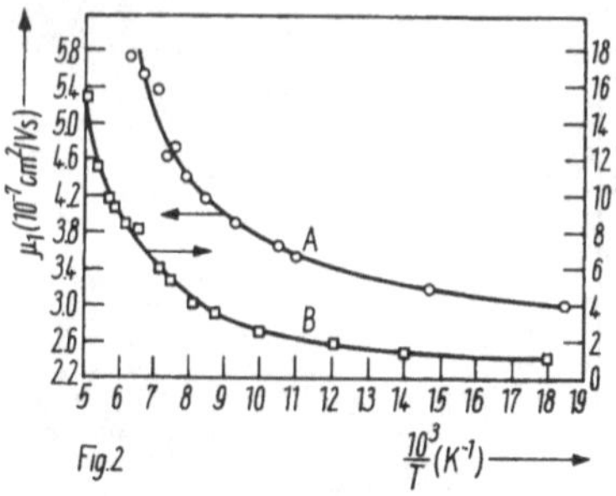

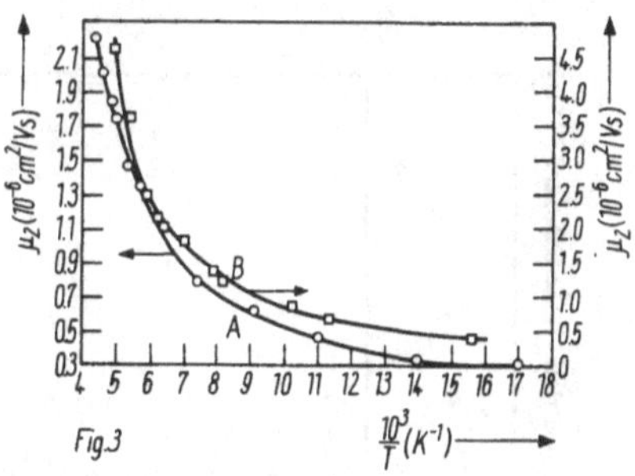

Fig. 2. Plot of mobility of charge carriers versus $10^3/T$ for $Mn_{0.75}Mg_{0.25}Fe_2O_4$ and $Mn_{0.1}Mg_{0.9}Fe_2O_4$

Fig. 3. Plot of mobility of charge carriers versus $10^3/T$ for $Mn_{0.5}Mg_{0.5}Fe_2O_4$ and $Mn_{0.25}Mg_{0.75}Fe_2O_4$

servations of the present investigation. A similar conclusion was arrived at by Morin /6, 7/ in the case of NiO and α-Fe_2O_3 and by Jonker /3/ in the case of cobalt ferrite.

In such cases, in the equilibrium lattice there is little overlap between the wave functions of ions on the adjacent octahedral sites, with the result that the electrons or holes are not free to move through a crystal but are fixed at the metal ions. In the presence of lattice vibrations, however, the ions occasionally come close enough together for transfer to occur with a high degree of probability. Thus the conduction is induced only by the lattice vibrations, and as a consequence the carrier mobilities show an increase with increasing temperature.

Using the conductivity data and the values of electron and hole mobilities, the energy gap (2Δ) for $MnFe_2O_4$ has been calculated using the expression for minimum value of conductivity /3/:

$$\sigma_{min} = 2\ Ne\sqrt{\mu_1 \mu_2}\ \exp(-\Delta/kT) \quad .$$

The activation energies involved in the lattice deformation for electrons and holes, q_1, and q_2, respectively, have been calculated from the observed Seebeck coefficients at room temperature using the well-known equations /3/. Two other important parameters viz., kinetic energy terms (α and β) and the lattice frequencies (ν_1 and ν_2) have also been calculated for the

$Mn_{1-x}Mg_xFe_2O_4$ system following the procedure adopted by Jonker /3/ and Elwell et al. /4/ and are included in Table 1 along with those for $Co_{3-x}Fe_xO_4$ and $NiFe_2O_4$.

The authors are grateful to Prof. G. Sivarama Sastry, Head of Department of Physics, Osmania University, for his interest in this work.

References

/1/ P. VENUGOPAL REDDY and T. SESHAGIRI RAO, J. less- common Metals 75, 255 (1980).

/2/ P. VENUGOPAL REDDY, T.S. RAO, and S.M.D. RAO, J. less-common Metals 79, 1 (1981).

/3/ G.H. JONKER, J. Phys. Chem. Solids 9, 165 (1959).

/4/ D. ELWELL, B.A. GRIFFITHS, and R. PARKER, Brit. J. appl. Phys. 17, 587 (1966).

/5/ G.I. EPIFANOV, Solid State Physics, Izd. Mir, Moscow 1979 (p. 180).

/6/ F.J. MORIN, Phys. Rev. 93, 1195 (1954).

/7/ F.J. MORIN, Phys. Rev. 93, 1199 (1954).

/8/ A.A. SAMOKHVALOV and A.G. RUSTAMOV, Soviet Phys. - Solid State 7, 961 (1965).

(Received January 6, 1983)

phys. stat. sol. (a) 77, K69 (1983)

Subject classification: 13.4 and 16; 22.2.2

Laboratorio de Fisica del Estado Sólido, Facultad de Ciencias, Valladolid (a) and Centre d'Etudes d'Electronique des Solides, Université des Sciences et Techniques du Languedoc[1] (b)

On the Thermal Evolution of Photoconductivity in Bulk Fe-Doped Semi-Insulating InP

By

J. JIMENEZ (a), M.A. GONZALEZ (a), V. CARBAYO (a), and J. BONNAFE (b)

Semi-insulating bulk InP is a very interesting substrate for microwave and optoelectronic devices. Semiconducting InP is converted into a high-resistivity material by doping with Fe and other transition metals, which are introduced in the lattice as substitutional impurities on In sites /1/. Deep Fe acceptor levels compensate the shallow residual donors and enhance the resistivity of InP to 10^7 to 10^8 Ωcm at room temperature. As well as semi-insulating GaAs /2/, high-resistivity InP exhibits a great variety of deep levels, which control its photoelectronic properties. In recent years these levels have been characterized by several experimental techniques, as DLTS /3/, photoluminescence /4/, photoconductivity /5/, EPR /6/, and TSC /7/.

In this short note we report the influence that carriers thermally emitted from traps have on the temperature evolution of the extrinsic photoconductivity of Fe-doped InP.

The measurements were performed with monocrystalline platelets tailored from MCP (Material Chemical Products) ingots, grown by the Czochralski technique and intentionally doped with Fe. Sample dimensions were typically 4x2x0.3 mm^3, and they were provided with silver past contacts, annealed for some minutes at 200 ^oC. In order to neglect the electrostatic inhomogeneities in the sample, electrical and photocurrent measurements were carried out in the linear part of the current-voltage characteristics, the applied electric field was in the 10 V/cm range.

The activation energy obtained from dark current measurements has found to be 0.66 eV, which agrees with the values reported by other authors /5, 8/,

1) Pl. E. Bataillon, 34060 Montpellier Cédex, France.

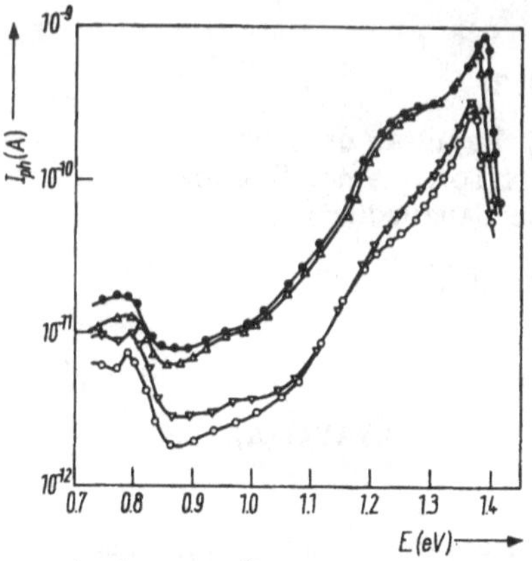

Fig. 1. Extrinsic photoconductivity spectra at several temperatures; ● 80 K, △ 124 K, ▽ 189 K, o 215 K

and it corresponds to the thermal ionization of the ^5E level of the Fe^{2+} substitutional impurity /5/.

The photoconductivity was measured at continuum monochromatic light illumination, between 77 and 300 K, in the 0.7 to 1.4 eV spectral range (lower excitation energies are not available with our experimental equipment). The measurement procedure has been described in a previous paper /9/.

In Fig. 1 are shown the extrinsic photoconductivity spectra for InP: Fe at several temperatures. The spectrum at liquid nitrogen temperature is characterized by the following features:

1) a band between 0.7 and 0.78 eV, which decreases for energies above 0.78 eV;

2) a weak structure is observed in the 0.95 eV zone;

3) a threshold can be observed at around 1.1 eV, followed by a band between 1.2 and 1.3 eV. Above these energies the intrinsic band-gap emission is observed.

The falling of the photoconductivity for energies higher than 0.78 eV, could be ascribed to a self-quenching mechanism /6/, which can be understood as follows: the 0.78 eV band is generated by the photoexcited electrons from the ^5E level of Fe^{2+} /6/ to the conduction band by 0.78 eV photons. For more energetic incident photons ($\hbar\nu > 0.78$ eV), a hole emission from ^5E to the valence band occurs simultaneously with the electron photogeneration, and hence there is an increase in the recombination rate, together with the corresponding decrease of the carrier lifetime, and consequently the photocurrent falls.

The thermal evolution of the photoconductivity is observed in the spectra of Fig. 1. It is characterized by an increase of the 0.78 eV band self-quenching; in fact the self-quenching percent (defined as $I_{ph}(\hbar\nu > 0.78)/I_{ph}(0.78\ \text{eV}) \times 100$) rises from 54% at 84 K to 77% at 190 K. On the other hand, the higher-energy side of the spectrum, that is the 1.2 eV band, is thermally quenched.

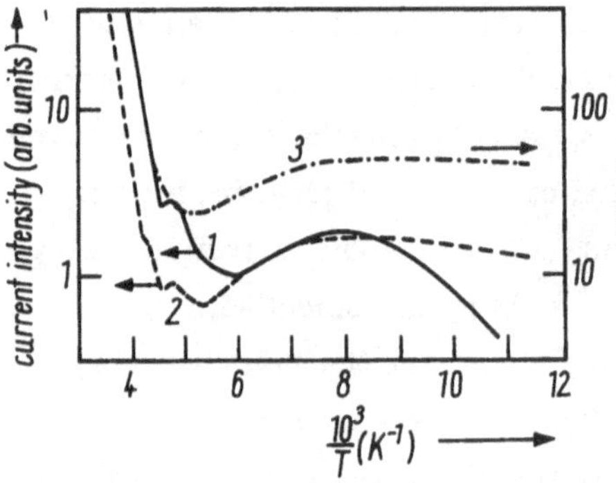

Fig. 2. Thermal evolution of TSC ($h\nu_{exc} \approx$ 0.79 eV) (1), photoconductivity (0.79 eV) (2), and 100 minus self-quenching percent (100-Q) (3) of the photoconductivity

In order to explain the former behaviour we show in Fig. 2 the thermally stimulated current (TSC), the thermal evolution of the photoconductivity for 0.79 eV photon excitation, and the evolution of 100 minus the self-quenching percent (100-Q), versus $10^3/T(K^{-1})$. The TSC spectrum was obtained by warming in the dark at a heating rate of 10 K/min, after 5 min of 0.79 eV monochromatic light excitation at liquid nitrogen temperature. A similar shape of these three curves can be observed in Fig. 2, this fact could explain the self-quenching behaviour. In the TSC spectrum a current peak can be observed, whose maximum temperature lies at 130 K, with an activation energy E_t of 0.24 eV ($E_t \approx 23\ kT_m$ /7/). In this way the thermal evolution of the 0.78 eV self-quenching band could be understood with the aid of a simple level sketch (Fig. 3), as follows: when the temperature increases, simultaneously with the photogenerated electrons and holes from level ^5E (Fe^{2+}) (transitions 1 and 2), there are electrons which are thermally emitted from level A (transition 3), which originates from the 0.24 eV TSC peak; the arrival of these electrons at the conduction band rises the electron density in this band, as well as their lifetimes, hence during the thermal emptying of the trap A, the self-quenching level remains unchanged (Fig. 2, curve 3), but when the trap A begins to exhaust its population, which occurs for temperatures above the TSC maximum (T > 130 K), an increase of the self-quenching is observed (Fig. 2, curve 3), with a maximum value of

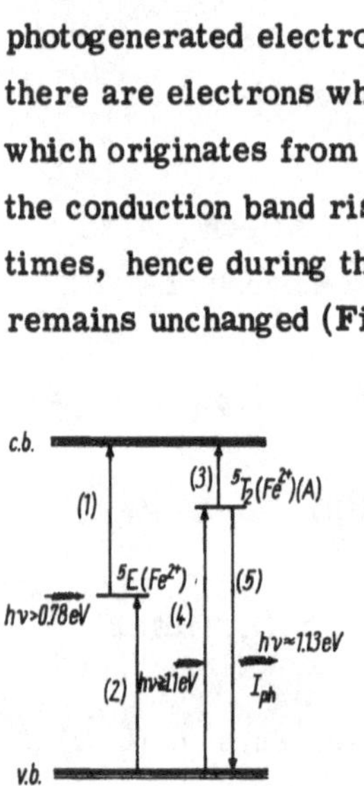

Fig. 3. Single level sketch of Fe-doped InP

77% at 190 K; later, when the equilibrium electrons are thermally emitted ($T > 190$ K, 0.66 eV activation energy) the electron lifetime increases (thermal equilibrium electrons are now dominant) and the self-quenching is not observed.

On the other hand, the thermal quenching of the 1.2 eV photoconductivity band can be ascribed to the thermal emptying of trap A. The activation energy for electrons emitted from trap A ($\approx$ 0.24 eV), is in agreement with the activation energy observed by Phil Won Yu /10/ for the thermal quenching of the 1.10 eV photoluminescence band. This luminescence band, that corresponds to our 1.2 eV photoconductivity band, whose threshold we have found at 1.1 eV (Fig. 1), is related to the radiative decay (transition 5) of the electrons photo-excited from the valence band to the $^5T_2$ excited (transition 4) level of the Fe^{2+} impurity. In this way thermal quenching can be explained as follows: for temperatures higher than 130 K electrons are thermally liberated from the $^5T_2$ level (trap A) to the conduction band (transition 3) and they recombine with photogenerated holes ($h\nu > 1.10$ eV); the lifetime of holes is shortened, and the photoconductivity is partially quenched. This mechanism explains also the thermal quenching of the 1.1 eV photoluminescence band /10/, observed by Phil Won Yu. In fact the deexcitation of the $^5T_2$ levels to ^5E levels via thermally stimulated electron emission into the conduction band, avoids a direct electron decay from the $^5T_2$ level to the valence band, and hence the 1.10 eV luminescence band cannot be observed at $T > 130$ K. In orther words, the recombination channel of excited electrons is changed with increasing temperature.

References

/1/ G.K. IPPOLITOVA, E.M. OMELYANOVSKII, N.M. PAULOV, A.Y. NASHELSKII, and S.V. YAKOBSON, Soviet Phys. - Semicond. 11, 733 (1977).

/2/ A.L. LIN, E.M. OMELYANOVSKII, and R.H. BUBE, J. appl. Phys. 47, 1852 (1976).

/3/ A. MAJERFELD, O. WADA, and N.N.M. CHOUDHURY, Appl. Phys. Letters 33, 957 (1978).

/4/ L.A. DEMBEREL, A.S. POPOV, D.B. KUSHEV, and N.N. ZHELEVA, phys. stat. sol. (a) 52, 341 (1979).

/5/ S. FUNG, R.J. NICHOLAS, and R.A. STRADLING, J. Phys. C 12, 5145 (1979).

/6/ W.H. KOSCHEL, U. KAUFMANN, and S.G. BISHOP, Solid State Commun. 21, 1069 (1977).

/7/ J. BONNAFE, M. CASTAGNE, J. ROMESTAN, M. DE MURCIA, and J.P. FILLARD, Electronic Letters 16, 313 (1980).

/8/ D.C. LOOK, Phys. Rev. B 20, 4160 (1979).

/9/ J. JIMENEZ, M.A. GONZALEZ, L.F. SANZ, and J. BONNAFE, phys. stat. sol. (a) 73, K189 (1982).

/10/ PHIL WON YU, Solid State Commun. 34, 183 (1980).

(Received March 24, 1983)

phys. stat. sol. (a) 77, K75 (1983)
Subject classification: 20.1; 22.2.4

Institut für Hochfrequenztechnik, Technische Universität Braunschweig[1)]

Optical Absorption and Refractive Index near the Bandgap for InGaAsP

By
W. KOWALSKY, H.-H. WEHMANN, F. FIEDLER, and
A. SCHLACHETZKI

Introduction Glass fibres for optical communication systems display minima of attenuation in the wavelength ranges around 1.3 and 1.5 μm and zero dispersion near the former wavelength. Therefore, semiconductors for these wavelengths are of prime interest. Particularly promising for suitable light sources and detectors is the alloy system InGaAsP with compositions lattice-matched to InP /1/. A knowledge of the optical absorption coefficient α and the refractive index n, measured with these alloys, is of special importance not only for the devices mentioned before, but also for possible waveguide and coupler applications. Unfortunately, most of the relevant measurements were either taken with rather different compositions /2/ or, limited by the measurement technique, only in the region of fundamental absorption /3/. Only few results were reported in the range of small absorption /4 to 7/.

It is the intention of this note to present measurements of α and n taken near the bandgap with two different compositions of interest and to provide a comparison with closely related compositions. In addition, several interpolation schemes /3, 5/ will be discussed with respect to their accuracy to calculate n for a given composition.

Experimental The crystals investigated here were grown by liquid-phase epitaxy using a standard graphite sliding-boat. Step-cooling was applied. The epitaxial films were deposited on either Fe-doped semi-insulating (s.i.) or on heavily Sn-doped InP substrates, supplied by MCP, England, in (100) orientation. The composition of the melt to obtain a particular composition of the solid was taken from /8/. By in-situ back-melting a possibly degraded surface layer of the substrate was removed to a depth of around 10 μm to improve the conditions of crystal growth. All samples were not intentionally doped, and no particular efforts were made to reduce the background impurity level.

---

1) Postfach der TU 3329, D-3300 Braunschweig, FRG.

Table 1

epitaxial layer	substrate	77K $N(10^{16}cm^{-3})$	77K $\mu(cm^2/Vs)$	300K $N(10^{16}cm^{-3})$	300K $\mu(cm^2/Vs)$	d (µm)
Q(1.2)	n^+InP	-	-	-	-	1.2
T	s.i.InP	2.7	16200	3	8800	2.8
T	s.i.InP	2.3	16300	2.6	8900	3.9
T	s.i.InP	2.2	15900	2.5	8500	7

The notation used in this paper is Q(1.2) for the quaternary $In_{0.82}Ga_{0.18}As_{0.41}P_{0.59}$ and Q(1.3) for $In_{0.73}Ga_{0.27}As_{0.64}P_{0.36}$, where the figures in parentheses give the bandgap wavelengths. T denotes the ternary $In_{0.53}Ga_{0.47}As$ alloy. The mobility and the carrier concentration N of the epitaxial layers grown on s.i. substrates were determined with the van der Pauw technique at 77 K and at room temperature. All films were found to be N-type. The measured electrical data as well as the film thickness d, obtained microscopically from the cleavage plane, are compiled in Table 1.

The sample preparation and the optical set-up were similar as reported in /7/. The substrate side was polished on a Pellon disc using a 0.4% Br-isopropanol solution in N_2 atmosphere. Mechanical surface profiles are also given in /7/. A current-stabilized tungsten lamp was used for illumination in the optical studies. Two different monochromators were used: a Bausch & Lomb monochromator for the wavelength range from 0.7 to 1.6 µm with a grating of 675 lines/mm and a dispersion of 15 nm and a Schoeffel GM 250 monochromator for the wavelength range from 0.5 to 1.8 µm with a grating of 590 lines/mm, a dispersion of 5 nm, and a blaze wavelength of 750 nm. To avoid higher orders, also passing the monochromator, a long-pass filter was employed. The radiation was focused on the sample at an angle of incidence of less than 7°. Thus multireflections between sample and lens were avoided. Either the reflected or the transmitted beam was detected by a Ge detector. Since its diameter is only 1 mm, a bi-convex lens is necessary to focus the radiation. The short-circuit output current of the detector was converted by an operational amplifier (CA 3140) into a voltage and detected with lock-in techniques (PAR 186). Because the surfaces of the epitaxial layers were of high optical quality, the total scattering

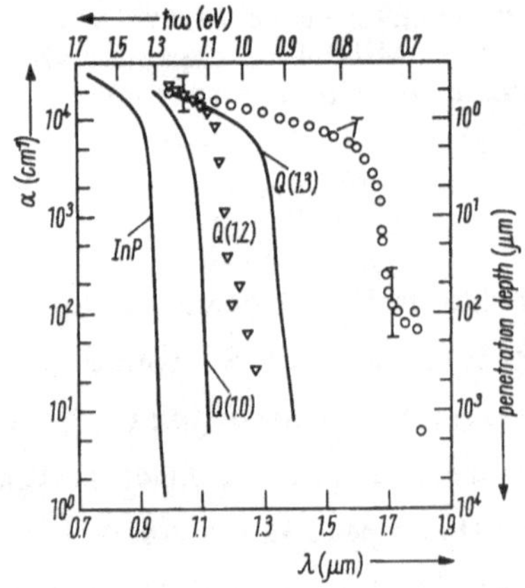

Fig. 1. Measured absorption coefficient α in dependence on wavelength λ. InP from /10/, Q(1.1) and Q(1.3) after /9/ and /7/, respectively. Data points: present measurements, T = 300 K

loss was measured as less than 0.5% so that no correction was necessary.

From the measured transmission and reflectance α and n were obtained by recursion formulas neglecting interference effects, but taking multireflections into account. Interference effects were indeed never observed and are very unlikely to occur because of epitaxial-layer thickness inhomogeneity and limited dispersion of the monochromators.

Discussion of the experimental results In Fig. 1 the measured absorption coefficient α in dependence on the wavelength λ is depicted for a number of compositions of particular importance in the present context. The solid curves are taken from the literature whereas the data points resulted from the present investigations. Q(1.0) marks the composition $In_{0.88}Ga_{0.12}As_{0.25}P_{0.75}$ /9/. The vertical bars represent the estimated experimental error, taking thickness inhomogeneities, wavelength dispersion, and instabilities of the light source into account. The error is relatively small in the range of high absorption since the reflected light originates only from the surface. Towards the bandgap the experimental accuracy decreases due to the larger contribution to the stray light by multireflections at the surfaces. Not shown are the experimental results by Burkhard et al. /3/ which complement the data of Fig. 1 in the short-wavelength region.

The measurements of Fig. 1 clearly depict the sharp decline of α at the bandgap, common to direct semiconductors. This allows the determination of the bandgap energy E_g by the standard formula /11/

$$\alpha\hbar\omega \sim (2m_{eff})^{3/2}(h\omega - E_g)^{1/2}/n ,$$

where m_{eff} is the effective mass and $\hbar\omega$ the photon energy. The evaluation of the present measurements results in $E_g = 0.75$ eV, corresponding to $\lambda_g = 1.65$ μm, for the ternary T and $E_g = 1.06$ eV or $\lambda_g = 1.17$ μm for the quater-

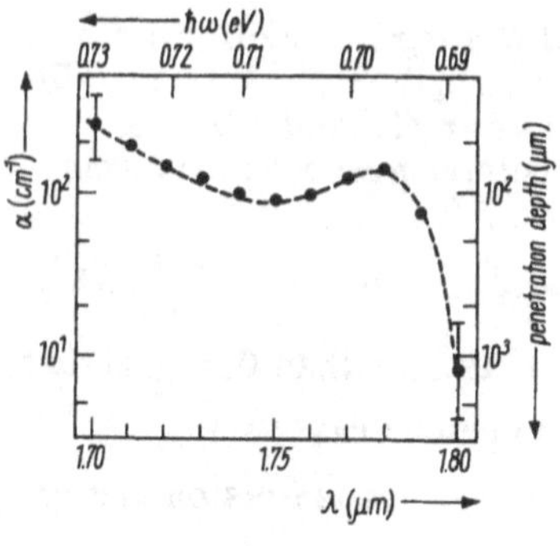

Fig. 2. Absorption coefficient α versus wavelength λ of the ternary alloy T on expanded abscissa showing the absorption anomaly near the bandgap

nary Q(1.2). The data in Fig. 1 can be interpreted without recourse to exciton absorption which becomes observable at lower temperatures /6/ and still more reduced carrier concentration.

The measurements for Q(1.2) and T both revealed a peak in the steep slope of the absorption edge. For the case of the ternary T this peak is displayed with high resolution in Fig. 2. The most likely explanation for this anomaly is in both cases a relatively deep acceptor level at roughly 50 meV from the valence-band edge. This figure can be inferred from a fit of the experimental data to the impurity-absorption formula /12/. A donor can be excluded since the carrier concentration does not change appreciably between liquid nitrogen and room temperature. The acceptors Cd and Zn were recently studied by photoluminescence /13/. But in the present case a contamination by Si /6/ is more probable. However, electron states below the conduction-band minimum and electrically inactive colour centers cannot be fully discarded. Photoconductivity and photoluminescence measurements, presently under way, will hopefully give a further clarification.

Fig. 3 presents the wavelength dependence of n for the alloys Q(1.2) and T as well as for Q(1.3) which was taken from /7/. The measured results join closely the data of Burkhard et al. /3/ in the short-wavelength range and are in good agreement with measurements taken by Chandra et al. /4/ (both not shown here) with similar compositions.

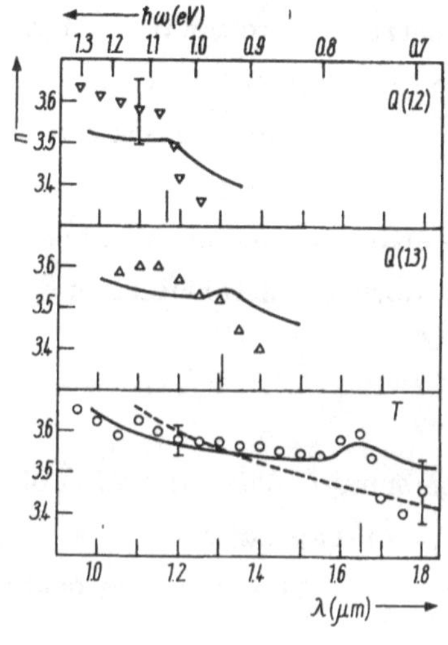

Fig. 3. Wavelength dependence of the refractive index. Solid curves calculated after /3/. Broken curve calculated after /5/; Δ experimental data taken from /7/, ∇ , o present experiments

The wavelengths of the bandgap, as obtained from the procedure previously described, are marked on the abscissa. The solid curves are derived from the weighted sum of the polarizabilities of the four constituent binary alloys, all taken at the same distance from the respective bandgap /3/. This physically plausible interpolation scheme gives the best fit to the experiments (less than 3% discrepancy). The broken curve results from a single-oscillator model with parameters adopted from the two binary constituents of the ternary alloy /5/. Below the bandgap wavelength an empirical modification of the oscillator model /14/ was employed. Although this procedure is less accurate than the interpolation of the polarizabilities after /3/, it has the advantage of only few experimentally determined input parameters. This makes estimates very convenient, particularly since an extension to quaternary compositions is possible /15/. It should be mentioned that an interpolation based on Sellmeier expressions /16/ is only applicable at some distance from the bandgap which is confirmed by a comparison with the present experiments.

Conclusion The optical absorption coefficient α and the refractive index n were measured for a quaternary alloy with bandgap-equivalent wavelength of 1.17 μm and for the ternary alloy, both lattice-matched to InP. α in dependence on the wavelength is characteristic of a direct semiconductor, but with a relative maximum on the order of 50 meV away from the bandgap in the small-absorption range. This peak is tentatively assigned to an acceptor impurity. The refractive index is fitted best by an interpolation scheme given by Burkhard et al. /3/ which adds the polarizability of the constituent components of the system InGaAsP.

The measurements reported here were partly supported by the Deutsche Forschungsgemeinschaft and the Stiftung Volkswagenwerk. The authors gratefully acknowledge this support. Clarifying discussions with Prof. F.-R. Keßler, Braunschweig, and Prof. P. Grosse, Aachen, were very helpful. The careful work of U. Heydecke during the crystal growth is deeply appreciated.

References

/1/ Y. SUEMATSU (Ed.), Optical Devices and Fibers, Japan. Ann. Rev. Electron., Comp. & Telecommun, 1982, Ohmsha-Tokyo, North-Holland, Amsterdam 1982.

/2/ Y.M. BURDUKOV, N.V. ZOTOVA, and K.A. KHALILOV, Soviet Phys. - Semicond. 4, 138 (1970).

G.A. SİKHARULIDZE, L.G. SAKVARELIDZE, and S.G. KONNIKOV, Soviet Phys. - Semicond. 5, 1928 (1972).
D.I. BILENKO and V.D. TSIPORUKHA, Soviet Phys. - Semicond. 13, 922 (1979).
/3/ H. BURKHARD, H.W. DINGES, and E. KUPHAL, J. appl. Phys. 53, 655 (1982).
/4/ P. CHANDRA, L.A. COLDREN, and K.E. STREGE, Electronics Letters 17, 6 (1981).
A.R. CLAWSON, W.Y. LUM, and H.H. WIEDER, J. Soc. Photo-Opt. Instrum. Engn. 132, 2 (1978).
/5/ T. TAKAGI, Japan. J. appl. Phys. 17, 1813 (1978).
/6/ Y.-S. CHEN and O.K. KIM, J. appl. Phys. 52, 7392 (1981).
/7/ W. KOWALSKY, A. SCHLACHETZKI, and F. FIEDLER, phys. stat. sol. (a) 68, 153 (1981).
/8/ L.W. COOK, M.M. TASHIMA, and G.E. STILLMAN, J. electronic Mater. 10, 119 (1981).
/9/ H.H. WIEDER, A.R. CLAWSON, and G.E. MC·WILLIAMS, Appl. Phys. Letters 31, 468 (1977).
/10/ B.O. SERAPHIN and H.E. BENNETT, in: Semiconductors and Semimetals, Vol. 3, Ed. R.K. WILLARDSON, A.C. Beer, Academic Press, New York/London 1967 (p. 499).
/11/ K. SEEGER, Semiconductor Physics, Springer, Wien/New York 1973 (p. 339).
/12/ K. SEEGER, ibid. (p. 356).
/13/ T. NISHINO, Y. YAMAZOE, K. OHTSUKA, Y. FUJIWARA, and Y. HAMAKAWA, in: GaAs and Related Compounds, Japan 1981; Inst. Phys. Conf. Ser. No. 63, Bristol 1982 (p. 149).
/14/ M.A. AFROMOWITZ, Solid State Commun. 15, 59 (1974).
/15/ K. UTAKA, Y. SUEMATSU, K. KOBAYASHI, and H. KAWANISHI, Japan. J. appl. Phys. 19, L137 (1980).
/16/ J. BUUS and M.J. ADAMS, Solid State and Electron. Dev. 3, 189 (1979).

(Received January 27, 1983)

phys. stat. sol. (a) 77, K81 (1983)
Subject classification: 15; 22.1.3

V.T. Division, Central Electronics Engineering Research Institute, Pilani[1)]

Thickness Dependence of the Thermoelectric Power of Tellurium Films[2)]

By
A.K. SHARMA

Recently, Chaudhuri et al. /1/ have reported the thermoelectric power (TEP) of tellurium films (500 to 1 700 Å thick) in the temperature range 298 to 453 K. Their results show an increase in the TEP with film thickness, contrary to others /2, 3/. Goswami and Ojha /2/ have studied the TEP in a few annealed tellurium films in the temperature range 78 to 450 K. They have found that thinner films had higher TEP than thicker ones. Ozkaplan et al. /3/ have also found in general similar results for tellurium films (100 to 2000 Å thick) in non-annealed, quenched, and annealed states. Studies by Phahle /4/ have been limited to the low temperature range (90 to 1 70 K) on annealed tellurium films. In this note the thickness dependence of the TEP of tellurium films (1660 to 4990 Å thick) is being reported under non-annealed and annealed conditions. Generally the TEP of Te films is higher at lower thicknesses, however, it increases with thickness from 1660 to 2120 Å in the non-annealed state.

Thermoelectric power measurements on various tellurium films (1660 to 4990 Å thick) have been done with respect to silver films (always 19000 Å thick). First, tellurium films have been obtained by the vacuum deposition technique onto glass substrates in controlled atmosphere at $3x10^{-5}$ Torr with 25 Å/s deposition rate by resistive heating of pure tellurium powder from molybdenum boat. Then silver electrodes onto Te films have been vacuum deposited using suitable s-steel masks. The silver layer on Te film prevents lateral diffusion of silver into tellurium /5/ at the overlapping portion and to form the Ag_2Te compound. The length to breadth ratio for Te films has been 13:1. For TEP measurements, a temperature difference upto 100 °C along the film has been produced by heating one end and maintaining the other end at

---

1) Pilani 333031, India.

2) Work done at Thin Films Laboratory, Department of Physics, University of Jodhpur, Jodhpur 342 001, India.

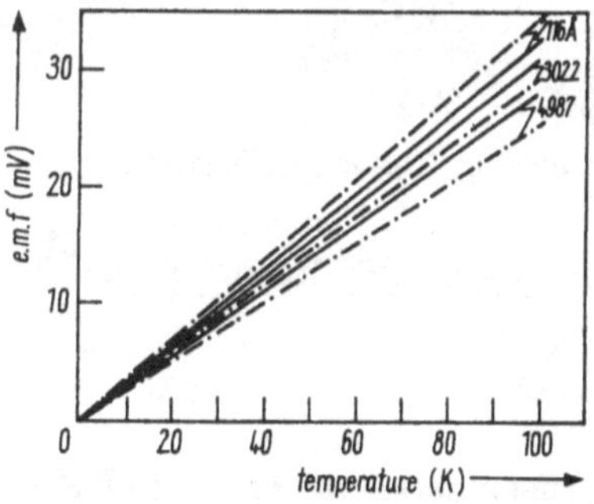

Fig. 1. Thermoelectric e.m.f. versus temperature difference in K; ——non-annealed, ----annealed films

room temperature (33 °C) by a large copper heat sink. The generated thermoelectric e.m.f. has been measured by a potentiometer (1 μV resolution) under non-annealed and annealed (120 °C in air for 1 h) conditions.

Fig. 1 shows the generated thermoelectric e.m.f. in a few tellurium films under non-annealed and annealed conditions for a temperature difference upto 100 °C. It indicates that the TEP does not vary with temperature (33 to 133 °C) in the present experiment and that the Fermi energy is pinned near the centre of the band gap at these temperatures. The polarity observed of the generated thermo e.m.f. suggests p-type characteristics of the Te films.

Fig. 2 shows the thickness dependence of the TEP of Te films under non-annealed and annealed conditions. The TEP of various films has been found to be in general higher at lower thicknesses under non-annealed and annealed conditions as observed by others /2, 3/. On the contrary, the TEP increases from 1660 to 2120 Å under non-annealed condition. It is also observed that annealing of Te films at 120 °C for 1 h in air causes the TEP to increase below 2900 Å thickness and to decrease above this thickness. It is well known that post-deposition annealing reduces the density of lattice defects or carrier concentration with slight change in grain size /4, 6/. Ozkaplan et al. have concluded that in the TEP of tellurium films the charge carrier concentration is more dominant than the carrier mobility, and hence annealing reduces the TEP, which has also been observed by us above 2900 Å film thickness. Below 2900 Å thickness in our case, the surface-space-charge layer (SSCL) which exists on the surface of the semiconductor, extends to a depth of 10 to 10000 Å /7/, possesses different concentrations of charge carriers, mobility /8/, and potential barriers on the grain boundaries /9/, plays a dominant role in increasing the TEP after annealing. Recently, Datta and Chaudhuri /10/ have studied the effect of SSCL on the thermoelectric effect for a non-degenerate semiconductor. The surface-space-charge region arises from the trapping of holes or electrons at the surface due to allowed levels corresponding to localised surface states

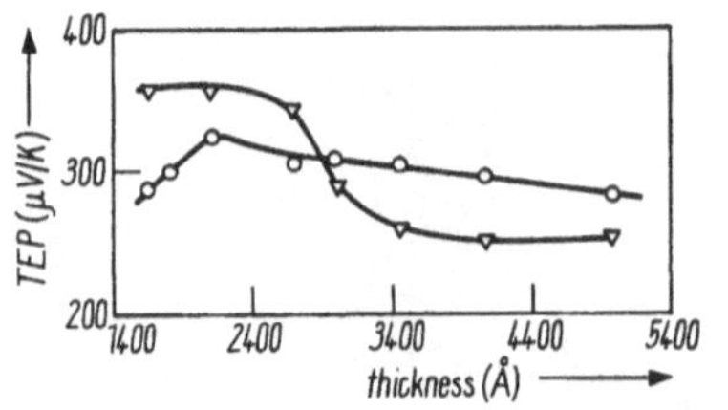

Fig. 2. Thickness dependence of thermoelectric power; -○- non-annealed, -▽- annealed films

in the forbidden gap, deliberates the introduction of impurities in the films or the sensitivity of the films to their environment. In our experiment, Te films are non-crystalline and deposited in a residual vacuum of $3x10^{-5}$ Torr, i.e. in the presence of $1.03x10^{6}$ gas molecules ($\approx 2.16x10^{15}$ oxygen molecules) striking per unit area of the substrate per second. As tellurium is susceptible to oxygen, chemisorption of oxygen molecules at the surface-ambient interface or even at the substrate-film interface causes to form the SSCL /11/. It has been found that annealing of Te films upto 100 °C for 3 h in air reduces the surface potential to zero due to the formation of surface adsorption layer of TeO_2 /12/. Also, the effective conductivity reduces. Hence, according to the well known relation

$$\alpha = A - B \log \sigma$$

(α THP, σ conductivity) TEP increases, which has been observed by us for Te films of thickness <2900 Å when these have been annealed in air for 1 h at 120 °C. The conductivity of various Te films, obtained separately under similar ambient conditions and deposition parameters is shown in Table 1. The conductivity at lower thicknesses does not change appreciably after annealing.

The TEP method together with conductivity data also permits us to estimate the carrier mobility in low mobility films /13/. The carrier concentration and mobility have been estimated in our Te films under non-annealed and annealed conditions from the following formulas:

$$\alpha = \frac{K_B}{e}\left[A + \ln\left(\frac{N_h}{n}\right)\right], \quad /13/,$$

where the scattering factor A = 4 /1/, the effective density of states in the valence band $N_h = 2(2\pi m^* K_B T/h^2)^{3/2}$; $m^* = 0.45\ m_o$ /14/, and $\mu = \sigma/n\,e$. The results are shown in Table 1.

Table 1

thickness (Å)	conductivity σ ($10^{-3}\,\Omega^{-1}\,cm^{-1}$)		carrier concentration n (10^{18} cm^{-3})		mobility μ ($10^{-3}\,cm^2\,V^{-1}\,s^{-1}$)		$n\mu$ (10^{16} $cm^{-1}\,V^{-1}$)	
	non-ann.	annealed	non-ann.	annealed	non-ann.	annealed	non-ann.	annealed
1660	3.72	4.37	15.4	6.7	1.5	4.0	2.31	2.68
2110	11.48	15.14	10.0	6.7	7.18	14.12	7.18	9.46
2710	25.70	39.87	12.2	7.88	13.16	31.6	16.65	24.94
3020	31.54	53.7	11.9	1.50	16.56	223.7	19.7	33.55
3470	52.48	77.62	12.8	2.13	25.62	227.7	37.79	48.5
4080	79.43	104.71	14.2	2.34	32.75	279.6	46.51	65.42

The product of carrier concentration n and mobility μ decreases with thickness and annealing in Table 1, which suggests that potential barriers at microcrystals in Te films decrease with film thickness and post-deposition annealing /15/. Thus we conclude that at lower thicknesses, the influence of SSCL is more dominant in the thermoelectric properties of Te films.

References

/1/ S. CHAUDHURI, B. CHAKRABATI, and A.K. PAL, Thin Solid Films 82, 217 (1981).

/2/ A. GOSWAMI and S.M. OJHA, Thin Solid Films 16, 187 (1973).

/3/ H. OZKAPLAN, M.A. DINNO, and M. SCHWARTZ, J. appl. Phys. 50, 1533 (1979).

/4/ A.M. PHAHLE, Thin Solid Films 41, 235 (1977).

/5/ D.B. JOHNSON and L.C. BROWN, J. appl. Phys. 40, 1711 (1969).

/6/ K. OKUYAMA, T. YAMASHITA, M. CHIBA, and Y. KUMAGAI, Japan. J. appl. Phys. 16, 1571 (1977).

/7/ J. BARDEEN, Phys. Rev. 71, 717 (1947).

/8/ R.L. PETRITZ, Phys. Rev. 110, 1254 (1958).

/9/ J.C. ANDERSON, Adv. Phys. 19, 311 (1970).

/10/ S.K. DATTA and A.K. CHAUDHURI, J. Phys. D 14, L141 (1981).

/11/ A. KUBOVÝ and M. JANDA, phys. stat. sol. (a) 35, 471 (1976).

/12/ A. KUBOVÝ and M. JANDA, phys. stat. sol. (b) 37, K127 (1976).

/13/ R.H. BUBE, Ann. Rev. Mater. Sci. 5, 201 (1975).

/14/ I.N. TIMCHENKO and S.S. SHALYT, Soviet Phys. - Solid State 4, 2642 (1963).

/15/ Y. YASUOKA and T. MIYATA, Japan. J. appl. Phys. 18, 1397 (1979).

(Received March 24, 1983)

phys. stat. sol. (a) 77, K87 (1983)

Subject classification: 18.2; 22.8.2

Hungarian Optical Works, Budapest (a), Institute of Physics, Czechoslovak Academy of Sciences, Prague[1] (b), and Central Research Institute for Physics, Budapest[2] (c)

Temperature Hysteresis of Demagnetized Domain Structures

By

G. VÉRTESY (a), I. TOMÁŠ (b), Z. VÉRTESY (c), and M. BALASKÓ (c)

The domain pattern being strongly dependent on the basic magnetic parameters is an important characteristic of magnetically ordered materials. It is sensitive to shape and size of the samples which sometimes renders it complicated for a theoretical description. In case of a simple geometry, however, the calculation can be performed rather easily. Such is the case with magnetic bubble layers where the lowest energy equilibrium period p_o of stripe domains /1/ and the collapse field of bubbles /2/ are commonly used for the calculation of the characteristic material length and the saturation magnetization, respectively.

In order to measure p_o the sample has to be brought to its equilibrium state, i.e. to be demagnetized, when the width d_1 of the domains magnetized "up" is equal to the width d_2 of those magnetized "down". This is usually achieved in an ac field with a slowly decreasing amplitude normal to the surface of the specimen and $p_o = d_1 + d_2$ is then measured by an optical method. If the stripe structure is not well "shaken up" by the ac field its parameters are generally different from the equilibrium ones and reflect the specimens latest magnetic and temperature history.

We have measured the stability of non-equilibrium demagnetized ($p \neq p_o$, $d_1 = d_2$) domain structures by the way of recording the temperature dependence of the stripe domain period. The studied samples were thin films of magnetic garnets grown by liquid phase epitaxy on GGG {111} substrates. The layers were placed in a sample holder in which the temperature was changed from -190 up to +200 ^{o}C. The period was measured using the first-order diffracted light from the domain pattern.

---

1) Na Slovance 2, 18040 Prague 8, Czechoslovakia.

2) P.O.B. 49, 1525 Budapest, Hungary.

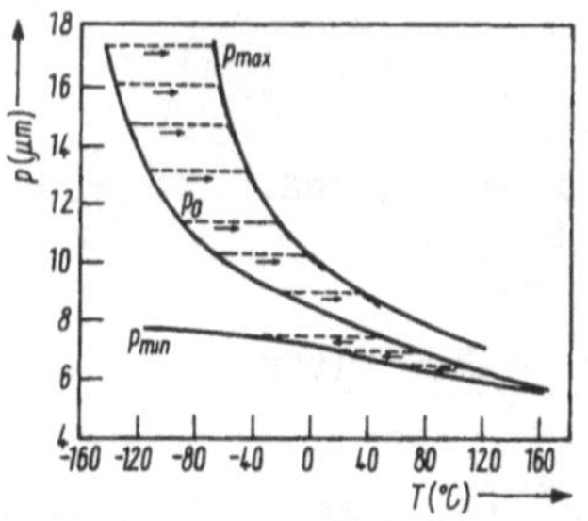

Fig. 1. Temperature hysteresis of the stripe domain period of the demagnetized sample. See text

The results can be described as a temperature hysteresis of the stripe domain period typically shown in Fig. 1. Here p_0 shows the temperature dependence of the equilibrium anhysteretic domain period measured after "shaking up" the structure in a normal ac field. If the temperature of the sample is changed without the ac demagnetizing field, hysteresis appears. For instance, starting from an equilibrium p_0 at $T = -140\ ^{\circ}C$ the domain structure remains unchanged until $T \approx -66\ ^{\circ}C$. At about this point the structure becomes unstable and any further increase of temperature brings about a rather smooth decrease of the period. If the starting point is the equilibrium period p_0 at a higher temperature, the new hysteresis curve usually overlaps the lower temperature hysteresis a bit. The curve connecting the critical breakdowns of period is called p_{max} in Fig. 1, as it represents the maximum demagnetized stable period ($d_1 = d_2$) which can be achieved at any given temperature. Similar hysteresis can be observed while decreasing the temperature from an equilibrium p_0 giving rise to the p_{min} curve which describes the temperature dependence of the minimum demagnetized stable period.

The curves in Fig. 1 show, on the one hand, that at a fixed temperature T_0 the sample can have zero magnetic moment with a domain structure period p within the limits $p_{min}(T_0) \leqq p \leqq p_{max}(T_0)$ (vertical cross section in Fig. 1) and, on the other hand, that the sample having an equilibrium period p_0 at a temperature T_0 would keep the same period in an appreciable range of temperatures $T(p_{min}) \leqq T \leqq T(p_{max})$ (horizontal cross section in Fig. 1). In devices which make use of domain structures of ferromagnetic materials the stability and parameters of the pattern play an important role. As it follows from Fig. 1 both the period of the stripe domain structure and its stabilitity limits can be conveniently influenced by a proper thermal and magnetic treatment. A demagnetized non-equilibrium domain structure tailored in such a way to a specific demand makes it possible to use a pattern being stiff and stable in a broad temperature range even in materials with a rather low value of domain wall coercivity. A similar behaviour of a close packed bubble array has been lately reported by Chodosov et al. /3/.

The coercivity which is supposed to be the principal cause of the hysteresis can be measured by quite a number of methods each of them giving somewhat different results. These differences are caused not only by the respective method of measurement used but to a large extent by the fact that different methods measure closely related but nevertheless somewhat different quantities.

The coercivity of garnet layers can be understood either classically as the normal field necessary to bring the sample from a remanent state to its zero net magnetic moment /4/ or the minimum normal field needed for moving the domain walls from the equilibrium while decreasing the width of one set of domains at the sake of the other /5 to 7/ or the minimum field difference across the diameter of a bubble necessary for the bubble to move /8/ and the like. In each case a characteristic quantity is obtained and it is just the ultimate use of the parameter which determines the most suitable choice of a method.

In each of those coercivity measurements actually the "friction" is registered opposing the sample's endeavour under the external condition to reach its minimum energy state by the way of shifting the domain walls in well defined directions as the magnetic pressure on one side of the walls is larger than that on the other side: the "up" domains grow while the "down" ones decrease, the bubble walls are shifted along the field gradient etc.

This is in contrast to our situation of a parallel stripe domain structure in zero external field if $d_1 = d_2$ but $p \neq p_0$. Now too, the energy E(p) is larger than the minimum value $E(p_0)$ but the only way how to decrease it is via a change of the period p. The pressure in both "up" and "down" domains is the same. Keeping the stripes in an infinite sample strictly parallel to each other there is no preferred direction of motion of the domain walls. The period could be changed by nucleation or annihilation of walls only. In reality the rather high energy barrier of the creation or extinction of domain walls is usually circumvented via changing the average period p by the winding and/or unwinding of the existing labyrinth pattern. However, the high stiffness of a well stabilized labyrinth domain structure under the above conditions apparently reflects (Fig. 1) the original indecisionof the domain walls as for "ewhere to move" even in this case.

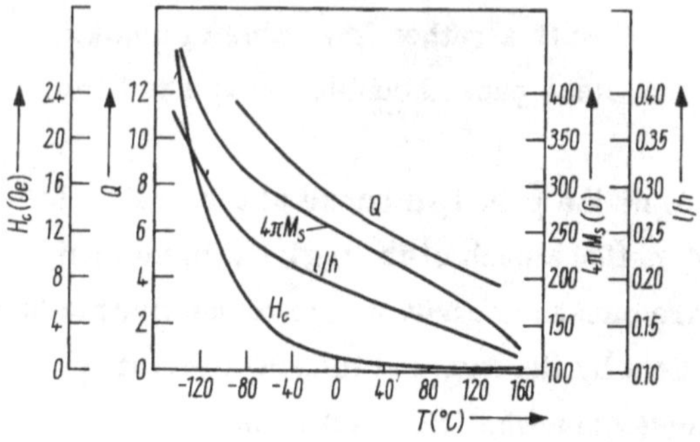

Fig. 2. Material parameters of the sample $Lu_{0.24}Sm_{0.27}Y_{1.52}Ca_{0.96}Fe_{4.01}Ge_{0.99}O_{12}$. Q quality factor, $4\pi M_s$ saturation magnetization, l characteristic length, H_c coercivity, h thickness of the sample

We have measured the effect of the temperature hysteresis of demagnetized stripe domain structures in a number of specimens with different composition and thickness and its connection with the coercivity and/or other material characteristics is presently studied. The sample studied in thepresent note is a garnet layer 2.8 m thick, having chemical composition $Lu_{0.24}Sm_{0.27}Y_{1.53}Ca_{0.96}Fe_{4.01}Ge_{0.99}O_{12}$. The magnetic parameters of the sample measured in a wide range of temperatures are given in Fig. 2.

References

/1/ Z. MÁLEK and V. KAMBERSKÝ, Czech. J. Phys. 21, 416 (1958).

/2/ A.A. THIELE, Bell Syst. tech. J. 48, 3287 (1969).

/3/ E.F. CHODOSOV, A.O. CHREBTOV, and YU.A. SIRYUK, Zh. tekh. Fiz., Pisma 8, 363 (1982).

/4/ D.J. CRAIK, P.V. COOPER, and G. MYERS, J. Phys. D 6, 872 (1973).

/5/ G.P. VELLA-COLEIRO, W.B. VENARD, and R. WOLFE, IEEE Trans. Magnetics 16, 625 (1980).

/6/ R.M. JOSEPHS, AIP Conf. Proc. 10, 286 (1972).

/7/ J.A. SEITCHIK, W.D. DOYLE, and G.K. GOLDBERG, J.appl. Phys. 42, 1272 (1971).

/8/ G.P. VELLA-COLEIRO and W.J. TABOR, Appl. Phys. Letters 21, 7 (1972).

(Received April 14, 1983)

phys. stat. sol. (a) 77, K91 (1983)

Subject classification: 1.6; 22.1.2

A.V. Shubnikov Institute of Crystallography, Academy of Sciences of the USSR, Moscow[1]

Investigation of the X-Ray Scattering Intensity for the Laue-Case Diffraction under Total-External-Reflection Conditions

By

A.L. GOLOVIN and R.M. IMAMOV

A diffraction scheme based on a new principle that allows one to obtain concrete information on the structure of thin surface layers with a thickness of the order of tens of nanometers was proposed in /1/. This scheme combines the requirements of diffraction according to the Laue geometry with conditions where the incident and diffracted beams are reflected from the crystal surface at small angles. The measurement of the angular dependence of the intensity of a diffracted beam which has suffered specular reflection requires a high precision and performance of the experimental set-up - i.e. high collimation of the incident beam - since the measurements should be carried out in a small range of deviations α from the exact Bragg condition. The proposed technique also takes into account the requirements limiting the vertical beam divergence which makes possible measurements at small angles of incidence of an X-ray beam on a crystal.

The theory of diffraction scattering of X-rays by ideal crystals under total-external-reflection conditions has been developed in /2/. It was shown that at small angles of incidence there exist definite directions in which the intensity of the specularly reflected diffracted wave is close to the incident wave intensity. Analytical expressions were obtained for the shape, intensity, and width of the reflection curve. It turned out that for diffraction curve measurements only high collimation through the angle of incidence of X-rays on the crystal Φ is sufficient. Thus, when measuring the intensity of the specularly reflected diffracted wave as a function of Φ_h, the angle of emergence of an appropriate wave from the entrace surface of the crystal, or of Φ, one can establish the dependence of the diffraction curves on the parameters α. This dependence is based on the rigid relation between Φ, Φ_h, and α, i.e., $\Phi^2 = \Phi_h^2 + \alpha$ /2/.

---

1) prospekt Leninski 59, 11733 Moscow, USSR.

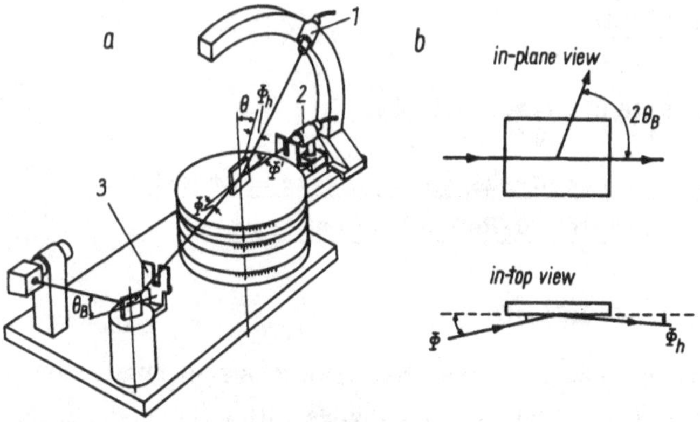

Fig. 1. X-ray experimental arrangement scheme: a) general view of the experimental arrangement; 1 and 2 detectors for recording the diffracted and specularly reflected waves, respectively, 3 slit for "dissecting" the $K\alpha_1$ line out of the CuKα X-ray spectrum; b) scheme of beam trajectory

This is very important for the experimental performance of measurements. In such a way, the measurements through the angles Φ or Φ_h with an accuracy of 30" which may, undoubtedly be achieved /3/, corresponds to an accuracy of the measurement through $\delta\Theta = \delta\alpha/2 \sin\Theta_B \approx 0.1''$ $(\delta\alpha = -2\Phi_h \delta\Phi_h)$.

To solve this problem, we propose a new experimental scheme which allows one to measure the intensity of specularly reflected waves under total-external-reflection conditions depending on the angle Φ or Φ_h. The measurement of the dependence of specularly reflected diffracted waves on the angle of incidence Φ in the integral regime, i.e. when recording the intensity of waves emerging at different angles from the entrance surface of the crystal, Φ_h, is the simplest one. These curves may be used to obtain some preliminary information on the structure of a surface layer. The results of such measurements on silicon samples are given below.

Fig. 1 shows the X-ray scheme of the experimental arrangement. The X-ray beam from the source (1.2 kW) is directed on to the crystal monochromator at the Bragg angle: silicon (220 reflection, CuKα radiation). The beam reflected from the crystal monochromator is incident on the investigated crystal (silicon) at a small glancing angle, Φ, experiencing the effect of Laue diffraction on the planes (220) perpendicular to the entrance surface of the crystal. The diffracted waves with different angles of emergence from the

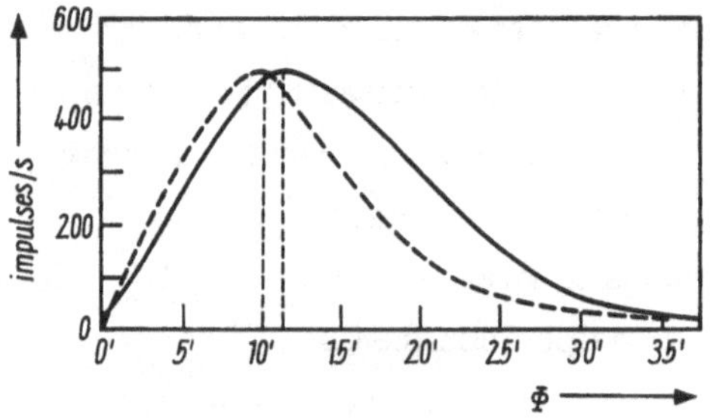

Fig. 2. Typical diffraction pattern from an ideal crystal; —— experiment (solid line), – – – theoretical curve

entrance surface of crystal, Φ_h, are recorded by detector (1). Between the crystal monochromator and the crystal under investigation there is a slit (3) which "dissects" the line $K\alpha_1$ out of the $CuK\alpha$ X-ray spectrum. The size of this slit determines the front width of the beam falling on the crystal (10 to 50 μm). The X-ray beam divergence in the horizontal plane is defined by the reflection curve from the first crystal and is measured to be $\approx 0.1'$. Apparently, the small value of the glancing angle (the angle between the crystal surface and the diffraction plane) may be easily obtained in real experiments.

In the vertical plane, collimation is not provided and the X-ray beam divergence is limited by the construction of the spectrometer. In the experimental arrangement the vertical divergence is limited to about $\approx 1^{\circ}$.

The specularly reflected wave is recorded by detector (2) with a slit of about 10 μm width which allows the transmitted and specularly reflected waves to be recorded separately.

The diffraction curves are measured on a goniometer fitted in the three-crystal spectrometer TRS. A typical diffraction pattern from a perfect Si crystal that shows the dependence of the specularly reflected wave intensity on the angle Φ (i.e. for all Φ_h in the 30' interval) is given in Fig. 2 (solid line). The same figure shows (dashed line) the theoretical curve which corresponds well with the experimental curve in the shape, size, and angular position of the intensity maximum. Comparison of the curves in Fig. 2 made it possible to reveal some peculiarities in the experimental curve: a more gentle sloping and some displacement of the angular position of the maximum in the right part. Evidently, this is due to the non-ideality of the near-surface layer of the crystal and its insufficient account in the theory of plane-polarized radiation falling on the crystal. Thus, the possibility of measuring the dependence of the specularly reflected diffracted wave intensity on the angle of incidence, has been experimentally realized. For this purpose a 1.2 kW source is sufficient. The next stage in the development of the technique of this experiment should be the measurement of the dependence of the specularly reflected diffracted wave intensity in the differential regime. However, the singling out of an appreciably narrow range of Φ_h angles would require a more powerful source.

The authors express their sincere thanks to A.M. Afanasev and M.V. Kovalchuk for unfailing attention to the work and helpful comments.

References

/1/ W.C. MARRA, P. EISENBERGER, and A.Y. CHO, J. appl. Phys. 50, 6927 (1979).

/2/ A.M. AFANSEV and M.K. MELKONYAN, to be published in Acta cryst., A 39 (1983).

/3/ E.K. KOVEV and YU.A. MATVEEV, Fiz. tverd. Tela 23, 587 (1981).

(Received January 17, 1983)

phys. stat. sol. (a) 77, K95 (1983)
Subject classification: 1.2; 14; 20

Department of Solid State Cryophysics, Institute of Experimental Physics, University of Wrocław[1)]

Some Electrical and Optical Effects Observed in Polycrystalline Ice at 200 K[2)]

By
K. BIEDRZYCKI, J. CHRZANOWSKI, and B. SUJAK

Phase transformations and associated effects in ice are still under study for their great importance from the point of view of hydrogen bond understanding. A number of papers have been already devoted to the phase transformation in ice and the associated electrical phenomena /1 to 7/.

Three modifications of ice are usually obtained and examined /8, 9/. In the very low temperature range isothermally prepared ice films are mainly amorphous. Above ≈120 K, almost within the region 120 to 150 K, the amorphous ice spontaneously transforms to the cubic form. Between ≈150 and 200 K thin ice layers condensed within this temperature range, usually exist in a cubic modification. Even ferroelectric behaviour of cubic ice has been suggested /1, 2/. Above this temperature range the hexagonal modification of ice layers can be formed only. The phase transformation to the hexagonal form at about 200 K, appearing when cubic ice is warmed-up, is accompanied by various effects as described in several papers /2 to 6, 10/.

In this short note some experimental results concerning effects appearing near 200 K in polycrystalline ice samples are presented.

In Fig. 1 thermally stimulated dc current curves plotted versus reciprocal temperature are shown for an ice layer of thickness 30 μm. The temperature dependence of the measured current I in the studied temperature range is exponential: $I \sim \exp(-E_T/k_B T)$, where I current intensity, k_B Boltzmann constant, T absolute sample temperature, E_T thermal activation energy for dc electrical conductivity.

---

1) ul. Cybulskiego 36, 50-205 Wrocław, Poland.

2) Sponsored in part by the Polish Academy of Sciences within the Main Scientific Project.

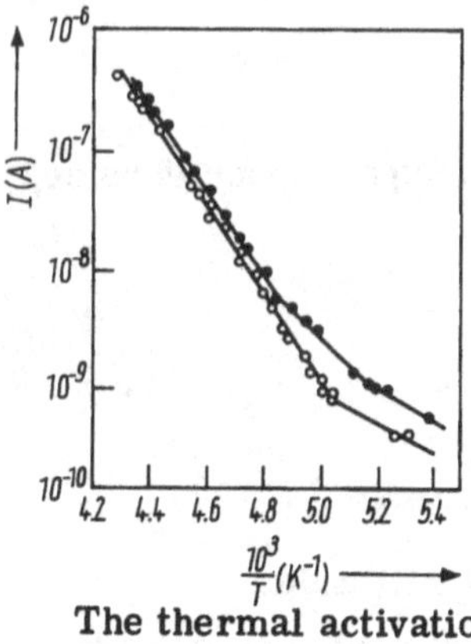

Fig. 1. Thermally stimulated dc current curve for a 30 μm thick ice polycrystalline sample plotted against the reciprocal temperature

The ice layers studied were always obtained by deposition at a vacuum of about 10^{-6} Torr, onto a substrate cooled down to LN_2 temperature. The deposited ice layers have been then heated up to a respective temperature with a constant rate of 3 K/min.

The thermal activation energy for the electrical conductivity is changing its value significantly at about 200 K from E_T^{I} = 0.35 eV (cubic form) to E_T^{II} = 0.95 eV (hexagonal form), Fig. 1 (open circles). In some cases a third activation energy of 0.52 eV has been observed, Fig. 1 (full circles).

In the case of condensation of ice layers onto a substrate cooled down to 192 to 202 K, the isothermally formed ice thin films show a positive electric surface charge (Fig. 2). The surface charge density is estimated to be $\approx 10^{-10}$ C/cm^2 at 195 K. Outside the temperature range 192 to 202 K only a negatively charged ice surface has been always observed.

As is known, only the hexagonal modification of ice is energetically stable. The cubic → hexagonal transition is irreversible. Therefore, effects associated to that transformation in ice should not be observed at 200 K during cooling or heating of the hexagonal form. This supposition is supported by the hysteresis processes appearing in the current and surface charge measurements. However, some of our optical measurements, described below, suggest additional processes, independent of the above discussed phase transformation, and appearing just at 200 K in the hexagonal form of ice.

A light emission by hexagonal polycrystalline ice samples when polarized by an alternating voltage, in the temperature range 190 to 210 K, has been observed and examined as a function of temperature. The studied samples were formed by water freezing in a dielectric vessel with a

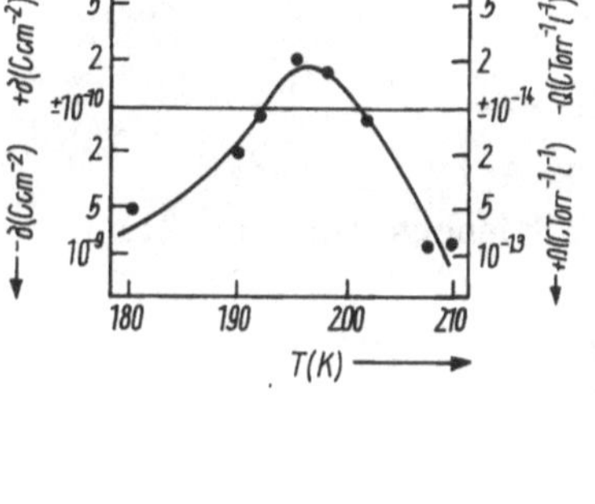

Fig. 2. Electric surface charges observed on thin ice samples formed isothermally within the range 180 to 210 K

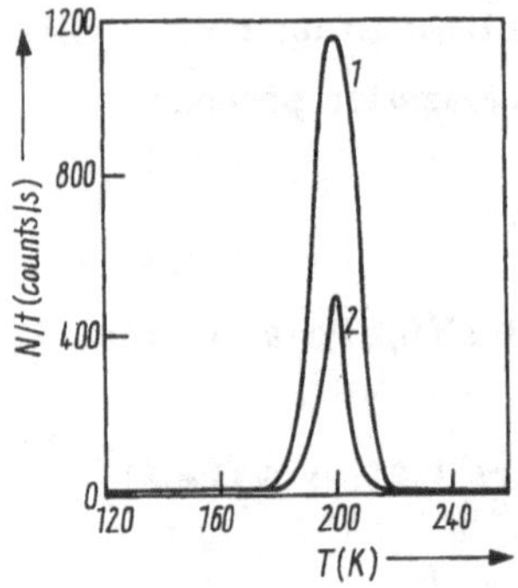

Fig. 3. Typical temperature dependence of the light emission intensity for the studied ice samples

metallic bottom. The thickness of the used samples was about 1 mm. To the upper ice surface a chromium-nickel grid electrode was put (pressed) on. Between the grid and the metallic bottom of the vessel a polarizing voltage was applied. The observed light emission was registered by an appropriate arrangement with a photomultiplier situated above the upper ice sample surface. The used photomultiplier worked in the pulse regime.

In Fig. 3 a typical temperature dependence of the registered light emission intensity from the studied ice samples is shown. The ice samples were polarized by sinusoidal voltage of amplitude about 600 V and frequencies 40 (curve 1) and 4000 Hz (curve 2), respectively. As is shown (Fig. 3), a maximum of the observed light emission intensity appears near 200 K. The maximum light intensity depends on the frequency of polarizing voltage and decreases for the highest frequency used. The observed maximum light intensity at 200 K did not change significantly during multiple cooling and re-heating of the same sample. The light emission near 200 K was observed only when the sample was warmed-up with simultaneous polarization of the sample by an alternating voltage with amplitude higher than 400 V. The observed maximum of the light emission for various ice samples was usually situated within the range 190 to 210 K. It is worth noting that Perez et al. /10/ observed a maximum of internal friction in hexagonal ice at about 190 K. Also von Hippel et al. /6/ observed a relaxation process in ice at 193 K. This observed effect has been assigned to H_2O molecule orientation at the grain boundaries.

The temperature of the maximum of the light emission intensity coincides with the temperature at which the surface charge of the ice layer changes its sign in the case of the isothermally grown ice thin films. The observed change of the thermal activation energy for the dc electric conductivity also appears at the very same temperature.

It seems, therefore, that the observed effect of light emission may be a consequence of some relaxation processes additionally stimulated by the external alternating voltage not strictly connected with the known cubic → hexagonal phase transformation in ice. The decrease of the light emission intensity for higher frequencies of the applied voltage suggests relaxation processes of rather great time constants.

References

/1/ O. DENGEL, U. ECKENER, H. PLITZ, and N. RIEHL, Phys. Letters 9, 291 (1964).

/2/ G. CUBIOTTI and R. GERACITANO, Phys. Letters A 24, 179 (1967).

/3/ J. CHRZANOWSKI, Ph.D. Thesis, Wrocław 1977.

/4/ J. CHRZAŃOWSKI and B. SUJAK, Thin Solid Films 79, 101 (1981).

/5/ G.P. JOHARI and S.J. JONES, J. chem. Phys. 62, 4213 (1975).

/6/ A. VON HIPPEL, R. MYKOLAJEWYCZ, A.H. RUNK, and W.B. WESTPHAL, J. chem. Phys. 57, 2560 (1972).

/7/ J. CHRZANOWSKI and B. SUJAK, submitted to Thin Solid Films.

/8/ F.V. SCHALLCROSS asnd G.B. CARPENTER, J. chem. Phys. 26, 785 (1957).

/9/ L.G. DOWELL and A.P. RINFERT, Nature 188, 1144 (1960).

/10/ J. PEREZ, J. TATIBOUET, R. VASOILE, and P. GOBEN, in: Mekhanizmy vnutrennego treniya v tverdykh telakh, Izd. Nauka, Moscow 1974.

(Received January 14, 1983)

Pre-Printed Titles

Papers to be published
in "physica status solidi (a) - applied research"
Vol. 77, No. 2

==

Original Papers

A.N. ANAGNOSTOPOULOS, C. MANOLIKAS, and D. PAPADOPOULOS, Thessaloniki
Comparison of the Structure and the Electric Properties of $ZnIn_2S_4$(III) - and $CdInGaS_4$ - Layered Crystals

J. CLAVERIE, G. CAMPET, D. CONTE, G. LE FLEM, and P. HAGENMÜLLER, Talence
Influence of Covalent Bonding on the Photoelectrochemical Properties of Some Perovskite-Type Related Compounds

M.S. BRODIN, N.I. VITRIKHOVSKII, A.A. KYPEN, S.G. SHEVEL, and N.I. YANUSHEVSKII, Kiev
Spatial and Spectral Characteristics and a New Model of Laser Generation for CdS-Type Single Crystals under One-Photon Excitation

J.L. DORMANN, A. TOMAS, and M. NOGUES, Meudon
Cation Ordering in $LiFe_5O_8$ Studied by Mössbauer Spectroscopy and X-Ray Crystallography

L. NANAI, I. HEVESI, L. MICHAILOVITS, I. APOSTEL, M. DINESCU, I.N. MIHAILESCU, and A.V. SIDORIN, Szeged, Moscow, and Bucharest
Damage of V_2O_5 Single Crystals by Powerful Pulsed CO_2 Laser Irradiation

R.Z. VALIEV, V.YU. GERTMANN, and O.A. KAIBYSHEV, Ufa
Non-Equilibrium State and Recovery of Grain Boundary Structure (II)

A.J. COHEN and G.G. JANEZIC, Pittsburgh, Pennsylvania
Relationships among Trapped Hole and Trapped Electron Centers in Oxidized Soda-Silica Glasses of High Purity

V.M. ORERA, V.D. RODRIGUEZ, R. ALCALÁ, and Y. CHEN, Zaragoza
Thermoluminescence of X-Irradiated CaO and CaO:Li Single Crystals

Z. GIERAK, J.W. MOROŃ, and A. LEHR, Katowice
Recovery of Structural Defects in Molybdenum

C.M. VAN BAAL, Delft
The Influence of Texture on the X-Ray Determination of Residual Strains in Ground or Worn Surfaces

N. KURAMOCHI, H. MIZUBAYASHI, and S. OKUDA, Ibaraki
Dislocation Breakaway at Low Temperatures in Nb

H. RATAJCZAK, P. SLANČO, T. TIMA, A. ZENTKO, E. DEMKÖVICSOVÁ, and M. KONČ, Poznań and Košice
Electrical Resistivity and Hall Effect in Thin Amorphous UFe Films

K.N. REDDY, MD. IQBAL AHMED, U.V. SUBBA RAO, and V. HARI BABU, Secunderabad and Hyderabad
Thermoluminescence and Optical Absorption Studies of Z_1-Centres in NaCl Crystals Doped with Terbium

B. RHEINLÄNDER, G. OELGART, R. REULKE, H. HAEFNER, and R. STEGMANN, Leipzig and Berlin
Degradation of Yellow Light-Emitting $GaAs_{0.1}P_{0.9}$:N Diodes

J. ZHANG and J.C. THUILLIER, Dijon
The Effect of Annealing Temperature on the Electron Concentration of $Hg_{0.3}Cd_{0.7}Te$

M.A. ANGADI and P.V. ASHRIT, Trinidad and Dharwad
The Effect of Diposition Parameters on the Electrical Properties of Thin Ytterbium Films

P.K.C. PILLAI, G.K. NARULA, A.K. TRIPATHI, and R.G. MENDIRATTA, New Delhi
A Study of Absorption Currents in Polycarbonate: Polypropylene Blend

H. SODOLSKI, Gdańsk
DC Conductivity and Electric Relaxation Currents in a Polyester Polymer

R.N. LITOVSKII, V.S. LYSENKO, A.N. NAZAROV, T.E. RUDENKO, S.B. KASCHIEVA, and N.R. NEDEV, Kiev and Sofia
The Influence of Defect Surface Layers on the Capacitive Properties of MOS Structures

I.E. TÜRE, F. POULIN, A.W. BRINKMAN, and J. WOODS, Durham
Electron Traps and Deep Levels in Cadmium Selenide

J. WOLLBRANDT, U. BRÜCKNER, and E. LINKE, Berlin
Investigations of Mechanically Induced Excited States on Cleavage Planes of Ionic Crystals (I)

SH. CHADRAABAL, A.S. POPOV, and D.B. KUSHEV, Sofia
Photoconductivity in n-Type InP:Fe

P. GAWORZEWSKI and K. SCHMALZ, Berlin
Oxygen-Related Donors Formed at 600 ^{o}C in Silicon in Dependence on Oxygen and Carbon Content

B. BRIDGE, N.D. PATEL, and D.N. WATERS, Uxbridge, Middlesex
On the Elastic Constants and Structure of the Pure Inorganic Oxide Glasses

M.M. CHANDRA and M. PRASAD, Bangalore
Schottky Barrier Characteristics at Low Temperatures

K.E. MIRONOV, V.K. OGORODNIKOV, V.D. ROZUMNYI, and V.I. IVANOV-OMSKII, Leningrad
Closed System LPE Growth of $Cd_xHg_{1-x}Te$

B. PISTOULET, D. COQUILLAT, J.C. TEDENAC, G. BRUN, and M. MAURIN, Montpellier
Hole Effective Mass and Impurity Levels in Undoped AgTlTe. Optimum Thermoelectric Figure of Merit

A. SZEKERES and S. ALEXANDROVA, Sofia
RF Annealing of Defects Induced in SiO_2 by Oxygen Plasma

M.K. AGARWAL, P.D. PATEL, and O. VIJAYAN, Gujarat
Electrical Studies on $(Mo/W)Se_2$ Single Crystals

R.N. MUKHERJEE, D.N.S. SRIVASTAWA, and S.C. SEN, Kanpur
Ultraviolet Absorption of Tl^+ Ion in K_2O-KBr-B_2O_3 and Na_2O-NaBr-B_2O_3 Glasses

D. PANKNIN, E. WIESER, R. KLABES, and H. SYHRE, Dresden
Dose Dependence of the Flash Lamp Annealing of Arsenic-Implanted Silicon

V.E. ANTONOV, T.E. ANTONOVA, I.T. BELASH, E.G. PONYATOVSKII, and V.I. RASHÚPKIN, Chernogolovka
The Pd-Pt-H System: Phase Transformations at High Pressure and Superconductivity

A.I. PROKHVATILOV and A.P. ISAKINA, Kharkov
Lattice Parameters, Thermal Expansion Coefficients, and Vacancy Density in Solid CD_4

D.W. FIELD, Brisbane
A Reexamination of Diffraction Data in Zinc to Determine Anharmonic Components in the One-Particle Potential Model

G.A. ANDREEV and V.A. KLIMOV, Leningrad
Preferential Precipitation on Extended Edge Dislocations in NaCl Crystals

J.C. LOU, M.S. LIN, J.I. CHYI, M.D. LIN, and S. SCHE, Hsinchu and Hsinfon
Electrical and Optical Characteristics of $GaAs_{0.6}P_{0.4}$ LEDs Fabricated by Zn Semi-Closed Diffusion Method

O.V. PRESNYAKOVA, V.I. ZAITSEV, and V.A. FOMCHENKO, Donetsk
Creep of Aluminium Single Crystal under High Pressure

E.W. KREUTZ, Darmstadt
Size Effects on the Magnetoresistance of p-InSb

P.A. BEZIRGANYAN and V.G. ASLANYAN, Erevan
About a Feature of X-Ray Diffraction in Imperfect Crystals

N.P. KALMYKOVA, B.T. KOLOMIETS, V.KH. SHPUNT, and E.A. SMORGONSKAYA, Leningrad
On the Influence of the Non-Homogeneous Structure on the Current-Voltage Behaviour of Chalcogenide Vitreous Semiconductors

A.S. ALEKSANDROV, V.F. ELESIN, and M.P. KAZEKO, Moscow
The Influence of Defects on the Low-Temperature Resistivity of Intermetallic Compounds with A-15 Structure

T. KACZOR, S. NOWAK, and J. ŻMIJA, Radom and Warszawa
Description of Coloration and Discoloration Processes in Inorganic Photochromic Materials

G.D. BAGRATISHVILI, R.B. DZHANELIDZE, and D.A. JISHIASHVILI, Tbilisi
Nonstoichiometric Germanium Nitride - A New Insulating Material for MIS Microelectronics (I)

Papers to be published
in "physica status solidi (b) - basic research"
Vol. 117, No. 2

Review Article

M. CARDONA, Stuttgart
Vibrational Spectra of Hydrogen in Silicon and Germanium

Original Papers

I. SILDOS, O. SILD, and I. DOLINDO, Tartu
Manifestation of Spin-Rotation Coupling in Intersystem Crossing in a Rotating Impurity Molecule

H. LUCHIAN, Iasi
Magnetic Symmetry in Several Cubic Compounds

P.K. ACHARYYA, A.K. TEWARI, S.K. ROY, and D.L. BHATTACHARYA, Serampore and Calcutta
Anharmonic Study on the Debye-Waller Factor for a Vacancy-Trapped Mössbauer Impurity in Solid Solution of Aluminium

A.S. AGABEKYAN, Erevan
Resonant Coherent Energy Transfer

H. BERGNER, V. BRÜCKNER, F. KERSTAN, and W. NOWICK, Jena
Comparative Studies of Fast Recombination Processes in Amorphous Silicon Films

Z. DZIUBA, Warsaw
Impurity Band Conduction in HgTe (II)

I. MERTIG, E. MROSAN, R. ZELLER, and P.H. DEDERICHS, Dresden and Jülich
Electronic Properties of Dilute Copper Alloys (II)

V.S. PATIL, R.G. PILLAY, P.N. TANDON, and H.G. DEVARE, Bombay
Hyperfine Field Investigations in Some Ru and Rh Based Heusler Alloys

A.V. EFANOV and M.V. ENTIN, Novosibirsk
Optical Orientation and Polarized Luminescence in Silicon

V.A. MURALEV, Moscow
Channeling of Ultrarelativistic Electrons in Crystals

A.V. MALAKHOVSKII and G.G. VASILEV, Krasnoyarsk
Covalency as Source of Intensity for d-d Transitions in Octahedral Complexes

A. JEZIERSKI, Poznań
Application of CPA to the Magnon Hamiltonian

M.D. KAPLAN, Kishinev
The Dynamical Electron-Phonon Mode Coupling in Jahn-Teller Crystals with Antiferroelastic Ordering

E.L. ALBUQUERQUE, Natal
Surface-Ripple Brillouin Scattering from a Glass-Metallic Liquid Interface

O.J. ŻOGAŁ, Wrocław
Knight Shifts of Protons in Some Rare-Earth Dihydrides

S. ASANO, N. YAMASHITA et Y. OGAWA, Okayama
Luminescence et structure vibrationelle des luminophores CaS:Ce^{3+} et CaSe:Ce^{3+}

L. SCHWARTZ and A. PAJA, Waltham and Cracow
On the Electrical Conductivity of Disordered Alloys in the Muffin-Tin Model

TH. HARBICH and G. MAHLER, Stuttgart
Bistability of the Electron-Hole Plasma Density in a Strong Electromagnetic Field

A.K. AHUJA, S. AULUCK, and S.K. JOSHI, Roorkee
Density of States and Optical Conducitivity of Ferromagnetic Nickel

A.P. FEDOTOV and V.F. SHABANOV, Krasnoyarsk
Deuteron Dynamics in Crystal of Sodium Trideuterium Selenite

H. HÜLSING and A. KASTEN, Karlsruhe
Magnetic Properties of Dysprosium Oxisulphate ($Dy_2O_2SO_4$)

T. AMAND, A. CORNET, M. PUGNET, J. COLLET, M. BROUSSEAU, and B.S. RAZBIRIN, Toulouse and Leningrad
Picosecond Study of the Exciton Gas Cooling

I.M. KANSKAYA, S.I. KOKHANOVSKII, R.P. SEYSYAN, and A.L. EFROS, Leningrad
Direct Determination of Indium Antimonide Energy Band Parameters from Diamagnetic Exciton Spectra

V.N. KASHCHEEV, Riga
Central Peak in Ising Model with Transverse Field (II)

S. LIGENZA, M. ŁUKASIAK, Z. KUCHARSKI, and J. SUWALSKI, Świerk
Study of the $3d^5$ Shell Deformation of Fe^{3+} Ions in $ZnFe_2O_4$ by High-Field Mössbauer Spectroscopy

H. GRIMM and H. KRONMÜLLER, Stuttgart
Investigation of Structural Defects in the Amorphous Ferromagnetic Alloy $Fe_{40}Ni_{40}P_{14}B_6$

J. PRAKASH and RAHUL, Gorakhpur
A Method for the Evaluation of Dielectric Relaxation Parameters in Electronic Excited State

T. SOMA, Y. KONNA, and H. MATSUO KAGAYA, Akita
Local Heine-Abarenkov Potential with Volume-Dependent Parameters of Al

K.P. GHATAK and A.N. CHAKRAVARTII, Calcutta
Carrier Statistics in Superlattices of Small-Gap Semiconductors in the Presence of a Quantizing Magnetic Field

J.C. UPADHYAYA and M.L. SHARMA, Maiduguri
Model Potential Calculation of Phonon Dispersion in Tb and Ho

A.E. NIKIFOROV, S.YU. SHASHKIN, M.L. LEVITAN, and T.H. AGAMALYAN, Sverdlovsk
Cooperative Jahn-Teller Orbital Ordering in $KCuF_3$ and K_2CuF_4 Crystals

O. BÍLEK, L. SKÁLA, and L. KÜNNE, Prague and Jena
Tight-Binding Cluster Model for C, Si, and Ge (I)

N.M. PLAKIDA, A.V. BELUSHKIN, I. NATKANIEC, and T. WASIUTYŃSKI, Dubna and Krakow
The Internal Soft Mode Phase Transition in Solid Biphenyl

R. BORROMEI, E. VACALLI, and G. INGLETTO, Parma
Electronic Spectrum of the Co(II) Ion in the Columbite $CoNb_2O_6$ Host Lattice at 4.2 K

A. LEHMANN, L. SCHUMANN, and K. HÜBNER, Rostock
Optical Phonons in Amorphous Silicon Oxides (I)

NGUYEN VAN HUONG, Warsaw
Theory of Free-Carrier Absorption in n-Type Lead Chalcogenides Due to Electron-Electron Scattering

K. REGIŃSKI, Warsaw
Some Extensions of the Shell Model of Ionic Crystals

S. MRÓZ, A. MRÓZ, A. GRUDNIEWSKI, and E. BĄK, Wrocław and Kielce
Effective Debye Temperature of the (110) Ni Face for Different Azimuths

Y. ITOH and C. HAMGUCHI, Osaka
Resonant Brillouin Scattering in HgI_2 Investigated by Acoustic Domain Injection Method

V.YA. PRINZ and S.N. RECHKUNOV, Novosibirsk
Influence of a Strong Electric Field on the Carrier Capture by Nonradiative Deep-Level Centers in GaAs

H. ZIMMERMANN and R. BOYN, Berlin
On the Configuration Coordinate Diagram of ZnS:Fe

G. JAANDEL, A. WARIN et G. MORLOT, Nancy
Analyse du spectre d'absorption infrarouge de $BaF_2:Y^{3+}$

19. Magnetic Resonance

20. Optical Properties
    - 20.1 Spectra. Optical Constants (X-Ray Spectra See 20)
    - 20.2 Lasers
    - 20.3 Luminescence (X-Ray Spectra See 20)

21. Metals. Alloys
    - 21.1 Transition Metals (Group Sc to Zn) and Their Alloys
        - 21.1.1 Fe and Fe Alloys
    - 21.2 Alkali Metals
    - 21.3 Alkali-Earth Metals
    - 21.4 Rare-Earth Metals
    - 21.5 Actinides
    - 21.6 Noble Metals
    - 21.7 Semi-Metals (Semiconducting Alloys 22.7)

22. Semiconductors and Ionic Crystals
    - 22.1 Elements
        - 22.1.1 Germanium
        - 22.1.2 Silicon
        - 22.1.3 Group VI Elements
    - 22.2 III–V Compounds
        - 22.2.1 Arsenides
        - 22.2.2 Phosphides
        - 22.2.3 Antimonides
        - 22.2.4 Mixed Crystals
    - 22.3 IV–IV, II–V, and III–VI Compounds
        - 22.3.1 Mixed Crystals
    - 22.4 II–VI Compounds
        - 22.4.1 Sulphides
        - 22.4.2 Selenides
        - 22.4.3 Tellurides
        - 22.4.4 Mixed Crystals
    - 22.5 Halides
        - 22.5.1 Silver Halides
        - 22.5.2 Alkali Halides
        - 22.5.3 Alkali-Earth Halides
        - 22.5.4 Mixed Crystals
    - 22.6 Simple Oxides (CdO, ZnO See 22.4)
        - 22.6.1 Mixed Crystals
    - 22.7 Semiconducting Intermetallic Compounds
        - 22.7.1 Mixed Crystals
    - 22.8 Tertiary and Higher Compounds
        - 22.8.1 Salts of Oxy-Acids (Carbonates, Phosphates, Silicates Including Mica etc.)
        - 22.8.2 Spinels, Garnets, Ferrites
    - 22.9 Organic Semiconductors

23. Solidified Gases

Attention

The preceding classification scheme is closely akin to the scheme used by Physics Abstracts to classify the various fields of solid state physics. A paper will either be classified under one or more of the subsections or, if none of these are relevant, it will just appear under the general section heading (i. e. sections and subsections are equivalent in this sense).

The *subject classification* given at the head of each paper begins with the main subject matter of the paper, and this is followed, if necessary, by figures referring to subsidiary subject matter and the type of solids investigated.

The papers published in this issue are listed in the order of their subject classification at the end of the table of contents.

Manuscripts and letters for physica status solidi (b) — basic research and physica status solidi (a) — applied research

should be sent to

DDR-1020 Berlin, Neue Schönhauser Str. 20

or

Internationaler Buchversand GmbH, 1000 Berlin 19, Eschenallee 21.

The object of "physica status solidi (b)" and "physica status solidi (a)" is to publish rapidly important and topical results concerning fundamental research in the field of solid state physics and aspects of application and experimental techniques of fundamental solid state research, respectively. The issues of "physica status solidi (b)" and "physica status solidi (a)" will be published on the 1st and the 16th day of each month, respectively. Either series will contain the following categories:

1. Review articles written by arrangement with the editors. They will be remunerated.
2. Original papers selected from those which have not been submitted for publication elsewhere, after being reviewed. (The latest date for acceptance of contributions is the 8th and the 23rd day of each month for "physica status solidi (b)" and "physica status solidi (a)", respectively, 50 days before the delivery of the journal.)
3. Short notes published in the off-set part of the journal. They are, if necessary after being reviewed, selected by the editorial staff. (The latest date for acceptance of manuscripts for this part is the 12th and the 27th day of each month for "physica status solidi (b)" and "physica status solidi (a)", respectively, 16 days before the delivery of the journal.)
4. Pre-printed titles of papers to be published in the next issue of both series.

Editorial Information

Manuscripts should be submitted preferentially in English. Besides, papers in German, French, and Russian are admitted. The final decision on the incorporation of the papers in the series (b) or (a) is taken by the editorial staff after consultation of the referee. Authors are asked to observe the following rules:

1. Manuscripts have to be written as concisely as possible. As most, 20 and 4 pages (28 lines per page; abstracts, tables, references, figures, and captions of figures included, where 3 figures are counted as about one page) are admitted for original papers and short notes, respectively. All manuscripts should be type-written on one side only. Two copies are required. Badly arranged manuscripts are returned.
2. The author's name, initials, and institution should be given.
3. One abstract of about ten lines in English and one in German, French, or Russian are required for original papers.
4. Because of lack of space conclusions (if any) must not have the character of a summary (i. e. substantially a repetition of the contents of the abstract).
5. Formulae should not be written too small and not with pencil. Separate lines for formulae are desirable. Indistinct writing will lead to a return of the manuscript.
6. Drawings should be enclosed on separate sheets of paper, not exceeding the maximum and minimum sizes of 20×30 and 10×15 cm^2, respectively. Photographs should be submitted, on white paper, only for review articles and original papers. Duplicates are required.
7. Captions for the figures should be submitted on separate sheets of paper, and special instructions with regard to the size of illustrations should be given on these.
8. References should be numbered in order of appearance and should be given, on a separate sheet, in the form used in physica status solidi.
9. SI-units should be used.
10. Each paper should contain a proposal for the subject classification according to the classification scheme appearing on the second and third cover of each issue.

The authors will be sent galley proofs of review articles and original papers. Proof reading of short notes and pre-printed titles will be done by the editorial staff.

For each original paper and short note a total of 75 reprints is available free of charge. The number of reprints free of charge per review article is 25. Requests for additional reprints (against payment) should be given together with the submitted manuscripts. **The deadline for orders of additional reprints is the 15th and the 30th of each month** for "physica status solidi (b)" and "physica status solidi (a)" respectively, two weeks before the delivery of the journal.

FSC
www.fsc.org
MIX
Papier | Fördert
gute Waldnutzung
FSC® C083411

Kolibri 360 GmbH
Ferdinand-Jühlke-Straße 7
99095 Erfurt, Deutschland
produktsicherheit@kolibri360.de